Lecture Notes in Civil Engineering

Volume 113

Lecture Notes in Civil Engineering (LNCE) publishes the latest developments in Civil Engineering—quickly, informally and in top quality. Though original research reported in proceedings and post-proceedings represents the core of LNCE, edited volumes of exceptionally high quality and interest may also be considered for publication. Volumes published in LNCE embrace all aspects and subfields of, as well as new challenges in, Civil Engineering. Topics in the series include:

- Construction and Structural Mechanics
- Building Materials
- Concrete, Steel and Timber Structures
- Geotechnical Engineering
- Earthquake Engineering
- Coastal Engineering
- Ocean and Offshore Engineering; Ships and Floating Structures
- Hydraulics, Hydrology and Water Resources Engineering
- Environmental Engineering and Sustainability
- Structural Health and Monitoring
- Surveying and Geographical Information Systems
- Indoor Environments
- Transportation and Traffic
- Risk Analysis
- Safety and Security

To submit a proposal or request further information, please contact the appropriate Springer Editor:

- Mr. Pierpaolo Riva at pierpaolo.riva@springer.com (Europe and Americas);
- Ms. Swati Meherishi at swati.meherishi@springer.com (Asia—except China, and Australia, New Zealand);
- Dr. Mengchu Huang at mengchu.huang@springer.com (China).

All books in the series now indexed by Scopus and EI Compendex database!

More information about this series at http://www.springer.com/series/15087

Harvinder Singh · Puneet Pal Singh Cheema ·
Prashant Garg
Editors

Sustainable Development Through Engineering Innovations

Select Proceedings of SDEI 2020

Springer

Editors
Harvinder Singh
Department of Civil Engineering
Guru Nanak Dev Engineering College
Ludhiana, Punjab, India

Puneet Pal Singh Cheema
Department of Civil Engineering
Guru Nanak Dev Engineering College
Ludhiana, Punjab, India

Prashant Garg
Department of Civil Engineering
Guru Nanak Dev Engineering College
Ludhiana, Punjab, India

ISSN 2366-2557 ISSN 2366-2565 (electronic)
Lecture Notes in Civil Engineering
ISBN 978-981-15-9553-0 ISBN 978-981-15-9554-7 (eBook)
https://doi.org/10.1007/978-981-15-9554-7

This Springer imprint is published by the registered company Springer Nature Singapore Pte Ltd.
The registered company address is: 152 Beach Road, #21-01/04 Gateway East, Singapore 189721, Singapore

Preface

The exploding world population, limited resources, and the consequent global climate changes are leading to major concerns across the countries, such as power shortage, limited space in and around the cities, extreme weather conditions, environmental pollution, and rapid depletion of the natural resources. Technological innovation is a 'must' for sustainable development in the prevailing scenario. The best way, and the challenge today, is to learn from past mistakes and correct what can be corrected with due haste. Under this backdrop, Guru Nanak Dev Engineering College, Ludhiana (an autonomous institute, under UGC Act) in association with the Institution of Engineers (India), Indian Geotechnical Society: Ludhiana Chapter and Testing and Consultancy Cell of the college organized an international congress on 'Sustainable Development through Innovations' on September, 17–19, 2020, at Ludhiana (Punjab), India, with an aim to invite original research and practice papers from all engineering domains to provide innovative inputs to meet the needs of growing world population and with a focus on different methods and technologies that could minimize environmental damage by designing products and processes to enable the use of wastes from one process as inputs to another and provide better livable conditions and mobility to peoples.

The congress aimed to provide a unique opportunity to highlight recent advances, new directions, and opportunities for sustainable and resilient approaches to design and protect infrastructure and the environment. Accordingly, call for papers was flashed to invite original research and practice papers on all engineering and construction-related aspects of the resource management cycle, such as minimization of waste through the eco-friendly re-use, and disposal of residual wastes. About 15 speakers from academia and industry delivered expert lectures during the congress. In addition, 200 papers were received from various researchers/academician/industry experts on the congress themes, and out of these, about 65 papers were recommended by a panel of reviewers for the presentation and publication in the congress proceeding.

The organizing committee extends a sense of deep appreciation to Springer Nature, who is publishing the congress proceeding and TEQIP-III for financially supporting the congress. We also express our sincere gratitude to all members of the advisory committee, organizing committee, editorial/technical committee, reviewers, and the student volunteers for their untiring efforts, besides the several individuals who made their sincere efforts for making this event a great success.

The success of any event in general and technical event in particular depends on the contributions and involvement of a maximum number of practicing engineers/academicians/researchers in their field of excellence, but also coming together of various organizations/individuals to support such activities because the aim of such activities is to take the information from the laboratory to the field.

The organizing committee received financial support from several other organizations, agencies, and individuals; our salutations are to them for supporting the event.

Ludhiana, India

Dr. Harvinder Singh
Dr. Puneet Pal Singh Cheema
Dr. Prashant Garg

Contents

About the Editors

Dr. Harvinder Singh is currently Professor at Guru Nanak Dev Engineering College, Ludhiana (Punjab). He is engaged in teaching, research, and consultancy in structural engineering. He has authored numerous technical publications, one laboratory manual, and one monograph on 'Reinforced Concrete Steel Fiber Concrete' and has edited four books. He is UGC Research Awardee, State Technical Auditor, Chartered Engineer and Structural Consultant involved in the planning and design of various infrastructure projects along with member of several professional bodies. He is an editorial board member and reviewer for several international journals published by ACI, ICE, fib and ASCE. His current research interests are mathematical modelling and limit analysis of engineering structures, largely directed towards solving practical problems and improving the related design practices.

Dr. Puneet Pal Singh Cheema is currently an Assistant Professor, Guru Nanak Dev Engineering College, Ludhiana (Punjab). He is engaged in teaching, research, and consultancy in environmental engineering at the civil engineering department of the college. His major areas of research interests include decentralized wastewater management, natural wastewater treatment, stormwater management and groundwater recharge. He has several papers in respected international journals and also as book chapters. Dr. Cheema has guided 20 M. Tech theses during the last 9 years and is also involved with government and the private sector in executing various consultancy assignments.

Dr. Prashant Garg is currently working as Assistant Professor, Guru Nanak Dev Engineering College, Ludhiana, Punjab. He has obtained his bachelors from the University of Roorkee and double masters in Highways Engg. and Transportation Planning from School of Planning & Architecture (SPA), New Delhi. He is the State Technical Auditor. His major area of research includes soil structure interaction, pavement analysis and design. He has guided about 70 M. Tech students. He has published around 50 papers in National and International Journal and Conferences. Apart from Academic activities, he is actively providing services as consultant in the field of highways and Geotechnical field.

Integration of Decentralised and Centralised Water Systems to Address Current Water Servicing Challenges

Ashok Sharma and Stephen Gray

1 Introduction

Conventional urban water supply systems transport treated fresh water to urban areas from distant water catchments via large-scale pipe networks. Generally, surface water is the main source of water, which is collected in large dams to meet the seasonal variability in demand. Similarly, wastewater collection system collects wastewater from residential, commercial and industrial developments for treatment and finally for disposal into receiving waters. Centralised stormwater systems are provided to quickly transport the urban runoff away from developments to prevent flooding and associated losses. These systems have been implemented over 100 years and have provided considerable benefits to modern society, via the provision of reliable services and increased health benefits. Due to rapid urbanisation and population growth, aging infrastructure and upsizing needs of water infrastructure, the long-term viability of conventional centralised solutions to provide water services is being questioned. Moreover, due to stringent conditions from regulatory authorities for wastewater disposal requiring minimisation of contaminant loads before discharging to water receiving environment. If the urban water services are provided only using centralised approaches, huge investment would be required to upgrade these systems in the future.

Population growth and increased urbanisation resulted in the change of densities within the existing urban developments and also have expanded the boundaries of the urban areas. This trend is expected to continue in the future. Thus, there is a need for expansion and upsizing of existing water, wastewater and stormwater systems or implementation of alternative water systems under Integrated Urban Water

A. Sharma (✉) · S. Gray
Institute for Sustainable Industries & Liveable Cities, Victoria University, Melbourne, Australia
e-mail: Ashok.Sharma@vu.edu.au

H. Singh et al. (eds.), *Sustainable Development Through Engineering Innovations*, Lecture Notes in Civil Engineering 113,
https://doi.org/10.1007/978-981-15-9554-7_1

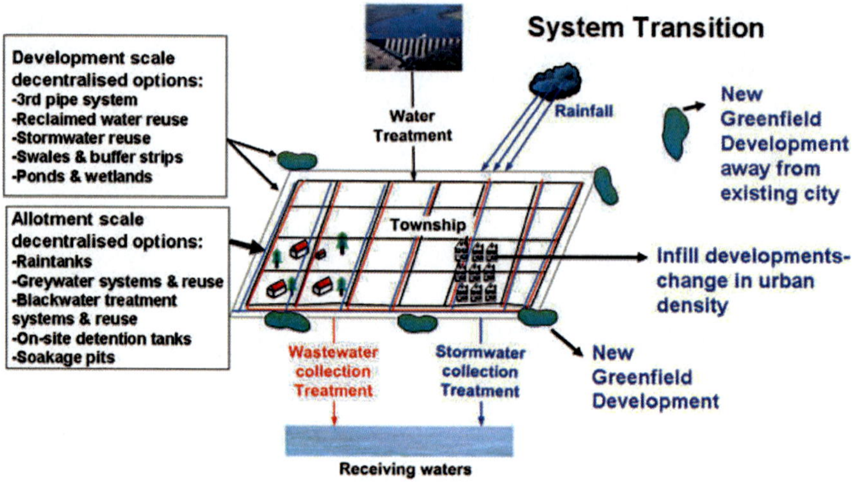

Fig. 1 Centralised/decentralised systems and urbanisation [11]

Management (IUWM) concepts. IUWM includes all aspects of the water cycle under fit for purpose concept to optimise water management solutions.

Decentralised systems can offer an alternative approach to provide municipal water, wastewater and stormwater services to urban and peri-urban areas, which can be integrated with centralised water systems or as standalone servicing solution, where the provision of centralised systems is not feasible due to technical, economical or environmental constraints. These decentralised systems can support existing centralised system to continue servicing their areas in a sustainable manner with no or limited capacity expansion.

The change in urban developments and the application of decentralised solutions are depicted in Fig. 1.

Under centralised systems approach, significant amount of high-quality water is used for toilet flushing and municipal waste is transported through large sewer network to the wastewater treatment plant and final disposal. A decentralised approach offers the opportunities to use local water sources and waste streams with a 'fit for purpose' approach that matches the quality of source water to the quality requirements of end-uses. Separate collection and treatment of various waste streams and recovery of valuable water, nutrients and energy are also possible through these systems [15].

Sharma et al. [12] presented a definition of decentralised systems, which was developed around the current recognition of the role of decentralised water, wastewater and stormwater systems in mitigating the environmental impact of urban development covering various development scales and application of alternative water resources including rainwater, wastewater and stormwater, based on a 'fit for purpose' concept.

The new water management and servicing approaches under water sensitive urban design (WSUD) is an aspect of IUWM, which focuses on the planning and configuration of developments to minimise environmental impacts and move towards more sustainable systems. WSUD approaches include decentralised systems. Some of the objectives of WSUD include protection of natural ecosystems, integration of water resources for provision of water services and reduction of peak stormwater flows. The Australian Intergovernmental Agreement on a National Water Initiative incorporated the concepts of WSUD into its urban water reform agenda and defined WSUD [2] as:

> The integration of urban planning with the management, protection and conservation of the urban water cycle that ensures urban water management is sensitive to natural hydrological and ecological processes.

A number of structural and non-structural tools can be used to achieve IUWM and WSUD objectives, the selection of which will be dependent on a large number of factors including; the type of development, catchment conditions, climate, customer acceptance and allocation of financial resources. Some examples of these tools are provided in Table 1.

Various studies have been conducted on the assessment of alternative water servicing options for infill and greenfield developments, but only a few of these studies have outcomes available in published literature. Manios and Tsanis [7] highlighted the importance of wastewater reuse in water resources management. Dixon et al. [4] highlighted that the society must move toward achieving efficient and appropriate water use for a sustainable urban future and described the reuse of domestic greywater and rainwater as a step in this direction. Hardy et al. [6] highlighted that the process of increased urbanisation can significantly change the input and output

Table 1 Decentralised/WSUD tools for adoption [5]

No.	WSUD approach/tools
1	Demand reduction
2	Rain gardens, green roofs and infiltration systems
3	Rainwater tanks
4	Pervious pavement
5	Urban water harvesting/reuse
6	Gross pollutant traps
7	Bioretention systems for streetscapes
8	Swales and buffer strips
9	Sedimentation basins
10	Constructed wetlands
11	Wastewater management
12	Siphonic roofwater systems

flows and quality from an area and can create a highly inefficient system in terms of water resources. However, inputs and outputs can be minimised through the efficient use of water resources.

This paper describes the role of decentralised systems in the transition of centralised systems to a more sustainable state using some examples and discusses some of the complexities in the implementation of decentralised systems.

2 Implementation of Decentralised Systems

The decentralised systems are planned along with centralised water systems or as stand-alone option to meet certain technical, economic and environmental objectives. The following three examples describe the planning studies conducted for the implementation of decentralised servicing options towards achieving desired objectives.

2.1 *Existing Development—Canberra Case Study*

The planning study was conducted for the township of Woden in Australian National Capital Canberra [3, 9]. The township was developed in the early 1960s and 1970s and is located in the South East region of Canberra. The total population at the time of the study was reported at 32,611 people, living in 13,890 dwellings over an area of 2990 ha with an average occupancy rate of 2.35 per dwelling. The residential area was divided into 13 suburbs for study purposes, with an overall average household block size of 1013 m^2 (377 m^2 min–5228 m^2 max). Urban water services are provided by conventional systems for water supply, wastewater collection & disposal and stormwater drainage. The stormwater from the township flows into Yarralumla Creek via lined open channels.

2.1.1 Study Objective

The following objectives were developed to incorporate both population growth and environmental issues for the case study sites:

1. A 25% reduction in household potable water demand
2. Peak stormwater flows for individual catchments no greater than pre-development.
3. Suspended solids, nitrogen and phosphorus loads in stormwater flows at pre-development levels.
4. A 20% reduction in wastewater discharged per household.

The first objective of a 25% reduction in potable household water demand was identified as the critical objective, with all others being of secondary importance.

2.1.2 Development of Servicing Options

Limited number of alternative water system options can be developed for existing developments due to various constraints. Dual pipe systems were considered too expensive to retrofit in existing suburbs due to economic considerations [1, 10] Considering the constraints, limitations and additional cost of reinstatement works in existing developments, it was reasonable to consider that such an option would be expensive.

The following on-site options were considered in existing development at allotment scale:

- Option A—Rainwater tanks for garden irrigation & toilet supply,
- Option B—Irrigation of gardens using greywater from the laundry and bathroom,
- Option C—Both rainwater tanks and greywater for irrigation of gardens.

Three stormwater ponds were proposed for storage of water for the irrigation of 166 ha of public open space. These ponds would also operate as a device for stormwater treatment. Two ponds 3 ha × 1.5 m deep and a third pond 4 ha × 1.5 m deep were considered. In order to achieve stormwater quality objectives, on-site detention tanks (OSD) at allotment scale, and gross pollutant traps (GPT), bioretention systems, buffers and swales at sub-catchment scale were considered as stormwater treatment options. Figure 2 depicts the water supply options adopted at allotment scale.

2.1.3 Analysis of Servicing Options

Based on the lot sizes, optimal rainwater tank sizes were estimated using Aquacycle (https://toolkit.ewater.org.au/Tools/Aquacycle). Thus, considering volumetric reliability of 40%, the optimum tank size of 3.5 kL, 7 kL, 14 kL and 20 kL for allotment sizes of 350 m^2, 700 m^2, 1000 m^2, and 5228 m^2 were considered respectively. The rate of water supply adopted for the analysis is provided in Table 2. Two demand management approaches were considered as evident from Table 2 [9]. The water balance analysis for all the options, adopting low rate of demand management, is provided in Table 3 [9].

The reductions in water, wastewater and stormwater flows were linearised to calculate the uptake rate of various decentralised systems (Table 3). The analysis showed that a 25% reduction in potable water usage could be achieved with the following combinations of water servicing options:

- 100% of the houses with raintanks.
- 75% of houses with both raintanks and greywater reuse.

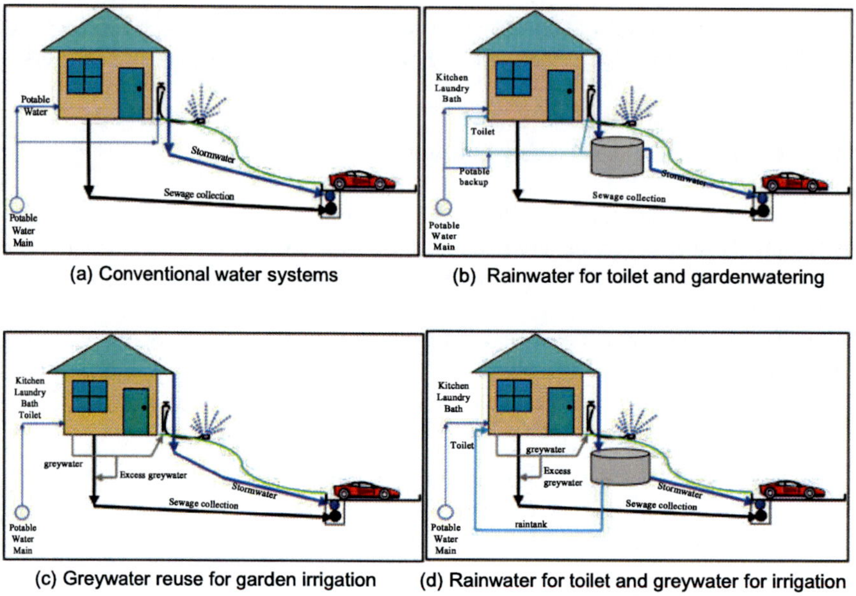

Fig. 2 Water supply options assessed for Woden existing development [3]

Table 2 Typical in-house water usage

Water demand	Typical water usage per-person	
	Low rate of demand management	High rate of demand management—Woden
	L/c/day	L/c/day
Kitchen	23	22
Bathroom	77	54
Laundry	50	43
Toilet	70	53
Total	220	172

Table 3 Outcome of water-saving options in Woden development

Water/wastewater/stormwater	Base case	Rain tanks in use (Option A)		Greywater for garden irrigation (Option B)		Raintanks and greywater in use (Option C)	
	ML/yr	ML/yr	% reduction	ML/yr	% reduction	ML/yr	% reduction
Water	4765	3649	24%	4166	13%	3160	34%
Wastewater	2836	2836	0%	2256	20%	2258	20%
Stormwater	4875	3758	23%	4858	0%	3850	21%

- 60% of houses with both raintanks and greywater reuse and remaining 40% of houses with greywater reuse only.

With high demand management, the uptakes rates for the various options are reduced to:

- $\geq$60% of the houses with raintanks
- $\geq$45% of houses with both raintanks and greywater reuse
- $\geq$30% of houses both raintanks and greywater reuse for garden irrigation and $\geq$40% of houses with greywater reuse.

The modelling indicated that peak stormwater flows would be reduced to pre-development levels in all allotment types except one (allotment size 5228 m^2) if rainwater tanks were provided as estimated. OSD tank sizes were estimated at allotment scale to reduce peak stormwater flows of 5 year average recurrence interval (ARI) to pre-development level.

Stormwater treatment trains alone could not provide stormwater quality equivalent to the pre-development level. Thus, the combined impact of raintanks and stormwater use from ponds for POS irrigation, GPT and ponds was modelled and found to predict contaminant loads of less than the pre-development case [3, 9]. Environment ACT amended planning controls to facilitate installation of raintanks and greywater systems and commenced a Government-funded rebate scheme for raintanks as a result of this study.

2.2 *Greenfield Kalkallo Development in Melbourne*

Kalkallo is a greenfield urban development in the north of Melbourne, approximately 30 km from the Melbourne CBD. It is anticipated that the development would house in the order of 86,000 residents and cover a total area of 3062 ha. Residential (1910 ha), industrial (840 ha), commercial (280 ha) and community (32 ha) areas are incorporated in this development. Alternative means of water servicing were considered for long-term ecological sustainability of the development. An environmental and economic assessment combined with various analysis means like water balance, life cycle costing and life cycle assessment (LCA) were considered to evaluate a range of alternative water servicing scenarios [10].

The options considered for servicing the development are listed in Table 4 [10]. The water resources used for household end-uses under various servicing options are also listed in this table.

The household water demand used for water balance analysis is listed in Table 5 [10]. Usual demand management (UDM) for Option A, improved demand management (IDM) for Options B and C and highest demand management (HDM) for Option D were considered.

Based on water balance analysis, the water supply-demand; piped water supply; stormwater runoff; wastewater flow; stormwater, wastewater, rainwater greywater

Table 4 Urban water servicing scenarios- application of various water resources

Scenario	Residential			Commercial and industrial		Demand management level (Table 5)
	Kitchen, bath, laundry	Toilet	Garden	Toilet	Garden	
A	Piped water					UDM
B		Reclaimed Water				IDM
C		Recycled Stormwater				IDM
D	Rainwater	Greywater				HDM

Table 5 End-use breakdown for water demand for Kalkallo residential lots

Usage	Usual demand management (UDM) (L/HH/day) (Option A)	Improved demand management (IDM) (L/HH/day) (Options B and C)	Highest demand management (HDM) (L/HH/day) (Option D)
Kitchen	48	48	41
Bathroom	267	195	156
Toilet	69	69	44
Laundry	111	78	48
Outdoor	175	175	88
Total	670	565	377

reuse quantities are listed in Table 6 [10]. It can be seen from Table 6 that Options A, B and C could achieve close to 100% water reliability, however, the water reliability for Option D was around 90% at highest demand management in comparison to other options, which would be significantly reduced under usual/improved demand management considerations.

A rainwater tank volume of 8.5 kL was adopted for Option D with volumetric reliability of ~0.9. This was adopted in preference to a rainwater tank (37 kL) with volumetric reliability of 1. This would need an average 242 ML/yr of water imported (external supply) for indoor residential use.

2.2.1 Conceptual Design of Water, Wastewater and Stormwater Infrastructure

The conceptual design of water, wastewater and stormwater infrastructure was conducted for life cycle costing (LCC) and life cycle assessment (LCA) of various servicing options. It also provided information for the estimation of embedded energy in developing the infrastructure and greenhouse gas emissions due to operational energy.

Table 6 Summary of average yearly water demand estimation for Kalkallo scenarios

Scenario	Total demand (ML/yr)	Demand met (ML/yr)	Piped water use (ML/yr)	Stormwater runoff, not reused (ML/yr)	Wastewater to sewer, not reused (ML/yr)	Stormwater/rainwater use (ML/yr)	Greywater/wastewater reuse (ML/yr)
A	14,439	14,439	14,439	9731	11,420	0	0
B	11,789	11,614	8289	9731	5682	0	3325
C	11,789	11,614	8291	6407	9006	3324 (SW)	0
D	8068	7694	4108	7301	5422	[a]2335 (RT)	[b]1251 (GW)

[a]Raintank supply short by 242 ML/yr for in-house usage. [b]Greywater supply short by 132 ML/yr for garden irrigation. *SW*—Stormwater, *RT*—Raintank water, *GW*—Greywater and *RW*—Reclaimed water

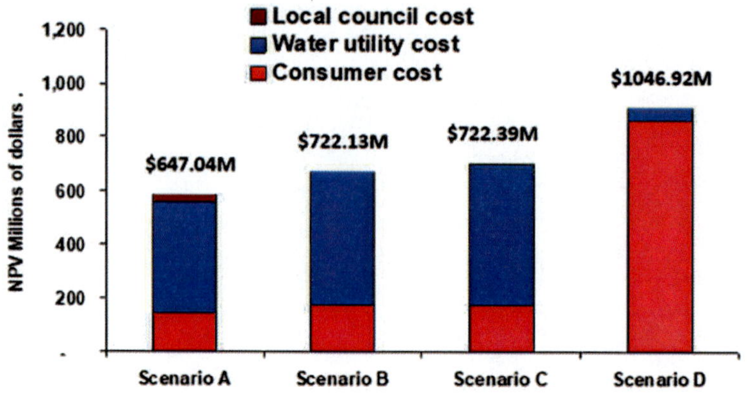

Fig. 3 Life cycle cost of various servicing options [10]

2.2.2 Life Cycle Costing (LCC)

Based on the conceptual design of infrastructure required, life cycle costing of various servicing options was conducted. The life cycle costs of Options A, B, C and D were estimated to $647.03 M, $722.13 M, $722.39 M and $1046.92 respectively, which are shown in Fig. 3. It can be concluded from cost considerations only that the centralised system's cost was minimum and self-sustained system's cost highest. There is a huge shift in cost to consumer (household) in moving from centralised to decentralised system.

2.2.3 Greenhouse Gas Emissions from Servicing Options

Greenhouse gas emissions from various servicing options were estimated based on embedded energy required in manufacturing the infrastructure and their construction on-site, and energy used in the operation of systems. It can be noticed from Fig. 4 that the embedded energy contribution from infrastructure is 10–15% of total GHG emissions and 85–90% contribution is from the operational energy required for servicing.

2.2.4 Total Community Cost of Options

A concept of community cost was developed to compare various servicing options based on combined economic and environmental considerations (Fig. 5). The community cost is the total cost of infrastructure for providing water, wastewater and stormwater services and the cost of avoidance or damage to the environment. The cost of environmental effects due to greenhouse gas emissions, eutrophication of receiving waters and use of fresh water resources was estimated (2005 data).

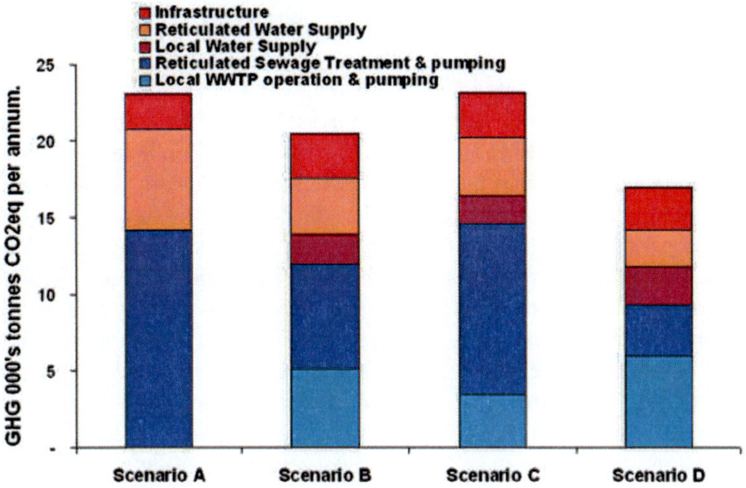

Fig. 4 GHG emissions CO_2 eq per annum [10]

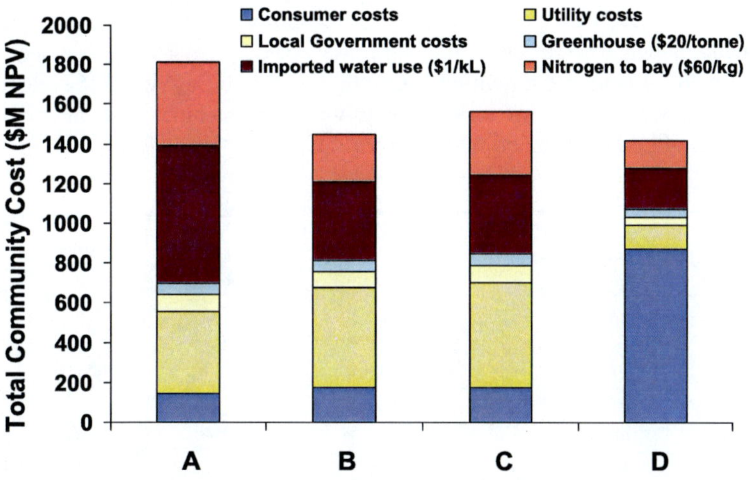

Fig. 5 Total community cost of servicing options [10]

It is highlighted that significant uncertainties are involved in their estimations due to limited data. As depicted in Fig. 5, community cost is highest for Option A and lowest for Option D. Option D was not considered finally due to the limited reliability of water supply. Option B (local wastewater recycling through dual pipe) was considered the best with least community cost. The water utility has constructed stormwater harvesting and treatment facilities for ultimate reuse of stormwater for potable application based on regulatory approval.

2.3 Infill Development—Melbourne Cases Study

The concept of Transit Oriented Developments (TOD) is encouraging infill developments in close proximity to public transport and mix of retail, health and recreation facilities. The proposed infill development "Box Hill Principle Activity Centre (PAC)" is an example of such a concept in Melbourne. Twenty high-rise residential/commercial buildings are proposed in a radius of 2 km as shown in Fig. 6a [8]. Each building is assumed to have 64 residential lots with an average population of two persons per lot, four retail lots with an average population of four persons per lot and four commercial lots with an average population of eight persons per lot. The development will significantly increase the population density and demand for the water infrastructure. It would require upsizing of existing water and wastewater infrastructure in the area. Due to this development, there will be an additional environmental impact from increased load on freshwater resources and increased wastewater flows and contaminant loads to the receiving environment. It was considered to explore alternative means of water and wastewater servicing to minimise the load on freshwater resources and receiving environment.

The following options were considered for assessment:

- Option A—Base case to reflect existing conventional services with adoption of demand management as per White Paper (WDM).
- Option B—Local reuse within an individual building envelope (localised reclaimed water reuse for toilet supply, washing machines and outdoor) with adoption of highest demand management (HDM).

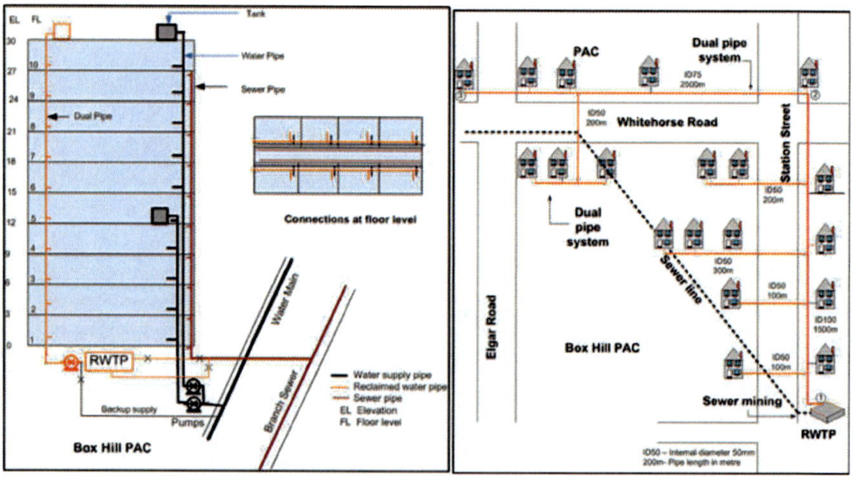

(a) Building scale reuse of wastewater (b) Sewer mining from trunk sewer

Fig. 6 Water servicing for high-rise buildings [8]

Table 7 Water usage end-use data (L/person/day)

Usage	As per White paper DM	As per highest DM
Residential	98	75
Commercial	8	7

Table 8 Water balance results

Option	Piped water (ML/yr)	Total wastewater generated (ML/yr)	Wastewater reuse (ML/yr)	Wastewater to sewers (ML/yr)	Demand management
A	98.3	93.0	0.0	93.0	WDM
B	46.3	70.9	27.9	43.0	HDM
C	61.3	93.0	35.6	57.4	WDM

Stormwater runoff for all the options = 10.3 ML/yr

- Option C—Centralised reclaimed water reuse for toilet supply, washing machines and outdoor from sewer mining with adoption of demand management as per White Paper (WDM).

The conceptual plan for water and wastewater services in a high-rise building with building scale wastewater reuse and layout of recycled water supply from sewer mining are shown in Fig. 6.

2.3.1 Water Balance Analysis

Water balance analysis of the options was conducted using Aquacycle software. The per-person water demand is listed in Table 7.

The piped water demand, wastewater generated, wastewater reuse, wastewater discharge to sewers and stormwater runoff generated are listed in Table 8 [8]. It can be seen that minimum piped water demand is in Option B, which also discharges minimum wastewater to sewers.

2.3.2 Life Cycle Costing

Based on the conceptual design of water servicing infrastructure, life cycle costing of various water servicing options was conducted, which is listed in Table 9 [8]. It can be seen that conventional (centralised) option was least expensive overbuilding scale recycled and sewer mining options from cost considerations.

Table 9 Water and wastewater servicing LCC

Option	Water supply and dual pipe system cost ($M)	Wastewater system cost ($M)	Total cost (NPV) ($M)
A	2.55	7.06	9.61
B	8.57	1.76	10.33
C	6.49	3.66	10.15

Table 10 Impact of environmental factors

Impact	Unit	Option A	Option B	Option C
Greenhouse gas emissions	Kg CO_2	212,200	130,700	178,100
Eutrophication	Kg PO_4	1734	1411	1511

2.3.3 Environmental Assessment

Environmental factors like GHG emissions and eutrophication were estimated for each option and are listed in Table 10 [8]. It can be seen from the table that Option B (building scale recycling) performed best.

The servicing option B (building scale wastewater reuse) with high water demand management was the best from potable water and contaminant load considerations, and from environmental factors. To achieve the most environmentally sustainable result requires an additional life cycle cost investment of 9% above the conventional system.

3 Post-implementation Assessment of Developments Designed with IUWM Approaches

Six developments designed based on IUWM and WSUD concepts were investigated to evaluate the effectiveness of WSUD in achieving implementation objectives, and to identify the knowledge gaps and other impediments that are hindering greater mainstream uptake of WSUD approaches in South Australia [13, 14]. These developments are Mawson Lakes, Springbank Waters, Harbrow Grove Reserve, Lochiel Park, Christie Walk and Mile End (Fig. 7).

Based on the post-implementation assessment of the above developments, the following issues were identified [13, 14].

Design intent of WSUD elements is often constrained by poor implementation by construction contractors. Need for oversight during the construction phase.

Cost remains a barrier for the broader adoption of WSUD approaches. In many cases, these schemes are financed under one-off capital funding opportunities to demonstrate more sustainable practices.

Fig. 7 Developments designed with IUWM concepts [14]

Those WSUD schemes that receive funds for capital can often neglect to consider the ongoing costs and human resources (skills) needed for the O&M.

There's still reluctance from local councils to assume responsibility for the O&M of some WSUD elements due to uncertainties on the cost burden.

In many cases, there has been no validation and monitoring of WSUD approaches, so there is no way of determining if they are achieving the sustainability objectives they were designed to (mains water savings, improvements in stormwater quantity and reduction in pollutant discharge).

The multi-functionality of some WSUD elements (e.g. flood mitigation, stormwater management, alternative water source, landscape amenity) can mean there are competing objectives.

4 Conclusions

It can be concluded that integration of decentralised systems with centralised systems can play a major role in the transition of current urban water systems to a more sustainable state and can address challenges faced by climate variability, population growth and urbanisation. The load on freshwater resources, urban water infrastructure and receiving environments can be reduced with the implementation of decentralised systems in conjunction with centralised systems. As these systems are new and their integration with existing centralised systems is complex, wide knowledge gaps exist in their planning, design, implementation, operation, management, governance, reliability, resilience and risk. Large-scale research is required to reduce these knowledge gaps for the increased uptake of decentralised systems.

References

1. Alegre N, McIntosh B, Thomas JS, Hardwick I, Riley S (2004) Strategic options for sustainable water management at new developments: the application of a simulation model to explore potential water savings. Water Sci Technol 50(2):9–15
2. COAG (2004) Intergovernmental agreement on a national water initiative. Commonwealth of Australian Governments. National Water Commission, Canberra, Australia
3. Diaper C, Sharma A, Gray S, Mitchell G, Howe C (2003) Technologies for the Provision of Infrastructure to Urban Developments—Canberra case study. CSIRO Technical Report, CMIT(C) 2003-183, September 2003
4. Dixon A, Butler D, Fewkes A (1999) Water saving potential of domestic water reuse systems using greywater and rainwater in combination. Water Sci Technol 39(5):25–32
5. DPLG (2009) Water sensitive urban design technical manual for the greater adelaide region. Department of Planning and Local Government, Government of South Australia, Adelaide
6. Hardy MJ, Kuczera G, Coombes PJ (2005) Integrated urban water cycle management: the urban cycle model. Water Sci Technol 52(9):1–9
7. Manios T, Tsanis IK (2006) Evaluating water resources availability and wastewater reuse importance in the water resources management of small Mediterranean municipal districts. Resour Conserv Recycl 47(3):245–259
8. Sharma A, Grant A, Tjandraatmadja G, Grant T, Opray L, Pamminger F (2007) Towards sustainable water and sewerage servicing options for a multi-story infill development. In: Ulanicki et al. (eds) Water management challenges in global change. Taylor and Francis, London
9. Sharma AK, Gray S, Diaper C, Howe C, Liston P (2008) Assessing integrated water management options for urban developments—Canberra case study. Urban Water J UK 5(2):147–159
10. Sharma A, Grant A, Grant T, Pamminger F, Opray L (2009) Environmental and economic assessment of urban water services for a Greenfield development. Environ Eng Sci 26(5):921–934
11. Sharma A, Burn S, Gardner T, Gregory A (2010) Role of decentralized system in the transition of urban water systems. Water Sci Technol Water Supply 10(4):577–583. https://doi.org/10.2166/ws.2010.187
12. Sharma A, Tjandraatmadja G, Cook S, Gardner T (2013) Decentralised systems—definition and drivers in the current context. Water Sci Technol 67:2091–2101
13. Sharma AK, Pezzaniti D, Myers B, Cook S, Tjandraatmadja G, Chacko P, Chavoshi S, Kemp D, Leonard R, Koth B, Walton A (2016) Water sensitive urban design: an investigation of current systems, implementation drivers, community perceptions and potential to supplement urban water services. Water 8(7):272. https://doi.org/10.3390/w8070272
14. Tjandraatmadja G, Cook S, Sharma ACP, Myers B, Pezzaniti D (2014) Water sensitive urban design impediments and potential: contributions to the SA urban water blueprint: Post-implementation assessment and impediments to WSUD; Goyder Institute for Water Research Adelaide, South Australia. http://www.goyderinstitute.org/uploads/WSUD_Task1_Final_web.pdf
15. Wilderer PA (2001) Decentralised versus centralised wastewater management. In: Lens P, Zeeman G, Lettinga L (eds) Decentralised sanitation and reuse, concepts, systems and implementation. Integrated environmental technology series. IWA Publishing, London, UK

Design Strategies for an MBR System for a Complex Influent

Harpreet Rai and Avneet Kaur

1 Introduction

Port Rowan is a town of around 1500 people located in the Long Point Region of Lake Erie in Southwest Ontario. Because of its location and being a hot spot for species diversity, it is a prime tourist location and attracts large numbers of visitors, especially during the summer and fall.

The town had an existing wastewater treatment system that consists of two facultative lagoons with hydraulic retention times in excess of three months. The lagoons were originally designed to treat domestic sewage of low to medium strength. However, with growth of the town as a tourist destination, the lagoons also received large and highly fluctuating batch loads (100–600 m^3/d) of septage and holding tank waste in addition to an average sewage flow of 600 m^3/d. Lagoons were discharged twice a year, during summer and fall, into a nearby creek, which eventually empties into Lake Erie.

As a result of overloading caused by the concentrated loads of septage and holding tank waste, TP and NH_4-N levels in the receiving creek had exceeded the background levels. In addition, odors caused by septage and holding tank loads (hauled wastes) were big cause of concern for the County as evidenced by frequent complaints by the residents. Furthermore, the town also expected to treat an additional load of landfill leachate in future. As a consequence of stringent effluent criteria for NH_4-N and P imposed by the Ministry of Environment (MOE) and inability of the lagoon system to handle current and additional future loads, an MBR-based wastewater treatment plant (WWTP) was chosen as the preferred option for upgrade of the existing system.

H. Rai (✉)
R.V. Anderson Associates Limited, London, ON, Canada
e-mail: hrai@rvanderson.com

A. Kaur
Department of Civil Engineering, Guru Nanak Dev Engineering College, Ludhiana, Punjab, India

H. Singh et al. (eds.), *Sustainable Development Through Engineering Innovations*, Lecture Notes in Civil Engineering 113,
https://doi.org/10.1007/978-981-15-9554-7_2

However, owing to the widely fluctuating and high strength streams of hauled wastes and leachate, the design of the MBR system faced several challenges. This paper discusses the design and operational strategies adopted to handle the unique challenges faced by the proposed system.

2 Design Issues

The major issues to be resolved for design and operation of the MBR system included the following:

- High hydraulic peaking factor for the sewage flow during wet weather
- Large fluctuating loads of TSS, BOD and TKN from hauled wastes and leachate with potential to destabilize the biological system (Table 1)
- Potential inhibition of the biological system by constituents of the hauled wastes and leachate.
- Stringent effluent TP and NH_4-N criteria (Table 2).

Table 1 Characteristics of hauled wastes and leachate

Stream	Value	Parameter				
		Flow (m^3/d)	CBOD (mg/L)	TSS (mg/L)	TKN (mg/L)	TP (mg/L)
Holding waste	Average	70	1300 ± 750	1320 ± 2800	175 ± 100	30 ± 30
	Maximum	550	3460	12,600	375	133
Septage	Average	10	3350 ± 2150	10,300 ± 10,000	450 ± 300	145 ± 200
	Maximum	80	6800	27,500	1200	738
Leachate	Average	50	150	100	250	2
	Maximum	100	150	100	250	2

Table 2 Effluent criteria for discharge

Parameter	Units	Value	Basis
BOD_5	mg/L	10	Monthly average
TSS	mg/L	10	Monthly average
NH_4-N	mg/L	1.0—Summer 3.0—Winter	Monthly average
TP	mg/L	0.1	Monthly average
pH	Dimensionless	6.5–7.5	Monthly average
Fecal coliform	CFU/100 mL	100	Monthly average

3 Design and Operational Strategies

3.1 Design Criteria

Three base loading scenarios emerged from analysis of yearly flow and characteristics data for raw sewage, hauled wastes and leachate. The loading scenarios were based on a future average flow of 950 m^3/d of typical raw sewage, combined with the hauled wastes and leachate loads based on characteristics given in Table 1. Further, as the combined stream is pre-treated in primary clarifier, the loadings reflect the combined primary effluent stream going into the MBR system. Table 3 summarizes the flow, temperature and loadings under the three scenarios.

Scenario #1 represents the annual average loadings to the MBR system. The minimum temperature for the average loading is 10 °C.

Scenario #2 represents the maximum month loadings to the MBR system, which normally occur in the months of April to June. The minimum temperature for the maximum month loading is 13 °C.

Scenario #3 represents the peak loadings to the MBR system, which occur in the months of May to June. The minimum temperature for the peak loading is 15 °C.

Based on the loading scenarios and the hydraulic peak of the raw sewage flow, the following general design criteria were considered for design and operation of the MBR system:

- Ability to meet effluent criteria for N and P under all loading conditions and minimum temperatures;
- Optimize the tank sizes to ensure desired treatment level under all operating conditions (Scenario #1, 2 and 3);
- Minimize the capital cost for MBR modules by optimum utilization of the existing lagoon system to equalize raw sewage hydraulic peaks;
- Minimize operational costs and
- Incorporate flexibility in operation to counter potential inhibition of microbes by hauled wastes and leachate.

Table 3 MBR system influent loading scenarios

Characteristic	Units	Average Scenario #1	Max. month Scenario #2	Peak Scenario #3
Flow	m^3/d	1110	1280	1680
BOD_5	kg/d	177	239	400
TSS	kg/d	186	289	575
TKN	kg/d	49	80	141
TP	kg/d	7	12	25
Alkalinity	kg-$CaCO_3$/d	100	100	100
Minimum temperature	°C	5	10	15

3.2 *Membrane Modules and MBR Tank*

The design peak hydraulic flow for the WWTP was 4000 m^3/d. Normally, the MBR surface area and hence the number of membrane modules are designed to handle the peak flow. Since, the peak flow is four times the annual average flow, designing MBR system for peak flow would mean providing MBR surface area 400% higher than required for average flow. This would mean not only a much higher capital cost but also operating the membrane system at 25% of its capacity for most of the year.

However, the above scenario was avoided by providing an off-line equalization tank. The process control for raw sewage flow was designed in a way that once the inflow to the biological system reaches a preset value for the day, the excess flow is diverted to the equalization tank. Further, since the primary clarifier is designed to handle the peak flow, inflow of suspended solids to the equalization tank is minimized under all flow conditions. In addition, odors being a big issue for the County, this strategy significantly reduces any potential odors while allowing the membrane modules to be designed for a lower, more reasonable peak flow value of 2500 m^3/d. Furthermore, since the minimum temperature at average flow conditions is 5 °C, membrane area would be sufficient to handle the peak loading flow of 1680 m^3/d at 15 °C as membrane permeability is higher at higher temperature. Refer to Fig. 1 for process flow details.

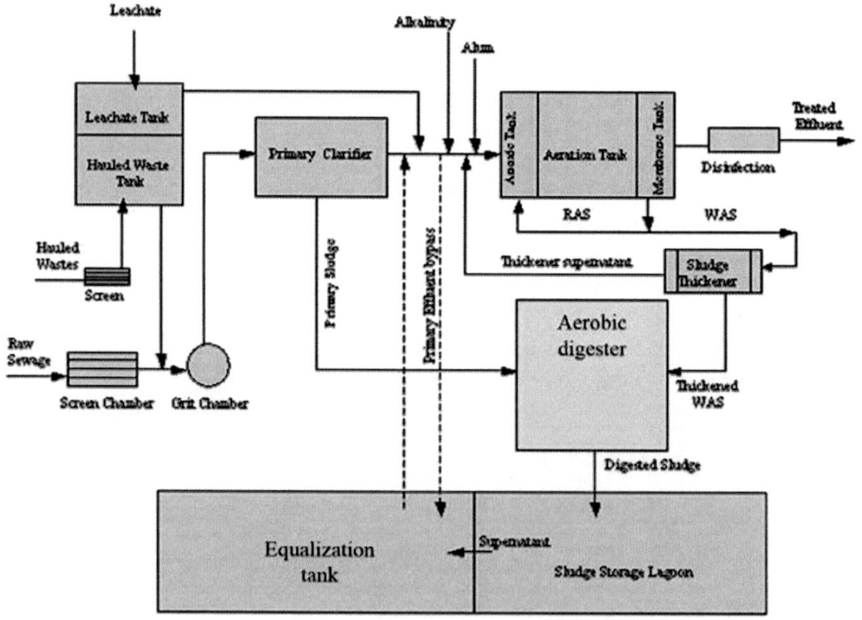

Fig. 1 Process flow diagram of the port rowan WWTP

Table 4 Aeration volumes for different loading scenarios

Parameter	Units	Loading		
		Average	Max. month	Peak
Aeration tank size	m^3	600	500	800
Aeration tank MLSS	mg/L	6500	6500	6500

3.3 Desktop Evaluation

Desktop evaluation was done in order to calculate the required aeration volumes for three scenarios of average, max month and peak loading. Preliminary calculations showed that the aeration volume required for average and maximum month conditions was nearly the same. However, the volume requirement to tackle peak loading is about 25% higher than the average and max month conditions in spite of a minimum temperature of 15 °C. Table 4 gives the aeration volumes required to meet the effluent criteria under the three loading scenarios.

3.4 BioWin Simulations

The program projection using BioWin was performed to calculate major parameters of the MBR plant. Based on the preliminary desktop calculations, BioWin simulations were run for the three loading scenarios using the configuration shown in Fig. 2. The volumes used for swing, oxic and MBR tanks were 200, 600 and 200 m^3, respectively. The swing zone was configured as anoxic for the average and peak loadings and as

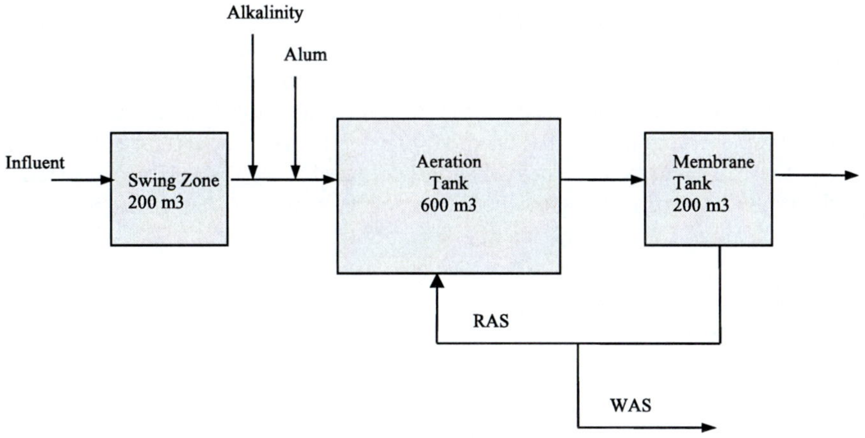

Fig. 2 Configuration of the MBR system for BioWin configuration

oxic for peak loadings. The results of the BioWin simulations are summarized in Table 5.

3.5 Discussion

By operating the swing zone as anoxic for average and maximum loadings and as oxic under peak loadings provides following advantages:

- Although there is no limit for total nitrogen in the CofA, operating swing zone in anoxic mode provides for denitrification capability, should a total nitrogen limit be imposed in future.
- Operation of swing zone in anoxic mode under average and maximum month loading scenarios reduces operational costs by oxygen and alkalinity credit generated by denitrification. Denitrification accounts for credits of 10–15% of daily oxygen demand and 15–25% of daily alkalinity demand under average and maximum month loading conditions (Table 5).
- Flexibility to operate the swing zone as oxic provides enough aeration volume for complete nitrification even under peak loading. As such, the ability of the biological system to assimilate peak day loads from hauled wastes was significant benefit from treatment and odor control viewpoints.
- The sizing of the tanks and configuration of the system offers sufficient flexibility to operate the system under a wide range of SRTs. This is of particular importance under maximum and peak loading scenarios when 50–80% of the nitrogen and BOD loading is contributed by the hauled wastes and can result in significant inhibition of nitrification. In case the nitrification does get inhibited, the biological system can be operated at higher SRTs to compensate for inhibition by increasing the MLSS in the membrane tank. However, as the MBR system performance is limited by MLSS concentration (maximum design concentration of 12,000 mg/L), the MLSS could be controlled by increasing the recirculation rate, if required, at higher SRTs under maximum and peak loadings (Table 5).
- The MBR tank volume based on the desktop evaluation was around 100 m^3. However, oversizing the MBR tank to 200 m^3 not only provided space for future addition of membrane cassettes to handle additional flows, but also helped keeping the MLSS concentration below the recommended maximum design concentration of 12,000 mg/L, under peak loading operation.

4 Conclusions

Design and operation of a WWTP receiving abnormally high flows of hauled loads (hauled wastes and leachate) can pose several potential issues including odor problems, large loading fluctuations and potential inhibition of the biological system by the constituents of the hauled loads.

Table 5 Simulation analysis under different loading scenarios

Parameter	Average loading $T_{min} = 10$ °C			Max month loading $T_{min} = 13$ °C			Peak loading $T_{min} = 15$ °C		
SRT (d)	12d	24d	36d	8d	20d	36d	5d	10d	15d
Swing mode	Anoxic			Anoxic			Oxic		
Effluent NH_4-N mg/l	1.5	0.3	0.3	0.9	0.2	0.1	0.2	0.1	0.1
$MLSS_{mem}$ mg/l	3800	7300	10,000	3600	8000	12,000	6000	9600	12,000
Qr/Q (%)	100	100	100	100	100	150	100	150	150
Air requirement m^3/h	270	330	350	550	730	800	1270	1350	1320
OTR kg O_2/h	9	11	12	15	18	20	26	27	27
Effluent NO_3-N mg/l	16	19	19	26	29	29	75	80	82
Denitrification oxygen credit (% of total oxygen demand)	17	10	9	19	14	12	–	–	–
Denitrification alkalinity credit (% of total alk. demand)	25	17	17	27	23	23	–	–	–

Incorporating flexibility in design of the MBR system to enable it to handle peak organic and nutrient loads along with potential inhibition by the hauled loads is vital to ensure desired treatment levels under diverse operating conditions caused by the hauled loads. Detailed analysis of the historical flow and characteristics data of the hauled loads can play an important role in establishing distinct loading scenarios and workable design criteria for the biological system.

In addition, providing a swing zone ahead of the aeration tank not only enhances flexibility to the MBR system to handle peak nutrient loads but also significantly reduces the operational costs by way of alkalinity and oxygen credit, when swing zone is operated in the anoxic mode under average and maximum loading conditions. Furthermore, utilizing existing storage to equalize hydraulic peaks can offer major savings in capital cost by designing the membrane system for average rather then peak hydraulic load.

Sustainable Rural Water Management Model

Puneet Pal Singh Cheema, Ashish Dehal, and Akepati Sivarami Reddy

1 Introduction

Household water supply and sanitation conditions in the rural human settlements (villages) continue to be inadequate, in spite of the long-standing efforts at the government and community's level. Adequate sanitation and water supply are development priorities, yet the ambition of international policy on drinking water and sanitation is not there. Globally, eight out of ten people suffer from of unimproved sanitation facilities, and six out of seven people defecate in the open in rural areas (Bartram and Cairncross [2]. Under millennium development goals (MDGs), the increase in the access to improved water supply and sanitation facilities in rural areas of India from 1990 to 2015 was 64% to 93% and 6% to 28%, respectively [12]. Health problems, especially in rural areas, are often related to sanitation conditions of the surroundings. Any form of sanitation system in developing countries should have the main objective of controlling pathogenic materials, which can be achieved by employing the on-site treatment options, like waste stabilization ponds, constructed wetlands, etc. Apart from acting as a barrier to the spread of pathogenic organisms, the on-site sanitation options have an added advantage of dispersing rather than concentrating the wastes, an important consideration if the sanitation facilities are not well maintained [5].

Poor technology choice is the most common reason for failure of the water supply and sanitation projects in rural areas. For increased success, stakeholders should be sensitized about the sustainability criteria associated with the water supply and sanitation projects [1]. Montgomery and Elimelech [8] identified three components,

P. P. S. Cheema (✉) · A. Dehal
Department of Civil Engineering, Guru Nanak Dev Engineering College, Ludhiana, India
e-mail: ppsc@gndec.ac.in

A. S. Reddy
Formerly at School of Energy and Environment, Thapar Institute of Engineering and Technology, Patiala, India

H. Singh et al. (eds.), *Sustainable Development Through Engineering Innovations*, Lecture Notes in Civil Engineering 113,
https://doi.org/10.1007/978-981-15-9554-7_3

i.e. community demand, local financing and cost recovery and dynamic operation and maintenance which are essential for the sustainability of any water supply and sanitation system. Integration of water supply, sanitation and watershed development has also been proposed by the researchers as a possible solution to the drinking water and sanitation problems. Kakade et al. [6] suggested that sustainable water management should be an integral part of any watershed/catchment development plan.

Traditionally, most of the villages in India have village ponds as an important source of water for rural human settlements. Village ponds played the role of a harvesting structure by collecting and storing the storm water and grey water from the pond catchment, which was then used by the inhabitants after being naturally treated in the pond. Presently, untreated wastewater (grey and black water) and polluted storm water runoff from dumps and cattle sheds are being discharged into the ponds. The ponds have become the source of disease vector, posing serious health hazard and creating poor living environment. Studies by Shivashankara and Puneeta [10] and Toor et al. [11] revealed that renovation of village ponds has positive effects on the environment, the rural economy and the groundwater regime. Further, if sufficiently treated, the village pond water can be used for irrigation purposes. Rejuvenation of village ponds through waste stabilization ponds (WSPs) has been attempted in Punjab, India. Periodical maintenance checks like removal of floating scum and duckweed from the surface of pond and removal of screenings and grit would be essential for the operation of ponds at maximum efficiency [7].

To find a solution for the water supply and sanitation problems in villages, it is imperative to adopt the integrated water resource management (IWRM) approach that will integrate water supply and sanitation sector in the rural areas. Moriarty et al. [9] presented that the application of IWRM is important in water supply and sanitation (WATSAN) sector. Where water is scarce or increasing competition among various water users (i.e. WATSAN, agriculture, industry, etc.) create a situation of conflict among water users, IWRM offers an integrated approach for managing the multiple uses of water without creating a situation of conflict. Cheema and Kaur [4] emphasized that a more realistic approach for the implementation of IWRM is to focus on the local levels instead of basin or national level. The present work is an effort towards the development of a sustainable water resource management system by adopting IWRM approach for villages (human settlements) in the state of Punjab, India. Focus of the study was to consider the village ponds at the centre of the sustainable rural water management model. The study has taken into consideration the various problems, issues and concerns related to the rural water supply and sanitation. Efforts have also been made to ensure that the developed integrated water resource management system is technologically and economically feasible, water conserving, and socially acceptable.

2 Water Management in Rural Human Settlements

2.1 Rural Water Supply System

A typical rural water supply system in India usually includes the following:

- A protected dug well or tube well or some other source of water (a stream or a lake),
- Pumping (used to lift water from the source to the treatment plant, to the water reservoir, or to the water distribution system, or to pump the water through the water treatment system and then to the distribution system. Often hand pumps are used to lift water or water is lifted manually),
- Water treatment plant (usually to remove TSS and/or turbidity from water and/or to disinfect the water. Occasionally, the treatment could be to reduce TDS levels or fluorides or any specific metal or metalloid species like cadmium, arsenic, selenium, etc.),
- Water reservoir (overhead reservoir and/or underground reservoir to ensure continued supply of water),
- Piped distribution network and individual household taps. Frequently users go to the source or to the communal faucets to fetch the water.

2.1.1 Water Service Levels

Water service levels are classified on the basis of the method by which the water is made available to the consumers:

- **Level-I**: This level provides a protected well or a developed spring with an outlet, but *without a distribution system*. The users go to the source to fetch the water. This is adapted where affordability is low and the houses in the intended service area are not crowded. A Level-I facility normally serves an average of 15 households within a radius of 250 m.
- **Level-II**: This type of system is composed of a source, a reservoir, a piped distribution network and communal faucets. Usually, one faucet serves four to six households within a radius of 25 m. It is generally suited for areas where houses are clustered in sufficient density to justify a simple piped system. The consumers still go to the supply point to fetch the water.
- **Level-III**: This system includes a source, a reservoir, a piped distribution network and individual household taps. It is generally suited for densely populated urban areas where the population can afford individual connections.

2.1.2 Rural Water Use

Water for rural human settlements of Punjab is mostly obtained from the local sources. Mostly, the water demands are met from the groundwater sources. Only, rarely water from a nearby river or stream or lake or pond is used.

Water consumption in rural human settlements mainly includes:

- human consumption,
- cattle and other domesticated animal consumption,
- religious, educational and other institutional consumption,
- agricultural use within the residential premises.

The fraction of water used for the non-consumptive purposes in the rural human settlements is relatively less. Open defecation and urination are quite common in rural human settlements. Toilets used in rural human settlements are mostly squat toilets (not the flush toilets), and washing clothes is mostly shifted away from the residential homes. Instead of bathtubs or showers, bathing involved use of one or a few buckets of water. Consequently, per capita day wastewater generated in rural areas is relatively very less.

Water consumption by cattle and other domesticated animals is relatively lesser. Mostly, the water consumed is for drinking (a consumptive use of water). Bathing of the animals is mostly shifted away from the residential homes. Cleaning the animal sheds is the major non-consumptive water use in case of cattle and other domesticated animals.

2.2 Rural Wastewater or Sewage

In line with relatively lower water consumption rates, wastewater generation rates are lower. Most often toilet flushings may not be contributing significantly to the wastewater generated. But, cattle and other domesticated animal contribution can be significant. The toilet facilities most often have on-site systems for the handling and management of the toilet flushings. Return factor for rural areas is relatively lower. Greater fraction of the used water is lost to evaporation and percolation moistening the soil.

Though volumes are lower, strength of wastewater from rural human settlements is usually higher. BOD, COD, TSS and grit levels could be higher in these wastewaters.

2.2.1 Sewerage Systems, Sewage Treatment and Disposal

Conventional underground sewerage systems are failures in the rural human settlements. Relatively lower population densities and lower per capita sewage generation rates, and also high costs make these systems not feasible. Small bore and simplified sewerage systems are often used in some of the human settlements as alternatives to the conventional sewerage systems.

Rural human settlements usually either have no sewerage systems or have poorly defined sewerage systems. The wastewater generated is mostly discharged or thrown out into roadside gutters or open drains, which in turn carry the wastewater to one or more points (usually village ponds). The high grit and suspended solid content in the sewage usually results in the siltation and clogging of the gutters and roadside drains.

Village ponds primarily function as primary facultative ponds or vegetated ponds in treating the received wastewater. From the ponds, the wastewater is lost or disposed off through evapo-transpiration and percolation. Sometimes, the pond water is drawn out (through pumping) and used mostly for irrigation within the premises. Smaller size, encroachments and weed infestation often limit the wastewater treatment potential of the village ponds.

2.3 Storm Water in Rural Human Settlements

Storm water flow rates are relatively lower due to low reaction of impervious area in rural areas. The combined runoff coefficient value is also lower for the rural areas. As a consequence, relatively lesser surface runoff is generated.

Storm water is relatively more polluted in rural areas. Unlined surfaces may be eroded, and grit and silt load of the storm water may be higher. Open defecation and urination, cattle dung, manure pits, loose soil and garbage dumps all these contribute to the higher pollution loads.

Rural human settlements usually have no defined storm water drainage system. Storm waters often flood many small depressions. The roadside gutters and drains usually carry/convey the storm water to the village ponds. For all the practical purposes, villages can be considered to have combined sewerage system carrying both sewage and storm water.

Storm water usually results in the siltation of the village ponds. The village ponds are usually flooded and overflow during the larger rainfall events and spread the pollution beyond the ponds to the surrounding areas. A contemporary rural water management system is shown in Fig. 1.

3 Sustainable Rural Water Management

3.1 Water Supply and Sanitation

Reliable supply of safe drinking water (treated and disinfected water satisfying the prescribed drinking water standards) and ensuring access of the safe drinking water

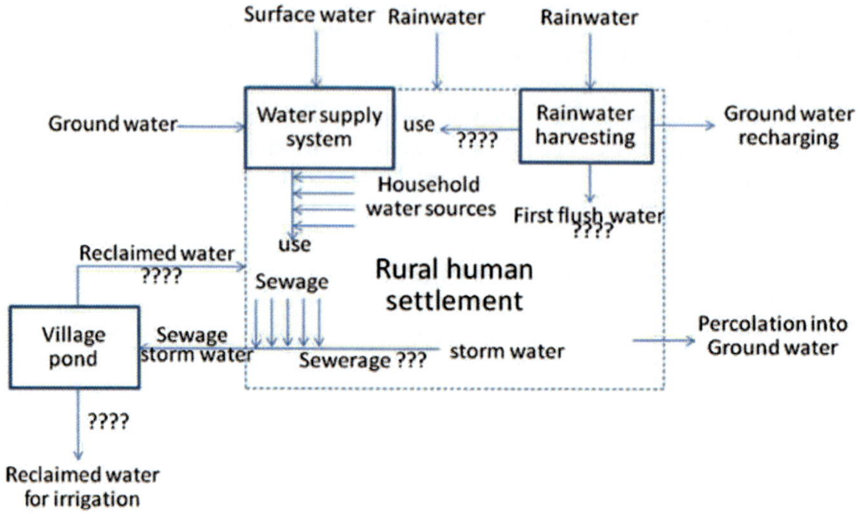

Fig. 1 Contemporary rural water management system

to all along with the minimum acceptable sanitation facilities to all and elimination of open defecation and urination practice is the prime motive of any sustainable rural water management system.

Ecological sanitation systems, which are water conserving and minimal black water generating, can prove to be the most appropriate. If not for social and security reasons, community toilets can suit the best in rural human settlements. The open defecation and urination can be avoided. Being a centralized facility, handling the toilet flushings becomes feasible. An on-site treatment facility comprising of a septic tank and a baffled anaerobic reactor can treat the wastewater from the community toilets to the level desired for disposal into the combined sewerage system being used in the rural human settlements. Salient features of an ecological sanitation system are:

- Not dumping black water into the sewerage system,
- Handling and management of cattle dung, urine and other wastewater generated at the cattle sheds separately and not allowing these wastes into the sewerage system,
- Avoiding storm water runon and runoff from the manure pits and garbage dumps.

A sustainable model of rural water management has been depicted in Fig. 2.

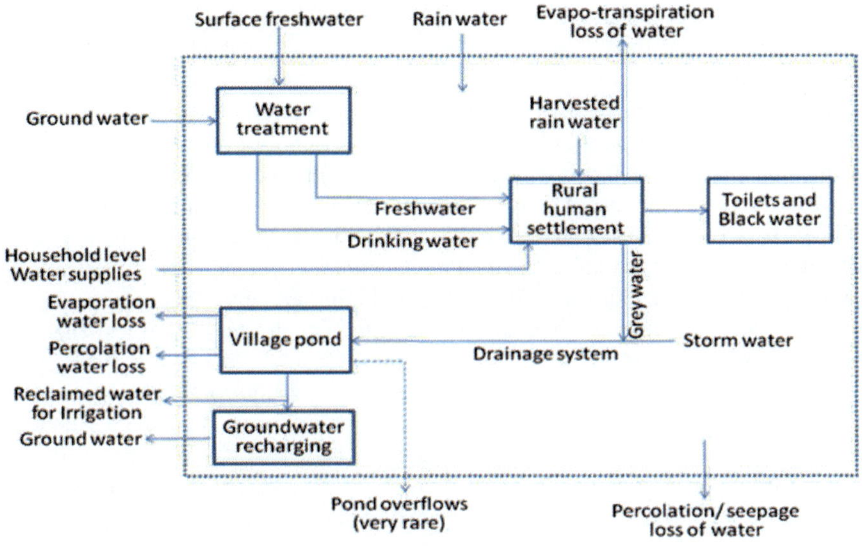

Fig. 2 Sustainable rural water management model

3.2 Domestic Sewage and Storm Water

Average daily domestic sewage generation rates and their average characteristics and storm water generated from the 90 percentile rainfall event and its characteristics assessed are required in the design of the storm water and domestic sewage treatment and disposal system.

Storm water generated by the rural human settlement or village for the 90 percentile rainfall event may be quantified and used in the design of storm water and sewage treatment and disposal system. Minimum size of a rainfall event (or the threshold rainfall event) that generates storm water for the village watershed may be identified, and all the rainfall events (larger than the threshold rainfall event) of the last 10 years for the region may be analysed to identify the 90 percentile rainfall event. Effective rainfall corresponding to the 90 percentile rainfall event and volume of storm water generated by the 90 percentile rainfall event may then be estimated.

Domestic sewage generation may be quantified through assuming (a) the per capita water consumption, (b) the sewage return factor and (c) the forecasted population of the village (or the rural human settlement) at the end of the design period. The domestic sewage generation rate may come out as very insignificant in comparison with the quantity of the storm water generated from the 90 percentile rainfall event.

Sewage samples collected at the inlet of the village pond and composited over a day (24 h) may be analysed for the characterization of the sewage from village in question. BOD, COD, total nitrogen, total phosphorus, MPN (coliform count), TSS, alkalinity, chloride and sulphate can be the parameters for the sewage characterization. This characterization of the domestic sewage may be carried out for all the 7 days. Mean parameter values represent average domestic wastewater characteristics. Mean ±2 times the standard deviation may provide the worst case and the best case domestic sewage characteristics.

Storm water may be characterized through composite sampling of the storm water received at the village pond over the period of the rainfall event. Characterization of the storm water usually involves testing of the samples for BOD, COD, total nitrogen, total phosphorus, MPN (coliform count), TSS, alkalinity, chloride and sulphate parameters. Characteristics of the storm water mainly depend on the size of the rainfall event being sampled and on the antecedent dry period of the rainfall event. Rainfall events are random, and their sizes are highly variable. Characterization of the storm water on the basis of a few storm water samples collection and analysis may thus not be acceptable.

Storm water characteristics can thus be assessed through predictive regression models. The models may be developed through multiple regressions of parameter values against the rainfall event size and antecedent dry period. The developed models may in turn be used to assess the characteristics of the storm water generated from a 90 percentile rainfall event with 90 percentile antecedent dry period.

3.3 *Sewerage System*

A combined sewerage system comprising of roadside drains may prove sufficient for the rural human settlements. Triangular drains with rounded bottom may be appropriate to ensure self-cleansing velocities even with just the domestic sewage flow. Catch basins if needed may be introduced at critical locations in order to prevent entry of grit and floating material into the drains.

3.4 *Sewage and Storm Water Treatment System*

The village pond and its premises may be used for the grey water and storm water treatment and disposal (shown in Fig. 3). Free water surface constructed wetland system may be added to the existing village pond mainly for the primary treatment and sludge stabilization. If needed, as part of the village pond may be transformed into the constructed wetland system. The village pond may be allowed to function as a secondary treatment unit (a facultative pond). Within the village pond, a two-stage filtration system (of an up-flow multi-grade roughing filter and a slow sand filter) may be created for the tertiary treatment. Provisions may be made within the pond to store the treated water for reuse (as and when required) and to dispose off the excess treated water through groundwater recharging.

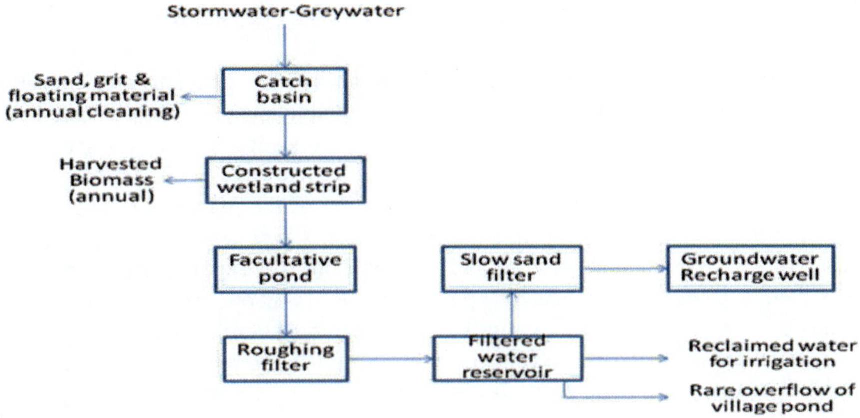

Fig. 3 Storm water and grey water treatment scheme

3.4.1 Constructed Wetland System

Constructed wetland system is supposed to act as a primary treatment unit and also as a sludge stabilization unit. Sheet flow of the wastewater within the wetland system will be ensured through dividing the wetland system into two or three wetland cells connected in series and terracing them. Each of the cells will be designed for the discrete settling of organic solids of 1.05 specific gravity and 20 micron size. The cells will have enough capacity to store the settled and stabilized solids. Outlets of the wetland cells will be designed to avoid wash down of suspended solids.

3.4.2 Village Pond or Facultative Pond

The pond will be designed as a secondary facultative pond and sized on the basis of the surface organic loading rates. Daily average temperature of the ambient air for the coldest month of the monsoon season will be used as the basis for designing the design surface organic loading. The pond will be assumed to have certain minimal water level. Maximum water level in the pond will be decided on the basis of the volume of storm water generated from a 90 percentile rainfall event.

Rate of accumulation of storm water in the pond will be influenced by the storm water storage capacity of the constructed wetland system and also by the groundwater recharge rate. No reuse of treated effluent is assumed during heavy rainfall events. For the 90 percentile rainfall event, water level in the pond should not cross the designed maximum water level of the pond.

3.4.3 Two-Stage Filtration System

The two-stage filtration system is supposed to remove the suspended solids concentration and to reduce the turbidity of water to acceptable level (5 NTU). Facultative pond water is believed to be turbid and has relatively high TSS levels specially because of the algal blooming. A slow sand filter which can significantly disinfect the water can probably not be used for the filtration of the pond water directly. A pre-filtration of the pond water in a roughing filter is probably needed. Filtration rates will be maintained at $\leq$2 m/h and $\leq$0.5 m/h in the roughing filter and slow sand filter, respectively. Whenever the pond water level is above the minimum, water should pass through the up-flow multi-grade roughing filter and then through the slow sand filter and stored in the filtered water sump. The filters should be positioned hydraulically in such a way that all the time the filters will remain submerged in water.

3.4.4 Groundwater Recharge System

Only filtered water would be made to enter the recharge well. Capacity of the recharge well will be equal to that of the filtration system. Water will be recharged into the first pervious layer provided it is having sufficient storage capacity. The recharge well can have provision for the flushing of the recharge well and removing the sludge accumulated at the bottom.

Area for the discharge of water may be fixed on the basis of the head available (water column height in the recharge well) and hydraulic conductivity of the pervious layer of the earth. Water will be discharged above the maximum mound height expected above the groundwater table [3].

The village pond-based treatment scheme is shown in Fig. 4.

4 Conclusions

For tackling the rural water supply and sanitation problems in villages of Punjab, implementation of IWRM at village level has been recommended. IWRM will constitute prohibition of household level private water supplies, ecological sanitation approach for water conservation, discharging only grey water into village drains, segregation of black water and cattle dung—cattle shed wastewater for separate on-site handling and management, design and creation of grey water—storm water drainage system, modification/restructuring the existing village ponds into village pond systems and disposal of the treated pond effluents through reuse as reclaimed water and through groundwater recharging. Pond-based sustainable water treatment system was proposed as viable solution for treatment of wet weather flows of pond catchments. The pond systems were sized to retain the wet weather flows in

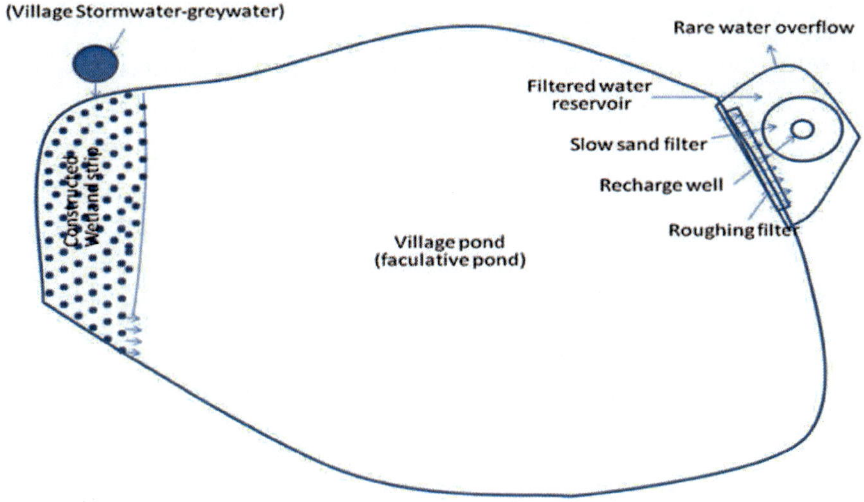

Fig. 4 Village pond system

the system till they are slowly treated and disposed off. The designed village pond systems can be mostly accommodated within the existing village ponds having very small additional land requirements.

References

1. Barnes R, Ashbolt N, Roser D, Brown P (2014) Implementing sustainable water and sanitation projects in rural, developing communities. Waterlines 33:71–88
2. Bartram J, Cairncross S (2010) Hygiene, sanitation, and water: forgotten foundations of health. PLoS Med 7:e1000367
3. Bouwer H (2002) Artificial recharge of groundwater: hydrogeology and engineering. Hydrogeol J 10:121–142
4. Cheema PPS, Kaur P (2018) Advocating integrated water resources management approach for rural human settlements. Int J Hydrol 2(3):278–279
5. DFID (1998) Guidance manual on water supply and sanitation programmes. Loughborough University, WEDC
6. Kakade B, Neelam G, Petare K, Doreswamy C (2003) Revival of a traditional water management system: an innovative farm pond network approach. Nat Resour Manag 6:2–12
7. MDWS (2012) Study of waste water treatment technologies adopted in Ludhiana and Muktsar Districts of Punjab. Ministry of Drinking Water and Sanitation, Govt. of India
8. Montgomery MA, Elimelech M (2007) Water and sanitation in developing countries: including health in the equation. Environ Sci Technol 41:17–24
9. Moriarty P, Butterworth J, Batchelor C (2004) Integrated water resources management and the domestic water and sanitation sub-sector. IRC Thematic Overview Paper, IRC International Water and Sanitation Centre, Delft, The Netherlands
10. Shivashankara G, Puneeta H (2004) Quality status of water in village ponds for irrigation. Nat Environ Pollut Technol 3:243–247

11. Toor A, Khurana M, Sidhu B, Khera JS, Brar KK (2011) Suitability of village pond waters for irrigation: a case study from district Ludhiana, India. Environ Monit Assess 172:571–579
12. WHO U (2015) Progress on sanitation and drinking water: 2015 Update and MDG Assessment. Geneva World Health Organisation

Utilization of Wastes and Energy-Efficient Materials Leading to Sustainable Construction

Jaspal Singh

1 Introduction

A large amount of carbon dioxide is emitted during the production of cement causing significant impact on environment. Cement industry is a major contributor in the emission of CO_2 as well as in using up high levels of energy resources in the production of cement. Industrial wastes contain some toxic elements which are released into the environment under natural weathering conditions. These toxic elements cause pollution of soils, surface waters and groundwater. Industrial wastes, such as copper slag, blast furnace slag, granulated phosphorus slag and steel making slag, are being used as supplementary cement replacement materials. The substitute/replacement of cement with waste materials not only helps to reduce the use of natural resources but also helps to mitigate the environment pollution. In other words, environment and economic considerations demand greater utilization of the wastes. It is worth mentioning that the conversion of industrial waste into cement needs only approximately 10% of the energy required for the production of Portland cement. Sources of natural aggregates are also depleting, and so, replacement of the same will augment in preparation of sustainable concrete. Published literature has supplemented that waste materials such as fly ash, coal bottom ash, copper slag, blast furnace slag, sugarcane bagasse ash, waste glass, hypo sludge, rice husk ash and corn cob ash can be used as a substitute/replacement of fine aggregates. An appropriate disposal of these wastes and by-products is a serious burden on our society. Furthermore, there is a lack of disposal sites to handle such wastes without causing harmful effects on the environment. Therefore, researchers from all over the world are advocating on

J. Singh (✉)
Department of Civil Engineering, Punjab Agricultural University, Ludhiana, India
e-mail: jaspalsingh@pau.edu

Guru Angad Dev Veterinary and Animal Sciences University, Ludhiana, India

H. Singh et al. (eds.), *Sustainable Development Through Engineering Innovations*, Lecture Notes in Civil Engineering 113,
https://doi.org/10.1007/978-981-15-9554-7_4

utilization of agro and industrial wastes. The concrete industry is one of the promising options where these wastes could have good perspective.

2 Utilization of Wastes

2.1 *Ashes*

2.1.1 Fly Ash/Alkali-Activated Fly Ash

Fly ash (FA) is a by-product from burning ground coal in electric power generating plants. Fly ash particles are generally spherical in shape and range in size from 0.5 to 300 μm. India is a resourceful country for fly ash, and it is estimated that 130–145 million tons of fly ash is generated by 70 major thermal power plants of which only 6–10% is utilized by cement, construction and road industries. The spherical shape of fly ash also improves the consolidation of concrete, which also reduces permeability.

Alkaline activation is a chemical process in which fly ash is mixed with an alkaline activator to produce a paste capable of setting and hardening within a reasonably short period of time. The strength, shrinkage, acid and fire resistance of the resulting materials depend on the activation process variables. The alkaline activation of fly ash is consequently of great interest in the context of new and environmentally friendly binders with properties similar to or that improved on the characteristics of conventional materials. Concrete made with alkali-activated fly ash (with no OPC) performs as well as traditional concrete and even better in some respects, exhibiting less shrinkage and a stronger bond between the matrix and the reinforcing steel. In addition to its excellent mechanical properties, the activated fly ash is particularly durable and highly resistant to aggressive acids, the aggregate-alkali reaction and fire.

Study on alkali-activated fly ash-based geopolymer mortar

The effect of nano-metakaolin on the compressive strength and microstructure properties of geopolymer mortar as partial replacement (0, 2, 4, 6, 8 and 10%) of fly ash in fly ash-based geopolymer mortar alkali activated with 12M, 14M and 16M molarity of NaOH alkaline activator solution has been studied.

Alkali-activated fly ash

Low calcium (class F) fly ash obtained from the Guru Nanak Dev thermal plant, Bathinda, was the source material. Two alkaline activators, viz. sodium hydroxide (NaOH) solution, prepared by dissolving sodium hydroxide pellets in water and sodium silicate in liquid form having Na_2O and SiO_2 in the range of 7.5–8.5%, 25–28%, respectively, were used.

Table 1 Chemical properties of nano-metakaolin

S. no.	Chemical composition	Values (%)
1.	SiO_2	51–54
2.	Al_2O_3	41–44
3.	K_2O	1–2
4.	Loss on ignition	1–2
5.	Fe_2O_3	0.35
6.	Na_2O	0.13
7.	MgO	0.065
8.	CaO	0.016

Nano-metakaolin

The nano-metakaolin (NMK) acquired from Sheetal industries, Gujarat, having the properties as shown in Table 1 was used.

Sample preparation

NaOH solution was prepared one day prior of casting. Both the alkaline activators (NaOH and Na_2SiO_3) were mixed together in a beaker. To prepare geopolymer mortar, fly ash and nano-metakaolin were mixed dry for five minutes. Later on, appropriate quantity of alkaline activator solution as per AAS/FA ratio was added slowly and mixed thoroughly until homogeneous paste was obtained. After that, sand was added and mixed thoroughly for another 3 min. The prepared geopolymer mortar was then poured into standard size 70.6 mm× 70.6 mm × 70.6 mm cubic moulds. After that, vibrator was used to remove entrapped air or voids from prepared moulds.

Alkali activation of fly ash-based geopolymer mortar was prepared using three different molarities 12M, 14M and 16M of sodium hydroxide (NaOH) along with Na_2SiO_3 solution in the ratio of 1:2.5. Locally available river sand, passing through 2.36 mm sieve, was used as fine aggregate. The ratio of fly ash to sand and alkaline activator to fly ash was 1:3 and 0.30, respectively. Total eighteen series of geopolymer mortar were prepared by partially replacing fly ash with different proportion of nano-metakaolin (0–10% @ 2% increment) by weight of fly ash. Mix proportion of these mixes is given in Table 2.

Compressive strength testing

The similar testing procedure for the compressive strength of geopolymer mortar was adopted as of cement mortar recommended by Bureau of Indian standards. Before testing, the specimens were cured at 70 °C for 24 h temperature curing followed by 28 days ambient curing. Three cubes of each mix were casted to find out the compressive strength.

Figure 1 shows the 28 days compressive strength of fly ash-based geopolymer mortar samples, alkali activated using three different molar concentrations 12M, 14M and 16M at different replacement levels of nano-metakaolin. At 0% replacement level of NMK, 28 days compressive strength of 14M and 16M was found to be 1.13 and 1.22 fold of 12M compressive strength. Similarly, for 2% replacement, 28 days

Table 2 Mix proportions of all the mixes

Mix	NaOH molarity (M)	NMK (%)	FA (g)	NMK (g)	AAS (ml)	NaOH (ml)	Na_2SiO_3 (ml)	Extra water (ml)
GM 1	12	0	2400	0	960	274.2	685.8	227.02
GM 2		2	2352	48	960	274.2	685.8	227.02
GM 3		4	2304	96	960	274.2	685.8	227.02
GM 4		6	2256	144	960	274.2	685.8	227.02
GM 5		8	2208	192	960	274.2	685.8	227.02
GM 6		10	2160	240	960	274.2	685.8	227.02
GM 7	14	0	2400	0	960	274.2	685.8	242.54
GM 8		2	2352	48	960	274.2	685.8	242.54
GM 9		4	2304	96	960	274.2	685.8	242.54
GM 10		6	2256	144	960	274.2	685.8	242.54
GM 11		8	2208	192	960	274.2	685.8	242.54
GM 12		10	2160	240	960	274.2	685.8	242.54
GM 13	16	0	2400	0	960	274.2	685.8	256.96
GM 14		2	2352	48	960	274.2	685.8	256.96
GM 15		4	2304	96	960	274.2	685.8	256.96
GM 16		6	2256	144	960	274.2	685.8	256.96
GM 17		8	2208	192	960	274.2	685.8	256.96
GM 18		10	2160	240	960	274.2	685.8	256.96

compressive strength of 14M and 16M was found to be 1.09 and 1.16 times the 12M. For 4% and 6% replacement of NMK, 28 days compressive strength of 14M and 16M was found to be 1.09 and 1.16 fold of 12M. Similar results were observed for 8% and 10% replacement of NMK.

Irrespective of molar concentration, the compressive strength of geopolymer mortar increases up to 4% replacement level of NMK with fly ash. There is reduction in compressive strength with more than 4% of NMK replacement. The main factor that is responsible for increase in compressive strength of fly ash-based geopolymer mortar with different percentage levels of NMK is contribution of silica and alumina

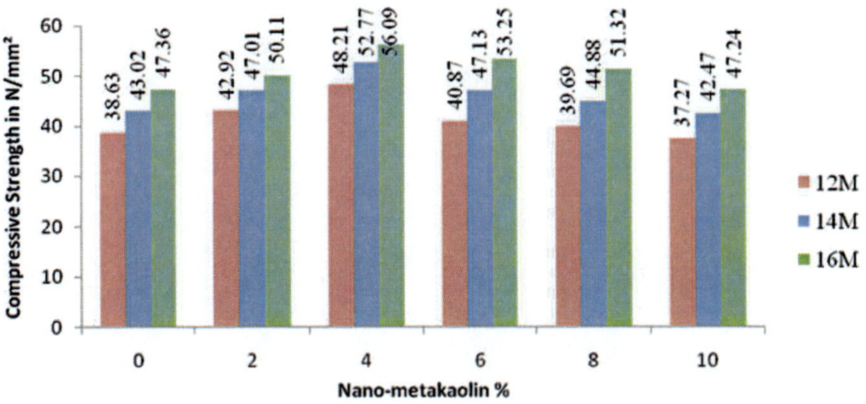

Fig. 1 Compressive strength of fly ash-based geopolymer mortar having different replacement levels of NMK

oxide of NMK in the geopolymerization reaction. However, the significant increase in compressive strength at 4% NMK is attributed to homogeneous dispersion of nano-metakaolin leading to the formation of nucleation material for geopolymer formation.

X-ray diffraction test analysis

The powder samples of GM 1, GM 7, GM 13 and GM 15 were analysed using XRD test. The XRD test results were represented in the form of peaks of different geopolymerization reaction products. The crystalline phases of all the samples were indexed using powder diffraction files (PDFs) on the basis of International Centre of Diffracted Data (ICDD). The XRD test results of GM 1, GM 7, GM 13 and GM 15 mixes are shown in Figs. 2, 3, 4, and 5.

The XRD test patterns of GM 1, GM 7, GM 13 and GM 15 indicate that quartz (SiO_2) with the mixed phase of mullite (Al6Si2O13) is the major phase present in all the samples and easily detected in the 2θ range of 26°–32°. The intensity of quartz peak at 26.6° in mixes GM 1, GM 7, GM 13 and GM 15 was 22121.46, 29671.47, 39,671.47 and 42,079.98 cps, respectively. Test results show that as the molar concentration of NaOH increases from 12M to 16M, intensity of quartz and mullite goes on increasing. From the test results, it is observed that the increase in molar concentration of NaOH and addition of 4% of NMK with 16M concentration of NaOH marginally affected the intensity of quartz peaks. The increase in molar concentration of NaOH from 12M to 16M resulted in enhancement in filler components intensities that consequently contributed to surge in compressive strength. Furthermore, with the addition of NMK, geopolymerization reaction increased significantly resulting in improvement of compressive strength further. Thus, microstructure study is in conformity with compressive strength test results.

Outcome from the experimental study elucidates that the compressive strength of fly ash-based geopolymer mortar is directly dependent on the molar concentration

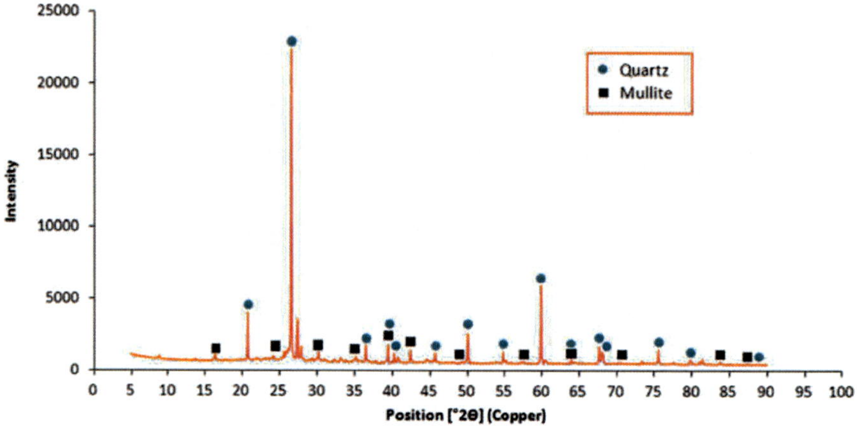

Fig. 2 XRD pattern of GM 1

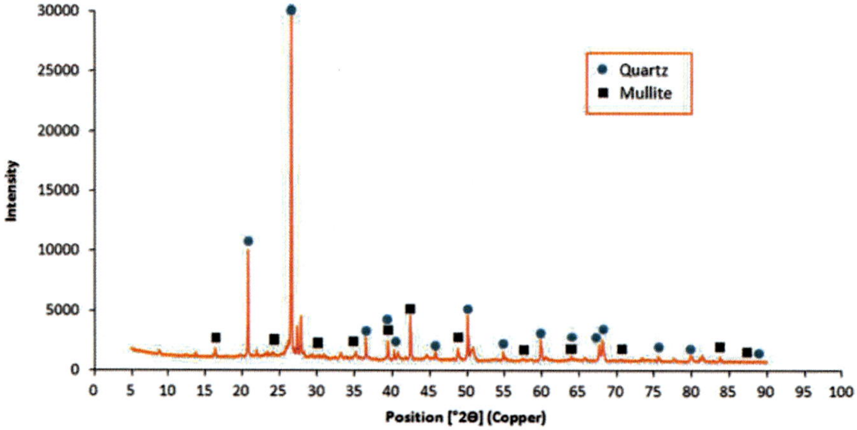

Fig. 3 XRD pattern of GM 7

of alkaline activator solution. The optimum compressive strength of 56.09 N/mm^2 is observed for 16M molarity of NaOH containing 4% proportion of NMK. The XRD test results also show that combination of fly ash-based geopolymer mortar having 4% nano-metakaolin replacement and 16M molar concentration of NaOH gives better results due to increase in the intensities of filler components. XRD test also depicts that with the addition of NMK, geopolymerization reaction increased significantly which in turn enhances the compressive strength.

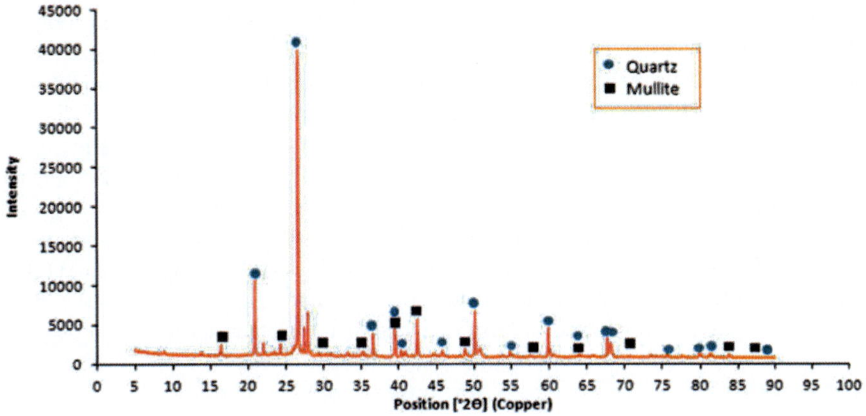

Fig. 4 XRD pattern of GM 13

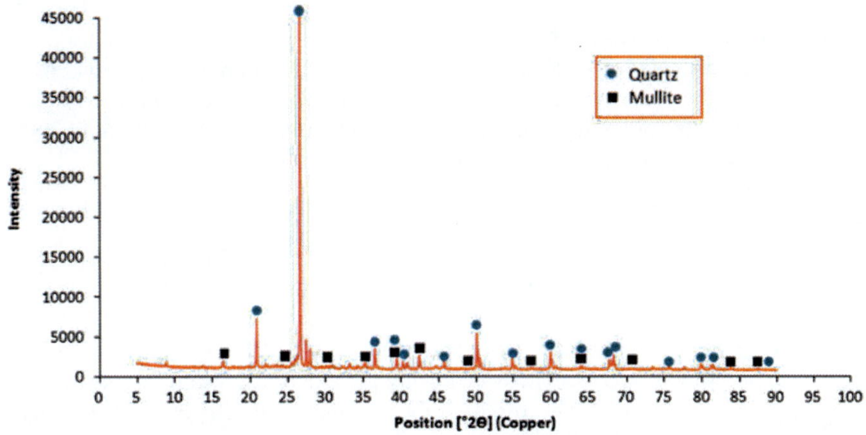

Fig. 5 XRD pattern of GM 15

2.1.2 Rice Husk Ash/Alkali-Activated Rice Husk Ash

The effects of nano-metakaolin on mechanical and microscopic properties of rice husk ash-based geopolymer were considered through compressive strength and scanning electron microscope (SEM). The nano-metakaolin was replaced with rice husk ash in geopolymer at a rate of 2, 4, 6, 8 and 10% by weight. Geopolymer is formed by alkali-activating the alumino-silicate material. The product obtained from geopolymerization reaction has tetrahedral Al^{3+} sites charge-balanced by alkali metal cations. The sodium silicate, sodium hydroxide, potassium hydroxide, potassium silicate and sodium carbonate may be used as alkaline activator. Based upon the chemical composition of source material, combination of two alkaline activators or one alkaline activator can be used for geopolymerisation. By geopolymerisation, wastes which are rich in alumino-silicate can be converted into cementitous material.

Rice husk ash, an agricultural waste, contains silica which exists dominatingly in an amorphous form. The amorphous silica present in rice husk ash has properties as cementitious binders to perform pozzolonic activity. The chemical composition of rice husk ash, in general, depends on the kind of paddy, geographical condition, climate and temperature used for burning the rice husk ash. Globally, an annual total production of rice paddy is 700 million ton approximately. About 22% of rice paddy is converted into rice husk, and 25% of rice husk is translated into rice husk ash.

Study on alkali-activated rice husk ash-based geopolymer mortar

The influence on microstructure and compressive strength of rice husk ash-based geopolymer by incorporating the nano-metakaolin was investigated. The microstructure was characterized by scanning electron microscope.

Alkali-activated rice husk ash

Rice husk ash, collected from Kissan rice mill in Faridkot district, Punjab, was used for synthesis of the geopolymer composite. The sodium hydroxide solution was used as alkaline activator for geopolymerization. The sodium hydroxide pellets with 98% purity were used to prepare the sodium hydroxide solution.

Nano-metakaoline

The nano-metakaolin was obtained from Sheetal Enterprises, Vadodara, Gujarat. It was found in the hydrate aluminium silicate (AI_20_3 $2Si0_2$ $2H_2O$) form.

Sample preparation

All the geopolymer mixes were activated using the alkaline activator of 14M molar concentration in the proportion of 0.60 to the rice husk ash. The ratio of rice husk ash to sand and water to solid was fixed at 1:3 and 0.35, respectively. Six mixes were prepared using nano-metakaolin of 0, 2, 4, 6, 8 and 10% of rice husk ash. The mix proportions are given in Table 3. At first, alkaline activator and rice husk ash (with/without) nano-metakaolin were mixed for 10 min. After that, appropriate amount of water and sand were added and mixed thoroughly.

The mixtures were poured into 70.6 mm × 70.6 mm × 70.6 mm cube moulds and subjected to thermal curing at 70 °C for 24 h. After de-moulding, the samples were cured at ambient temperature for testing.

Compressive strength testing

Compressive strength test was conducted on 70.6 × 70.6 × 70.6 mm^3 specimen at age of 3, 7, 14 and 28 days using universal testing machine, the values of which are given in Table 4.

The compressive strength improved with the 2, 4 and 6% replacement as compared to control sample. Development of compressive strength showed the good dispersion

Table 3 Mix proportions

Mix	Alkaline activator/binder ratio	Water/solid ratio	Nano-metakaolin (%)	Molar concentration of alkaline activator	Rice husk ash (g)	Nano-metakaolin (g)	Alkaline solution (g)	Fine aggregates (g)	Water (l)
N0	0.60	0.35	0	14M	2400	0	1440	7200	185.38
N2	0.60	0.35	2	14M	2352	48	1440	7200	185.38
N4	0.60	0.35	4	14M	2304	96	1440	7200	185.38
N6	0.60	0.35	6	14M	2256	144	1440	7200	185.38
N8	0.60	0.35	8	14M	2208	192	1440	7200	185.38
N10	0.60	0.35	10	14M	2160	240	1440	7200	185.38

Table 4 Compressive strength of geopolymer sample

Mix	NMK (%)	Compressive strength (N/mm^2)			
		3 days	7 days	14 days	28 days
N0	0	28.79	33.52	36.49	39.95
N2	2	30.20	34.52	37.90	40.76
N4	4	31.40	36.3	39.46	42.25
N6	6	36.80	40.67	45.67	46.67
N8	8	33.30	37.23	40.26	43.48
N10	10	29.13	33.55	35.60	38.73

of nano-metakaolin. Nano-metakaolin also escalated the geopolymerization reaction by contributing the silica and alumina content. The ultrafine particle of nano-metakaolin increased the density of the geopolymer. Furthermore, increasing the proportion of nano-metakaolin resulted in decrease of compressive strength. This reduction may be attributed to the poor dispersion and agglomeration due to high content of nano-metakaolin.

The compressive strength was increased when the rice husk ash was replaced by nano-metakaolin up to 6% in geopolymer mortar. Beyond that there was reduction in compressive strength. This may be attributed to the dispersion properties of nano-metakaolin which assist in formation of denser matrix. But higher dosage of nano-metakaolin prompts the formation of agglomerate and hinders the intact in the reactive product.

Scanning Electron Microscope (SEM) Analysis

To observe the microstructure, SEM analysis was conducted on remnants of specimen after the 28 days compressive strength testing. The analysis was performed using scanning electron microscope (Hitachi S-3400 N) device at an accelerating voltage of 15 kV with secondary electron detector. Firstly, the specimen were stubbed on aluminium stub using double-sided carbon disc that was sputter coated with gold plasma in ion sputter coater model Hitachi E-1010. Then, microscopic images were collected using in SEM (Hitachi S-3400 N) voltage in secondary electron imaging mode.

Figure 6a–f exhibits the SEM images of rice husk ash, rice husk ash-based geopolymer without and with 6% of nano-metakaolin. The geopolymer seemed to be less dense matrix as compared to geopolymer containing nano-metakaolin. The dense matrix of geopolymer with nano-metakaolin has closely packed particles that result in high compressive strength.

2.1.3 Sugarcane Bagasse Ash (SCBA)

According to Food and Agriculture Organization (FAO), India is the second largest producer of sugarcane in the world and produces 340 million tons of sugarcane every

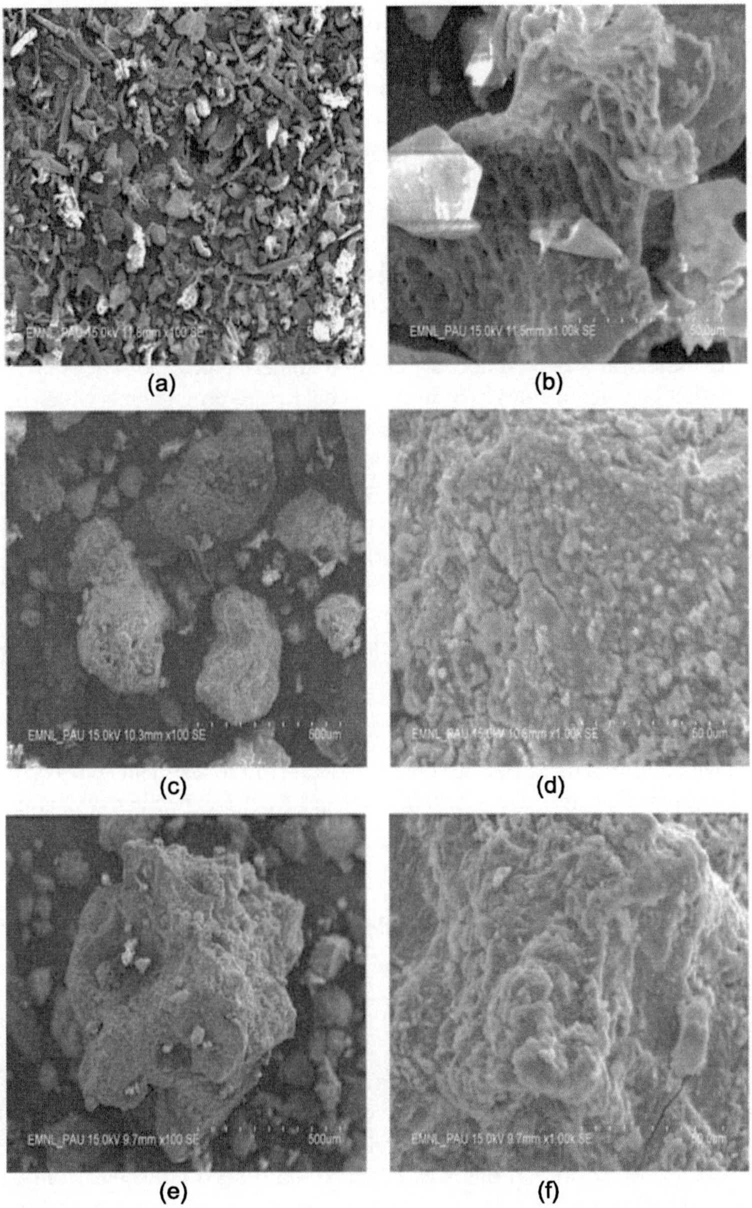

Fig. 6 SEM images of rice husk ash, rice husk ash-based geopolymer without and with 6% of nano-metakaolin

year. The fibrous matter that remains after crushing and juice extraction of sugarcane is known as sugarcane bagasse. This sugarcane bagasse comprises 50% cellulose, 25% hemicelluloses and 25% lignin. It is estimated that daily, in India, about 44,000 ton sugarcane bagasse ash is produced from cogeneration process from sugar mill industries leaving about 10 million ton of unutilized ash. Nowadays, this bagasse is also reutilized as a biomass fuel in boilers for vapour and power generation in sugar mills. This improved compressive strength depends on both physical and chemical effects of the SCBA. The physical effect (also called filler effect) relates to shape, size and texture of the SCBA particles, and the chemical effects relate to the ability of SCBA to participate in the pozzolanic reaction with calcium hydroxide by providing reactive siliceous compounds.

Among industrial wastes, a non-combustible material, coal bottom ash (CBA) is produced after burning of coals in furnaces of coal fired thermal plants which is either disposed off in an open area or mixed with water and pumped into water bodies. Both direct disposal and mixing of CBA with water are a source of pollution. CBA is well graded, and majority of its grains sizes are similar to grains sizes of river sand. These physical properties of CBA augment in partial replacements to fine aggregates in production of concrete.

Study of sugarcane baggase ash

Mixes of concrete were prepared at different replacement levels of SCBA at replacement levels of 5, 10, 15 and 20% vis-a-vis CBA at replacement levels of 10, 20, 30 and 40%. Test results demonstrate that workability of concrete decreases with addition of SCBA in concrete; however, there was an increase in compressive strength of concrete with addition of SCBA. An optimum value of replacement levels of cement and fine aggregates, respectively, with SCBA and CBA has been arrived at in order to make sustainable concrete.

Cement

Ordinary Portland Cement (OPC) of Grade 43 was used in this experimental investigation. The physical properties of OPC determined conformed to the requirements of BIS specifications.

Aggregates

The coarse aggregates were collected from Pathankot quarry. Specific gravity, water absorption and fineness modulus of coarse aggregates were experimentally determined as 2.65, 0.61 and 6.57%, respectively. A mixture of two sizes of 10 mm and 20 mm size in 50:50 proportions was taken. Natural sand collected from Chakki River (Pathankot) was used as fine aggregates. The specific gravity, water absorption and fineness modulus of fine aggregates were 2.71, 1.21 and 2.67, respectively.

Sugarcane bagasse ash

When this bagasse is burned under controlled temperature, ash is formed which contains high levels of SiO_2 and Al_2O_3, enabling its use as a supplementary cementitious material (SCM). The use of SCBA as SCM not only reduces the production of

Table 5 Chemical properties of SCBA

S. no.	Chemical component	% of Chemical component
1.	SiO_2	78.34%
2.	Fe_2O_3	3.61%
3.	Al_2O_3	8.55%
4.	CaO	2.15%
5.	Na_2O	0.12%
6.	K_2O	3.46%
7.	Ignition loss	0.42%

cement which is responsible for high energy consumption and carbon emission but also can improve the compressive strength of cement-based materials like concrete and mortar. The collection of the ash was carried out during the boiler cleaning operation. The ash was ground before it was used as a cement replacement material. The chemical properties of SCBA are given in Table 5.

Coal bottom ash

The CBA obtained from Guru Hargobind Thermal Plant situated at Lehra Mohabbat (Bathinda District) was used in this study. The physical properties of CBA are given in Table 6.

Mix design of concrete by BIS recommendations

Concrete mix for M20 grade of concrete was designed, and proportion of different materials obtained is as shown in Table 7.

Preparation of trial mixes

Based on the concrete mix design, four trials mixes as shown in Table 8 were prepared. Two trials mixes were prepared with water–cement ratio of 0.55, and other two mixes were prepared with water–cement ratio of 0.50. Nine cubes were cast for each mix and were tested at 3, 7 and 28 days.

The mix M_{R2} was chosen as the control mix because its average cube strength was very close to the target mean strength of concrete among all mixes and this mix

Table 6 Physical properties of CBA

Colour	Grayish or shiny black
Particles shape and texture	Spherical, irregular and porous
Specific gravity	1.78
Water absorption	9.64%

Table 7 Proportion of different materials

Water	Cement	Fine aggregates	Coarse aggregates
186 L	338.18 kg	747.23 kg	1142.87 kg
0.55	1	2.21	3.38

Table 8 Quantities per cubic metre for trial mixes

Mix no.	Water–cement ratio	Water (L)	Cement (kg)	Sand (kg)	Coarse aggregates (kg)	Average cube strength at 7 days (N/mm^2)	Average cube strength at 14 days (N/mm^2)	Average cube strength at 28 days (N/mm^2)	Slump (mm)
M_{R1}	0.55	186	338.18	747.23	1142.87	18.01	23.65	28.40	35
M_{R2}	**0.55**	**197.16**	**358.47**	**728.30**	**1113.77**	**17.72**	**23.57**	**27.73**	**110**
M_{R3}	0.50	186	372.00	716.74	1143.52	19.72	25.40	31.90	28
M_{R4}	0.50	197.16	394.32	698.20	1113.95	18.23	23.74	29.85	86

also had good workability characteristics. Based on mix M_{R2}, the mix proportions of other mixes calculated are as shown in Table 9.

Workability of concrete

Workability of each mix measured in terms of slump is given in Fig. 7. From results, it was observed that the workability of concrete mixes decreased with the increase in content of SCBA and CBA. The reason for decrease in workability is due to the porous structure of CBA's particles which cause the higher water absorption. In addition, rough and angular shape of SCBA's particles increased the interparticle friction which further decreased the slump values of concrete.

Table 9 Mix proportions of different concrete mixes

Mix	SCBA (%)	CBA (%)	Cement (Kg/m^3)	SCBA (Kg/m^3)	Fine aggregates (Kg/m^3)	CBA (Kg/m^3)	Coarse aggregates (Kg/m^3)	Water (L/m^3)
D1	0	0	358.47	0	728.20	0	1113.77	197.16
D2	5	0	340.55	17.92	728.20	0	1113.77	197.16
D3	10	0	322.62	35.85	728.20	0	1113.77	197.16
D4	15	0	304.70	53.77	728.20	0	1113.77	197.16
D5	20	0	286.78	71.69	728.20	0	1113.77	197.16
D6	0	10	358.47	0	655.38	72.82	1113.77	197.16
D7	5	10	340.55	17.92	655.38	72.82	1113.77	197.16
D8	10	10	322.62	35.85	655.38	72.82	1113.77	197.16
D9	15	10	304.70	53.77	655.38	72.82	1113.77	197.16
D10	20	10	286.78	71.69	655.38	72.82	1113.77	197.16
D11	0	20	358.47	0	582.56	145.64	1113.77	197.16
D12	5	20	340.55	17.92	582.56	145.64	1113.77	197.16
D13	10	20	322.62	35.85	582.56	145.64	1113.77	197.16
D14	15	20	304.70	53.77	582.56	145.64	1113.77	197.16
D15	20	20	286.78	71.69	582.56	145.64	1113.77	197.16
D16	0	30	358.47	0	509.74	218.46	1113.77	197.16
D17	5	30	340.55	17.92	509.74	218.46	1113.77	197.16
D18	10	30	322.62	35.85	509.74	218.46	1113.77	197.16
D19	15	30	304.70	53.77	509.74	218.46	1113.77	197.16
D20	20	30	286.78	71.69	509.74	218.46	1113.77	197.16
D21	0	40	358.47	0	436.92	291.28	1113.77	197.16
D22	5	40	340.55	17.92	436.92	291.28	1113.77	197.16
D23	10	40	322.62	35.85	436.92	291.28	1113.77	197.16
D24	15	40	304.70	53.77	436.92	291.28	1113.77	197.16
D25	20	40	286.78	71.69	436.92	291.28	1113.77	197.16

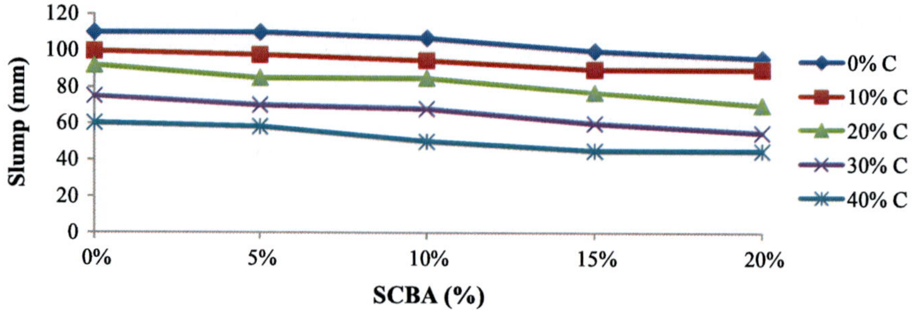

Fig. 7 Slump values of concrete with different replacement levels of SCBA and CBA

Compressive strength of concrete

Twenty-five concrete mixes were prepared each with 0.55 w/c ratio by replacing the cement with SCBA (0–20% @ increment of 5%) and fine aggregates with coal bottom ash (0–40% @ increment of 10%). To investigate the effect of SCBA and CBA on compressive strength, 225 cubes of 150 mm × 150 mm × 150 mm were prepared by varying percentage of SCBA and CBA.

The compressive strength of all concrete mixes was measured at the age of 14, 28 and 60 days. The effect of both waste materials on compressive strength at curing ages of 14, 28 and 60 days is illustrated by Figs. 8, 9, 10, 11, and 12. It is evident that with increasing the SCBA content (up to 15%), the improvement in compressive strength of concrete mixes is continuous. It was also observed that the gain in compressive strength of SCBA concrete mixes at curing age of 60 days was more than that of 14 and 28 days of curing. After 60 days of curing, compressive strength of SCBA concrete mixes exceeded by 4.3, 6.5 and 2.7% while decreased by 0.5% as compared to control concrete mix when cement was replaced by 5, 10, 15 and 20%, respectively.

On the other hand, there was a significant loss in compressive strength of concrete when fine aggregates were replaced with CBA. The incorporation of bottom ash produced relatively low strength concrete as compared to reference concrete due to its

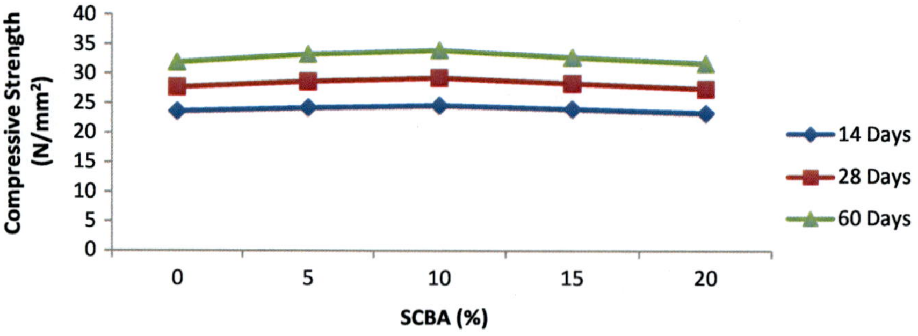

Fig. 8 Compressive strength of concrete with different replacement levels of cement with SCBA for 0% CBA

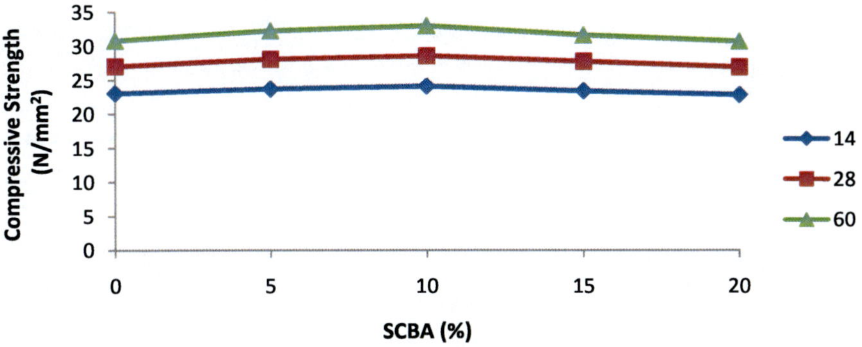

Fig. 9 Compressive strength of concrete with different replacement levels of cement with SCBA for 10% CBA

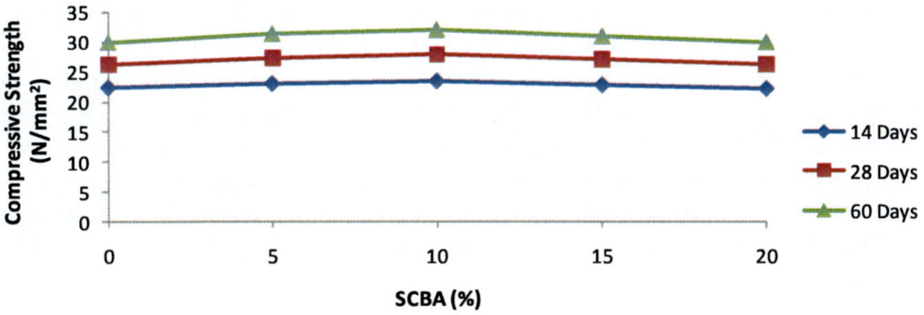

Fig. 10 Compressive strength of concrete with different replacement levels of cement with SCBA for 20% CBA

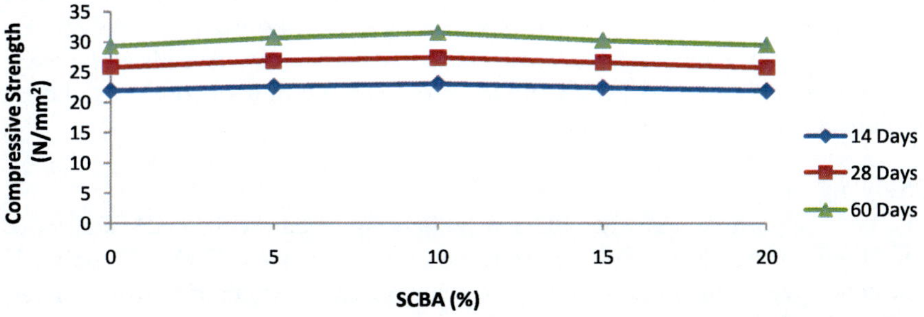

Fig. 11 Compressive strength of concrete with different replacement levels of cement with SCBA for 30% CBA

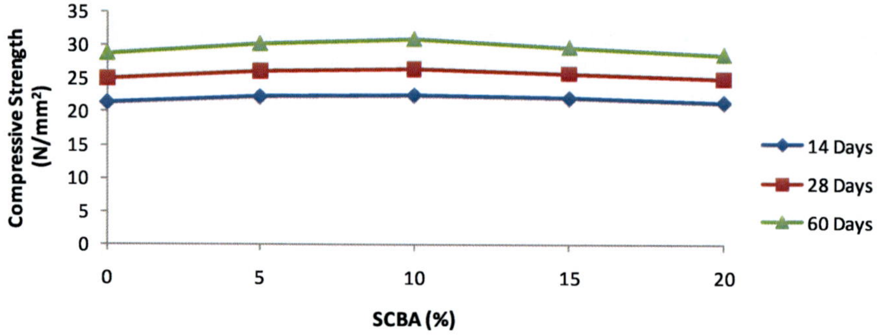

Fig. 12 Compressive strength of concrete with different replacement levels of cement with SCBA for 40% CBA

porous structure and higher water absorption. The compressive strength decreases at every replacement level of CBA with fine aggregates. Though there was no improvement in strength at any curing period, yet, it can be seen that the loss in compressive strength was minor up to 10% of CBA. Beyond 10%, there was large reduction in compressive strength. From the above, it can be concluded that as combination, replacing cement with SCBA up to 10% and fine aggregates with CBA up to 10% gives better result in strength characteristics as compared to reference concrete, and so, the combination of 10% SCBA and 10% CBA is recommended.

2.1.4 Corn Cob Ash

Corn cob is available in abundant in northern region of our country. Generally, burning of corn cobs in fire is the easiest and economical way to dispose of corn cobs adopted by the farmers of Punjab state that produces lot of carbon dioxide emission causing global warming as well as health problems also. The burning of corn cobs in open air pollutes the land, air and water.

Study on corn cob ash

The corn cob ash (CCA) was obtained by burning of corn cobs under controlled conditions for 6 h in muffle furnace at temperature of about 700 °C. The study was conducted to find the workability and compressive strength of cement concrete containing CCA at different replacement levels (0, 5, 10, 15, 20%).

The scanning electron microscope (SEM) test was conducted on corn cob ash in the Electron Microscopy and Nano Science lab of PAU Ludhiana. The SEM test was conducted at magnification of ×500SE, ×1000SE and ×1500SE. The bean-shaped

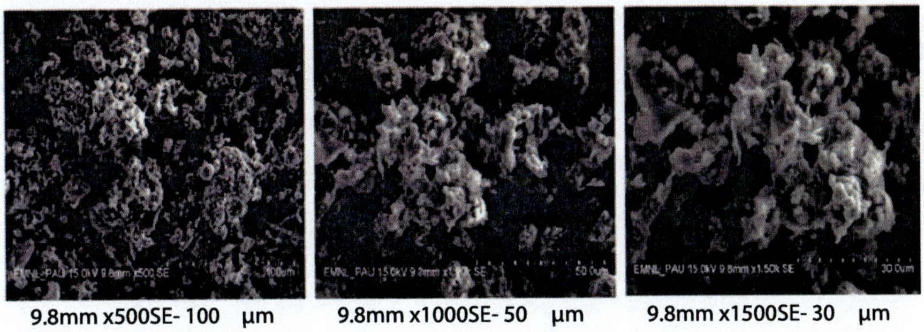

Fig. 13 SEM images of corn cob ash

grain is about 100 μm, 50 μm and 30 μm at magnification of ×500SE, ×1000SE and ×1500SE, respectively. The SEM images of corn cob ash are shown in Fig. 13

Compressive strength of concrete

Concrete cubes were tested for determining compressive strength. The test was carried out at 7, 28, 60 and 90 days of curing. Figure 14 shows the graphical representation of average compressive strength results.

It is noted that the average strength gain at 7 days curing is 69.09% when compared to 28 days average compressive strength of five mixes. The compressive strength is considered 100% at 28 days curing. The average strength calculated of all five mixes at 60 days of curing age and at 90 days of curing age is 104.79% and 107.40%, respectively. It is clearly shown that the strength increases with increase in curing age until completion of all calcium silicate hydrate reaction.

The increase in strength for the mix containing CCA replacement as 20% was noted as 10.03% and 15.16% at curing age of 60 days and 90 days, respectively, when compared to 28 days curing age. That means with the increase in CCA content, the later age strength increases as it takes more time to produce calcium silicate hydrate

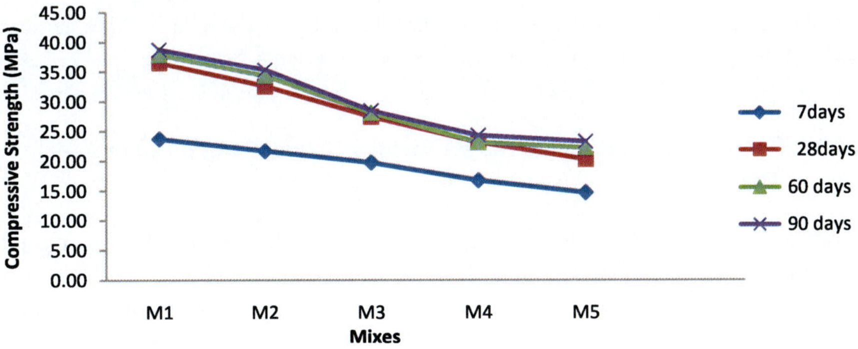

Fig. 14 Average compressive strength of the mixes

gel. This is due to the reason that increase in silica content present in CCA will slow down the rate of reaction. Hence, we can say that with increase in CCA content, the later age strength increases.

With reduction in cement content to 20%, the compressive strength reduces about 40%. Similarly, at 15% replacement, the reduction in strength was noted as 37.38% at curing age of 90 days. So, it is not recommended to replace cement beyond 10% with CCA.

2.2 *Slags*

2.2.1 Copper Slag

Copper slag is one of the materials that is considered as a waste material which could have a promising future in cement industry as partial substitute of cement. It is a by-product obtained during the smelting and refining of copper. To produce every ton of copper, approximately 2.2–3 tons copper slag is generated, and approximately 24.6 million tons of slag is generated from the world copper industry.

Copper slag has a specific gravity varying from 3.40–3.70, which is higher than that for OPC (3.15) which may result in production of dense concrete. Copper slag usually has a good content of silicon dioxide (SiO_2), and hence, it exhibits pozzolanic properties. Due to its pozzolanic properties, it reacts with the calcium hydroxide (CH), which is produced during cement hydration. Silicon dioxide combines with CH to produce additional cementing compound calcium silicate hydrate (C–S–H), which is responsible for holding concrete together.

Copper slag can be used as an iron adjustment material in cement clinker production. Researchers have also used the tailings of copper slag to produce cement clinker. The performance of the cement using copper slag tailing was even better than that produced using traditional clay, limestone and mill scale. Various investigations have been conducted to determine the suitability of copper slag as a partial replacement of cement. The use of copper slag as mineral admixture in concrete seems feasible and shows better pozzolanic properties than fly ash. Copper slag up to 15% cement replacement is recommended by most of the research workers.

2.2.2 Granulated Phosphorus Slag/Activated Granulated Phosphorous Slag

Phosphorus slag is a by-product during the production of elemental phosphorus and is composed mainly of SiO_2 and CaO. The air-cooled phosphorus slag does not exhibit cementitious properties and can be crushed for use as ballast or aggregate. The glass content of granulated phosphorus slag may reach 98% due to the high viscosity of the molten slag. The replacement of gypsum with Na_2SO_4 can activate the potential activity of phosphorus slag and increase the early strength of Portland phosphorus slag cement very significantly. The latent cementitious properties of granulated phosphorus slag can be very effectively activated using NaOH or Na_2SiO_3. But, Na_2SiO_3 is more effective than NaOH at 28 and 90 days when 3% Na_2O (by mass of slag) is used. The other interesting fact is that the presence of a soluble phosphate does not affect the setting and strength development of alkali-granulated phosphorus slag cements. The addition of Na_3PO_4 does not affect the times of setting of the alkali-granulated phosphorus slag cements. Strength testing by researchers has indicated that addition of Na_3PO_4 increases the strength of alkali-granulated phosphorus slag cements, which may be explained as being due to the composite activation of Na_3SO_4 and NaOH or Na_2SiO_3.

2.2.3 Iron Blast Furnace Slag/Ground Granulated Blast Furnace Slag

Iron blast furnace slag is produced when iron ore is reduced by coke at about 1350–1550 °C in a blast furnace. Typically, about 220–370 kg of blast furnace slag are produced per metric ton of pig iron. The granulated slag is produced by quenching the liquid slag with large amount of water to produce sand-like granulates. Granulates normally contain more than 95% of glass. Normally, they are ground to fine powder, called ground granulated blast furnace slag (GGBFS).

GGBFS is used as cement replacement at levels from 20 to 80%, which is varied depending on the application. According to the Bureau of Indian Standards (BIS), the slag content in the Portland slag cement can be up to 65% in mass. Portland slag cement is obtained by mixing Portland cement clinker, gypsum and granulated slag, in suitable proportions and grinding the mixture to get a thorough and intimate mix between the constituents. In addition, it has low heat of hydration and is relatively better resistant to soils and water containing excessive amounts of sulphates of alkali metals, alumina and iron, as well as to acidic waters, and therefore, used for marine works.

2.3 Energy-Efficient Aspects/Materials Leading to Sustainable Construction

2.3.1 Compressed Earth Blocks

Compressed earth blocks (CEBs) are comprised of mixed dry subsoil, clay and waste aggregates, such as building rubble, compressed with machine press or hydraulic compactor at high pressures. CEBs are used in construction since they have a compressive strength that exceeds the requirements for regular cement blocks. Most importantly, they are sustainable since the earth blocks use only 8% cement, which helps to reduce carbon emissions. Because only soil slurry is used for bonding instead of cement mortar, the construction process is also faster, and low technology can be used to manufacture them. CEBs also have better insulation properties than regular concrete blocks. CEBs can withstand loads, rain and snow, and most importantly, they are biodegradable, i.e. the humans can break them down thus giving back soil to the earth.

2.3.2 Trombe Walls

Trombe walls can be installed to possibly heat the building. Trombe walls consist of external walls made out of glass or plastic glazing panes and internal walls with high heat capacity. In between these two walls, there is air that separates them. The system works when the external wall absorbs heat through UV waves during the day and transfers the hot air to the insides of the building through internal vents during the night. For a Trombe wall to work effectively, it must be built in a fashion to maximize the total heat gain from the sun along with minimization of losses that are likely to occur during cold times. The economics of Trombe wall is dependent upon many factors such as climatic conditions, cost of materials and manpower and type of construction. The adoption of Trombe walls results in high energy conservation and reduction of CO_2 emissions.

In a school at Ladakh, Trombe walls have been used as a passive solar design thin order to optimize thermal performance. These walls are perfect for places such as Ladakh which has 320 days of sunlight.

2.3.3 Bottle Bricks

Bottle bricks are plastic bottles stuffed with inorganic landfill trash (plastic bags, wrappers, etc.) until they become compressed just like bricks. They are used to build houses, school buildings and other structures. We have enormous saving of building embodied energy by using plastic bottles instead of bricks as well as reduction of CO_2 emission. Generally, the structures made up of bottle houses are bioclimatic in design. Reusing the plastic bottles as the building material can have substantial effects on saving the building embodied energy and making the structure sustainable.

2.3.4 Earth Bags

Woven polypropylene bags are filled with locally available inorganic material and moist subsoil, called earth bags. The bags are designed to have internal stability and resilience against earthquakes due to its properties of self-interlocking and consolidating moist fill inside. The bags are also connected by barbed wire and mesh for friction and tension resistance between bags. An education centre in Nepal built using earth bags just five days prior to the occurrence of Nepal earthquake (2015) was completely safe.

3 Salient Points/Conclusions

Today, numerous challenging and exciting buildings are being designed and constructed in our society. Examples of such are tall and slender buildings or structures with long spans. Most likely, this trend will further develop, and in parallel, new and increasing demands regarding long-term sustainability will be put on the built environment.

Environmental sustainable building technologies are gaining the attention of the construction industry. Factors for these sustainable technologies include appropriate site selection, use of environmental-friendly building materials, energy-efficient building practices, use of water-efficient technologies, use of renewable energy resources, etc.

Due to the concerted efforts of various Govt and non-government agencies, safe disposal and effective utilization of fly ash have gained importance. There is greater acceptance of fly ash products as the agencies are taking positive initiatives in creating awareness among the people. Even fly ash can be used in manufacturing of several products like extender/pigment in emulsion paint formulations, wood substitute in combination with organic fibre, lightweight aggregate, tiles, etc. These products are competitive to conventional material and provide several indirect benefits like reduced pollution, waste recycling, conservation of soil, etc.

It has been established that coal ashes are not only used as substitute materials in concrete, but also when used as additives, lower the slope of virgin compression curve, thereby reducing the consolidation settlement of the parent soil.

The utilization of industrial wastes like fly ash and rice husk ash not only finds application in cement/concrete, but they also improve the soil expansiveness. These are very good alternatives to other expensive techniques like use of geotextiles and stone columns, thereby reducing the cost of construction of roads.

For meeting the need of infrastructure, adoption of fast-track cost-effective steel construction and other materials will have to be resorted to. Steel concrete composite technology, pre-engineered building system and space frame structures are going to become popular in future.

Research and development of sustainable or environmental-conscious construction materials have been carried out in the recent years. It is established that concrete polymer composites are highly sustainable and promising construction materials.

The use of marble powder and coal ash as a partial or full replacement of fine aggregate in self-compacting concrete makes the concrete construction more sustainable and environmental friendly. Use of other waste materials such as blast furnace slag, red mud, waste glass, cinder, coconut husk, banana leaves, jute fibres, rubber from automobile tires is well documented by research.

As far as possible, locally available materials should be preferred so as to minimize the energy spent in transportation of the building materials. Energy consumed in transportation should be considered as total energy spent on transporting materials starting from the place of manufacturing. Effort should always be made to minimize the embodied energy which includes energy required to extract raw materials from nature, energy used to transport raw materials to the manufacturing unit and energy used in manufacturing activities to provide a finished product.

Building materials can be manufactured using waste materials or recycled materials. Use of recycled materials helps the environment and economy in several ways. Waste materials that would have landed in landfills after its useful life can be reprocessed for use in other products. To be professionally successful in the future, engineers need to develop thorough understanding of the possibilities and limitations of different materials and also possess knowledge with respect to building elements and structural load bearing systems. A time has come when the Institutes may start offering course in sustainable engineering especially for those persons who want to shape and construct the buildings of the future.

Reference

1. Singh M, Siddique R (2014) Strength properties and micro-structural properties of concrete containing coal bottom ash as partial replacement of fine aggregate. Constr Build Mater 50:246–56

Degradation and Decolourization of Methyl Orange Dye Using Fe-TiO_2 Hybrid Technology (Photo-Fenton and Photocatalysis) in Fixed-Mode

Lavneet Kumar, Ina Thakur, Anoop Verma, B. S. Bhatia, and Charanjit Kaur Mangat

1 Introduction

Azo dyes such as methyl orange (MO) have widely utilized in multiple commercial applications due to their chemical stability and flexibility. Generally, azo dyes are characterized by one or more azo bond (−N55N−) related to an aromatic structure. The degradation of such dyes is very difficult because of their stability in nature. Therefore, it creates pollution in the environment [1, 2]. In the past literature, a lot of different technologies were adopted for the decomposition of residual azo dyes such as membrane filtration [3], biodegradation [4], sorption [5], and electrochemical degradation [6]. However, these technologies generally require long resident time and frequently generate secondary pollutants. In recent years, the advanced oxidation process (AOP) such as photocatalysis [7], photo-Fenton [8] has been substantially used to decompose the organic pollutants. In an AOP, the organic pollutant oxidized via highly oxidizing radicals such as OH radicals, $O_2^{\bullet -}$, $HO_2\bullet$, etc. produced because of the occurrence of series of reactions generated as intermediate products. Titanium-based photocatalysis and the photo-Fenton process have drawn significant attention in past years [7, 8]. The mechanism of the studies mentioned above is well documented in the literature [9]. However, high doses H_2O_2 along with the formation of vast quantities of iron sludge really restricting the applications of photo-Fenton process at a large scale [10]. The implementation of TiO_2 photocatalysis is always favoured in the fixed form to avoid the additional step of post-treatment separation of catalyst.

L. Kumar (✉) · I. Thakur · A. Verma
School of Energy and Environment, Thapar Institute of Engineering and Technology, Patiala 147004, India
e-mail: lavneetdubey72@gmail.com

B. S. Bhatia · C. K. Mangat
RIMT University, Mandi Gobindgarh, Punjab, India

H. Singh et al. (eds.), *Sustainable Development Through Engineering Innovations*, Lecture Notes in Civil Engineering 113,
https://doi.org/10.1007/978-981-15-9554-7_5

Yet, there are many drawbacks such as increment in treat time, $e^{-}h^{+}$ recombination, mass transfer limitations, etc. that impede its commercial application [10].

With the mentioned implications, the proposed technology (a combination of photo-Fenton and photocatalysis in the fixed form) has a potential to overcome most of the problems faced by the technologies for the treatment of wastewater [11]. The main emphasis was given on the utilization of waste material such as foundry sand (FS) and fly ash (FA) as an iron source. FS and FA were mixed with clay (support material) to form spherical beads and coated with TiO_2, thus facilitates the effect of the dual process. The leaching of iron from the composite in the presence of H_2O_2 under acidic conditions leads to the photo-Fenton process, whereas the coating of TiO_2 in the presence of light caused surface excitation of an electron from balance band (VB) to the conduction band (CB) leads to photocatalysis. This technology eliminates the problem of electron-hole recombination in photocatalysis via reaction of electron present in the CB with the ferric ion leaching out of the composite beads leads to the formation of ferrous ion, thus starts photo-Fenton simultaneously. Moreover, the systematic leaching of iron in the system eliminates the formation of iron sludge. The utilization of waste material such as FS and FA also proves the cost-effectiveness of the technology.

In the present investigation, the degradation and decolourization of MO dye were demonstrated using Fe-TiO_2 composite beads in the batch reactor incorporating dual effect. In this study, three types of Fe-TiO_2 catalysts; FS + clay, FA + clay, and FS + FA + clay were made, and an increase in the efficiency and decrease in treatment time of MO was evaluated. Then, these composite beads were also checked and assessed for the durability study.

2 Chemicals and Reagent

Methyl orange azo dye was bought from the local dyeing shop in Patiala, Punjab, India. The rest of the other chemicals used is same as explained in our previous study [10].

3 Analytical Determination

The sample analysis was executed using UV–vis spectrophotometer, LABINDIA, model-T60U. The morphological analysis of different types of Fe-TiO_2 composite beads was analysed using SEM, JSM-6510LV, JEOL (Japan), while their elemental composition was confirmed through EDS, INCAX-act, Oxford Instruments (UK). The crystalline structure of iron and TiO_2 on the surface of the composite beads was confirmed by performing X-ray diffraction (XRD) using X'Pert pro diffractometer (PANalytical) with Cu Kα wavelength ($\lambda = 1.5405$) in the range of $2\theta = 20 - 80°$.

4 Composite Material for the Dual Effect Study

Dual-effect study of photo-Fenton and photocatalysis was investigated using three types of composite beads made up of FA + clay (1:2), FS + clay (1:2), and FA + FS + clay (1:1:2). FA and FS are the natural sources of iron, and the same has been confirmed by SEM/EDS analysis, as shown in Fig. 1a–c. These spherical beads were made manually and later baked for 2 h at 800 °C. Then, the beads were cooled at room temperature and cured for 48 h to attain the strength. The vernier calliper was used to measure the size of the beads. The diameter of the beads was measured from at least four different directions, and average was 15.20 mm. The surface of the composite beads was twice coated by the TiO_2 using the dip-coating method. After the surface coating of the beads, they were muffled at 350 for 1.5 h for the proper binding of TiO_2 on the surface of the beads. The SEM-EDS analysis confirmed the uniform coating of TiO_2. The EDS analysis confirmed the presence of TiO_2 and iron in different composite beads, as shown in Fig. 1a–c, thus confirmed the occurrence of dual process in the system.

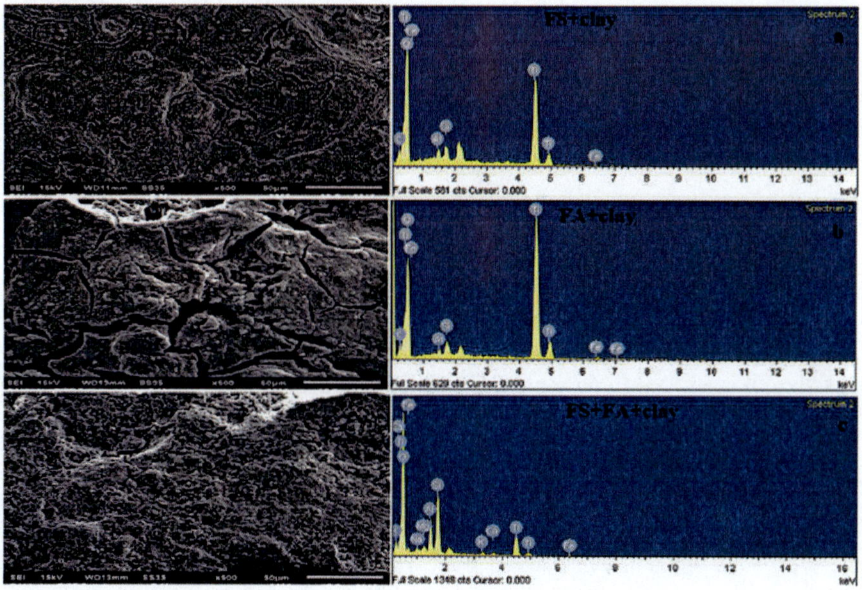

Fig. 1 SEM/EDS images of different composite beads

5 Experimental Analysis

The photocatalytic analysis of MO was performed under a laboratory-scale reactor using a 1000 ml glass reactor (diameter = 18 cm × height = 5 cm). Three different reactors were used, and the surface of each glass reactor was covered with the different composite beads. The stock solution was prepared using MO (50 mg L^{-1}) in 200 mL of double-distilled water. After that, 300 mg L^{-1} of H_2O_2 was added to the reactor, and the pH of the solution was maintained at 3.5. Air spargers having airflow of 3–3.5 L min^{-1} was kept under the reactors during the experiment to maintain the homogeneity of the system. Then, the reactors were placed under a Wooden UV chamber having UV tubes (36 W × 7, UV-A, Philips) fixed under the roof of the chamber. The average intensity provided to the system was 20 W m^{-2} and was measured using Eppley radiometer (model no. −33013). For the degradation and decolourization study, the samples were withdrawn from the reactor at a regular interval of time. Each experiment was repeated three times, and the average of the three was reported in the results.

6 Results and Discussion

6.1 *Adsorption and Photolytic Studies*

The blank experiment in a batch reactor was carried out using bare beads (without TiO_2 coating) in the presence of UV light (photolysis), which showed a small amount of degradation (15%) and decolourization (29%) of MO dye. In the adsorption studies (TiO_2 coated beads only), 8% and 12% of compound and colour removal were observed, respectively, as shown in Fig. 2. This corresponds to the development of the compound's monolayer over the catalyst surface. This caused hindrance to all the active sites and led to no more degradation and decolourization of MO further. When only H_2O_2 was added in the system, only 14 and 15% of degradation and

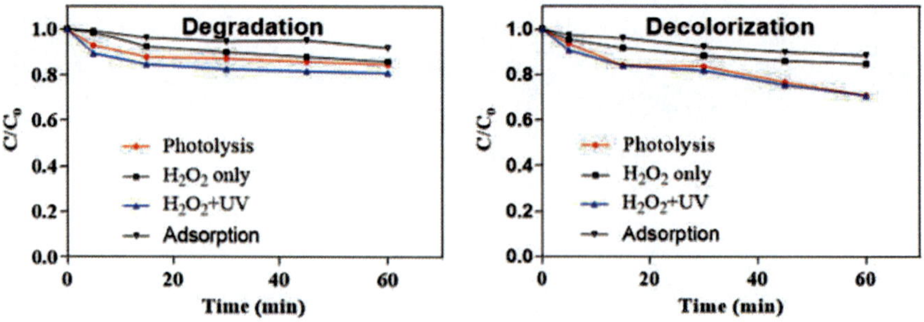

Fig. 2 Preliminary studies for the degradation and decolourization of MO under different conditions

decolourization of MO were observed, respectively. This might be due to the minimal amount of OH° production. When H_2O_2 was added in the reactor with UV light, the results showed a slight increase in the degradation (20%) and decolourization (21%) of MO after 60 min of treatment time, as shown in Fig. 2.

6.2 Dual-Effect Studies

To confirm the efficacy of hybrid effect using TiO_2-coated clay beads with FA and FS, efforts were made to study the degradation of MO at the acidic condition (acetate buffer). All experiments were performed under UV-A irradiations with 50 mg L^{-1} of MO, having an H_2O_2 concentration of 300 mg L^{-1}. Initially, photocatalysis was performed using FS/clay, FA/clay, and FS/FA/clay beads separately with TiO_2 coating, keeping the pH at neutral condition. The best result was obtained with FS/FA/clay beads showing 53% (0.0138 min^{-1}) and 57% (0.0261 min^{-1}) of degradation and decolourization of MO after 60 min of treatment time, as shown in Fig. 3a1–c2. The photo-Fenton process was also used to study the degradation and decolourization of composite beads (FS/clay, FA/clay, and FS/FA/clay) beads without TiO_2 coating. The results showed the maximum degradation (38%) having 0.0076 min^{-1} of reaction rate and decolourization (58%) with 0.0139 min^{-1} of reaction rate with FS/FA/clay beads, as shown in Fig. 3a1–c2.

For the dual process, TiO_2-coated clay beads are with FA and FS separately; the maximum degradation (58%) and decolourization (83%) rate were observed with FA/clay beads. But when FS and FA together mixed with clay to form composite beads (FS/FA/clay) and coated with TiO_2, it showed the highest increase in the degradation (66%) and decolourization (89%) with increase in reaction rate in 60 min of treatment time, as shown in Fig. 3a1–c2. Actually, at lower pH (3.5), leaching of the iron from the beads containing FA and FS takes place; hence, the support of simultaneous effect of the photo-Fenton process along with photocatalysis leads to the production of more hydroxyl radicals.

Moreover, during the dual process, the rate constants observed were in the following order: FS/FA/clay/TiO_2 beads (0.0187 min^{-1}) > FA/clay/TiO_2 beads (0.0143 min^{-1}) > FS/clay/TiO_2 beads (0.0130 min^{-1}) as shown in Fig. 3a1–c2. In FA/FS/clay/TiO_2 composite beads, iron content is high as compared to the other two composite beads, which lead to an increase in the production of oxidative species in the photo-Fenton reaction. This technology eliminates the problem of electron-hole recombination in photocatalysis via reaction of electron present in the CB with the ferric ion leaching out of the composite beads leads to the formation of ferrous ion, thus starts photo-Fenton simultaneously. The results obtained from this study might help in designing the fixed-bed pilot-scale reactors (using waste FS, FA) incorporating a dual effect for the treatment of industrial waste.

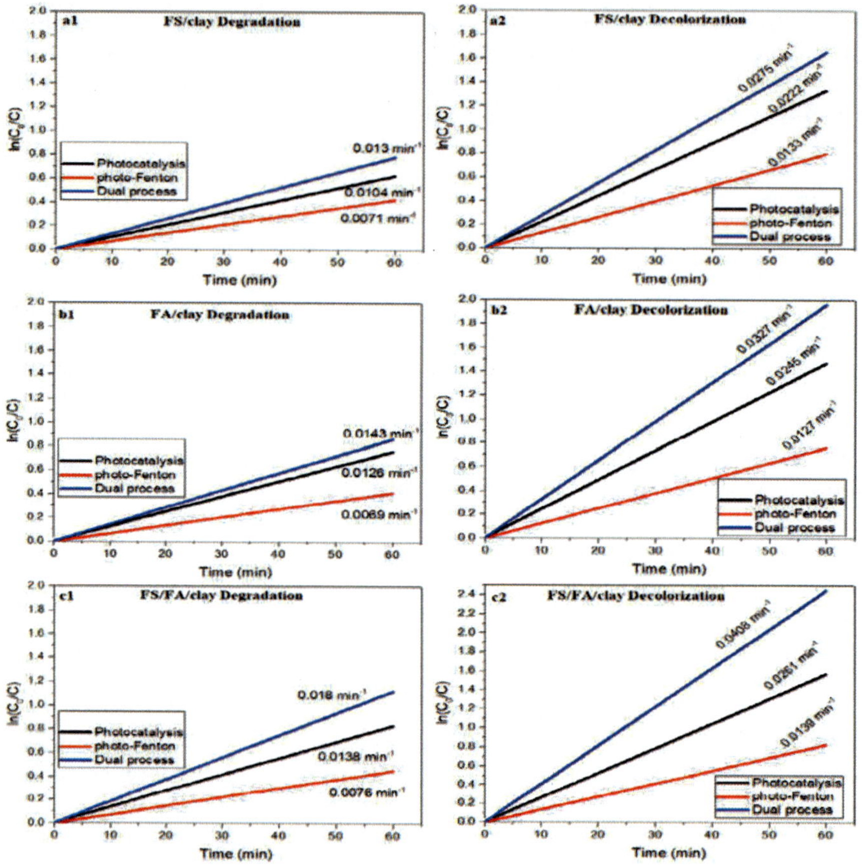

Fig. 3 Degradation and decolourization rate constant using FS/clay, FA/clay, and FS/FA/clay with TiO_2 coating

6.3 *Durability Studies*

The practical viability of the dual process in a commercial application depends on the durability and stability of the composite material. In this study, composite beads (FS/FA/Clay) coated with TiO_2 were recycled 50 times. The clay used in this study showed excellent binding properties for the immobilization of catalysts during the degradation and decolourization of MO. Only 4% (degradation) and 3% (decolourization) decrease in the activity of the catalyst were observed even after 50 times of recycling as shown in Fig. 4a. This minimal decrease can be attributed to the scrapping of the catalyst during washing after each experiment. The TiO_2 was perfectly intact over the surface of the composite beads after 50 cycles, and the same can be confirmed by the SEM/EDS image of recycled composite beads, as shown in Fig. 4b. The peaks of iron and titanium oxide in EDS data indicate the presence of

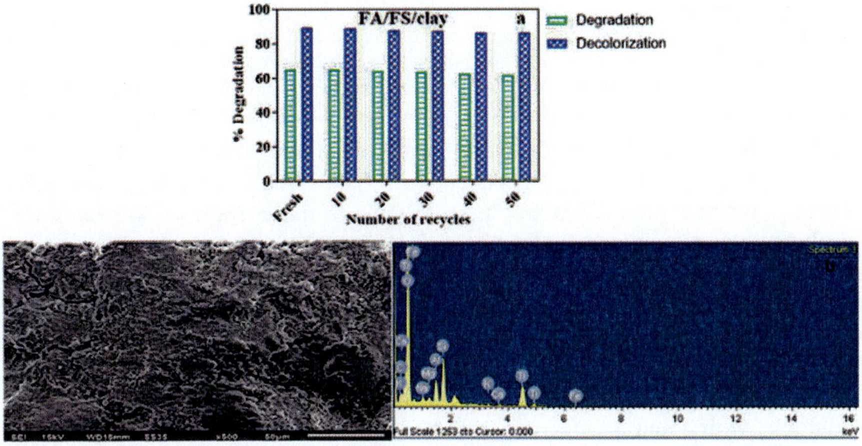

Fig. 4 **a** Durability studies of MO, **b** SEM/EDS images of recycled FS/FA/clay coated with TiO_2

Fe with TiO_2 in the recycled composite beads. Hence, it demonstrates the potency of composite beads for dual effect studies. The XRD analysis of freshly coated and recycled composite beads also confirm the intactness of TiO_2 and iron in the system, as shown in Fig. 5. No change in the crystalline planes of the coated catalyst was observed even after 50 recycles. The anatase (A) and rutile (R) phase of the TiO_2 catalyst can be seen in Fig. 5. The presence of Ti-O-Ti and Fe-O-Ti bonds in the beads even after recycling as shown in our previous FTIR study [10] also confirms the potency of the technique, thus proves the significant possibility of this technology for its commercial application.

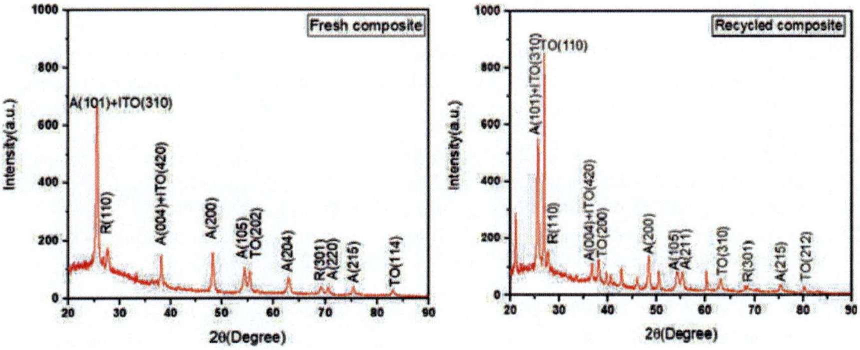

Fig. 5 XRD pattern of FS/FA/clay coated with TiO_2

7 Conclusion

The current study investigates the efficacy and stability of Fe-TiO_2 composite for the degradation and decolourization of MO dye in double-distilled water, pH 3.5 under UV-A irradiation. The degradation and decolourization of MO dye were found to be 66% and 89%, respectively, with FS/FA/clay composite with TiO_2 coating, UV-A radiation (25 W/m^{-2}), and 60 min of treatment time. The considerable increase in first-order rate constant in dual process (degradation = 0.187 min^{-1} and decolourization = 0.0408 min^{-1}) as compared to photo-Fenton (degradation = 0.0076 min^{-1} and decolourization = 0.0139 min^{-1}) and photocatalysis (degradation = 0.0138 min^{-1} and decolourization = 0.0261 min^{-1}) confirms the feasibility of the dual process. Hence, this in situ fixed-bed hybrid technique offers a viable technical solution for treating contaminated water and waste water. The beads were found to be durable, with 4 and 3% decrement in the activity of composite for degradation and decolourization efficiency.

Acknowledgements Authors are thankful to sophisticated analytical instrumentation facility, Punjab University, Chandigarh, India, for extending their facilities for the characterization of samples.

References

1. Li Y, Li X, Li J, Yin J (2006) Photocatalytic degradation of methyl orange by TiO_2-coated activated carbon and kinetic study. Water Res 40:1119–1126. https://doi.org/10.1016/j.watres.2005.12.042
2. Youssef NA, Shaban SA, Ibrahim FA, Mahmoud AS (2016) Degradation of methyl orange using Fenton catalytic reaction. Egypt J Pet 25:317–321. https://doi.org/10.1016/j.ejpe.2015.07.017
3. Cui P, Chen Y, Chen G (2011) Degradation of low concentration methyl orange in aqueous solution through sonophotocatalysis with simultaneous recovery of photocatalyst by ceramic membrane microfiltration. Ind Eng Chem Res 50:3947–3954. https://doi.org/10.1021/ie100832q
4. Thao TP, Kao HC, Juang RS, Lan JCW (2013) Kinetic characteristics of biodegradation of methyl orange by *Pseudomonas putida* mt2 in suspended and immobilized cell systems. J Taiwan Inst Chem Eng 44:780–785. https://doi.org/10.1016/j.jtice.2013.01.015
5. Allouche F-N, Yassaa N, Lounici H (2015) Sorption of methyl orange from aqueous solution on Chitosan biomass. Procedia Earth Planet Sci 15:596–601. https://doi.org/10.1016/j.proeps.2015.08.109
6. Li SH, Zhao Y, Chu J, Li WW, Yu HQ, Liu G (2013) Electrochemical degradation of methyl orange on Pt-Bi/C nanostructured electrode by a square-wave potential method. Electrochim Acta 92:93–101. https://doi.org/10.1016/j.electacta.2013.01.012
7. Nguyen CH, Fu CC, Juang RS (2018) Degradation of methylene blue and methyl orange by palladium-doped TiO_2 photocatalysis for water reuse: efficiency and degradation pathways. J Clean Prod 202:413–427. https://doi.org/10.1016/j.jclepro.2018.08.110
8. Yusuf AA, Rehman F (2018) Photo-oxidative degradation of methyl orange in aqueous medium by photo-Fenton reaction

9. Thakur I, Örmeci B, Verma A (2020) Inactivation of E. coli in water employing Fe-TiO_2 composite incorporating in-situ dual process of photocatalysis and photo-Fenton in fixed-mode. J Water Process Eng 33:101085. https://doi.org/10.1016/j.jwpe.2019.101085
10. Bansal P, Verma A (2018) In-situ dual effect studies using novel Fe-TiO_2 composite for the pilot-plant degradation of pentoxifylline. Chem Eng J 332:682–694. https://doi.org/10.1016/j.cej.2017.09.121
11. Talwar S, Sangal VK, Verma AK (2019) In-situ dual effect of novel Fe-TiO_2 composite for the degradation of phenazone. Sep Purif Technol 211:391–400. https://doi.org/10.1016/j.seppur.2018.10.007

Landslide Hazard Assessment Along Lahru to Chamba, Himachal Pradesh, India, Using Certainty Factor Approach

Desh Deepak Pandey, R. S. Banshtu, and Kanwarpreet Singh

1 Introduction

Loss of life and property due to landslides triggered by natural and anthropogenic activities is increasing environmental risk worldwide. Taxonomy showing types of landslides based on the slope forming materials like rock debris, and earth depicting various types of mass movements like fall, topple, slide, flow, etc. has been described by various researchers. The stability of the slope is generally controlled by the shear strength of the slope forming materials and gradient of the slope. The expert's knowledge, field practice, and landslide history/Inventory are helpful for selecting landslide causative factors (LCF's) of the study area. The influence of rainfall has been analyzed in terms of thresholds for triggering landslides by various researchers [1, 4, 5, 7, 14, 15, 19, 23, 27]. A few of the Indian states are affected due to landslide activities that get reactivated during every monsoon cum rainfall season. The state of Himachal Pradesh is occupied by Himalayan mountains in many districts like Chamba, Kangra, Mandi, Kullu, Bilaspur, and Shimla where there is always a danger of rockfall, debris flows, and slides along the road corridors because of a natural climatic condition like rainfall, seismicity, geology, etc. [8] and external human interference like road cut slopes, toe cutting due to river flow, deforestation,

D. D. Pandey (✉) · R. S. Banshtu
Civil Engineering Department, National Institute of Technology, Hamirpur, Himachal Pradesh, India
e-mail: deepak2171985@gmail.com

R. S. Banshtu
e-mail: banshtu.rajeshwar@gmail.com

K. Singh
Civil Engineering Department, Chandigarh University, Mohali, Punjab, India
e-mail: kanwarpreet.e9750@cumail.in

H. Singh et al. (eds.), *Sustainable Development Through Engineering Innovations*, Lecture Notes in Civil Engineering 113,
https://doi.org/10.1007/978-981-15-9554-7_6

etc. Some of the authors suggested geology, soil, landuse landcover (LULC), topography as major landslide contributing factors [1, 3, 12, 23, 26, 27]. The slope failure activities affect the economy of the state as well as the country which sometimes leads to life loss. The delineation of the LHZ maps based on the LCF's is helpful for suggesting precautionary control measures in the areas highly prone to landslides.

The mountainous terrain of district Chamba in Himachal Pradesh is suffering from landslide disaster mainly in the region where the slope gradient is greater than 40°, which is harming lives and million dollars property across the region. An earthquake-triggered landslides caused the death of around 500 people in Chamba in the year 1905. The constructional activities of infrastructural megaprojects are also disturbing hilly terrains in district Chamba for the last 10 years. Hence, the landslide hazard assessment is essential in the region for future development and construction. Therefore, a vulnerable road stretch in the district Chamba, Himachal Pradesh has been selected for landslide hazard zonation (LHZ) mapping from Lahru to Chamba district headquarters along (NH-154A) in the present study.

Previous literature is putting minor reflection on LHZ assessment in the present study area using various techniques that are applicable in ArcGIS software. Numerous methods like the analytical hierarchy process (AHP) by Saaty [24], landslide hazard evaluation factor (LHEF) known as [13] were utilized for analyzing slope failure activities from 1970 to 1990 [2] which are widely applied by various researchers. The statistical and deterministic approaches have been employed by numerous researchers for LHZ mapping using ArcGIS [10, 16, 20]. Distribution of landslides and LCF's have been correlated with historical events by using statistical methodologies recently by various researchers [22, 23, 27]. The certainty factor approach was adopted for LHZ modeling in the present study to categories the area into various hazard zones varying from very low hazard to very high hazard zone.

2 Study Area

Study zone covers 27,355.47 m^2 region encompassed between 80.84 km NH-154A from Lahru to Chamba by means of Banikhet (76° 10′ E, 32° 35′ N) and 54.90 km along the state highway 28 from Lahru to Chamba by means of Jot (75° 51′ E, 32° 26′ N). The region has a dendritic drainage design and can be effortlessly distinguished on utilizing Survey of India 1986 toposheets (43P/14, 43P/15, 52D/2, and 52D/3) at the size of 1:50,000. The examination territory has high elevation extending from 656 feet close to Lahru to 2757 feet in Dalhousie. Greatest temperature in summer is 30 °C though the least is 14 °C while in winter season temperature fluctuates between 2 and 11 °C. Precipitation of 186.42 mm is recorded during the year 2018 in this locale (Indian meteorological division). Atmosphere conditions and physiography are the main drivers of slant disappointments in the investigation zone. Mountain inclines are to a great extent influenced by human impedances for advancement and impromptu development in spite of the fact that precipitation and seismic exercises additionally assume fundamental jobs for activating landslides. Street development started a colossal translational slide because of the development of road close toChowari [25].

3 Landslide Inventory

The landslide inventory map gives data about area, timing, and sorts of landslides in a specific territory. It is set up from the authentic information of individual landslide location, satellite pictures, field studies, and elevated photos [18]. In the current research, 242 landslides were distinguished by utilizing chronicled records of Border Road Organization, National Highway Authority of India (NHAI), Public Works Department (PWD), LISS IV imagery, and Google Earth (Fig. 1). With the assistance of handheld GPS, field surveillance of the landslides was affirmed and refreshed. The reason for most extreme quantities of landslides is associated with thrusts and

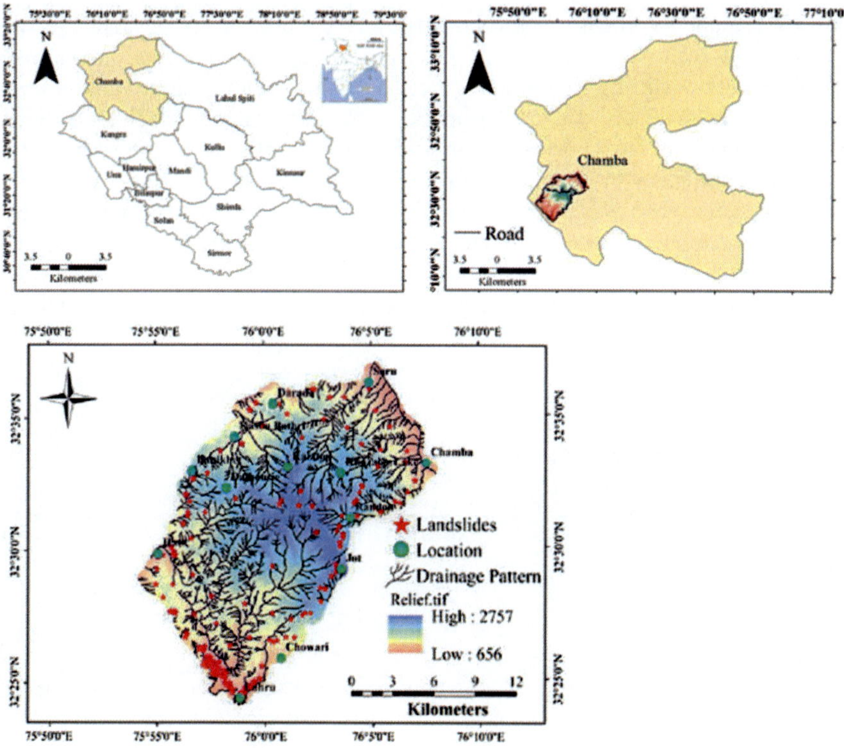

Fig. 1 Study area Lahru to Chamba, Chamba District, and Himachal Pradesh, India

lineaments. Rockfall close to the Chowari town was activated by the seismic exercises. Banikhet landslide occurred because of overwhelming precipitation which is an activating element for 75% of the landslides in the zone.

4 Methodology

Several researchers have used the certainty factor (CF) model for the identification of landslide hazard zones [11, 17]. The CF approach is one of the conceivable proposed idealness capacities to deal with the issue of consolidating various information layers and the heterogeneity and vulnerability of the information. Equation 1 is used for the analysis of CF weight values.

$$\mathrm{CF}_{ij} = \begin{cases} \frac{f_{ij}-f}{f_{ij}(1-f)} = f_{ij} \geq f \\ \hline \frac{f_{ij}-f}{f_{if}(1-f_{ij})} = f_{ij} \leq f \end{cases} \tag{1}$$

where CF_{ij} denotes certainty factor weight values f_{ij} probability of landslides in each class i of each parameter j and f prior probability having a total number of landslide events occurring in the study area. Certainty factor weight values were obtained into the range of −1.0 to 1.0 from Eq. 1. Negative values of each class show the low probability of landslides although positive values indicate a high probability of landslides frequency in each class of each parameter [21]. Density of landslides in each parameter class was analyzed for the calculation of certainty factor weight values. Thematic layers of certainty factor were developed from CF weight values (Table 1) by using Eq. 1. The, pairwise comparison matrix tool in excel has been used for the combination of CF values. The combination of two CF values was shown in the form of X and Y in Eq. 2, shown below:

$$Z = \begin{cases} X + Y - xy \\ \dfrac{X + Y}{1 - \min(|X|, [Y[)} \\ X + Y - XY \end{cases} \tag{2}$$

$X, Y \geq 0;$

X, Y Opposite site ;

$X, Y < 0.$

Landslide hazard index (LHI) values were obtained by using pairwise comparison method to summation of all CF values. LHI values were classified into five groups by using natural break classification method of GIS [6, 9, 28].

Table 1 Analysis of weight values for each initiating factors for landslides with landslide inventory

FID	Factor classes	Ncl_Pix	Nsl_Pix	F_{ij}	F	$F_{ij}(1-F)$ or $F(1-F_{ij})$	CF
Slope gradient	0–15	597,082	0	0	0.00144	0	0
	15–30	519,152	764	0.001472	0.00144	0.00144183	0.01
	30–45	608,522	918	0.001509	0.00144	0.00150857	0.042836
	45–60	638,258	1039	0.001628	0.00144	0.00162787	0.112980
	60–80	372,533	1229	0.003299	0.00144	0.00329424	0.563124
Aspect	North East	263,231	414	0.001573	0.00144	0.99855605	0.000129
	East	79,738	24	0.000301	0.00144	0.00144352	−0.791793
	South East	364,841	768	0.002105	0.00144	0.99855605	0.000662
	South	322,945	55	0.000170	0.00144	0.0014437	−0.882205
	South West	337,058	1333	0.003955	0.00144	0.99855605	0.002515
	West	335,608	7	2.086E−05	0.00144	0.00144392	−0.985576
	North West	369,206	1302	0.003526	0.00144	0.99855605	0.002086
	North	662,920	47	7.09E−05	0.00144	0.00144385	−0.950963
Lithology	Alluvium	95,316	0	0	0.00144	0	0
	River	9769	0	0	0.00144	0	0
	Dalhousie granite	75,902	0	0	0.00144	0	0
	Chamba formation	177,997	574	0.003225	0.00144	0.99855605	0.001784
	Dharmsala formation	326,207	14	4.292E−05	0.00144	0.00144389	−0.970319
	Ghar formation	30,528	62	0.002031	0.00144	0.99941303	0.001443
	Kalhel formation (Lilang group)	767,634	200	0.000261	0.00144	0.00144357	−0.819778
	Khokan formation	342,393	676	0.001974	0.00144	0.99946961	0.001443
	Mandi formation	829,363	376	0.000453	0.00144	0.00144330	−0.686340
	Pindru formation	19,345	17	0.000879	0.00144	0.00144268	−0.391751
	Upper Siwaliks	61,093	2031	0.033244	0.00144	0.96819956	0.001398

(continued)

Table 1 (continued)

FID	Factor classes	Ncl_Pix	Nsl_Pix	F_{ij}	F	$F_{ij}(1-F)$ or $F(1-F_{ij})$	CF
Soil	Loamy skeletal soil with loamy surface	488,350	51	0.000104	0.00144	0.00144380	−0.927772
	Loamy skeletol soil	970,307	443	0.000457	0.00144	0.00144329	−0.684127
	Rocky outcross	131,122	0	0	0.00144	0	0
	Loamy soil	508,273	235	0.00046	0.00144	0.00144	−0.68012
	Sandy skeletol soil	542,058	3221	0.00594	0.00144	0.99856	0.00450
	Coarse loamy soil	95,437	0	0	0.00144	0	0
Weathering	Slight erosion	970,855	18	0.0000	0.0014	0.0014	−0.9872
	Moderate erosion	94,921	538	0.0057	0.0014	0.9986	0.0042
	High erosion	996,302	2687	0.0027	0.0014	0.9986	0.0013
	Very high erosion	673,469	707	0.0010	0.0014	0.0014	0.2733
Land use and cover	Khajiar lake	208	0	0	0.00144	0	0
	River	26,971	0	0	0.00144	0	0
	Snow cover	2296	0	0	0.00144	0	0
	Forest	1,266,924	190	0.0001	0.00144	0.0014	−0.8963
	Agriculture land	535,376	1096	0.0020	0.00144	0.9986	0.0006
	Rocky terrain	148,181	352	0.0024	0.00144	0.9986	0.0009
	Shrub	267,686	679	0.0025	0.00144	0.9986	0.0011
	Settlement	92,669	243	0.0026	0.00144	0.9986	0.0012
	Sparse vegetation	376,062	1064	0.0028	0.00144	0.9986	0.0014
	Road	19,174	326	0.0170	0.00144	0.9986	0.0156
Rainfall	1437–1547	89,859	582	0.0065	0.00144	0.9986	0.0050

(continued)

Table 1 (continued)

FID	Factor classes	Ncl_Pix	Nsl_Pix	F_{ij}	F	$F_{ij}(1 - F)$ or $F(1 - F_{ij})$	CF
	1628–1691	629,317	1527	0.0024	0.00144	0.9986	0.0010
	1548–6127	951,594	347	0.0004	0.00144	0.0014	−0.7477
	1692–1752	1,064,777	1494	0.0014	0.00144	0.0014	−0.0283
Lineament density	Very low	266,883	508	0.0019	0.00144	0.9986	0.0005
	Low	711,703	621	0.0009	0.00144	0.0014	−0.3961
	Moderate	999,906	1350	0.0014	0.00144	0.0014	−0.0651
	High	757,055	1471	0.0019	0.00144	0.9986	0.0005
Drainage density	Very low	254,298	40	0.0002	0.00144	0.0014	−0.8912
	Low	783,874	863	0.0011	0.00144	0.0014	−0.2378
	Moderate	830,755	1199	0.0014	0.00144	0.0014	−0.0005
	High	866,620	1848	0.0021	0.00144	0.9986	0.0007
Relative relief	656–1190	681,747	1510	0.0022	0.00144	0.9986	0.0008
	1190–1581	868,767	1468	0.0017	0.00144	0.9986	0.0002
	1581–2041	696,372	857	0.0012	0.00144	0.0014	−0.1479
	2041–2757	488,661	200	0.0004	0.00144	0.0014	−0.7168
Curvature	Flat	152,199	0	0	0.00144	0	0
	Convex	1,283,651	1094	0.0009	0.00144	0.0014	−0.4101
	Concave	1,299,697	2856	0.0022	0.00144	0.9986	0.0008

5 Results and Discussion

Based on previous records and interpretation of various images and ground truth surveys, 242 landslides were mapped (Fig. 2a). Lineament structures like joints and fractures in rock discontinuities along a bedding plane and steep terrain gradient helps in rock slides. Heavy rainfall in the region triggers the debris slope failure. Rotational landslides are mainly associated with the undercutting of toes of slopes through haphazard road cuttings, mining activities, and infrastructural projects.

Weight values are analyzed as mentioned in (Table 1). Terrain of slopes between 45 and 80° has highest weight values between 0.56 and 0.11 followed by 15–45° (−0.01 to 0.42) while terrain slopes between 0 and 15° have none evidence of landslips (Fig. 2g). Terrain slopes facing southwest and northwest directions (aspect) have maximum numbers of landslides resulting in high weight values 0.0025–0.002 followed by southeast and northwest direction whereas east, west, north and south directions have least number of landslides (Fig. 2i). Terrain slopes and aspects are correlated with relative relief and curvature of the region. Relative relief between 656 and 1581 m (0.007–0.008) from mean sea level with concave curvature (0.008) has obtained highest weight values (Fig. 2k). Soil properties also play a very important role in the instigation of landslides (Fig. 2d). There are six categories of soil found

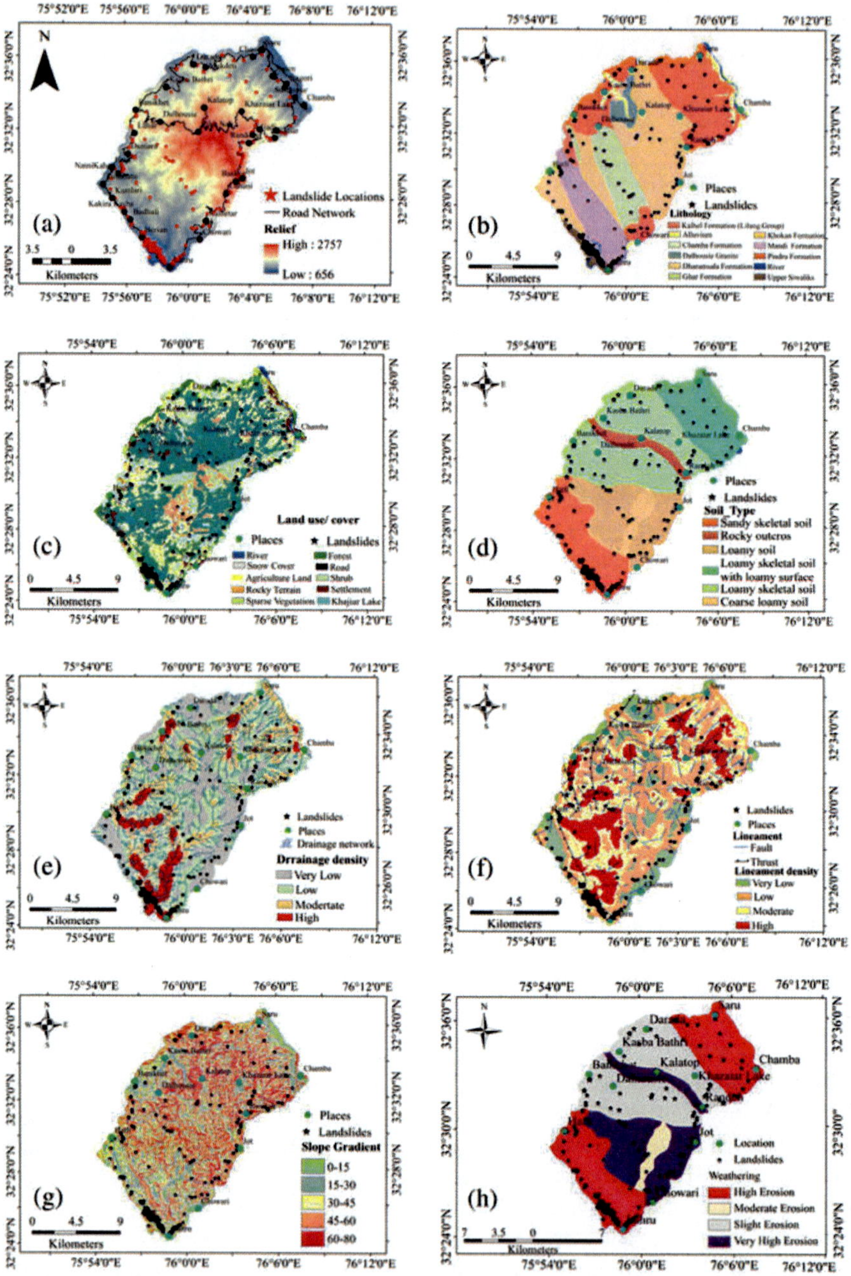

Fig. 2 **a** Landslide inventory map. **b** Lithology map. **c** Land use and cover map. **d** Soil map. **e** Drainage density map. **f** Lineament density map. **g** Slope gradient map. **h** Weathering map. **i** Aspect gradient map. **j** Rainfall map. **l** Curvature map. **k** Relative relief map

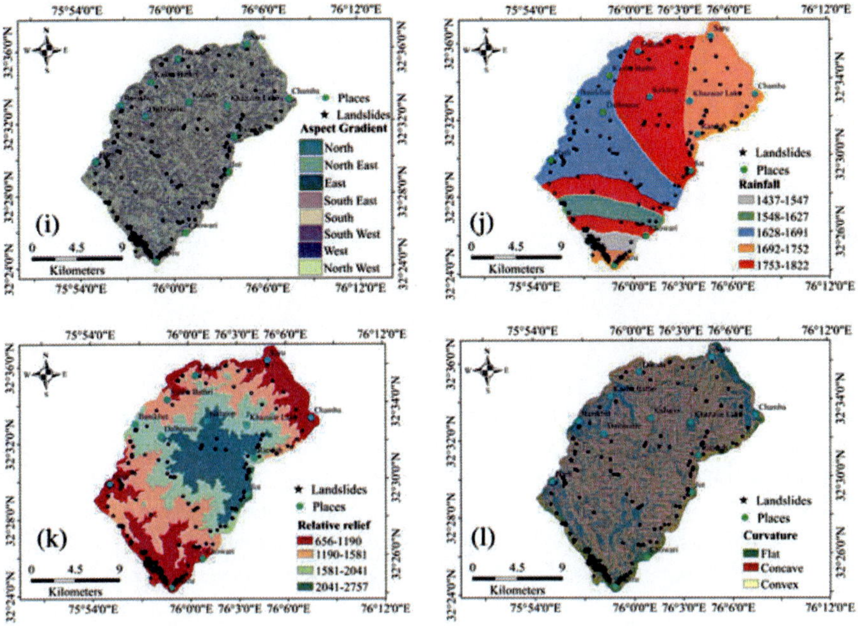

Fig. 2 (continued)

in region, in which landslides, sandy skeletal soil type have obtained highest weight value of 0.0045 while coarse loamy soil and rocky outcrops obtained zero weight values due to no evidence of landslides. Out of ten geological formations, Dalhousie granite, Chamba formations, and alluvium (Fig. 2b) have stable slopes due to no evidence of landslides while Upper Siwalik (1.0014) has highest weight values due to highest number of landslips trailed by Dharamsala formation sandstones (0.0014). Combination of thrust, joint, and fault planes (lineaments) in bedding planes are disastrous combination for instigation of landslips (Fig. 2f). High density of lineament obtained high weight value of 0.0005 while very low has −0.89. Lineaments of the region have play a very important role in formation of drainage networks. High density of drainage networks obtained high weight value of 0.0007 trialed by moderated density of drainage networks −0.0005 (Fig. 2e). Anthropogenic activities like settlement (0.0012) and road networking (0.016) have maximum weight values due to high number of landslips associated with these land uses while agricultural land has the least weight value of 0.0006. Natural land cover like sparse vegetation shrub and forest has high weight value although water bodies i.e. snow cover, Khajjiarlake, and river have zero weight value (Fig. 2c).Very High Erosion (0.27) with rainfall between 1437 and 1547 mm (0.0050) have high weight value followed by moderate and rainfall between 1628 and 1691 mm (Fig. 2h, j).

Landslide Hazard Index (LHI) map was obtained from the combination of all weight values of designated thematic layers in GIS environment (Fig. 3). Reclassify

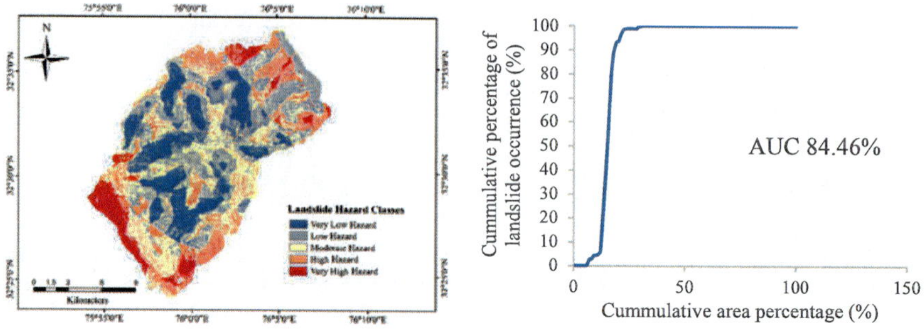

Fig. 3 Landslide hazard zonation map and validation

tool of GIS has been used for the classification of LHI values into very low hazard, low hazard, moderate hazard, high hazard, and very high hazard zone classes. Very high hazard zone class 10% area, high hazard zone class 20% area, moderate hazard zone class 26% area, low hazard zone class 27% area, and very low hazard zone class 17% area is recognized in landslide hazard zonation map.

6 Conclusion

Himalayan region of India is prominent evidence of most disastrous natural hazards like landslides. Assessment of landslide hazard was essential for Chamba district, Himachal Pradesh, India before any construction such as road (SH-28, NH-154A, link road), reservoirs, etc. In the present study, inventory of different types of landslides was compared with all factors to analyze the responsible factor for occurrences of landslides using certainty factor model. Natural (weathering processes and rainfall) and anthropogenic activities (road, settlement, and agriculture) activities are responsible for occurrences of landslides. Very high hazard zone was associated with Upper Siwaliks which is highly fragmented due to high density of lineaments and drainage networks. Very high and high hazard zones are also affected by high mode of weathering processes due to rainfall, drainage networks, and wind. Higher altitude regions are less affected due to low density of drainage, lineament, and highly dense vegetation cover. Weights for each factor class were attributed during the process of modeling. Very low 44% of the area is low-risk zones for the occurrence of landslides while 30% area is falling in high risk of landslides and 26% area is falling in the moderate zone. The accuracy of the hazard zonation map was evaluated using area under the curve method. The prediction rate of the landslide hazard zonation map was evaluated as 84.46% based on the area under the curve method.

Acknowledgements The authors are thankful to the Chamba district, Public works department, Himachal Pradesh for providing database of landslide related and Department of Civil Engineering, National Institute of Technology, Hamirpur, Himachal Pradesh for providing the essential equipment's to conduct the study.

References

1. Achour Y, Boumezbeur A, Hadji R, et al (2017) Landslide susceptibility mapping using analytic hierarchy process and information value methods along a highway road section in Constantine, Algeria. Arab J Geosci 10:194
2. Aleotti P, Chowdhury R (1999) Landslide hazard assessment: summary review and new perspectives. Bull Eng Geol Env 58(1):21–44. https://doi.org/10.1007/s100640050066
3. Anbalagan R (1992) Landslide hazard evaluation and zonation mapping in mountainous terrain. Eng Geol 32(4):269–277
4. Brooks SM, Crozier MJ, Glade TW, Anderson MG (2004) Towards establishing climatic thresholds for slope instability: use of a physically-based combined soil hydrology-slope stability model. Pure Appl Geophys 16:881–905
5. Chung MC, Tan CH, Chen CH (2016) Local rainfall thresholds for forecasting landslide occurrence: Taipingshan landslide triggered by Typhoon Saola. Landslides 14(1):19–33
6. Constantin M, Bednarik M, Jurchescu MC, Vlaicu M (2011) Landslide susceptibility assessment using the bivariate statistical analysis and the index of entropy in the Sibiciu Basin (Romania). Environ Earth Sci 63:397–406
7. Crosta GB, Frattini P (2003) Distributed modelling of shallow landslides triggered by intense Rainfall. Natural Hazards Earth Syst Sci 3:81–93
8. Cruden DM, Varnes DJ (1996) Landslide types and processes, special report. Transp Res Board National Academy Sci 247:36–75
9. Devkota K, Regmi A, Pourghasemi HR, Yoshida K, Pradhan B, Ryu I, Dhital M, Althuwaynee O (2018) Landslide susceptibility mapping using certainty factor, index of entropy and logistic regression models and their comparison at a landslide prone area in Nepal Himalaya. Nat Hazards 65:135–165. 10.1007/S11069-012-0347-6
10. Ercanoglu M, Gokceoglu C (2004) Use of fuzzy relations to produce landslide susceptibility map of a landslide prone area (West Black Sea Region, Turkey). Eng Geol 75:229–250. https://doi.org/10.1016/j.enggeo.2004.06.001
11. Gökçeolu C, Sonmez H, Nefeslioglu HA, Duman TY, Can T (2005) The 17 March 2005 Kuzulu landslide (Sivas, Turkey) and landslide-susceptibility map of its near vicinity. Eng Geol 81:65–83
12. Hutchinson JN (1995) Landslide hazard assessment, In: Landslides: proceedings of the sixth international symposium, Christchurch, vol 1, pp 1805–1842
13. I.S. (1998) Preparation of landslide hazard zonation maps in mountainous terrain—guidelines (Part 2—Macrozonation), vol 14496, 2nd edn. BIS, New Delhi, pp 1–19
14. International Association of Engineering Geology (IAEG), Bull Eng Geol Environ 63(4)
15. Iverson RM (2000) Landslide triggering by rain infiltration. Water Resour Res 36(7):1897–1910
16. Kanungo DP, Arora MK, Sarkar S et al (2009) Landslide susceptibility zonation (LSZ) mapping—a review. J South Asia Disaster Stud 2(1):81–105
17. Kanungo DP, Sarkar S, Sharma S (2011) Combining neural network with fuzzy, certainty factor and likelihood ratio concepts for spatial prediction of landslides. Nat Hazards 59:1491–1512
18. Kayastha P, Dhital MR, De Smedt F (2013) Application of the analytical hierarchy process (AHP) for landslide susceptibility mapping: a case study from the Tinau watershed, west Nepal. Comput Geosci 52:398–408. https://doi.org/10.1016/j.cageo.2012.11.003

19. Montrasio L, Valentino R, Losi GL (2012) Shallow landslides triggered by rainfalls: modeling of some case histories in the Reggiano Apennine (Emilia Romagna Region, Northern Italy). Nat Hazards 60:1231–1254
20. Nandi A, Shakoor A (2009) A GIS-based landslide susceptibility evaluation using bivariate and multivariate statistical analyses. Eng Geol 110:11–20. https://doi.org/10.1016/j.enggeo.2009.10.001
21. Pourghasemi HR, Mohammady M, Pradhan B (2012e) Landslide susceptibility mapping using index of entropy and conditional probability models in GIS: Safarood Basin, Iran. Catena 97:71–84. https://doi.org/10.1016/j.catena.2012.05.005
22. Pourghasemi HR, Pradhan B, Gokceoglu C et al (2013) Application of weights-of-evidence and certainty factor models and their comparison in landslide susceptibility mapping at hazard watershed, Iran. Arab J Geosci 6:2351–2365. https://doi.org/10.1007/s12517-012-0532-7
23. Wang Qiqing, Guob Yinghai, Li Wenping, He Jianghui, Zhiyong Wu (2019) Predictive modelling of landslide hazards in Wen County, north western China based on information value, weights-of-evidence, and certainty factor. Geomatics Nat Hazards Risk 10(1):820–835
24. Saaty TL (1980) The analytic hierarchy process. McGraw Hill, New York. International, Translated to Russian, Portuguese, and Chinese, Revised editions, Paperback
25. Singh KC, Thakur VC (1986) Crenulation cleavage in slate of the Chamba syncline: associated microstructures and origin. J Geol Soc India 28:311–314
26. Varnes DJ (1984) Landslide hazard zonation: a review of principles and practice. United Nations Educational, Scientific and Cultural Organization, p 63
27. Chen W, Pourghesmi HR, Kornejady A, Xie X (2018) GIS-based landslide susceptibility evaluation using certainty factor and index of entropy ensemble with alternating decision tree models, natural hazards GIS-based spatial modeling using data mining techniques, pp 225–251. https://doi.org/10.1007/978-3-319-73383-8_10
28. Xu C, Xu X, Lee YH, Tan X, Yu G, Dai F (2012) The 2010 Yushu earthquake triggered landslide hazard mapping using GIS and weight of evidence modelling. Environ Earth Sci. https://doi.org/10.1007/s12665-012-1624-0

Decision-Making Rating System for Prioritization of Municipal Solid Waste Dumps for Closure Based on Aesthetics

Himanshu Yadav, Lalit Kumar, and V. P. Singh

1 Introduction

Rapid urbanization and population growth result in generation of large quantity of waste. In India, most of the municipal solid waste was started to dump into a place in the outskirts of the city, with the development of the city the dumps come within the domain of the city. Open dumping is the most common method for the disposal of the municipal waste of the city in developing nations. Municipal solid waste dumps are hazardous to the environment by various ways like groundwater contamination, surface water contamination, air pollution, soil contamination, explosion and bird hit to the aircrafts [1–3]. Various dumpsites from India and other countries of world are given in Table 1. It is necessary to close these dumpsites because it is not feasible to manage the landfills in sustainable way, and there is limited resources to upgrading these dumpsites. A number of approaches are available for risk assessment to the environment due to the MSW dumpsites. Hazard rating systems are generally used as they are simpler and require less data [4]. These rating systems are based on mathematical scoring algorithm. Various input parameters are identified arranged in mathematical structured form to give rating. There are numerous rating systems available based on the migration route considered for the contaminants. The study of various rating systems shows that there are three types of systems (i) additives, (ii) multiplicative and (iii) additive–multiplicative [5].

H. Yadav (✉)
Indian Institute of Technology Guwahati, Guwahati, Assam, India
e-mail: yadav.himanshu1234@gmail.com

L. Kumar · V. P. Singh
Motilal Nehru National Institute of Technology Allahabad, Prayagraj, Uttar Pradesh, India
e-mail: lalitkgupta2015@gmail.com

V. P. Singh
e-mail: vps15783@mnnit.ac.in

H. Singh et al. (eds.), *Sustainable Development Through Engineering Innovations*, Lecture Notes in Civil Engineering 113,
https://doi.org/10.1007/978-981-15-9554-7_7

Table 1 Various dumpsites of India and World

A. Indian dumpsites			
Site	Area (ha)	Height (m)	References
Indore	8	16	[5]
Nagpur	21.5	5	
Jaipur	54	5	
	49.4	2	
Chennai	81	8	
	81	6.4	
Ahmedabad	28	24	
Kolkata	21.4	24	
Mumbai	120	15	
Delhi	13	65	[10]
	28.3	60	
Gurgaon	4	35	
Lucknow	4.15	9.7	[9]
Prayagraj	0.6	8.5	[7]
	5.5	10	
B. Other countries dumpsites			
Site	Area (ha)	Height (m)	Reference
Ibaden, Nigeria	14	4	[5]
Dandora, Kenya	53	56	
Surjani, Pakistan	202	10.6	
Lahore, Pakistan	25.5	33	
Santo Domingo Dominican Republic	128	17	

2 Development of New Rating System

This study focuses on the development of novel decision-making tool for prioritization of dumpsites based on aesthetics. It is considered important because it creates the first impression about the environmental condition and presence of the dumpsite in the city affects the tourism of the city. Various factors affecting the aesthetics are identified and arranged in a mathematical scoring system to find a rating range from 0 to 1000. The rating system is applied to the hypothetical sites of varying characteristics. Later, the system is applied to the five dumpsites from Indian cities to evaluate the score. The parameters that are important for aesthetics of waste site are categorized into two categories:

(i) related to the dumpsite characteristics

(ii) related to the surrounding characteristics.

Area of dumpsite, height of dumpsite, presence of leachate, presence of vegetation cover and the present status of site considered the important parameters of dumpsite characteristics. Distance from airport, distance from national highway (NH) or state highway (SH), population density of area, distance from important monument or tourist spot, and relative position of site with respect to humans are considered the parameters related to the humans population (surroundings). All parameters are not equally important, and some parameters (like waste area and height) are more important than other parameters. For considering different importance of different parameters, additive–multiplicative algorithm is used. Scores related to site characteristics are calculated using addition of some parameters to the multiplication of other parameters, score related to surrounding characteristics is calculated after addition of parameters, and overall score is calculated by multiplication of site and surrounding scores. The overall score or rating can be evaluated using Eq. 1. In this rating system, scores are scaled on a range of 0–1000, and maximum overall rating is limited to 1000. Generally, rating systems are based on source–pathway–receptor methodology, but in case of aesthetics rating system there is no need of pathway. So, source–receptor methodology is used.

$$S_O = S_{\text{Site}} * S_{\text{Surrounding}} \tag{1}$$

where S_O = the overall score due to aesthetics of MSW dumpsites, S_{Site} = score due to the parameters related to the site characteristics, and $S_{\text{Surrounding}}$ = score due to the parameters related to the surrounding of the dumpsite.

2.1 Site Score

Site score represents the score related to the site characteristics and is given by

$$S_{\text{Site}} = A * H * (S_L + S_{\text{VC}} + S_{\text{CS}})/6 \tag{2}$$

where A = area of dumpsite (in hectares), H = height of dumpsite (m), S_L = leachate presence score; S_{VC} = vegetation cover presence score, and S_{CS} = current status of site score.

The maximum value of site score is limited to 1000. If the site score obtained from Eq. 2 is greater than 1000, then value of site score is to be taken as 1000.

Leachate presence score considers the presence of leachate around the dumpsite. If leachate is present around the site throughout the year, then the score is taken as 2. For seasonal presence of leachate around the dumpsite, the score is taken as 1, and for absence of leachate the score is taken as 0.

Vegetation cover score is taken as per Table 2.

Table 2 Vegetation cover score for aesthetics rating system

Vegetation cover	Score
None	2
Poorly established (Sparse, Root Zone < 6")	1.5
Established (Good, Root Zone = 6-12")	1
Well established (Lush, Root Zone > 12")	0.5

Table 3 Score due to current status of site for aesthetic rating system

Type of waste dumpsite	Presence of final cover and daily cover	Only final cover is present but daily cover is absent	Absence of final and daily cover
Old waste only (No fresh waste is dumped)	0.2	1	1.5
Receives fresh waste	0.8	1.5	2

Current status score of site indicates whether site accepts fresh waste or not and presence of daily cover and final cover. The score can be taken as per Table 3.

2.2 Surrounding Score

This score represents the parameters related to the surrounding of the dumpsite and is given by

$$S_{\text{Surrounding}} = \left(S_{\text{Airport}} + S_{\text{Highway}} + S_{\text{PD}} + S_{\text{RP}} + S_{T}\right)/10 \tag{3}$$

where S_{Airport} = score representing distance from airport; S_{Highway} = score representing distance from national highway (NH) or state highway (SH); S_{PD} = population density score; S_{RP} = relative position of dumpsite score, and S_T = score representing distance from tourist spot and monuments.

Distance from airport is considered as an important factor in aesthetics rating as the number of people from other cities and countries visits the place and if they find out the dumpsite in the first vision that will create a negative impact of the place. Dumpsite should be at least 20 km away from the airport, so that it is not visible to the human population visiting the place. Score representing distance from the airport can be taken as 1 if distance between airport and dumpsite is greater than 20 km, and it should be taken as 2 if distance between airport and dumpsite is less than 5 km, otherwise score can be calculated using Eq. 4.

$$S_{\text{Airport}} = 1 + \left(\frac{4}{3} - \frac{D}{15}\right) \tag{4}$$

where D = distance from the airport in km.

Distance from highway is considered in rating as humans travelling through highway can see the dumpsite. Dumpsite should be at least 2 km away from the highway. Score representing distance from national highway (NH) or state highway (SH) can be taken as 0 if distance between highway and dumpsite is more than 2 km, otherwise score can be calculated using Eq. 5.

$$S_{\text{Highway}} = 2 - D_{\text{NH/SH}} \tag{5}$$

where $D_{\text{NH/SH}}$ = distance from national highway (NH) or state highway (SH) in km.

Population density within 3 km of dumpsite is defined by percentage of built-up area, more than 50% being dense, between 20 and 50% medium and less than 20% indicating sparse population. Rating based on population density is given in Table 4.

Relative position of dumpsite score indicates the score due to the topography of the area. Dumpsites at the elevated area are not aesthetically pleasant as compared to the plain area. Waste dumpsites in low-lying area are not visible to the humans or effect of height does not seem to be effective. The relative position scores are given in Table 5.

Distance from tourist spot and monument is important; if dumpsite is visible from that point then, it affects the tourism of the city. Presence of tourist spot and monument is considered within the radius of 2 km from the dumpsite. Score can be taken as 2 if any of tourist spot or monument is present within radius of 2 km, and otherwise the score is taken as 1.

Table 4 Population density score for aesthetic rating system

Population density	Score
Sparse (5–20%)	1
Medium (20–50%)	1.5
Dense (>50%)	2

Table 5 Relative position score for aesthetic rating system

Relative position of dumpsite with respect to humans	Score
Best scenario (low lying)	1
Medium scenario (plain)	1.5
Worst scenario (elevated)	2

Table 6 Site characteristics of hypothetical waste sites

Parameters	W_1	W_2	W_3	W_4	W_5
Area (ha)	5	10	15	20	30
Height (m)	5	15	20	30	40
Leachate presence	No	No	Seasonal	Yes	Yes
Vegetation cover	Well Established	Established	Established	Poorly established	None
Presence of daily cover and final cover	Both Present	Both Present	Final cover present; daily cover absent	Final cover present; daily cover absent	Both Absent
Type of dumpsite	Old	New	Old	New	New
Distance from airport	20	15	12	10	5
Distance from highway	3	2	1.5	1	0.5
Population density(within 3 km)	Sparse	Medium	Medium	Dense	Dense
Relative position of dumpsite with respect to humans	Low lying	Low lying	Plain	Elevated	Elevated
Presence/absence of tourist spot or monuments (within 2 km)	Both tourist spot and monument absent	Only tourist spot present	Only tourist spot present	Only monument present	Both tourist spot and monument present

3 Application of System to Hypothetical Sites

The aesthetic rating system is applied to the five dumpsites having varying characteristics. The characteristics are given in Table 6, rating increases from W_1 to W_5. Dumpsite W_1 represents best characteristics and W_5 as worst. Best and worst values of the parameters are selected based on various literatures.

The scores obtained from rating system for hypothetical sites vary from 1 to 950 (Fig. 1), which means full range of the system (0-1000).

4 Application of System to the Indian MSW Sites

New system was applied to the six Indian sites, and data for the sites is collected with the help of existing literature [4, 6–9] and field visit to the site. Site characteristics are provided in Table 7.

The results from the application of aesthetic rating system to the Indian sites are shown in Fig. 2. The sites have the rating from 26 to 706. Sites A and E have the

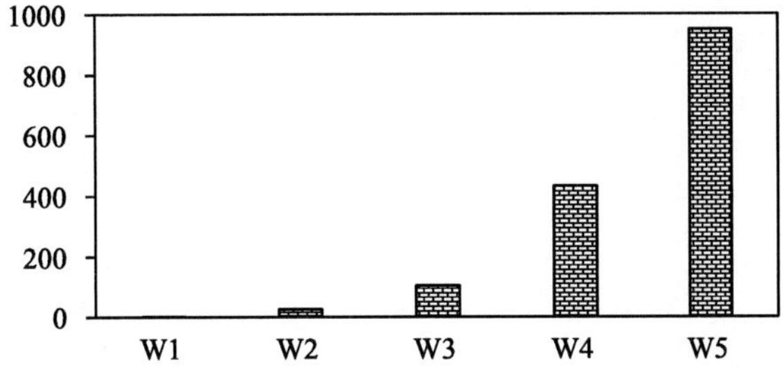

Fig. 1 Scores for aesthetic rating system for hypothetical sites

Table 7 Site characteristics for waste sites from Indian cities

Parameters	Site A	Site B	Site C	Site D	Site E
City	Prayagraj	Delhi	Delhi	Kolkata	Lucknow
Area (ha)	5.5	29	16	21.4	4.15
Height (m)	10	60	40	24	9.7
Leachate presence	Seasonal	Yes	Yes	Yes	Seasonal
Vegetation cover	Poorly established	None	None	None	None
Presence of daily cover and final cover	None	None	None	None	None
Type of dumpsite	Old	Active	Active	Active	Active
Distance from airport (km)	15	25	21	12.5	15.2
Distance from highway (km)	1.2	0.44	0	0.38	0
Population density(within 3 km)	Dense	Dense	Dense	Sparse	Medium
Relative position of dumpsite with respect to humans	Low lying	Plain	Plain	Plain	Plain
Presence/absence of tourist spot or monuments (within 2 km)	Only tourist spot present	Both absent	Only monument present	Only tourist spot present	Only monument present

lowest rating, and Site A and Site E are smaller sites as compared to others. Site B has large area and height, so it has the highest score.

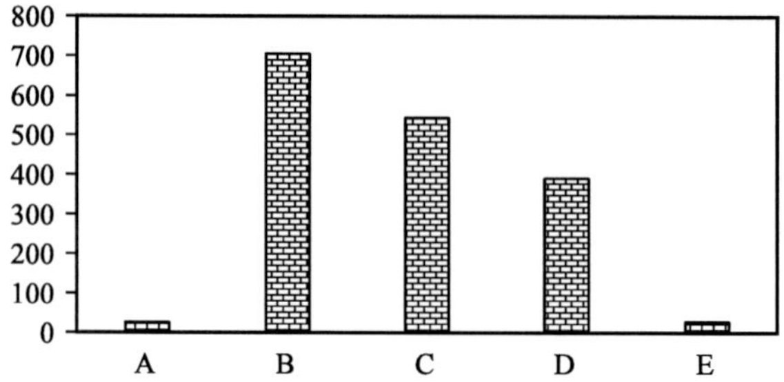

Fig. 2 Aesthetic rating for different Indian sites

5 Conclusions

A new system is developed for rating dumpsites based on aesthetics. This system can be used to the prioritization of dumpsites for planning of closure of them. This system is easy to understand and requires less or minimum data. Various parameters related to the aesthetics are found out, and mathematical system is developed. System was applied to the five hypothetical sites of varying characteristics and score obtained from 1 to 950. The system is applied to the five Indian dumpsites of different characteristics. Site B from Delhi has the highest, and site A from Prayagraj and site E of Lucknow have the lowest score.

References

1. Powrie W, Richards D, Beaven R (2015) Geotechnical hazards associated with closed municipal solid waste landfill sites. In: International symposium on geohazards and geomechanics (ISGG2015). IOP Conf Ser Earth Environ Sci
2. Baxter A (2001) Bird control on landfill sites–is there still a hazard to your aircraft? In: 2001 bird strike committee-USA/Canada, third joint annual meeting, Calgary, AB, pp 48–55
3. Yadav H, Kumar P, Singh VP (2019) Hazards from the municipal solid waste dumpsites: a review. In: Proceedings of the 1st international conference on sustainable waste management through design. ICSWMD 2018, Lecture notes in civil engineering, vol 21. Springer, Cham
4. Datta M, Kumar A (2015) Hazard rating of MSW dumps and geoenvironmental measures for closure. In: 50th Indian geotechnical conference, Pune, India
5. Kumar A, Datta M, Nema AK, Singh RK (2017) Closure of municipal solid waste dumps–site rating for odour impact. Environ Eng Protect J
6. Kumar A, Datta M, Nema AK, Singh RK (2015) An improved rating system for assessing surface water contamination potential from MSW landfills. Environ Model Assess. Springer International Publishing, Switzerland
7. Yadav H, Singh VP (2020) Risk assessment due to municipal solid waste dumpsites and geoenvironmental measures for closure. Int J Environ Waste Manag 26(2):190–211
8. Kumar A, Tomar RK (2019) A new decision support system for assessing bird-hit hazard from municipal waste dumps. In: Proceedings of national conference: advanced structures, materials and methodology in civil engineering (ASMMCE 2018), pp 261–265

9. Archana A, Ali D, Yunus M, Dutta V (2014) Assessment of status of municipal solid waste management (MSWM) in Lucknow–capital city of Uttar Pradesh, India. IOSR J Environ Sci Toxicol Food Technol (IOSR-JESTFT) 8(5)(Ver. II):41–49
10. Mohit S, Datta M, Gupta SK, Sreekrishnan, Ramana GV (2019) Comprehensive assessment of the leachate quality and its pollution potential from six municipal waste dumpsites of India. Bioresour Technol Rep 6(2019):198–206

A Preliminary Study of Bioremediation on Oil-Contaminated Soil Using Bacteria and Organic Manure

Surya Muthukumar, P. Dharuneeswar, John Jesuran, Jayakrishnan, Yamini Jayaprakash, Sakthipriya, and Amritha Velayudham

1 Introduction

Soil contamination happens due to the excessive usage of fertilizers, pesticides, petroleum products, and chemicals [1]. The percolation of oil into the ground takes place during oil exploration, transportation, buried pipeline leakages, petroleum productions, spillage from vehicles, discharge from the coastal oil wells, etc. The evolution of petroleum fractions has consistently induced and forced various sectors to pollute the adjacent environment bodies with used oils since these do not possess a regulated disposal methodology. The introduction of these vicious materials to the environment discharges huge amount of hydrocarbon [2] into the environmental bodies. This is considered to be a major pollution sources for the soil in the world, especially in oil-producing countries [3–5]. This menace is found to be very severe in poorly regulated countries since oil spillage is not considered as pollution for a longer period. This had resulted in continuous accumulation and turned existing problem into a hazard to the environment [6–8]. This soil contamination leads to a cyclic disaster since the oil-contaminated soil leads to loss of vegetation productive soil and groundwater pollutions which ultimately affect the local economic conditions of the people [9]. Remediation of these contaminated soils is basically expensive [10]. To eliminate the toxicity in soil with high efficiencies, low-cost bioremediation techniques are adopted [11, 12]. Bioremediation refers to the transformation of the contaminants to potentially less harmful entities with the help of bacteria and microorganisms. These organisms infest on the nutrient available as pollutants on the site and break to degrade the pollutants to potentially less pollutant state. The contaminated sites can be treated using physicochemical methods like soil flushing,

S. Muthukumar (✉) · P. Dharuneeswar · J. Jesuran · Jayakrishnan · Y. Jayaprakash · Sakthipriya · A. Velayudham
Department of Civil Engineering, Amrita School of Engineering, Amrita Vishwa Vidyapeetham, Coimbatore, India
e-mail: m_surya@cb.amrita.edu

H. Singh et al. (eds.), *Sustainable Development Through Engineering Innovations*, Lecture Notes in Civil Engineering 113,
https://doi.org/10.1007/978-981-15-9554-7_8

soil washing, and electrokinetic remediation and by biological remediation methods. Bioremediation techniques have been observed to be effective techniques that stimulate the biodegradation of contaminated soils [13]. It has been observed that bioremediation of oil-contaminated soil is the most suitable technique for remediation of oil contaminants [14]. Bioremediation of oil-contaminated soil includes the technology of adding nutrients to the soil to stimulate the microbial population called biostimulation and to degrade the contaminants called bioaugmentation. Most commonly used degrading bacteria to remove the petroleum contamination in soil are bacillus, pseudomonas, Acinetobacter, Azamines, etc [15, 16].

In this study, pseudomonas fluorescence and bacillus sublitis, where bacillus sublitis is a combination of three bacteria, namely bacillus licheniformis, bacillus circulans, and bacillus megaterium, is used. Pseudomonas fluorescence is a rod-shaped bacterium which is gram-negative, and it is found everywhere, in virtually all environment on earth that supports life. This type of bacteria has multiple flagella and can survive at room temperature. This bacterium can surpass the other soil microorganisms by giving a competitive advantage as useable for iron. This bacterium can produce compounds antagonistic to other soil microorganisms like phenazine and antibiotics of hydrogen cyanide (HCn). The growth of bacteria has more involvement in reducing the oil and grease content from the soil, and bacillus subtilis is a gram-positive, spore-forming bacterial species. It is the best-characterized bacteria and is used as a model organism. It is a rod-shaped bacterium, and as it produces endospores it can survive the high-temperature environmental condition. In the soil, the natural environment of bacillus subtilis of the bacterium can survive environmental conditions including drastic differences in oxygen tension. Bacillus subtilis is strongly linked to the other bacterium bacillus licheniformis, bacillus circulans, and bacillus megatherium. Though many research are conducted on biodegradation using seven families of bacteria together for efficient results, here an attempt has been made with the abovementioned two families.

Compost is a source of nutrients that improve the soil retention and preservation and promotes aeration. The technique of compost bioremediation is applied with dual purpose of biostimulation and soil fertilization. The compost supplies nutrients to the microorganisms in the soil which enables to degrade the target contaminants [16]. Tests were also done to study the remediation rate of petroleum hydrocarbon using agrowaste and swine wastewater. This technique showed that this bioremediation method is economical and utilization of all the materials. This process could offset the disposal of wastes from agriculture without the influence of any chemical surfactant [17]. A study [18] shows that the bioremediation of oil-contaminated soil can be achieved by the use of cow dung and goat manure, which resulted in the positive degradation of contaminated soil. It is also found that the microbes present in these manures can remove the contaminants. Cow dung is a composite of cellulose, protein, hemicellulose, and minerals like nitrogen, potassium, magnesium, calcium, sulfur, etc. It includes a microbial composition of nearly sixty species of bacteria, fungi, and about hundred species of protozoa and yeasts. The bacterial species in cow dung include bacillus, corynebacterium, and lactobacillus [19]. Rice husk which is a natural fiber can be used as a bulking agent in bioremediation of hydrocarbon polluted agricultural land [20]. The bulking agents will improve the permeability and porosity of the contaminated soil and increase the rate of the remediation process. The hydrocarbon degraders from the rice husk will have enzymes that metabolize hydrocarbon as a source of energy. The rice husk mostly contains fungi due to its lignocellulosic content and results in logarithmic microbial multiplication. The increased microbial activities lower the concentration of hydrocarbon within a maximum of 4 weeks. Goat manure is a good biostimulant that increases the activities of the hydrocarbon clastic bacteria cause a decrease in the hydrocarbon content in the soil. The nutrients present in the manure stimulate microbial growth and produce enzymes to breakdown the hydrocarbon present in the soil.

However, this study reports on various bioremediation techniques used such as bacterial biodegradation (BA), cow dung–rice husk (CR) combination, and goat manure–rice husk (GR) combination and to compare the remediation potential for these three different stimulants with respect to the chemical and geotechnical aspects of the remediated soil within standard stipulated time frames.

2 Materials

2.1 *Sample Location*

The uncontaminated soil samples are collected from Amrita University, Coimbatore, India. The sample is taken from a depth of 15 feet below the ground level to eliminate the interference of organic matter. The sample is collected from an open excavation by core cutter method and sealed to avoid the loss of moisture content.

2.2 *Compost and Bacteria*

The soil is contaminated artificially by used engine oil. The specific gravity of the oil is 0.79 at 27 °C with a kinematic viscosity of 67.56 mm^2/s.

The process of bioremediation is done using a hydrocarbon-degrading family of bacteria. Pseudomonas fluorescence and bacillus subtilis were bacillus subtilis which is a combination of three bacteria, namely bacillus licheniformis, bacillus circulans, and bacillus megaterium.

Organic manure, such as cow dung and goat manure with rice husk, is a bulking agent. The compost and bacteria used for bioremediation were obtained from Tamil Nadu Agricultural University, Coimbatore, India.

3 Experimental Program

3.1 *Bioremediation Experimental Design*

The soil sample was sieved through a 4.75 mm sieve to ensure the removal of coarser particles. The soil was artificially contaminated by 3% (w/w) with engine oil. The contaminated soil was air-dried, grind, and sieved using 4.75 mm. The bioremediation experiment was conducted in contaminated soil divided into three parts by different treatment methods.

(a) BA (bacterial remediation): The bacteria mixture composed of three isolates of 2 ml was added to the 3 kg soil mass and kept for remediation for 30 days.
(b) CR (cow dung with rice husk)
(c) GR (goat manure with rice husk).

The ratio of the compost with the bulking agent in the contaminated soil was kept as 1:1:3. The prepared sample was stored in separate labeled containers and maintained at room temperature for 30 days for the development of microorganisms to destroy the oil content.

To analyze the influence of oil content on the strength of soil, the density behavior of the contaminated soil was tested for different compositions of oil. The shear strength of the uncontaminated soil and the remediated soil using various methods were obtained to assess the reusability of the remediated soil.

3.2 *Tests Conducted*

3.2.1 Preliminary Tests

Preliminary tests were conducted on the uncontaminated soil samples to determine their physical properties such as sieve analysis, specific gravity, consistency limits, and plasticity index in accordance with respective IS standards, and the results are shown in Table 1.

3.2.2 Determination of Physiochemical Properties of Soil

The soil samples were tested for different physiochemical properties such as pH, moisture content, total nitrogen content, phosphate content, potassium content, total organic carbon content, electrical conductivity, organic content, and oil and grease content in accordance with IS standards.

Table 1 Properties of uncontaminated soil

Properties of uncontaminated Soil	Method	Value (%)
Sand (0.075-4.75 mm)	IS 2720-(Part 4)—1985	53.12
Silt and clay (<0.075)	IS 2720-(Part 4)—1985	46.88
Natural moisture content (%)	IS 2720-(Part 2)—1973	21
Organic content (%)	IS 2720-(Part 22)—1972	3.1
Specific gravity	IS 2720-(Part 3)—1980	2.69
Liquid limit	IS 2720-(Part 5)—1985	42
Plastic limit	IS 2720-(Part 5)—1985	20

3.2.3 Soil Compaction Test

The test was carried out according to IS 2720-Part-8, where the maximum dry density of the soil and optimum moisture content of the soil were determined by plotting a graph between dry density and water content.

3.2.4 Unconfined Compression Test

The shear strength of the soil was identified according to IS 2720-10.

4 Results and Discussion

4.1 Physiochemical Properties of Various Methods of Bioremediation

The results of the physiochemical properties of the soil samples are given in Table 2. It is observed that after it has been adopting the remediation technique, there is a drastic reduction in the oil and grease content in a span of 30 days. It observed a decrease in the oil content of 95 and 92% for cow dung–rice husk (CR) and goat manure–rice husk (GR). This decrease is mainly due to the presence of pseudomonas and bacillus bacteria present in the manure, where these microorganisms tend to digest the oil as a possible source of nutrient. These bioremediation techniques also tend to promote the plant growth in the remediated site due to an increase in the concentration of N, PO_4, K, organic carbon, and organic matter.

Table 2 Physiochemical properties of various methods of bioremediation

Parameters	Units	Method	Contaminated soil	BA (30 days)	CR (30 days)	GR (30 Days)
Oil and grease	mg/kg	FAO MANUAL	81	32	4	6
Moisture	%	IS 15106:1999	11.2	9.30	8.20	9.21
N	mg/kg	IS 14684:1999	930	1927	916	1502
PO_4	mg/kg	IS 10158:1982	238	409	1093	4053
K	mg/kg	USEPA 3050B	183	124	119	125
Total organic carbon	%	IS: 2720	7.7	8.9	16.8	14.9
EC@25 °C	µm hos/cm	IS: 14767:2000	180	70	120	230
Organic matter	%	IS: 2720	13.3	15.3	28.9	25.6
pH@25 °C	–	IS: 2720	8.33	9.30	8.23	8.02

BA—Bacterial remediation technique
CR—Cow dung–rice husk remediation technique
GR—Goat manure–rice husk remediation technique

4.2 *Compaction Test*

The compaction test results of various composition of oil are shown in Fig. 1. It is observed that an increase in oil concentrations decreases the optimum moisture

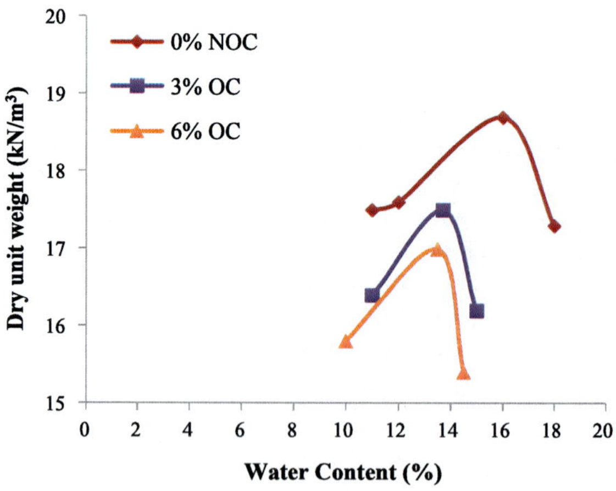

Fig. 1 Compaction test for different concentrations of oil

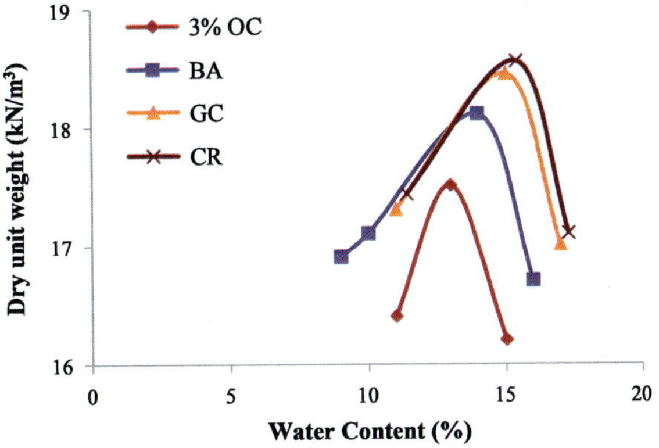

Fig. 2 Compaction tests for bioremediated soils

content and the maximum dry unit weight of the soil. This behavior is observed since the oil is hydrophobic in nature, which wraps the particles and thus restricts the interaction of water with the clay particles, thus reducing the amount of water required to attain its maximum unit weight and thus decreasing the moisture content. The oil present induces small structural transformations, thus increasing the interlayer expansion within the clay particle. Therefore, using similar compaction effort as used for compacting uncontaminated soil, the soil particles are less packed which leads to the decrease in dry unit weight of the contaminated soil. As oil contamination in soil increases by 3 and 6%, the optimum moisture content of the soil reduces by 14% and 17%, respectively. According to the compaction curve in Fig. 2, the compaction test results for the remediated soils clearly indicate that the optimum moisture content, and the maximum dry unit weight of the remediated soil increases by 3.4, 5.3, and 6.1% for bacterial remediation BA, cow dung–rice husk CR, and goat manure–rice husk GR remediation techniques, respectively, due to the decrease in oil content by bacterial digestion.

4.3 *Unconfined Compression Test*

Figure 3 shows the axial stress–strain variation for different oil contents. The unconfined compression strength (UCS) of the soil is found to be decreasing drastically with the increase in oil contamination concentrations. The UCS value of the uncontaminated soil is 130 kN/m^2, whereas the strength of 3 and 6% oil-contaminated soil is 85 and 70 kN/m^2. The increase in oil content decreases the UCS of the soil, due to a decrease in the compaction of soil which ultimately decreases the UCS of the soil mass. It is also evident that the axial strain increases with the increase in oil content, due to the wrapping of soil particles by oil, which in turn increases the interparticle sliding effect.

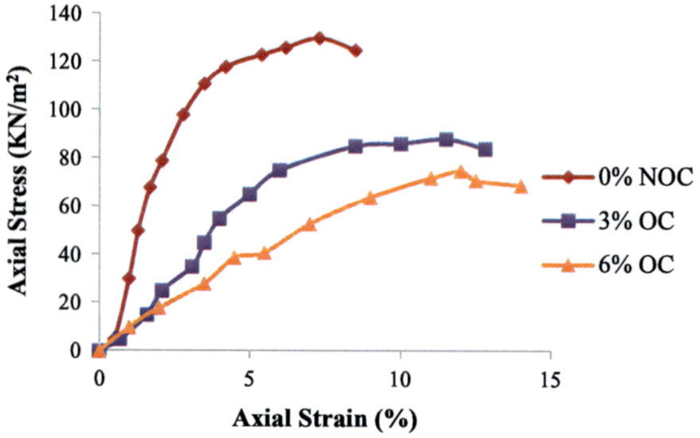

Fig. 3 Variation of axial stress and strain for different oil content

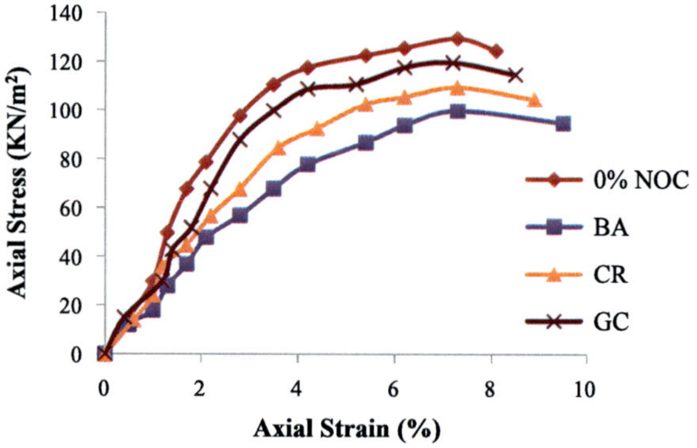

Fig. 4 Variation of axial stress and strain for various bioremediation methods

According to Fig. 4, it is evident that the UCS values of the remediated soil had increased drastically due to the reduction of oil content by the remediation methods as listed in Table 3. The remediation using CR and BR has shown an improved shear strength value as the oil content has reduced and due to the presence of the bulking agent. These fibers can influence the improvement of shear strength. But the quantity of materials mixed in these methods is high, they are in the ratio of 1:1:3. Therefore, it is difficult to handle for larger quantity remediation.

Table 3 Unconfined compression test for various conditions

Soil type	q_u (kN/m^2)	C (kN/m^2)
Uncontaminated soil	130	65
Contaminated soil (3% oil)	85	42.5
Contaminated soil (6% oil)	70	35
Bacterial biodegradation (BA)	98	49
Cow dung–rice husk (CR)	105	52.5
Goat manure–rice husk (GR)	110	55

5 Conclusions

The present study analyzed the amount of oil and grease content in the contaminated soil and the effect of the three different remediation techniques.

The following conclusions are obtained from the study:

- The bioremediation techniques with bacteria and organic manure showed an improved result in the removal of oil contaminants in the soil.
- The oil and grease content in contaminated soil is initially 81 mg/kg. The obtained percentage removal of oil contaminants is 60.49, 92.59, and 95.06% for the remedial combinations bacteria (BA), cow dung–rice husk (CR), and goat manure–rice husk (GR).
- It is found that bacterial remediation is found to be economical in comparison with the other methods.
- The unconfined compression strength of the uncontaminated soil has decreased by 34.61 and 46.15% for 3 and 6% oil contamination.
- The UCS value of the bioremediated soil has increased by 15.29, 23.53, and 29.41% for the remedial combinations bacteria (BA), cow dung–rice husk (CR), and goat manure–rice husk (GR).

6 Future Studies

The oil concentration used in the current study is in the range of 1–6% tested for three different remediation methods. Tests can be done for higher concentration of contaminants for long degradation period. The growth of bacteria, its propagation, and metabolism for different oil concentration can be studied. Microbial dynamics can also be analyzed in future.

References

1. Li X, Liu L, Wang Y, Luo G, Chen X, Yang X et al (2013) Heavy metal contamination of urban soil in an old industrial city (Shenyang) in Northeast China. Geoderma 192:50–58
2. Butler CS, Mason JR (1997) Structure-function analysis of the bacterial aromatic ring–hydroxylating dioxygenases. Adv Microbial Physiol 38:47–84
3. Adu AA, Aderinola OJ, Kusemiju V (2015) comparative effects of spent engine oil and unused engine oil on the growth and yield of Vigna Unguiculata (Cowpea). Int J Sci Technol 4(3):105–118
4. Abdulsalam S, Adefia SS, Bugaje IM, Ibrahim S (2012) Bioremediation of soil contaminated with used motor oil in a closed system. J Bioremediat Biodegrad 3(12):3–9
5. Stephen E, Ijah UJJ (2011) Comparison of glycine max and *sida acuta* in the phytoremediation of waste lubricating oil polluted soil. Nat Sci 9(8):190–193
6. Abioye OP, Aziz AA, Agamuth P (2010) Biodegradation of used motor oil in soil using organic waste amendments enhanced biodegradation of used engine oil in soil amended with organic wastes. Water Air Soil Pollut 209:173–179

7. Onuoha SC, Olugbe VU, Uraku JA, Uchedu DO (2011) Biodegradation potentials of hydrocarbon degraders from wastelubricating oil-spilled soils in Ebonyi state, Nigeria. Int J Agric Biol 3(4):586–590
8. Ebenezer LA (2013) Bioremediation of hydrocarbon contaminated soil using compost, NPK fertilizer and cattle bile as amendment materials. Kwame Nkrumah University of Science and Technology
9. Akpan EE, Ogboi Kingsley C, Nwadinigwe CA (2013) Bioremediation of hydrocarbon polluted soil in the lowland forest ecosystem in the Niger Delta through enhanced natural attenuation process (ENAP). Int J Appl Sci Technol 3(8)
10. Bushnell LD, Haas HF (1941) The utilization of hydrocarbons by microorganisms. J Bacteriol 41:653–673
11. Nie M, Zhang X, Wang J, Jiang L, Yang J, Quan Z et al (2009) Rhizosphere effects on soil bacterial abundance and diversity in the Yellow River Deltaic ecosystem as influenced by petroleum contamination and soil salinization. Soil Biol Biochem 41(12):2535–2542
12. Rimmer DL, Vizard CG, Pless-Mulloli T, Singleton I, Air VS, Keatinge ZAF (2006) Metal contamination of urban soils in the vicinity of a municipal waste incinerator: one source among many. Sci Total Environ 356(1–3):207–216
13. McLaughlin B (2001) Soil Remediation. Enginr Sci Rev 2:69–77
14. Harder E (2004) Bioremediation of engine oil. Little Flower Academy, Dallas, Texas
15. Bello RA Bioremediation of Engine Oil Contaminated Site O. Ogunbayo
16. Wu M, Guo X, Wu J, Chen K (2020) Effect of compost amendment and bioaugmentation on PAH degradation and microbial community shifting in petroleum-contaminated soil. ECSN, p 126998
17. Zhang C, Wu D, Ren H (2020) Bioremediation of oil contaminated soil using agricultural wastes via microbial consortium, pp 1–8
18. Abdulyekeen KA, Muhammad IM, Giwa SO, Abdulsalam S (2016) Bioremediation of used motor oil contaminated soil using elephant and horse dung as stimulants
19. Gupta KK, Aneja KR, Rana D (2016) Current status of cow dung as a bioresource for sustainable development. Bioresour Bioprocess
20. Nwogu TP, Azubuike CC, Ogugbue CJ (2015) Enhanced bioremediation of soil artificially contaminated with petroleum hydrocarbons after amendment with Capra aegagrus hircus (Goat) Manure. Biotechnol Res Int 2015:1–7

Application of Brick Kiln Dust for Sustainable Construction

Hemant Sood, Gaurav Gupta, and Pardeep Kumar Gupta

1 Introduction

Clay bricks (solid-fired) form the primary building blocks employed for construction in India, where India is already producing the largest number of bricks among all nations in the world, next to China. Approximately, 0.25 trillion bricks are manufactured annually in about 140,000 brick kilns in India, accounting for 11.3% of the global brick production [7]. An annual growth of 2–5% in the production of bricks has been witnessed in the recent past [5, 12]. Brick kiln dust, a blend of coal ash, burnt soil, wood ash and brickbats, is a by-product generated during the process of brick manufacturing in brick kilns, i.e., during combustion of wood and coal, for the firing of bricks. The predominant exercise of discarding the waste in undeveloped ground has led to its recognition as the largest source of air pollution in India, after fly ash [13]. The escalation in production of the waste has resulted in shortage of terrestrial land for its disposal, and growing volume of the waste is becoming a hazardous threat to ecosystem. The leachate originating from the waste comprises toxic and heavy metals elements that pose a potential danger of polluting surface water, groundwater and land [8].

H. Sood · G. Gupta (✉)
Department of Civil Engineering, National Institute of Technical Teachers' Training and Research, Chandigarh, India
e-mail: gaurav007gupta@gmail.com

H. Sood
e-mail: sood_hemant@yahoo.co.in

P. K. Gupta
Department of Civil Engineering, Punjab Engineering College (Deemed to be University), Chandigarh, India
e-mail: p_gupta_2000@yahoo.com

H. Singh et al. (eds.), *Sustainable Development Through Engineering Innovations*, Lecture Notes in Civil Engineering 113,
https://doi.org/10.1007/978-981-15-9554-7_9

Table 1 Physical characteristics of brick kiln dust

Physical property	Results	
	Gupta et al. [4]	Riaz et al. [9]
Specific gravity	2.20	2.42
Gravel (%)	0.0	0.0
Sand (%)	72.7	70.9
Silt (%)	27.7	29.1
Clay (%)	0.0	0.0

In spite of a steady mount in accretion of the waste, the concern associated with the influence of the waste on the ecosystem has unfortunately been ignored so far. Consequently, application of the waste in construction-based activities is crucial to practice sustainable construction and effective management of the waste.

2 Characteristics of Brick Kiln Dust and Its Application

For the current section, the literature related to the characteristics of the waste has been presented along with its application. Important engineering characteristics such as physical, chemical, mineralogical, microstructural and toxicity are discussed below to get an insight of the governing mechanism prevalent in areas where the waste shall be applied to attain effectual application such as construction projects.

2.1 Physical Characteristics

The physical characteristics of the waste reported by various scholars have been reported in Table 1.

The particle size distribution of brick kiln dust reported in the literature shows that the waste is non-plastic in nature and is composed of sand and silt size particles with absence of clay size particles.

2.2 Chemical Characteristics

The chemical composition of the waste, obtained by X-ray fluorescence (XRF) method, depicts that the waste chiefly comprises quartz, alumina and iron with a low amount of calcium [4, 6, 9] The chemical composition of brick kiln dust, determined by various researchers, has been reported in Table 2. Loss on ignition values (LOI) was determined to be in the range of 1–2%. LOI being less than 5% ensures no peril of sudden self-ignition of the waste.

Table 2 Chemical composition of brick kiln dust

Elements	Results		
	Gupta et al. [4]	Riaz et al. [9]	Metwally et al. [6]
SiO_2 (%)	50.6	68.25	60.42
Al_2O_3 (%)	19.4	2.3	24.71
Fe_2O_3 (%)	11.4	8.1	5.01
CaO (%)	5.9	2.1	1.57
MgO (%)	1.72	2.72	1.89
TiO_2 (%)	0.93	–	0.85
P_2O_5 (%)	2.94	–	–
SO_3 (%)	3.66	–	2.23
MnO (%)	0.06	–	0.04
Na_2O (%)	0.87	0.1	1.79
K_2O (%)	2.23	–	–
LOI (%)	1.1	1.2	1.49

2.3 Mineralogical Characteristics

The diffractogram of the waste has been obtained through X-ray diffraction (XRD) method, reported by scholars, which shows the existence of crystalline phases of quartz (SiO_2), hematite (Fe_2O_3) and alumina (Al_2O_3) as major peaks accompanied by a minor quantity of calcite as carbonates ([1, 4, 6, 9]). The calcite peaks were not distinct in the waste. Presence of high amount of quartz, alumina and hematite and low amount of calcite was corroborated by the results of XRF analysis where the content of calcium oxide (CaO) was reported to be less than 5% in the literature. Low presence of CaO was due to decomposition of the majority of the carbonate during combustion of bricks. Being low in the content of CaO, the waste is expected not to possess any cementitious properties.

2.4 Microstructural Characteristics

The scanning electron microscopy (SEM) micrographs of brick kiln dust, reported by scholars, show that the waste particles were found to have a rough surface, and they mostly comprise irregular, spherical and flaky-shaped particles, justifying the presence of variable-sized particles due to varying particles size distribution ([4, 6, 9]; Ahmad 2017). The SEM images of brick kiln dust have been presented in Fig. 1a, b at 30× and 100× magnification, respectively, Gupta et al. [4].

Fig. 1 SEM images of brick kiln dust at a magnification of **a** 30× and **b** 100×, Gupta et al. [4]

2.5 *Toxicity*

Mondal et al. [8] reported high concentrations of potential toxic elements present in the waste samples, collected from two different locations in India. The literature highlights the potential threat of hazardous contamination of groundwater and surface water bodies by leaching of heavy metals due use of the waste in embankments. Therefore, it is essential to assess the concentration of hazardous elements in the leachate of the waste before its application in any type of construction project.

Gupta et al. [4] carried out toxicity characteristic leachate procedure to acquire leachate of the waste and evaluated the concentration of toxic elements. Results as stated in Table 3 show that zinc, arsenic and lead were recognized in the leachate of the waste whereas erstwhile toxic elements were found below the detection limit of 0.1 mg/L of atomic absorption spectrophotometer.

Overall, the concentrations of heavy metals were discovered to be noteworthily under the permissible limits for hazardous waste, as prescribed by United States Environmental Protection Agency (USEPA). Therefore, it was determined that brick kiln dust is safe for application in construction projects to attain sustainability.

Table 3 Concentration of heavy metals in the waste

Heavy metals	Concentration (mg/L)	USEPA limits for hazardous wastes (mg/L)
Chromium	BDL[a]	5.0
Zinc	0.85	Not reported
Lead	0.11	5.0
Arsenic	0.13	5.0
Cadmium	BDL*	1.0
Nickel	BDL*	Not reported
Copper	BDL*	Not reported
Mercury	BDL*	0.2

[a]BDL: Below detection limit

Physical, chemical, mineralogical, microstructural and toxicity characteristics of the waste reported in literature show that the waste possesses a number of favorable characteristics, including high sand content and absence of plasticity, for stabilization of weak soils for the subgrade of flexible pavements. Prediction of high demand of bricks for construction in future necessitates the need for sustainable management of the waste.

2.6 *Application of Brick Kiln Dust for Soil Stabilization*

Weak soils have been reported to be stabilized by the waste for the effective application of the waste. Gupta et al. [2] added the waste to soils of different plasticity, in the range of 0–40%, by the dry mass of the soil. The researchers depicted the benefit of stabilization in enhancing strength and stiffness characteristics of soils, reducing carbon footprint, pavement thickness cost associated during the life cycle of pavements [2–4]. The consequence of the stabilization on compaction, strength, plasticity and stiffness of soil was reviewed.

2.6.1 Compaction Parameters of Brick Kiln Dust Stabilized Soils

Stabilization of soils with the waste results in alteration of the compaction parameters of the soil: optimum moisture content (OMC) and maximum dry density (MDD) [2–4]. With a raise in the dosage of brick kiln dust, it was reported that MDD decreases whereas OMC increases. The reason for reduction in MDD, as cited by the authors, was lesser specific gravity of the particles of the waste than the soil. The reason for increase in OMC, as cited by the authors, was higher water absorption by the particles of brick kiln dust than the soil, due existence of un-burnt carbon in the particles of the waste.

2.6.2 Strength Parameters of Brick Kiln Dust Stabilized Soils

Stabilization of soils of varying plasticity with the waste results in alteration of bearing strength of the soils, determined through California bearing ratio (CBR) test [2–4]. With raise in the dosage of the waste, it was reported that CBR rises for all the soils, till the optimum content of the stabilizer is attained, past that the CBR falls. The optimum content of the waste of 30%, added by the dry mass of the soil, for all the soils was reported. The reason for increase in CBR, as cited by the authors, was reduction in plasticity of plastic soils and improvement in particle size distribution of non-plastic soils due to insertion of sand and silt size particles of the waste.

2.6.3 Stiffness Parameters of Brick Kiln Dust Stabilized Soils

Stiffness parameters such as resilient modulus of a fine-grained soil stabilized with the optimum dose of 30% brick kiln dust, added by the dry mass of the soil, were determined through advanced cyclic triaxial testing [4]. Resilient modulus of the soil increased by 47.2 and 54.4%, for samples tested under unsoaked and four-days water-soaked conditions, respectively.

2.6.4 Plasticity Parameters of Brick Kiln Dust Stabilized Soils

Stabilization of soil with the waste results in reduction of the plasticity parameters of the soil such as plasticity index, plastic limit and liquid limit [4]. All the plasticity parameters of the soil reduce with rise in contents of the waste. The reason for fall in plasticity, as cited by the authors, was non-plastic nature of the particles of the waste.

2.7 Application of Brick Kiln Dust in Other Engineering Applications

The waste has been utilized in several engineering applications other than soil stabilization such as to increase the growth and yield of plants, for pyrolysis of polyethylene (high-density) in order to obtain a beneficial oil which is competent enough for use a substitution of petro-fuels, as a low-cost adsorbent and for replacement of fresh clay soil that is used in manufacturing of bricks.

Rizvi and Khan [10] utilized the waste for increasing the yield and growth performance of eggplant. A soil was replaced by the waste in the range of 5–50%, added by the dry mass of the soil. The growth and yield of the plant were substantially escalated with rise in the dosage of the waste in the altered soil. Though, at larger dosage of the waste, in the range of 40–50%, growth and yield parameters mitigated considerably. The examination of the results shows that the optimal dosage of the waste was depicted as 30% for the improved yield and development of eggplant. Also, in a similar study, Rizvi and Khan [11] further studied the impact of the waste on development and penetration of *Meloidogyne javanica* in the roots of eggplant. The authors reported that the waste was capable of suppressing the penetration of *Meloidogyne javanica* in the roots of eggplant. Overall brick kiln dust emerged as an effective fertilizer that increased the yield of eggplant.

Ahmad et al. [1] utilized the waste for pyrolysis of polyethylene ((high density) to obtain beneficial oil that is competent enough for use a substitution of petro-fuels. The authors reported that ZnO catalyst containing 5% brick kiln dust was found to be appreciably active. The resultant oil depicted fuel values close to standard petro-fuels.

Riaz et al. [9] utilized the waste in brick construction. The waste was applied for replacement of good quality soil used for brick manufacturing, in the contents of 5, 10, 20 and 25%, by the dry mass of the soil. On the basis of the tests performed, the authors reported that the bricks containing the waste have a porous structure and are apt for use in medium weather resistance and for the purpose of insulation. These bricks reported greater absorption of water than the control bricks; therefore, it is essential to soak the bricks in water before use. The waste containing bricks were discovered to be tougher to efflorescence than the bricks containing only clay soil and conversely less resistant to sulfate attack than the control bricks. The authors reported reduced compressive strength for bricks containing brick kiln dust than the control bricks. It was reported that the strength decreases with rise in contents of brick kiln dust. However, replacement of fresh clay soil with up to 15% brick kiln dust resulted in brick manufacturing conforming to second class of bricks, where the control bricks also conformed to second class bricks. The bricks containing 25% brick kiln dust reported flexure strength of more than 1 MPa, meeting the criteria of least flexure. Overall replacement of 25% of fresh clay soil with brick kiln dust, for brick manufacturing, emerged as a useful and an economical method for brick construction.

Metwally et al. [6] used fixed bed column technique for application of the waste as an economical adsorbent in aqueous solution for elimination of strontium, cesium and lead ions. The column performance was 80.1, 73.6 and 68.5% for lead, cesium and strontium ions, respectively. Results show that the waste was determined to be an effectual adsorbent. The process was reported to be easy, eco-friendly, environmentally efficient and economical.

2.8 Application of the Waste for Sustainable Construction Practices

Sustainable construction can be practiced by application of the waste in activities and construction projects. Gupta et al. [3] reported diminution of greenhouse gas emissions and thickness of layers of flexible pavements, on application of the waste in stabilization of soils forming the subgrade of flexible pavements. It was highlighted that the stabilization process leads to lesser consumption of bitumen and natural aggregates. Gupta et al. [3] reported that inclusion of brick kiln dust for stabilization of a fine-grained soil forming the subgrade of flexible pavements designed to cater to the traffic for 50 years results in lesser cost and associated greenhouse gas emissions.

3 Conclusions

Brick kiln dust was identified as a major pollutant to the environment. Brick kiln dust imposes environmental threats of leaching of heavy metals into the water and land bodies. Therefore, there is an urgent for application of the waste in construction which is expected to yield sustainable construction. Brick kiln dust has been effectively utilized in stabilization of weak soils and many other engineering applications. Following conclusions were drawn from the review of the literature.

1. Physical characteristics of brick kiln dust reported in the literature show that the waste comprises of sand and silt size particles and is non-plastic in nature.
2. Chemical characteristics of brick kiln dust reported in the literature show that the waste is rich in silica, alumina and iron whereas the waste is poor in calcium.
3. Mineralogical characteristics of brick kiln dust reported in the literature show that the waste is rich quartz, alumina and hematite whereas a minute quantity of calcite, present as carbonates, has been reported to be present.
4. Microstructural characteristics of brick kiln dust reported in the literature show that the waste has a rough surface, and it mostly comprises irregular, spherical and flaky-shaped particles.
5. Toxicity characteristics of brick kiln dust reported in the literature show that the waste possesses high concentrations of potential toxic elements. Heavy metals such as zinc, arsenic and lead were present in leachate of the waste. Overall, the concentrations of heavy metals were found within permissible limits for hazardous waste, as prescribed by USEPA.
6. Brick kiln dust has been utilized in engineering applications other than soil stabilization for increasing the growth and yield of plants, for pyrolysis of polyethylene (high density) to obtain beneficial oil that is competent enough for use a substitution of petro-fuels, as a low-cost adsorbent and for replacement of fresh clay soil for manufacturing of bricks.
7. Stabilization of soil with brick kiln dust results in decrease of MDD and plasticity and increase of OMC, with increase in contents of the waste.
8. Stabilization of soils with brick kiln dust results in increased bearing strength of soils of different plasticity at an optimum dosage of 30% content of the waste, added by the dry mass of the soils.
9. Stabilization of soil with brick kiln dust results in increased stiffness, measured as resilient modulus, determined through cyclic triaxial test. Resilient modulus

of the soil increased by 47.2 and 54.4%, for samples tested under unsoaked and four-days water-soaked conditions, respectively.

10. Sustainable construction can be practiced by application of the waste in stabilization of soils forming the subgrade layer of flexible pavement. The stabilization process leads to reduction of cost of pavements construction, lowering of pavement thickness and lessening of release greenhouse gas emissions associated with pavement construction.
11. No study has been executed to find the performance of the waste on permeability, split tensile strength, shear strength and unconfined compressive strength of soils. Also, studies to determine the effect of the waste on stiffness, measured as resilient modulus, of soils of different plasticity have not been performed.

References

1. Ahmad I, Khan MI, Khan H, Ishaq M, Khan R, Gul K, Ahmad W (2017) Pyrolysis of HDPE into fuel like products: evaluating catalytic performance of plain F metal oxides impregnated waste brick kiln dust. J Anal Appl Pyrol 124:195–203
2. Gupta G, Sood H, Gupta PK (2018a) brick kiln dust waste management through soil stabilization. In: International conference on sustainable waste management through design. Springer, Cham, pp 422–431
3. Gupta G, Sood H, Gupta PK (2018b) Life-cycle cost analysis of brick kiln dust stabilized perpetual pavements for lowering greenhouse gas emissions in India. In: ASCE India conference 2017 American Society of civil engineers. Indian Institute of Technology Delhi Institution of Engineers, India
4. Gupta G, Sood H, Gupta P (2020) Performance evaluation of pavement geomaterials stabilized with Pond Ash and Brick Kiln dust using advanced cyclic triaxial testing. Materials 13(3):553
5. Lalchandani D, Maithel S (2013) Towards cleaner brick kilns in India. In: A win–win approach based on Zigzag firing technology. Greentech Knowledge Solutions Pvt. Ltd, New Delhi
6. Metwally SS, El-Sherief EA, Mekhamer HS (2020) Fixed-bed column for the removal of cesium, strontium, and lead ions from aqueous solutions using brick kiln waste. Sep Sci Technol 55(4):635–647
7. Mills SJ (2010) Coal use in the new economies of China, India and South Africa. IEA Clean Coal Centre, London
8. Mondal A, Das S, Sah RK, Bhattacharyya P, Bhattacharya SS (2017) Environmental footprints of brick kiln bottom ashes: geostatistical approach for assessment of metal toxicity. Sci Total Environ 609:215–222
9. Riaz MH, Khitab A, Ahmed S (2019) Evaluation of sustainable clay bricks incorporating brick kiln Dust. J Build Eng 24:100725
10. Rizvi R, Khan AA (2009) Response of eggplant (Solanum melongena L.) to fly ash and brick kiln dust amended soil. Biol Med 1(2):20–24
11. Rizvi R, Khan AA (2011) Influence of brick-kiln dust on penetration and development of Meloidogyne javanica in eggplant roots and its impact on plant growth and yield
12. Singh AL, Asgher MS (2005) Impact of brick kilns on land use/land cover changes around Aligarh city, India. Habitat Int 29(3):591–602
13. Upadhyay E (2002) Impact of particulate air pollutants on Brassica Juncea L. and Linumusitatissimum L. Doctoral dissertation, Aligarh Muslim University

Effect of Municipal Solid Waste Incinerator Ash and Lime on Strength Characteristics of Black Cotton Soil

Kapil Kumar Gautam, Ravi Kumar Sharma, and Abhishek Sharma

1 Introduction

Black cotton soil (BCS) is prone to welling and shrinkage due to their property to react with moisture. This swelling and shrinkage of black cotton soil cause distress to the structures founded over them (Nelson and Miller 1992, [8, 9]). There is differential settlement of the structures founded over these soils, and hence, construction is often neglected on these soils. However, if it is not possible to avoid the construction on these soils, then these soils are stabilized prior to construct structures over them. The stabilization using lime is a very old practice to improve the geotechnical characteristics of poor soil. When lime gets added to BCS, its engineering properties are improved depending on soil characteristics, temperature, and period of curing [1]. Lime reacts with clay minerals of BCS instantaneously, thus satisfying its moisture affinity before executing any pozzolanic reactions, and the amount of this lime is often termed as 'fixed-lime' [4]. When an adequate lime amount is added to expansive soils such that the soil mass attains a pH value 10 or above, the pozzolanic reactions occur resulting in the development of gel-like compounds that continue to crystalline with curing period [5]. The reactions that occur within the soil mass due to lime reduce the thickened double layer formed owing to adsorption of water by clay minerals resulting in flocculation and agglomeration of particles, thus modifying characteristics of the soil mass. A lot of researchers found that lime is highly

K. K. Gautam · R. K. Sharma · A. Sharma (✉)
National Institute of Techonolgy Hamirpur, Hamirpur, Himachal Pradesh, India
e-mail: abhishek1@nith.ac.in

K. K. Gautam
e-mail: er.kapilgautam@gmail.com

R. K. Sharma
e-mail: rksnithp61@gmail.com

H. Singh et al. (eds.), *Sustainable Development Through Engineering Innovations*, Lecture Notes in Civil Engineering 113,
https://doi.org/10.1007/978-981-15-9554-7_10

Table 1 Geotechnical characteristics of black cotton soil

Property	Value/Description
Specific gravity, G	2.71
Soil type	CH
DFS (%)	40.167
Liquid limit (%)	69.62
Plastic limit (%)	30.00
Plasticity index (%)	39.62
Maximum dry density, MDD (gm/cc)	1.57
Optimum moisture content, OMC (%)	20.00
Unconfined compressive strength at 28 days, UCS (kPa)	385.00

effective in gaining the desirable properties of soil ([7], Bhardwaj and Sharma 2020). As per manual of Government of India (2016), the waste production in India is much massive than its population, and incineration of these wastes remains the best alternative to restricted landfills. Show et al. [10] found incinerator ash as an efficient stabilizer and reported a drop in plasticity and compressibility with an increase in permeability of the soft soil. Poran and Ahtchi-Ali [6] found municipal solid waste (MSW) incinerator ash suitable for replacing with conventional materials for fills in embankments, and mixing the waste ash with lime or cement improves permeability, shear strength, compaction properties, and CBR of soil. MSW incinerator ash stabilized BCS reported a drop in swelling and a rise in shear strength characteristics and internal friction angle of soil particles [3]. This study deals with the effect of addition of MSWIA to poor clayey soil to improve its geotechnical characteristics.

2 Materials

2.1 Soil

The soil was bought from Rajkot, Gujarat, India, and was categorized as a highly plastic clay (CH), conforming to Unified Soil Classification System (USCS) as mentioned in Table 1 along with other calculated geotechnical characteristics.

2.2 Lime

Lime used for investigation was obtained from a regional store in Hamirpur, Himachal Pradesh, India. The chemical structure of lime supplied by manufacturer is represented in Table 2.

Table 2 Chemical structure of lime

Chemical composition	Content (%)
Silicon dioxide (SiO_2)	45.40
Aluminium oxide (Al_2O_3)	18.50
Calcium oxide (CaO)	10.90
Ferric oxide (Fe_2O_3)	8.69
Potassium oxide (K_2O)	2.51
Sodium oxide (Na_2O)	2.30
Titanium dioxide (TiO_2)	2.10
Sulphur trioxide (SO_3)	1.86
Magnesium oxide (MgO)	1.42
Phosphorus trioxide (P_2O_3)	0.82
Loss of ignition	5.36

2.3 *Incinerator Ash*

The incinerator ash (IA) was obtained from the MSW Incinerator Ash plant in Chandigarh, India. The chemical composition of incinerator ash supplied by manufacturer is tabulated in Table 3.

Table 3 Chemical composition of incinerator ash

Chemical composition	Content (%)
Silicon dioxide (SiO_2)	55.37
Calcium oxide (CaO)	19.39
Aluminium oxide (Al_2O_3)	9.2
Ferric oxide (Fe_2O_3)	4.93
Sulphur dioxide (SO_2)	1.53
Organic carbon	0.96
Chloride (Cl)	0.44
Potassium oxide (K_2O)	0.43
Magnesium oxide (MgO)	0.41
Sodium oxide (Na_2O)	0.24
Nitrogen (N)	0.12
Barium (Ba)	0.11
Lead (Pb)	0.07
Phosphorus pentoxide (P_2O_5)	0.07
Chromium (Cr)	0.051
Zinc (Zn)	0.02
Copper (Cu)	94.4 mg/kg
Cadmium (Cd)	7.45 mg/kg

3 Method

Differential free swell index, Atterberg's limits, and UCS tests were conducted on combinations of lime and IA with BCS as per the standard codes and are mentioned along with the test names in the following section of this paper. The BCS obtained in the form of huge lumps was first manually pulverized, and debris such as particles of glass and plastic was separated. Thereafter, the natural properties of BCS were computed before beginning with stabilization tests. Results were plotted for BCS combination containing 3, 6, 9, and 12% lime and 8, 16, 24, and 32% IA.

4 Results

4.1 Differential Free Swell

DFS was calculated conferring to IS: 2720 (Part 40)-1977. The DFS of BCS decreased from 40.167 to 10% on adding 9% lime (Fig. 1), attributing to the calcium-saturation of high potential swell minerals due to cation exchange reactions and development of cementitious matrix resulting in reduced affinity for water (Bhardwaj and Shamra 2020). Furthermore, DFS reduced to 0% on adding 24% IA to BCS (Fig. 2), attributing to the presence of multivalent cations of silica, calcium, alumina, and iron that aid flocculation of soil particles owing to cation exchange reactions (Table 3). Consequently, flocs formations lead to a decreased average surface area of BCS particles that lower the moisture affinity of the clay minerals. Due to this lowered affinity, water absorption in clay minerals reduced, and the swelling potential of BCS gets reduced [2].

Fig. 1 DFS of BCS at various lime contents

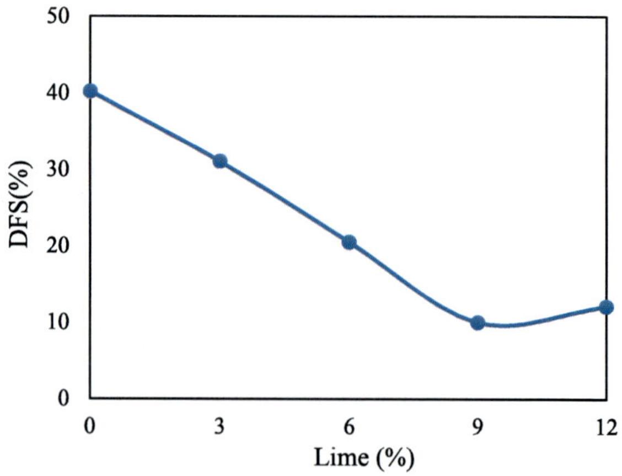

Fig. 2 DFS of BCS at various IA contents

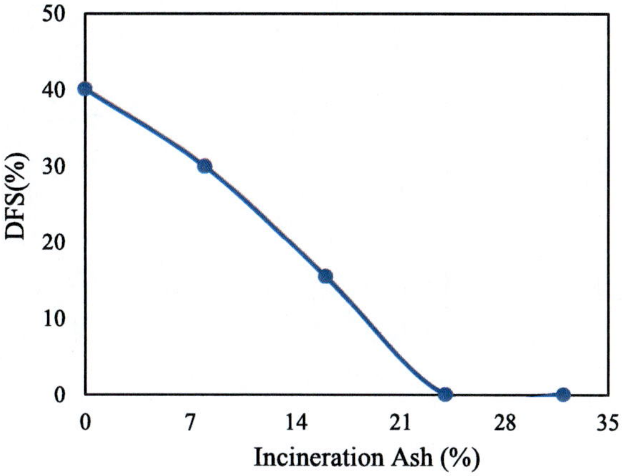

4.2 Atterberg's Limits

The Atterberg's limits were found as per the specifications of ASTM D-4318-00. The plasticity index reduced from 39.62 to 15.8% on increasing lime content in BCS to 12% (Fig. 3). The obtained plasticity may be because of formation of poorly developed cementitious products as a result of lime-modification reactions with soil [11]. Plasticity index of IA stabilized BCS decreased to 12.3% on up to 32% replacement of soil particles by IA (Fig. 4). The reduced plasticity index is related with the presence of calcium oxide and silica dioxide in IA responsible for physicochemical reaction. The combined lime-IA mixtures further reduced the plasticity to 2.1% for BCS mixture containing 9% lime and 24% IA (Fig. 5).

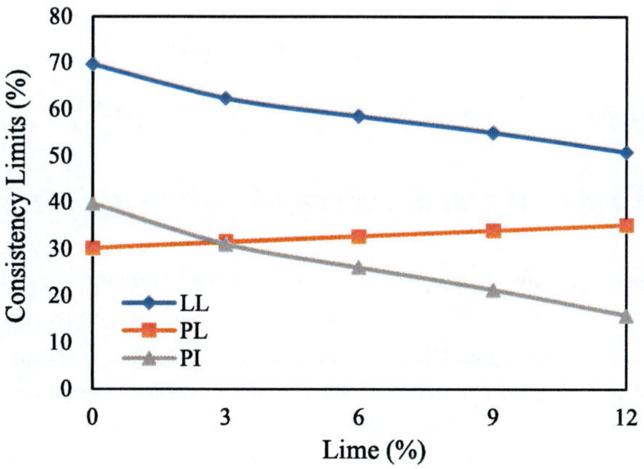

Fig. 3 Consistency limits of BCS at various lime contents

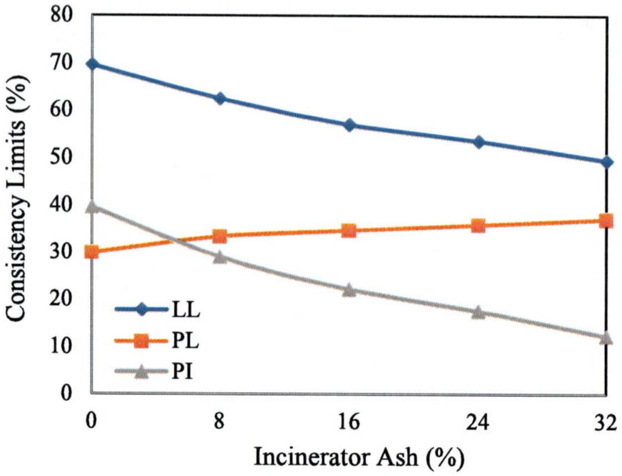

Fig. 4 Consistency limits of BCS at various IA contents

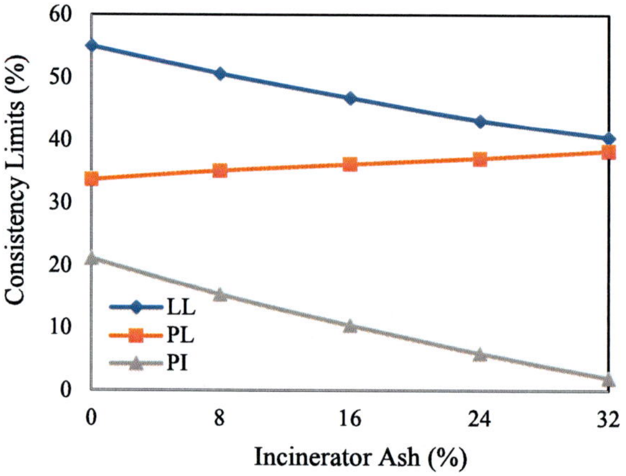

Fig. 5 Consistency limits of BCS at various IA contents with 9% lime

4.3 Unconfined Compressive Strength

The UCS tests were achieved in agreement with ASTM D 2166: 13 on various mixes of BCS along with IA and lime. The maximum strength development of BCS was 378 kN/m^2 which was observed for 28 days curing. A similar trend of the highest strength gain, i.e., at 28 days curing period, was observed for every specimen of BCS combinations with lime and IA due to their time-dependent pozzolanic reactions. The increase in the strength up to 714 kN/m^2 for 9% lime modified soil at 28 days of curing (Fig. 6) is attributed to the further development of lime-modification reactions that lead to the crystallization of cementitious products [11]. IA mixing to BCS increased

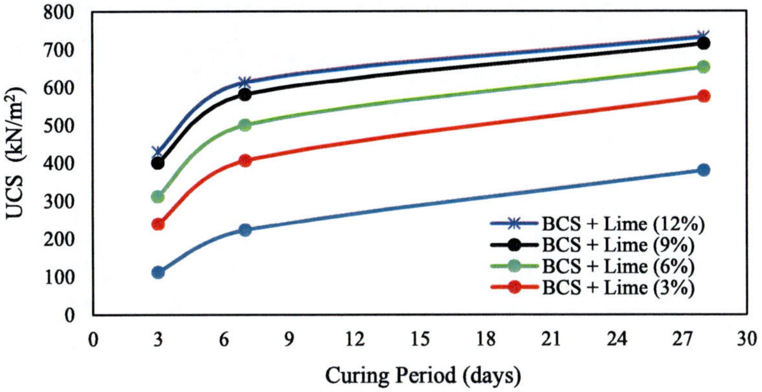

Fig. 6 UCS of BCS with curing days at various lime contents

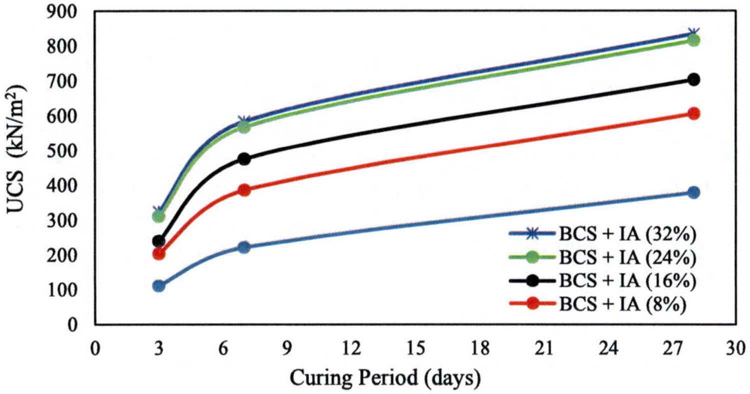

Fig. 7 UCS of BCS with curing days at various IA contents

the UCS to 815 kN/m^2 at 24% IA content (Fig. 7). The rise in strength is attributed to better packing of IA and soil particles leading to a coherent composition that sustains higher compression loading. A remarkable increase in strength to 1085 kN/m^2 was achieved for BCS stabilized with 9% lime and 24% IA (Fig. 8).

5 Conclusion

Based on the above computed results, following conclusion on optimum mixes can be drawn:

1. The differential free swell index of the BCS decreased 75.10% on 9% lime addition and further reduced by 100% on adding 24% incinerator ash.

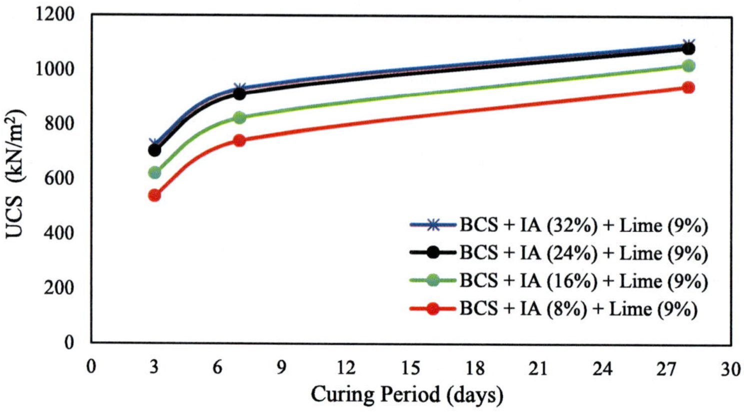

Fig. 8 UCS of BCS with curing days for 24% IA and various lime contents

2. The liquid limit and plasticity index reduced by 20.99 and 46.49% on 9% lime addition and 23.01 and 55.57% on adding 24% incinerator ash to BCS. Further improvement of plasticity index to 71.69% and reduction in liquid limit by 21.45% was obtained on adding 24% incinerator ash to 9% lime stabilized BCS.
3. The UCS for BCS, along with different combinations, was maximum for 28 days as compared to 3 and 7 days curing period. For an individual mix of 9% lime and 24% incinerator ash, UCS increased 1.88 times and 2.156 times, respectively. The highest optimum strength increment of 2.87 times was obtained for 9% lime and 24% incinerator ash.

Based on the results obtained from the experimental study, reduced swelling and remarkable strength gain in BCS reveal that lime and MSW incinerator ash together make an efficient stabilizer when added in optimum quantities in high plasticity clay soils like black cotton soil. This study is helpful in protecting the environment from the harm caused due to dumping of waste incinerator ash.

Acknowledgements The author remains thankful to NIT Hamirpur for providing with every possible facility for the successful accomplishment of the research.

References

1. Bell FG, Coulthard JM (1990) Stabilization of clay soils with lime. Municipal Eng 7(3)
2. Cokca E (2001) Use of class C fly ashes for the stabilization of an expansive soil. J Geotechn Geoenviron Eng 127(7):568–573
3. Ferreira C, Ribeiro A, Ottosen L (2003) Possible applications for municipal solid waste fly ash. J Hazard Mater 96(2–3):201–216
4. Hilt GH, Davidson DT (1960) Lime fixation in clayey soils. Highway Research Board Bulletin (262)

5. Little DN (1998) Evaluation of structural properties of lime stabilized soils and aggregates. National Lime Association, vol 1
6. Poran CJ, Ahtchi-Ali F (1989) Properties of solid waste incinerator fly ash. J Geotech Eng 115(8):1118–1133
7. Sharma RK, Hymavathi J (2016) Effect of fly ash, construction demolition waste and lime on geotechnical characteristics of a clayey soil: a comparative study. Environ Earth Sci 75(5):377
8. Sharma A, Sharma RK (2019) Effect of addition of construction–demolition waste on strength characteristics of high plastic clays. Innov Infrastruct Solutions 4(1):27. https://doi.org/10.1007/s41062-019-0216-1
9. Sharma A, Sharma RK (2020) Strength and drainage characteristics of poor soils stabilized with construction demolition waste. Geotech Geol Eng
10. Show KY, Tay JH, Goh AT (2003) Reuse of incinerator fly ash in soft soil stabilization. J Mater Civil Eng 15(4):335–343
11. Thyagaraj T, Rao SM, Sai Suresh P, Salini U (2012) Laboratory studies on stabilization of an expansive soil by lime precipitation technique. J Mater Civil Eng 24(8):1067–1075

Sustainable Use of Agri-Waste in the Production of Green Hybrid Composites

Amrinder Singh Pannu, Sehijpal Singh, Vikas Dhawan, and Jai Inder Preet Singh

1 Introduction

Since few decades, the design and development of the materials and their products became vital issue with the introduction of concept of green environment. Presently, the agenda of sustainable development is significant within industrial and public sectors. For the sustainable development, cleaner production of materials, goods and services is the only way where energy and resources are used in an efficient way so that in production very small amount of emissions and wastes are produced. Composite materials are playing very important role in sustainable development because of their easier production, less waste and emissions that result into fulfillment of agenda of green environment. Moreover, using different reinforcements and matrices, one can manage the waste also which in turn results not only into environmental protection but also boosts the economy of the country through waste management. So in making of biodegrade composite, various matrix materials like wheat gluten, whey proteins, polyglycolic acid, polylactic acid, etc., are being used in the development of composite products. Reinforcements in form of fibers are being extracted from plants like sisal, agave, banana, etc., where stems and leaves are important source

A. S. Pannu (✉)
Mechanical Engineering Department, Guru Nanak Dev Engineering College, Ludhiana, Punjab 141006, India
e-mail: erpannu@yahoo.com; mech@gndec.ac.in

S. Singh
Guru Nanak Dev Engineering College, Ludhiana, Punjab 141006, India

V. Dhawan
Academics, Research and Development, SGT University, Gurugram, Haryana 122505, India

J. I. P. Singh
School of Mechanical Engineering, Lovely Professional University, Phagwara, Punjab 144411, India

H. Singh et al. (eds.), *Sustainable Development Through Engineering Innovations*, Lecture Notes in Civil Engineering 113,
https://doi.org/10.1007/978-981-15-9554-7_11

of fiber. Also, India being an agricultural country, cultivation of wheat, rice, corn, etc., is being done on large scale. So, the waste left behind after the cultivation could be very important in development of the composite materials. In terms of applications, many renowned automotive companies are using fibers like flax and hemp in making roofs of their cars. Similarly, in the door cladding, seatback linings, floor panels, etc., hemp and sisal are being used. On the other hand, cotton fibers are not only being used for cushioning of seat but also to make the vehicles sound proof [1, 2]. Other very important field is biomedical where composites are being used in tissue engineering and showed outstanding potential [3, 4]. In the transportation and marine sector, hull reinforced with flax has been used widely for the specialized catamaran [5]. Moreover, the flax-reinforced PLA is being used to produce the canoe for boats (www.jeccomposites.com). Manufacturing processes like extrusion, resin transfer molding, injection molding, pultrusion are being used to fabricate the bio-composites [6]. Altun et al. [7] used the injection molding machine to manufacture composite based on wood flour and PLA, whereas Hu et al. [8] used the compression molding process to develop composite based on jute and PLA. Single-screw extrusion and twin-screw extrusion are being used to process the composites based on PLA and other natural reinforcements. This process is advantageous in terms of functional versatility and compounding capabilities. Torres et al. [9] used single crew extruder to produce the thermoplastic based on natural fiber reinforcements along with material characterization.

2 Development of Composite

In present scenario, protecting the environment is main concern, so eco-friendly and sustainable materials could be of utmost importance to stop the degradation of environment. In terms of matrix, as polylactic acid (PLA) is very common and emerging biodegradable material with good mechanical properties, so its use in the development of composite will aid in the enhancement of the composite properties. Similarly, banana and rice husk are abundantly available in India due to the large area under its cultivation which could be used as reinforcement in making of composites.

2.1 Treatment of Materials

To develop a composite with good mechanical properties, treatment of the reinforcement is must which in turn will increase the interfacial bonding between reinforcement and matrix. The components present in the rice husk and banana are shown in Table 1.

After drying the banana waste and rice husk in open sunlight, fibers were chopped down in mixer grinder to a size of 600 μm (0.6 mm) having aspect ratio of 2.5 as shown in Fig. 1a and 2b. To remove the unwanted particles and dust from the fibers

Table 1 Chemical composition of rice husk and banana [10, 11]

Fiber	Lignin (wt%)	Cellulose (wt%)	Hemicellulose (wt%)
Rice husk	26–31	25–35	18–21
Banana	5–10	60–65	6–19

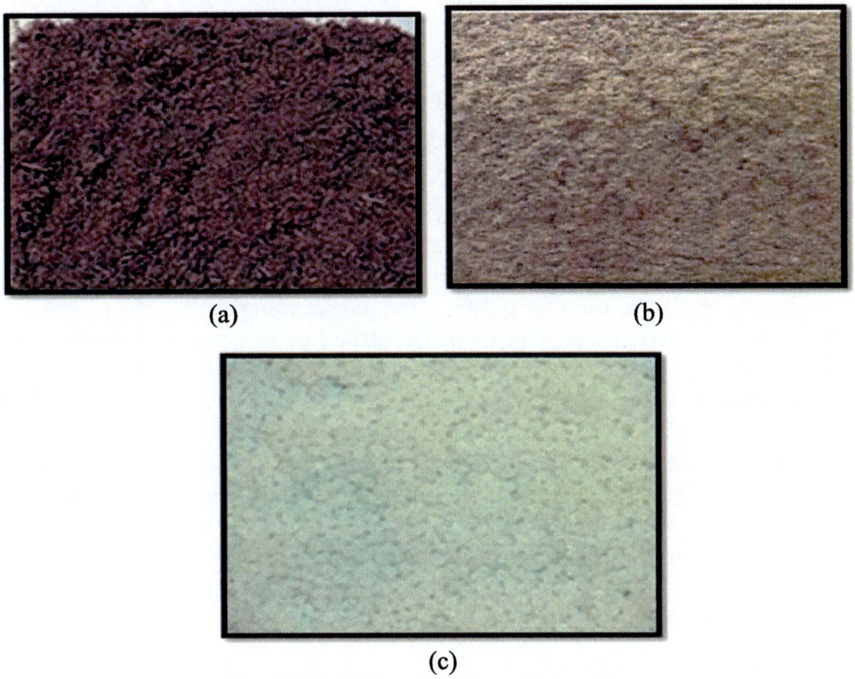

Fig. 1 **a** Chopped banana fibers; **b** chopped rice husk; **c** PLA granules (3052D grade)

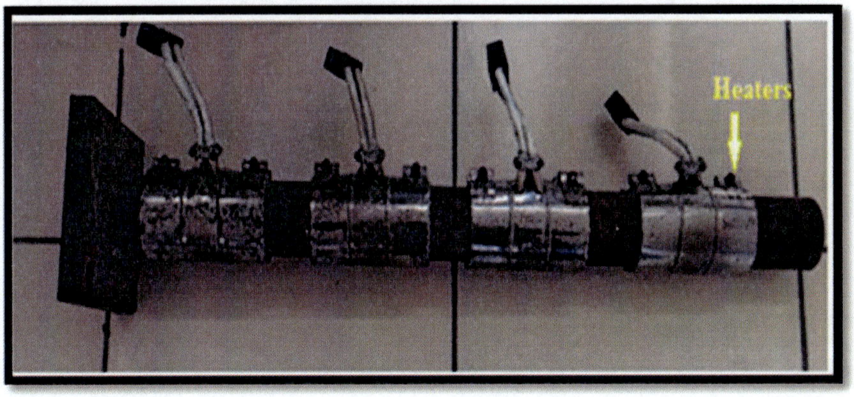

Fig. 2 Extrusion die

(rice husk and banana waste), washing has been done using distilled water for 1 h at temperature of around 60 °C. Thereafter, they are left to be dried in open air for 2 days. To increase the bonding between fibers and matrix, lignin removal is must, so NaOH treatment of 10% [12–15] was given to the fibers followed by washing and drying. To reduce the moisture content from the PLA pallets of 3052D grade (shown in Fig. 1c), heat was given to pallets at temperature of 70 °C in an oven for approximately 2 h.

2.2 *Fabrication and Testing of Composite*

Extrusion technique was used to fabricate the composite rod. Die as shown in Fig. 2 was used to develop the rod using different fiber volume ratio, i.e., 0, 10, 20, 30 and 40% (50/50 banana and rice husk in each ratio) in PLA matrix. Single-screw extruder (as shown in Fig. 3) was used for the extrusion of composite rod.

Fibers were mixed manually in a small container using coconut oil to stick the fibers properly and uniformly with the PLA granules, and then, this mixture is transferred to the hopper. Desired temperature range has been set by controller unit. Screw rotates inside the barrel. Thermocouples and heaters were installed throughout the barrel to control the temperature at different points as per requirement. Samples were made as per ASTM D7205 as shown in Fig. 4. Five samples at each volume ratio were developed to get average strength.

Mechanical testing in terms of tensile, flexural and impact was done to check the strength of the composite rod developed. For the tensile testing, samples according to ASTM D7205 are made with diameter of 12.7 mm and 45 inches of length. Five

Fig. 3 Single-screw extruder

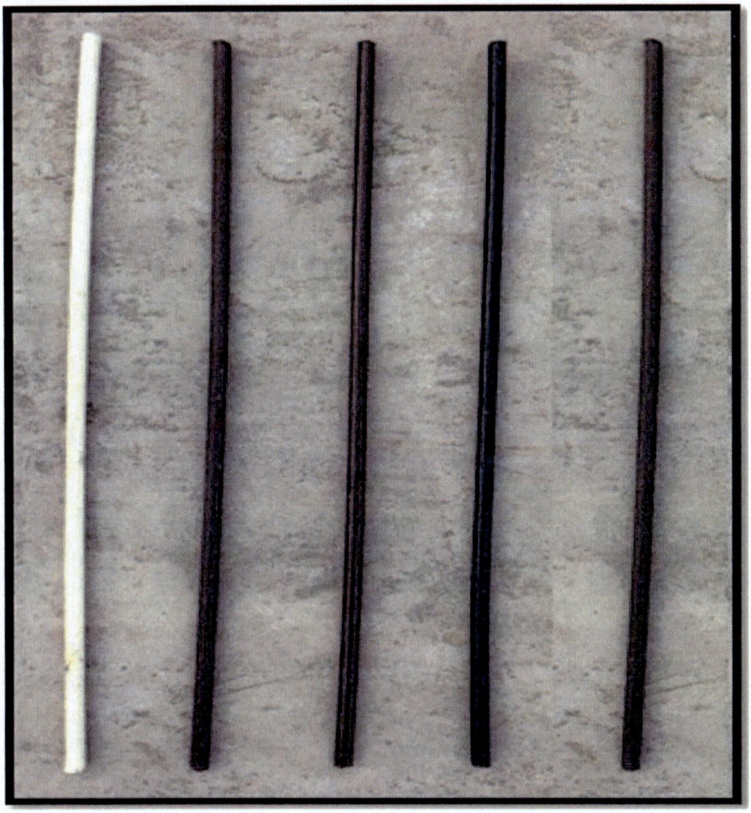

(a) (b) (c) (d) (e)

Fig. 4 Developed rods **a** pure PLA; **b** 10% fiber fraction; **c** 20% fiber fraction; **d** 30% fiber fraction; e 40% fiber fraction

samples of each rod with a specific fiber ratio were tested on computerized UTM of capacity 100 kN (Model HEICO-HLC 693.40) with strain rate of 0.001 mm/min, round sample with gauge length of 15 inches and at room temperature (around 25 °C). To check the flexural strength, samples were made as per ASTM D4476. Three point flexural testing is used to find the bending limit of material required to fail the specimen. Five specimens of each type of rod were tested on UTM of capacity 10 kN (Model-Shanta Engineering). Similarly, the impact energy of the specimen was measured using the Izod testing machine (Model 0701, Avery Birmingham). Five samples for each volume fraction were prepared as per ASTM D-256.

3 Results and Discussion

During mechanical testing, the results shown in Fig. 5 were obtained. It showed the variation of mechanical strength of the hybrid composite with the increase in the fiber volume ratio. With the increase in fiber ratio, the strength of hybrid composite rod increases. Trend of increase in strength continues up to 40% fiber fraction where 20% is banana waste and 20% is rice husk.

It has been observed that the tensile strength of 44.56 MPa is observed in 40% fiber fraction rod while for neat PLA it is 15.82 MPa. During the flexural testing of specimen, the flexural strength comes out to be 38.59 MPa for 40% reinforcement, and for pure PLA rod, it is 13.64 MPa. In case of impact testing, the value touched 5.12 kJ/m^2 for 40% reinforcement, and for neat PLA, it has been observed as 1.98 kJ/m^2. All these strength values are maximum at fiber fraction ratio of 40% (where 20% is banana fiber and 20% rice husk) the reason being good bonding between the fibers and matrix. Shih and Huang [16] and Pannu et al. [17] had also reported the increase in mechanical strength of composite with the increase in reinforcement of banana fibers in PLA. So from the above results, overall there has been an increase of 181.66% in tensile strength, 182.91% in flexural strength and 158.58% in impact strength with reinforcement of fibers in polylactic acid. From these results, it has been observed that, with the use of waste (banana and rice husk) as reinforcement, the agenda of waste management is fulfilled. In the sustainable development, cleaner production, reduction of waste and emission are prime requirement. So during the extrusion of composite rod, the reinforcement is agricultural waste that is completely biodegradable has been utilized, and PLA is also very common biodegradable matrix that can be easily manufactured from corn starch. This is a closed process of manufacturing the composite. Moreover, utilizing the waste coming from agricultural and industrial sector, it will not only increase the income of the farmers but also boost up the economy of the country.

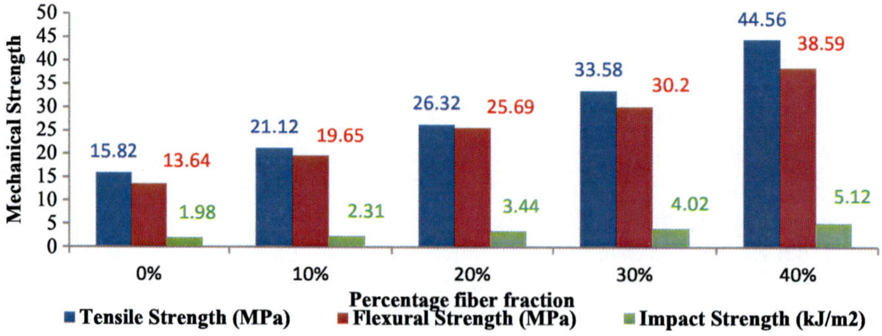

Fig. 5 Mechanical strength of the specimens

4 Conclusion

The fully biodegradable hybrid composite was made successfully using banana waste and rice husk as reinforcement in PLA matrix. During testing of developed hybrid composite, significant improvement has been observed in rice husk-/banana-reinforced hybrid composite as compared to pure PLA sample. The following conclusions were drawn from the mechanical testing of the developed composite:

- Fulfilled the goal of sustainable development by using waste material in manufacturing of fully biodegradable composite.
- Extrusion has emerged as efficient method for developing composite products like rod where any type of chopped fiber can be used.
- Waste management issue has been addressed by using banana waste and rice husk as reinforcement.
- Tensile and flexural strength increased by around 2.5 times at 40% fiber reinforcement.
- In case of impact testing, strength observed was around 2 times at maximum reinforcement as compared to pure PLA.

Acknowledgements The author is highly thankful to Principal, Guru Nanak Dev Engineering College, Ludhiana, to carry out this research work and using testing facilities of the institute. Guidance given by Supervisors (Dr. Sehijpal Singh and Dr. Vikas Dhawan) is of great importance in completing the entire research work. Author is also grateful to organizers of SDEI International Engineering Congress for giving platform to present the findings at this level. In the end, the author is also thankful to all the staff members, colleagues and friends for giving support/advice whenever needed.

References

1. Singh B, Verma A, Gupta M (1998) Studies on adsorptive interaction between natural fiber and coupling agents. J Appl Polym Sci 70(9):1847–1858
2. Winandy JE, Stark NM, Clemons CM (2004) Considerations in recycling of wood-plastic composites. In: 5th Global wood and natural fibre composites symposium, 27–28 April 2004, Kassel, Germany, p 9
3. Ngiam M, Liao S, Patil AJ, Cheng Z, Yang F, Gubler MJ, Chan CK et al (2009) Fabrication of mineralized polymeric nanofibrous composites for bone graft materials. Tissue Eng Part A 15(3):535–546
4. Qian J, Xu W, Yong X, Jin X, Zhang W (2014) Fabrication and in vitro biocompatibility of biomorphic PLGA/nHA composite scaffolds for bone tissue engineering. Mater Sci Eng C 36:95–101
5. Van Rijswijk K, Brouwer WD, Beukers A (2001) Application of natural fibre composites in the development of rural societies. Delft University of Technology, Delft
6. Faruk O, Bledzki AK, Fink HP, Sain M (2012) Biocomposites reinforced with natural fibers: 2000–2010. Prog Polym Sci 37(11):1552–1596

7. Altun Y, Doğan M, Bayramlı E (2013) Effect of alkaline treatment and pre-impregnation on mechanical and water absorbtion properties of pine wood flour containing poly (lactic acid) based green-composites. J Polym Environ 21(3):850–856
8. Hu RH, Ma ZG, Zheng S, Li YN, Yang GH, Kim HK, Lim JK (2012) A fabrication process of high volume fraction of jute fiber/polylactide composites for truck liner. Int J Precis Eng Manuf 13(7):1243–1246
9. Torres FG, Ochoa B, Machicao E (2003) Single-screw extrusion of natural fibre reinforced thermoplastics (NFRTP) study of the material transport processes. Int Polym Process 18(1):33–40
10. Arjmandi R, Hassan A, Majeed K, Zakaria Z (2015) Rice husk filled polymer composites. Int J Polym Sci
11. Mohanty AK, Misra M, Drzal LT (2001) Surface modifications of natural fibers and performance of the resulting biocomposites: an overview. Compos Interfaces 8(5):313–343
12. Kumar R, Choudhary V, Mishra S, Varma IK (2008) Banana fiber-reinforced biodegradable soy protein composites. Frontiers Chem China 3(3):243–250
13. Thiruchitrambalam M, Alavudeen A, Athijayamani A, Venkateshwaran N, Perumal AE (2009) Improving mechanical properties of banana/kenaf polyester hybrid composites using sodium laulryl sulfate treatment. Mater Phys Mech 8(2):165–173
14. Alavudeen A, Rajini N, Karthikeyan S, Thiruchitrambalam M, Venkateshwaren N (2015) Mechanical properties of banana/kenaf fiber-reinforced hybrid polyester composites: effect of woven fabric and random orientation. Mater Des 1980–2015 66:246–257
15. Ndazi BS, Karlsson S, Tesha JV, Nyahumwa CW (2007) Chemical and physical modifications of rice husks for use as composite panels. Compos Part A Appl Sci Manuf 38(3):925–935
16. Shih YF, Huang CC (2011) Polylactic acid (PLA)/banana fiber (BF) biodegradable green composites. J Polym Res 18(6):2335–2340
17. Pannu AS, Singh S, Dhawan V, Singh J (2017) Development of natural fiber reinforced fully biodegradable composite rod. IJMSE 8(1):53–59

Optimization Model in Sustainable Development: Multiobjective Programming Approach

Kailash Lachhwani, Naveen Jha, Deepam Goyal, Abhishek Dwivedi, and Rajeev Kumar Dang

1 Introduction

In manufacturing sector, social, economic and environmental are three dimensions, which form basis of sustainability, and these three dimensions constitute three bottom line (TBL) [6]. In recent years, sustainability or sustainable development has achieved a greater visibility based on all kinds of dimensions. Large number of sustainability issues have been explored by various researchers and presented as journal articles. Such issues can be categorized as sustainability assessment, engineering awareness and sustainability as service. Other important issues which have been discussed are value creation, renewable energy and green supply chain. Besides these, some important topics related to manufacturing which have been addressed are efficient product design, maintenance processes, manufacturing and remanufacturing processes. Focus has also been given on reuse, recycling and life cycle engineering. But along with such issues, qualification requirements for sustainability engineering along with role of information and communication technologies have also been explored as sustainability issues. At last, need was felt to deliberate on providing

K. Lachhwani (✉)
Department of Applied Science, National Institute of Technical Teachers' Training and Research, Chandigarh, India
e-mail: kailashclachhwani@yahoo.com

N. Jha
Department of Mathematics, Govt. Engineering College, Bharatpur, Rajasthan, India

D. Goyal
Institute of Engineering and Technology, Chitkara University, Patiala, Punjab, India

A. Dwivedi
Department of Mathematics, MCM DAV College, Kangra, Himachal Pradesh, India

R. K. Dang
UIET, PUSSGRC, Hoshiarpur, Punjab, India

H. Singh et al. (eds.), *Sustainable Development Through Engineering Innovations*, Lecture Notes in Civil Engineering 113,
https://doi.org/10.1007/978-981-15-9554-7_12

adequate environment for initiating entrepreneurship and providing education for same and making it as society policy.

2 Background

Many researchers have worked and published papers in the field of sustainability on such diverse theoretical topics in [1, 3–5, 7]. Nowadays, optimization theory for the sustainable development in the complex problems plays a key role and becomes important area of research. But still, work on optimization of factors for sustainability and their inter alia relationships have not been addressed in details. Time and cost are two parameters which need urgent attention while designing optimization algorithm. Many researchers have worked and published papers in the field of sustainability on diverse topics, but explicit focus on optimization has been scarce. Some important contributions are as follows: Gunasekaran and Spalanzani made a review of certain techniques and tools of various manufacturing and services from sustainability point of view and business development [4]. Kannegiesser and Gunther focused on environment-related issues, and three models related to sustainable optimization were presented [6]. Garetti and Taisch [3] deliberated on importance of sustainable manufacturing for sustainable development.. Recently, Gabrie [2] discovered optimization model of sustainability index by minimization of total development cost and time using mathematical differentiation. The aim of this paper is to show mathematical formulation of the sustainable development problem as multiobjective optimization model with recent available optimization techniques useful for the sustainable development.

In Sect. 3 of paper, we briefly describe the components of sustainability development, associate development costs and time, etc. Multiobjective optimization model formulation with different costs along with solution by elementary optimization technique (linear weighted aggregated technique) is discussed in Sect. 4 in details. The proposed multiobjective optimization model is illustrated on real problem of manufacturing industry as a case study in Sect. 5. Significance of proposed model and optimization techniques are discussed in Sect. 6. Concluding remarks are presented in last section.

3 Cost Components of Sustainability

In this section, we discuss various costs associated with an enterprise, and further, this section is bifurcated on the basis of cost and time for sustainability optimization. Various aspects, dimensions, and related model have been presented. Prominent aspects of each pillar of sustainability as suggested by Gabrie [2] and [1] are presented in Table 1.

Table 1 .

Economic sustainability	Social sustainability	Environmental sustainability
Globalization issues	CSR management	Environmental management
Competitive issues	Human rights	Use of resources
Innovation	Societal commitments	Pollution and dangerous outputs
New product design	Customers issues	Natural environmental
Production and other costs	Ethical business practices	

3.1 Sustainability Cost

3.1.1 Infrastructure Cost

The infrastructure cost is one time cost which is incurred for accomplishing initial setting up of infrastructure. Infrastructure sustainability cost is defined as $C(s)$ as suggested by [2]:

$$C(S_c) = f_1\{S_c, T\} \tag{1}$$

where $C(S_c) = f_1\{S_c, T\}$ is the objective function of sustainable cost S_c and time T that is to be optimized.

3.1.2 Sustainability Development Cost

The sustainability development cost includes all those costs which are proportional to sustainability development in the way of development to be updated and upgraded. Various aspects of costs for development towards sustainability are some of these [2]:

C_1 = Cost towards improving globalization
C_2 = Cost towards innovative process
C_3 = Cost towards applying competitive manufacturing strategies
C_4 = Cost related to improving work management
C_5 = Cost of resources
C_6 = Cost for elimination of pollution and other dangerous outputs

and may be many other costs based on the functions and nature of an enterprise. The development cost will be the sum of these costs involved, and costs may be some functions of time.

Hence, the total development cost

$$C(D_c, T) = \sum (c_1 + c_2 + c_3 + c_4 + c_5 + \ldots) = \sum_i c_i = f_2\{D_c, T\} \quad (2)$$

3.2 *Sustainability Time*

3.2.1 Infrastructure Time (Setup Time)

This time is basically time needed for setting up initial infrastructure development, and this may be treated as function to be optimized given by [2] as

$$T(S_t) = f_3\{S_c, T\} \quad (3)$$

3.2.2 Development Time (Carrying Time)

This time includes time associated with various development related activities. Some development time for sustainability is as suggested by [2]:

t_1 = time for improvement and/or enhancement globalization
t_2 = time for implementation and solving any emerging issue
t_3 = time for innovation design processes
t_4 = time towards changing organization management, etc.

The development time for sustainability can be expressed as the summation of all these times as

$$T(D_t) = \sum (t_1 + t_2 + t_3 + t_4 + t_5 + \ldots) = \sum_i t_i$$

Also, total development time involves sustainability development cost and time. So, the development time may be considered as function of development cost and time consumed as

$$T(D_t) = \sum (t_1 + t_2 + t_3 + t_4 + t_5 + \ldots) = \sum t_i = f_4\{D_c, T\} \quad (4)$$

4 Multiobjective Optimization Model for Sustainability Development

In complex situation where more than one objective function is required to be optimized, we here propose the multiobjective optimization model as follows:

Optimize

$$\begin{aligned}&\{C(S_c), C(D_c), T(S_t), T(D_t)\}\\&= \{f_1\{S_c, T\}, f_2\{D_c, T\}, f_3\{S_c, T\}, f_4\{D_c, T\}\}\end{aligned} \tag{5}$$

Subject to the given constraints.

Here, this multiobjective programming model may be linear or nonlinear depending upon the complexity or the relationship between the variables involved. There are many optimization techniques are available in the literature for the solution of complex multiobjective programming problem. One of the classical methods is linear aggregated weighted multiobjective method.

4.1 *Linear Aggregated Weighted Multiobjective Method*

In this method, we convert this multiobjective programming problem (5) into single objective problem by summation of weighted objectives as

Optimize

$$\begin{aligned}F &= w_1 f_1\{S_c, T\} + w_2 f_2\{D_c, T\}\\&\quad + w_3 f_3\{S_c, T\} + w_4 f_4\{D_c, T\}\end{aligned} \tag{6}$$

Subject to the given constraints.

Where w_1, w_2, w_3, w_4 represent the relative importance of the objective functions. Some of other techniques for solution of problem (5) are

(a) Goal programming approach
(b) Fuzzy goal programming approach
(c) Search methods

- Genetic Algorithm (GA)
- Artificial Immune Systems (AIS)
- Geometric Programming (GP)
- Tabu Search (TS)
- Differential Evolution (DE)
- Particle Swarm Optimization (PSO), etc.

5 Case Study

Let us consider the case of manufacturing industry producing farm equipment components. Assume, in this organization, the sustainable development cost of equipment is dependent upon the total development cost D_c and total time T in the form of following relationship as

$$D(S_c) = f_2\{D_c, T\} = 2.5D_c^2 - 1000T^3$$

Also, sustainable development time has linear relationship with total cost and total time as

$$f_4\{D_c, T\} = D_c + 1000T$$

and infrastructure development cost a constant equals \$1000 and infrastructure development time is given as

i.e.

$$f_1\{S_c, T\} = 10{,}000$$
$$f_3\{S_c, T\} = 250\,T$$

Further, assuming the organization requires only ten to twenty years as total available time with the total cost should be in between \$5000 and \$10,000. Assuming all the objectives of equal importance then in order to find the optimum values of all objectives.

Using the linear aggregated weighted model for the multiobjective programming model [2], we have

$$\text{Minimize } 10{,}000 + 2.5D_c^2 - 1000T^3 + 250T + D_c - 1000T$$

Subject to,

$$10 \le T \le 20$$
$$5000 \le D_c \le 10{,}000$$
$$T, D_c \ge 0$$

Solving this nonlinear programming technique (or by LINGO software—as given in figures—Screenshot Figs. 1 and 2), we have the optimal solution as

Total cost D_c = \$5000 and total time T = 20 with the optimum value of each objective function as

$$f_1\{S_c, T\} = 10{,}000$$

Fig. 1 Description of problem in LINGO

$$f_2\{D_c, T\} = 54{,}500{,}000$$
$$f_3\{S_c, T\} = 5000$$
$$f_4\{D_c, T\} = 25{,}000$$

6 Significance of Optimization Technique in Sustainable Development

Using multiobjective optimization model, complex problem related to sustainability development of any enterprises specifically manufacturing industries can be solved effectively with optimum results. In this continuation, the significance of optimization techniques in current area may be summarized as

- Tool for scientific analysis
- Solution for various complex problems
- Proper deployment of resources
- Minimization of time and cost wastage

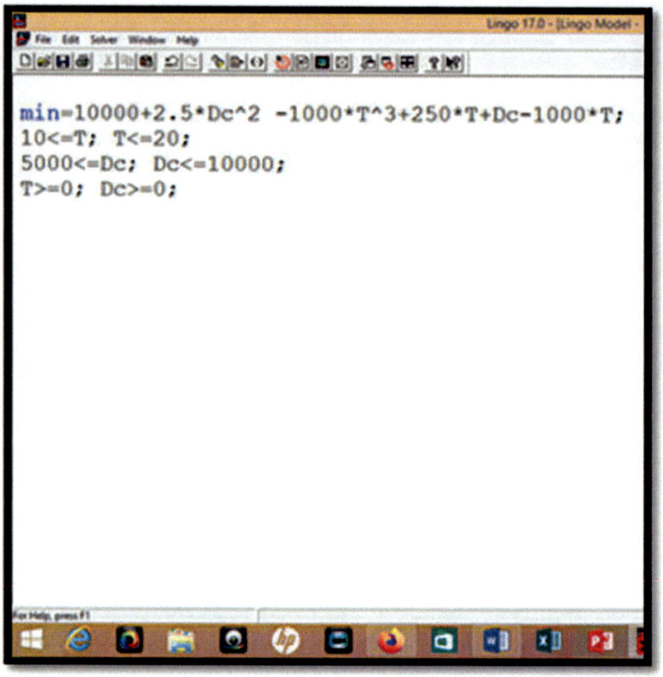

Fig. 2 Solution window of problem in LINGO

- Selection of optimum strategy
- Optimum resource allocation
- Facilitation in decision-making.

7 Concluding Remarks

The concluding remarks are presented as follows:

- The present paper proposes a unique multiobjective optimization model on sustainability costs in view of sustainable development in any enterprise with numerical applicability of proposed model for manufacturing industry.
- The proposed model can be used for minimization of sustainability cost/time and maximization of productivity of an organization.
- The aim of this study is to suggest a unique optimization model in complex area of sustainable development in which there is lack of such mathematical modelling.
- The present study will be useful for future researchers working in the areas of sustainable development and cost minimization problem of any enterprise in general and manufacturing industries in particular.

References

1. Garbie IH (2014) An analytical technique to model and assess sustainable development index in manufacturing enterprises. Int J Prod Res 52(16):4876–4915. https://doi.org/10.1080/00207543.2014.893066
2. Garbie IH (2015) Sustainability optimization in manufacturing enterprises. Procedia CIRP 26:504–509. https://doi.org/10.1016/j.procir.2014.07.085
3. Garetti M, Taisch M (2012) Sustainable manufacturing: trends and research challenges. Prod Plann Control 23(2–3):83–104. https://doi.org/10.1080/09537287.2011.591619
4. Gunasekaran A, Spalanzani A (2012) Sustainability of manufacturing and services: Investigations for research and applications. Int J Prod Econ 140(1):35–47. https://doi.org/10.1016/j.ijpe.2011.05.011
5. Kannegiesser M, Günther H-O (2013) Sustainable development of global supply chains—part 1: sustainability optimization framework. Flexible Serv and Manuf J 26(1–2):24–47. https://doi.org/10.1007/s10696-013-9176-5
6. Kannegiesser M, Günther HO (2014) Sustainable development of global supply chains—part 1: sustainability optimization framework. Flexible Serv Manuf J 26(1–2):24–47. https://doi.org/10.1007/s10696-013-9176-5
7. Liu S, Leat M, Smith MH (2011) State-of-the-art sustainability analysis methodologies for efficient decision support in green production operations. Int J Sustain Eng 4:236–250. https://doi.org/10.1080/19397038.2011.574744

A Sustainable Resource Allocation Techniques for Fog Computing

Jagdeep Singh and Parminder Singh

1 Introduction

In today's world, fog computing plays an essential role in sustainable development where the number of Internet of things (IoT) devices is increased day by day. Fog computing has various characteristics that can support sustainable development in computing areas like save the bandwidth, support mobility, reduce latency, data security, low energy consumption, and heterogeneous in nature [6, 25]. Today, IoT is fast growing in many fields such as health care, smart city, agriculture, industrial automation, smart energy grid, smart vehicles, and transportation [2, 19]. Resource allocation is an important area where it needs to work to make the overall system more efficient concerning energy consumption, latency, and quality of service (QoS) [9].

In fog computing environments, the storage, network, and computing capacity of fog devices are limited, so the proper utilization of resources is required to improve the performance. In fog, resources are dynamic in nature due to different variety of IoT devices, nodes, and their accessible resources. In general, fog devices are responsible for doing their own application computation. Available resources are dynamic and need analysis to predict the performance and capability of fog devices [1]. The state of the resources may vary according to the demand of their own application, so the

J. Singh (✉)
Lovely Professional University, Phagwara, Punjab, India
e-mail: jagdeepmalhi@gndec.ac.in

Department of Information Technology, Guru Nanak Dev Engineering College, Ludhiana, Punjab, India

P. Singh
School of Computer Science and Engineering, Lovely Professional University, Phagwara, Punjab, India
e-mail: parminder.16479@lpu.co.in

H. Singh et al. (eds.), *Sustainable Development Through Engineering Innovations*, Lecture Notes in Civil Engineering 113,
https://doi.org/10.1007/978-981-15-9554-7_13

predication of available resources plays a significant rule for fog device. However, the purpose of fog is to utilize the idle resources ready for use on available fog devices with fog computation always using as a secondary priority. The resource allocation, resource scheduling, and load balancing in fog computing are more challenging as compared to the conventional resource allocation in the cloud environment [10].

1.1 Resource Allocation in Fog Computing

The method assignment concerns distinct in fog and cloud computing networks. A set of geographical distribution diversified fog devices for IoT services are used effectively by authors. Resource allotment issues are becoming a challenge in the fog network. Resource allotment concern is sometimes considered as a double matching issue. Fog nodes and cloud servers are integrated for IoT users, whereas IoT users and fog nodes are combined for cloud networks [16]. Pairing in fog nodes is essential between fog nodes and cloud computing. The main objective of double matching course allotment is to find out optimum cloud computing for entire IoT smart users to get main functions. Resource allotment methods are of two types: auction-based and optimized-based. The auction-based allotment of resources is based on market pricing method. It expands the requirement and supply of fog nodes with the help of a bid [20]. With regards to optimized ways, resource allotment is used as a double matching concern because cloud computing servers and fog devices are joined for IoT users [24]. It finds an optimum way for allotment of fog servers accurately with distinct QoS requirements [27].

2 Literature Review

In the fog computing platform, the challenges based on resource allocation have been examined in several computing paradigms like fog, edge, and cloud. The resource allocation strategies adopted in the other computing paradigms do not precisely use in the fog computing. The reason behind this is the dynamic and heterogeneous characteristics of devices with respect to resources and users in the fog environment. In the past few years, some researches concentrate on only available fog resources in fog computing platforms, and others focus on cloud-fog resource allocation [21]. In this section, we summarize the work done by the researcher on the topic of resource allocation techniques in fog environment. Table 1 shows the brief summary of the research work done by various authors in resource allocation techniques.

3 Resource Allocation Techniques

The various authors proposed different resource allocation techniques as discussed below. The authors [24] get the dynamic load balancing for the different types of computing devices in fog computing environment, the dynamic resource allocation technique called DRAM has been developed to balance the load in the research paper. In fog computing, the IoT applications can run with the help of edge computing node.

Table 1 Brief summary of research work done by authors

Author and year	Proposed algorithms/method applied	Evaluation process	Main feature	Drawbacks/weakness
[24]	Dynamic Resource Allocation	CloudSim	• Improving throughput • Resource utilization	• Missing service migration • Delay and energy consumption have not evaluated
[26]	Dinkelbach algorithm, Alternating direction method of multiplier	MATLAB	• Energy efficient and time saving.	• Low scalability • CPU availability not considered
[28]	Privacy preserving resource allocation scheme	Not mentioned	• Enhanced security and privacy	• Lack of an appropriate simulation technique
[11]	Double matching strategy based on deferred acceptance algorithm (DADMS)	MATLAB	• Improve the efficiency and performance	• Lack of implementation • Energy consumption has not evaluated
[7]	Clustering algorithm for resource allocation and security based on priority	iFogSim	• Low latency with high security. • Reduces the complexity	• Scalability not evaluated • Cost analyzing not done
[3]	Efficient resource allocation algorithm (ERAA)	Cloud Analyst/CloudSim	• Improves QoS in fog environment • Low processing time	• Not analyzing cost
[17]	Multi-criteria-based resource allocation policy	CloudSim	• Reduces processing time and SLA violation	• Not considered failure of the fog devices • Not analyzing cost
[4]	logistic regression, node procession	MATLAB and Hadoop	• Improves the QoS of both user's and provider	• Low scalability • Not evaluating time and cost

(continued)

Table 1 (continued)

Author and year	Proposed algorithms/method applied	Evaluation process	Main feature	Drawbacks/weakness
[13]	Nash bargaining solution (NBS), bicooperative voting game (BVG)	Not mentioned	• Quality of experience • Resource utilization • Task failure probability	• Need to work on privacy and energy issues • Lack of simulation
[5]	Deep reinforcement learning, adaptive learning-based task offloading algorithm (ALTO)	Simulation vehicular network	• Reduce perception reaction time (PRT)	• Lack of an appropriate simulation
[14]	Energy-efficient resource allocation and computation offloading	Two tier collaborative computation offloading scheme	• Decrease the overall system cost	• Lack of an appropriate simulation • Not evaluating time with respect to cost
[23]	Blockchain-based resource allocation model	MATLAB	• Improve latency and performance	• Complexity and cost increased
[18]	Using resource ranking in hybrid and hierarchical Fashion	CloudSim Toolkit	• Improve data processing time, instance cost, network delay	• Energy utilization has not evaluated • Low scalability
[12]	Symmetric/asymmetric encryption algorithm	Codes are written by Java	Based on the security of resource allocation	• High computational cost
[15]	Efficient resource allocation algorithm (ERAA)	CloudSim	• Enhanced quality of service (QoS) • Low processing time	• Not evaluating cost and scalability • Complexity is high
[8]	Predictive offloading and resource allocation (PORA), fine-grained queue model	Trace-driven simulations	• Near optimal power consumption • Exploits predictive information and	• Low scalability • Presence of prediction errors

(continued)

Table 1 (continued)

Author and year	Proposed algorithms/method applied	Evaluation process	Main feature	Drawbacks/weakness
[22]	User association decision approximation, joint optimization iterative algorithms	MATLAB	• Low complexity • Lower delay tolerance • Efficient with few and limited computing capacity of fog nodes	• Scalable to user specific tasks

The DRAM technique has applied that is related to the dynamic and static source scheduling. Thus, the experimental results and comparative analysis are taken place to confirm the validity of the projected procedure. In future work, the negative impact of service migration would be analyzed. Furthermore, a design for the corresponding process to balance the positive and negative implications of service migration would be proposed.

Yu et al. [26] have investigated green fog computing by expanding framework utility function by consideration of energy efficient with the restriction of power in the present research study. To deal with large-scale mixed integer nonlinear programming issues, the authors have designed a method to solve issues in a parallel and distributed way. The outer part of the current issue is related to the benders decomposition to split the continuous and integer variables from the master problem and sub-problems, respectively. Authors have also used the Dinkelbach method to convert the fractional programming into an identical explainable form. It can be controlled and maintained by the alternating direction method of multiplier (ADMM) algorithm. The main issue authors have proposed is a centralized branch and bound framework to handle complexity. The authors have also discussed the properties and performance of the framework. In the final stage, the simulation results demonstrate the suggested algorithm is time saving and energy efficient.

Daoud et al. [7] proposed a security prototype that is based on collaboration between fog computing and IoT. The prototype merges an effective access control procedure related to a monitoring method ensuring cooperation between various resources and distinct parts of the operation. An inclusive scheduling procedure and resource allocation procedure applying a security mechanism are proposed to enhance the overall system's performance. The experimental results show the iFogSim simulation assessment, which proved proposed methods reduce latency with privacy and high security. It also minimizes the complex structure of administration and management of the frameworks.

Akintoye and Bagula [3] focus on the stage of quality of service (QoS) in the fog computing that can be measured the resources used in the framework. The fog environment is used to extend the effectiveness of the cloud computing and cut back the amount of data that has to move to the cloud servers for the objective of storage and data processing. In the present paper, efficient resource allocation architecture and algorithms are suggested, and implementation is done using a tool called CloudSim. Also, evaluation of the proposed working framework is done using CloudSim. The recommended framework processing time is less in comparison with other techniques.

In this paper, [17] proposed multi-criteria-based resource allocation policy with resource reservations to cut back time of processing, entire delay and violation. The fog computing's relevant characteristics are resource constraint, mobility, different device heterogeneity, and dynamic changes in user requirements. The experimental outcome clearly shows that the proposed policy performs excellently compared to the existing policy, mitigating the total delay by 51%. The proposed framework cut back SLA violation and processing time that is profitable for running time-sensitive applications in fog computing.

Bashir et al. [4] proposed the resource prediction and allotment issue for the Internet of things (IoT) and the fog environment. User data is developed against the lower layer to assist in the fog nodes. TOPSIS is used to rank various fog node models that are based on the available resources such as networking, RAM, and CPU. The workload of every fog node is calculated by the logistic machine learning method, and output is transferred to fog broker. Improvement in QoS for provider's side and user's side can be taken place with the help of the proposed network.

In this paper, [13] presented a resource allotment method for the SFIoT. This proposed way contains two steps: (i) smart devices take votes for deciding the resource allotment that depends on the BVG framework. (ii) Every application, which depends on NBS, totally depends on the distribution of the partitioned computation capacity. The two game frameworks proceed in a step by step way according to the SFIoT system information. The proposed model can effectively increase the QoE, resource allotment, and system compared to CFIC and SSEC methods. In future scope, the authors will investigate energy and privacy issues for the SFIoT system. Furthermore, the authors will add a blockchain to the network.

In this research work, [18] dynamic differences from the user side are explored in the fog environment. The assessment was undertaken in the simulation environment with the help of exploring the CloudSim simulation toolkit. The output revealed that by using the proposed network, as compared to lately offered resource and latency aware systems, there was declination in considerable processing time delay and cost. It is discovered that proposed networks reduce the cost and average of time processing by 15% and 12%, respectively, when compared with present remedial measures. Future scope of this work includes dynamics differences in resource and failure handling in a fog computing, for example, design a multiplex environment for simulation with big data IoT application and communication and resource failures.

Jiang et al. [12] developed a secure computing resource allocation framework for open fog computing, where resource allocations and computing requests are handled by the server. The behavior of fog devices and fog server is maintained by cloud audit center. In this paper, author works on resource allocation technique for sub-tasks priority with the availability of computing service devices. Furthermore, the author also does work on the security of the framework for the allocation of resources in fog computing. To check the performance of the proposed framework was tested and analyzed through experiments on different task amounts, different data amounts, and the different number of service devices. As a result of these experiments shown, some tasks require high computational cost that is easily identified.

In the above-studied paper, most of the authors tired best to make efficient resources allocation techniques as per the framework required. Those techniques can be improved in terms of cost, energy utilization, application scalability, security, reduced failure rate, and decrease waiting time for resources. So that proposed techniques may be used for other fog applications for better performance. Moreover, the proper implementation and experimentation of any resource allocation technique are most important in the simulation environment or real test bed that is missing. In future work, the authors can work out on optimal resource allocation techniques that can cover the drawbacks of presented techniques, and experiments can be done on an appropriate simulation environment for more effective results.

4 Conclusion

In this research article, the different techniques are studied on the topic of resource allocation in the fog computing platform for suitable development. The different researchers proposed and implemented various techniques as per the requirement of an application to increase the performance in terms of latency, bandwidth, energy consumption, and security. The comparison of the proposed algorithms, the method applied, the evaluation process, the main feature, and drawbacks/weaknesses is made, so the other researchers can work further to develop more efficient techniques. Finally, the future direction is discussed for continuation research to find better techniques for resource allocation in the fog computing.

References

1. Aazam M, Zeadally S, Harras KA (2018) Offloading in fog computing for IoT: review, enabling technologies, and research opportunities. Future Gener Comput Syst 87:278–289. https://doi.org/10.1016/j.future.2018.04.057
2. Ahmed A, Arkian H, Battulga D, Fahs AJ, Farhadi M, Giouroukis D, Gougeon A, Gutierrez FO, Pierre G, Souza PR, Tamiru MA, Wu L (2019) Fog computing applications: taxonomy and requirements. 1–16. http://arxiv.org/abs/1907.11621
3. Akintoye SB, Bagula A (2019) Improving quality-of-service in cloud/fog computing through efficient resource allocation. Sensors (Switzerland) 19(6). https://doi.org/10.3390/s19061267
4. Bashir H, Lee S, Kim KH (2019) Resource allocation through logistic regression and multicriteria decision making method in IoT fog computing. Trans Emerg Telecomun Technol 1–14. https://doi.org/10.1002/ett.3824

5. Chen X, Leng S, Zhang K, Xiong K (2019) A machine-learning based time constrained resource allocation scheme for vehicular fog computing. China Commun 16(11), 29–41. https://doi.org/10.23919/JCC.2019.11.003
6. Chiang M, Zhang T (2016) Fog and IoT: an overview of research opportunities. IEEE Internet Things J 3(6):854–864. https://doi.org/10.1109/JIOT.2016.2584538
7. Daoud WB, Obaidat MS, Meddeb-Makhlouf A, Zarai F, Hsiao KF (2019) TACRM: trust access control and resource management mechanism in fog computing. Human-Centric Comput Inform Sci 9(1). https://doi.org/10.1186/s13673-019-0188-3
8. Gao X, Huang X, Bian S, Shao Z, Yang Y (2020) PORA: predictive offloading and resource allocation in dynamic Fog computing systems. IEEE Internet Things J 7(1):72–87. https://doi.org/10.1109/JIOT.2019.2945066
9. Ghobaei-Arani M, Souri A, Rahmanian AA (2019) Resource management approaches in Fog computing: a comprehensive review. J Grid Comput. https://doi.org/10.1007/s10723-019-09491-1
10. Hong C-H, Varghese B (2019) Resource management in Fog/Edge computing. ACM Comput Surv 52(5):1–37. https://doi.org/10.1145/3326066
11. Jia B, Hu H, Zeng Y, Xu T, Yang Y (2018) Double-matching resource allocation strategy in fog computing networks based on cost efficiency. J Commun Netw 20(3):237–246. https://doi.org/10.1109/JCN.2018.000036
12. Jiang J, Tang L, Gu K, Jia W (2020) Secure computing resource allocation framework for open Fog computing. Comput J. Oxford University Press. https://doi.org/10.1093/comjnl/bxz108
13. Kim S (2019). Novel resource allocation algorithms for the social internet of things based Fog computing paradigm. Wirel Commun Mob Comput. https://doi.org/10.1155/2019/3065438
14. Li Q, Zhao J, Gong Y, Zhang Q (2019). Energy-efficient computation offloading and resource allocation in fog computing for Internet of Everything. China Commun 16(3):32–41. https://doi.org/10.12676/j.cc.2019.03.004
15. Mani SK, Meenakshisundaram I (2020) Improving quality-of-service in fog computing through efficient resource allocation. Comput Intell 1–21. https://doi.org/10.1111/coin.12285
16. Mouradian C, Naboulsi D, Yangui S, Glitho RH, Morrow MJ, Polakos PA (2018) A comprehensive survey on Fog computing: state-of-the-art and research challenges. IEEE Commun Surv Tutorials 20(1):416–464. https://doi.org/10.1109/COMST.2017.2771153
17. Naha RK, Garg S (2019). Multi-criteria-based dynamic user behaviour aware resource allocation in Fog computing. 1(1):1–31. http://arxiv.org/abs/1912.08319
18. Naha RK, Garg S, Chan A, Battula SK (2020) Deadline-based dynamic resource allocation and provisioning algorithms in Fog-cloud environment. Future Gener Comput Syst 104:131–141. https://doi.org/10.1016/j.future.2019.10.018
19. Naha RK, Garg S, Georgakopoulos D, Jayaraman PP, Gao L, Xiang Y, Ranjan R (2018) Fog computing: survey of trends, architectures, requirements, and research directions. IEEE Access 6:47980–48009. https://doi.org/10.1109/ACCESS.2018.2866491
20. Ni L, Zhang J, Jiang C, Yan C, Yu K (2017) Resource allocation strategy in Fog computing based on priced timed petri nets. IEEE Internet Things J. https://doi.org/10.1109/JIOT.2017.2709814
21. Patil-Karpe S, Brahmananda SH, Karpe S (2020) Review of resource allocation in Fog computing. Smart Innov Syst Technol 159:327–334. https://doi.org/10.1007/978-981-13-9282-5_30
22. Tong S, Liu Y, Cheriet M, Kadoch M, Shen B (2020) UCAA: user-centric user association and resource allocation in Fog computing networks. IEEE Access 8:10671–10685. https://doi.org/10.1109/ACCESS.2020.2965218
23. Wang H, Wang L, Zhou Z, Tao X, Pau G, Arena F (2019) Blockchain-based resource allocation model in fog computing. Appl Sci (Switzerland) 9(24). https://doi.org/10.3390/app9245538
24. Xu X, Fu S, Cai Q, Tian W, Liu W, Dou W, Sun X, Liu AX (2018) Dynamic resource allocation for load balancing in Fog environment. Wirel Commun Mob Comput 2018:1–15. https://doi.org/10.1155/2018/6421607

25. Yousefpour A, Fung C, Nguyen T, Kadiyala K, Jalali F, Niakanlahiji A, Kong J, Jue JP (2019) All one needs to know about fog computing and related edge computing paradigms: a complete survey. J Syst Architect 98:289–330. https://doi.org/10.1016/j.sysarc.2019.02.009
26. Yu Y, Bu X, Yang K, Wu Z, Han Z (2019) Green large-scale fog computing resource allocation using joint benders decomposition, Dinkelbach algorithm, ADMM, and branch-and-bound. IEEE Internet Things J 6(3):4106–4117. https://doi.org/10.1109/JIOT.2018.2875587
27. Zhang H, Zhang Y, Gu Y, Niyato D, Han Z (2017) A hierarchical game framework for resource management in Fog computing. IEEE Commun Mag 55(8):52–57. https://doi.org/10.1109/MCOM.2017.1600896
28. Zhang L, Li J (2018). Enabling robust and privacy-preserving resource allocation in fog computing. IEEE Access 6(c):50384–50393. https://doi.org/10.1109/ACCESS.2018.2868920

Sensor-Based Optimization of Energy Efficiency in Internet of Things: A Review

Meera Sharma, Manish Kumar Singla, Parag Nijhawan, and Arvind Dhingra

1 Introduction

Internet of things is the latest technological example. This includes processes such as computer technology, international, a few computers, the Internet, sensor technology, and telecommunications. Power as an integral part of the Internet for things where millions of devices are connected to the Internet and performance is limited. The next step to extend the life of the sensor is to save energy. This paper discusses the energy savings of the network. Today, buildings are having difficulty analyzing service expectations and service needs. Customers are concerned that the process of self-verification is often overlooked and that the system provides people with experience. Users can be accessed by phone or personal settings. The user experience is controlled by language, text, programming, and communication. Energy suddenly explodes and regains danger. The services provided by IoT provide customers with the best local and long-term services needed for the platform. As such, personal communication is multi-factorial for the exchange process and technical handling. Most of them work and have a balanced and efficient system. Other users can interact

M. Sharma (✉) · M. K. Singla · P. Nijhawan
Electrical and Instrumentation Engineering Department, Thapar Institute of Engineering and Technology, Patiala 147004, India
e-mail: taurusmeera@gmail.com

M. K. Singla
e-mail: msingla0509@gmail.com

P. Nijhawan
e-mail: parag.nijhawan@rediffmail.com

A. Dhingra
Electrical Engineering Department, Guru Nanak Dev Engineering College, Ludhiana 141006, India
e-mail: arvinddhingra@gmail.com

H. Singh et al. (eds.), *Sustainable Development Through Engineering Innovations*, Lecture Notes in Civil Engineering 113,
https://doi.org/10.1007/978-981-15-9554-7_14

with the management system. The outcome of this has been multisensory networks which are coupled to actuators, information sinks, and automated balancing systems. Out of these, some operate automatically and have self-regulating and self-balancing systems, while others depend on the consumer's communication and control [1].

Residential and commercial real estate is one of the most productive places in the world. This is mainly discussed in developed countries where from 20 to 40% of total energy is understood [2]. The major goals prioritized for multi-disciplinary researches in energy policy drafting and building engineering are ensured energy efficiency along with the carbon footprint reduction globally. In order to reduce the amount of energy available to improve current performance, this step focuses on current energy consumption. Given the growing demand for heat, ventilation, and air-conditioning (HVAC) technology to provide thermal comfort, it is clear that one of the reasons behind low energy consumption is to address the driving forces. But so far, many approaches have provided several solutions for the power consumption problem for structures with completely different properties and also rejected previous proposals [3]. As a result, there is a lack of analysis that shows the steps needed to find a solution to determine the energy efficiency of a building.

Houses with certain properties have different electronic information, so to use energy; you need to carry out a groundwork classification of the core contributors. Cities face many challenges related to housing, climate, transport, and infrastructure development, which makes it difficult for cities to understand their objectives and integrate into urban communities. Recently, cities have begun to use improved technology particularly in smart cities described as information and communication technologies (ICT). It provides the basis of decision-making and the importance of information technology to support the development of competitive advantage in urban areas. Particularly smart cities cover all aspects and regulators that unite sustainable development and efficiency and use a wide range of regulatory capabilities and stable and healthy cities [4]. However, it is generally believed that achieving energy saving in buildings is a complex and complex process [5]. If cities are to have the tools and mechanisms to use advanced ICT equipment to significantly reduce energy consumption and carbon dioxide emissions, a comprehensive, open, and integrated approach is needed [6]. In fact, despite the large amount of energy data in cities, systems and verification tools need to be collected, coordinated, and analyzed for people supporting energy management [7].

2 Energy Efficiency in IoT

Energy efficiency encapsulates various aspects of a system in IoT stated as shown in Fig. 1 [8]:

Sensor nodes use their functions to perform tasks such as collecting, transferring, and processing data. Acquisition depends on the type of monitoring. Communication consumes more energy than other activities. It includes meters and reception. Data

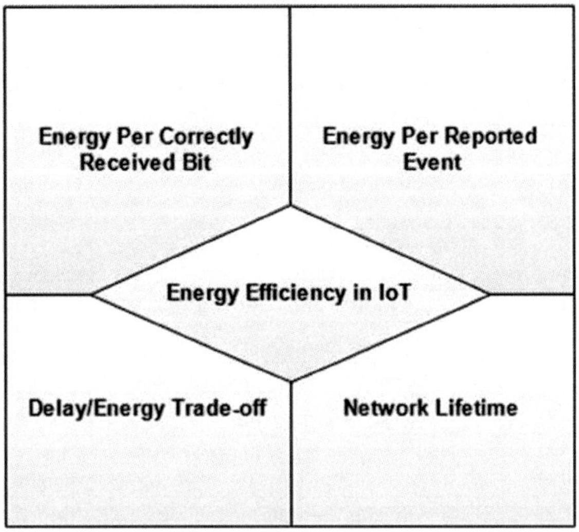

Fig. 1 Schematic of energy efficiency

processing is the process of selecting internal filtering to connect the sensor data cycle from source to receiver.

2.1 A Study on Issues and Methods Used in Energy Efficiency

On the basis of above-discussed literature, various constraints related to energy efficiency in IoT can be listed as per the following.

Idle Listening
The vital source of energy consumption is the node, when it is in the active mode. While waiting for the data to be transmitted, in between when the packets are neither send nor received, idle listening occurs.

Collision
Sometimes there is useless data due to if multiple data packets are received at the nodes, causing collision, thereby increasing latency and unnecessarily consuming energy.

Over Hearing
During the time of data conveyance, when neighbor nodes are interfered with the high-density sensor nodes, lead to over hearing. This is the root cause of exhaustion of energy resources while processing information.

Table 1 Issues and methods used in energy efficiency

References	Issues	Methods used
[5]	Encryption and authentication	SSL protocol with maximum bandwidth
[7]	Energy in wireless sensor networks	Topology control, routing, and data link protocols
[8]	Spectrum efficiency	Optimization-constraints, variables and algorithms
[9]	Security and privacy key issues	Encryption and communication mechanism and cryptographic algorithms
[10]	MAC-Insufficient energy efficiency	Heterogeneous MAC duty-cycle
[11]	Energy consumption	Automatic repeat request scheme (ARQ)
[12]	Node energy, Throughput and Latency	Improving routing protocols
[13]	Sleep and wake up periods	The PaderMAC protocol
[14]	Node sleeping scheduling schemes	Cluster-based energy-efficient mobility-centric node scheduling scheme
[15]	Energy efficiency during routing	Survey on routing protocols
[16]	Green, reliability, and security	Energy-efficient activity scheduling scheme

Reduction of Protocol Overhead

Energy sources are exhausted because of the protocol header information. Various techniques to overcome this depletion overhead are adaptive transmission; cross-layering along with optimized flooding.

Traffic Fluctuation

There could be high delays and congestion due to the traffic. It could further escalate in case of the network is functioning at its highest capacity.

Brief studies of various issues and methods have been compiled in Tables 1 and 2 represents the comparison between different wireless technologies.

3 Layers of the Base Architecture of a Smart Building Management System

Our energy-saving project includes the first phase of receiving the sensors (Architecture Part 1 and Data Set (Part 2) in Fig. 2). Energy-saving requirements and services for any home or environment entitle the monitoring phase patterns. These parametric patterns can be building occupancy, behavior patterns, electrical devices, electronic, environmental, and other processes. The information in the technical document requires careful references in specific contexts that can help assess and verify the building management system. Once the data is collected, methods are used to determine the appropriate building strength control (layer 3). In addition to the

Table 2 Comparison between different wireless technologies [17–19]

Technology	Range (km)	Data rate	Power usage (years)	Security	Installation cost	Example application
LoRA (Long range)	≤50	0.3–38.4 kbps	8–10	High	Low	Smart buildings
Narrowband-IoT	≤50	≤100 kbps	1–2	High	Low	Smart grid
Long-term evolution for machine-type communications	≤200	0.2–1 mbps	7–8	High	Moderate	Smart meter
Sigfox	≤50	100 bps	7–8	High	Moderate	Smart buildings
Weightless LPWAN	≤5	100 kbps	Very long	High	Low	Smart meter
Bluetooth	≤50	1 mbps	Few months	High	Low	Smart home appliances
Zigbee	≤100	250 kbps	5–10	Low	Low	Smart metering in renewable energies
Satellite	Very long >1500	100 kbps	high	High	Costly	Solar and wind power plants

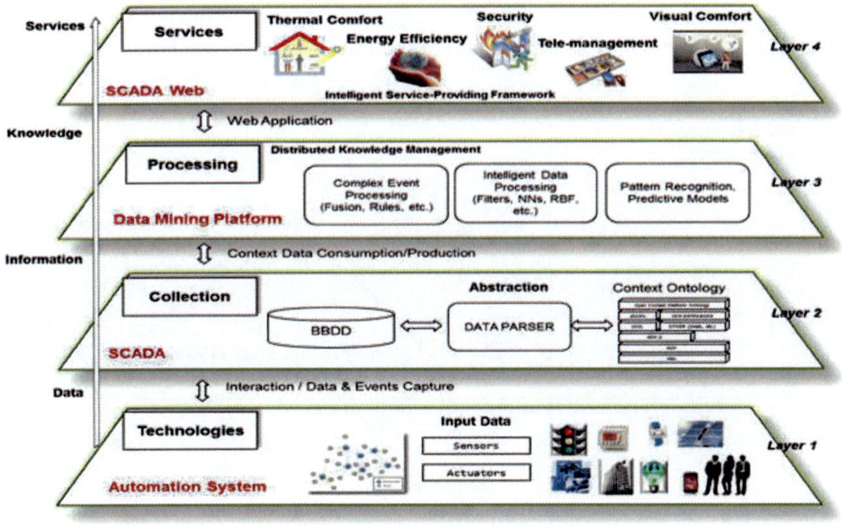

Fig. 2 Architecture with different layers [20]

energy efficiency of the building, the warmth and comfort of the screen, security, and other services are also provided. Available in the same way supplementary services except the thermal and visual comfort could be provided (layer 4) [21].

4 IoT Wireless and Sensor Energy Requirements

IoT devices are ready to access the real impact of the solar industry. These devices reduce energy costs which are true, but do not require confirmation of the application. Information intelligence is partially obtained through sensible data storage for IoT. Information centers have difficulty securing their operations as it requires large amount of energy. Communication and transfer IoT and machine data management are of a major concern [21]. Electricity bills are needed to deny IoT devices and infrastructure. It must be designed in terms of disposal, replacement, processing, and disposal. Some IoT systems are fragile and have significant damage.

4.1 Optimization of Smart Home Energy Requirements

The optimization of energy requirements has four main components as shown in Fig. 3 [22]. Optimization of energy initiates by managing the life cycles of sensors. A surveillance camera or a light detector is certain peripherals on which the power consumption depends. The sensor nodes are quite dependent on the life cycle phases, optimal settings and the relevant data which is transmitted, making the system stable henceforth. Knowing more about the fact of increasing energy consumption, humans

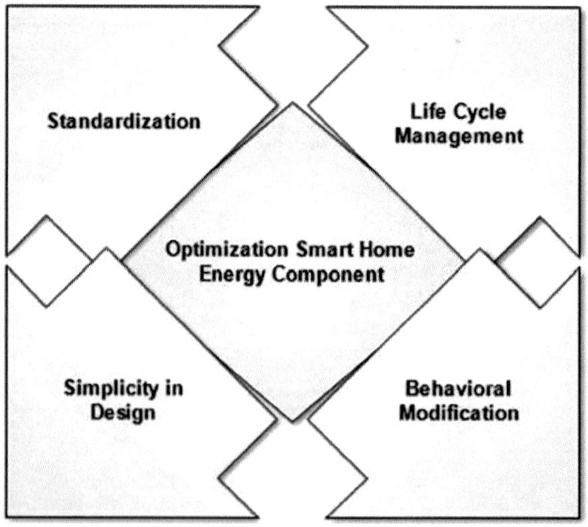

Fig. 3 Block diagram of smart home energy component

are looking for new ways. In a sense, the Internet to meet the expectations of customer service is the essence of many retailers. Hence, the modification of human behavior is required, power consumption optimization is required.

The goal of general positioning is to balance systems and between systems. Setting energy costs can be important. Standardization is an important aspect of simple design. Modern IoT systems are complex multidimensional systems with unbalanced, asymmetric forms, and energy optimization problems that are not yet fully resolved [23].

Cities are faced with a number of challenges associated with accommodation, atmosphere, transport, and infrastructural development, making difficult for urban communities and cities to realize their objectives. In recent years, cities have been turning to advanced technologies to become smart cities. This term is used to describe information and communication technological (ICT) solutions for cities and to highlight ICT importance and potential in helping the city to develop competitive advantages. More specifically, smart cities are comprised of cities that work in frugal and sound ways, by incorporating every one of its substructure and administrations into unified whole and utilizing insightful gadgets for observing and control, in order to guarantee maintainability and effectiveness [24].

In the monitoring phase, before considering the specific environment of the building, before decisive measures are proposed to reduce energy consumption, data are collected and analyzed from various sources. Functions buildings with different functions have different energy consumption characteristics, so it is necessary to determine the energy consumption characteristics of the main participants.

The intelligent management system should provide automation tools and appropriate measures to meet the basic requirements of building convenience and energy savings. Home automation systems receive data from corridors and sensors installed in the home. Use this data (light, temperature, humidity, etc.) to handle things like HVAC, light or safety. By capturing the boundaries of the system and the environment, this expanded resource can be used to consciously conserve energy. Feedback is a learning tool that can be used to save energy. According to a careful measurement analysis, smart metering that provides real-time information about the use of internal fuel, energy-monitoring technology can reduce energy consumption from 5 to 15% [25].

5 Conclusion

By supporting behavioral changes among occupants of the smart buildings/home, ICT-based solutions through the Internet of things can play an important role in saving energy. The proposed IoT system allows energy end users to determine the amount of energy being consumed. On behalf of users specific functions and other complementary peers, an understanding of individual recommendations for energy conservation and redistribution can be drafted for consumers. However, not only must supply and demand be met, but also accurate estimates of overall usage can better meet energy demand. The main purpose of the system is that it must be flexible, easy to use, and easy to customize, comprehensive maintenance (all licenses) and at personal level (users) ICT platform. Using sensor data as a starting point to measure real-time information on system, data entry, data behavior, set system parameters, and much more. This makes it difficult for information collected by the system and puts energy users (building occupancy) in their hands (e.g., end users know how to improve building performance through orders). The city government can monitor and control energy in the building space. Semantic network technology can play an important role in the rapid development of smart city infrastructure with various replication research areas. The coalition of Web Services, Semantic Web, and energy management may help city authorities towards a path-breaking shift of the future of cities.

References

1. Santiago S, Arockiam L (2016) Energy efficiency in internet of things: an overview. Int J Recent Trends Eng Res (IJRTER) 2(6):475–482
2. Jumira O, Zeadally S (2013) Energy efficiency in wireless networks. Wiley
3. Santiago S, Arockiam L A methodology for an energy efficient machine to machine (M2M) communications
4. Haller S, Serbanati A, Bauer M, Carrez F (2013) A domain model for the internet of things. In: 2013 IEEE international conference on green computing and communications and IEEE internet of things and IEEE cyber, physical and social computing. IEEE, pp 411–417
5. Venckauskas A, Jusas N, Kazanavicius E, Stuikys V (2015) Energy efficient protocol for the internet of things. J Electr Eng 66(1):47–52
6. Kumar C, Paulus R (2014) A prospective towards M2M Communication. J Convergence Inform Technol 9(2):102
7. Chen F, Guo L, Chen C (2012) A survey on energy management in the wireless sensor networks. IERI Procedia 3:60–66
8. Wu G, Yang C, Li S, Li GY (2015) Recent advances in energy-efficient networks and their application in 5G systems. IEEE Wirel Commun 22(2):145–151
9. Suo H, Wan J, Zou C, Liu J (2012) Security in the internet of things: a review. In: 2012 international conference on computer science and electronics engineering, vol 3. IEEE, pp 648–651
10. Beaudaux J, Gallais A, Noël T (2013) Heterogeneous MAC duty-cycling for energy-efficient Internet of Things deployments. Netw Sci 3(1–4):54–62
11. Kim K, Lee J, Lee J (2014) Energy efficient and reliable ARQ scheme (ER-ACK) for mission critical MM/IoT services. Wirel Pers Commun 78(4):1917–1933
12. Talwar M (2015) Routing techniques and protocols for internet of things: a survey. Proc NCRIET-2015 Indian J Sci Res 12(1):417–423
13. Autenrieth M, Frey H (2014) PaderMAC: energy-efficient machine to machine communication for cyberphysical systems. 7:243–254
14. Al-Kahtani MS (2015) Efficient cluster-based sleep scheduling for M2M communication network. Arab J Sci Eng 40(8):2361–2373
15. Katkar PS (2015) A survey on energy efficient routing protocol for wireless sensor networks
16. Lu R, Li X, Liang X, Shen X, Lin X (2011) GRS: the green, reliability, and security of emerging machine to machine communications. IEEE Commun Mag 49(4):28–35
17. Kabalci Y, Kabalci E, Padmanaban S, Holm-Nielsen JB, Blaabjerg F (2019) Internet of Things applications as energy internet in smart grids and smart environments. Electronics 8, 972

18. GSMA (2019) Security features of LTE-M and NB-IoT networks. Technical report. GSMA, London, UK
19. Fraire JA, Céspedes S, Accettura N (2019) Direct-to-satellite IoT—a survey of the state of the art and future research perspectives. In: Proceedings of the 2019 international conference on ad-hoc networks and wireless, Luxembourg, 1–3 October 2019, pp 241–258
20. Kumar V, Fensel A, Fröhlich P (2013) Context based adaptation of semantic rules in smart buildings. In: Proceedings of international conference on information integration and web-based applications and services, pp 719–728
21. Pérez-Lombard L, Ortiz J, Pout C (2008) A review on buildings energy consumption information. Energ Build 40(3):394–398
22. Moreno M, Úbeda B, Skarmeta AF, Zamora MA (2014) How can we tackle energy efficiency in IoT based smart buildings? Sensors 14(6):9582–9614
23. Kramers A, Höjer M, Lövehagen N, Wangel J (2014) Smart sustainable cities—exploring ICT solutions for reduced energy use in cities. Environ Model Softw 56:52–62
24. Giffinger R, Fertner C, Kramar H, Kalasek R, Pichler-Milanović N, Meijers E (2007) Smart cities: ranking of European medium-sized cities. Vienna, Austria: Centre of Regional Science (SRF), Vienna University of Technology. www.smart-cities.eu/download/smartcitiesfinalreport.pdf
25. Corrado V, Ballarini I, Madrazo L, Nemirovskij G (2015) Data structuring for the ontological modelling of urban energy systems: the experience of the SEMANCO project. Sustain Cities Soc 14:223–235

Effect of Microbes on the Unconfined Compressive Strength of Dredged Sediments

K. M. N. Saquib Wani, B. A. Mir, and Ishfaq Rashid Sheikh

1 Introduction

Soil stabilization techniques have been gaining a lot of attention from researchers all over the world. There are a lot of soil stabilization techniques available at hand which include chemical stabilization, soil nailing, use of geosynthetics, grouting, use of sand drains, etc. [1–9]. There are many problems associated with the use of chemicals like cement, lime, resins, etc. as they contribute to the carbon emission into the atmosphere and alter groundwater quality [10]. Researchers are thus more focused on sustainable new eco-friendly materials and technologies which can have multiple advantages like reduction in carbon emissions, no health hazards, can assist in waste disposal, and above all are economic.

One such sustainable soil stabilization technique is the introduction of microbes into the soil, which cements the particles together with the inclusion of a mixture of cementation solution. The basic principle that follows this process is when bacteria are added to the soil they release urease enzyme which increases the pH of the soil in turn making the environment feasible for precipitation. The calcium from supplied source then forms calcium carbonate which is responsible for the strength gain in soils. This is a natural process that can be seen in ancient caves and formations but the time factor of millions of years can be reduced in the laboratory by using specific conditions and urease-producing bacteria. The following chemical reactions take place during the process of stabilization.

$$CO(NH_2)_2 + 2H_2O \rightarrow 2NH_4^+ + CO_3^{2-} \quad (1)$$

$$Ca^{2+} + CO_3^{2-} \rightarrow CaCO_3 \quad (2)$$

K. M. N. S. Wani (✉) · B. A. Mir · I. R. Sheikh
National Institute of Technology Srinagar, Srinagar 190006, India
e-mail: sakibwani_02phd17@nitsri.ac.in

H. Singh et al. (eds.), *Sustainable Development Through Engineering Innovations*, Lecture Notes in Civil Engineering 113,
https://doi.org/10.1007/978-981-15-9554-7_15

A lot of work has been done on stabilization of soils using biological means. In the last few years, this technique has emerged as an eco-friendly sustainable, and viable solution for soil improvement. Many studies depicting the rise in UCS values of bio-treated soils have been reported [11–15]. High calcium carbonate precipitation at varying cementing solution molarities was also noticed by many researchers [16, 17]. Overall this technique is gaining a lot of praise for being an environmentally friendly option available to engineers with promising results.

In this research, a nonpathogen bacteria namely *Bacillus sphaericus* has been used to induce calcium carbonate precipitation into the weak dredged sediments. Cementing solution molarity of 0.25, 0.5, and 0.75 has been used along with curing of 3 and 7 days. The testing has been carried out at 0.9 $\gamma_{d(max)}$, to study the UCS characteristics. At last, the enhancement in strength values of the treated soil has been concreted by the precipitation of calcium carbonate as determined by acid washing technique.

2 Materials and Testing Methodology

2.1 Materials

2.1.1 Dredged Sediments Used in This Research

Soil was collected from two distinct locations on the Doodh-Ganga tributary of river Jhelum (34°01′43.5″ N 74°48′25.5″ E). As dredging was taking place on the day of sample collection, it made it easier for the team to collect disturbed samples. However, undisturbed samples were collected at a depth of 2 m below the top surface using core cutters, dolly, and hammer. The soil was light in color with shades of grey as it contained a mixture of silt, sand, and clay. The soil was devoid of any organic matter. The grain size distribution of the two sites is shown in Fig. 1. Table 1 gives the different properties of the two soils.

2.1.2 Microbe Used in This Research

A nonpathogen bacteria's namely *B. sphaericus* has been used to induce calcium carbonate precipitation into the weak dredged sediments. It is a Gram-positive, mesophilic, rod-shaped bacterium commonly found in soil. It can form resistant endospores that are tolerant of high temperatures, chemicals, and ultraviolet light and can remain viable for long periods of time. The bacterium was procured from a local laboratory and cultivated in sterile conditions in nutrient broth following standard procedures. In this research, an optical density of 1.0 (~10^7 cells per ml) has been used. We can calculate the cells of the bacterium by the basic empirical equation given by Ramachandran et al. [18].

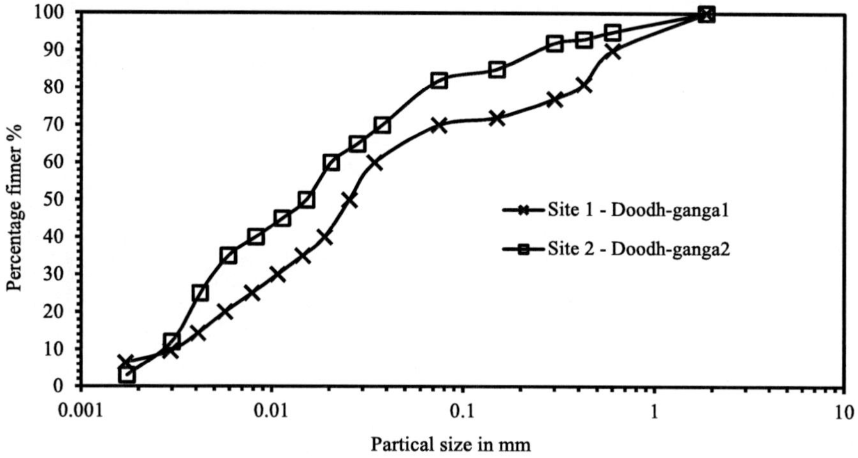

Fig. 1 Grain size distribution of two site samples

Table 1 Different property of dredged sediments from two distinct sites

Properties	Site DG[a]1	Site DG[a]2
Natural water content (%)	39	35
Bulk unit weight (kN/m^3)	18.5	19.0
In situ dry unit weight (kN/m^3)	13.8	14.9
Specific gravity (G)	2.62	2.67
Percentage finer than 75 μm	70	82
Clay (%)	7	8
Silt (%)	63	74
Sand (%)	30	18
Maximum dry unit weight (kN/m^3)	16.9	17.8
OMC (%)	20	17
In situ UCS, q_u (kPa)	24	29
UCS at OMC, q_u (kPa)	49	61
In situ cohesion by DST, c_u (kPa)	22	27
Cohesion by DST at OMC, c_u (kPa)	34	40
In situ angle of internal friction by DST, Φ_u (°)	21	26
Angle of internal friction by DST at OMC, Φ_u (°)	28	23
CBR, un-soaked (%)	6.0	8.0
CBR, soaked @ 96 h (%)	2.0	3.0

[a]*DG* = Doodh-Ganga

Table 2 Constituents of the cementation media implied for the bio-cementation process

Chemical	Chemical concentration (g)		
	0.25*M*	0.5*M*[a]	0.75*M*[a]
Urea	15.02	30.03	45.05
$CaCl_2 \cdot 2H_2O$	36.75	73.50	110.3
Distilled water	1000	1000	1000
pH	6.5	6.5	6.5
NB[b] per L	4.0	4.0	4.0

[a]*M* = molarity
[b]*NB* = nutrient broth

$$C = 8.59 \times 10^7 \times R^{1.3627} \tag{3}$$

where R = reading at OD_{600}; and C = concentration (cells per ml).

2.1.3 Cementing Solution Used in This Research

A cementing solution of three different molarities was used viz. 0.25, 0.5, and 0.75. The composition of each substituent is shown in Table 2. Many researchers have used different cementing solution molarities and some have even used a constant amount of nutrient broth in order to provide nourishment to the cells [19, 20]. We have used a constant 4 g/L of nutrient broth for all the experiments.

2.1.4 Testing Methodology

Initially, the soil samples from the two locations were tested in the lab for all basic properties following standard codal provisions [21–27]. The weakest soil amongst the two site locations was chosen based on the different strength parameters, Table 1. Testing included both in situ samples and disturbed samples. Finally, the weakest soil site, i.e., Site DG1 was chosen for further research. UCS tests were done on two different sized samples viz. 38 and 50 mm dia. samples. After the testing, small pellet samples were chosen from the failure plane for determining the calcium carbonate precipitation using gravimetric analysis or simple acid washing technique.

3 Results and Discussions

3.1 *Effect of Microbes on the UCS of Bio-Cemented Soils*

The testing had been carried out at 0.9 $\gamma_{d(max)}$ since the soil contained 30% sand particles, it had enough pore spaces to let the bacteria and cementing solution move uniformly through the matrix. The authors in their previous works have used a density of 85% owing to the fact that the soil contained fine particles in excess [11–13]. The soil was mixed with bacteria and supplied with cementing solution at three different molarities (0.25, 0.5, and 0.75), the treatment duration varied from 40 to 50 h followed by a resting period of 24 h in order to extract the treated samples from the molds [19, 28]. The samples were tested immediately after this period and for each combination of cementing solution molarity two samples were kept for curing for 3 and 7 days, respectively in desiccators. A noticeable rise in UCS was seen at 7 days of curing amongst all treatment conditions. A UCS rise of approximately 300% was seen for both the sample sizes with the maximum enhancement from 280 to 764 kPa being seen at a molarity of 0.5 in the 50 mm dia. sample. A UCS increase in 38 mm dia. sample from 140 to 642 kPa was also note worthy, Fig. 2a–f. All the testing was carried at a strain rate of 1.2 mm/min according to standard IS codes [27].

It can be noticed from the stress versus strain plots that as the molarity and curing increases the strain values get reduced indicating a stiff sample and a brittle failure. At higher cementing solution molarities it is seen that the strength values get slightly reducing owing to the salt accumulation as has been previously highlighted by many researchers. Another factor that leads to this type of result is attributed to the nonuniform distribution of calcium carbonate as might have been due to a cluster of bacteria in a single area or change in pH and temperature. Figure 3a, b shows the extruded samples before UCS testing and after testing. In these samples, a clear defined failure plane can be seen which is not always the case, as the precipitate tends to make the sample fail at a plane of more precipitate deposition [13].

3.1.1 Effect of Microbes on the Calcium Carbonate Precipitation of Bio-cemented Soils

Simple acid washing technique was adopted to determine the amount of precipitate in the treated samples. This was a confirmatory test for the experimental results as treatment with an acid liberated carbon dioxide, in turn, increasing the weight of the sample after drying and thus the difference between the two weights, i.e., before treatment and after treatment gave us approximately the amount of precipitate formed which was in turn responsible for the strength enhancement. Table 3 gives the values of different calcium carbonate precipitation percentages at different test conditions. It is clear that as the curing and molarity of the cementing solution increased the precipitate percentage increased. Researchers established studies that clearly showed that at lower molarity of cementing solution, strength imparted was more. This was

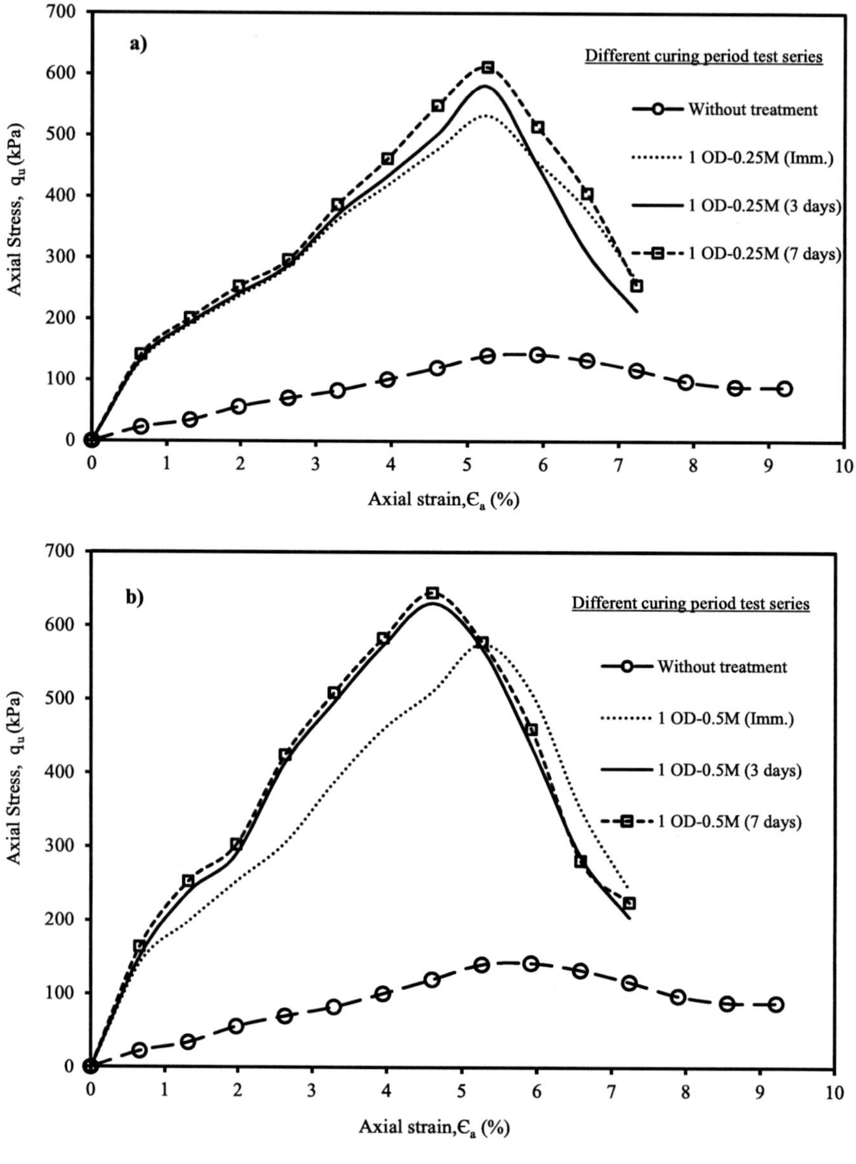

Fig. 2 **a–c** Stress versus strain plots for 38 mm diameter UCS samples. **d–f** Stress versus strain plots for 50 mm diameter UCS samples

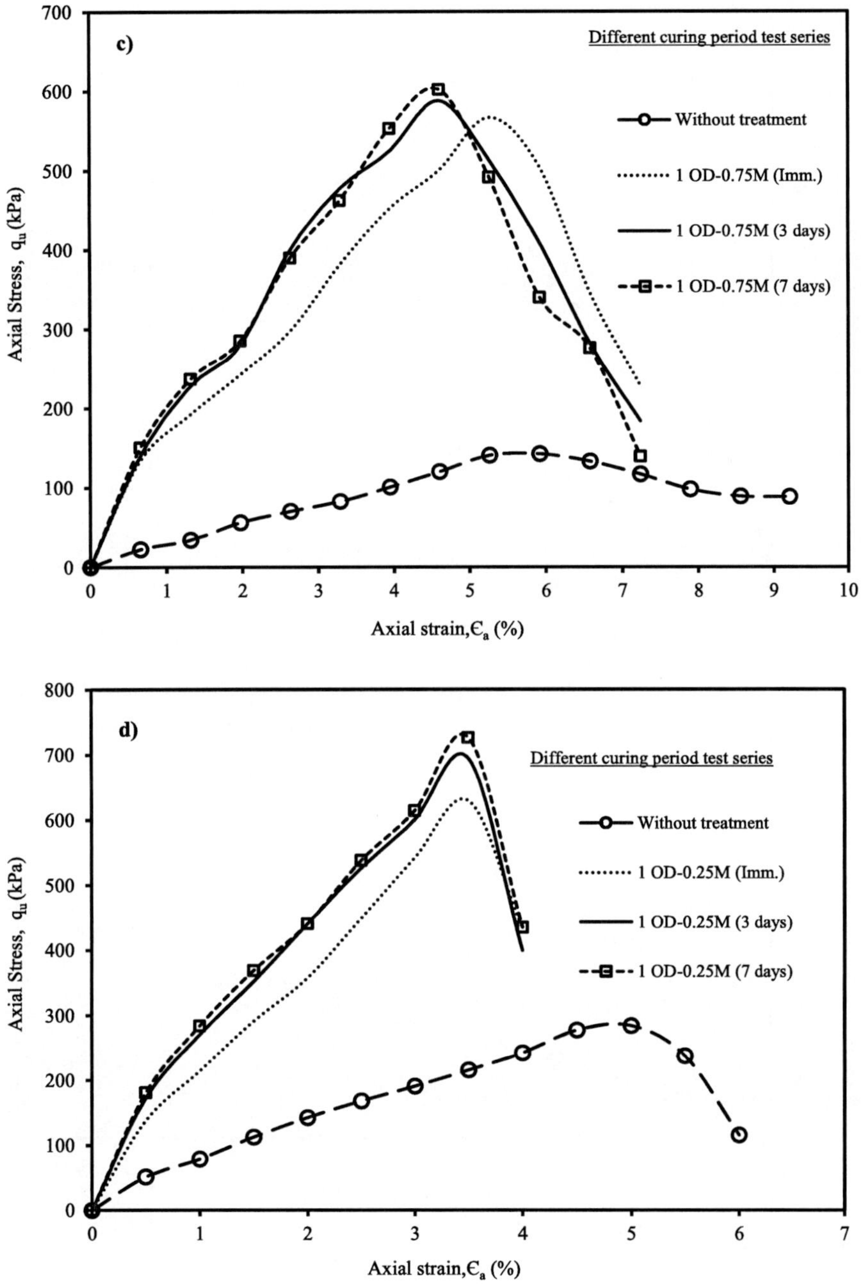

Fig. 2 (continued)

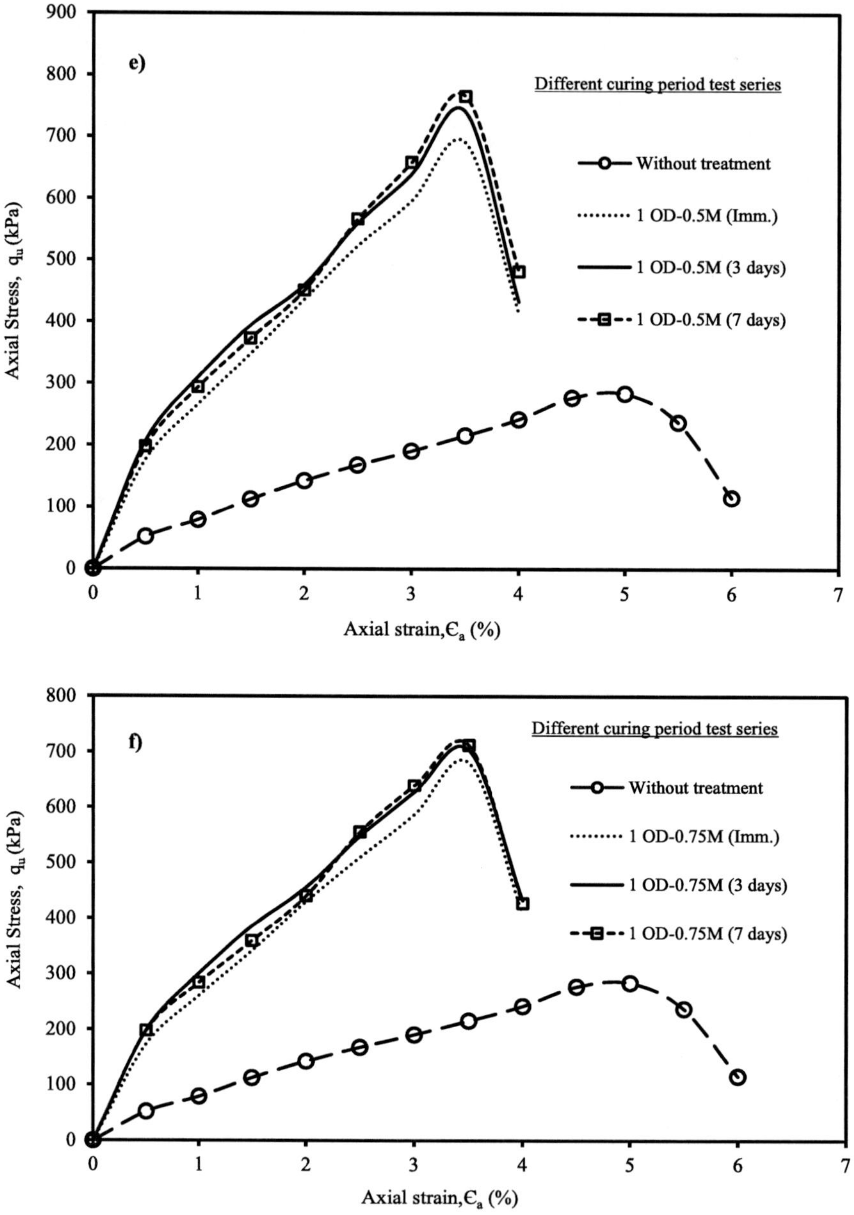

Fig. 2 (continued)

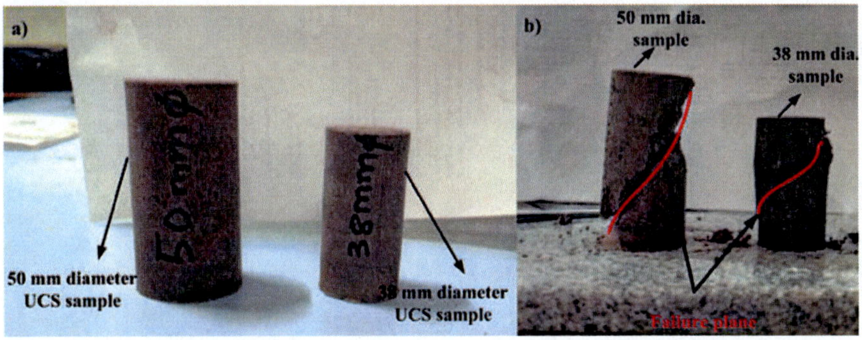

Fig. 3 **a, b** Extruded UCS samples before and after testing

Table 3 Values of different calcium carbonate precipitation percentages at different test conditions

38 mm diameter sample				
Molarity of urea	Molarity of calcium chloride	Curing (days)	UCS (kPa)	$CaCO_3$ (%)
0	0	0	140	0
0.25	0.25	0	530	5.5
0.25	0.25	3	580	6
0.25	0.25	7	610	6.4
0.5	0.5	0	575	5.9
0.5	0.5	3	629	6.5
0.5	0.5	7	642	6.8
0.75	0.75	0	565	6.2
0.75	0.75	3	587	7
0.75	0.75	7	601	7.5
50 mm diameter sample				
Molarity of urea	Molarity of calcium chloride	Curing (days)	UCS (kPa)	$CaCO_3$ (%)
0	0	0	280	0
0.25	0.25	0	630	6.1
0.25	0.25	3	695	6.5
0.25	0.25	7	725	6.9
0.5	0.5	0	690	6.4
0.5	0.5	3	741	7
0.5	0.5	7	764	7.4
0.75	0.75	0	680	6.8
0.75	0.75	3	704	7.4
0.75	0.75	7	712	8.1

attributed to a more uniform distribution whereas at higher molarities strength was less but the precipitate percentage was more [29–32]. These findings clearly support our experimental data.

4 Conclusions

Based on the results and discussions, the following conclusions can be made:

1. Use of microbes in soil stabilization proved very effective as the UCS values were enhanced considerably.
2. Different sized UCS sample testing proved vital in establishing the fact that this technique can be employed in situ as the strength increased by about 300%, which is almost comparable to cement stabilized soils.
3. The acid washing technique provided conclusive evidence about the formation of calcium carbonate which was responsible for bio-cementation of the UCS samples.
4. This technique is a sustainable green environment friendly technique and can be employed in the field after proper validation which would in turn reduce the carbon emission into the atmosphere.

5 Future Scope of Work

This technique can prove very beneficial in order to replace the age-old conventional soil stabilization techniques which are very detrimental to the environment. Future work related to the use of different urease-producing bacteria keeping an account of other factors like humidity, temperature, pH, even distribution of calcite, predatory organisms, etc., can prove very beneficial in applying this technique in the field. Tests like consolidated tri-axial tests, effect of pore water on the strength of bio-cemented soils, etc. can prove handy when applying this technique in situ. Most important consideration can be given to the sites where urease-producing bacteria's are available naturally, this would reduce the cost of this technique. Researchers should focus on cheaper cementing solutions and natural deposits for calcium chloride.

Acknowledgements The authors would like to thank the Division of Geotechnical Engineering and Department of Civil Engineering, National Institute of Technology—Srinagar, India.

Funding There was no direct funding available for this research. The first and the third author receive a doctoral fellowship from Ministry of Education (MoE)-India.

Conflict of interest The authors declare that there is no conflict of interest in this research.

References

1. Degirmenci N, Okucu A, Turabi A (2007) Application of phosphogypsum in soil stabilization. Build Environ 42(9):3393–3398. https://doi.org/10.1016/j.buildenv.2006.08.010
2. Morrison WR (1971) Chemical stabilization of soils—laboratory and field evaluation of several petrochemical liquids for soil stabilization
3. Sheikh IR, Shah MY (2020) Experimental study on geocell reinforced base over dredged soil using static plate load test. Int J Pavement Res Technol 1–10. https://doi.org/10.1007/s42947-020-0238-2
4. Ajayi MA, Grissom WA, Smith LS, Jones EE (1991) Epoxy-resin-based chemical stabilization of a fine, poorly graded soil system. Transp Res Rec 1295:95–108. http://onlinepubs.trb.org/Onlinepubs/trr/1991/1295/1295-012.pdf
5. Phear A, Dew C, Ozsoy B, Wharmby NJ, Judge J, Barley AD (2005) Soil nailing-best practice guidance, No. C637
6. Turner JP, Jensen WG (2005) Landslide stabilization using soil nail and mechanically stabilized earth walls: case study. J Geotech Geoenviron Eng 131(2):141–150. https://doi.org/10.1061/(ASCE)1090-0241(2005)131:2(141)
7. Karol RH (2003) Chemical grouting and soil stabilization, revised and expanded, vol 12. CRC Press, Boca Raton
8. Rutledge PC (1958) Study of deep soil stabilization by vertical sand drains
9. Johnson SJ (1970) Foundation precompression with vertical sand drains. J Soil Mech Found Div 96(1):145–175
10. DeJong JT, Mortensen BM, Martinez BC, Nelson DC (2010) Bio-mediated soil improvement. Ecol Eng 36(2):197–210. https://doi.org/10.1016/j.ecoleng.2008.12.029
11. Wani KS, Mir BA (2019) Influence of microbial geo-technology in the stabilization of dredged soils. Int J Geotech Eng 1–10. https://doi.org/10.1080/19386362.2019.1643099
12. Wani KS, Mir BA (2019) Effect of biological cementation on the mechanical behaviour of dredged soils with emphasis on micro-structural analysis. Int J Geosynth Ground Eng 5(4):32. https://doi.org/10.1007/s40891-019-0183-9
13. Wani KS, Mir BA (2020) Unconfined compressive strength testing of bio-cemented weak soils: a comparative upscale laboratory testing. Arab J Sci Eng. https://doi.org/10.1007/s13369-020-04647-8
14. Van Paassen LA, Ghose R, van der Linden TJ, van der Star WR, van Loosdrecht MC (2010) Quantifying biomediated ground improvement by ureolysis: large-scale biogrout experiment. J Geotech Geoenviron Eng 136(12):1721–1728. https://doi.org/10.1061/(ASCE)GT.1943-5606.0000382
15. Martinez BC, DeJong JT (2009) Bio-mediated soil improvement: load transfer mechanisms at the micro- and macro-scales. In: Advances in ground improvement: research to practice in the United States and China, pp 242–251. https://doi.org/10.1061/41025(338)26
16. Whiffin VS, Van Paassen LA, Harkes MP (2007) Microbial carbonate precipitation as a soil improvement technique. Geomicrob J 24(5):417–423. https://doi.org/10.1080/01490450701436505
17. Chen KB, Bin Kassim KA (2015) Microbial induced cementation on tropical residual soil. https://engineering.utm.my/civil/wp-content/uploads/sites/29/2016/12/Microbial-Induced-Cementation-on-Tropical-Residual-Soil.pdf
18. Ramachandran SK, Ramakrishnan V, Bang SS (2001) Remediation of concrete using micro-organisms. ACI Mater J Am Concr Inst 98(1):3–9
19. Cheng L, Shahin MA, Cord-Ruwisch R, Addis M, Hartanto T, Elms C (2014) Soil stabilisation by microbial-induced calcite precipitation (MICP): investigation into some physical and environmental aspects. In: 7th international congress on environmental geotechnics: ICEG2014. Engineers Australia, p 1105
20. Stocks-Fischer S, Galinat JK, Bang SS (1999) Microbiological precipitation of $CaCO_3$. Soil Biol Biochem 31(11):1563–1571

21. Bureau of Indian Standards (1970) IS 1498: Classification and identification of soils for general engineering purposes. Bureau of Indian Standards, New Delhi
22. Bureau of Indian Standards (1980) IS 2720-Part 3(1): Determination of specific gravity of fine grained soils. Bureau of Indian Standards, New Delhi
23. Bureau of Indian Standards (1985) IS 2720-Part 4: Determination of grain size distribution. Bureau of Indian Standards, New Delhi
24. Bureau of Indian Standards (1985) IS 2720-Part 5: Determination of Atterberg limits. Bureau of Indian Standards, New Delhi
25. Bureau of Indian Standards (1980) IS 2720-Part 7: Determination of water content-dry density relation using light compaction. Bureau of Indian Standards, New Delhi
26. Bureau of Indian Standards (1979) IS 2720-Part 16: Laboratory determination of CBR. Bureau of Indian Standards, New Delhi
27. Bureau of Indian Standards (1986) IS 2720-Part 10: Laboratory tests procedure for unconfined compressive strength test. Bureau of Indian Standards, New Delhi
28. Cheng L, Shahin MA, Mujah D (2016) Infuence of key environmental conditions on microbially induced cementation for soil stabilization. J Geotech Geoenviron Eng 143(1):0401608
29. DeJong JT, Soga K, Banwart SA, Whalley WR, Ginn TR, Nelson DC, Mortensen BM, Martinez BC, Barkouki T (2011) Soil engineering in vivo: harnessing natural biogeochemical systems for sustainable, multi-functional engineering solutions. J R Soc Interface 8(54):1–15
30. DeJong JT, Fritzges MB, Nüsslein K (2006) Microbially induced cementation to control sand response to undrained shear. J Geotech Geoenviron Eng 132(11):1381–1392
31. Harbottle MJ, Lam MT, Botusharova S, Gardner DR (2014) Self-healing soil: biomimetic engineering of geotechnical structures to respond to damage
32. Wani KMNS, Mir BA (2020) Effect of microbial stabilization on the unconfined compressive strength and bearing capacity of weak soils. Transp Infrastruct Geotechnol. https://doi.org/10.1007/s40515-020-00110-1

Waste Material a Sustainable Remediation for Settlement Caused by Pseudo-static Loading at Nearby Footing

Aashim Gupta, Prashant Garg, and Charnjeet Singh

1 Introduction

The phenomenon of foundation interference becomes an important aspect to be considered in closely built structures and the effect becomes further sensitive under dynamic condition caused by machine vibration operating at high frequency. The main objective which is to be considered by a civil engineer while constructing a structure is the safety of that structure and also the safety of nearby structure. So the proper study of behaviour of machine foundation and its effects on nearby structure is important. It has been noticed that in many cases ignoring the safety and care while construction leads to the failure of its structure or the nearby structure.

Dynamic loading basically comes under category of machine loading like motors and generators which consists of unbalanced moving masses which create different types of loads on soil. So, main purpose while constructing a structure is that the structure itself should be safe, and also, no harm shall be caused to the nearby structure due to the construction of new structure. So proper study of the behaviour of machine foundation and itself effect on nearby structure is important. The studies conducted by Liang [1], Wong and Luco [2], Ghosh [3], Ghosh and Kumari [4], and Ghosh [5] explain the interference effect of nearby foundations and ground anchors under the earthquake loading. In most of the research, it was found the vibration generated due to the presence of any source affects the behaviour of foundation located in vicinity. The magnitude of damage depends upon the soil properties, load intensity, material

A. Gupta (✉) · P. Garg · C. Singh
Guru Nanak Dev Engineering College, Ludhiana, India
e-mail: aashimgupta789@gmail.com

P. Garg
e-mail: prashant.ced@gmail.com

C. Singh
e-mail: charnjeet1989@gmail.com

H. Singh et al. (eds.), *Sustainable Development Through Engineering Innovations*, Lecture Notes in Civil Engineering 113,
https://doi.org/10.1007/978-981-15-9554-7_16

and dimension of footing. The waves due to vibration move up to a certain depth and beyond which the effect is negligible; but in horizontal direction, significant influence can be observed. Therefore, creating empty space or space filled with different material up to a certain depth may be advantageous to damp the vibration and finally reduce the adverse effect on the nearby footing. Also, economy should be considered in construction and foundation design but there should be no compromise with the safety of the structure and nearby structure.

Tyre wastes are preferred in many different engineering applications due to their convenient engineering properties such as thermal insulation, permeability, compressibility, stiffness and also high damping. This study is a new approach to use tire wastes as energy absorption material due to its enhanced damping and stiffness properties as compared to recycled aggregates. The use of waste tyre shredded is taken into consideration as its disposal is a problem and also the reuse of tyre is limited to certain extent. The stone column of recycled aggregates derived from constructional waste was used in the study. Both materials are extract of waste, so study is done to examine the use of these in creating damping effect to counteract the pseudo-static loading as sustainable remediation.

2 Methodology of Work

Plaxis 2D is well known and reliable finite element method geotechnical-based software. Many established consultancy firms are using it effectively. In spite of this for the present study, the software is proposed to be validated with the experimental work, conducted by Swain and Ghosh [6]. In the present study, the effect of vibration generated by a pump operating with different frequency was observed in terms of the displacement of secondary footing. A blanket of three different materials was proposed to be installed in between the primary and secondary footing to dampen the effect of vibration. The length and thickness of blanket were varied to determine the optimum length and thickness, and the variables used for the study are listed in Table 1. The soil and footing properties as shown in Table 2 and size of both footing were kept constant. The weight of pump on the primary footing was also kept constant.

3 Numerical Modelling

The primary footing and secondary footing were modelled using plane strain model. Appropriate boundary condition was assigned. Properties to the footing (plate element) and soil were assigned. Mohr–Coulomb model was selected for the analysis as this model can provide the first order accuracy only using elastic parameters. Medium size mesh was generated for the triangulation of the whole soil mass. Initial

Table 1 Variables for the study

Parameter	Symbol	Value	Units
Spacing between primary and secondary footing	s	750	mm
		1000	
Length of blanket	L	500	mm
		1000	
		1500	
Operating frequency of pump	f	3.5	Hz
		6.5	
		8.0	
Thickness of blanket	t	100	mm
		200	
		300	
Damping material	–	Coarse aggregates	
		Shredded waste tyre	

Table 2 Soil and footing properties

S. No.	Parameter	Property
Footing parameters		
1	Primary footing	0.55 m × 0.55 m × 0.2 m
2	Secondary footing	0.65 m × 0.65 m × 0.2 m
3	Material	M20 concrete
4	Density of material	24 kN/m^3
5	Clear spacing	0.25 m (S/B = 0.45)
6	Static loading	10.75 kN/m^3
7	Motor speed	1200–3000 rpm (3.5–8.0 Hz)
Soil parameters		
1	Soil type	Clayey silt up to depth 2 m
2	Liquid limit	35%
3	Plasticity index	17%
4	γ_{sat}	17.5 kN/m^3
5	γ_{dry}	16.0 kN/m^3
6	E_s	20,000 kN/m^3
7	K_x	1.15×10^{-6} m/s
8	K_y	1.15×10^{-6} m/s
9	K_s	25,000
10	υ	0.35

condition was simulated using K_0 value method. Calculation was performed using dynamic load on the different models with and without blanket.

4 Validation of Work

The results of numerical model are compared with the experimental work to validate the result of software. The various soil and footing parameters determined by Swain and Ghosh [6] in the laboratory and mentioned in their research publication are given in Table 2. The experimental setup used for the experimental study by Swain and Ghosh is also shown in Fig. 1.

The model was validated for three dynamic loading as mentioned in Table 3. The results obtained by the numerical study and experimental study were given in Table 3 for all three different loadings generated due to different operating frequency of pump. From Table 3, it is observed that the resonant frequency, i.e. at which maximum displacement was occurred is in the good agreement with that of obtained from numerical study. The difference was varied from 2 to 8% only. Similarly if the displacement obtained from numerical study is compared with the experimental result, it was observed that maximum error is 7% in case of displacement at primary footing and about 18% in case of displacement at secondary footing. Therefore, it may be concluded that Plaxis 2D may be used to study the effect of vibration on the nearby footing/structure efficiently and economically with sufficient accuracy.

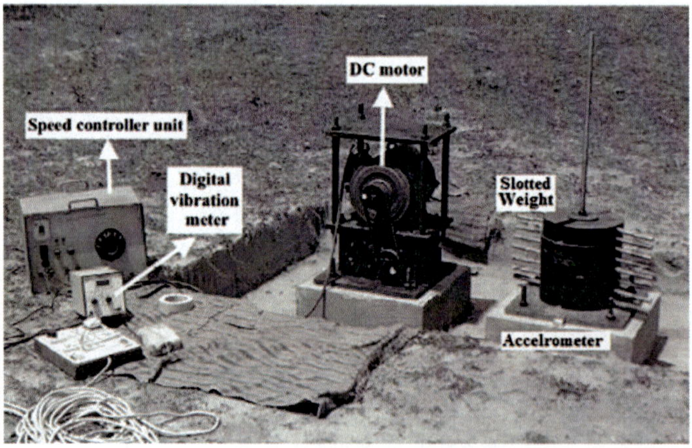

Fig. 1 Experimental setup (Swain and Ghosh)

Table 3 Result of experimental and numerical study

S. No.	Force (m.e)	Experimental results Displacement (10E-6m)				Numerical results Displacement (10E-6m)			
		Resonant frequency (Hz)	Resonant frequency (rpm)	Primary footing	Secondary footing	Resonant frequency (Hz)	Resonant frequency (rpm)	Primary footing	Secondary footing
1	0.017	6.35	2300	**128**	**121**	6.5	2450	**117.5**	**100**
2	0.025	6.15	2200	**210**	**170**	6.5	2450	**205**	**155**
3	0.033	6.05	2100	**272**	**220**	6.5	2450	**262**	**180**

5 Results and Discussion

To observe the adverse affect of dynamic load on the nearby footing, the study was carried out for three different frequency and distance between source (Primary footing) and secondary footing (effected footing) were taken as 750 and 1000 mm. The effect on the secondary footing in terms of its vertical settlement was shown in Table 4. Various remedial measures were taken, and their comparative study was carried out.

5.1 Effect of Stone Blanket of Recycled Aggregates Between Two Footings

In this case, two different models are created for two different spacing between primary footing and secondary footing for 0.75 and 1 m. Length of stone column is taken as 0.5 m, 1 m and 1.5 m, respectively. Thickness of stone column is taken as 0.1, 0.2 and 0.3 m. Three frequencies are taken for computation, i.e. 3.5, 6.5 and 8 Hz. The geometric model generated on Plaxis 2D was shown in Fig. 2. Properties of

Table 4 Displacement of secondary footing

Frequency (Hz)	Displacement (mm)	
	If c/c spacing between footing(s) as 0.75 m	If c/c spacing between footing(s) as 1.00 m
3.5	0.9	0.7
6.5	2.2	2.0
8	2.85	2.6

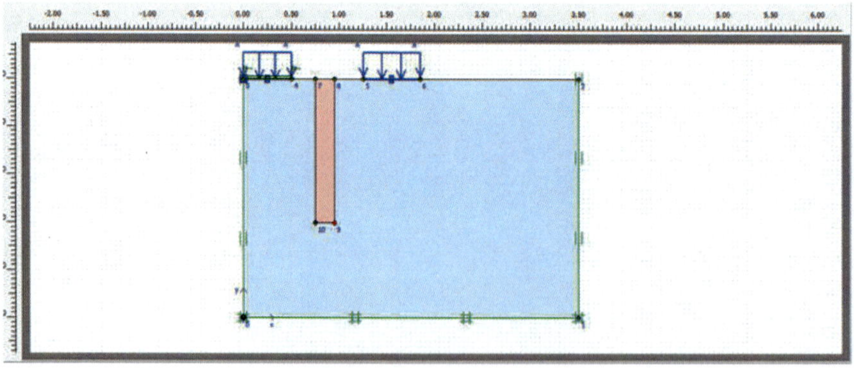

Fig. 2 Geometrical model of footings with stone column

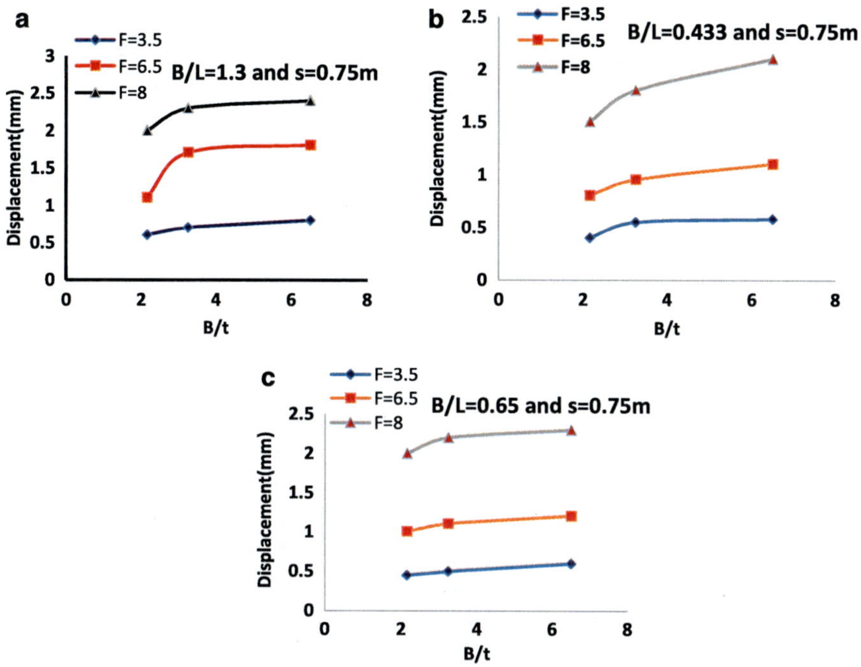

Fig. 3 **a** Effect of thickness on displacement. For c/c spacing = 750 and length 500 mm. **b** Effect of thickness on displacement. For c/c spacing = 750 and length 1500 mm. **c** Effect of thickness on displacement for c/c spacing = 750 and length 1000 mm

stone column: unsaturated unit weight = 20 kN/m^3, $\nu = 0.330$, $E = 38{,}000$ kN/m^2, $c = 10$ kN/m^2, $\phi = 30°$, $\Psi = 8°$. The result obtained from Plaxis with stone blanket represented in Fig. 3 for different thickness of stone blanket and for the length of stone blanket of 500, 1000 and 1500 mm. In these curves, the spacing between primary and secondary footing is kept same, i.e. 750 mm. Three curves were shown in figure which represents the displacement observed for three different frequencies of motor kept at primary footing. It was observed that as the thickness of stone blanket increases, the displacement at secondary footing decreases irrespective of frequency. Maximum decrease in displacement at secondary footing was observed when the frequency was 6.5 Hz. If the thickness of stone blanket increases from 100 to 300 mm, a reduction of 40–75% in displacement was observed.

It is also observed that as the magnitude of frequency effect a lot on the displacement at secondary footing. It increases with the frequency of source. The similar results were also observed when the length of stone blanket is 1000 and 1500 mm as represented in Fig. 3b and c.

Figure 4a–c represents the settlement as obtained from the Plaxis 2D. In these cases, the c/c spacing between primary and secondary footing was kept as 1000 mm. The length and thickness of stone blanket were varied. Figure 4a shows the effect

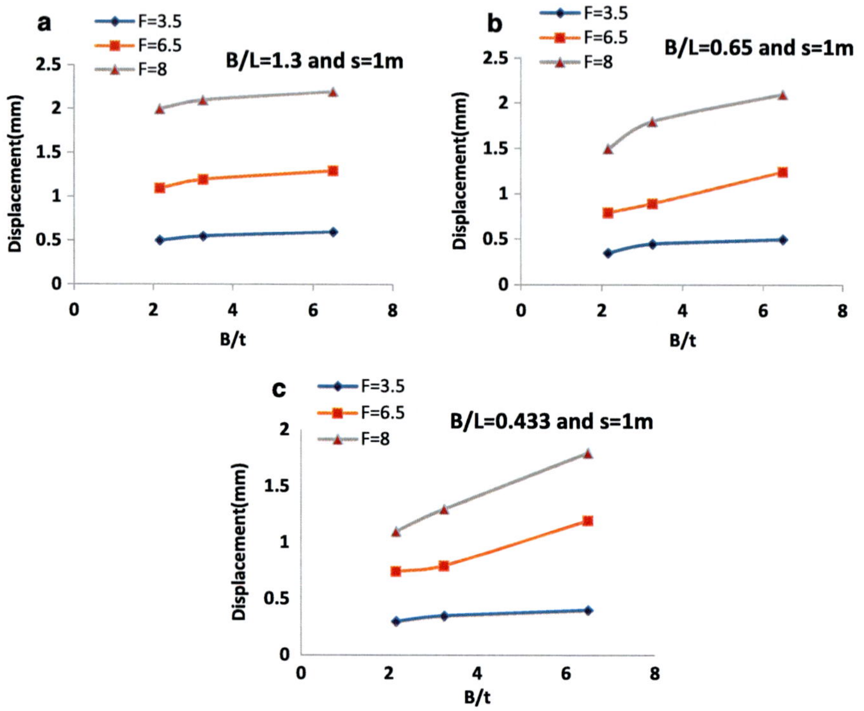

Fig. 4 **a** Effect of thickness on displacement. For c/c spacing = 1000 and length 500 mm. **b** Effect of thickness on displacement. For c/c spacing = 1000 and length 1000 mm. **c** Effect of thickness on displacement for c/c spacing = 1000 and length 1500 mm

stone blanket thickness on the displacement of secondary footing for motor running on primary footing with different frequencies as 3.5, 6.5 and 8 Hz. The result obtained is nearly same as for above. It was also observed that length of stone blanket also has a considerable impact on the settlement of secondary footing. As the length of blanket increases from 500 to 1500 mm, the settlement of secondary footing decreases about 80–100% for different thickness of stone blanket.

5.2 *Effect of Installing Tyre Scrap Layer Between Footings*

A blanket of shredded waste tyre is used instead of recycled aggregates with the properties: unsaturated unit weight = 6.5 kN/m^3, $\nu = 0.1$, $E = 900$ kN/m^2, $c =$ 14 kN/m^2, $\phi = 24°$. The results obtained from Plaxis 2D with stone blanket were represented in Fig. 6 for different thickness of stone blanket and for the length of stone blanket of 500, 1000 and 1500 mm. The spacing between primary and secondary footing was kept same, i.e. 750 mm. It was observed that as the thickness of stone

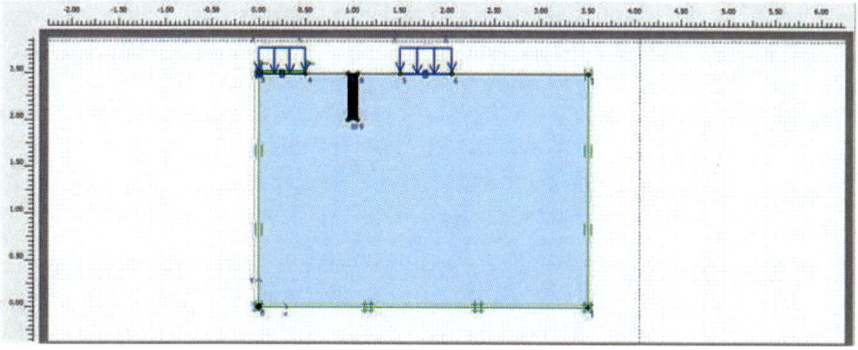

Fig. 5 Geometrical model of footings with the shredded waste tyre blanket

blanket increases, the displacement at secondary footing decreases irrespective of frequency. Maximum decrease in displacement at secondary footing was observed when the frequency was 6.5 Hz. If the thickness of shredded tyre blanket increases from 100 to 300 mm, a reduction of 60–105% in displacement was observed (Fig. 5).

It is also observed that as the magnitude of frequency has great effect on the displacement of secondary footing. It increases with the frequency of source. The similar results were also observed when the length of stone blanket is 1000 and 1500 mm as represented in Fig. 6b and c.

Figure 7a–c represents the settlement as obtained from the Plaxis 2D. In these cases, the c/c spacing between primary and secondary footing was kept as 1000 mm. Figure 7 shows the effect shredded tyre blanket thickness on the displacement of secondary footing for motor running on primary footing with different frequencies as 3.5, 6.5 and 8 Hz. It was also observed that length of shredded tyre blanket also has a considerable impact on the settlement of secondary footing. As the length of blanket increases from 500 to 1500, the settlement of secondary footing decreases about 100–140% for different thickness of shredded tyre blanket.

6 Conclusions

The study was carried out to observe the effect of two different waste extracted materials to dampen the vibrations. The blanket in between the primary and secondary footing using different materials was used. The thickness and length of the blanket were also varied. The following conclusions were drawn from the present study:

1. With increase in the operating frequency, the displacement on nearby footing increases up to certain limit and that is called resonant frequency. With further increase in frequency, displacement starts decreasing.
2. The displacement on secondary footing decreases with the increase in the thickness and length of blanket irrespective of material used as shown in Figs. 3 and 4.

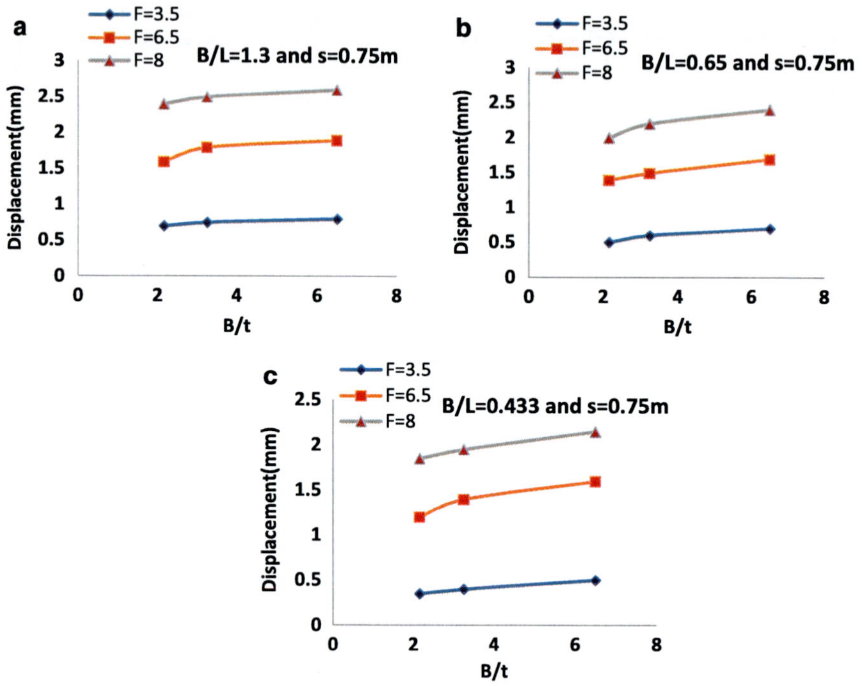

Fig. 6 **a** Effect of thickness on displacement. For c/c spacing = 750 mm and length 500 mm. **b** Effect of thickness on displacement. For c/c spacing = 750 mm and length 1000 mm. **c** Effect of thickness on displacement for c/c spacing = 750 mm and length 1500 mm

The optimum thickness and length of blanket depend upon the material used to damp, soil properties and spacing between the source of vibration and secondary. In case of recycled aggregates and waste shredded tyre reduction in settlement was observed 80–100% and 100–140%, respectively, if the length of blanket is increased from 500 to 1500 mm.

3. When the coarse recycled aggregates are used in the blanket, then 40–75% reduction in displacement of secondary footing was observed and when shredded tyre waste is used as blanket, then 60–105% reduction in displacement of secondary footing was observed.
4. It may be concluded that waste tyre is more effective to dampen the effect of vibration on any structure while length of blanket depends upon soil type.

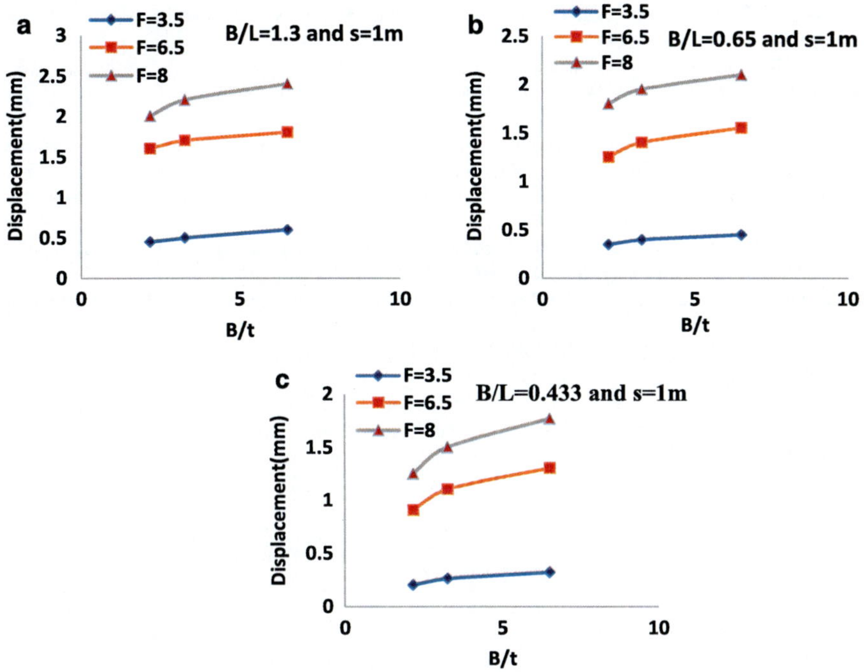

Fig. 7 **a** Effect of thickness on displacement. For c/c spacing = 1000 and length 500 mm. **b** Effect of thickness on displacement. For c/c spacing = 1000 and length 1000 mm. **c** Effect of thickness on displacement for c/c spacing = 1000 and length 1500 mm

Acknowledgements The authors express their heartiest appreciation for all those at Guru Nanak Dev Engineering College, Ludhiana, who rendered help and support in this research project. This help is greatly appreciated.

References

1. Liang CV (1974) Dynamic response of structures in layered soils. Ph.D. thesis, Massachusetts Institute of Technology, Cambridge, USA
2. Wong HL, Luco JE (1986) Dynamic interaction between rigid foundations in a layered half space. Soil Dyn Earthq Eng 5(3):149–158
3. Ghosh P (2011) Seismic interference effect of two nearby square footings. Geotechnical Special Publication, ASCE (GSP 211), pp 352–361
4. Ghosh P, Kumari R (2012) Seismic interference of two nearby horizontal strip anchors in layered soil. Nat Hazards 63(2):789–804
5. Ghosh P (2013) Numerical studies on seismic interference of two nearby embedded shallow footings. Disaster Adv 6(9):19–30
6. Swain A, Ghosh P (2016) Experimental study on dynamic interference effect of two closely spaced machine foundations. Can Geotech J 53(2). ISNN: 0008-3674

7. Hadi K (2016) Dynamic behavior of machine foundation on two-layer soil system. Int J Eng Dev 19. ISNN 1813-7822
8. IS 2974 (Part-I): 1987. Design and construction of machine foundation (First Revision)
9. IS 2974 (Part-III): 1992. Design and construction of machine foundation for rotary type of machine
10. Ishihara K, Ansal AM (2008) Dynamic behavior of soil, soil amplification and soil–structure interaction. J Earthq Eng Greece 19(3). ISNN 355-381
11. Kumar P, Ghosh P (2009) Interference effect of two nearby strip footings on reinforced sand. Contemp Eng Sci 2(12). ISNN 577-592
12. Nikam S, Bhagat SR (2014) Dynamic behavior of underground structures during earthquake: a critical review. Int J Eng Res 5(Special 1). ISSN: 2319-6890
13. Ramesh HN, Kumar HA (2005) Effect of static and cyclic loading on behavior of fiber reinforced sand. IOSR J Eng. ISSN 2278-8719
14. Tan SA, Tjahyono S, Oo KK (2008) Simplified plane-strain modeling of stone-column reinforced ground. ASCE1090-0241
15. Srinivasulu P (2001) Handbook of machine foundations. Book on design & construction of machine foundation by Tata McGraw Hill publishers, New Delhi
16. Vivek P, Ghosh P (2012) Dynamic interaction of two nearby machine foundations on homogeneous soil. GeoCongress 2012 © ASCE 2012

Performance-Based Evaluation: A Quantitative Tool for Sustainable Earthquake-Resistant Design

Krishna Murari, Harvinder Singh, and Savleen Takkar

1 Introduction

Earthquake forces are most destructive forces among all natural hazards and severely affect the built environment. Moreover, building damage contributes in worldwide environmental instability owing to this people faces various environmental challenges like those that depletion of non-renewable natural resources, global warming, excessive amounts of toxic material and non-ecological waste. Important research and development programme have been undertaken to minimize the global scale environmental impact of manmade structures. A range of quantitative tools, like performance-based evaluation and some advance structural system concepts such as seismically isolated structures, structures that rock on spread footing or pile foundations and self-centring structures reduce post-earthquake damage can contribute in sustainable development of infrastructure.

The earthquake-resistant design (ERD) of structures has developed gradually, initializing in the early twentieth century coming down to the performance-based design (PBD) concept in today's scenario. By this latest design philosophy, namely "performance-based design" or "displacement-based design", a designer is enabled to design any structure with anticipated performance, and, this is not the case with the "force-based design" (FBD) concept, because the latter does not explicitly contemplate upon the predictable performance of a structure. In this paper, performance of the code-conforming RC buildings and their seismic vulnerability are examined.

K. Murari (✉) · H. Singh · S. Takkar
Guru Nanak Dev Engineering College, Ludhiana 141006, India
e-mail: krishnamurarigne@gmail.com

H. Singh
e-mail: hvs1221@gmail.com

S. Takkar
e-mail: savleentakkar0605@gmail.com

H. Singh et al. (eds.), *Sustainable Development Through Engineering Innovations*, Lecture Notes in Civil Engineering 113,
https://doi.org/10.1007/978-981-15-9554-7_17

(a) Bihar- Nepal Earthquake (1934)

(b) Assam Earthquake (1950)

(c) Latur Earthquake (1993)

(d) Bhuj Earthquake (2001)

Fig. 1 Damage during past earthquakes

2 Motivation

The seismic zone map of Indian subcontinent emphasizes upon the fact that more than 60% of the land comes under severe to moderate earthquake conditions (IS 1893-Part 1, 2016) and approximate habitat requirement is 20–25 lakhs of buildings per year. Past earthquakes (Fig. 1) in India have led to an increase in the seismic zoning factor for many parts of the country. In addition, ductility has become an issue for all those buildings, designed and detailed using earlier versions of the Indian codes. Under such circumstances, assessing the capacity of existing building as per the existing codes in practice has become imperative. In order to enhance the performance of existing buildings to the present level of ductile design prescribed by present codes and find the retrofit or design a rehabilitation system, there is an urgent need to assess accurately the actual lateral load resistance and the potential failure modes.

3 Performance-Based Evaluation

PBD is continuously under development and used for the design of new structures and evaluation and retrofitting of existing structures. It allows realistic and reliable design and construction of structure keeping in mind the economic loss, occupancy and risk to life, which can happen because of the future seismic hazard. Defining parameter for PBSD is its performance objective, i.e. the acceptable level of damage

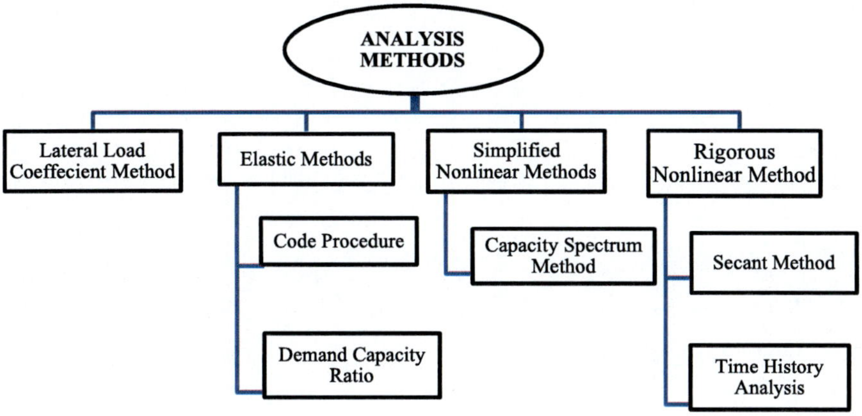

Fig. 2 Available analysis methods

selected for a specified earthquake intensity level. A building may be designed based on one or multiple performance objectives. Let us take an example of residential building; it could be designed for two performance objectives; first is to be fully operational (no damage and continuous service) based on an earthquake of low intensity and high probability of occurrence, and second is to achieve collapse prevention (extensive damage while not endangering the lives of its occupants) for a severe but low probability earthquake event. The selected performance objectives will depend on the intended use of the structure; for example, safety-critical buildings, such as hospitals and fire halls, are required to remain operational (light damage, most operations can resume immediately) after a severe earthquake event. Available methods of analysis are shown in Fig. 2. Selection of methods for analysis depends on various factors like, "what performance level you are hoping to achieve", the configuration of the structure, "how accurate you need to be". Lateral load coefficient method and elastic method give the linear response while nonlinear methods simulate more realistic behaviour of structures. Besides, this ATC 40, FEMA 356, FEMA 440 and ASCE/SEI 41-06 present some simplified nonlinear procedures, making an ease for practical design purpose [1–4].

4 Seismic Vulnerability of Generic Buildings

Vulnerability can be defined in many different ways. In simple words, seismic vulnerability is sensitiveness of any structure towards damage cause by any ground shaking of a given intensity, Seismic vulnerability and risk assessment of RC buildings in earthquake prone areas is the first step in any disaster mitigation plan. Prediction of cost of repair as a ratio of the cost of replacement, for the seismic hazard at particular

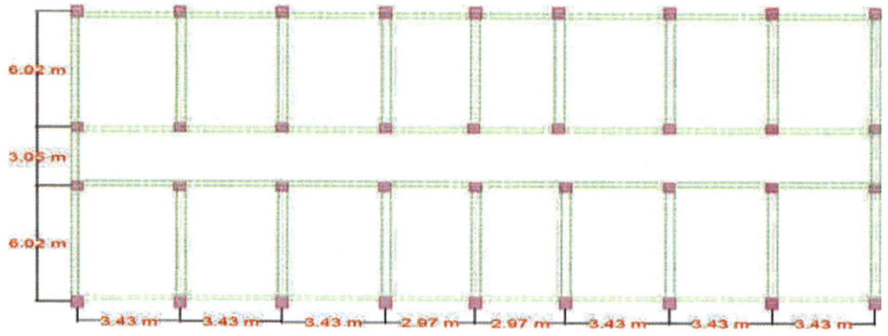

Fig. 3 Typical floor-level plan

site, is main goal of vulnerability assessment, which ultimately gives the loss estimation. There are two main categories; namely, empirical and analytical methods are available for vulnerability assessment. There is also a hybrid method, which uses both of these methods.

HAZUS methodology deals with wide range of different types of losses with nearly all aspects of built environment. It is developed by National Institute of Building Science (NIBS) for FEMA to reduce seismic hazard in USA and has been used, in some from or other, all over the world, for loss assessment of urban areas. HAZUS methodology has been used for the generation of curves (*Hazus Earthquake Model User Guidance*, 2019), According to this, the probability of being or exceeding a given damage state is calculated as

$$P[\mathrm{ds}/S_{\mathrm{d}}] = \Phi\left[\frac{1}{\beta_{\mathrm{ds}}}\ln\left(\frac{S_{\mathrm{d}}}{\bar{S}_{\mathrm{d,ds}}}\right)\right] \tag{1}$$

where $\bar{S}_{\mathrm{d,ds}}$ = median spectral displacement for damage state ds, Φ = normal cumulative distribution function, and β_{ds} = standard deviation of the threshold spectral displacement (Table 1).

5 Results and Discussion

Capacity curves (Figs. 4 and 5) and vulnerability curves (Figs. 6 and 7) are shown for 4-storey and 9-storey buildings (plan shown in Fig. 3) and description of buildings as given in Table 1 designed as per relevant Indian Standards for gravity load only and for special moment resisting frame (SMRF) case as per the same standards using computer program SAP2000v19.0.0 [5–10]. Capacity curves of structures are plot of base shear versus roof displacement when it is subjected to lateral loads. These curves can be used for assessment of the actual behaviour of structures under given ground motion intensity, whereas vulnerability curves are lognormal curves with a

Table 1 Description of buildings

No. of stories	4, 9
Typical floor height	3.43 m
Seismic zone	IV
Soil type	Medium stiff soil
Importance factor	1.5
Grade of concrete	M25, M30
Grade of steel	Fe 500
Density of RC	25 kN/m^2
Live load at corridor	4 kN/m^2
Live load at rooms	3 kN/m^2
Member loading at roof level	5 kN/m
Member loading at other level	15 kN/m

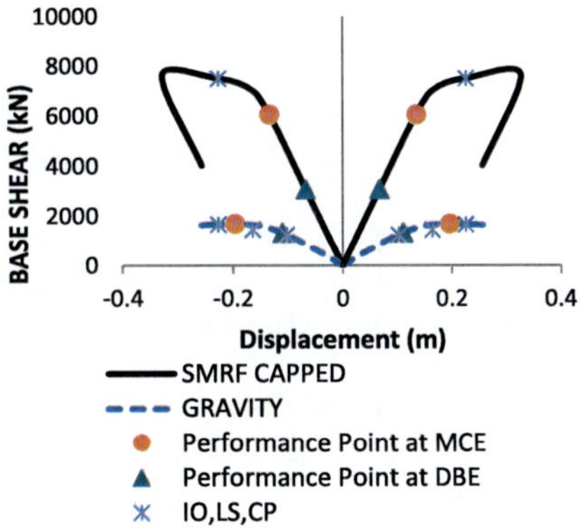

Fig. 4 Capacity curves for 4-storey buildings designed for gravity and as SMRF, as per relevant IS codes

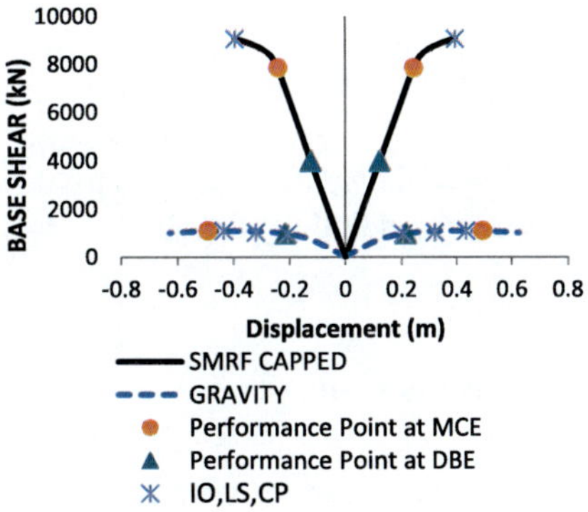

Fig. 5 Capacity curves for 9-storey buildings designed for gravity and as SMRF, as per relevant IS codes

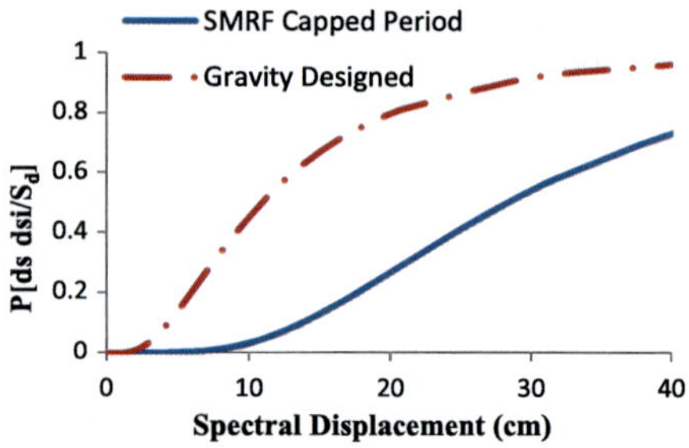

Fig. 6 Vulnerability for moderate damage state threshold (as per HAZUS) of 4-storey buildings

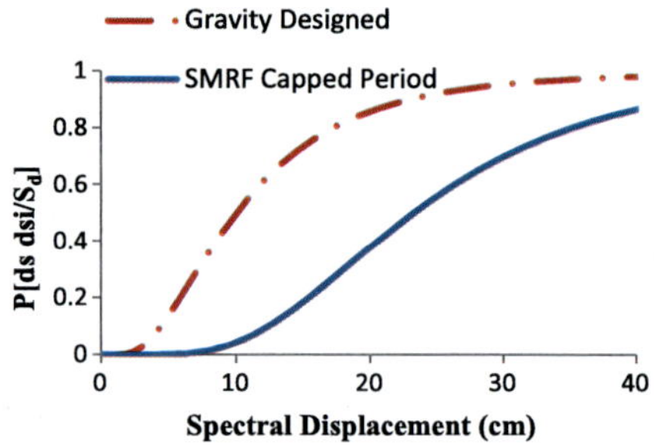

Fig. 7 Vulnerability for moderate damage state threshold (as per HAZUS) of 9-storey buildings

logarithmic standard deviation, to give the probability of being in a particular damage level.

It can be observed from the capacity curves, the seismic performance of buildings in all the cases considered for which seismic forces are taken into account shows better performance, i.e. immediate occupancy (IO) performance level. Moreover, graph shows that after considering the seismic forces in design, there is increase in strength and ductility capacity. The relative increase depends upon the building height and design period of vibration.

Fragility curves/vulnerability curves show probability of damage for different damage grade. It can be observed from the above shown fragility curves in Figs. 6 and 7, seismic vulnerability greatly depends on the design level and increases with the increase in building height.

6 Conclusion

Current earthquake-resistant design method adopted in most of the national codes intended for life safety with assumption of insuring minimal probability of collapse of structures while subjected to numerous cycles of inelastic deformation during seismic event. However, this approach results in substantial damage of structural and non-structural elements, which further leads to profound impact on economy and society. In addition, there is no prescribed method mentioned for quantification of damage and prediction of actual behaviour.

The following major conclusions are listed below:

- Performance-based design approach provides better confidence in achieving a specified strain or drift performance level or multiple performance objectives under a specified seismic activity also quantifies damage produced by earthquake and hence can contribute in sustainable development of infrastructures.
- In comparison to deterministic framework, probabilistic framework of performance-based seismic design gives better insight into expected performance and associated risk.
- Seismic performance of buildings in all the cases considered for which seismic forces are taken into account shows better performance, i.e. immediate occupancy (IO) performance level.
- Seismic vulnerability greatly depends on the design level and increases with the increase in building height.

References

1. ATC-40 (1996) Seismic evaluation and retrofit of concrete building. Applied Technology Council, California
2. ASCE 41-13 (2013) Seismic evaluation and retrofit of existing buildings. American Society of Civil Engineers, Virginia, USA
3. FEMA (2003) HAZUS-MH technical manual. Federal Emergency Management Agency, Washington, DC
4. Goel RK (2008) Evaluation of current nonlinear static procedures for reinforced concrete buildings. In: 14th world conference on earthquake engineering, Dec, pp 57–80
5. BIS (1987) IS: 875 (Part 1)—1987—Indian Standard Code of Practice for Design Loads (Other than Earthquake) for Buildings and Structures, Part 1: Dead Loads—Unit Weights of Building Materials and Stored Materials (Second Revision). Bureau of Indian Standards, New Delhi
6. BIS (1987) IS: 875 (Part 2)—1987—Indian Standard Code of Practice for Design Loads (Other than Earthquake) for Buildings and Structures, Part 2: Imposed Loads (Second Revision). Bureau of Indian Standards, New Delhi
7. BIS (2000) IS 456: 2000—Indian Standard Plain and Reinforced Concrete—Code of Practice (Fourth Revision). Bureau of Indian Standards, New Delhi
8. BIS (2016) IS 1893 (Part 1): 2016—Indian Standard Criteria for Earthquake Resistant Design of Structures, Part 1: General Provisions and Buildings (Sixth Revision). Bureau of Indian Standards, New Delhi
9. BIS (2016) IS 13920: 2016—Indian Standard Ductile Detailing of Reinforced Concrete Structures Subjected to Seismic Forces—Code of Practice. Bureau of Indian Standards, New Delhi
10. CSI. Analysis reference manual for SAP2000. Computers and Structures Inc., Berkely, CA

Building Information Modeling (BIM)-Based Sustainable Management of a Construction Project

Satinder Kaur Khattra, Hardeep Singh Rai, and Jagbir Singh

1 Introduction

The advent of information technology in the construction industry has prompted a gradual shift from manual paper-based processes to computer-aided design and production. In this transfer, there has been an overwhelming curiosity in the application of building information modeling (BIM). BIM and sustainability are comparatively new ideas inside the AEC industry. There has been an expanding awareness of BIM use in sustainable building projects because of its amazing treatment of multidisciplinary data inside a single model. Despite the advantages of BIM and sustainable data management in the design stage [3], there are a few difficulties to BIM acceptance in sustainable development such as the absence of practical tools. The integration process requires significant effort and time to such an extent that sustainable data evaluation remains an after design stage process for most of the cases. The typical approach to representing a model on a paper by hand has been replaced by the 3D model in BIM-based workflow. "Building Information Model" refers to a digital model (as opposed to a physical model) that is capable of holding semantic information relating to building components. BIM is the structured procedure to create, store, exchange, and share building-related information between the different stages of the building project [2]. Significantly after such a large number of years since the advancement of BIM, industry cannot clarify the BIM by a solitary exceptional definition. Indeed, even it is does not appear to be achievable in the current situation on account of the significance of BIM continues changing and consistently has been explained depending upon the undertaking or reason that uses

S. K. Khattra (✉)
I.K.G. Punjab Technical University, Kapurthala Road, Jalandhar, Punjab, India
e-mail: satinderkhattra14@gmail.com

H. S. Rai · J. Singh
Department of Civil Engineering, Guru Nanak Dev Engineering College, Ludhiana, Punjab, India

H. Singh et al. (eds.), *Sustainable Development Through Engineering Innovations*, Lecture Notes in Civil Engineering 113,
https://doi.org/10.1007/978-981-15-9554-7_18

it and besides various procedures about how it should be applied in the structure, advancement and support of structures [13]. According to "US National Building Information Model Standard (NBIMS) Project Committee" has been defined BIM as:

> Building Information Modeling (BIM) is a digital representation of physical and functional characteristics of a facility. A BIM is a shared knowledge resource for information about a facility forming a reliable basis for decisions during its life cycle; defined as existing from earliest conception to demolition. [18]

Despite all the complex acronyms and technical jargon, BIM is essentially a process about collaborative design by allowing multiple project team members to work on a single virtual representation of a physical building known as BIM model. Although BIM has been an emerging trend in the architecture, engineering, and construction (AEC) domain, however the ultimate success of BIM depends on the ability to capture all relevant data in the BIM model and to effectively communicate data between the various applications. BIM workflow can be divided into two phases. The first phase refers to the geometric modeling that is used in order to describe information regarding the building while the second refers to the way that this model is used to share or exchange this information [20]. Although BIM is designed to help with such collaborations, the lack of integration and interoperability is still causing problems in data transfer, which is considered as the main barrier for a wide adoption of BIM in the AEC industry [17]. Successful use of BIM tools in collaboration depends on whether the information produced by the various participants in different stages of the project can be successfully shared/exchanged through the entire building life cycle. The power of BIM to develop a structure earlier to development of the real structure gives a compelling way to check its constructability in reality and to determine any uncertainty during the project. It allows architect, structural engineer, contractor, and the user to collaborate on a single platform, helping them to work more efficiently by limiting the waste of resources, improve energy performance, and so on.

2 Industry Foundation Classes (IFC)

The adoption of BIM brings great benefit to the AEC industry, including visual advantages and direct use of the model in all kinds of computer-based analysis. However, it has been found that a number of firms that follow BIM are not reaping the full benefits because of difficulties such as the lack of communication between the various professionals in design and construction. Successful implementation of BIM is directly measured by the integration and exchange of wealth of information contained in BIM model among different parties and different stages of the project which is technically known as interoperability. Interoperability is defined as a process to share common information of objects such as geometry and properties among heterogeneous software tools. Interoperability can be seen as a method which permits diverse software and people to work together, in which a software tool exports its model in a standard format for sharing and exchange of data and the other one imports this model for further use [14]. For example, a structural engineer can import a 3D structural frame from architectural model to be used for design and analysis purposes.

The construction industry is characterized by an extremely fragmented procedure with several diverse and autonomous participants who only communicate for the duration of project only. A plenty of diverse software/tools are utilized for preparing domain specific models, and therefore, homogeneous standards are hard to impose. Most of the software programs used in development of models are usually developed to work as standalone applications and are not typically designed to share data with other programs. Proprietary software applications develop their own data exchange formats specific to their domain which restricts the user to exchange data by selecting his desired BIM software. Hence, the data sharing is not seamless through these commercial software. If the user got the freedom of data exchange and sharing by successfully using a common sharing platform and effective interoperability, then it saves a lot of time and improves communication among various stakeholders.

To enhance interoperability between BIM software and to cater the increasing need for collaboration, the transformation of explicit information (geometric as well as non-geometric) is carried out with the help of a widespread file format, known as Industry Foundation Classes (IFC) (Thein). IFC—essential concept of BIM is an important neutral, open and object-based, industry standard for the exchange of BIM model information. The IFC data model transfers 3D geometry, properties and relationships of "intelligent" building elements among all stakeholders to provide them a clearer visualization of the project and escalating their capability to make quicker, more conversant decisions. This open information attempt, to break down the silos of information, started in 1996 when a group of 12 US companies invited by Autodesk, met up to look at the potential for making diverse software applications cooperate and established the International Alliance for Interoperability [11], presently known as buildingSMART International [16]. BuildingSMART is an international, open, vendor free and non-profit association mainly focused to enhance the exchange of information between software applications working in the construction industry to develop global, standards, rules, norms, to support the extensive use of BIM. BuildingSMART defines their most widely used exchange format as follows:

> The Industry Foundation Classes (IFC) represent an open specification for Building Information Modeling (BIM) data that is exchanged and shared among the various participants in a building construction or facility management project... [8].

IFC, a continuously developing standard by buildingSMART International, is frequently referred to as the excellent approach by which BIM interoperability can be realized. In this comprehensive data model, object-oriented approach is used to describe both the semantic information and geometry of a building model. Exclusively designed as a medium to exchange the information among model-based software systems in the AEC industries, IFC is now supported by the major software vendors as well as by several downstream analysis applications. For the past few years, IFC has emerged as a major neutral standard to improve interoperability among the individuals, domain-specific applications that are used to plan, design, and construct the buildings by encapsulating information about all aspects of a building throughout its life span. The advantage of using the IFC model is that information can be shared in a system which facilitates and promote interoperability.

3 IFC Structure

IFC is actually a schema not a format. A schema refers to a structured framework at the back data organization. IFC schema described using EXPRESS data definition language. EXPRESS is a standard data modeling language formulated with STEP (STandard for the Exchange of Product model data), utilized to properly define the semantics of data. The IFC schema characterizes a model incorporated with number of entities, attributes and their relationships in a logical manner. STEP is universally acknowledged neutral file format of almost all commercial computer-aided software. When building model is saved in IFC file format, it organizes the IFC units in a tree structure as shown in Fig. 1. The substance of the IFC Model is attached to a particular convention which will be followed in this work and is denoted hereafter in italicized text. Various entities in IFC model are prefixed with *Ifc,* for instance *ifcWall, ifcSlab,* etc. The order the structure is on higher to lower level that is *ifcproject* > *ifcsite* > *ifcbuilding* > *ifcbuildingstorey* > *ifcbuildingelement.* This IFC tree structure will be followed for the geometric exchange in this paper.

All components of the building and their spaces besides their interrelationships are expressed in detail to establish all the needed information to analyze and process

Fig. 1 IFC tree structure

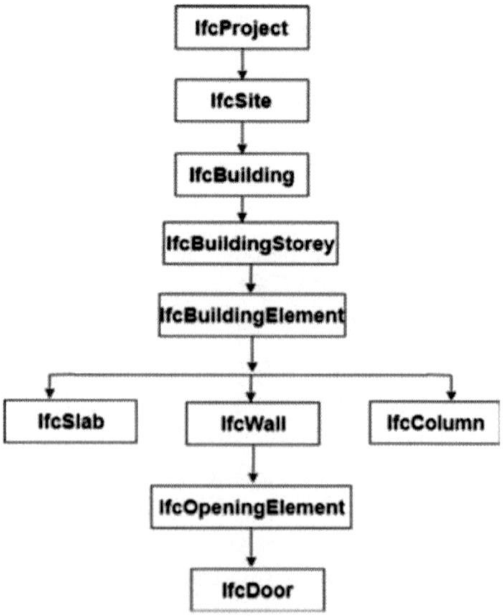

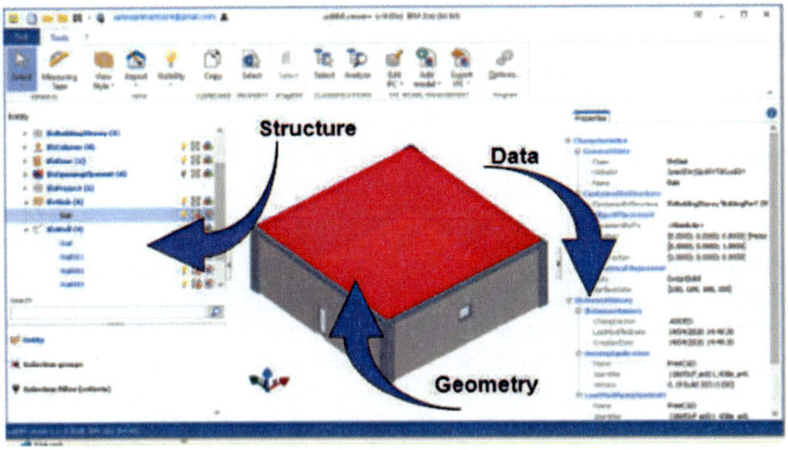

Fig. 2 IFC models with building *geometric and non-geometric data*

every element of the building. It can be utilized for almost any type of data exchange condition in the life span of a building. The IFC a neutral platform is very much significant for implementing BIM concepts. One of the major benefits of IFC is that it is an open standard and one and all have complete access to the data contained in it. In this manner, it is perfect for sharing of information between various software applications. Although IFC a most widely used ISO standard for exchange of information, its complex nature is a great challenge to get IFC data into diverse business applications. Despite the rich quality of IFC model, information instance files using IFC schema are not clearly understandable like a bar graph. IFC4, the current version of schema comprises more than 750 entities and above 350 types [5]. In this schema, entities, attributes, and their relations are used to represent the building elements as shown in Fig. 2. For instance, *ifcbeam* is used to represent standard beam element. The attribute *ifcOwnerHistory* is used to describe owner history attached to the beam. Also, *ifcUnit* is another attribute that represents the units associated with the beam element. ifcName is used to represent the name property of the beam.

As the complete IFC model by and large methods a coordinated model, that incorporates a large amount of data about a BIM-based building project created by different task members. So the size of the IFC model document ought to be expanded [9]. Most of the time, it is difficult and irrelevant to transfer the complete building information model among the project participants. The project participants need information related to their specific domain such as design, construction, mechanical, and energy performance or electrical and plumbing (MEP)-related data. Along these lines, it is increasingly productive to utilize IFC model containing just fundamental data to do explicit exercises [9] than to use a complete model; yet, it is not easy to generate and manage the BIM model information according to the characteristics of the definite tasks that will be carried out [15, 23]. Chen et al. [4] proposed an IFC-based web server to automatically extract geometric data for structural analysis from a 3D object-oriented model. Case studies were used to validate the process for extraction of information related to columns and beams only.

4 Scope and Significance of the Study

This paper focuses on the implementation of a framework for BIM and sustainable integration of data based on IFC schema to automatically extracting a partial BIM model from the original complete BIM model corresponding to the user's choice of building elements by repeatedly iterating through IFC data model depending on the rules specified in the algorithm. The objective of this research is to provide an efficient sustainable data approach to extract subset model containing entities, attributes and their relations associated with the requested building elements. The framework was implemented to utilize the BIM technology to sustain a collaborative process between the team members by recognizing entities associated to building elements, which clients needed to perform their domain related work. The information retrieval process is capable of exchange and manages certain specific data stored in the model efficiently by utilizing only the minimum valid IFC file data. The main idea of creating a partial model is to reduce the size and complexity of an IFC model, either to satisfy the special domain application requirements or to suit the data transformation requirements [1, 24].

5 Research Methodology

In the first step, a BIM tool is adopted to create building model and then model is exported as *.ifc* file. Based on the exported IFC model, for extraction of information from the architectural model, that is pertinent to a specific domain, a Python code, integrated with the Ifcopenshell library, has been developed to extract the target information that takes an IFC file as input. The reading of IFC files was accomplished by incorporating the Ifcopenshell (IfcOpenShell) into the IFC Parser program. The program is utilized to develop the geometric information extraction algorithm, which has been developed to reuse the rich content of the IFC model. The program recognizes different strings in the IFC file which can be queried to extract the required geometric data. The information is directly extracted from an IFC model and saved to text file which can be further utilized in various downstream applications such as for creating structural elements or energy performance evaluation of building in initial design stages or for creating fire simulation models. A complete BIM model can be used to extract two types of partial models. In first kind of partial extraction not addressed in this work, information on a certain view or characteristic of the building from the total model is retrieved. The subsequent extracted model is a "more slender" model, which despite everything comprises of all the structure components yet just contains the pertinent data. For instance, a model with just the geometries or a model with just the data identified with green structure rating. The model view definition (MVD) proposed by buildingSMART is an instance of such extractions. The second category of partial model, which is undertaken in this study, is the creation of a subset models with certain specific entities with user's choice of building elements in the original BIM model.

Even though an IFC model is text-based file and can be easily opened and readable in any text editor. But to generate a logically complete partial model with meaningful entities, attributes and with their topological relationships [6], to make it useful in the downstream application is not possible by simply copy and paste of IFC file. As the building elements in an IFC file may refer to a number of other elements and/or might be referred to by some other elements. The IFC model has been defined in EXPRESS language which is not a programming language rather it is a modeling language. Therefore, information in IFC model about building objects and their attributes needs to be mapped to a programming language to support the generation of partial model. Python language is employed in this study for this purpose. For extraction of information from the IFC model, that is relevant to the requested model, an IFC Parser tool has been developed that extracts BIM objects directly from an exported IFC model primarily written in Python. IFC files are not readable by Python. The reading of IFC files was accomplished by incorporating the Ifcopenshell into the IFC Parser program. Technically visualization of model could be achieved by opening the IFC model using IfcOpenshell in the program. Python is a great object-oriented, open source, flexible, and easy to learn programming language. Python is a progressively composed language, so the variables are automatically defined. The basic syntax rules of the language allow the developer to concentrate on the issues and not on the syntax to understand. It is easy to write and user-friendly. Also, Python can be effortlessly coordinated with other programming languages. Its enormous and vigorous standard library makes Python score over other programming languages. The standard library permits you to look over a wide scope of modules as per your requirements. The data visualization libraries provided by language help to visualize and present data in a more appealing and effective way.

6 Test Model and Criteria

The implementation of partial model extraction algorithm is explained by using a sample project. A 3D BIM-based model was prepared using FreeCAD (freecadweb). FreeCAD is a general purpose parametric 3D computer-aided design (CAD) modeling tool which meets all the requirements in order to be used as an open-source platform. It is extremely customizable, scriptable and extensible. New functions and open-source libraries can be added without changing the main core system. It includes an inbuilt Python interpreter, and thus, Python code can be used in the software. It can be considered to be the only open-source modeling software equivalent to the current proprietary industry tools like Revit or ArchiCAD. In this research, an entity related to the slab among many building elements such as ifcbeam, ifccolumn, ifcdoor was utilized for the evaluation of developed algorithm. Slab element in the IFC model is directly connected with the IfcSlab. Extraction of different strings and entities related to slab element from IFC model was carried out accordance to the hierarchy shown in Fig. 3.

An outline of the extraction procedure making use of a sample model, starting from the original BIM model and ending with output as the extracted partial model is shown in Fig. 4.

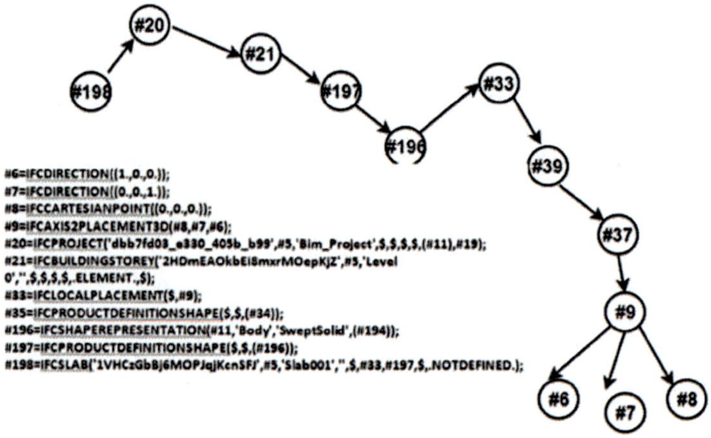

Fig. 3 Entity relationships for IFCSLAB

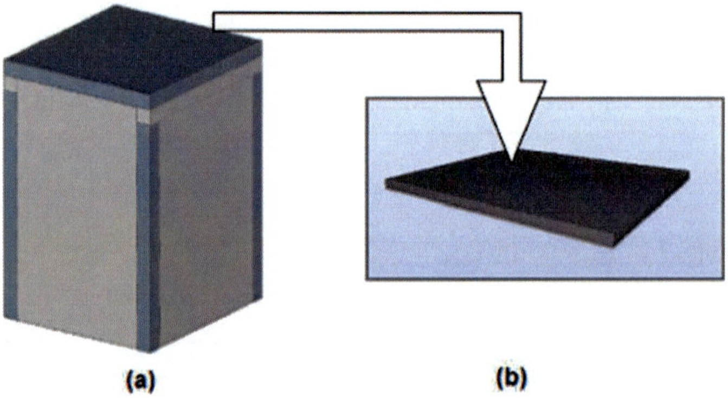

Fig. 4 Extraction algorithm overview

7 BIM Advances Sustainable Design

BIM-based information-sharing framework discussed in this paper can provide an environment that creates coordination among involved stakeholders through the life cycle of sustainable construction. Data sharing and information management using BIM platform helps to improve project performance by integrating building design data. Extraction and analysis of data suitable to different users' needs provide the designers with more competent ways of designing and allow them to create an improved design with all this information that will guide to better decision making and less likely to select products or design alternatives that harms the environment. High-quality and more precise decisions in turn will maximize the impact of the sustainability of the building [12]. Similarly, such object-oriented, intelligent models carry a wealth of information for many other different aspects of sustainable design. For example, schedules of building elements can be produced directly from the model to find percentages of material reuse, recycling, or salvage. Collaborative environment between BIM-based design tools and improved workflow reduces the time required for delivering a facility [22]. This can be of huge benefit as it allows initial designers to make immediate decision on which elements and design they can choose from which then allows for a more high quality and more accurate decisions which in turn will maximize the impact of the sustainability of the building and also cost.

8 Conclusions

- There is no doubt that BIM is a revolutionary concept and contain sufficient data needed for supporting sustainable design.
- BIM can essential be utilize to optimize the design and improve collaboration among various stakeholders enabling them to make more rational design decisions.
- However to achieve the full potential of BIM and to keep pace with the growing technology, BIM interoperability and data exchange is still a research challenge. As most of building models come only in proprietary formats, so they can only be used in the BIM authoring tool for which they have been created.
- To address this issue, in the proposed approach, the geometric information needed for downstream applications was automatically extracted from an IFC-based model and stored in text file.
- The proposed framework is appropriate for multi-participants BIM applications. This research is expected to impact the overall interoperability of applications in the building information modeling field helping to present more sustainable buildings in future.
- It is sure that BIM will be all the more broadly utilized assisting with giving more sustainable buildings and infrastructure in future. The conceptual design

and construction work process and even operational management will be arranged and controlled with the assistance of BIM.

Acknowledgements The primary author would like to mention and acknowledge the Department of Research, Innovation & Consultancy of I.K. Gujral Punjab Technical University, Kapurthala Road, Jalandhar, for giving her an opportunity and providing all the required assistance to write this research paper. This study is a part of a PhD research work carried out at I.K. Gujral Punjab Technical University, Jalandhar (Punjab).

References

1. Beetz J, Van Leeuwen J, De Vries B (2009) IfcOWL: a case of transforming EXPRESS schemas into ontologies. Ai Edam 23(1):89–101
2. Bryde D, Broquetas M, Volm JM (2013) The project benefits of building information modelling (BIM). Int J Project Manag 31(7):971–980
3. Bynum P, Issa RR, Olbina S (2013) Building information modeling in support of sustainable design and construction. J Constr Eng Manag 139(1):24–34
4. Chen PH, Cui L, Wan C, Yang Q, Ting SK, Tiong RL (2005) Implementation of IFC-based web server for collaborative building design between architects and structural engineers. Autom Constr 14(1):115–128
5. Cheng JC, Deng Y, Das M, Anumba C (2014) Evaluation of IFC4 for the GIS and green building domains. In: Computing in civil and building engineering (2014), pp 2216–2223
6. Daum S, Borrmann A (2014) Processing of topological BIM queries using boundary representation based methods. Adv Eng Inform 28(4):272–286
7. Eastman C, Lee JM, Jeong YS, Lee JK (2009) Automatic rule-based checking of building designs. Autom Constr 18(8):1011–1033. https://www.freecadweb.org/
8. IFC (2015) Industry Foundation Classes, buildingSMART International. http://www.buildingsmart-tech.org/specifications. Accessed 29 May 2015
9. Hwang YS (2004) Automatic quantity takeoff from drawing through IFC model. Archit Inst Korea 20(12):89–97
10. IfcOpenShell (2018) About ifcopenshell. http://www.ifcopenshell.org/. Accessed 23 Feb 2020
11. International Alliance for Interoperability (IAI) (1995) http://www.iai-international.org
12. Jalaei F, Jrade A (2015) Integrating building information modeling (BIM) and LEED system at the conceptual design stage of sustainable buildings. Sustain Cities Soc 18:95–107
13. Jongeling R, Kim J, Fischer M, Mourgues C, Olofsson T (2008) Quantitative analysis of workflow, temporary structure usage, and productivity using 4D models. Autom Constr 17(6):780–791
14. Lai H, Deng X (2018) Interoperability analysis of IFC-based data exchange between heterogeneous BIM software. J civil Eng Manag 24(7):537–555
15. Lee SK, Kim KR, Yu JH (2014) BIM and ontology-based approach for building cost estimation. Autom Constr 41:96–105
16. Liebich T (2013) IFC4—the new buildingSMART standard. In: IC meeting. bSI Publications, Helsinki, Finland
17. Lin JR, Hu ZZ, Zhang JP, Yu FQ (2016) A natural-language-based approach to intelligent data retrieval and representation for cloud BIM. Comput-Aided Civil Infrastruct Eng 31(1):18–33
18. NBIMS 2007. http://www.facilitiesinformationcouncil.org/bim/publications.php
19. Thein V (2011) Industry Foundation Classes (IFC). BIM interoperability through a vendor-independent file format

20. Vanlande R, Nicolle C, Cruz C (2008) IFC and building lifecycle management. Autom Constr 18(1):70–78
21. Won J, Lee G, Cho C (2013) No-schema algorithm for extracting a partial model from an IFC instance model. J Comput Civil Eng 27(6):585–592
22. Wong KD, Fan Q (2013). Building information modelling (BIM) for sustainable building design. Facilities
23. Yang D, Eastman CM (2007) A rule-based subset generation method for product data models. Comput-Aided Civil Infrastruct Eng 22(2):133–148
24. Zhang L, Issa RR (2013) Ontology-based partial building information model extraction. J Comput Civil Eng 27(6):576–584

Electrochemical Oxidation of Amido Black 10B Under Amperostatic Conditions with Vertically Oriented Graphite/Platinum Electrodes

Rajvir Kaur and Harpreet Kaur

1 Introduction

Azo colourants constitute an important class of synthetic coloured organic compounds (60–70%), which are characterized by the presence of one or more azo bonds (R_1-N=N-R_2) and aromatic rings mostly substituted by sulfonate groups [1]. Azo colorants are all around us as the important part of our everyday colourful life. They are used widely in substrates such as; textile fibers, leather, plastics, papers, hair, mineral oils, waxes, foodstuffs and cosmetics [2, 3]. The strongly coloured effluent created environmental and aesthetic problems [4]. AB 10B is an amino acid staining diazo dye, which is applicable to all kinds of natural and synthetic fibers like; wool, cotton, silk, polyesters, rayon and acrylic [5]. The dye is used for staining of protein in biochemical research, mainly to stain for total protein on transferred membrane blots [2]. High toxicity of AB 10B is responsible for various health problems [5, 6]. Various alternative techniques [7–11] such as; electrochemical coagulation, photocatalytic decolorization, adsorption, fenton oxidation process and microbiological decomposition have been proposed from time to time for the removal of different type of dyes from effluents. Most of the dyes are not oxidized by conventional physical and biological treatments because of their complex molecular structures and large sizes [12]. The removal of complex structure of dyes from wastewater always remains a major problem. Electrochemical method has been received great attention due to its unique features [13–15] like; versatile, high energy efficiency, high degradation

R. Kaur (✉)
Department of Applied Science, Guru Nanak Dev Engineering College, Ludhiana, Punjab 141006, India
e-mail: srawrajvir021@gmail.com

H. Kaur
Department of Chemistry, Punjabi University, Patiala, Punjab 147002, India
e-mail: preetjudge@yahoo.co.in

H. Singh et al. (eds.), *Sustainable Development Through Engineering Innovations*, Lecture Notes in Civil Engineering 113,
https://doi.org/10.1007/978-981-15-9554-7_19

efficiency, amenability of automation, does not cause secondary pollution and safer as it operates under mild conditions. Electrochemical oxidation provides a direct and economically cheap route for the degradation of dye from aqueous as process provides the environmental protection with maximum energy resource management. In the electro-oxidation containing chloride ions as supporting electrolyte, active chlorine species have been generated in situ during the reaction, by the following reaction mechanism [14, 16, 17]:

At anode

$$2Cl^- \rightarrow Cl_2(aq) + 2e^- \tag{1}$$

At cathode

$$2H_2O + 2e^- \rightarrow 2OH^- + H_2(g) \tag{2}$$

In the bulk solution

$$Cl_2(aq) + H_2O \rightarrow HOCl + Cl^- + H^+ \tag{3}$$

$$HOCl \rightarrow H^+(g) + ClO^- \tag{4}$$

The electro-generated chlorine compounds act as the main oxidizing species in the degradation of dye present in the effluent. Literature reveals [11] that very few studies are available, which elucidate the degradation mechanism of dye into a small fragment during degradation process. On the basis of above-said considerations, the aim of the present work is to study the electrochemical oxidation of hazardous dye AB 10B from aqueous solution using graphite anode and platinum cathode. Chemical oxygen demand, kinetics and energy consumption parameters of the process have also been calculated. The intermediates have been identified by using mass spectroscopic techniques.

2 Experimental Technique

2.1 Materials

AB 10B (4-Amino-5-hydroxy-3-[(4-nitrophenyl)azo]-6-(phenylazo)-2,7-Naphthalene disulfonic acid, disodium salt) dye with molecular formula $C_{22}H_{14}N_6Na_2O_9S_2$ (CAS No. 1064-48-8, Molecular weight 616.49 g/mol, purity) has been obtained from sd fine-chem limited, Mumbai. All other chemicals such as; potassium chloride, sodium chloride, potassium dichromate, ferrous ammonium sulphate and sulphuric acid have been used as procured. Water has been purified by adding potassium permanganate with a pellet of potassium hydroxide followed by fractional distillation.

2.2 *Instrumentation*

The electrochemical experiments have been carried out by using a digital DC Electrophoresis power supply (Perfit, India) as shown in Fig. 1. The electrochemical cell consists of an undivided cell with two vertically oriented electrodes, i.e., graphite as anode with total effective electrode area of 16.623 cm^2 and platinum as cathode with 8.240 cm^2. The electrodes have been connected to a digital DC power supply (0–100 mA, 1–300 V) under amperostatic operational conditions. The potential across the electrodes has been adjusted so that a constant current of 30 mA passed through the cell. Potassium chloride has been used as a supporting electrolyte. The solution has been stirred thoroughly during the process with the help of a magnetic stirrer.

The electrolytic cell can be represented as:

$$C_{(+)}\left|\begin{array}{c}\text{Aqueous AB 10B dye solution}\\ +\text{KCl}\end{array}\right|Pt_{(-)}$$

where, $C_{(+)}$ and $Pt_{(-)}$ represents graphite anode and platinum cathode, respectively.

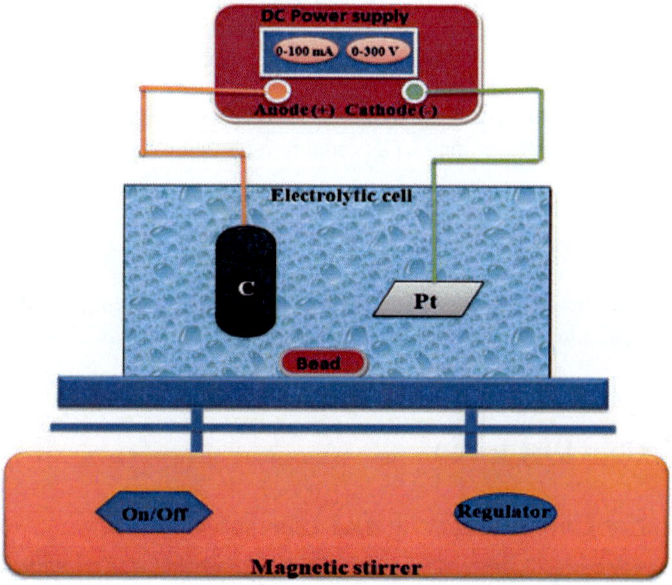

Fig. 1 Schematic diagram of the experimental setup

The change of *p*H in aqueous solution has been recorded with the help of digital *p*H meter Elico Private Limited, Hyderabad (India) to desired level by adding 0.1 N HCl or 0.1 N NaOH solutions.

2.3 Electrochemical Degradation Analysis

All the experiments have been performed by taking 100 mL solution of fixed concentration of AB 10B dye followed by the addition of KCl as supporting electrolyte at room temperature and thoroughly stirring the solution with the help of magnetic stirrer. The electrodes have been vertically installed inside the undivided electrolytic cell with a distance of 5 cm. 2 mL sample has been taken out from the reaction mixture periodically using a micro-pipette. The amount of dye degrade has been measured with the help of Shimadzu-1800, UV-Visible spectrophotometer at the maximum wavelength of AB 10B dye ($\lambda_{max} = 618.00$ nm). The effect of different parameters has been studied by varying the current density, electrolysis time, initial dye concentration, amount of supporting electrolyte, *p*H and comparison of different electrodes. The percentage of dye degraded has been calculated [14] by using the following relation as:

$$\%\text{degradation of dye} = [(C_0 - C_t)/C_0] \times 100 \tag{5}$$

where, C_0 and C_t are the initial dye concentration and concentration of dye at time t in solution (mg/L), respectively.

Electrochemical method degrades the organic pollutants in the wastewater and thereby causes a reduction in COD and a decrease in the colour. The chemical oxygen demand (COD) has been measured by volumetric analytical method as reported in the literature [18]. The percentage COD reduction has been calculated by using the following relation:

$$\text{COD reduction (\%)} = [(\text{COD}_0 - \text{COD}_t)/\text{COD}_0] \times 100 \tag{6}$$

where, COD_0 and COD_t are the initial COD value and COD value at time t after electrolysis (mg/L), respectively.

3 Results and Discussion

3.1 Effect of Current Density and Electrolysis Time

Current density is defined as the current applied by the projected area of the electrodes. Anode has been taken as graphite of projected area, 16.623 cm^2. In

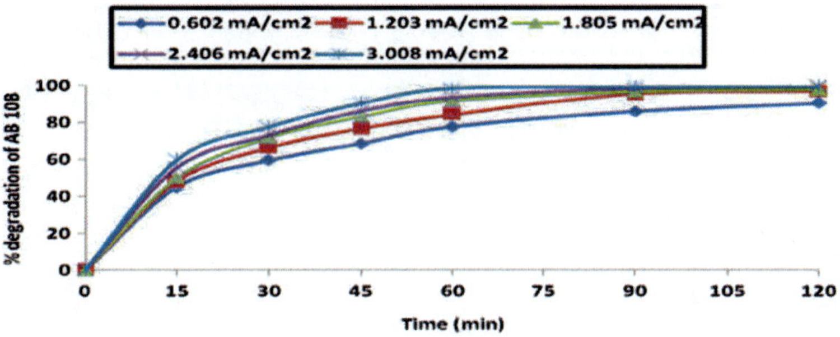

Fig. 2 Effect of current density with electrolysis time on electrochemical degradation of AB 10B (conditions: $C_0 = 100$ mg/L, KCl = 0.5 g, $pH = 7.0$)

order to study the effect of current density and electrolysis time, the experiment has been conducted at different current densities (0.602, 1.203, 1.805, 2.406 and 3.008 mA/cm^2) for the electrochemical degradation of AB 10B dye in an aqueous solution. It is clear from the results (Fig. 2) that percentage degradation of AB 10B increases with an increase in electrolysis time for different current densities. It may be due to reason [17, 19] that the rate of generation of oxidants (active chlorine compounds) increases with an increase in current density, which ultimately increases the degradation of dye. It is interesting to note that the rate of degradation is quite high initially (in first 60 min) and after this time rate remains constant.

3.2 Effect of Initial Dye Concentration of AB 10B

In order to study the effect of initial concentration of AB 10B with electrolysis time, experiments have been carried out at different concentrations of dye (Fig. 3). The degradation efficiency decreases from 99.00 to 94.20% with an increase in the concentration of AB 10B from 50 to 150 mg/L at 120 min. During electrolysis, by increasing the dye concentration the molecules of dye tend to associate to give large molecules of low diffusivity, which lowers the rate of dye diffusion towards the anode with a consequent decrease in percentage degradation of dye [17].

3.3 Effect of Amount of Supporting Electrolyte

It is evident from Fig. 4a that the addition of KCl to the wastewater increases the conductivity of aqueous solution, which increases the rate of degradation of AB 10B dye. This may be due to the fact that with an increase in concentration of KCl more hypochlorite ions/hypochlorous acid is produced through the liberation of chlorine. These generated chlorine compounds react with the dye contaminants

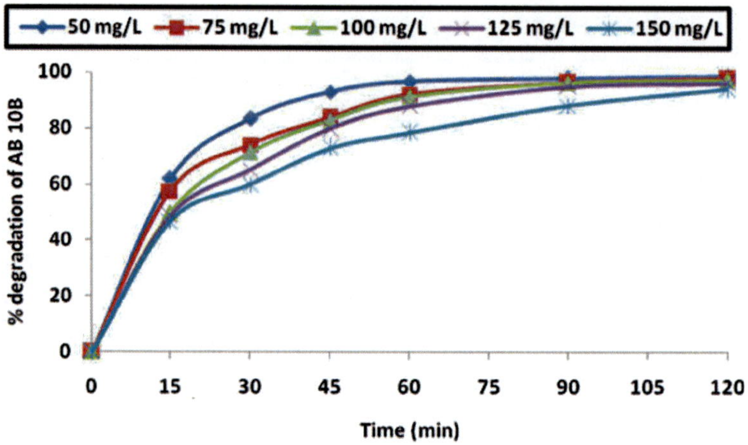

Fig. 3 Effect of initial dye concentration with electrolysis time on electrochemical degradation of AB 10B (conditions: current density = 1.805 mA/cm^2, KCl = 0.5 g, pH = 7.0)

in an aqueous solution and degraded these into smaller molecules like; CO_2 and H_2O [20]. Secondly, the increase in the concentration of chloride ion decreases the anodic activity towards oxygen evolution. A comparative study of KCl with NaCl as supporting electrolytes has been carried out at optimum conditions. Figure 4b shows the comparative behaviour of supporting electrolytes KCl and NaCl towards the degradation of AB 10B dye.

3.4 Effect of pH

Electrochemical experiments have been performed in *p*H range of 3.0–9.0 at optimized conditions of the degradation of AB 10B. It is clear from Fig. 5 that the rate of dye degradation is more at initial stages of electrolysis time at different *p*H values of dye solution and at final stage affect of degradation is negligible. It may be due to the reason that the increase in $HOCl/Cl_2$ amount in acidic medium, having the $HOCl/Cl_2$ system higher standard potential than that of ClO^-/Cl^- [21].

3.5 Effect of Cathode and Anode Materials

It has been observed that when platinum has been used as anode as well as cathode then there is 100.00% degradation of AB 10B dye. Similar results have been obtained, when cathode material is replaced with graphite. When anode is replaced with graphite then there is decrease in percentage of dye degradation at initial but at

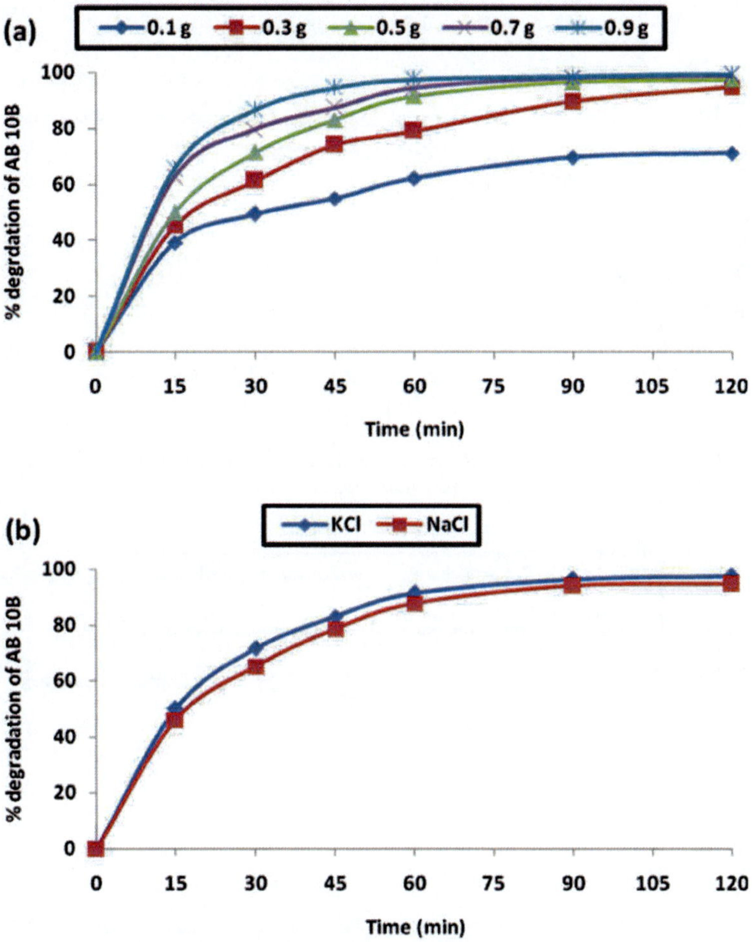

Fig. 4 **a** Effect of amount of KCl with electrolysis time on electrochemical degradation of AB 10B (conditions: current density = 1.805 mA/cm^2, C_0 = 100 mg/L, pH = 7.0). **b** Comparison study of amount of KCl with NaCl on electrochemical degradation of AB 10B (conditions: current density = 1.805 mA/cm^2, C_0 = 100 mg/L, pH = 7.0)

final stage, the degradation is quite high (Fig. 6). From this, it can be concluded that yet the efficiency of degradation at graphite electrode is low at middle stage but it degrades 97.50% of dye. Thus it can be used as an economical electrode material for the degradation of wastewater.

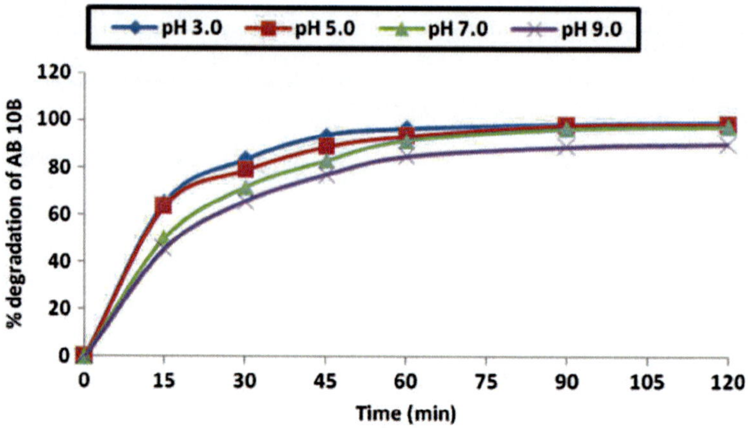

Fig. 5 Effect of *p*H with electrolysis time on electrochemical degradation of AB 10B (conditions: current density = 1.805 mA/cm^2, C_0 = 100 mg/L, KCl = 0.5 g)

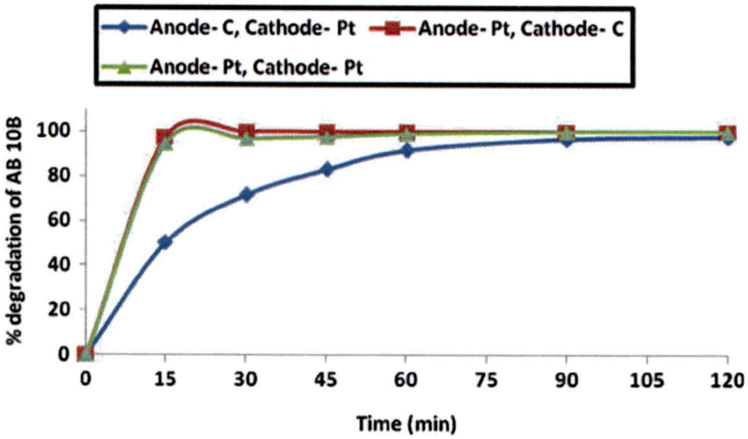

Fig. 6 Comparison study of different electrodes on electrochemical degradation of AB 10B (conditions: current density = 1.805 mA/cm^2, C_0 = 100 mg/L, KCl = 0.5 g, *p*H = 7.0)

3.6 Kinetic Studies

Kinetic first-order rate equation is as follows:

$$\ln(C_0/C_t) = k_1 t \tag{7}$$

where, C_0 and C_t are the initial dye concentration and concentration of dye at time t in solution (mg/L), respectively. k_1 is the first-order kinetic rate constant (min^{-1}).

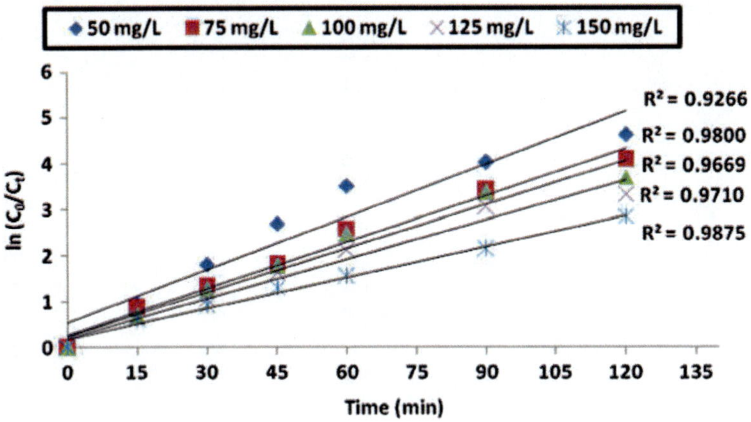

Fig. 7 Kinetic first-order plot at a different initial dye concentration of AB 10B dye

Table 1 Values of rate constant k_1 for first-order kinetic rate equation at different initial concentrations of AB 10B

Initial dye concentration (mg/L)	50	75	100	125	150
Rate constant (min^{-1})	0.0381	0.0334	0.0317	0.0285	0.0223

It is clear from Fig. 7 that the experimental data well fitted to kinetic first-order rate equation with good values of correlation coefficient for all initial concentrations of AB 10B dye. The values of rate constants have been obtained from the linear plot of $\ln(C_0/C_t)$ verses t and are given in Table 1 for the variable concentration of dye (50–150 mg/L).

3.7 Analysis of COD

COD allows the feasibility of waste in terms of the total quantity of oxygen required for the oxidation of organic matter to CO_2 and H_2O. This study has been taken as an effective parameter to measure the efficiency of electrochemical degradation process for the treatment of hazardous AB 10B dye from aqueous solution. The volumetric analytical method has been applied for COD measurement. It is clear from the results (Table 2) that percentage reduction in COD increases with the increase in the current density and amount of supporting electrolyte. This may be due to reason that the electro-generated chlorine compounds play an important role in the electrochemical degradation process [21]. But the reduction percentage in COD value decreases with an increase in initial dye concentration and *p*H of dye solution due to the reason that the dye molecules tend to associate with large molecules of low diffusivity, which may lower the dye diffusion to the anode.

Table 2 Percentage reduction in COD at different operating conditions for electrochemical degradation of AB 10B

	Current density (mA/cm^2)					Amount of KCl (g)					*pH*				Initial dye concentration (mg/L)				
	0.602	1.203	1.805	2.406	3.008	0.1	0.3	0.5	0.7	0.9	3.0	5.0	7.0	9.0	50	75	100	125	150
COD reduction %	68.99	79.67	82.00	84.00	85.00	65.50	75.64	82.00	84.50	89.90	85.97	84.00	82.00	79.00	88.00	85.00	82.00	81.00	80.00

3.8 Electric Energy Consumption

The electric energy consumption during electrochemical process has been calculated in terms of kWh/kg dye degraded by using the following equation [14]:

$$\text{Energy consumption} = VIt \times 10^3/[60(C_0 - C_t) \times \text{treated volume}(L)] \quad (8)$$

where V is the cell voltage (V), I is the current (A) and t is the time of electrolysis (min). Electric energy consumption values have been calculated at various current densities and amount of KCl and are presented in Table 3. The maximum electric energy consumption is 21.5384 kWh/kg for 100 mg/L dye concentration at 1.805 mA/cm^2 current density with 0.5 g of KCl. With this very small electrical energy consumption, almost complete degradation of dyes has been achieved after 120 min of electrolysis. In India, ~6.43 rupees are required to remove 1 kg of dye from wastewater during this process.

3.9 Fourier Transform Infrared Spectroscopy (FT-IR) Analysis

FT-IR spectra of AB 10B dye before and after electrolysis have been recorded in the region of 4000–400 cm^{-1}. FT-IR spectrum of AB 10B consists of peaks [22] at 1047 cm^{-1} due to S=O stretching vibrations of sulfonic acid, 1225 cm^{-1} due to C–N stretching vibrations, 1331 cm^{-1} due to C–N bending vibrations, 1454 cm^{-1} due to C–H bending vibrations, 1572 cm^{-1} due to N=N stretching vibrations, 2924 cm^{-1} due to O–H stretching vibrations and 3431 cm^{-1} due to N–H stretching vibrations of aromatic primary amine (Fig. 8a). FT-IR spectrum obtained after electrolysis of AB 10B consist of no peak at 1572 cm^{-1} due to azo group. The peaks due to other functional groups are very weak (Fig. 8b), which supports the electrochemical degradation of AB 10B dye.

3.10 Liquid Chromatography–Mass Spectrometry Studies (LC–MS)

The intermediate compounds formed during the electrochemical degradation of dye has been identified by MS/MS. The sample has been collected after 60 min of electrolysis. Based on the LC–MS identification, the following pathway (Fig. 9) is proposed for the electrochemical degradation of AB 10B. The structures of various intermediates as predicted from LC–MS fragmentation analysis are given below in Fig. 9 with a base peak at 112 and indicating that the entire dye has been decomposed to low molecular weight fragments.

Table 3 Electric energy consumption values at various parameters during electrochemical process

	Current density (mA/cm^2)					Amount of KCl (g)				
	0.602	1.203	1.805	2.406	3.008	0.1	0.3	0.5	0.7	0.9
Energy consumption (kWh/kg dye degraded)	5.9172	12.4352	21.5384	32.3232	45.1808	37.9214	25.3432	21.5384	18.1818	15.0754

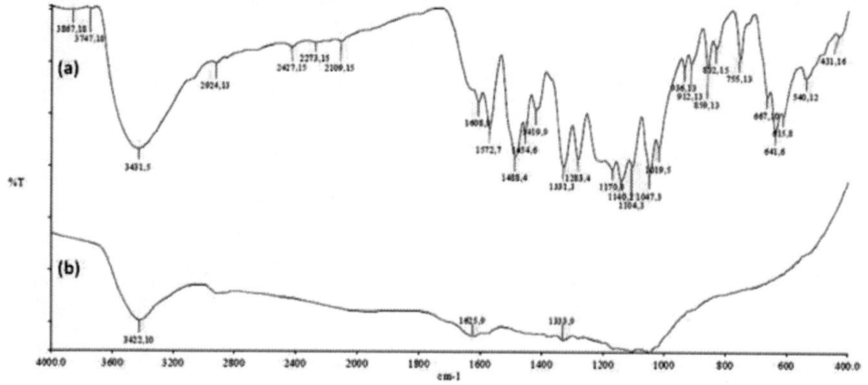

Fig. 8 FT-IR spectrum of AB 10B dye: **a** before electrochemical degradation and **b** after electrochemical degradation process

4 Conclusions

The present work is devoted to the electrochemical oxidation of toxic dye, AB 10B at graphite anode and platinum cathode, under the optimum operational electrolytic conditions, which influence the degradation of dye. The process of degradation is found to be of first-order kinetic. LC–MS and FT-IR spectral studies confirmed that the proposed electrochemical degradation process could be effective method for the degradation of AB 10B in short treatment time. The study shows that electrochemical oxidation is a superior technology for treatment of AB 10B with this 82.00% reduction in COD and 97.50% dye degradation has been achieved with maximum power management.

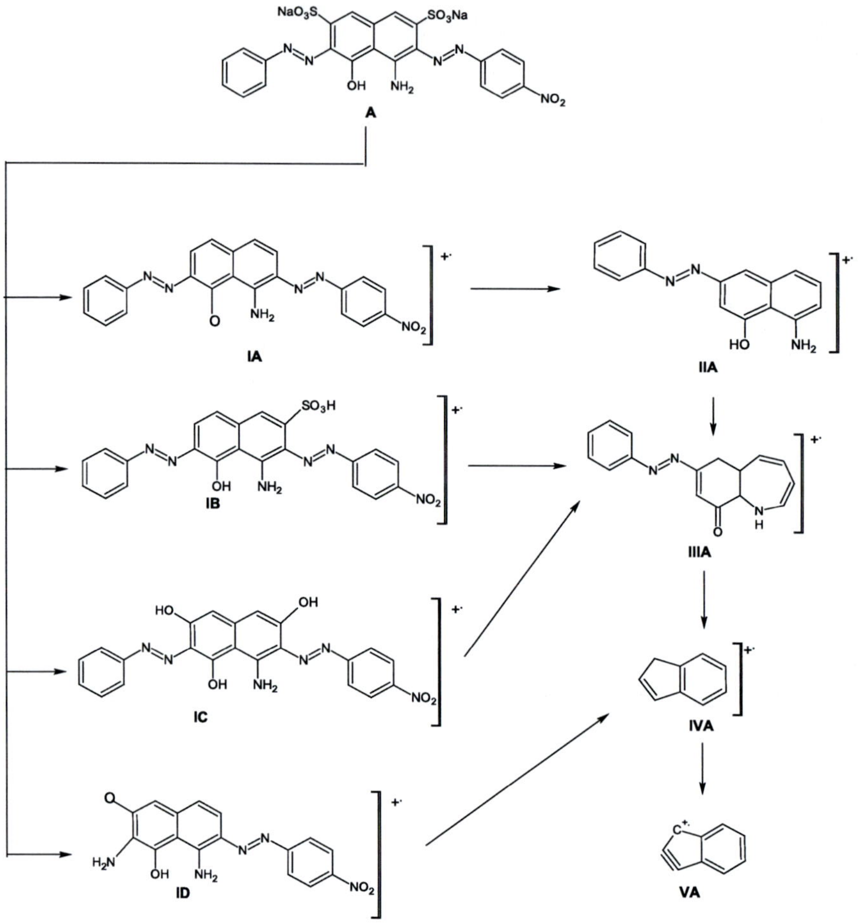

Fig. 9 Proposed pathways of electrochemical degradation of AB 10B dye

Acknowledgements Authors acknowledge sincere thanks to UGC, New Delhi for awarding the UGC-BSR Fellowship for carrying out the research work successfully.

References

1. Kaur R, Kaur H (2019) Adsorptive removal of Amido black 10B from aqueous solution using stem carbon of Ricinus communis as adsorbent. Asian J Chem 31(5):1071–1076. https://doi.org/10.14233/ajchem.2019.21813
2. Sun JH, Sun SP, Wang GL, Qiao LP (2007) Degradation of azo dye Amido black 10B in aqueous solution by Fenton oxidation process. Dyes Pigm 74:647–652. https://doi.org/10.1016/j.dyepig.2006.04.006

3. Kaur H, Swati, Kaur R (2014) Kinetic and Isotherm studies of Congo red adsorption from aqueous solution by biowaste material. Chem Sci Trans 3(4):1300–1309. https://doi.org/10.7598/cst2014.922
4. Kaur R, Kaur H (2017) Calotropis procera as effective adsorbent for removal of malachite green dye: a comprehensive study. Desalin Water Treat 78:253–262. https://doi.org/10.5004/dwt.2017.20548
5. Kaur R, Kaur H (2017) Adsorption of amido black 10B from aqueous solution using weed waste as adsorbent: characterisation, equilibrium, kinetic and thermodynamic studies. Asian J Chem 29(2):441–446. https://doi.org/10.14233/ajchem.2017.20242
6. Mittal A, Thakur V, Gajba V (2013) Adsorptive removal of toxic azo dye Amido Black 10B by hen feather. Environ Sci Pollut Res 20(1):260–269. https://doi.org/10.1007/s11356-012-0843-y
7. Jovic M, Stankovic D, Manojlovic D, Andelkovic I, Milic A, Dojcinovic B, Roglic G (2013) Study of the electrochemical oxidation of reactive textile dyes using platinum electrode. Int J Electrochem Sci 8:168–183
8. Kaur R, Kaur H (2017) Calotropis procera an effective adsorbent for removal of congo red dye: isotherm and kinetics modelling. Model Earth Syst Environ 3(9):1–13. https://doi.org/10.1007/s40808-017-0274-3
9. Kaur H, Kaur R (2014) Removal of Rhodamine-B dye from aqueous solution onto Pigeon Dropping: adsorption, kinetic, equilibrium and thermodynamic studies. J Mater Environ Sci 5(6):1830–1838
10. Katwal R, Kaur R, Kaur H (2017) Photodegradation of congo red, methylene blue and methyl red dyes using electrochemically synthesized Al_2O_3 nanocatalyst. Asian J Chem 29(5):1095–1097. https://doi.org/10.14233/ajchem.2017.20428
11. Singh S, Srivastva VC, Mall ID (2013) Mechanism of dye degradation during electrochemical treatment. J Phys Chem C 117:15229–15240
12. Mittal A, Kurup L, Mittal J (2007) Freundlich and Langmuir adsorption isotherms and kinetics for the removal of Tartrazine from aqueous solutions using hen feathers. J Hazard Mater 146:243–248. https://doi.org/10.1016/j.jhazmat.2006.12.012
13. Martinez-Huitle CA, Brillas E (2009) Decontamination of wastewater containing synthetic organic dyes by electrochemical methods: a general review. Appl Catal B 87:105–145. https://doi.org/10.1016/j.apcatb.2008.09.017
14. Kaur R, Kaur H (2016) Electrochemical degradation of congo red from aqueous solution: role of graphite anode as electrode material. Portug Electrochim Acta 34(3):185–196. https://doi.org/10.4152/pea.201603185
15. Kaur A, Kaur R, Kaur H (2016) Electrochemical degradation of Rhodamine B dye. Moroccan J Chem 4(1):93–100
16. Kariyajjanavar P, Narayana J, Nayaka YA, Umanaik M (2010) Electrochemical degradation and cyclic voltammetric studies of textile reactive azo dye Cibacron navy WB. Portug Electrochim Acta 28(4):265–277. https://doi.org/10.4152/pea.201004265
17. El-Ashtoukhy ESZ (2013) Removal of indigo carmine dye from synthetic wastewater by electrochemical oxidation in a new cell with horizontally oriented electrodes. Int J Electrochem Sci 8:846–858
18. Clesceri LS, Greenberg AE, Trussell RR (1989) Standard methods for the examination of water and wastewater, 17th edn
19. Kariyajjanavar P, Narayana J, Nayaka YA (2013) Degradation of textile dye C.I. Vat Black 27 by electrochemical method by using carbon electrodes. J Environ Chem Eng 1:975–980. https://doi.org/10.1016/j.jece.2013.08.002
20. Sendhil J, Muniswaran PKA (2012) Electrochemical treatment of Acid green V dye solution in a tubular flow reactor. Int J Eng Res Dev 1(6):67–74
21. El Hajj Hassan MA, El Jamal MM (2012) Kinetic study of the electrochemical oxidation of methylene blue with Pt electrode. Portug Electrochim Acta 30(5):351–359. https://doi.org/10.4152/pea.20120535
22. Pavia DL, Lampman GM, Kriz GS (2001) Introduction to spectroscopy, 3rd edn

Performance Review of Polyester Fibre Reinforced Soil for Sustainable Construction

Gaurav Gupta, Hemant Sood, and Pardeep Kumar Gupta

1 Introduction

Soil reinforcement is an effectual method for enhancing the mechanical characteristics of soils such as shear strength and bearing strength. Chemically stabilized soils are weak in tension and strong in compression. To increase the tensile strength of the soil, soil reinforcement is done by placing discrete elements that are high in tensile strength such as fibres or geosynthetics, entitled the reinforced soil. Fibre reinforcement forms a matrix of soil with embedded reinforcement and it possesses novel properties. To sum up, fibre reinforced soil is an amalgamated material containing high tensile resistance elements with improved engineering properties. Fibres have also been efficient in improving the shear strength of soil, limiting post-failure loss of strength [6]. The key motive behind reinforcement of soil is enhancement of soil's mechanical properties such as ductility, strength and stiffness and reduction of settlement and lateral deformation in the soil. The ensuing paragraphs elaborate on the concept of soil reinforcement and discuss the potential of polyester fibre for soil reinforcement.

G. Gupta (✉) · H. Sood
National Institute of Technical Teachers' Training & Research, Chandigarh, India
e-mail: gaurav007gupta@gmail.com

H. Sood
e-mail: sood_hemant@yahoo.co.in

P. K. Gupta
Punjab Engineering College (Deemed to Be University), Chandigarh, India
e-mail: p_gupta_2000@yahoo.com

H. Singh et al. (eds.), *Sustainable Development Through Engineering Innovations*, Lecture Notes in Civil Engineering 113,
https://doi.org/10.1007/978-981-15-9554-7_20

2 Randomly Distributed Fiber-Reinforced Soil

Reinforcement of soil can be effectively performed with randomly oriented discrete elements, i.e. fibres, to improve the geotechnical characteristics of soil. Reinforcing soil with fibres produces a composite material wherein, fibres are implanted in a matrix of soil. Fibres for soil reinforcement can be broadly classified as natural or synthetic, depending upon the source or origin of the fibres.

Natural fibres that have been recognized to be effectual for soil reinforcement are jute, sisal, bamboo and coir fibres. Usage of natural fibres is beneficial because these are low in cost, environmental friendly and available in bulk quantities. The drawbacks of utilization of natural fibres for soil reinforcement are that they are biodegradable, non-reproducible and show poor resistance to the alkali environment. Also, natural fibres undergo a gradual loss of strength and deterioration when used for soil reinforcement, where the rate of loss is dependent on the type of natural fibre used. The non-durable characteristic of natural fibres limits their application to soil reinforcement.

The limitations in the employment of natural fibres can be surmounted by the use of synthetic fibres like polyester and polypropylene fibres. The benefits of synthetic fibres over natural fibres are that these fibres are much less prone to chemical and biological degradation when exposed to various environmental conditions. Also, synthetic fibres can be manufactured as per the required and anticipated dimensions. The drawback of the use of synthetic fibres is that their manufacturing is an energy-intensive process resulting in greenhouse gas emissions. Also, as these fibres are manufactured in an industry, the fibres come at a cost and therefore the overall economic feasibility of use of fibre for soil reinforcement may become questionable, depending upon the economics of the fibre selected.

3 Factors Concerning the Properties of Soils Reinforced by Fibres

The betterment of the geotechnical properties of the soil achieved by fibre reinforcement is dependent upon factors such as classification of the soil, material of the fibre, proportion of fibre in the soil, orientation and dispersion of fibre in the soil matrix, compaction and moisture of soil-fibre mixture. For the soils reinforced with fibre, the gain in the strength of these soils relies critically on the development of the interfacial shear resistance at the in interface of soil and fibre. This interfacial shear resistance is dependent on matrix suction, interface morphologies friction and bonding force. The above listed key parameters are therein affected by the moisture content of the soil-fibre mix, shape and size of the soil particles, and addition of chemical stabilizers to the soil as under load application the fibre at soil-fibre interface may be forced to undergo rotation, due to tensile force acting on the fibre, resulting in plastic deformation of the fibre. Soil stabilization with lime or cement restricts the rotation of fibre under load application due to added bonding forces developed through cementation of soil-fibre at their interface.

4 Potential of Polyester Fibre for Soil Reinforcement

Polyester (chemical name: polyethylene terephthalate) fibre is a synthetic fibre manufactured as a reinforcing agent for soils and concrete. Polyester is produced from a chemical reaction using acid and alcohol. The synthetic fibre is composed of purified terephthalic acid and monotheluene glycol. Polyester material is thermoplastic, that is, the material can be reformed after being melted at high temperatures. During the production of polyester fibres, the manufacturers melt the polyester pellets and push them across small holes, resulting in release of continuous filaments that solidify to form fibres, where the size and dimensions of the small holes govern the diameter and shape of the fibres. The fibres produced are solid polymers with an absence of void spaces inside the fibres.

'Recron 3S' Polyester fibre manufactured by Reliance India Limited has been widely used for soil reinforcement. According to the developer (Reliance India Limited), the polyester fibres have an exclusive triangular cross-section that provides a 40% larger surface area for bonding in comparison with other shapes. These polyester fibres are also engineered to keep the fibre uniformly dispersed and dimensionally straight, so as to safeguard against curling, balling and bunching. The properties of 'Recron 3S' Polyester fibre, as specified by the developer (Reliance India Limited) have been listed in Table 1. Photograph of the fibre, as obtained by SEM (scanning electron microscopy), has been presented in Figs. 1 and 2 at 150× and 2000× magnification, respectively [6]. The current work intends to review the impact of reinforcing action of polyester fibre on the geotechnical properties of weak soils, for use in the subgrade layer of flexible pavements. Unified soil classification system (USCS) for soil classification was followed in the present paper.

Table 1 Properties of polyester fibre

Material	Polyester
Effective diameter	25–40 μm
Length	6 and 12 mm
Specific gravity	1.34–1.40
Shape/cross section	Triangular
Tensile strength	4–6 MPa
Young's modulus	>5000 MPa
Melting point	240–260 °C

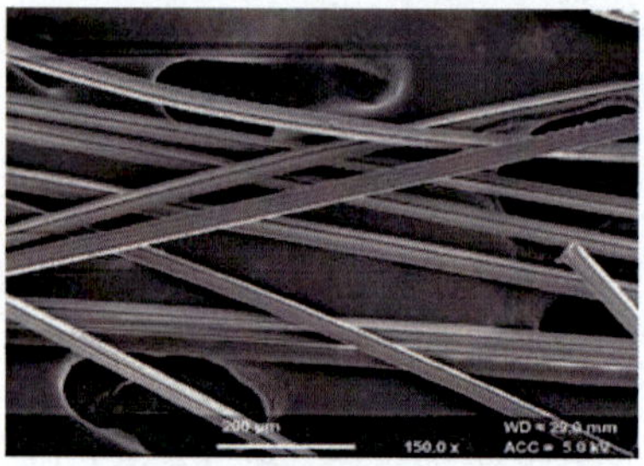

Fig. 1 SEM image of polyester fibre at 150× magnification [6]

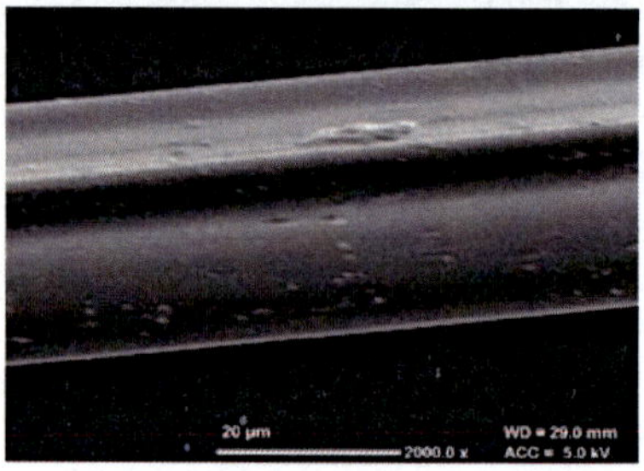

Fig. 2 SEM image of polyester fibre at 2000× magnification [6]

4.1 Compaction Parameters

In the current section, the impact of polyester fibre reinforcement on compaction parameters of soil such as optimum moisture content (OMC) and maximum dry density (MDD) were reviewed. It was seen that some studies reported that reinforcement with polyester fibre failed to have a significant consequence on the MDD and OMC of the soil whereas other studies reported that with rise in content of the fibre, the OMC of the soil increases due to sticking of water at the soil-fibre interface or absorption of water by the fibres, whereas the MDD decreases primarily due to lesser specific gravity of the fibre than the soil.

Kumar et al. [9] reinforced an inorganic soil (USCS classification: SM) with 0.1, 0.2, 0.3 and 0.4% content of polyester fibre, added by the dry mass of the soil and performed Proctor compaction tests under modified effort, to ascertain the impact of fibre reinforcement on MDD and OMC of the soil. The length of the fibre adopted was 30 mm. Results depict that addition of fibre, failed to have a significant effect on the MDD and OMC of the soil.

Kaniraj and Havanagi [8] prepared mixtures of fly ash and a silty soil and a fly ash and a sandy soil, containing equal contents of fly ash and the soil and reinforced each one them with 1.0% content of polyester fibre, added by the dry mass of the mixture. The authors performed Proctor compaction tests under standard effort. The length of the fibre adopted was 20 mm. Results depict that addition of the fibre did not have a note-worthy impact on MDD and OMC of the soil.

Kumar et al. [10] reinforced a black cotton soil (USCS classification: CL) with 0.5, 1.0, 1.5 and 2.0% content of polyester fibre, added by the dry mass of the soil and performed Proctor compaction tests under standard effort. The lengths of the fibre adopted were 3, 6 and 12 mm. Results depict that addition of fibre, failed to have a significant impact on MDD and OMC of the soil.

Das and Singh [4] reinforced a lateritic soil (USCS classification: CL) with 0.125, 0.25, 0.375 and 0.5% content of polyester fibre, added by the dry mass of the soil and performed Proctor compaction tests under standard effort. The length of the fibre adopted was 12 mm. Results depict that with rise in content of polyester fibre, the reinforced soil specimens reported lowering of MDD and rise of OMC of the soil. MDD of the soil decreased from 15.0 to 14.8 kN/m^3 and the OMC increased from 22.0 to 22.8% on addition of 0.5% dose of polyester fibre. The cause cited by the authors for reduction in MDD was lesser specific gravity of the fibres than the soil, whereas, the cause for rise in OMC was increased absorption of water by the fibres.

Gupta et al. [6] used a clay soil (USCS classification: CL) and a sandy soil (USCS classification: SM) and two mixtures of the soils, where each soil was mixed with 4% content of a commercial chemical stabilizer marketed under the name of RBI Grade 81(Road Building International Grade 81). The length of the fibre adopted was 12 mm. Each soil and mixture was reinforced with 0.25, 0.50, 0.75 and 1.0% content of polyester fibre, added by the dry mass of the soil/soil-RBI mixture. The authors performed Proctor compaction tests under modified effort. Results for both the soils and their mixtures show that with rise in content of polyester fibre, the reinforced samples showed lowered MDD and increased OMC. The cause cited by the authors for reduction in MDD was lesser specific gravity of the fibres than the soil whereas the cause for rise in OMC was entrapping of water at soil-fibre interface.

Sarli et al. [14] reinforced a loess soil (USCS classification: ML-CL) with 0.5, 1.0 and 1.5% content of polyester fibre, added by the dry mass of the soil and performed Proctor compaction tests under standard effort. The length of the fibre adopted was 35 mm. Results depict that with rise in content of polyester fibre, the reinforced samples showed lowered MDD and increased OMC. MDD of the soil decreased from 18.7 to 18.0 kN/m^3 and the OMC increased from 13.0 to 15.1% on addition of 1.5% content of polyester fibre, added by the dry mass of the soil.

Divya et al. [5] reinforced a mixture of fine-grained soil (USCS classification: ML) containing 5% content of bentonite, with 0.5% content of polyester fibre, added by the dry mass of the stabilized soil and performed Proctor compaction tests under standard effort. The length of the fibre adopted was 90 mm. Results depict that, the reinforced samples showed reduction in MDD and rise in OMC. The MDD decreased from 15.93 to 15.81 kN/m^3 and the OMC increased from 18.0 to 18.6% on addition of 1.5% content of polyester fibre.

Gupta et al. [7] reinforced a fine-grained soil (USCS classification: CL) with 0.25, 0.50, 0.75 and 1.0% content of polyester fibre. The length of the fibre adopted was 12 mm. The authors performed Proctor compaction tests under modified effort. Results depict that with rise in content of polyester fibre, the reinforced samples showed a reduction in MDD and a rise in OMC. The cause, as cited by the authors, for reduction in MDD was lesser specific gravity of the fibres than the soil whereas the cause for rise in OMC was absorption of water by the fibre in the soil.

4.2 Strength Parameters

In the current section impact of polyester fibre reinforcement on bearing strength, unconfined compressive strength (UCS) and shear strength characteristics like angle of internal friction and cohesion were reviewed. It was seen that the addition of polyester fibre enhances all the strength parameters of soil. Addition of fibres altered the brittle performance of chemically stabilized soils to a ductile one and reduced post-failure loss in strength of the specimen.

Kumar et al. [9] reinforced an inorganic soil (USCS classification: SM) with 0.1, 0.2, 0.3 and 0.4% content of polyester fibre, added by the dry mass of the soil and performed California bearing ratio (CBR) tests (under soaked condition), UCS tests and direct shear tests, to ascertain the impact of fibre reinforcement on bearing strength of the soil. Results depict that, with rise in content of polyester fibre, the reinforced samples showed increased CBR, UCS and shear strength. Angle of internal friction and cohesion improved from 32.0° to 41.0° and 15.0 to 17.0 kPa, respectively, on reinforcement of the soil with 0.4% content of polyester fibre. For the effect of the same dosage, UCS improved by 28% and CBR improved from 21.0 to 27.0%.

Kaniraj and Havanagi [8] reinforced mixtures of Indian fly ash and a silty soil, and Indian fly ash and a sandy soil, containing equal contents of the fly ash and the soil, with 1.0% content of polyester fibre, added by the dry mass of the fly ash-soil mixture. These mixes were also stabilized with 3% content of cement, added by the dry mass of the mixtures. The authors performed direct shear tests and unconsolidated undrained triaxial compression tests on un-cemented samples whereas only UCS tests were performed on both cemented samples and un-cemented samples, to ascertain the impact of fibre reinforcement on strength of the mixtures. The cement stabilized samples were cured for 7, 14, 28 and 56 days. The length of the fibre adopted was 20 mm. Results depict that addition of fibre, resulted in improved UCS, cohesion and angle of internal friction for all the non-cemented-mixtures and cemented-mixtures. Regarding unconsolidated undrained triaxial compression test results, on reinforcement with the fibre, angle of internal friction and cohesion showed a rise from 30.2° to 40.0° and 25.8 to 93.3 kPa, respectively, for the un-cemented mixture of the silty soil and fly ash. Whereas, angle of internal friction and cohesion improved from 30.4° to 32.9° and 16.5 to 160.0 kPa, respectively, for the un-cemented mixture of the sandy soil and fly ash. Regarding direct shear tests, on reinforcement with the fibre, cohesion increased and angle of internal friction decreased from 14.7 to 32.5 kPa and 36.0° to 35.1°, respectively for the un-cemented blend of the silty soil and fly ash. Whereas, angle of internal friction and cohesion improved from 9.8 to 17.4 kPa and 33.0° to 39.4°, respectively, for the un-cemented mixture of the sandy soil and fly ash. Regarding UCS tests, on reinforcement with the fibre, UCS improved from

47.1 to 304.0 kPa for un-cemented mixture of the silty soil and fly ash. Also, for the cement stabilized mixtures, on reinforcement with the fibre, 56 days cured-UCS improved from 498.5 to 943.4 kPa for the blend of the silty soil and fly ash, and 367.8 to 724.7 kPa for the blend of the and fly ash and sandy soil.

Consoli et al. [3] reinforced, a uniform fine sand (USCS classification: SP) and a mixture of the soil containing 7% content of cement, with 0.5% content of polyester fibre, added by the dry mass of the soil and performed triaxial compression tests under consolidated drained conditions. The authors determined the impact of fibre reinforcement on strength of the soil. The lengths of the fibre adopted were 12 and 36 mm. Results depict that fibre reinforcement of the un-cemented soil increased the deviator stress at failure from 311.0 to 383.0 kPa and 420.0 kPa, calculated at a confining pressure of 100 kPa, on reinforcement with 12 mm and 36 mm lengths of the fibre, respectively. Also, fibre reinforcement of the cemented soil increased the deviator stress at failure from 1308.0 to 1465.0 kPa and 1564.0 kPa, calculated at a confining pressure of 100 kPa, on reinforcing the cemented soil with 12 mm and 36 mm lengths of the fibre, respectively. It was clearly established that with rise in fibre length, deviator stress at failure increases. Fibre reinforcement of the un-cement soil increased the angle of internal friction from 36.0° to 37.0° and 40.0°, on reinforcing the soil with 12 mm and 36 mm lengths of the fibre, respectively. Also, fibre reinforcement of the cemented soil increased the angle of internal friction from 38.0° to 46.0° and 52.0°, on reinforcing the cemented soil with 12 mm and 36 mm lengths of the fibre, respectively. It was clearly established that with rise in fibre length, the angle of internal friction increases. Also, it was concluded that reinforcement of the soil with polyester fibre converted the brittle nature of the cemented soils to a more ductile one, where the impact was noted to be more prominent for the longer fibres.

Kumar et al. [10] reinforced a mixture of, black cotton soil (USCS classification: CL) stabilized with 8% content of lime and 20% content of fly ash, with 0.5, 1.0, 1.5 and 2.0% content of polyester fibre, added by the dry mass of the soil and performed UCS and split tensile strength (STS) tests to ascertain the impact of fibre reinforcement on strength characteristics of the soil. The samples were subjected to 28 days curing, followed by 4-hours soaking in water. The lengths of the fibre adopted were 3, 6 and 12 mm. Results depict that with rise in content and length of polyester fibre, UCS and STS increases. As compared to stabilized soil without fibres, on addition of 1.5% content of fibres of length 6 mm, UCS rises by 74% whereas, on addition of 1.0% content of fibres of length 12 mm, UCS rises by 100%. Similarly, on inclusion of 1.5% dosage of fibres of length 6 mm, STS increases by 100%, whereas on addition of 1.0% content of fibres of length 12 mm, STS increases by 135%. Rise in the ratio of STS by UCS with rise in fibre content shows that the role of fibre inclusion was more significant in enhancing the tensile strength than the compressive strength of the soil.

Changizi and Haddad [2] reinforced an expansive soil (USCS classification: CL) and mixtures of the soil containing 0.5, 0.7 and 1.0% content of nano-SiO_2 and 0.1, 0.3 and 0.5% content of polyester fibre, added by the dry mass of the soil. The length of the fibre adopted was adopted as 20 mm. The authors performed UCS tests and direct shear tests to ascertain the influence of fibre reinforcement, on UCS and shear strength of the soil. Results depict that, with rise in content of polyester fibre, the reinforced soil samples showed an increase in UCS and shear strength. Cohesion and angle of internal friction increased from 38.0 to 64.0 kPa and 13.5° to 23.3°, respectively, on inclusion of 0.5% content of polyester fibre to the unreinforced soil. Also, inclusion of the fibre to the SiO_2-soil mixtures containing 1% content of nano SiO_2, increased shear strength and UCS of the mixes. For soils-SiO_2 mixtures containing 1% nano SiO_2 reinforced with the fibre, cohesion and angle of internal friction increased from 38.0 to 113.0 kPa and 13.5° to 38.5°, respectively, on inclusion of 0.5% content of polyester fibre to the unreinforced soil.

Nguyen et al. [13] reinforced an inorganic soil (USCS classification: SC) with 0.5, 1.0 and 1.5% content of polyester fibre, added by the dry mass of the soil and performed direct shear tests to ascertain the influence of fibre reinforcement on shear strength of the soil. The length of the fibre adopted was adopted as 20 mm. Results depict that with rise in content of polyester fibre content to up to 1%, the reinforced samples showed an increase in shear strength. By not taking into account continual change in the shear area of the specimen, the authors reported that on reinforcing the soil with 1% content of the fibre, the shear strength of the soil increases from 201.2 to 271.0 kPa, measured at a normal stress of 200 kPa. The shear strength of the soil reinforced with 1.5% content of fibre falls to 269.0 kPa. Taking into account continual change in the shear area of the specimen, the authors reported that on reinforcing the soil with 1% content of the fibre, the shear strength of the soil increases from 227.9 to 329.0 kPa, measured at a normal stress of 200 kPa. The shear strength of the soil reinforced with 1.5% content of fibre falls to 303.7 kPa.

Das and Singh [4] reinforced a lateritic soil (USCS classification: CL) with 0.125%, 0.25%, 0.375% and 0.5% content of polyester fibre, added by the dry mass of the soil and performed CBR and UCS tests, to ascertain the impact of reinforcement with fibre on the strength of the soil. The length of the fibre adopted was 12 mm. Results depict that, with rise in content of polyester fibre, the reinforced samples showed an in increase in UCS and CBR. UCS improved from 212.0 to 603.9 kPa, on addition of 0.5% content of polyester fibre whereas, at the same dosage, CBR increased from 8.6 to 15.9% and 14.2 to 27.7% for soaked and unsoaked conditions, respectively.

Gupta et al. [6] used two inorganic soils, a clay soil (USCS classification: CL) and a sandy soil (USCS classification: SM) and two mixtures of the soils, each soil mixed with 4% content of RBI (Road Building International) Grade 81-a chemical stabilizer, added by the dry mass of the soil. All the soils and mixtures were reinforced with 0.25, 0.50, 0.75 and 1.0% content of polyester fibre, added by the dry mass of the soil/soil-RBI mixture. The length of the fibre adopted was 12 mm. The authors performed UCS tests and 4 days of water-soaked CBR tests. Results depict that with a rise in content of polyester fibre up to 0.5%, UCS and CBR of the reinforced soil increases. For the clay soil reinforced with 0.5% content of the fibre, UCS improved from 147.4 to 283.9 kPa whereas CBR improved from 3.2 to 6.7%. For the sandy soil reinforced with 0.5% content of the fibre, UCS improved from 249.7 to 407.5 kPa whereas CBR improved from 9.4 to 14.8%. Addition of fibre was efficient in imparting ductility to

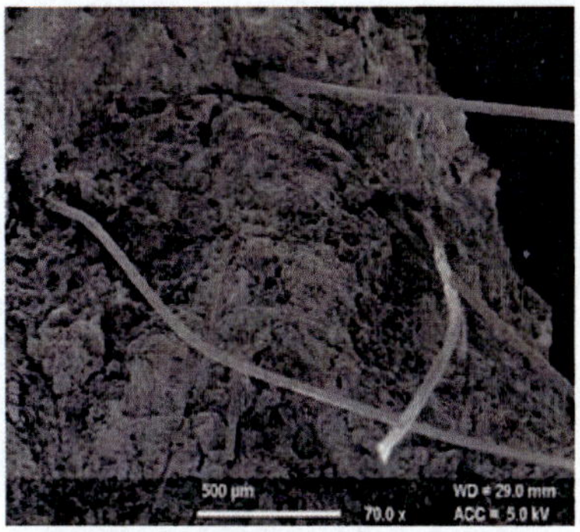

Fig. 3 SEM image of fibre-reinforced and RBI Grade 81 stabilized clay soil [6]

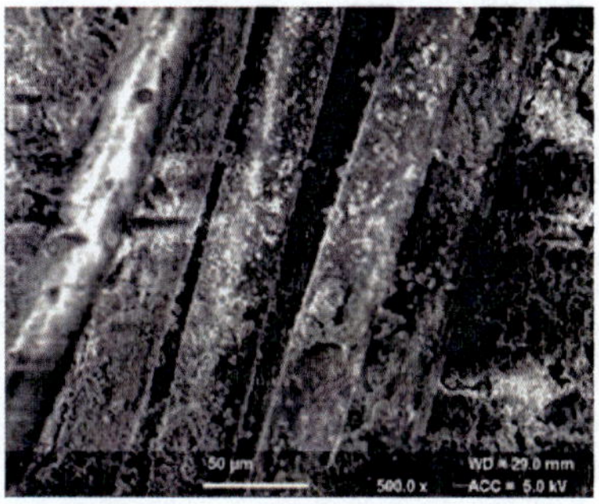

Fig. 4 SEM image of fibre reinforced and RBI Grade 81 stabilized sandy soil [6]

brittle samples of chemically stabilized soils. Also, addition of fibre reduced post-failure loss in strength of all the specimens tested. SEM images of fibre reinforced and RBI Grade 81 stabilized clay and sandy soils have been presented in Figs. 3 and 4, respectively.

Aouali et al. [1] reinforced mixtures of silty sand (USCS classification: SM) containing 0% and 20% content of silt, added by the dry mass of the soil, with 0.1, 0.25, 0.3 and 0.5% content of polyester fibre, added by the dry volume of the soil. The length of the fibre adopted was 10 mm. The authors performed direct shear tests to ascertain shear strength of fibre reinforced soil samples. Results depict that with rise in content of polyester fibre up to 0.25%, the reinforced samples showed an increase in shear stress and higher vertical displacement, calculated at 2 mm horizontal displacement. Also, lower shear stress and reduced vertical displacement was reported for soils containing higher amount of fines.

Mariri et al. [12] reinforced a mixture of, a fine-grained inorganic soil (USCS classification: CL) stabilized with 4% content of cement and varying content of zeolite in the range of 0–90% added by the dry mass of the soil, with 0.1%, 0.3% and 0.5% content of polyester fibre, added by the dry mass of the soil. the authors performed UCS tests to ascertain the impact of fibre reinforcement on strength characteristics of the soil. The length of the fibre adopted was 20 mm. The samples were cured for 7 and 28 days. Results depict that with rise in content of polyester fibre up to

0.5% content of the fibre, UCS of the fibre reinforced soil increased. As compared to stabilized soil without fibres, addition of 0.5% content of fibres, increased UCS of the soil by 1.32 times. It was reported that addition of fibres converted the brittle performance of the stabilized soils to a ductile one.

Lian et al. [11] reinforced a loess soil with 0.5, 0.75 and 1.0% content of polyester fibre, added by the dry mass of the soil. The lengths of the fibre adopted were 9 and 18 mm. The authors performed consolidated drained triaxial compression tests to ascertain the impact of fibre reinforcement on shear strength characteristics of the soil. Results depict that, with rise in content of polyester fibre up to 0.75%, the reinforced samples show rise in angle of internal friction and cohesion. Angle of internal friction and cohesion improved from 23.3° to 25.3° and 32.7 to 46.2 kPa, respectively, on addition of 0.75% content of polyester fibre of length 9 mm to the un-reinforced soil. Further on rise in fibre content to 1.0% added by the dry mass of the soil, angle of internal friction and cohesion decreased from 25.3° to 24.2° and 46.2 to 35.8 kPa, respectively. Deviator stress at failure improved from 332 to 406 kPa on addition of 0.75% content of polyester fibre of length 9 mm to the un-reinforced soil. Further on rise in fibre content to 1.0%, added by the dry mass of the soil, deviator stress at failure decreased from 406 to 364 kPa.

Sarli et al. [14] reinforced a loess soil (USCS classification: ML-CL) with 0.5, 1.0 and 1.5% content of polyester fibre, added by the dry mass of the soil and performed direct shear tests to ascertain the impact of fibre reinforcement on shear strength characteristics of the soil, such as angle of internal friction and cohesion. The length of the fibre adopted was 35 mm. Results depict that, with rise in content of polyester fibre, the reinforced samples showed an increased angle of internal friction and cohesion. Angle of internal friction and cohesion improved from 24.0° to 31.0° and 11.0 to 32.0 kPa, respectively, on addition of 1.5% content of polyester fibre.

Divya et al. [5] reinforced a mixture of fine-grained soil (USCS classification: ML) mixed with 5% content of bentonite, with 0.25, 0.50, and 0.75% content of polyester fibre of length 90 mm, added by the dry mass of the soil. With an aim to evaluate the impact of fibre length on tensile strength of soil, The length of the fibre adopted was varied as 30, 60 and 90 mm for sample prepared with 0.25 and 0.50% content of polyester fibre, added by the dry mass of the soil. Results depict that with rise in the length of content of the fibre, with rest of the other variables constant, the strain for crack initiation and energy absorption increases, depicting higher tensile strength of the soil samples reinforced with fibre. Longer fibres depicted higher tensile strength. The highest gain of 2.5 times, in strain at crack initiation and peak tensile strength was achieved on reinforcement with 0.75% dosage of polyester fibre of 90 mm length. Inclusion of fibres converted brittle failure characteristics of un-reinforced samples to ductile one. With rise in fibre length, post cracking behaviour got enhanced.

Gupta et al. [7] reinforced a fine-grained soil (USCS classification: CL) with 0.2%, 0.50, 0.75 and 1.0% content of polyester fibre, added by the dry mass of the soil. The authors performed soaked and un-soaked CBR tests. The length of the fibre adopted was 12 mm. Results depict that, with rise in content of polyester fibre up to 0.5%, all the reinforced samples showed an increase in CBR of the soil.

4.3 Stiffness Parameters

Stiffness parameters such as resilient modulus(M_R) of a fine-grained soil (USCS classification: CL) stabilized with polyester fibre of length 12 mm, was determined through advanced cyclic triaxial testing under soaked and unsoaked conditions [7]. On reinforcement of the soil with 0.5% content of the fibre, added by the dry mass of the soil, M_R of the soil increased from 49.0 to 58.9 MPa under unsoaked condition and 18.2 to 24.2 MPa under 4-day water-soaked condition.

5 Sustainable Construction with Polyester Fibre Reinforced Soil

Scholars have reported rise in CBR and UCS of soils on their reinforcement with polyester fibre. Rise in the strength of soils on reinforcement with polyester fibre, is expected to lower the required thickness of pavement layers, when fibre-reinforced soil is used as the subgrade layer of flexible pavements, saving bitumen and aggregates and thereby making the construction sustainable. Also, mitigation of greenhouse gas emissions, associated with pavement construction, is expected due to reduction in use of bitumen and aggregates in pavement construction.

6 Conclusions

Polyester fibre was found to perform efficiently for soil reinforcement. Ensuing conclusions were developed from the review of the literature.

1. Reinforcement with polyester fibre results in lowering of MDD of soils and rise of OMC of soils. In some cases, the impact of the reinforcement on MDD and OMC was established to be non-significant.
2. Reinforcement with polyester fibre results in rise in strength characteristics of soil, such as CBR, UCS, cohesion and angle of internal friction and stiffness characteristics such as resilient modulus.
3. Reinforcement with polyester fibre imparts ductility to all soil samples. Samples stabilized with chemicals or cement, etc., characterize brittle failure and therefore imparting ductility to them reduces the risk of a sudden catastrophic failure.
4. Reinforcement with polyester fibre results in reduction in post-failure loss of strength.
5. The effect of polyester fibre reinforcement on permeability of soil, release of greenhouse gas emissions, cost economics of project where polyester fibre reinforced soil is used, of use of fibre has not been studied so far and needs to be determined in the future. Also, the influence of polyester fibre on the MDD and

OMC, bearing strength, shear strength and UCS of soils of different plasticity have not been studied.

References

1. Aouali N, Benessalah I, Arab A, Ali B, Abed M (2019) Shear strength response of fibre reinforced Chlef (Algeria) silty sand: laboratory study. Geotech Geol Eng 37(2):1047–1057
2. Changizi F, Haddad A (2015) Strength properties of soft clay treated with mixture of nano-SiO_2 & recycled polyester fiber. J Rock Mech Geotech Eng 7(4):367–378
3. Consoli NC, Montardo JP, Donato M, Prietto PD (2004) Effect of material properties on the behaviour of sand–cement–fibre composites. Proc Inst Civil Eng Ground Improv 8(2):77–90
4. Das N, Singh SK (2019) Geotechnical behaviour of lateritic soil reinforced with brown waste & synthetic fibre. Int J Geotech Eng 13(3):287–297
5. Divya PV, Viswanadham BVS, Gourc JP (2014) Evaluation of tensile strength-strain characteristics of fiber-reinforced soil through laboratory tests. J Mater Civ Eng 26(1):14–23
6. Gupta G, Sood H, Gupta PK (2019) Geotechnical & geo-environmental properties of discrete polyester fibre-reinforced & RBI Grade-81-stabilized clay & sand. In: IOP conference series: earth & environmental science, vol 219, no 1. IOP Publishing, p 012019
7. Gupta G, Sood H, Gupta PK (2020) Mathematical modelling of resilient modulus response of fibre reinforced clay subgrade for pavement design. J Interdiscipl Math 23(1):247–255
8. Kaniraj SR, Havanagi VG (2001) Behavior of cement-stabilized fiber-reinforced fly ash-soil mixtures. J Geotech Geoenviron Eng 127(7):574–584
9. Kumar R, Kanaujia VK, Chandra D (1999) Engineering behaviour of fibre-reinforced pond ash & silty sand. Geosynth Int 6(6):509–518
10. Kumar A, Walia BS, Bajaj A (2007) Influence of fly ash, lime, & polyester fibers on compaction & strength properties of expansive soil. J Mater Civ Eng 19(3):242–248
11. Lian B, Peng J, Zhan H, Cui X (2019) Effect of randomly distributed fibre on triaxial shear behavior of loess. Bull Eng Geol Environ 1–9
12. Mariri M, Ziaie Moayed R, Kordnaeij A (2019) Stress-strain behavior of loess soil stabilized with cement, zeolite, & recycled polyester fiber. J Mater Civ Eng 31(12):04019291
13. Nguyen G, Hrubešová E, Voltr A (2015) Soil improvement using polyester fibres. Procedia Eng 111:596–600
14. Sarli JM, Hadadi F, Bagheri RA (2019) Stabilizing geotechnical properties of loess soil by mixing recycled polyester fiber & nano-SiO_2. Geotech Geol Eng 1–13

Urban Heritage Conservation for Sustainable Development: A Case of Kapurthala

Vivek Sehgal and Harsimran Kaur

1 Introduction

Urban heritage within sustainability means to tie together the past, present and future of the cities that meets the requirements of human in any era [6, 7]. Heritage is what has been handed over to generations by their ancestors [2]. It lies in everything right from the fairy tales told by their grandmothers to the buildings one live in or work in. As being heirs to this legacy, there is need to conserve the heritage in the same way as ancestors did. Even the same message was conveyed by the former president of India, Dr. A. P. J. Abdul Kalam, quoted as "we will be remembered only if we give to our younger generation a prosperous and safe India, resulting out of economic prosperity coupled with civilizational heritage" [4].

Kapurthala is one such city in Punjab with rich treasure of cultural heritage in the form of architecture and urban spaces that needs to be conserved in such a way that it becomes the reason for social integrity and economic prosperity for generations to come. But Kapurthala has been neglected and vandalized beyond its limit. Therefore, the paper aims to investigate the existing issues related to the city, and its surroundings and intervene in the form of designing the surrounding areas, recommending the strategies and framework for urban design guidelines.

This paper is structured in five sections. Following the introduction (Sect. 1), a focus on Kapurthala for the contextualization of the study is discussed in Sect. 2.

Vivek Sehgal and Harsimran Kaur equally contributed.

V. Sehgal (✉)
GNDEC School of Architecture, Ludhiana, Punjab, India
e-mail: hod_arch@gndec.ac.in

H. Kaur
Department of Architecture, Planning and Design, Indian Institute of Technology (BHU), Varanasi, Uttar Pradesh, India
e-mail: harsimran1601@gmail.com

H. Singh et al. (eds.), *Sustainable Development Through Engineering Innovations*, Lecture Notes in Civil Engineering 113,
https://doi.org/10.1007/978-981-15-9554-7_21

Existing conditions of the city are discussed in Sect. 3. Section 4 draws recommendations for the sustainable urban heritage, and Sect. 5 concludes the paper by summarizing the findings of the study.

2 Setting the Context: An Overview of Kapurthala and Its Heritage

Kapurthala also known as the "City of Palaces and Gardens," recording continuity of settlement and culture since eleventh century, is unique in the aesthetics, and blend of French, Greek, Roman and Indo-Saracenic (Mughal-Gothic) architectural styles and is, even today, a living example of these secular and aesthetic mix of the city. Kapurthala city spreads over 1947 hectares housing total population of over 1.01 lakhs as per 2011 census (Census of India, 2011). Kapurthala Municipal Council has been divided into 29 wards.

Kapurthala, a princely state in British India, is also quoted as "Paris of Punjab" (Fig. 1) because of the monuments which were constructed here during the regime of Jagatjit Singh (1872–1948). He was a frequent traveler across the globe, so he had the global exposure too. That is how he got all the monuments built which were inspired by different styles of architecture. The state introduced fresh innovations and improvements from each such visit. Modern sewage, water system and a telephone system were implemented in 1901. Many highlights were an improved justice system in 1904, legislative and police reforms from 1906 to 1910, a state assembly and

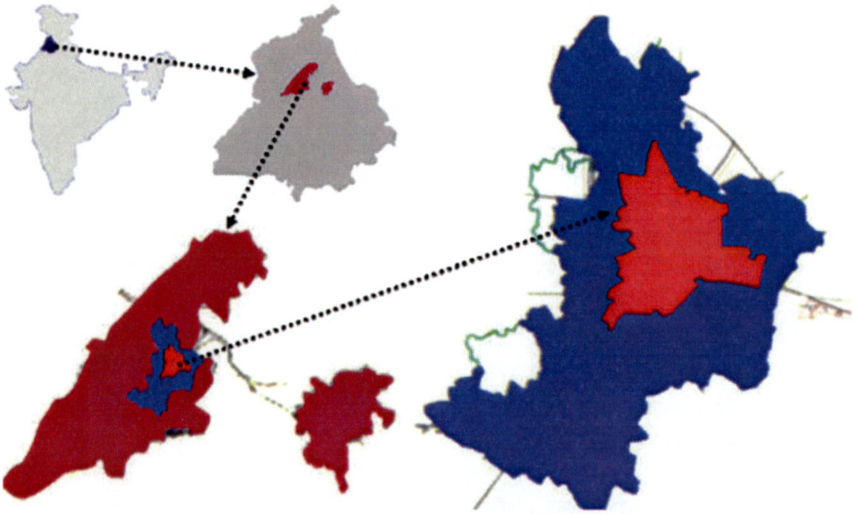

Fig. 1 Kapurthala location on India map. *Source* Prepared from Master plan Kapurthala, 2010–2031

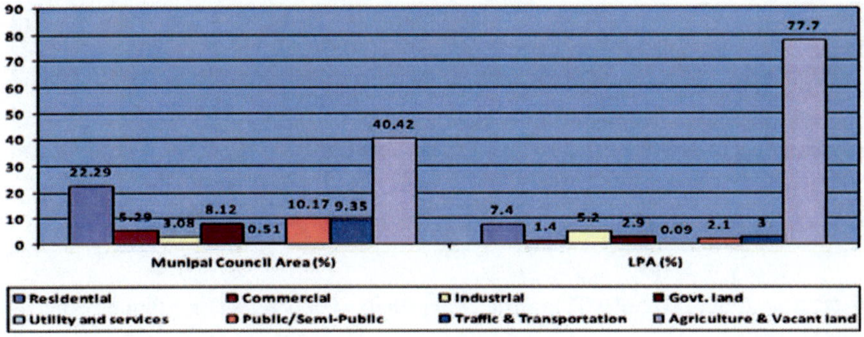

Fig. 2 Land-use map of Kapurthala with percentage distribution of various land uses. *Source* Master plan Kapurthala, 2010–2031

state council in 1916, free compulsory primary education from 1918, agricultural cooperative credit societies in 1920, and factories in the 1940s.

Land-use pattern: Most of the core area of the city is very densely populated with very less open spaces. The graph actually shows above 40% of open space in municipal council area. But the fact is that open land is actually trapped in some complexes like Jagatjit palace, Kamra Bagh, D.C. residence, etc., which actually is not accessible to the public. City is now expanding outwards, as is shown in master plan which shows considerable amount of residential development on Kapurthala-Jalandhar road (Fig. 2).

Sustainability and quality of life are closely linked to the quality and effectiveness of physical and social infrastructure in urban areas. Physical infrastructure is like the backbone of any city where as social infrastructure is all about better society and community development.

Traffic and Transportation: Intercity traffic is managed by bus stand and railway station which is located inside the city. Though there is no airport in close vicinity except one at Amritsar which is 75 km from here. There is no truck terminal present in the city. Autorickshaws and rickshaws are there in the name of public transport within the city. Some of the key issues include acute shortage of parking, absence of railway over bridge, leading to jam on rail crossings, inadequate road width within the urban area, due to encroachment on both sides of the roads.

According to Kapurthala Water Supply and Sewerage Board, "82% of total area of town (2008) is covered with sewerage system and rest area is served through septic tanks and independent institutional setup." According to the data available from Kapurthala Municipal Council, the total solid waste generated on daily basis in the town is 25 tons. Out of the total, 80% waste (20 t) is domestic waste and rest 20% (5 t) waste is from sabzi Mandi.

There is no system for storm water drainage, and water generally flows into open drains (Fig. 3). Kali Bein (Black Stream–rivulet of Punjab) is the main source of surface water in the town. But due to the water pollution caused by the industries, its

Fig. 3 Condition of neighborhood. *Source* Author

water is no longer potable. Therefore, Kapurthala is dependent on ground water for water supply.

Recreational/Open spaces: There are two parks and one stadium in the town. Parks are Shalimar Bagh and Kamra Bagh, both having area of 8 ha each. Then, there is Guru Nanak Stadium present in the city. Municipal Council Office and Fire Brigade are in the Shalimar Bagh. Guru Nanak Stadium on Jalandhar Road is used for all types of sports activities and government functions organized in the city.

Socio-cultural: Socio-cultural facilities include community hall, sports complex, stadium, playgrounds, public libraries, art gallery, cinema hall (550 seats), dharamshala's (rest houses). The key issues are the poor infrastructure in District Library and Community Hall, no multiplex or mall in a city; the only cinema hall present there is in critical condition, a need of community rooms, community halls, library, recreational club, drama, dance, musical center, meditation hall and spiritual centers.

2.1 Heritage Buildings in Kapurthala

City has a number of buildings which have historic, cultural and heritage importance. Built heritage includes museum and palaces, gardens, clock tower, temples, gurudwaras, community hall, clubs, schools, colleges and guest house. Efforts made by the last ruler of Kapurthala, Jagatjit Singh, transformed the image of the city into a colonial look alike. Beautiful gardens like Shalimar Garden and Kamra Garden are famous for their tree lined avenues and quality of plantation. The city has a very prominent urban axis (Jalandhar–Kapurthala Road) along which nearly all the major historical structures were built, with Jallowkhana at the beginning of this axis. All the heritage structures and the axis itself are shown in Fig. 4. However, architectural style of various heritage buildings inside the city has been shown in Fig. 5.

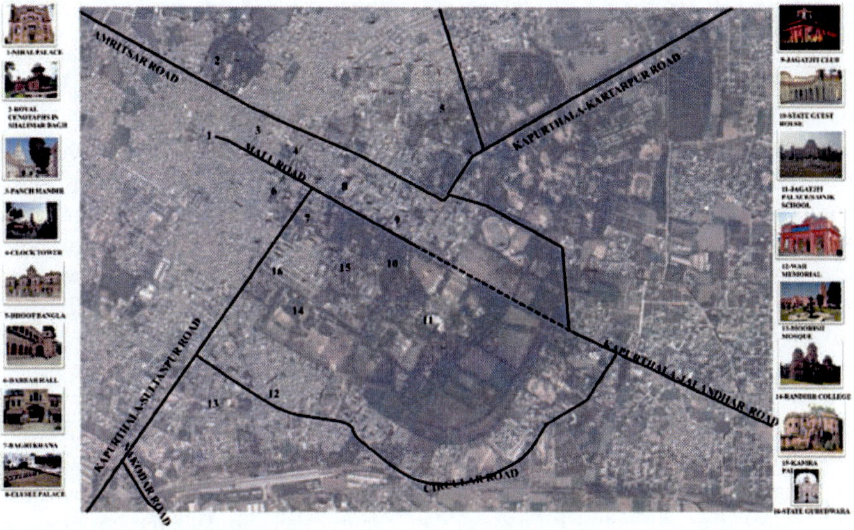

Fig. 4 Map showing the location of heritage sites and their approach routes. *Source* Primary survey done by author

Fig. 5 Architectural style of Kapurthala. *Source* Primary survey done by author

3 Issues and Challenges in Kapurthala

Kapurthala is an example of historic towns but the current scenario of the city does not reflect the magnificence of its past. This is due to shear neglect, continued dereliction and vandalism of some of its historic monuments and poor level of social and physical infrastructure. The city is no longer Paris of Punjab. Its rich cultural heritage is deteriorating and dying due to lack of awareness among the local residents. Although the city has abundant heritage resources, planning and management of heritage core areas has always been neglected in Kapurthala. Limited attempts to conserve the built heritage and lack of initiatives regarding marketing of Kapurthala heritage has resulted in decay and destruction of the rich heritage. This section is an attempt to map out the issues related to heritage management in Kapurthala which could be reversed back with various interventions.

Vandalism and neglect: There have been a total neglect for many public as well as private heritage buildings. Two such palaces, Jallowkhana and Gol Kothi, are the paragons of total institutional neglect. Gol Kothi has to be maintained by State Archaeology Department of Punjab [1], but it is in ruins and declared not fit to be occupied by local authorities. Jallowkhana is also in a very pathetic state and is in ruins too. It is even encroached by local neighborhoods. Ex-army headquarters which is also known as Bhoot Bangla is also in very pitiable plight.

Lack of green spaces: Most of the core area of the city is very densely populated with very less open spaces. There is 40% of open space in municipal council area [5]. But the fact is that green spaces are very unevenly spread and trapped in some complexes like Jagatjit palace, D.C. residence, etc., which actually is not accessible to the public.

High-density core areas: The core areas of the city are very densely populated. There are zig-zag narrow streets with open drains on either one or both sides. There is lot of encroachment in terms of vehicle parking and in the form of informal commercial activities by the shopkeepers too. For instance, Jallowkhana lies in ward 20 of the city which is a medium density ward but immediate wards surrounding it are highly dense in terms of population per hectare (Figs. 6 and 7).

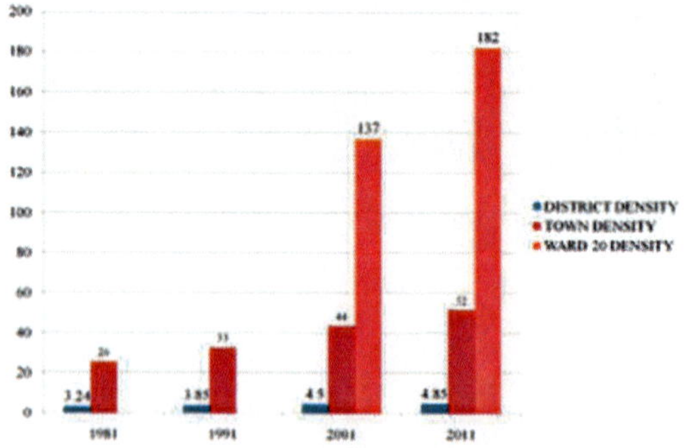

Fig. 6 Comparison of population density. *Source* Master plan Kapurthala, 2010–2031

Fig. 7 Urban fabric around Jallowkhana. *Source* Master plan Kapurthala, 2010–2031

Uneven distribution of various amenities: Schools, colleges, hotels, religious buildings, etc., are unevenly distributed throughout the city which increases the walkable distance from them. And when the distance increases, people prefer using automobiles than being pedestrians. Thus, adding to the number of vehicles on the road. So, uneven distribution of such amenities, along with other reasons like situation of bus terminal in the center of the city, is some reasons for a greater number of traffic jams in the city. The following are some of the ped-shed analysis showing walkability from Jallowkhana site (Fig. 8).

Some new insertions are unsympathetic to the scale and character of existing historic structures, and such haphazard developments have not been monitored by the local authorities. Resource shortages have resulted in no maintenance and thus environmental degradation. Additional damage has also occurred due to unconsidered additions, repairs and alterations.

4 Proposals and Recommendations for Sustainable Development of the Historic Town

The first step is to develop the entire building form and architectural style of palaces and gardens as they once existed. Nowadays, historic buildings are completely ruined and encroached by residents (Fig. 9). Residents have built their houses on the historic lands. However, the court ordered that the original position of Palaces such as Jallowkhana be restore but the local authorities took very little or no steps

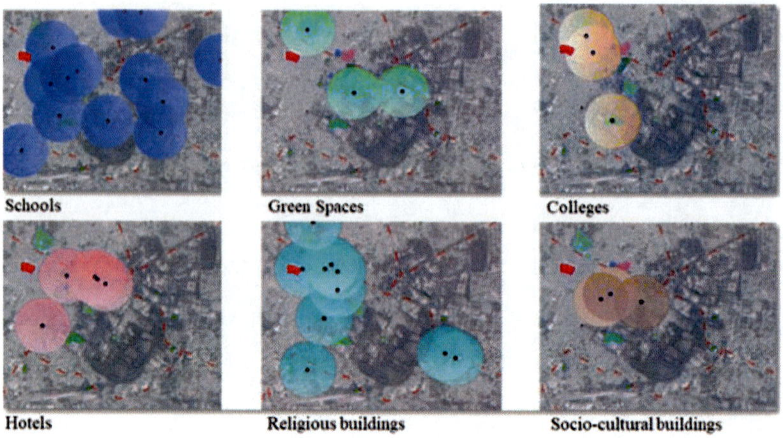

Fig. 8 Ped-shed analysis. *Source* Author

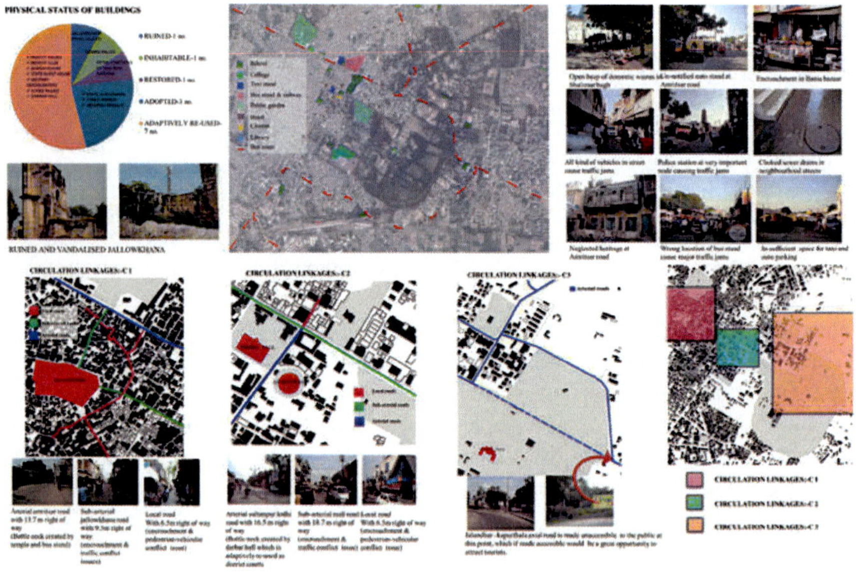

Fig. 9 Existing conditions of Kapurthala. *Source* Author

to protect their land and existing structure. Some of the urban design proposals and guidelines are suggested that can be adopted by the local government officials for the conservation of urban heritage and the integrated revitalization of the Kapurthala. In addition, by conducting a SWOT analysis, one can focus on the strengths, minimize weaknesses, and make the most significant possible advantage of opportunities available [3].

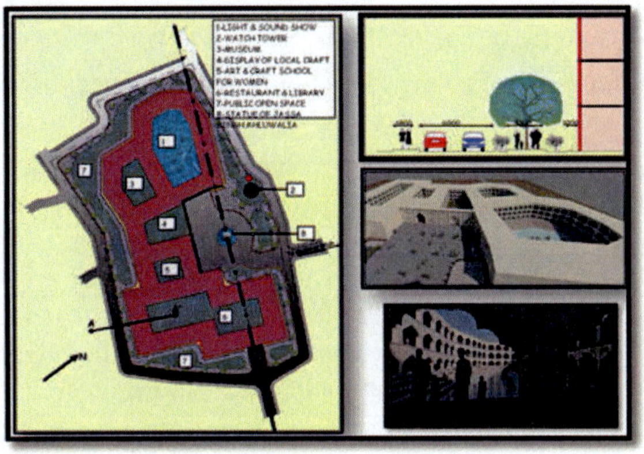

Fig. 10 Design proposal. *Source* Author

4.1 Urban Design Proposals and Guidelines

Walkways around the peripheral roads can be proposed and carved out from the historic structure's land itself. Further, ample open spaces can be left around the palace, so that people can sit here and relax. Huge space can be designed in front of the street, which makes its way from Shalimar Bagh because it was there in the original design. Copper water tank present in this open area is adaptively re-used. Its new use is of a new watchtower. Tourists can come here and see the whole area from here. Further, its function of supplying water is also proposed to be re-established, and it would supply water to the new development only. Some of the proposed spaces are library, restaurant, museum, art and craft school for women, light and sound shows, display areas for the retail of handicrafts, watchtower, statue of Jassa Singh Ahluwalia, etc. (Fig. 10).

Framing the urban design guidelines is very important for this part of the city. Now, as the proposal is to renovate and conserve the palaces, the surrounding areas also need to be intervened. So, guidelines for the façade regulations of the two main approach streets to the Jallowkhana are prepared. Some of the guidelines are listed below:

- Streets from Shalimar Bagh and Sadar bazaar need to be pedestrianized. The street width would be such that an emergency vehicle could move.
- As the corbels/brackets are very much used in the city buildings. So, the implementations of the corbels need to be there, which would support the roof projections.
- Typical arches which are used in many of the heritage buildings need to be made part of the façade of these two historical streets.
- Another proposal is to remove all the visual pollution by laying down underground cables for telephone and television. In addition to this, a standard hoarding design would be made for all the shops on the streets. The material of the hoarding would

be red sandstone, and the name of the shop would be carved in that with font height of 100 mm only.

- Land use of all the shops needs to be assessed. In order to promote tourism, local handicrafts and materials would be sold here.
- All the street lights would be operational from solar energy.

5 Conclusion

The findings and observations from the study highlight the fact that while the urban heritage of the city is currently under severe stress.

- There are tremendous opportunities to renew that area through sustainable urban renewal.
- To sustain the heritage value of the city, it requires a comprehensive set of visions to enhance same and inclusive economic development.
- Further explorations for improving urban design elements and community-based area renewal in the city may be carried out along with the recommendations provided in this paper.

Acknowledgements Authors would like to thank local government authorities of Kapurthala city for allowing visits to the site and conduct the surveys.

References

1. ASI Punjab (n.d.) Archaeological survey of India. Retrieved from Protected Monuments in Punjab website: https://asi.nic.in/. 1 June 2020
2. Goodey B (2001) Tourists in historic towns: Urban conservation and heritage management. URBAN DESIGN Int 6(2):113–114. https://doi.org/10.1057/palgrave.udi.9000030
3. Kaur H, Ahuja R, Praharaj S (2015) Revitalization of the historic urban quarters and upgrading tourism economy in the walled city of Diu, India revitalization of the historic urban quarters and upgrading tourism. In: National conference on rediscovering cities (NCRC), Oct. https://doi.org/10.13140/RG.2.1.1821.4489
4. Pokhriyal R (2016) Dreams that don't let you sleep: based on the life skills management of Dr. Ramesh Pokhriyal "Nishank"—Google Books
5. PUDA (2010) Punjab Urban Planning & Development Authority, Punjab, India. Retrieved from Draft master plan for Kapurthala lpa 2010–2031 website: https://www.puda.gov.in/. 1 June 2020
6. Tweed C, Sutherland M (2007) Built cultural heritage and sustainable urban development. Landsc Urban Plan 83(1):62–69. https://doi.org/10.1016/j.landurbplan.2007.05.008
7. Udeaja C, Trillo C, Awuah KGB, Makore BCN, Patel DA, Mansuri LE, Jha KN (2020) Urban heritage conservation and rapid urbanization: insights from Surat, India. Sustainability 12(6):2172

Liquefaction Behavior of Soil in Light of Sustainability

Sunita Kumari, Kallol Saha, A. K. Choudhary, and J. N. Jha

1 Introduction

Bihar is located in the high seismic zone of India which falls on the boundary of the tectonic plate joining the Himalayan tectonic plate near the Bihar–Nepal Border and has six subsurface fault lines moving toward the Gangetic planes in four directions. The state experienced major earthquakes in the past; the worst was the 1934 earthquake in which more than 10,000 people lost their lives, followed by 1988 earthquake and recent earthquake was the Sikkim earthquake in September 2011. The new and growing urban centers in the state where building codes and control mechanisms are not enforced, earthquake remains a major threat to cities. This could result devastation in social infrastructures such as schools and hospitals which are not built to be earthquake-resistant. This may lead to serve damage and loss of lives as well.

Bihar region is very well recognized as a zone of high seismic activity. Some of the densely populated districts like Darbhanga, Sitamarhi, Madhubani, and Supaul are situated in seismic Zone-V. Most of the districts situated in southwestern area of Bihar lie in Zone-III whereas the other remaining districts of Bihar including

S. Kumari (✉) · K. Saha
Department of Civil Engineering, National Institute of Technology Patna, Patna, Bihar 800005, India
e-mail: sunitafce@nitp.ac.in

K. Saha
e-mail: kallol001.saha@gmail.com

A. K. Choudhary
Civil Engineering Department, NIT Jamshedpur, Jamshedpur, Jharkhand 831014, India
e-mail: akchoudhary.ce@nitjsr.ac.in

J. N. Jha
MIT Muzaffarpur, Muzaffarpur, Bihar 842003, India
e-mail: jagadanand@gmail.com

H. Singh et al. (eds.), *Sustainable Development Through Engineering Innovations*, Lecture Notes in Civil Engineering 113,
https://doi.org/10.1007/978-981-15-9554-7_22

Patna lie in Zone-IV. An earthquake of estimated magnitude 8.4 on Richter scale occurred on January 15, 1934, in the border region of Nepal and Bihar state of India which leads to the widespread devastation due to liquefaction resulting huge loss of human life. Another earthquake of magnitude 6.6 on the Richter scale occurred in the same region on August 21, 1988, resulting in widespread damage and liquefaction of alluvial deposits available in that pocket of Bihar. Also, the evidence of liquefaction was also marked in Madhubani district of Bihar as a consequence of the Nepal earthquake of April 25, 2015.

However, research following the 1994 Northridge, 1999 Kocaeli, and 1999 Chi-Chi earthquakes has identified a significant number of cases where ground failure in silty and clayey soils containing more than 15% clay-size particles caused considerable damage to buildings. For example, Bray and Sancio [1] found that liquefaction in the silts of Adapazari, Turkey, was responsible for much of the damage observed in this city. These fine-grained soils typically had clay contents greater than 15%, but contrary to the often relied upon Chinese criteria, these soils liquefied, which led to significant ground failure.

The performance of fine-grained soils in Adapazari, Turkey, during the 1999 Kocaeli earthquake offers an exceptional opportunity for understanding their susceptibility to liquefaction and response under cyclic loading. Following the earthquake, a comprehensive program of subsurface investigations through energy measured standard penetration test and soil index testing was completed by Bray and Sancio [1]. Soils that were observed to have liquefied in Adapazari during the Kocaeli earthquake did not typically meet the Chinese criteria for liquefaction-prone fine-grained soils. Xenaki and Athanasopoulos [2] performed experimental investigation for liquefaction resistance of sand–silt mixtures. Most of the districts of Bihar falls under seismic Zones IV and V. Hence, evaluation of liquefaction potential of high seismic region of Bihar will provide a guide map for engineers and designers.

From the literature it is found that hazard mitigation has become a crucial and important activity to eliminate the risk to life and property and can lead to a path of sustainable development. Therefore, the objective of present paper is to find out liquefaction susceptibility using [3] method based on SPT. The outcome is compared with [4] method to check the effect of fine content, as uncertainty occurs in both the loading and resistance aspects of liquefaction problem. Also, probabilistic seismic hazard analysis has been also carried out using Liao et al. [5] method to determine the uncertainties in liquefaction resistance which include effect of fine content. The results of this study provide useful insights for projects encountering fine-grained soils because the liquefaction susceptibility of these soils is not well-understood.

2 Methodology

The present works consider bore-hole data for the four different districts of Bihar, i.e., Samstipur and Begusarai, which falls under seismic Zone-V. Analysis has been done using [3] method and uncertainty is determined using probabilistic approach as described in Liao et al. [5].

2.1 Idriss and Boulanger [3] Approach

This method provides an update on the semi-empirical field-based procedures for evaluating liquefaction potential of cohesion less soils during earthquakes. This update includes recommended relations for many parameters which are considered to capture the essential physics of the liquefaction phenomenon while keeping it as simplified as possible. These relations are then used in re-evaluations of the field case histories to derive revised deterministic SPT-based correlations.

2.2 Stress Reduction Coefficient (r_d)

The shear stresses induced at any point in a level soil deposit during an earthquake are primarily due to the vertical propagation of shear waves in the deposit. The most practical approach to express stress reduction coefficient (r_d) is a function of depth and earthquake magnitude (M).

$$\ln(r_d) = \alpha(z) + \beta(z)M \tag{1}$$

$$\alpha(z) = -1.012 - 1.126\sin\left(\frac{z}{11.73} + 5.133\right) \tag{2}$$

$$\beta(z) = 0.106 + 0.118\sin\left(\frac{z}{11.28} + 5.142\right) \tag{3}$$

In which, z is depth in meters and M is moment magnitude. These equations are applicable to a depth $z \leq 34$ m, whereas for $z > 34$ m, the following expression is applicable:

$$r_d = 0.12\exp(0.22M) \tag{4}$$

The uncertainty in r_d increases with increasing depth such that Eqs. (1)–(3) should only be applied for depths less than about 20 m.

2.3 Magnitude Scaling Factor (MSF)

The magnitude scaling factor, MSF, has been used to adjust the induced *CSR* during earthquake magnitude M to an equivalent *CSR* for an earthquake magnitude, $M =$ 7.5. The MSF is thus defined as:

$$\text{MSF} = \frac{\text{CSR}_M}{\text{CSR}_{M=7.5}} \tag{5}$$

Thus, MSF provides an approximate representation of the effects of shaking duration or equivalent number of stress cycles.

The MSF (MSF $\leq$ 1.8) relation produced by this re-evaluation is expressed as:

$$\text{MSF} = 6.9\exp\left(-\frac{M}{4}\right) - 0.058 \tag{6}$$

2.4 Overburden Correction Factor ($\mathbf{K}_\sigma$)

The overburden stress effects on CRR could be represented in either of two ways: (i) through the additional normalization of penetration resistances for relative state, thereby producing the quantities $N_{1\,60}$ or (ii) through a K_σ factor.

The recommended K_σ relations, as expressed by Idriss and Boulanger [3] are as follows:

$$K_\sigma = 1 - C_\sigma \ln\left(\frac{\sigma'_{\text{vo}}}{P_\text{a}}\right) \leq 1.0 \tag{7}$$

$$C_\sigma = \frac{1}{18.9 - 17.3D_\text{R}} \leq 0.3 \tag{8}$$

$$D_\text{R} = \sqrt{\frac{(N_1)_{60}}{46}} \tag{9}$$

$$C_\sigma = \frac{1}{18.9 - 2.55\sqrt{(N_1)_{60}}} \tag{10}$$

In which, D_R is defined as relative density of soil deposit. The K_σ values were restricted to 1 for the re-evaluation of the SPT liquefaction correlations presented later, although conceptually the K_σ values should be allowed to exceed 1.0 when $\sigma'_{\text{vo}}/P_\text{a}$ is less than unity.

2.5 *Overburden Correction Factor (C_N) for Penetration Resistance*

Idriss and Boulanger [3] obtained the following equation for C_N:

$$C_N = \left(\frac{P_a}{\sigma'_{vo}}\right)^{\alpha} \leq 1.7; \quad \alpha = 0.784 - 0.0768\sqrt{(N_1)_{60}} \tag{11}$$

2.6 *Effect of Fines Content on SPT*

The SPT penetration resistance is adjusted to an equivalent clean sand value as:

$$(N_1)_{60cs} = (N_1)_{60} + \Delta(N_1)_{60} \tag{12}$$

$$\Delta(N_1)_{60} = \exp\left(1.63 + \frac{9.7}{FC + 0.1} - \left(\frac{15.7}{FC + 0.1}\right)^2\right) \tag{13}$$

2.7 *Cyclic Resistance Ratio (CRR)*

The value of CRR for a magnitude $M = 7.5$ earthquake and an effective vertical stress atm $\sigma_{vo} = 1$ can be calculated based on cs $(N_1)_{60cs}$ using expression:

$$CRR = \exp\left\{\frac{(N_1)_{60cs}}{14.1} + \left(\frac{(N_1)_{60cs}}{126}\right)^2 - \left(\frac{(N_1)_{60cs}}{23.6}\right)^3 + \left(\frac{(N_1)_{60cs}}{25.4}\right)^4 - 2.8\right\} \tag{14}$$

2.8 *Cyclic Stress Ratio (CSR)*

The value of CRR for a magnitude $M = 7.5$ earthquake and an effective vertical stress $\sigma'_{vo} = 1$ atm can be calculated using the following expression:

$$(CSR)_{M=7.5,\sigma=1} = 0.65\left(\frac{\sigma_{vo} a_{max}}{\sigma'_{vo}}\right)\frac{r_d}{MSF}\frac{1}{K_\sigma} \tag{15}$$

At a depth below the ground surface, liquefaction will occur if shear stress induced by earthquake is more than the shear stress predicted by various approaches. By comparing the induced and predicted shear stresses at various depths, liquefaction zone can be obtained.

2.9 *Probabilistic Approach by Liao et al. [5]*

There are many potential sources of uncertainty in both the loading and resistance aspects of liquefaction problems, hence probabilistic seismic hazard analysis has been done to deal with them. This is based on the probabilistic characterization of the parameters shown by the laboratory tests which influence pore pressure generation. Haldar and Tang [6] characterized uncertainty in the parameters of the simplified cyclic stress approach. Fardis and Veneziano [7] used a similar approach with total stress and effective stress model. Chameau and Clough [8] described pore pressure generation probabilistically using experimental data and an effective stress model. Each of these methods can compute the probability of liquefaction due to a particular set of loading conditions. Its accuracy depends on the accuracy of the underlying liquefaction/pore pressure model and accuracy in determination of the uncertainty of the model parameters. An alternative group of methods are based on the in situ test-based characterization of liquefaction resistance. These methods use various statistical classification and regression analysis to assign probabilities of liquefaction to different combinations of loading and resistance parameters. For example, Liao et al. [5] worked on 278 case studies to produce an expression for the probability of liquefaction.

$$P_L = \frac{1}{1 + \exp\left[-(\beta_o + \beta_1 \ln(\text{CSR}) + \beta_2 (N_1)_{60})\right]} \tag{16}$$

In which, the value of parameters of equation are given in Table 1.

Table 1 Regression parameters for calculating P_L

Data	Number of cases	β_o	β_1	β_2
All cases	278	10.167	4.1933	0.24375
Clean sand cases only	182	16.447	6.4603	0.39760
Silty sand cases only	96	6.4831	2.6854	0.18190

3 Results and Discussion

Standard penetration test (SPT) resistance is commonly used in situ test parameter for characterization of liquefaction potential of soil. The SPT allows a sample to be retrieved (for identification, measurement of fines content, etc.) and has the largest case history database of any in situ test. The present works consider the borehole data for the two different districts of Bihar, i.e., Samstipur and Begusarai which fall under Zone-IV.

3.1 Soil Profile of Samastipur

From Fig. 1, it is observed that most of the boreholes or pits examined in study region have FOS less than 1. This indicates that most of the sites in study regions are susceptible to liquefaction. The respective graphs plotted between depth versus FOS also depicts this behavior for both the cases either [4] or [3]. It is also indicated that, at a particular depth factor of safety against liquefaction is high in case of [3] approach as compared to [4] approach. Thus, at a particular depth, the soil has less liquefaction resistance in case of Seed and Idriss [4] than Idriss and Boulanger [3]. This may be due to consideration of fine content. Hence, cost on remedial measure will decrease while considering [3].

For instance, at depth of 1.5 m below ground surface FOS against liquefaction by Seed and Idriss [4] method is 1.66 while the same FOS against liquefaction is 5.85 by Idriss and Boulanger [3] approach at same depth 1.5 m below surface. These values are almost higher or same for higher depth. This is because there is a remarkable difference between the two approaches. The analytical framework of Idriss and Boulanger [3] approach includes the effect of fines content on SPT penetration resistance value. Such provision is not observed in methodology of Seed and Idriss [4]. This leads to fact that the presence of fines enhances the resistance of soil to liquefaction.

Figure 2 shows the variation of probability of liquefaction verses depth for better comprehension of site by different approaches. It is observed that for same fine

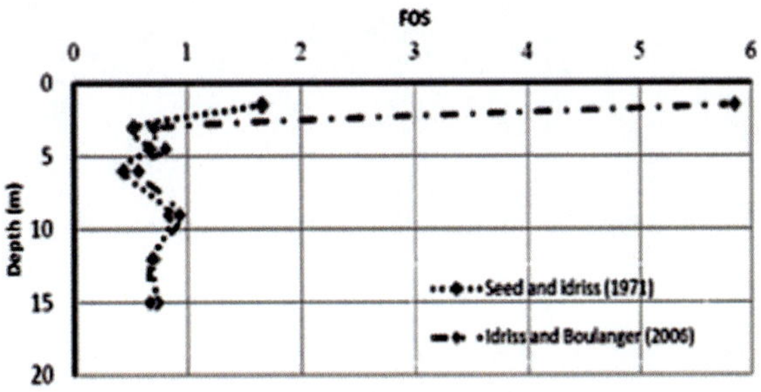

Fig. 1 FOS versus depth at Samastipur

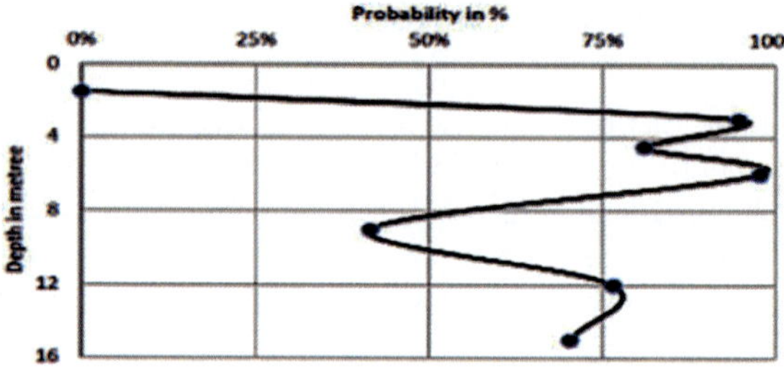

Fig. 2 Probability for liquefaction versus depth at Samastipur

content and depth <10 m, increasing the depth will result decrease in liquefaction resistance. But the trend is reversed for 10 m < depth < 15 m. Here, for same fine content, increasing the depth will result in decrease in liquefaction susceptibility or increase in liquefaction resistance. Now, in probabilistic approach by Liao et al. [5] for this site, it is observed that normalized SPT value $(N_1)_{60}$ is a measure of soil liquefaction resistance. As value of $(N_1)_{60}$ increases, probability of liquefaction decreases monotonically. For example, Fig. 3 shows that when normalized SPT value $(N_1)_{60}$ increases from 12 to 15, the probability of liquefaction decreases from 98 to 95%. These types of trend are seen at higher depth also. This accounts for the direct impact of $(N_1)_{60}$ on probability of liquefaction. The same general trend was observed with almost all sites under study region of Samastipur district.

Equation (14) also shows that $(N_1)_{60}$ has exponential relationship with CRR. Thus, increasing the $(N_1)_{60}$ will directly increase the CRR and thus increases the resistance of soil against liquefaction. An increment in resistance of soil to liquefaction will also lead to decrease in probability to liquefaction. However, this variation is slightly deviated at one point when the $(N_1)_{60}$ value changes in the range of 16–21. Effect of fine content on liquefaction resistance of soil is evaluated for [3].

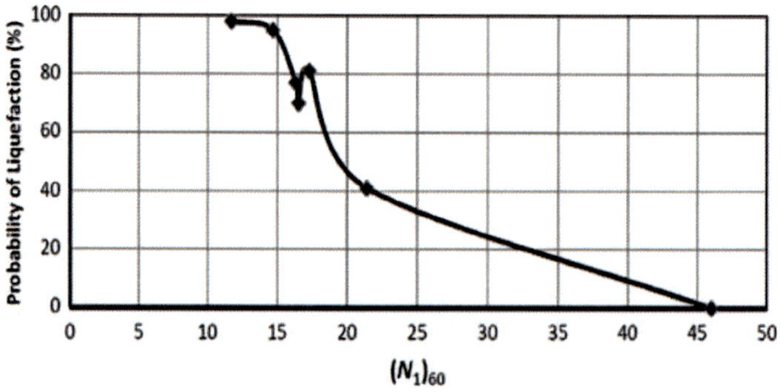

Fig. 3 Probability for liquefaction versus SPT at Samastipur

3.2 Soil Profile of Begusarai

Begusarai district comes under Zone-IV and the bore log of this district has been analysed for effect of fine content. Similar trend was observed as Samastipur district. Here also FOS against liquefaction, as shown in Fig. 4, evaluated by Idriss and Boulanger [3] is higher than that of [4] approach. Thus, at a particular depth, the soil has less liquefaction resistance in case of [4] approach than that of [3] approach. These values are almost higher or same when depth is increasing. For instance, at depth of 3 m below ground surface, FOS against liquefaction in case of Idriss and Boulanger is 1.95 whether the same FOS against liquefaction for Seed and Idriss [4] approach is 1.2. This may be due to remarkable difference between the methodology of [4, 3] approach. Similar trend for probability of liquefaction is seen in Fig. 4 also. Here, also increasing the normalized SPT value $(N_1)_{60}$ value results in decreasing the probability of liquefaction. Figure 5 shows the probability of liquefaction verses depth. Figure 6 shows that when $(N_1)_{60}$ increases from 25 to 28, the probability of liquefaction decreases from 22% to 10%.

The SPT $(N_1)_{60}$ of the soil deposit has exponential relationship with CRR. Thus, increasing the SPT $(N_1)_{60}$ value will directly increase the CRR and thus the resistance of soil to liquefaction. Increment in resistance of soil to liquefaction will lead to decrease in probability to liquefaction. A general trend was observed in [3] approach that regarding effect of same fines percentage with changing depths. This states that for same fine content and depth < 10 m, increasing the depth will result in increase in liquefaction susceptibility or decrease in depth decreases liquefaction resistance.

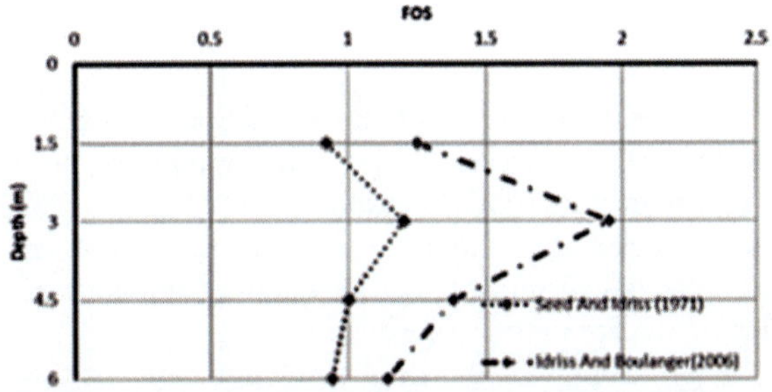

Fig. 4 Factor of safety (FOS) versus depth at Begusarai

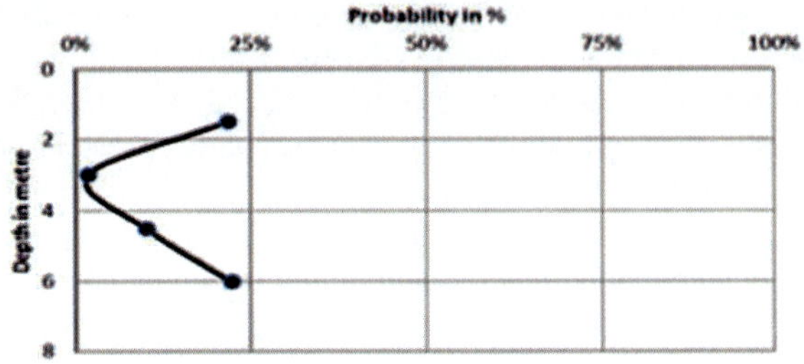

Fig. 5 Probability for liquefaction versus depth at Begusarai

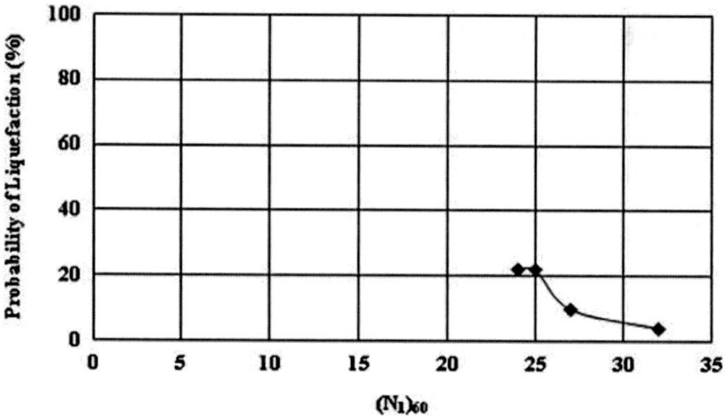

Fig. 6 Probability for liquefaction versus SPT at Begusarai

But the trend is reversed for 10 m < depth < 15 m. Here, for same fine content increasing the depth will result in decrease in liquefaction susceptibility or increase in liquefaction resistance due to overburden effect. This conclusion is validated for all districts except Begusarai (Zone-IV) due to presence of non-plastic fines. In both the earthquake-prone districts of Bihar, it was found that fines content <6% do not affect the *N*-values and hence in turn the CRR of the soil whereas fines content greater than 6% increases the resistance of soil toward liquefaction.

4 Conclusion

The present paper presents the evaluation of liquefaction potential of high seismic region of Bihar which will provide a guide map for engineers and designers. This guide map will provide guidelines for hazard mitigation that can save massive damages caused due to hazards, thus will result in cost efficiency and quick predictions for the sustainable development of human resources. The following findings are significant in the present study:

- Increasing the value of $(N_1)_{60}$ will result in decrease in probability of liquefaction since energy corrected normalized SPT value $(N_1)_{60}$ has direct influence on the calculation of cyclic resistance ratio (CRR) of soil.
- An apparent increase in CRR is observed with increased fines content.
- It was found that fines content <6% do not affect the *N*-values and hence in turn the CRR of the soil whereas fines content greater than 6% increases the resistance of soil toward liquefaction.
- For same fine content, increasing the depth will result in decrease in liquefaction susceptibility or increase in liquefaction resistance due to overburden effect.

Table 2 Soil profile for Samstipur

S. No.	Depth (m)	Water content (%)	Saturated unit weight (kN/m^3)	Fines (%)	SPT (*N*-value)
1	1.5	22.4	20.06	7	10
2	3.0	25.7	19.53	7	11
3	4.5	25.5	19.56	5	10
4	6.0	27.6	19.24	2	9
5	9.0	28.4	19.13	6	22
6	12.0	22.7	20.01	8	24
7	15.0	25.3	19.59	10	20

Table 3 Soil profile for Begusarai

S. No.	Depth (m)	Water content (%)	Saturated unit weight (kN/m^3)	Fines (%)	SPT (*N*-value)
1	1.5	22.4	20.06	5	10
2	3.0	23.8	19.53	10	18
3	4.5	25.5	19.56	7	18
4	6.0	26.6	19.24	6	19

Hence, evaluating the liquefaction susceptibility without considering fine content is conservative as well as uneconomical for the sustainable design of structures. This directs the saving of construction materials resulting saving of natural resources.

Appendix

See Tables 2 and 3.

References

1. Bray JD, Sancio RB (2006) Assessment of the liquefaction susceptibility of fine grained soil. J Geotech Eng 132(9):1165–1177
2. Xenaki VC, Athanasopoulos GA (2003) Liquefaction resistance of sand-silt mixtures: an experimental investigation of the effect of fines. Soil Dyn Earthq Eng 23(2003):183–194
3. Idriss IM, Boulanger R (2006) Semi-empirical procedures for evaluating liquefaction potential during earthquakes. J Soil Dyn Earthq Eng 26(2):115–130
4. Seed HB, Idriss IM (1971) Simplified procedure for evaluating soil liquefaction potential. J Soil Mech Found Div ASCE 92(6):1249–1273
5. Liao SSC, Veneziano D, Whitman RV (1988) Regression models for evaluating liquefaction probability. J Geotech Eng 114(4)

6. Haldar A, Tang WH (1979) Probabilistic evaluation of liquefaction potential. ASCE J Geotech Eng Div 105(2):145–163
7. Fardis MN, Veneziano D (1982) Probabilistic analysis of deposit liquefaction. J Geotech Eng Div ASCE 108(GT3):395
8. Clough GW, Chameau JL (1983) Seismic response of San Francisco waterfront fills. J Geotech Eng ASCE 109:491–506

Biotransformation of Industrial Wastes for Nutrient Rich Vermicompost—A Review of the Bioconversion Process by Earthworms

Vinay Kumar Badhwar, Sukhwinderpal Singh, and Balihar Singh

1 Introduction

Industrial solid waste management has been a rising concern with the growth of industrial sector and economy buildup of developing countries like India where standards of living are getting better each day resulting in destruction of green space and major demand for urbanization and adequate infrastructure leading to expanding industrialization. According to Central Statistics Office, Ministry of Statistics and Programme Implementation, Government of India, total 371, 336 polluting industries were in operation in India (*Sugar; Pulp & Paper; Pharmaceutical; Pesticides; Iron & Steel; Fertilizer; Distillery; Soap & Detergent; Cement; Aluminium; Copper; Zinc*), out of which 40,095 were large-scale and 331,241 were small-scale industries [1]. About 30% of 960 million tons solid waste generated in India comes from the industrial sector [2]. Different industries produce different levels of hydrocarbons, phenols, heavy metals, organic chemicals, and other effluents in different forms which pollute the overall environment through its unsafe disposal. Hazardous industrial wastes are destructive in all manners while non-hazardous industrial wastes are non-toxic by-products generated by commercial and industrial activities and are safe for recycling and other waste-to-energy processes. All the *agro-based industries, pulp & paper industry, tanneries, food processing industries & textile industries*, thus become a major source of raw materials as their by-products can be reused again.

V. K. Badhwar (✉) · S. Singh · B. Singh
Department of Civil Engineering, Guru Nanak Dev Engineering College, Ludhiana 141006, India
e-mail: vinaybadhwar.vb@gmail.com

S. Singh
e-mail: sukhwinderpalsingh92@gmail.com

B. Singh
e-mail: balihar1994@gmail.com

H. Singh et al. (eds.), *Sustainable Development Through Engineering Innovations*, Lecture Notes in Civil Engineering 113,
https://doi.org/10.1007/978-981-15-9554-7_23

In the current scenario of sustainable development where the population is increasing at an enormous rate and the natural resources are decreasing in an uncontrollable manner, the environment is facing the adverse effects of the overall human activities. The management of solid wastes is a major concern in the path of achieving sustainable needs that can be solved only by providing access to engineering innovations. The current methods available for solid waste management include landfilling, incineration, recycling, source reduction, and reuse [3]. The operation of incineration is least-preferred considering the toxic emissions, high capital costs, and Global Warming Potential. Landfilling has been the most dominant technique of disposing waste due to cheaper costs among all operations [4]. The problem of water & land contamination through leachate and greenhouse gas emissions arises in landfilling and is, therefore, not the best solution for managing waste [5, 6]. Recycling is slow and informal in developing countries like India which lacks in providing basic recycling facilities [7, 8]. As a result, reuse is the most feasible option for industrial waste management. Reuse can be done by both composting and vermicomposting. The cost-benefit analysis of reusing is much attractive as compared to other management technologies [7]. Therefore, the need of the hour is to adopt vermicomposting as a sustainable waste management strategy in bio-transforming industrial wastes into fertilizers.

Vermicompost is the bio-oxidized and stabilized product formed by the combined action of earthworms and micro-organisms. Vermicomposting process is an eco-friendly process that is best suited for industrial waste bioconversion [9]. The earthworm species *E. foetida* is extensively used in vermistabilization process because of its high potential for bioconversion of organic wastes into vermin casts [10].

The cocoon production, the increase in biomass, and the low mortality rate of earthworms in the vermistabilization of paper mill wastewater sludge is an indication of the suitability of *E. fetida* in industrial waste stabilization [11]. Higher heavy metal concentration levels occur in different industrial waste by-products due to the involvement of chemicals in different processes. In the final vermicompost samples, the heavy metals content decreased as compared to the initial *sewage sludge* while that in the earthworms increased due to the bioaccumulation factor (BAF) [12]. This clearly depicts the efficiency of vermiconversion process in effectively removing the heavy metals content. An increase of 6.6 times the initial value of Total Nitrogen Content occurred in the finally stabilized *tea coal factory ash* vermicompost sample which clearly shows its high nutrient value [13]. The macronutrients and micronutrients are readily available to plants in vermicompost [14]. Therefore, in these adulteration times, the excessive use of inorganic fertilizers should be countered while the demand for organic fertilizers such as vermicompost should be given preference in agriculture. The existing resources should be used adequately and industrial wastes need to be utilized in a manner such that sustainability could be achieved. The main objective of this review is to focus on the potential of stabilization of industrial wastes by using earthworms and implementing this technology on a large-scale for sustainable agricultural development.

2 Vermicomposting

In the process of waste conversion, the waste material to be stabilized passes through the gut of the earthworms, where it gets transformed and is excreted out in the form of worm manure. These castings are not only helpful in regulating the growth of plants but also act as pest repellents thereby enriching microbial activity. This process of valorization of waste material is known by different names such as vermicomposting/vermitechnology/vermiconverion/vermistabilization. This process can work within a mesophilic temperature range of 10–32 °C and both microbes and earthworms survive and work effectively in this range [15]. Vermicomposting is a closed-loop renewable technology for amendment of waste materials into rich nutrients source for soil fertility and also enhances the water retention capacity of soil. It also results in compositional change of waste, decrease in C:N ratio due to decrease in organic content, and gain in nutrient content by the earthworm activity [16]. The earthworms not only aerate the substrate but also decompose it thereby increasing their biomass and overall microbial activity of the process [17]. The agro-industrial organic residues need to be transformed for use in food production by the addition of vermistabilized humus to soil instead of expensive disposal of such wastes [18]. The researchers have been managing the industrial wastes using earthworms extensively [19–21].

Industrial wastes cannot be directly fed to earthworms due to containment of hazardous chemicals as it can drastically affect the vermicomposting process. Amendments are bound to be added to the industrial wastes for successful working conditions. Cow dung was used as a bulking agent to stabilize *bakery industrial sludge* in vermicomposting trials [22]. Three bulking agents cow dung (CD), biogas plant slurry (BGS) and wheat straw (WS) were added to stabilize *agro-industrial sludge* procured from *sugar mill* [23]. *Tea waste* was stabilized in a short period of 28 days with the condition that it should be mixed with cow dung [24]. 50:50 combination of sludge obtained from *beverage industry* and cow dung was ideally suggested as nutrient-rich end product in terms of its earthworm growth and nutrient content [25]. Therefore, it is evident that a bulky material is necessary for the degradation of industrial wastes if they are mixed in suitable ratios.

Many researchers have also reported a decrease in toxicity levels of the industrial waste in the vermicomposting process on the basis of heavy metal bioaccumulation in the gut of earthworms. Vermicompost made from 1:1 combination of *textile mill wastewater sludge* and cow dung was reported ideal due to reduced bioavailability of toxic heavy metals after vermistabilizing it using two epigeic earthworm species *Eudrilus eugeinae* and *Perionyx excavates* [26]. The earthworm also contains a protein called metallothionein which is known to bind heavy metals such as Cu^{2+}, Zn^{2+}, and Mn^{2+} [27]. Detoxification of industrial wastes is done by chloragocyte cells and gut microbes available in earthworms [28]. A considerable increase in heavy metal content was also noticed [29, 30]. It is, therefore, crucial to study the behavior of heavy metals and their bioaccumulation in the vermistabilization process. The industrial waste should be properly utilized by using earthworms in order to obtain end product that can sustainably be used as a crop nutrient and reduce the use of inorganic fertilizers [31].

3 Biotransformation of Different Industrial Wastes

Non-hazardous industrial wastes that are otherwise disposed of unhygienically by the industries, have been successfully used by researchers in biotransformation for the preparation of vermicompost using earthworms of different species. Vermicomposting process depends on a variety of factors such as type of substrate material to be degraded, its physicochemical properties, species and number of earthworms added, temperature, moisture content, amendment material, mixing ratio of different added materials, aeration process in the material. Bioconversion by red earthworm *Eisenia fetida* on *brewers's spent grain (BGS)* showed a decrease in C/N ratio whereas earthworm population increased at the end [32]. Increase in dehydrogenase activity also depicts the release of extra-cellular enzymes in the manure castings and the degrading ability of active microbial population [33]. Detoxification and recycling of *brick kiln coal ash* increased the nitrogen and phosphorus contents, thus decreasing the heavy metal content which resulted in less polluted vermicompost [34]. The vermiconversion process also involved the use of green plants such as *Tephrosea purpurea* (TEP) and *Gliricidia sepium* (GLS) that were utilized to stabilize *paper industry sludge.* PMS + CD + TEP/GLS (2:1:1) combination being ideal for sustainable agriculture [35]. The waste material maturity can also be indicated by dehydrogenase, urease, and phosphatases activities. The problem of *leachates* in landfills was also countered when *Eisenia fetida* were added to it which led accumulation of toxins in the tissues of earthworms enhancing their reproductive ability and detoxified the municipal leachate [36]. Potential of earthworms *Eisenia fetida* was also checked after transformation of *citronella bagasse* and *paper mill sludge* mixture (3:2) leading to nutrient content increase and bacterial population growth. Increase in ash content and decrease in humification index are indicative of better maturity of compost [37]. By-products of agricultural industry can be effectively converted into useful resource just like transformation of 2:1:1 ratio of *pressmud* of sugar industry with cow dung and green plants (*Gliricidia sepium* and *Leucaena leucocephala*) produced nutrient-rich manure [38]. Similarly, 60:40 combination of *distillery sludge waste* (DSW) and *tea leaf residues* (TLR) acquired favorable results in terms of casting activity of earthworms and depicted the preference of earthworms for less DSW content and more TLR content consumption when compared with 80:20 ratio [39]. S*ewage sludge derived biochar (SSDB)* was applied as an amendment in *sewage sludge* (SS) which increased the cocoon production with addition of SSDB and high bioaccumulation of heavy metals. The decrease in C/N ratio showed positive trends in good vermicompost transformation [40]. In a separate study, mushrooms were cultivated using *tea factory waste* (TFW), and the obtained product (mycotea waste) was completely transformed into vermicompost. The 1:1 mixture showed maximum cellulose reduction. High C/N ratio is said to be an outcome of high cow dung content [41].

Similarly, other industrial wastes that have successfully been used by various researchers include *milk industry sludge*; *apple pomace waste*; *soft drink industry sludge; herbal pharmaceutical industry waste*; *paper mill sludge, tannery sludge and distillery waste* each in combination with municipal solid waste [42–46]. Residues from *palm oil mill*; biomass of an oil yielding plant named *java citronella*; *biosludge from beverage industry* are some other by-products of industrial processes that were vermicomposted and proved to be beneficial for environment instead of harming when disposed of in a rendered manner [47–49]. The industrial wastes are stabilized and detoxified after bioconversion by earthworms and the end product brings nutrient value with it that adds to the effective yield of crops and soil conditioning. The vermitechnology is an environment-friendly technique of converting industrial wastes into resourceful manure and covers the ecological, social and economic dimensions of sustainable development.

4 Role of Earthworms in Bioconversion Process

Aristotle, the Greek philosopher, called earthworms "*intestines of the earth*". Earthworms were also denoted as "*ecosystem engineers*" [50]. Earthworm castings are soil aggregates that help in the organic matter decomposition in the soil and also create soil porosity by humus formation [51, 52]. Many exogenic and endogenic enzymes are available in the gut of the earthworms for conversion of organic minerals into exchangeable forms for growth of plants and soil structural development [53].

The addition of organic fertilizer into the soil is advantageous for C and N mineralization and biomass of earthworm along with the availability and assimilation of nutrients such as Mg, K and P [54]. Poly-aromatic hydrocarbons (PAHs) are bioaccumulated to a great extent by earthworms [55]. The microbial and biochemical soil activities are facilitated by earthworms on the substrate and the environmental risks are also reduced by partial degradation of excretions of earthworms [56]. The ingestion of organic wastes in the gut of earthworms occurs including microflora present in it which is expelled out in the form of vermin castings (faecal matter) [57]. High bioaccumulation of heavy metals in earthworms can be observed due to worm activity whereas it may not be seen in substrates in which earthworms are absent [58, 59]. Earthworm castings are humus-like substances, enriched with certain hormones, enzymes, microbes and other complexes that are transferred to the soil through the gut of earthworm [60]. The feeding regime of the earthworm is also dependent on the selection of substrate for the nutrient pool soil [61].

The growth of earthworms and complete decomposition of organic matter is depicted by high levels of humic acid present in vermicomposts thereby indicating "hormone inducing activity" [62]. 25% increase was seen in the yield of crops when the earthworms are present in the agricultural-based ecosystem [63]. Aerobic conditions are maintained by earthworms through continuous mixing of soil, eliminating contaminants present in it. Both earthworms and their vermicast play a vital role in sustainable agriculture and food production by increasing the nutritional quality of the harvested crop. Therefore, the cheapest and permanent solution to make modern-day agriculture more sustainable and productive is vermitechnology.

5 Conclusions

- The researches of almost all authors have indicated to comply with vermicomposting as a sustainable method to transform a variety of industrial wastes/sludges into useful end product. This reduces their toxicity and adds to the existing resources thus maintaining ecological balance.
- Adequate recycling of by-products of industries by the most effective vermitechnology should be implemented on a larger scale to decrease the load on the environment and control the existing resources from depletion.
- The investment cost of setting up a vermicompost system is lower as compared to other waste treatment methods. Additionally, it is considered clean and sustainable technology because waste is reused to produce organic fertilizer which could be applied to agricultural lands.
- Biotransformation leads to zero waste production without the misuse or burning of organic wastes. As a result, the concept of circular economy is better achieved by less exploitation of virgin resources. It also aims at increasing the efficiency of materials, energy use, recycling and reuse, accompanied by low waste generation.
- Vermicomposting is a sustainable method that works on the same principle of Waste-to-Energy (WTE) processes and is an energy-efficient process for waste to biomass conversion using engineering innovations.

Acknowledgements The authors acknowledge Dr. Puneet Pal Singh Cheema, Assistant Professor, Department of Civil Engineering, Guru Nanak Dev Engineering College, Ludhiana for his guidance and the necessary research facilities provided in the Environment Engineering Laboratory.

References

1. https://unstats.un.org/unsd/envaccounting/londongroup/meeting20/LG20_7_4.pdf
2. Pappu A, Saxena M, Asolekar SR (2007) Solid wastes generation in India and their recycling potential in building materials. Build Environ 42:2311–2320
3. Wu TY, Lim SL, Lim PN, Shak KPY (2014) Biotransformation of biodegradable solid wastes into organic fertilizers using composting or/and vermicomposting. Chem Eng Trans 39:1579–1584
4. Laner D, Crest M, Scharff H, Morris JWF, Barlaz MA (2012) A review of approaches for the long-term management of municipal solid waste landfills. Waste Manag 32:498–512
5. Romero C, Ramos P, Costa C, Márquez MC (2013) Raw and digested municipal waste compost leachate as potential fertilizer: comparison with a commercial fertilizer. J Clean Prod 59:73–78

6. Pozza SA, Penteado CSG, Criscuolo VG (2015) A greenhouse gas inventory in the municipal landfill of the City of Limeira, Brazil. Chem Eng Trans 43:2083–2088
7. https://sustainabledevelopment.un.org/content/documents/126GER_synthesis_en.pdf
8. Song Q, Li J, Zeng X (2015) Minimizing the increasing solid waste through zero waste strategy. J Clean Prod 104:199–210
9. Bhat SA, Singh J, Vig AP (2016) Management of sugar industrial wastes through vermitechnology. Int Lett Nat Sci 55:35–43
10. Garg P, Gupta A, Satya S (2006) Vermicomposting of different types of waste using *Eisenia foetida*: a comparative study. Bioresour Technol 97:391–395
11. Negi R, Suthar S (2013) Vermistabilization of paper mill wastewater sludge using *Eisenia fetida*. Bioresour Technol 128:193–198
12. Liu F, Zhu P, Xue J (2012) Comparative study on physical and chemical characteristics of sludge vermicomposted by *Eisenia Fetida*. Procedia Environ Sci 16:418–423
13. Goswami L, Sarkar S, Mukherjee S, Das S, Barman S, Raul P, Bhattacharyya P, Mandal NC, Bhattacharya S, Bhattacharya SS (2014) Vermicomposting of Tea Factory Coal Ash: metal accumulation and metallothionein response in *Eisenia fetida* (Savigny) and *Lampito mauritii* (Kinberg). Bioresour Technol 166:96–102
14. Atiyeh RM, Arancon N, Edwards CA, Metzger JD (2001) The influence of earthworm processed pig manure on the growth and productivity of marigolds. Bioresour Technol 81:103–108
15. Adhikary S (2012) Vermicompost, the story of organic gold: a review. Agric Sci 3:905–917
16. Bhat SA, Singh J, Vig AP (2017) Amelioration and degradation of pressmud and bagasse wastes using vermitechnology. Bioresour Technol 243:1097–1104
17. Lavelle P, Decaens T, Aubert M, Barot S, Blouin M, Bureau F (2006) Soil invertebrates and ecosystem services. Eur J Soil Biol 42:3–15
18. Suthar S (2006) Potential utilization of guar gum industrial waste in vermicompost production. Bioresour Technol 97(18):2474–2477
19. Elvira C, Sampedro L, Benitez E, Nogales R (1998) Vermicomposting of sludges from paper mill and dairy industries with *Eisenia ndrei*: a pilot scale study. Bioresour Technol 63:205–211
20. Garg VK, Kaushik P, Dilbaghi N (2006) Vermiconversion of wastewater sludge from textile mill mixed with anaerobically digested biogas plant slurry employing *Eisenia foetida*. Ecotoxicol Environ Saf 65:412–419
21. Ravindran B, Dinesh SL, John Kennedy L, Sekaran G (2008) Vermicomposting of solid waste generated from leather industries using epigeic earthworm *Eisenia foetida*. Appl Biochem Biotechnol 151:480–488
22. Yadav A, Garg VK (2019) Biotransformation of bakery industry sludge into valuable product using vermicomposting. Bioresour Technol 274:512–517
23. Suthar S (2010) Recycling of agro-industrial sludge through vermitechnology. Ecol Eng 36:1028–1036
24. Kaur S, Kour G, Singh J (2014) Vermicomposting of tea leaves waste mixed with cow dung with the help of exotic earthworm *Eisenia fetida*. Int J Adv Res Biol Sci 1(9):229–234
25. Singh J, Kaur A, Vig AP, Rup PJ (2009) Role of Eisenia fetida in rapid recycling of nutrients from biosludge of beverage industry. Ecotoxicol Environ Saf 73:430–435
26. Yuvaraj A, Karmegam N, Tripathi S, Kannan S, Thangaraj R (2020) Environment-friendly management of textile mill wastewater sludge using epigeic earthworms: bioaccumulation of heavy metals and metallothionein production. J Environ Manag 254:109813
27. Irvine GW, Summers KL, Stillman MJ (2013) Cysteine accessibility during $As3^{+}$ metalation of the α- and β-domains of recombinant human MT1a. Biochem Biophys Res Commun 433:477–483
28. Srivastava R, Kumar D, Gupta SK (2005) Bioremediation of municipal sludge by vermitechnology and toxicity assessment by *Allium cepa*. Bioresour Technol 96:1867–1871
29. Bhat SA, Singh J, Singh K, Vig AP (2017) Genotoxicity monitoring of industrial wastes using plant bioassays and management through vermitechnology: a review. Agric Nat. http://dx.doi.org/10.1016/j.anres.2017.11.002

30. Hait S, Tare V (2012) Transformation and availability of nutrients and heavy metals during integrated composting-vermicomposting of sewage sludges. Ecotoxicol Environ Saf 79:214–224
31. Bhat SA, Singh J, Vig AP (2017) Earthworms as organic waste managers and biofertilizer producers. Waste Biomass Valoriz. http://dx.doi.org/10.1007/s12649-017-9899-8
32. Saba S, Zara G, Bianco A, Garau M, Bononi M, Deroma M, Pais A, Budroni M (2019) Comparative analysis of vermicompost quality produced from brewers'spent grain and cow manure by the red earthworm *Eisenia fetida*. Bioresour Technol 293:122019
33. Benitez E, Sainz H, Nogales R (2005) Hydrolytic enzyme activities of extracted humic substances during the vermicomposting of a lignocellulosic olive waste. Bioresour Technol. 96:785–790
34. Mondal A, Goswami L, Hussain N, Barman S, Kalita E, Bhattacharyya P, Bhattacharya SS (2019) Detoxification and eco-friendly recycling of brick kiln coal ash using *Eisenia fetida*: a clean approach through vermitechnology. Chemosphere. S0045-6535(19)32710-9
35. Karmegam N, Vijayan P, Prakash M, John Paul JA (2019) Vermicomposting of paper industry sludge with cowdung and green manure plants using *Eisenia fetida*: a viable option for cleaner and enriched vermicompost production. J Clean Prod 228:718–728
36. Jóźwiak MA, Jóźwiak M, Kozłowski R, Żelezik M (2019) Zooremediation of leachates from municipal waste using *Eisenia fetida* (SAV). Environ Pollut 254:11287
37. Boruah T, Barman A, Kalita P, Lahkar J, Deka H (2019) Vermicomposting of citronella bagasse and paper mill sludge mixture employing *Eisenia fetida*. Bioresour Technol 294:122–147
38. Balachandar R, Baskaran L, Yuvaraj A, Thangaraj R, Subbaiya R, Ravindran B, Chang SW, Karmegam N (2019) Enriched pressmud vermicompost production with green manure plants using *Eudrilus eugeniae*. Bioresour Technol. S0960-8524(19)31808-5
39. Mahaly M, Senthilkumar AK, Arumugam S, Kaliyaperumal C, Karupannan N (2018) Vermicomposting of distillery sludge waste with tea leaf residues. Sustain Environ Res 2018(28):223–227
40. Krystyna M, Małgorzata G, Rafaela C, Agnieszka, R, Patryk W, Ślęzak E (2016) Biochar amendment for integrated composting and vermicomposting of sewage sludge—the effect of biochar on the activity of *Eisenia fetida* and the obtained vermicompost. S0960-8524(16)31564-4
41. Abbiramy KS, Ross PR, Paramanandham J (2015) Degradation of tea factory waste by mushroom cultivation and vermicomposting. J Environ Sci Eng 57(2):126–130
42. Singh S, Bhat SA, Singh J, Kaur R, Vig AP (2017) Earthworms converting milk processing industry sludge into biomanure. Open Waste Manag J 10:30–40
43. Hanc A, Chadimova Z (2014) Nutrient recovery from apple pomace waste by vermicomposting technology. Bioresour Technol. https://doi.org/10.1016/j.biortech.2014.02.031
44. Singh J, Kaur A (2013) Vermidegradation for faster remediation of chemical sludge and spent carbon generated by soft drink industries. J Environ Sci Sustain 1:13–20
45. Singh D, Suthar S (2012) Vermicomposting of herbal pharmaceutical industry waste: earthworm growth, plant-available nutrient and microbial quality of end materials. Bioresour Technol 112:179–185
46. Hemalatha B (2012) Recycling of industrial sludge along with municipal solid waste—vermicomposting method. Int J Adv Eng Technol 3:71–74
47. Singh RP, Embrandiri A, Ibrahim MH, Esa N (2011) Management of biomass residues generated from palm oil mill: vermicomposting a sustainable option. Resour Conserv Recycl 55:423–434
48. Deka H, Deka S, Baruah CK, Das J, Hoque S, Sarma H, Sarma NS (2011) Vermicomposting potentiality of *Perionyx excavatus* for recycling of waste biomass of java citronella—an aromatic oil yielding plant. Bioresour Technol 102:11212–11217
49. Singh J, Kaur A, Vig AP, Rup PJ (2010) Role of *Eisenia fetida* in rapid recycling of nutrients from bio sludge of beverage industry. Ecotoxicol Environ Saf 73:430–435
50. Jones CG, Lawton JH, Shachak M (1994) Organisms as ecosystem engineers. Oikos 69:373–386

51. Jouquet P, Dauber J, Lagerlof J, Lavelle P, Lepage M (2006) Soil invertebrates as ecosystem engineers: intended and accidental affects on soil and feedbackloops. Appl Soil Ecol 32:153–164
52. Lavelle P, Spain AV (2001) Soil ecology. Kluwer Academic Publishers, Dordrecht, p 654
53. Suthar S (2012) Vermistabilization of wastewater sludge from milk processing industry. Ecol Eng 47:115–119
54. Briones MJL, Barreal ME, Harrison AC, Gallego PP (2011) Earthworms and nitrogen application to improve soil health in an intensively cultivated kiwifruit orchard. Appl Soil Ecol 49:158–166
55. Jing YP, Liu MQ, Yin QP, Li HX, Hu F (2013) Effects of earthworms and ryegrass on the removal of fluoranthene from soil. Pedosphere 23:523–531
56. Coutino-González E, Hernández-Carlos B, Gutiérrez-Ortiz R, Dendooven L (2010) The earthworm *Eisenia fetida* accelerates the removal of anthracene and 9,10-anthraquinone, the most abundant degradation product, in soil. Int Biodeterior Biodegrad 64:525–529
57. Suthar S, Sajwan P, Kumar K (2014) Vermiremediation of heavy metals in wastewater sludge from paper and pulp industry using earthworm *Eisenia fetida*. Ecotoxicol Environ Saf 109:177–184
58. Rodriguez-Campos J, Dendooven L, Alvarez-Bernal D, Contreras-Ramos SM (2014) Potential of earthworms to accelerate removal of organic contaminants from soil: a review. Appl Soil Ecol 79:10–25
59. Suthar S, Singh S, Dhawan S (2008) Earthworm as bioindicators of metals (Zn, Fe, Mn, Cu, Pb and Cd) in soils: is metal bioaccumulation affected by their ecological categories. Ecol Eng 32:99–107
60. Tersic T, Gosar M (2012) Comparison of elemental contents in earthworm cast and soil from a mercury contaminated site (Idrija area, Slovenia). Sci Total Environ 430:28–33
61. Norgrove L, Hauser S (2000) Production and nutrient content of earthworm casts in a tropical agrisilvicultural system. Soil Biol Biochem 32:1651–1660
62. Atiyeh RM, Lee S, Edwards CA, Arancon NQ, Metzger JD (2002) The influence of humic acids derived from earthworms-processed organic wastes on plant growth. Bioresour Technol 84:7–14
63. Groenigen JWV, Lubbers IM, Vos HMJ, Brown GG, Deyn GBD, Groenigen KJV (2014) Earthworms increase plant production: a meta-analysis. Sci Rep 4:63–65. https://doi.org/10.1038/srep06365

Ageing-Friendly Neighbourhoods: A Study of Mobility and Out-of-Home Activity

Parshant Rehal, Prabhjot Singh Chani, Sonal Atreya, and Vivek Sehgal

1 Introduction

In a fast urbanising world, rapid population ageing country is becoming a present-day situation in the twenty-first century. With most of the developed countries facing this situation, 24% of a global ageing population is estimated to be in Asia of which the majority would be in India by 2050 [14]. The percentage of elderly population share is expected to be 12.6% by 2026. India is expected to be home to around 324 million elderly population constituting to be around 1/5th of its total population by 2050. Changing social structure and family bondages in the wake of the younger working population migrating to urban areas, elderly population often experiences social isolation and has to arrange for their material requirements of daily use on their own. Facing challenges related to mobility and independent functioning within the immediate housing and neighbourhood environments, the ageing population often struggles to maintain a healthy, independent lifestyle. Population among 60+ and above age groups has been facing acute challenges to liveability and independence in rapidly urbanising and globalising urban built and social environments.

P. Rehal (✉) · P. S. Chani · S. Atreya
Department of Architecture and Planning, Indian Institute of Technology Roorkee, Roorkee, Uttarakhand 247667, India
e-mail: prehal@ar.iitr.ac.in

P. S. Chani
e-mail: p.chani@ar.iitr.ac.in

S. Atreya
e-mail: sonal.atreya@ar.iitr.ac.in

V. Sehgal
GNDEC School of Architecture, Ludhiana, Punjab, India

H. Singh et al. (eds.), *Sustainable Development Through Engineering Innovations*, Lecture Notes in Civil Engineering 113,
https://doi.org/10.1007/978-981-15-9554-7_24

World Health Organization (WHO) envisioned the concept of 'age-friendly' or 'elder-friendly' cities with approaches of universal design, accessibility and sustainability to improve the quality of life and well-being of a larger population. India adopted the National Policy of Older Persons (NPOP) in 1999, followed by National Policy for Senior Citizens in 2011 for the well-being and welfare of elderly citizens.

Most of the research associated with ageing has been taken up in the fields of gerontology and geriatrics. A few other researches have been done in relation to demographic changes for economic and health factors. However, there are very few studies done in India on ageing concerning the built environment. It becomes imperative to study ageing and their built environment with a focus on active ageing and independent functioning. Mobility becomes the principal factor that can encourage successful and healthy ageing. Pedestrian-friendly environments improve overall social participation, social inclusiveness, thereby paving the way for independent living in performing activities of daily life (ADL). The research studies various factors which affect the independent functioning of ageing populations within out-of-home environments. The paper would be helpful to understand various challenges to the active ageing concerning mobility and would establish the need for age-friendly neighbourhoods.

2 Methodology

The two concepts, liveability and sustainability, are used interchangeably in the study. A literature review is undertaken to bring out various factors which are essential for the well-being, liveability and quality of life of the ageing population in context to the out-of-home environments. Past studies and reports are studied to establish the relation of these factors to the liveability of ageing persons. Mobility and availability of open spaces are seen as a significant factor which influences the social participation, inclusiveness and independent performing of ADL in out-of-home environment. A 'comparative analysis' is conducted between different neighbourhoods for identifying challenges to streamlined pedestrian mobility among the ageing population in out-of-home neighbourhood environments. The availability of open spaces is also analysed in the perspective of how pedestrian environments can be made more accessible, pleasant and comfortable for the ageing people.

3 Literature Review

The literature review is conducted to understand the various concepts which are associated with elderly specific liveability, their health and well-being. Factors are identified, which influenced the out-of-home activities of aged persons within their neighbourhoods. Quality of life of ageing persons is often described as a broad concept which includes income, physical health, psychological state, level of independence, social relationships, personal beliefs and relationship to salient features in the environment [7]. As people age, their quality of life is mostly determined by their ability to maintain autonomy and independent functioning. Liveability of the neighbourhood is often defined by the spatial relationship between their dwelling and amenities they (residents) wanted to live near [1]. Elderly specific liveability within neighbourhoods has been researched in relation to the available amenities in their neighbourhoods [16, 22]. Various models have been developed in different studies over the past to define satisfaction of elderly populations.

The term active ageing is adopted by World Health Organization (WHO) which defines it as 'the process of optimising opportunities for health, participation and security in order to enhance the quality of life as people age' [19]. Active ageing promotes respect for the ageing population's needs and lifestyle, as well as promoting inclusion of the ageing population [18–20]. Neighbourhood liveability for older people is often described in terms of higher densities which encourage ease of accessibility and proximity to different amenities in out-of-home environment [16]. Growing age is often associated with reduced activity in out-of-home environments and thereby reduces the life satisfaction among these populations [5]. Mobility and accessibility in out-of-home environments have been stressed upon by various researchers as an essential factor for the liveability of the ageing population [11, 12, 16, 17]. Mobility of urban areas needs to be planned more sustainably encouraging pedestrianisation and encouraging inclusiveness. Neighbourhood liveability specific for ageing people can also be determined through independent functioning while performing activities of daily life in terms of ease of accessibility and proximity to different amenities. However, different factors such as topography, terrain, ability to walk, presence of resting spaces and modes of transport can alleviate or hinder the accessibility of ageing residents within or outside the neighbourhoods [16]. Parks, open spaces and other microelements around the neighbourhood improve the social participation and interactions among the elderly [4].

There is a lack of understanding about how green neighbourhood spaces can actually improve physical activity among elderly population and also improve health and well-being. Studies suggest that the neighbourhood with good amenities and better walkable environments exhibit better social cohesion in comparison with neighbourhoods which are less walkability and mobility friendly [15]. Presence of open and green spaces in proximal spaces within the neighbourhood spaces helps in maintaining an active lifestyle and is considered to positively influence the health of the ageing population [2, 21, 23]. High-density neighbourhoods and residencies in the city are believed to give higher satisfaction to ageing residents in terms of all amenities within reach [13]. Proximity to the basic services and amenities including daily needs stores, pharmacy shops and even recreational spaces within neighbourhoods encourages ageing persons to walk. It benefits the health of ageing persons by helping them to maintain a moderate level of physical activity [6]. Seating spaces in public open spaces tend to improve mobility among aged persons and also improve the social interaction and walkability experiences [8]. There is a need to identify how green spaces and environment can be used to keeping ageing population healthier [3].

4 Mobility Issues for Ageing Persons in Out-of-Home Environments

Encouraging sustainable mobility, accessibility and pedestrianisation in and around the neighbourhoods becomes key for active ageing. Age-friendly neighbourhoods should provide for various amenities like shops, health clinics, chemists, restaurants and other social amenities in walkable limits of the ageing persons keeping in mind their abilities and constraints. Accessibility to the transit points in and around the neighbourhoods should be a pleasantly walkable zone. Higher traffic speeds often pose a threat to the elderly persons while walking or crossing the streets and roads. These age-friendly neighbourhoods should also provide green open and public spaces to provide enough opportunity for people to maintain their social connections. With a view of identifying various challenges to the independent functioning with respect to mobility and accessibility among ageing populations, various neighbourhoods of Delhi and Chandigarh are selected in urban areas.

Chandigarh has an elderly population of 0.7 lakhs in 2011. Chandigarh is among the few planned cities in India and has a walkability score of 0.9. The site study is conducted at various locations to understand the ease of walkability among ageing persons within the neighbourhoods around the transit points. Open spaces within the neighbourhoods are also observed to identify the challenges for walkability and accessibility issues.

5 Case-Based Findings

A number of challenges are identified for the ageing population to their free pedestrian movement within the neighbourhoods and around different transit points. Chandigarh performed reasonably well on providing the ageing residents with comfortable, safe out-of-home environments. A few challenges are also observed in the case of Chandigarh. In the case of Delhi, there is variation in the pedestrian condition quality along the selected paths connecting to different neighbourhoods.

5.1 Challenges for the Ageing Population Around Different Transit Points

Connaught place, Sansad Marg area and Bhikaji Cama place area have been selected in Delhi to identify the challenges to a safe, comfortable and pleasant walk among ageing persons.

Absence of pedestrian paths, sidewalks and unequal width of pedestrian paths: The pedestrian paths are non-continuous at various points along the Sansad Marg area (Fig. 1). At a few places, the pedestrian paths are of inadequate width and are insufficient for two persons to walk along. Walking with the help of walking aids or wheelchair becomes a difficult task at such places. Connaught place area has adequate width of walkways along the central park. Challenges faced by pedestrians at Sansad Marg area, Delhi, can be seen in Fig. 1.

Irregular height and absence of kerb ramp along pavements: The pedestrian paths being too high and without a comfortable kerb ramp to walk from the road to

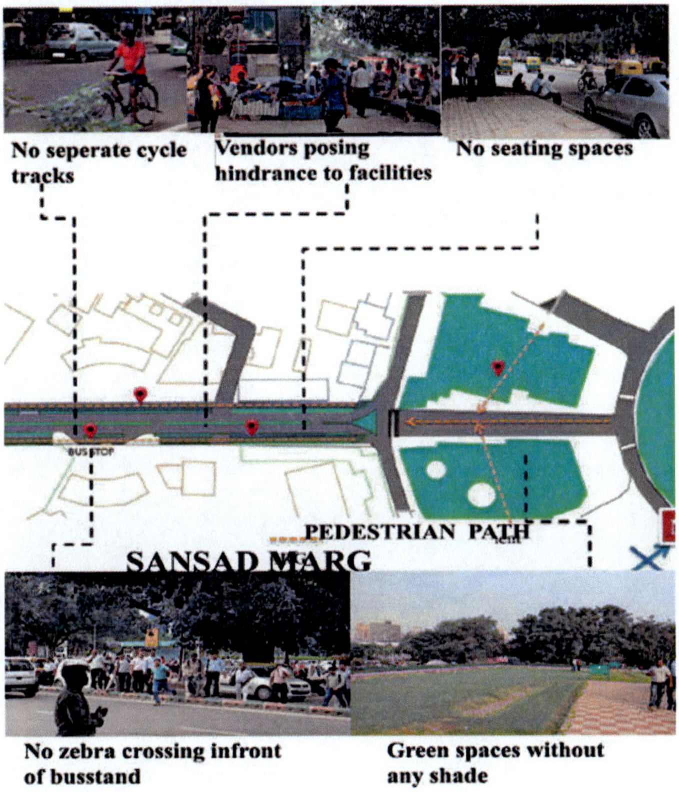

Fig. 1 Roadside and crossing at Sansad Marg, Delhi

footpath/sidewalks are often obstructive for aged persons forcing most of them to walk on the road only, thereby making them prone to accidents. Kerb cut has been constructed along the walkways around the central park in Connaught place but is found missing at a few points leading to other road junctions. The absence of Kerb cut at different places along the pedestrian paths in Bhikaji Cama place made it difficult for the elderly to climb up a relatively high pedestrian path. Universal accessibility measures have been implemented only at few selected places.

Crossing Junction/Road crossing points: Along Sansad Marg, absence of zebra crossing ahead of the bus stops forces people to cross the road riskfully (Fig. 1). It was observed that a few vehicles did not stop well behind the zebra crossing. Traffic light was also observed to be malfunctioning through it may not be a regular event (Fig. 2). Traffic speed calming measures like speed tables are absent ahead of the

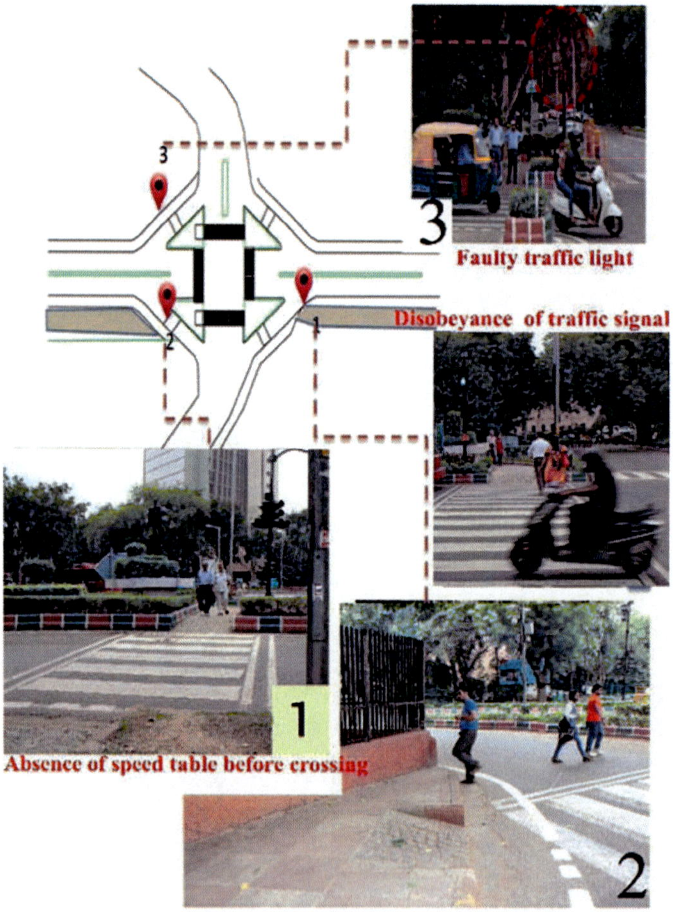

Fig. 2 Crossing junction at Sansad Marg area

zebra crossings not giving enough time for elderly to cross the roads. Speed tables are present in Connaught place area to calm down the traffic speed at most places (Fig. 3).

Encroachment of pedestrian walks, parked cars on sidewalks: At many places, the footpaths are either encroached by vendors (Fig. 1). The pedestrian paths are either too dirty to walk on or become a dumping place for garbage and are often occupied by parked vehicles (Fig. 4).

Insufficient amenities: The sidewalks are not provided with shade or plantation, making it challenging to walk in the extremely hot summers (Fig. 3). There are no intermediate resting spaces for elderly and children to sit at comfortable walking distances along the Sansad Marg (Fig. 1) and Bhikaji Cama place. No dustbins are provided at most of the places, and there are no urinals at adequate distances for the people (Fig. 5). Connaught place area is observed to provide maximum amenities

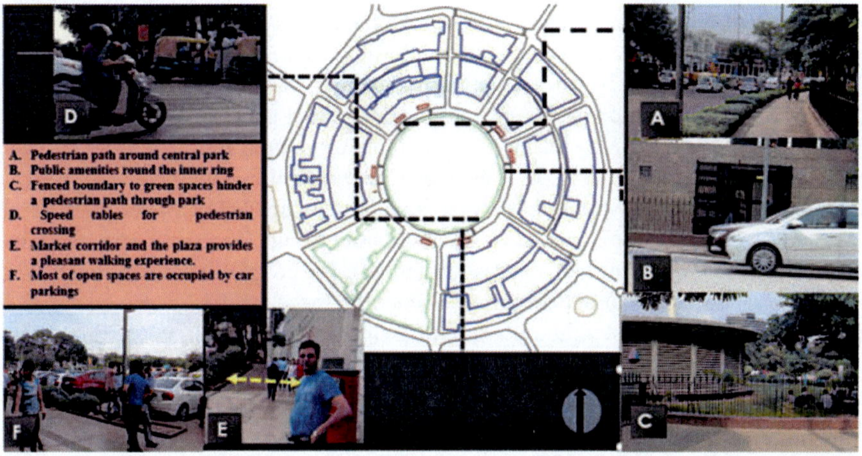

Fig. 3 Connaught place area, Delhi

Fig. 4 Inadequate width of sidewalks

Fig. 5 Absence of basic amenities along pedestrian poor upkeep and maintenance paths

to the pedestrian elderlies. A few drivers have been seen to stop the buses on roads often making it uncomfortable to walk through the busy road.

Lack of shades: Natural shade is present because of trees planted along the route on Sansad Marg and Bhikaji Cama place area. The Connaught place area did not have the thoughout shaded pavements around the central park.

Lack of roadside open spaces or roadside public spaces or well-connected public spaces: Ample green space has been provided along the pedestrian paths in Connaught place and Sansad Marg area. The open spaces or parks have been restricted in Connaught place area by fencing with railings (Fig. 3). These pleasant quality roadsides or street-side spaces are inviting for people to walk through and come again and again. Inevitably, the absence of these public spaces at various place spaces in various neighbourhoods in Delhi tends to play a negative role in the implementation of age-friendly environments measures. The vacant lands are often turned into garbage dumps, making it unpleasant to pass by. Also, the open spaces in different developments are not walkable through and thus elongate the walking distances, which becomes all the more difficult without the public realm. Even the open spaces within different neighbourhoods are not interconnected in Delhi to provide pedestrian access and also have localised barriers like boundary walls and gated entries.

Neighbourhoods within Chandigarh area lot more pedestrian-friendly for ageing persons when compared to Delhi, yet there are certain issues of poor maintenance, dumping of garbage along the pedestrian paths, localised hindrances, absence of zebra crossings on roads linking parks and schools and other facilities (Fig. 6). But pedestrian pathways inside the sectors have often been occupied as personal green space or parking space and are discontinuous at few points. A few electrical installations are risky to streamlined walkability of ageing persons. A few pedestrian paths are poorly lit and rendered them unsafe for ageing persons (Fig. 6). Along the paths, a few open wind storm drains posed both a risk and challenge to walkability (Fig. 6). Though the green spaces and parks are present, they are often not connected to each other and bounded by walls. Staircases at the park entries make most of the open spaces universally inaccessible for the aged persons (Fig. 6). Variation in

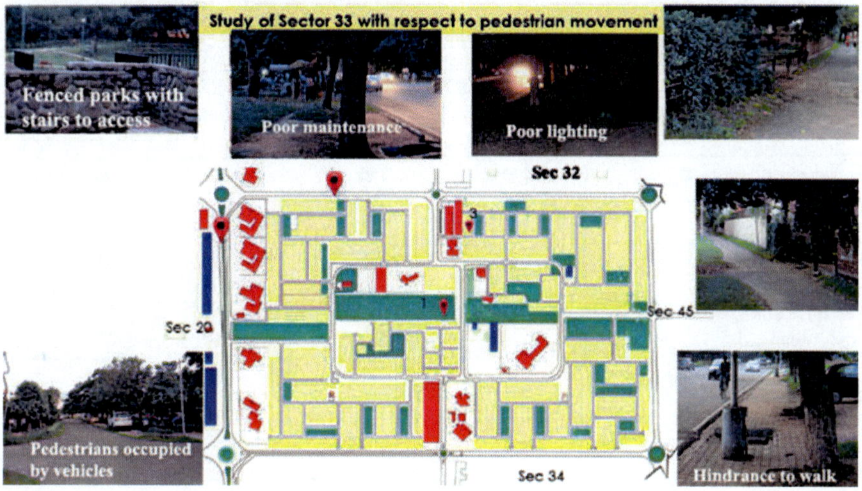

Fig. 6 Analysis of pedestrian networks in Sector 33, 34

maintenance levels of pedestrian quality is also visible in different sectoral V4 and V5 roads. Sector 17 being the heart of the city provided better amenities for pedestrians. Kerb cuts are not provided along the pedestrians and sidewalks. The inner pedestrian walks are usually occupied by vehicles in Sector 33 and 34.

Unfinished pedestrian surfaces have been observed in few places. The subway entry (towards 17 bus stand) is to be accessed by stairs and becomes inaccessible for ageing persons using walking aid or wheelchairs. Seating spaces are not provided anywhere in front of the subway entry (Fig. 7). The implementation of measures for universal accessibility for elderly seems to be missing at many places.

Table 1 shows a comparative analysis based on observation during site visits to Delhi and Chandigarh. Sectors 33 and 17 in Chandigarh have good infrastructure for the elderly, but maintenance is the key to improve the infrastructure. Connaught place area exhibits good pedestrian-friendliness, but provision of natural shade along pedestrian paths around the central park needs to be encouraged. Sansad Marg

Fig. 7 Local obstructions, maintenance and walking surface, crowded entry to subway, no seating provision

Table 1 Site observations in New Delhi and Chandigarh

Age-friendly features	New Delhi			Chandigarh	
	Connaught place (around central park)	Sansad Marg area	Bhikaji Cama place area	Sector 33	Sector 17
Presence of walkways and sidewalks	●	●	●	●	●
Pathways are well-shaded, by tree lines and other means	o	●	●	●	●
Pedestrian paths are continuous	●	o	o	o	●
Pathways are not congested	●	●	o	●	●
Over crowdedness at pathways	●	o	●	o	o
Height issues of pedestrian paths	o	●	●	●	o
Irregular width of pedestrian paths	o	●	●	o	o
Encroachment of walkways (by vehicle parking or construction materials)	o	o	●	●	o
Good upkeep and maintenance	●	●	o	o	●
Neatness on pedestrian paths	●	●	o	●	●
Presence of speed tables ahead of zebra crossings	●	o	o	o	●
zebra crossings and road crossings	●	●	o	o	●
Presence of kerb ramps along pedestrian paths	●	●	o	o	o

(continued)

Table 1 (continued)

Age-friendly features	New Delhi			Chandigarh	
	Connaught place (around central park)	Sansad Marg area	Bhikaji Cama place area	Sector 33	Sector 17
Presence of amenities along pedestrian paths (washrooms, seating spaces, etc.).	●	●	o	o	o
Faulty traffic light	o	●	o	o	o
Disobeyance of traffic signals	o	●	o	o	o
Restricted and fenced open spaces	●	●	●	●	●
Interlinked continuous green spaces for pedestrian movement	o	o	o	**o**	o

Note o means not available, ● means available, and **o** means partly available

exhibits good pedestrian user-friendly infrastructure, but minimising the height variation and provision of speed table ahead of zebra crossings are required along its full distance along with the provision of speed tables to reduce vehicular speeds along with the pedestrian crossings. Connaught place exhibits a continuous pedestrian network, while in Sansad Marg the pedestrian's paths reduce in width and even ends (Fig. 2) at a few places exhibiting discontinuity due to a boundary wall corner. In Sector 33, the continuity is broken by unfinished pedestrian path surfaces.

5.2 *Open Spaces for Age-Friendly Neighbourhoods*

Exciting and inviting public spaces, well-shaded sidewalks along with pleasant roadside spaces can encourage not only aged but all persons to walk rather than using personal vehicles. The adjacent open spaces also make the overall environment friendly for aged pedestrian users and bicyclists. Well-connected green open spaces provide safe and comfortable accessibility from home to transit stations, especially for elders and children who prefer resting spaces during a walk. Street-side spaces can also be used to generate some local activities which, on the one hand, provide facilities to the people and also adding to the safety of people. Besides the travel aspect, such interconnected open spaces also provide a crucial opportunity for people to socialise and recreation. Building adjacent softscapes and hardscapes/linear street-side spaces, open spaces with landscape features and benches also provide the pedestrian better connectivity and rest places while transits. Park, parklets in front of buildings adjacent to streets nurture a feeling of healthy living. Cycle tracks, waterfronts, jogger tracks, parks and green corridors are planned in most transit and age-friendly cities. Walkable interlinked open spaces in neighbourhoods thus reduce the overall ridership distances and cost for the individuals by encouraging people to walk. The examples from Curitiba, Portland and San Juan Puerto Rico Edmonton exhibit plenty of open public spaces which provide a pleasant experience of walking and cycling for the last mile connectivity and provide the socialising and recreational opportunities besides making the place a haven for ageing persons.

Public spaces should count around 45–50% of a city's land area, with 30–35% street area with 15–20% open space (UN-Habitat's Global Urban Observatories Unit) [9]. Open areas thus also include streets, sidewalks and also roadside and street-side spaces along with parks, parking lots and transit station adjacent areas. '*If the point of transportation planning is to get people places they want to be, then most transportation planning should start with place making*' [10]. Thus, these open spaces play an important role in the implementation of a smart, sustainable mobility plan for a city. An age-friendly neighbourhood essentially needs to look and provide for these open spaces for the users to encourage people to walk thus creating a sustainable built environment.

6 Conclusion

Sustainable urban development stresses upon walkability and creation of green open spaces within neighbourhood spaces which fall in line with the ideas of age-friendly neighbourhoods. Mobility becomes the critical factor which affects the quality of life of ageing persons in out-of-home environments. Safe environments in and around the neighbourhoods are essential for outdoor activity among ageing populations. The open spaces play an important role in the implementation of age-friendly neighbourhoods and add to the sustainability of cities. Interlinked green open spaces can provide a shorter, comfortable route for ageing residents of the neighbourhoods. Placemaking improves the overall imageability of the areas and streets are more inviting for all age group persons, especially the elderly ones. The ageing populations should be encouraged to walk and bicycle in order to maintain a healthy lifestyle.

Placemaking activities are essential to make the spaces in and around the neighbourhood and transit spaces peripheral areas more attractive and usable. Ageing persons can maintain a healthy lifestyle while performing their ADLs in out-of-home environments when provided with safe, comfortable and ambient neighbourhood environment. All these measures encourage ageing-in-place and lead to successful and healthy ageing.

Acknowledgements The authors would like to acknowledge the support of the Indian Institute of Technology (IITR) Roorkee, India, and Ministry of Human Resource Development (MHRD) for conducting this study.

References

1. Allen N, Haarhoff E, Beattie L (2018) Enhancing liveability through urban intensification: the idea and role of neighbourhood. Cogent Soc Sci 4(1). https://doi.org/10.1080/23311886.2018.1442117
2. Bergefurt L, Kemperman A, van den Berg P, Borgers A, van der Waerden P, Oosterhuis G, Hommel M (2019) Loneliness and life satisfaction explained by public-space use and mobility patterns. Int J Environ Res Public Health 16(21). https://doi.org/10.3390/ijerph16214282
3. Dalton AM, Wareham N, Griffin S, Jones AP (2016) Neighbourhood greenspace is associated with a slower decline in physical activity in older adults: A prospective cohort study. SSM Popul Health 2:683–691. https://doi.org/10.1016/j.ssmph.2016.09.006
4. de Vries S, van Dillen SME, Groenewegen PP, Spreeuwenberg P (2013) Streetscape greenery and health: stress, social cohesion and physical activity as mediators. Soc Sci Med 94:26–33. https://doi.org/10.1016/j.socscimed.2013.06.030
5. Kim S (2003) Analysis of elderly mobility by structural equation modeling. Transp Res Rec: J Transp Res Board 1854(1):81–89. https://doi.org/10.3141/1854-09
6. Levasseur M, Généreux M, Bruneau JF, Vanasse A, Chabot É, Beaulac C, Bédard MM (2015) Importance of proximity to resources, social support, transportation and neighborhood security for mobility and social participation in older adults: Results from a scoping study. BMC Public Health 15(1):1–19. https://doi.org/10.1186/s12889-015-1824-0
7. Netuveli G, Blane D (2008) Quality of life in older ages. Br Med Bull 85(1):113–126. https://doi.org/10.1093/bmb/ldn003
8. Ottoni CA, Sims-Gould J, Winters M, Heijnen M, McKay HA (2016) "Benches become like porches": built and social environment influences on older adults' experiences of mobility and well-being. Soc Sci Med. https://doi.org/10.1016/j.socscimed.2016.08.044
9. PPS.org (n.d.) From place to place: reinventing transportation planning with placemaking. Retrieved from: https://www.pps.org/article/from-place-to-place-reinventing-transportation-planning-with-placemaking. 16 June 2019
10. Scruggs G (n.d.) How much public space does a city need?—next city. Retrieved from: https://nextcity.org/daily/entry/how-much-public-space-does-a-city-need-UN-Habitat-joan-clos-50-percent. 20 June 2020
11. Shamsuddin S, Rasyiqah N, Hassan A, Fatimah S, Bilyamin I (2012) Walkable environment in increasing the liveability of a city. Procedia Soc Behav Sci 50:167–178. https://doi.org/10.1016/j.sbspro.2012.08.025
12. Shoval N, Wah H-W, Auslander G, Isaacson M, Oswald F, Edry T, Landau R, Heinik J (2018) Use of the global positioning system to measure the out-of-home mobility of older adults with differing cognitive functioning. https://doi.org/10.1017/S0144686X10001455
13. Temelová J, Dvořáková N (2012) Residential satisfaction of elderly in the city centre: the case of revitalizing neighbourhoods in Prague. Cities 29(5):310–317. https://doi.org/10.1016/j.cities.2011.11.015
14. UNFPA (2017) Caring for our elders: early responses caring for our elders: early responses. In: United Nations Population Fund 2017. 'Caring for Our Elders: Early Responses'—India Ageing Report—2017. UNFPA, New Delhi, India. https://doi.org/10.1002/adfm.201100401
15. van den Berg P, Sharmeen F, Weijs-Perrée M (2017) On the subjective quality of social interactions: influence of neighborhood walkability, social cohesion and mobility choices. Transp Res Part A: Policy Pract 106:309–319. https://doi.org/10.1016/j.tra.2017.09.021

16. Vine D, Buys L, Aird R (2012) The use of amenities in high density neighbourhoods by older urban Australian residents. Landsc Urban Plan 107(2):159–171. https://doi.org/10.1016/J.LANDURBPLAN.2012.05.013
17. Walford N, Samarasundera E, Phillips J, Hockey A, Foreman N (2011) Older people's navigation of urban areas as pedestrians: measuring quality of the built environment using oral narratives and virtual routes. Landsc Urban Plan 100(1–2):163–168. https://doi.org/10.1016/J.LANDURBPLAN.2010.12.006
18. WHO (2002) Active ageing: a policy framework, vol 11. Retrieved from: http://www.who.int/hpr/. 23 Nov 2018
19. WHO (2007) Global age-friendly cities: A guide. Retrieved from: https://www.who.int/ageing/publications/active_ageing/en/. 12 Dec 2018
20. WHO (2015) World report on ageing and health. Retrieved from: www.who.int. 23 Nov 2018
21. Xue F, Gou Z, Lau SSY (2017) Green open space in high-dense Asian cities: site configurations, microclimates and users' perceptions. Sustain Cities Soc 34(February):114–125. https://doi.org/10.1016/j.scs.2017.06.014
22. Yan B, Gao X, Lyon M (2014). Modeling satisfaction amongst the elderly in different Chinese urban neighborhoods. Soc Sci Med 118(C):127–134. https://doi.org/10.1016/j.socscimed.2014.08.004
23. Yung EHK, Conejos S, Chan EHW (2016) Social needs of the elderly and active aging in public open spaces in urban renewal. Cities. https://doi.org/10.1016/j.cities.2015.11.022

Study of Lateral Capacity of a Single Pile in Clay Overlying Sand

Amanpreet Kaur, Harvinder Singh, and J. N. Jha

1 Introduction

Pile foundations are deep foundations which are used to support civil engineering structures when it is not feasible to provide foundations at shallow depths due to the presence of weak soil. These foundations pass through different soil layers with varying strength properties affecting their performance. Use of pile foundations is common in structures such as dams and earth retaining structures, high rise buildings, transmission towers, and bridges. In these structures, foundations not only support vertical load but are also subjected to lateral loads being caused by various agents like seismic forces, wave action, earth pressure, wind action and traffic forces. Design of laterally loaded pile foundations is not only based on ultimate lateral load, but on lateral deflection which is an important criterion from the serviceability point of view.

Considering sustainability assessment of a geotechnical project, its cost and environmental effects are two most important factors. Concrete piles are usually preferred to steel piles as former are less expensive. However, as cement manufacturing industry is one of the major contributors to global warming, use of steel piles may prove to be an environmental friendly substitute for concrete piles. In pile foundations, embedment depth is a major factor which directly affects the cost of the project. So, to

A. Kaur (✉)
I.K. Gujral Punjab Technical University, Jalandhar, Punjab, India
e-mail: kaurpreetaman79@gmail.com

H. Singh
Department of Civil Engineering, Guru Nanak Dev Engineering College, Ludhiana, Punjab, India
e-mail: hvs1221@gmail.com

J. N. Jha
Muzaffarpur Institute of Technology, Muzaffarpur, Bihar, India
e-mail: jagadanand@gmail.com

H. Singh et al. (eds.), *Sustainable Development Through Engineering Innovations*, Lecture Notes in Civil Engineering 113,
https://doi.org/10.1007/978-981-15-9554-7_25

have an idea of an optimum embedment length of the pile which satisfies all aspects of strength, serviceability as well as economic requirements is much needed. In this paper, an attempt has been made to relate lateral capacity and lateral deflection to the optimum embedment length of a steel pile beyond which the effect of increase in length becomes insignificant. Taking this optimum embedment length into consideration in the design can help to reduce the cost.

Many experimental and numerical approaches have been reported to study the behaviour of laterally loaded piles in homogenous as well as layered soils [3–8, 10–12]. Ai et al. [2] suggested that pile behaviour is affected by both transverse isotropy and stratified character of soil and the pile's size and physical properties. Murugan et al. [9] reported that lateral load capacity increases with increase in diameter of the pile for the same length. Abdrabbo and Gaaver [1] stated that the effective depth of a flexible laterally loaded pile embedded in cohesionless soil is about 16 times the pile diameter. Erdal and Laman [5] reported that that lateral load capacity of piles in layered sand conditions decreases nonlinearly as the thickness of the upper loose sand layer increases. However, experimental research data incorporating relationship between lateral capacity and embedment length of laterally loaded piles in layered soil containing clay and sand layers is rather scanty.

This paper presents the results obtained from an experimental model study performed on a laterally loaded free-head single model pile embedded in two-layered soil containing a clay layer overlying a sand layer. The parameters varied in this study were embedment length of pile and thickness of clay layer. Effect of variation in these parameters on lateral response of pile is studied from the obtained lateral load–displacement curves and presented here in form of charts relating lateral capacity to two parameters, viz. ratio of pile length to pile diameter and ratio of clay layer thickness to pile length. From the charts, an optimum embedment length corresponding to maximum improvement in lateral strength is obtained.

2 Test Bed Preparation

Model tests were performed in a testing tank of dimensions 2 m × 2 m × 1 m. Selected tank size was large enough to eliminate the boundary effects. Two-layered test bed was prepared using inorganic clay of high plasticity (CH) at liquid limit in the upper layer and poorly graded dry sand (SP) of relative density 80% in the lower layer. Properties of sand and clay used in experiments are given in Tables 1 and 2, respectively.

The sand layer was prepared using rainfall technique in which dry sand was dropped from a predetermined height using a moveable spreader to achieve the required density corresponding to 80% relative density. To prepare clay for upper layer of test bed, predetermined quantity of water, i.e. 60%, by weight was added to the required quantity of air-dried clay which was then filled in the tank above the sand layer. Prepared soil bed was covered with moistened jute bags to minimize water loss in order to maintain the water content equal to liquid limit throughout the

Table 1 Properties of sand

S. No.	Property	Unit	Value
1	Maximum unit weight	kN/m^3	18.30
2	Minimum unit weight	kN/m^3	15.68
3	Unit weight used in test bed	kN/m^3	17.70
4	Specific gravity	–	2.64
5	Coefficient of uniformity, C_u	–	2.45
6	Coefficient of curvature, C_c	–	0.98
7	Cohesion, c_u	kN/m^2	0
8	Angle of internal friction, φ	°	34.6

Table 2 Properties of clay

S. No.	Property	Unit	Value
1	Wet unit weight	kN/m^3	18.74
2	Maximum dry unit weight	kN/m^3	15.40
3	Optimum moisture content	%	21.5
4	Specific gravity	–	2.70
5	Liquid limit	%	60
6	Plastic limit	%	30
7	Grain size distribution		
	Sand size	%	6
	Silt size	%	51
	Clay size	%	43

testing procedure. Model piles were pushed into the prepared soil bed using a guide frame to maintain vertical alignment of the piles.

3 Model Pile and Test Instrumentation

Stainless steel pipes of 12.7 mm outer diameter (D) and 0.4 mm thickness were used as model piles. Embedment length (L) of the pile was varied. Modulus of elasticity of the model piles was 1.5×10^5 N/mm^2 which was determined in a three point bending test. Lateral load was applied at head of the piles using static weights with the help of a pulley and string arrangement. Foil-type strain gauges of resistance 120 Ω, gauge factor 2.1 and 5 mm length were fixed at three different levels along the length of these piles. The readings of the strain gauges were used to measure strain from which bending moment produced in the piles was calculated. Linear variable differential transducers were used to measure lateral displacement of the pile head. Readings of strain gauges and LVDT were recorded in a computer system using an electronic data acquisition system. The schematic diagram of model test setup is shown in Fig. 1.

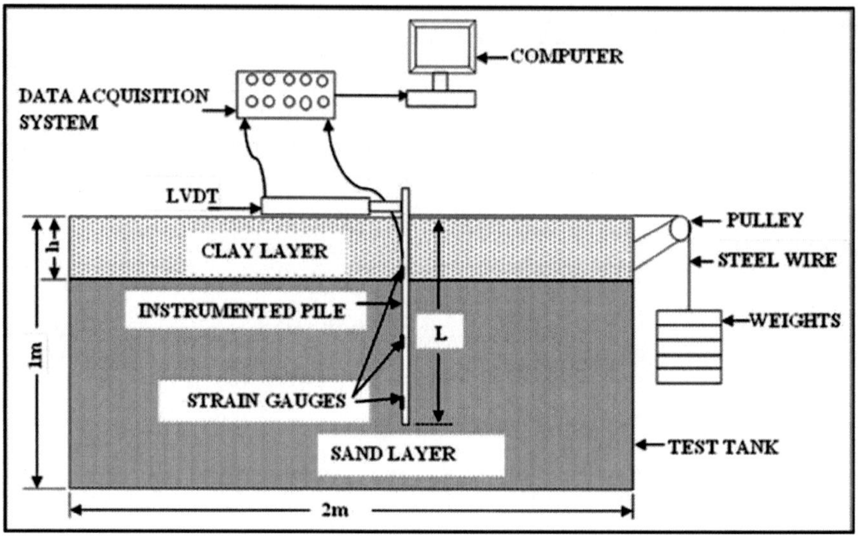

Fig. 1 Schematic diagram of test setup

4 Test Series Details

Results of total 12 tests performed on laterally loaded free-head single pile embedded in layered soil are being presented in this paper. Parameters varied in the test programme include embedment length of the pile and thickness of clay layer (*h*). Three different values of thickness of clay layer, viz. 150, 200 and 250 mm, were used in the experimental test programme. Test programme is classified into three series designated as CS1, CS2 and CS3 with different values of thickness of clay layer. Details of the parameters used in the test programme are given in Table 3. Effects of the two variable parameters on lateral performance of pile are being presented in this paper by comparing lateral load capacity variation with respect to variation in *L*/*D* ratio and *h*/*L* ratio.

5 Results and Discussion

From the observed results of model testing, comparison is made for lateral capacity variation for various values of *L*/*D* and *h*/*L*. The observed trends are presented in the form of charts in following sections.

Table 3 Parameters varied in test series

Test series designation	Clay layer thickness, *h* (mm)	Pile length, *L* (mm)	*L*/*D* ratio	*h*/*L* ratio
CS1	150	191	15	0.78
		254	20	0.59
		318	25	0.47
		381	30	0.39
CS2	200	254	20	0.78
		318	25	0.62
		381	30	0.52
		445	35	0.45
CS3	250	254	20	0.98
		318	25	0.78
		381	30	0.65
		445	35	0.56

5.1 Effect of Variation in L/D *Ratio*

L/*D* ratio was varied by changing embedment length of the pile while keeping its diameter constant. Typical lateral load–displacement curves obtained for different values of *L*/*D* ratio in series CS1 are shown in Fig. 2. Trends of the curves clearly show that lateral load capacity of a single pile increases with increase in *L*/*D* ratio. However, rate of this increase is not linear and reduces for longer piles. Similar trends were observed for curves obtained in series CS2 and CS3.

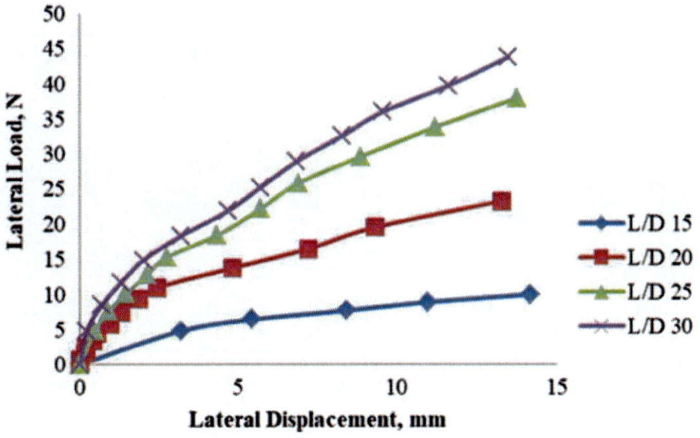

Fig. 2 Lateral load–displacement curves for series CS1

To analyse the relationship between lateral capacity and *L*/*D* ratio, lateral load carried by pile corresponding to 5 mm lateral displacement of pile head for various pile lengths are compared for all the three series. Figure 3 shows the obtained trends of variation of lateral capacity with respect to change in *L*/*D* ratio. It can be concluded from the observed data that the lateral capacity of a single pile increases with increase in *L*/*D* ratio for smaller pile lengths. But the rate of increase of lateral capacity decreases for longer pile lengths, and after obtaining an optimum value, this effect somehow diminishes and lateral capacity decreases with further increase in *L*/*D* ratio. Percentage variation in lateral capacity corresponding to 5 mm displacement with change in *L*/*D* ratio for all the three series is given in Table 4. It is clearly visible from the data presented that for the series CS2 and CS3, variation in lateral capacity

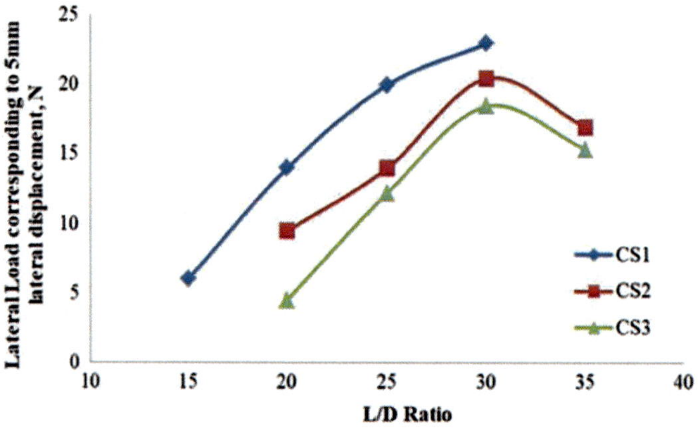

Fig. 3 Variation of lateral capacity w.r.t. change in *L*/*D* ratio

Table 4 Lateral load corresponding to 5 mm lateral displacement

Series designation	*L*/*D*	Lateral load, N	Variation in lateral load, %
CS1	15	6	–
	20	14	133.3
	25	20	42.9
	30	23	15.0
CS2	20	9.5	–
	25	14	47.4
	30	20.5	46.4
	35	17	−17.1
CS3	20	4.5	–
	25	12.2	171.1
	30	18.5	51.6
	35	15.4	−16.8

becomes negative representing reduction in lateral capacity when value of *L*/*D* ratio approaches 35. Also for series CS1, percentage variation reduces with increase in *L*/*D* ratio and is minimum for *L*/*D* equal to 30. It can be concluded that the optimum value of *L*/*D* ratio is about 30 for all the three series.

Initial increase in lateral capacity with increase in pile length may be due to increased surface friction with increase in embedment depth in dense sand. However, after attaining an effective embedment depth in sand, this effect diminishes and pile being laterally loaded shows a decrease in capacity with further increase in length. For pile with *L*/*D* 30, portion of pile length embedded in sand is about 18D, 14D and 10D in series CS1, CS2 and CS3, respectively, which shows that effective depth of pile in sand layer is dependent on thickness of overlying clay layer.

5.2 *Effect of Variation in Clay Layer Thickness*

Effect of change in clay layer thickness on lateral capacity of a single pile was studied by comparing the lateral load corresponding to a specific value of lateral displacement for a pile of a certain length for different values of *h*/*L* ratio. It was noticed that for all pile lengths, lateral capacity at a certain displacement level decreases with increase in thickness of clay layer. Comparing lateral load corresponding to 5 mm lateral displacement for different series, for pile with *L*/*D* ratio 30, decrease in lateral capacity is 10.9% when *h*/*L* varies from 0.39 to 0.52 and decrease in lateral capacity is 9.8% when *h*/*L* varies from 0.52 to 0.66. Similarly decrease in lateral capacity for pile with *L*/*D* ratio 25 is 30% for *h*/*L* varying from 0.47 to 0.63 and 12.9% for *h*/*L* varying from 0.63 to 0.79. For pile with *L*/*D* equal to 20, decrease in lateral capacity is 32.1% for *h*/*L* varying from 0.59 to 0.79 and 52.6% for *h*/*L* varying from 0.79 to 0.98. Typical curves relating the lateral capacity corresponding to 5 mm lateral displacement to *h*/*L* ratio for pile lengths with *L*/*D* ratio 20, 25 and 30 are shown in Fig. 4. These curves clearly show that relationship between variation in lateral load and variation in *h*/*L* ratio is nonlinear and to some extent it follows a power law in terms of *h*/*L* value. However, this relationship varies for different pile lengths.

6 Conclusions

The conclusions drawn from the observed results are summarized as following:

- Lateral load carrying capacity of a single pile embedded in layered soil containing soft clay layer over dense sand increases with increase in pile length. But rate of increase in lateral capacity decreases with increase in *L*/*D* ratio. Optimum value of *L*/*D* ratio is about 30 and further increase in length results in decrease in lateral capacity of pile.

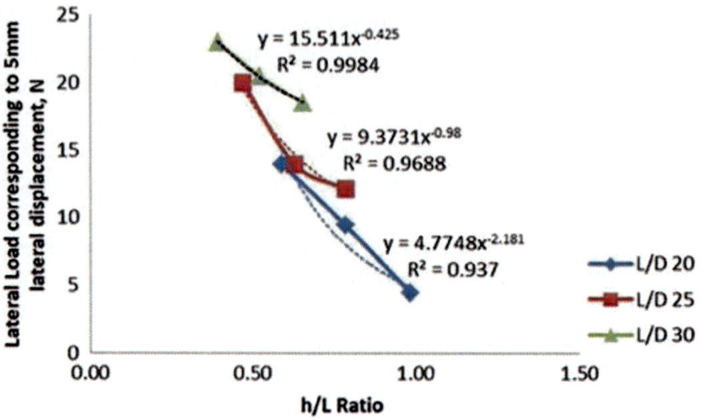

Fig. 4 Variation of lateral capacity for different pile lengths with change in *h*/*L* ratio

- Effective embedment depth of pile in sand layer is dependent on the thickness of overlying clay layer.
- Lateral load capacity of a single pile embedded in layered soil consisting of clay overlying sand decreases with increase in the thickness of clay layer. Relationship between lateral capacity and *h*/*L* ratio is nonlinear and follows a power law.
- Consideration of optimum length of pile and obtained relationship of lateral capacity with *L*/*D* and *h*/*L* ratio in design practices may help to reduce the construction cost while satisfying strength and serviceability requirements.

References

1. Abdrabbo FM, Gaaver KE (2012) Simplified analysis of laterally loaded pile groups. Alexandria Eng J 51:121–127
2. Ai ZY, Feng DL, Cheng YC (2013) BEM analysis of laterally loaded piles in multi-layered transversely isotropic soils. Eng Anal Bound Elem 37:1095–1106
3. Ashour M, Norris G (2000) Modeling lateral soil-pile response based on soil-pile interaction. J Geotech Geoenviron Eng Am Soc Civil Eng 126(5):420–428
4. Basu D, Salgado R, Prezzi M (2008) Analysis of laterally loaded piles in multilayered soil deposits. Publication FHWA/IN/JTRP-2007/23. Joint Transportation Research Program, Indiana Department of Transportation and Purdue University, West Lafayette, Indiana. https://doi.org/10.5703/1288284313454
5. Erdal U, Laman M (2011) Lateral resistance of a short rigid pile in a two-layer cohesionless soil. Acta Geotech Slov 2:19–43
6. Gupta BK, Basu D (2017) Analysis of laterally loaded short and long piles in multilayered heterogeneous elastic soil. Soils Found 57:92–110
7. Kim BT, Kim YS (1999) Back analysis for prediction and behavior of laterally loaded single piles in sand. KSCE J Civil Eng 3(3):273–288
8. Kim BT, Yoon GL (2010) Laboratory modeling of laterally loaded pile groups in sand. KSCE J Civil Eng 15(1):65–75

9. Murugan M, Natarajan C, Muthukkumaran K (2011) Behaviour of laterally loaded piles in cohesionless soils. Int J Earth Sci Eng 4(6):104–106
10. Phanikanth VS, Choudhury D (2013) Single piles in cohesionless soils under lateral loads using elastic continuum approach. Indian Geotech J 44(3):225–233. https://doi.org/10.1007/s40098-013-0072-4
11. Phanikanth VS, Choudhury D, Srinivas K (2012) Response of flexible piles under lateral loads. Indian Geotech J. https://doi.org/10.1007/s40098-012-0030-6
12. Sivapriya SV, Gandhi SR (2012) Experimental and numerical study on pile behaviour under lateral load in clayey slope. Indian Geotech J 43(1):105–114. https://doi.org/10.1007/s40098-012-0037-z

Stress–Strain Characteristics of Natural and Recycled Aggregate Concrete with Waste Foundry Sand and Additives

Rachit Sharma

1 Introduction

Construction waste volume has increased tremendously across the world over the last decade. Particularly in developing countries like India, the new zoning bylaws, urbanization and industrialization, reconstruction of old structures, and roadways for the ever-growing traffic demands have led to a vast quantity of demolition waste generation. To achieve the ever-increasing demands of road and structural construction, there is a shortage of natural aggregates which can be fulfilled by using construction waste aggregate. Therefore, the reuse of waste aggregate in concrete construction would be beneficial from economic and environmental perspectives in building a sustainable future.

The use of construction waste in concrete will result in reducing the strength of concrete as the percentage of replacement increases due to old mortar attached to it [9]. Construction waste aggregate concrete is more porous due to adhered mortar [21] and has a lower density as compared to natural aggregates concrete due to the lower specific gravity of recycled aggregates [5]. On full replacement of natural aggregates with construction waste aggregates in concrete strength, the reduction is not much, and applications involving low strength concrete could be possible [30]. Under static loading conditions, the stress–strain curve for construction waste aggregate was similar to natural aggregate concrete although there was a certain inconsistency in characteristic parameters between construction waste aggregate concrete and natural aggregate concrete [29].

Waste foundry sand is rejected material from iron castings. There are around 5000 foundries estimated in India [8]. Due to high reclamation cost, it is dumped in landfills, becoming a major threat to the environment. Enhanced mechanical properties

R. Sharma (✉)
National Institute of Technology Jalandhar, Jalandhar, Punjab, India
e-mail: rachitrooney10@gmail.com

H. Singh et al. (eds.), *Sustainable Development Through Engineering Innovations*, Lecture Notes in Civil Engineering 113,
https://doi.org/10.1007/978-981-15-9554-7_26

of waste foundry sand concrete were reported with 8–19% and 6.5–14.5% increment for compressive and split tensile strength, respectively [25]. Due to difference in the casting industries, there are discrepancies in reports for optimal replacement of waste foundry sand in concrete with 15% being the optimal replacement percentage [26] and 20% replacement level being optimal [19] for 0.5 water-binder ratio. Reusing the waste foundry sand in concrete applications will reduce the impact on natural resources along with negating the negative environmental factors associated with its disposal [7].

Due to the high carbon footprint from the massive production of cement [24], the cement is supplemented by certain cementitious material one such being silica fume. Construction waste aggregate concrete with 8% silica fume addition displayed better compressive strength performance whereas marginal improvements were found in split tensile strength [4]. By introducing fibers the brittleness of concrete can be compensated. A marginal reduction in workability was observed with the addition of chopped glass fibers however compressive, flexural strength, and shear toughness all improved [28].

2 Research Significance

In this study, to investigate the influence of construction waste on stress–strain curves of compressive and split tensile strength, the natural fine and coarse aggregates were replaced to an extent of 100%. Further, partial replacement of waste foundry sand for a replacement of 20% for natural fine aggregates was studied on natural and construction waste aggregate. To study the influence of additives on compressive and tensile stress–strain curves mixes were treated with superplasticizer (0.5%), silica fume (10%), and waste glass fiber (0.5%). The strength development with age was studied for each mix. The stress–strain curves will shed light on the plausible utilization of waste in the concrete applications for sustainable development of the construction sector.

3 Materials and Methods

3.1 Materials

Portland cement (IS: 8112-1989 grade 43) [14] was used as a binder in this study. Standard consistency of 31% and the initial and final settling time was recorded at 50 min and 520 min, respectively. Natural river aggregates tested as per IS 2386: 1963 [16] were used in the study. As per IS 383: 1970 standards [11], the coarse aggregates 20 mm downgraded and fine aggregates 4.75 mm downgraded were used.

Construction waste aggregates were obtained from crushed old concrete construction. The coarse construction waste aggregates tested as per IS 2386: 1963 [16] were crushed to a nominal size of 20 mm and fine aggregates 4.75 mm downgraded as per IS 383: 1970 requirements have been used. Waste foundry sand was obtained from local iron industries. To remove impurities, waste foundry sand was thoroughly washed and air-dried for 24 h before testing. The gradation of waste foundry sand was similar to natural fine aggregates. Polycarboxylic ether (PCE) based superplasticizer with a relative density of 1.10 g/cm^3 at 25 °C conforming to ASTM C494 Type F was used [2]. Waste glass fiber with an average fiber length of 4.2 mm for 0.5% addition of weight of cement was used in the study. The physical and chemical properties of silica fume were given by the manufacturer. Standard tap water conforming to IS: 456-2000 [12] was used for casting and curing purposes.

3.2 *Mix Proportions and Testing*

The concrete mix was designed as per IS: 10262-2009 [10] for 1 m^3 concrete volume and 0.5 water-binder ratio taking cement:fine sand:coarse aggregate:water (330 kg:660 kg:1320 kg:165 kg) and further substitution and addition by weight was performed using various waste materials. The construction waste aggregate was replaced for 100% natural fine and coarse aggregate, waste foundry sand for 20% replacement for natural sand, silica fume for 10% cement. Further, fiber addition of 0.5% weight was also exercised. Additionally, the effect of superplasticizer was observed for 0.9% addition of the weight of cement keeping water-binder ratio at 0.45. The mix design was made considering the aggregates to be in surface saturated dry (SSD) condition. A total of 10 mixes were prepared as per IS 456: 2000 [12] and their description was tabulated under Table 1.

Cube (150 × 150 × 150 mm) and cylinder (ϕ 150 × 300 mm) specimens were employed for making each mix. The concrete mix was prepared with proper batching of materials. Each batch was verified for consistency test immediately after mixing as per IS: 1199-1959 [15], filled in steel molds, and compacted under the vibration table. The steel molds were stored in the laboratory for 24 h before demolding and placed in water-filled curing tanks under room temperature.

3.2.1 Testing Procedure

The cylinder and cube specimens were tested for splitting tensile and compressive strength in surface dry condition after 3, 7 and 28 days of curing under computer-controlled compression testing machine as per IS: 516-1959 [13]. The respective stress–strain curves were plotted from the load-deflection plot of each specimen and the peak value of compressive stress was duly recorded.

Table 1 Mix description

Mix name	Mix description	Specimens count		Slump value (mm)
		Cube	Cylinder	
N	Cement: fine sand: coarse aggregate: water	9	9	75
CW	Cement: construction waste fine sand: Construction waste coarse aggregate: water	9	9	65
N WFS	Cement: fine sand (20% waste foundry sand replacement): coarse aggregate: water	9	9	70
CW WFS	Cement: construction waste fine sand (20% waste foundry sand replacement): construction waste coarse aggregate: water	9	9	60
N WFS + SF	Cement (10% silica fume replacement): fine sand (20% waste foundry sand replacement): coarse aggregate: water	9	9	65
CW WFS + SF	Cement (10% silica fume replacement): construction waste fine sand (20% waste foundry sand replacement): construction waste coarse aggregate: water	9	9	55
N WFS + SP	Cement: fine sand (20% waste foundry sand replacement): coarse aggregate: water + 0.9% superpalsticizer	9	9	70
CW WFS + SP	Cement: construction waste fine sand (20% waste foundry sand replacement): construction waste coarse aggregate: water + 0.9% superplasticizer	9	9	65
N WFS + F	Cement: fine sand (20% waste foundry sand replacement): coarse aggregate: water + 0.5% waste glass fiber	9	9	60
CW WFS + F	Cement: construction waste fine sand (20% waste foundry sand replacement): construction waste coarse aggregate: water + 0.5% waste glass fiber	9	9	50

4 Results and Discussion

4.1 Stress–Strain Characteristics

The stress–strain characteristics of concrete on complete replacement of coarse and fine aggregates with construction waste aggregate sand partial substation with waste foundry sand including superplasticizer, silica fume, and fiber at 3, 7, and 28 days are discussed.

4.2 Compressive Strength

4.2.1 Effect of Construction Waste Aggregates

The stress–strain characteristics of natural aggregate concrete indicate a consistent increase in the failure of stress and strain with age as shown in Fig. 1.

The design strength is attained after 28 days at a strain of more than 2% and stress–strain curves indicate an increase in stress till brittle failure immediately after attaining peak stress. The complete replacement of coarse and fine aggregates with construction waste aggregates shows the stress–strain curves steeply rising. The stress increases with age till brittle failure after reaching peak stress and failure stress is slightly less than the corresponding stress in case of natural aggregate concrete [18].

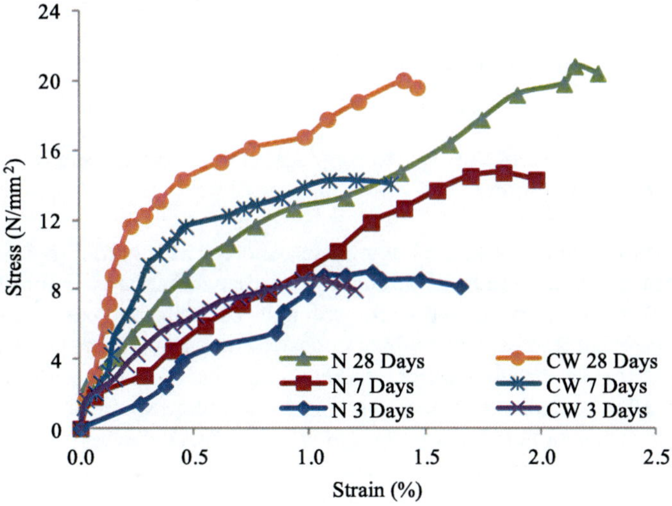

Fig. 1 Stress–strain characteristics of N/CW concrete

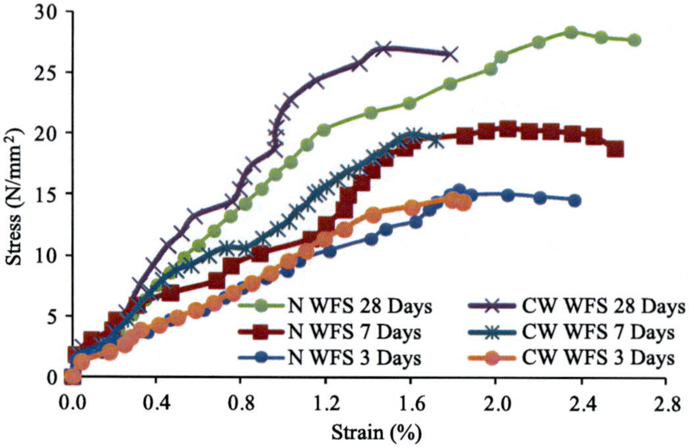

Fig. 2 Stress–strain characteristics of N/CW concrete with optimum WFS

4.2.2 Effect of Construction Waste Aggregates and Waste Foundry Sand

The partial substitution of fine aggregate with waste foundry sand in natural concrete indicates stress–strain curves with increased failure stress and strain as shown in Fig. 2. The failure strength of concrete at 28 days is higher than the control of concrete and brittle failure occurs at increased strain [26]. The construction aggregate concrete indicates increased failure stress and strain with age and the curves are closer to the curves for natural aggregate concrete. The failure strength and strain of construction aggregate concrete were lesser than those for natural concrete which indicate relatively brittle and porous concrete.

4.2.3 Effect of Construction Waste Aggregates and Waste Foundry Sand + Silica Fume

The stress–strain curves of natural aggregate concrete with partial replacement of sand with waste foundry sand and the addition of silica fume shown in Fig. 3 indicate a consistent increase in stress and strain with age. The failure of concrete occurs at increased stress but at lower strain which indicates compact concrete [1]. The stress–strain curves for construction waste concrete with waste foundry replacement and silica fume addition indicate increased strength and failure strain. The strength and failure strain is comparable for two types of concrete thus indicating improved characteristics of construction waste concrete. The waste foundry sand and silica fume fill the pores or interstices in construction waste concrete making it comparatively

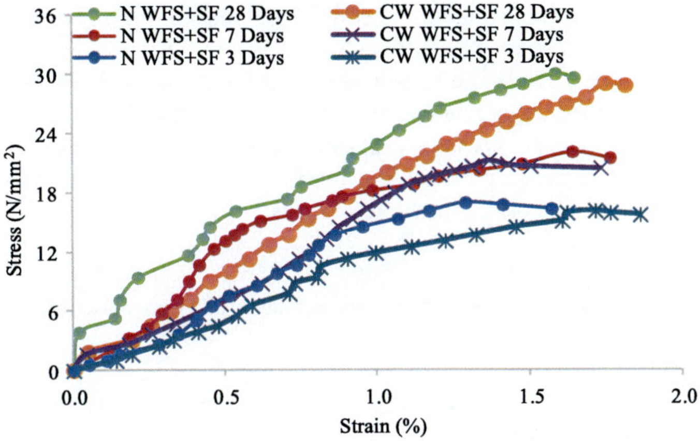

Fig. 3 Stress–strain characteristics of N/CW concrete with optimum WFS + silica fume

compact and attaining better stress–strain characteristics. Typical failure of construction waste concrete with waste foundry sand and silica fume at 7 days indicates a brittle mode of failure and cracks are vertically as well as diagonally along the weak inter-transition zones [6].

4.2.4 Effect of Construction Waste Aggregates and Waste Foundry Sand + Superplasticizer

The development of stress–strain characteristics of natural aggregate concrete with partial replacement of sand with waste foundry sand and addition of superplasticizer demonstrated in Fig. 4 reveal an increase in strength at 3, 7, and 28 days. The

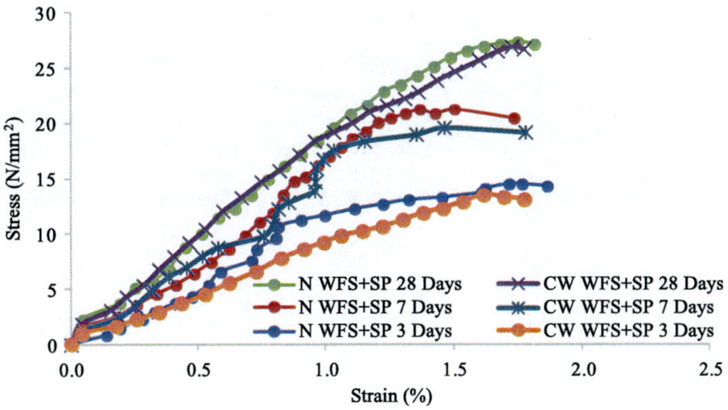

Fig. 4 Stress–strain characteristics of N/CW concrete with optimum WFS + superplasticizer

compressive strength is more than that of control concrete but slightly less than that of concrete with waste foundry sand [3]. The stress–strain characteristics of construction aggregate concrete are lower but comparable to those for natural concrete and the strength of construction aggregate concrete is lesser than that of concrete with waste foundry sand substituent [20]. The superplasticizer covers porous and rough-surfaced construction waste aggregate particles and thus improves their stress–strain behavior. The failure of concrete is brittle in all the samples.

4.2.5 Effect of Construction Waste Aggregates and Waste Foundry Sand + Fiber

The stress–strain curves of natural aggregate concrete with waste foundry sand substituent and fiber are illustrated in Fig. 5. There is a consistent development of strength with age in natural aggregate concrete and strength is higher than that of normal concrete excluding or including waste foundry sand, waste foundry sand with silica fume, or superplasticizer. The fibers interconnect the inter-transition layers together and thus form compact concrete mass and strength increases [23]. The stress–strain behavior of construction waste aggregate concrete improves appreciably with the addition of fiber and the strength increase with age. The strength of construction waste aggregate concrete with waste foundry sand and fiber is higher than that with other admixtures. The failure pattern of construction waste aggregate concrete with waste foundry sand and fiber at 28 days, the appearance of vertical cracks showed a change in crack pattern as compared to that in concrete without fibers.

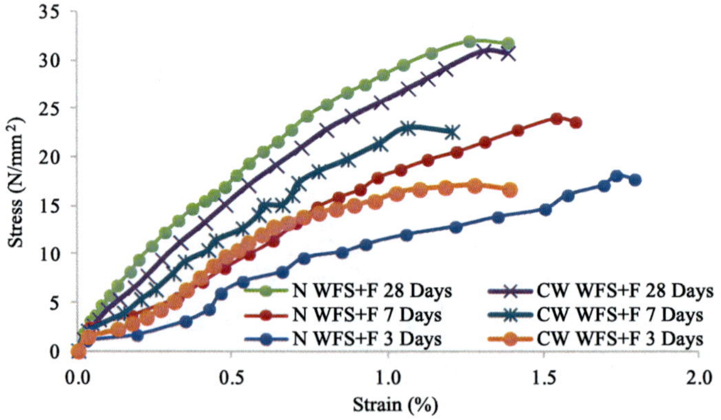

Fig. 5 Stress–strain characteristics of N/CW concrete with optimum WFS + glass fiber

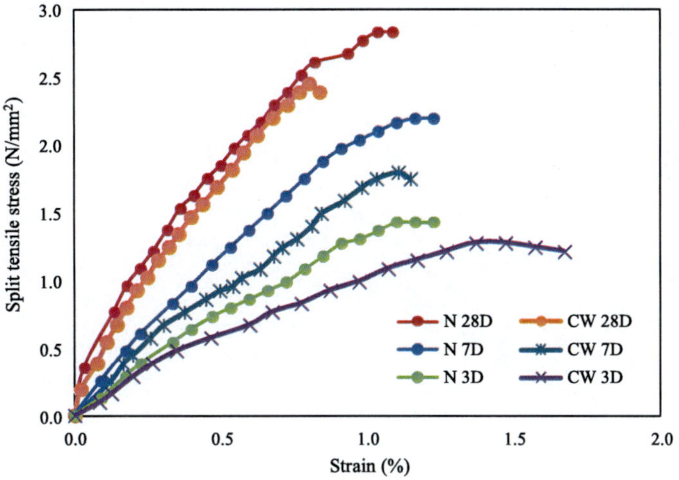

Fig. 6 Split tensile stress–strain characteristics of N/CW concrete at different age

4.3 Split Tensile Strength

4.3.1 Stress–Strain Characteristics of N/CW Concrete

The split tensile stress–strain characteristics of natural concrete and construction waste concrete at 3, 7, and 28 days are demonstrated in Fig. 6. The stress–strain characteristics of construction waste concrete are inferior to those of natural concrete at 3 and 7 days but improve at 28 days. The split tensile strength of CW concrete is lesser than that of NA concrete at 3, 7, and 28 days which is due to the porous and poor surface of construction waste aggregate particles forming improper joints over a large number of inter-transition layers [27]. Typical split tensile failure of cylindrical CW concrete sample at 7 days, indicates diagonal cracks in addition to vertical cracks revealing the failure along the poor surface of waste aggregate particles.

4.3.2 Stress–Strain Characteristics of N/CW + WFS Concrete

Figure 7 demonstrates the split tensile stress–strain characteristics of natural concrete and construction waste concrete containing optimum content of waste foundry sand at 3, 7, and 28 days. The stress–strain characteristics of CW concrete are poor as compared with those of NA concrete at 3 and 7 days but there is not much difference in characteristics of the two types of concrete at 28 days. The split tensile strength of CW concrete is less but comparable to that of NA concrete. The strength of NA and CW concrete improves due to waste foundry sand substituent which fills the voids in concrete making it compact [19].

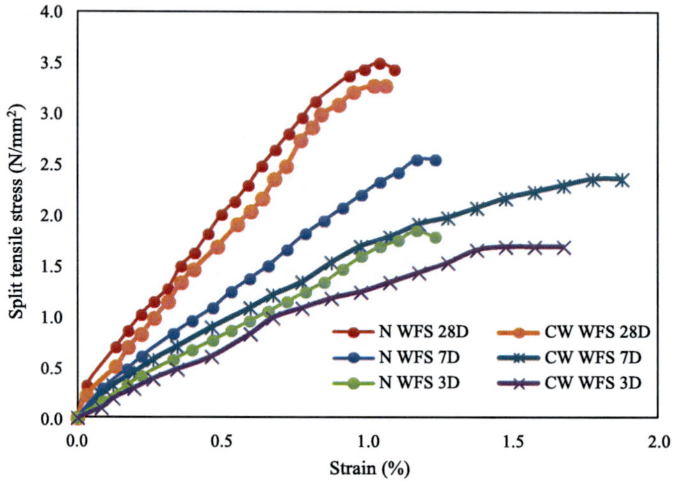

Fig. 7 Split tensile stress–strain characteristics of N/CW + WFS concrete at different age

4.3.3 Stress–Strain Characteristics of N/CW + WFS + SF Concrete

The addition of silica fume to natural or construction waste concrete containing optimum content of waste foundry sand substituent improves the stress–strain characteristics as shown in Fig. 8. The stress–strain characteristics of CW concrete at 3 and 7 days are inferior to those of NA concrete but are comparable at 28 days. The addition of silica fume improves the split tensile strength of concrete appreciably as

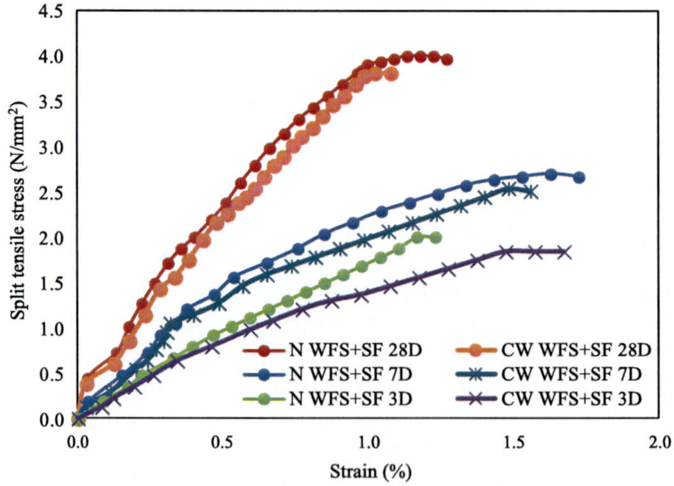

Fig. 8 Split tensile stress–strain characteristics of N/CW + WFS + SF concrete at different age

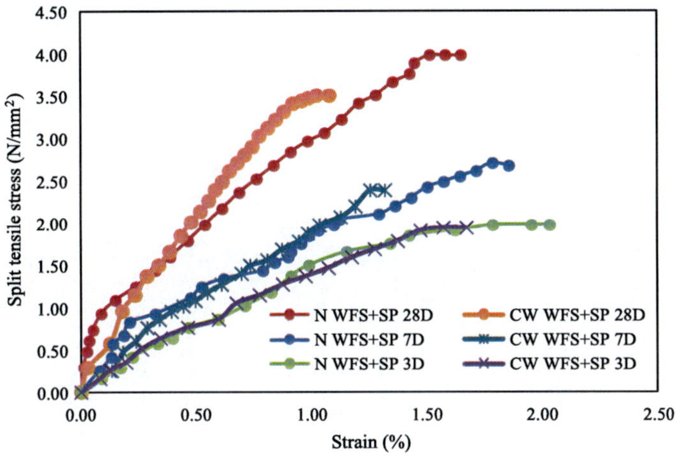

Fig. 9 Split tensile stress–strain characteristics of N/CW + WFS + SP concrete at different age

compared to that of control concrete. The waste foundry sand fills the large voids and silica fume fills microvoids in concrete thus transforming it into a compact mass and increasing the strength [22].

4.3.4 Stress–Strain Characteristics of N/CW + WFS + SP Concrete

The split tensile stress–strain characteristics of natural concrete or construction waste concrete containing optimum content of waste foundry sand substituent and superplasticizer are illustrated in Fig. 9. The stress–strain characteristics of NA/CW concrete improve slightly on the addition of superplasticizer but are inferior to those of NA/CW concrete containing waste foundry sand substituent and silica fume. The superplasticizer acts as a lubricant changing the surface characteristics of construction waste aggregate particles thus turning the stress–strain characteristics of CW concrete similar to those of NA concrete at 3 and 7 days. However, there is not much influence of superplasticizer on split tensile strength of CW concrete which is less than that of NA concrete [20].

4.3.5 Stress–Strain Characteristics of N/CW + WFS + F Concrete

In case fibers are added to natural or construction waste concrete containing optimum content of waste foundry sand substituent, its split tensile strength characteristics are improved significantly as compared to those of control concrete or concrete with other combinations of waste foundry sand and admixtures as shown in Fig. 10. The stress–strain characteristics of CW concrete are comparable to those of NA concrete, particularly at 28 days. The split tensile strength of CW concrete improves

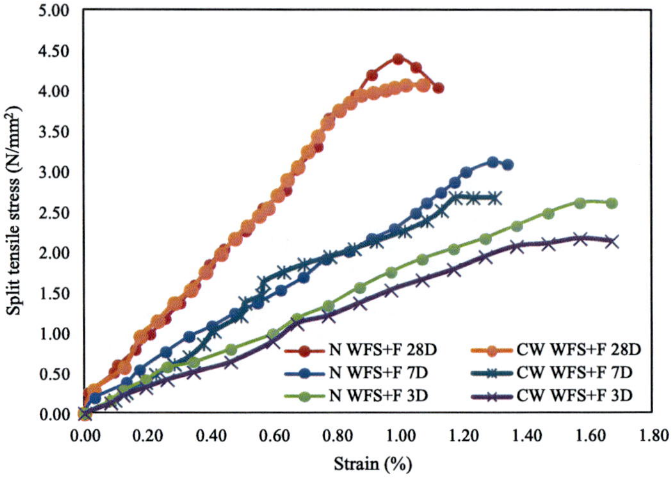

Fig. 10 Split tensile stress–strain characteristics of N/CW + WFS + F concrete at different age

significantly however it is less than that of NA concrete. The improvement in split tensile strength is due to the filling of voids by waste foundry sand particles and fibers connecting the layers in concrete mass [17].

5 Conclusions

The experimental study on stress–strain characteristics of natural and recycled concrete containing waste foundry sand and additives can be concluded as:

- The stress–strain characteristics of construction waste concrete are inferior to those of natural aggregate concrete; the strength and failure strain decrease in CW concrete. The optimum content of waste foundry sand substituting natural sand improves the stress–strain characteristics of NA concrete and CW concrete; the strength of the latter being less.
- The addition of silica fume with waste foundry sand improves the stress–strain characteristics of both the concrete appreciably but lesser in CW concrete. The failure pattern reveals the presence of diagonal and vertical cracks indicating brittle failure.
- The waste foundry sand and superplasticizer appreciate the stress–strain characteristics of NA concrete and CW concrete but the effect of superplasticizer is on strength is marginal.
- The waste foundry sand and fiber enhance the stress–strain characteristics of both the concrete significantly and strength is marginally less in CW concrete. Thus,

fibers bridge the gaps in between the layers in concrete and increase the strength and change the failure pattern.

- The construction, foundry, and fiber wastes can be employed as constituent materials in concrete solving the problem of unplanned disposal and conserve the environment.

References

1. Ahmadi M, Farzin S, Hassani A, Motamedi M (2017) Mechanical properties of the concrete containing recycled fibers and aggregates. Constr Build Mater 144:392–398
2. ASTM C494 (2017) Standard specification for chemical admixture for concrete. West Conshohocken, USA
3. Atis CD, Özcan F, Kılıc A, Karahan O, Bilim C, Severcan MH (2005) Influence of dry and wet curing conditions on compressive strength of silica fume concrete. Build Environ 40(12):1678–1683
4. Fonteboa BG, Abella FM (2008) Concretes with aggregates from demolition waste and silica fume. Materials and mechanical properties. Build Environ 43(4):429–437
5. Butler L, West JS, Tighe SL (2011) The effect of recycled concrete aggregate properties on the bond strength between RCA concrete and steel reinforcement. Cem Concr Res 41(10):1037–1049
6. Çakır Ö, Sofyanlı ÖÖ (2015) Influence of silica fume on mechanical and physical properties of recycled aggregate concrete. HBRC J 11(2):157–166
7. Du H, Tan KH (2017) Properties of high volume glass powder concrete. Cem Concr Compos 75:22–29. https://doi.org/10.1016/j.cemconcomp.2016.10.010
8. Global Casting Production expands (2019) Census of world casting production. American Foundry Society. Metal casting design & purchasing, pp 27–29
9. Hansen TC (1986) Recycled aggregate and recycled aggregate concrete, second state-of-the-art report, developments from 1945–1985. Mater Struct 19(3):201–246
10. IS: 10262-2009: Concrete mix proportioning—guidelines. Bureau of Indian Standards Manak Bhawan, New Delhi, India
11. IS: 383-1970: Specification for coarse and fine aggregates from natural sources for concrete. Bureau of Indian Standards Manak Bhawan, New Delhi, India
12. IS: 456-2000. Plain and reinforced concrete—code of practice. Bureau of Indian Standards Manak Bhawan, New Delhi, India (reaffirmed in 2011)
13. IS: 516-1959: Methods of tests for strength of concrete. Bureau of Indian Standards Manak Bhawan, New Delhi, India (reaffirmed in 1999)
14. IS: 8112-1989: Specification for 43 grade ordinary Portland cement. Bureau of Indian Standards Manak Bhawan, New Delhi, India
15. IS: 1199-1959: Methods of sampling and analysis of concrete. Bureau of Indian Standards Manak Bhawan, New Delhi, India
16. IS: 2386-1963: Methods of test for aggregate for concrete. Bureau of Indian Standards Manak Bhawan, New Delhi, India
17. Khan MI, Umair M, Shaker K, Basit A, Nawab Y, Kashif M (2020) Impact of waste fibers on the mechanical performance of concrete composites. J Text Inst. https://doi.org/10.1080/00405000.2020.1736423
18. Khatib JM, Hibbert JJ (2005) Selected engineering properties of concrete incorporating slag and metakaolin. Constr Build Mater 19(6):460–472
19. Manoharan T, Laksmanan D, Mylsamy K, Sivakumar P, Sircar A (2018) Engineering properties of concrete with partial utilization of used foundry sand. Waste Manag 71:454–460

20. Matias D, De Brito J, Rosa A, Pedro D (2013) Mechanical properties of concrete produced with recycled coarse aggregates—influence of the use of superplasticizers. Constr Build Mater 44:101–109
21. Poon CS, Shui ZH, Lam L (2004) Effect of microstructure of ITZ on compressive strength of concrete prepared with recycled aggregates. Constr Build Mater 18:461–468
22. Sahoo KK, Pradip Sarkar P, Davis R (2019) Mechanical properties of silica fume concrete designed as per construction practice. Proc Inst Civil Eng Constr Mater 172(1):20–28
23. Saravana RMK, Sumathi A (2017) Effect of fly ash in fiber reinforced concrete composites. Jordan J Civil Eng 11(1):30–39
24. Scrivener KL, John VM, Gartner EM (2018) Eco-efficient cements: potential economically viable solutions for low-CO_2 cement-based materials industry. Cem Concr Res 114:2–26. https://doi.org/10.1016/j.cemconres.2018.03.015
25. Siddique R, de Schutter G, Noumowe A (2009) Effect of used-foundry sand on the mechanical properties of concrete. Constr Build Mater 23(2):976–980
26. Siddique R, Singh G, Belarbi R, Ait-Mokhtar KK (2015) Comparative investigation on the influence of spent foundry sand as partial replacement of fine aggregates on the properties of two grades of concrete. Constr Build Mater 83(15):216–222
27. Tabsh SW, Abdelfatah AS (2009) Influence of recycled concrete aggregates on strength properties of concrete. Constr Build Mater 23(2):1163–1167
28. Tassew ST, Lubell AS (2014) Mechanical properties of glass fiber reinforced ceramic concrete. Constr Build Mater 51:215–224
29. Xiao J, Li J, Zhang C (2005) Mechanical properties of recycled aggregate concrete under uniaxial loading. Cem Concr Res 35:1187–1194
30. Yehia S, Helal K, Abusharkh A, Zaher A, Istaitiyeh H (2015) Strength and durability evaluation of recycled aggregate concrete. Int J Concr Struct Mater 9(2):219–239

Performance Evaluation for Use of Recycled Concrete Aggregates in Flexible Pavement

Pushpinder Singh and S. K. Singh

1 Introduction

Sustainable development is very necessary to balance our economic, environmental, and social needs considering the well-being of now and upcoming generations. Recycling or reuse of waste materials is an important part of sustainability because it reduces the pressure on natural resources. So, whenever possible, waste materials should be recycled and reused. Cement concrete has been used as one of the major construction materials for buildings, dams, bridges and other structures. That is why concrete waste constitutes the major part of solid waste generated during the construction and demolition of buildings and other structures. About 7.5–10 million tonnes of concrete waste is generated every year considering 25–30 million tonnes of annual generation of construction and demolition (C&D) waste and 30% proportion of concrete waste in C&D waste [1]. Shortage of landfills leads to disposal problems and promotes the illegal dumping of concrete waste in water bodies, on hill slopes, in roadside areas, etc. Also, it is undesirable to dispose of concrete waste considering the high strength of concrete. It is better to utilize concrete waste in some ways, so that sustainability can be achieved in construction and demolition practices.

Recycled concrete aggregates (RCA), obtained after crushing the segregated concrete waste, have different properties and composition as compared to the natural aggregates (NA) due to the attached cement mortar on the aggregate surfaces. A number of researches have been carried out for determining the viability of RCA in structural concrete [2–5]. RCA is recommended to be used in structural concrete as

P. Singh (✉)
Department of Civil Engineering, Guru Nanak Dev Engineering College, Ludhiana 141006, India
e-mail: pushpinder722@gmail.com

S. K. Singh
Civil Engineering Department, Punjab Engineering College, Chandigarh 160012, India
e-mail: sksingh@pec.ac.in

H. Singh et al. (eds.), *Sustainable Development Through Engineering Innovations*, Lecture Notes in Civil Engineering 113,
https://doi.org/10.1007/978-981-15-9554-7_27

a replacement of natural aggregates up to 25% [6], whereas, even higher replacing percentage i.e., 30–50% is proposed in structural concrete for the construction of bank protection, base/fill of drainage structures, rigid pavements, sidewalks, kerbs, gutters, etc. [7]. But, due to higher rate of concrete waste generation as compared to its utilization, there is a need for finding other alternatives for using the RCA.

The utilization of RCA in flexible pavements is very limited in India. Various researches have been conducted for determining their feasibility in subbase and base course [8–11] and surface course of flexible pavement [12–14]. Bennert et al. [15] investigated the use of reclaimed asphalt pavements (RAPs) and recycled concrete aggregates (RCA) for subbase and base applications and experimental results observed the higher value of resilient modulus for RAPs and RCA and lower value of permanent deformation in case of RCA as compared to RAPs. In another study, Barbudo et al. [16] concluded that segregation of concrete waste from other wastes in C&D waste (brick waste, ceramic waste, etc.) is necessary, as C&D waste aggregates consisting a higher percentage of concrete showed better performance in CBR, Los Angeles coefficient, shape and compaction tests.

Coarse aggregates form the basic structure of bituminous mixes and significantly influence the performance and behavior of surface course. Since the aggregates constitute the major portion of bituminous mixes, the use of RCA can be very advantageous in bulk utilization of waste material. Yang et al. [17] investigated the feasibility of coarse and fine RCA in dense-graded HMA mixes with different replacing percentages (20 and 40% for fine RCA; and 20, 40, and 60% for coarse RCA) and observed satisfactory performance in case of mixtures containing 40% coarse RCA or 20% fine RCA. Tahmoorian and Samali [11] concluded that all the properties of recycled concrete aggregates were appropriate, but the water absorption and wet/dry strength variation of RCA were coming out to be more than the virgin aggregates due to the attached mortar and cement paste.

Most of the studies investigated the performance of RCA in bituminous mixes by using them in some proportion or pre-treating them with some methods such as acid treatment, treatment with epoxy resins, etc. [3, 18]. The practical feasibility of some of the treatment methods is in question due to their higher cost and difficult application. Moreover, the use of RCA as full replacement of natural aggregates can be effective in terms of easiness in practical application and mass utilization of concrete waste. Therefore, in the present study, RCA has been investigated for full replacement of natural aggregates for utilization in flexible pavement. In a study by Mahendra et al. [18], RCA properties were evaluated after treating them with acid and thermal treatment, and experimental results observed better results in case of treated RCA. Although the characterization of thermally treated RCA has been conducted in different studies, the influence of using the thermally treated RCA in bituminous mixes has not been investigated yet. So, the performance of thermally treated concrete aggregates (TRCA) has also been investigated for use in binder course of flexible pavement. Additionally, the feasibility of RCA for the construction of subbase and base course has been determined in this study.

2 Materials and Methodology

Concrete waste was collected from a demolition site after the segregation from other waste materials such as brick waste, reinforcements, etc. Crushing and sieving of waste concrete were done for obtaining the aggregates of the required sizes. Locally available natural aggregates (NA) were obtained and used for preparing control mixes. Recycled concrete aggregate (RCA), obtained from crushing and sieving, were then treated by thermal treatment. Thermal treatment was done by heating the RCA in muffle furnace at 350 °C for 2 h. Due to the difference between coefficient of thermal expansion of cement mortar and natural aggregate, the excess weak mortar gets separated from the aggregate surfaces, thus, improving the aggregate properties. Characterization of NA, RCA, and TRCA aggregates was done by determining different aggregate tests. Wet aggregate impact test was conducted, as it is recommended whenever water absorption exceeds the permissible limit of 2% [19]. Marshall tests were conducted on the samples prepared by using NA, RCA, and TRCA to evaluate the optimum bitumen content and other design parameters. Moisture susceptibility of bituminous mixes was determined by retained stability test. Compaction and CBR tests were conducted for determining the adequacy of RCA in subbase and base course of flexible pavement.

2.1 Bitumen

Bitumen acts as a binder for aggregates in the bituminous mixes. In the present study, locally available binder corresponding to 80/100 penetration grade (viscosity grade 10) was used for preparing DBM mixes. The characterization results of binders are tabulated in Table 1.

Table 1 Characterization of bitumen

Test	Result	IS 73:2013 specifications
Penetration number	87.18	80 (Min)
Softening point (°C)	45.20	40 (Min)
Ductility (cm)	93	75 (Min)
Flash point (°C)	260	220 (Min)
Fire point (°C)	302	–
Specific gravity	1.02	–

Table 2 Characterization of aggregates and filler

Aggregate property	NA	RCA	TRCA	Test procedure
Aggregate impact value (%)	13.63	16.00	15.12	IS 2386 (Part 4):2002
Aggregate crushing value (%)	19.95	25.71	25.13	IS 2386 (Part 4):2002
Los Angeles abrasion value (%)	22.42	24.93	23.36	IS 2386 (Part 4):2002
Combined flakiness and elongation index (%)	18.56	28.23	25.62	IS 2386 (Part 1):2002
Water absorption (%)	1.37	4.19	2.91	IS 2386 (Part 3):2002
Wet impact value (%)	–	18.63	16.94	IS 5640:1998
Bulk specific gravity of coarse aggregate	2.62	2.38	2.45	IS 2386 (Part 3):2002
Apparent specific gravity of coarse aggregate	2.71	2.64	2.64	IS 2386 (Part 3):2002
Apparent specific gravity of fine aggregate	2.70	2.49	2.45	IS 2386 (Part 3):2002
Apparent specific gravity of filler	2.56	2.16	2.15	IS 2386 (Part 3):2002

2.2 Aggregates and Filler Material

In case of DBMmixes prepared with NA, locally available stone dust was used as a filler material, whereas in other two cases, concrete dust was used as filler material, which was obtained during crushing of waste concrete. Characterization of aggregates and filler material is given in Table 2.

3 Experimental Procedure

3.1 Marshall Mix Design

Marshall tests were performed in accordance with AASHTO T245-97 (2001) [20]. Grading 2 was adopted for DBM mixes as per the MORTH specifications for roads and bridges, which is given in Table 3.

Optimum bitumen content (OBC) was determined as per the recommendation of Asphalt Institute MS-2 Manual [21]. It suggests choosing the bitumen content at the median of the percent air voids limit, which is 4%. If all the calculated and measured mix properties at this bitumen content are satisfying the mix design criteria, then this is the OBC for the mix design. If other parameters do not satisfy the required specifications at bitumen content corresponding to 4% air voids, some adjustments can be made such as taking the OBC equal to the average of bitumen contents corresponding to maximum stability, maximum density, and 4% air voids.

Table 3 Composition of dense bituminous macadam layer as per MORTH specifications, 2013

IS Sieve	Cumulative % passing	Cumulative % passing adopted
37.5 mm	100	100
26.5 mm	90–100	95
19 mm	71–95	83
13.2 mm	56–80	68
9.5 mm	–	–
4.75 mm	38–54	46
2.36 mm	28–42	35
1.18 mm	–	–
600 μm	–	–
300 μm	7–21	13.5
150 μm	–	–
75 μm	2–8	5

3.2 Retained Stability Test

Retained stability tests were conducted in accordance with ASTM: D1075 (2011) [22] specifications to examine the performance of bituminous mixes against the moisture-induced damage. In this test, Marshall stability values of specimens prepared at optimum bitumen contents, are determined before and after the moisture conditioning process. Conditioning of specimens is done by immersing them in the water at 60 °C for 24 h and unconditioned specimens are considered as those, which are immersed in water at 60 °C for half an hour before testing. Retained stability is calculated by taking the ratio of Marshall stability of conditioned specimen to the Marshall stability of unconditioned specimen as given in Eq. 1.

$$\text{Retained stability (\%)} = \frac{\text{Marshall stability of Conditioned specimen}}{\text{Marshall stability of Unconditioned specimen}} \times 100 \tag{1}$$

3.3 Compaction Test

Two different procedures were adopted in accordance with IS:2720 (Part 7) (2011) [23] and Colorado procedure 23–13 (2017) [24] for determination of Optimum Moisture Content (OMC) and Maximum Dry Density (MDD). Wet Mix Macadam (WMM) gradation was assumed to be provided for the subbase and base course layer as given in Table 4. As per IS method, air dried material passing a 20 mm IS sieve is taken. The coarser fraction is replaced by equivalent weight of finer material by

Table 4 Composition of Wet Mix Macadam as per MORTH specifications, 2013

IS Sieve	Cumulative % passing (specification limits)	Cumulative % passing adopted
53 mm	100	100
45 mm	95–100	97.5
22.40 mm	60–80	70
11.20 mm	40–60	50
4.75 mm	25–40	32.5
2.36 mm	15–30	22.5
600 μm	8–22	15
75 μm	0–5	2.5
Pan	0	0

maintaining the specified proportion as per grading requirements. Due to the limitations of IS method observed while conducting the test such as poor workability and inadequate compaction, the OMC and MDD obtained by Colorado procedure were adopted. In Colorado procedure, the parameters used for the determination of optimum moisture content and maximum dry density of mix (OMC_{mix} & MDD_{mix}) are OMC of material finer than 4.75 mm (M_f), MDD of material finer than 4.75 mm (D_f), water absorption of material coarser than 4.75 mm (M_c) and bulk specific gravity of material coarser than 4.75 mm (D_c) taking M_f and M_c in percentage and D_f in kN/m^3. P_f is percentage of material finer than 4.75 mm and P_c is percentage of material coarser than 4.75 mm (percentage taken with respect to total weight) as per the WMM gradation. Water absorption and bulk specific gravity (M_c & D_c) are evaluated in accordance with AASHTO T85 (2001) [25], whereas OMC and MDD of material finer than 4.75 mm (M_f & D_f) are evaluated by using AASHTO T99 (2001) [26]. Equation 2 and 3 are given for determination of OMC_{mix} & MDD_{mix}.

$$OMC_{mix} = \frac{(M_f \times P_f) + (M_c \times P_c)}{100} \tag{2}$$

$$MDD_{mix} = \frac{(P_f \times P_f) + (0.90 \times P_c \times P_c)}{100} \tag{3}$$

3.4 California Bearing Ratio (CBR) Test

CBR test was conducted in accordance with IS:2720 (Part 16) (2002) [27]. The material used in this test shall pass a 19 mm IS sieve. Material which is larger than 19 mm is replaced by equal amount of material which passes a 19 mm sieve but is retained on a 4.75 mm sieve. The test was conducted after four days soaking of prepared specimens for taking the worst moisture conditions into account. The load values were recorded at penetrations 0.5, 1, 1.5, 2, 2.5, 3, 4, 5, 6, 7, 8, 9, 10, and

12.5 mm. CBR is determined based on either 2.5 or 5 mm penetration by taking the ratio of load corresponding to 2.5 or 5 mm penetration to standard load (1370 kg for 2.5 mm penetration and 2055 kg for 5 mm penetration).

4 Results and Discussion

4.1 Material Characterization

Physical requirements of coarse aggregates for use in different layers as per MORTH specifications are given in Table 5.

The aggregate properties of natural aggregates (NA), recycled concrete aggregates (RCA), and thermally treated concrete aggregates (TRCA) were determined in characterization process and comparison results are given in Fig. 1.

As observed in Fig. 1 and Table 5, all other aggregate properties of RCA and TRCA except water absorption were satisfying the permissible limits specified by the MORTH for using them in dense bituminous macadam (DBM) and wet mix macadam (WMM). The water absorption of RCA and TRCA was observed as 4.19 and 2.91%, more than the permissible limit i.e., 2%. But, the wet aggregate impact

Table 5 Physical requirements for coarse aggregates as per MORTH specifications, 2013

Aggregate property	DBM	WMM
Aggregate impact value (%)	Max. 27	Max. 30
Los Angeles abrasion value (%)	Max. 35	Max. 40
Combined flakiness and elongation index (%)	Max. 35	Max. 35
Water absorption (%)	Max. 2	Max. 2

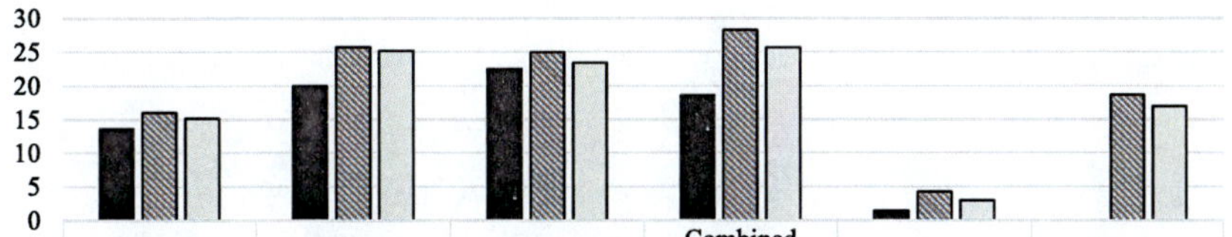

	Aggregate impact value (%)	Aggregate crushing value (%)	LA abrasion value (%)	Combined flakiness and elongation index (%)	Water absorption (%)	Wet aggregate impact value
■NA	13.63	19.95	22.42	18.56	1.37	
▨RCA	16	25.71	24.93	28.23	4.19	18.63
□TRCA	15.12	25.13	23.36	25.62	2.91	16.94

■NA ▨RCA □TRCA

Fig. 1 Comparison of aggregate properties of NA, RCA, and TRCA value for RCA (18.63%) and TRCA (16.94%) were lesser than the maximum permissible limits (max. 24%, 27%, and 30% for BC, DBM, and WMM, respectively). Thermal treatment had lesser effect on properties like aggregate impact value, aggregate crushing value, and LA abrasion value but the water absorption and combined elongation and flakiness index showed a significant reduction in their value (1.28% decrease in water absorption and 2.61% decrease in combined flakiness and elongation index). The decrease in water absorption and combined flakiness and elongation index could be attributed to the separation of weak mortar from the aggregate surfaces during the thermal treatment.

4.2 *Marshall Mix Design*

The determination of mixed design parameters is necessary in order to predict the performance of mix. Marshall stability value is indicative of the resistance towards the permanent deformation. It is largely dependent upon the internal friction provided by the aggregate skeleton and meagrely upon the cohesion provided by bitumen. Sufficient air voids are necessary to enable extra compaction of the bituminous layer during traffic loading and expansion of binder during hot climate. Voids filled with bitumen (VFB) is an essential parameter in order to predict the assurance of adequate bitumen film thickness around the aggregates. Voids in mineral aggregates (VMA) is another design parameter influencing the durability of the bituminous mixes. Marshall tests were conducted on prepared Marshall specimens and Figs. 2, 3, 4, 5 and 6 show the variation of Marshall mix design parameters with bitumen content for DBM mixes containing each type of aggregates.

As it can be observed from Figs. 2, 3, 4, 5 and 6, the optimum values for some mix design parameters achieved at greater bitumen contents in case of mixes containing RCA and TRCA in comparison to NA. As indicated in Fig. 4, maximum Marshall

Fig. 2 Air voids versus bitumen content

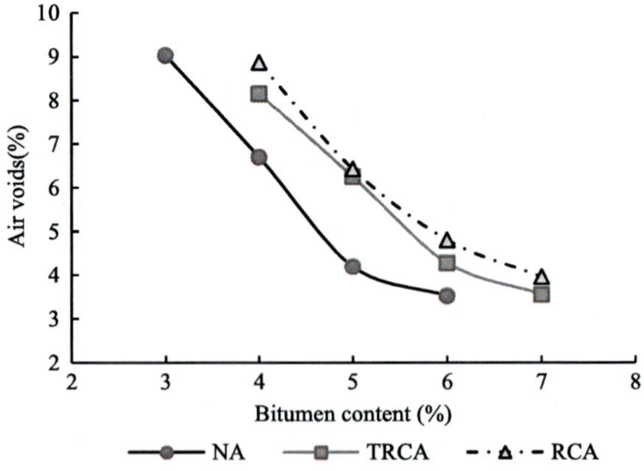

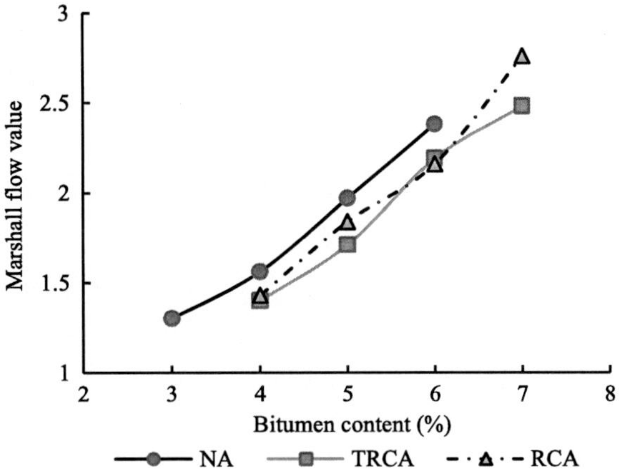

Fig. 3 Marshall flow value versus bitumen content

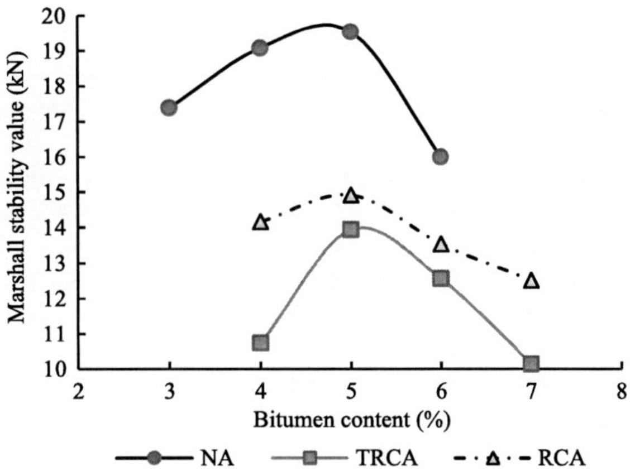

Fig. 4 Marshall stability value versus bitumen content

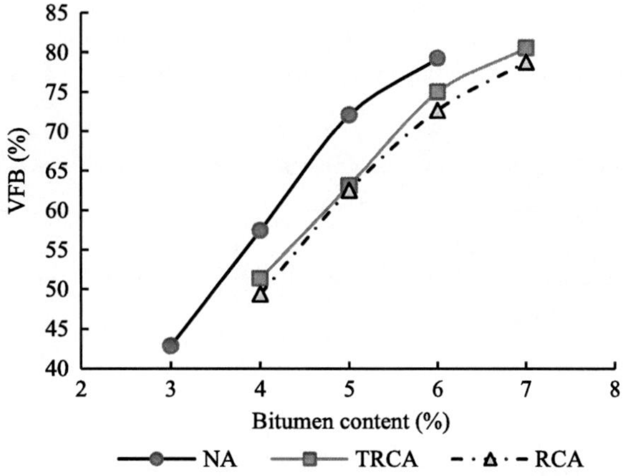

Fig. 5 VFB versus bitumen content

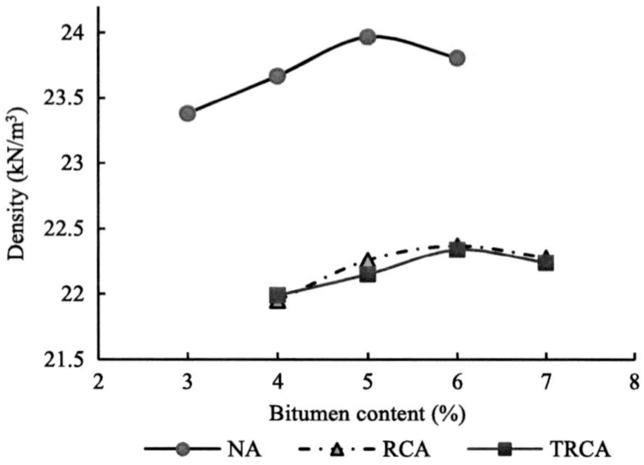

Fig. 6 Density versus bitumen content

stability value was obtained at almost similar bitumen content in case of RCA and TRCA. Similar Marshall flow value trends were observed with an increase in bitumen contents for each aggregate type albeit the shift in bitumen contents. There was a significant decrease in bulk density of mixes containing RCA and TRCA in comparison to NA. Overall, DBM mixes containing NA observed better results as compared to mixes containing RCA and TRCA. The mix design parameters at corresponding optimum bitumen contents for each type of aggregates are tabulated in Table 6.

As observed in Tables 6, there was an increase in optimum bitumen content in the case of DBM mixes containing RCA and TRCA as compared to NA, the reason being the higher bitumen content required to achieve 4% air voids. Although, there was a significant decrease in Marshall stability and density of mixes containing RCA and TRCA as compared to mixes containing NA, these were satisfying the permissible requirements for use in DBM. The utilization of TRCA in DBM mixes improved all other mix properties except Marshall stability value as compared to RCA. Extreme heating of aggregates could be the reason for decrease in Marshall stability in the case of TRCA. Marshall flow value at optimum bitumen content was observed as minimum in the case of NA. VFB and VMA were satisfying the permissible requirements in all three cases. Although, NA showed better results, DBM

Table 6 Marshall mix design parameters corresponding to optimum bitumen content (DBM mixes)

Aggregates used	NA	RCA	TRCA	MORTH limits, 2013
Optimum binder content (%)	5.15	5.95	5.83	Min. 4.5
Marshall stability value (kN)	19.20	13.60	12.90	>9
Marshall flow value	2.00	2.15	2.12	2–4
G_m (kN/m^3)	23.97	22.37	22.32	–
V_v (%)	4.00	4.85	4.60	3–5
VMA (%)	15.78	17.59	17.06	Min. 11–13
VFB (%)	73.00	72.50	73.00	65–75

mixes containing RCA and TRCA were also meeting the required specifications for mix design parameters.

4.3 Retained Stability Test

Resistance towards moisture damage is another factor that affects the durability characteristics of the mix. It was necessary to determine the moisture susceptibility of bituminous mixes prepared with RCA and TRCA due to their higher water absorption. Therefore, Marshall specimens were prepared at optimum bitumen content for each type of aggregates, and retained stability tests were conducted. Tables 7 showed that the retained stability for the mixes with natural aggregates had a higher value as compared to the other two mixes. Thermal treatment of aggregates increased the retained stability value. Lower value of retained stability in case of RCA mixes could be due to more water absorption of RCA resulting in more permeable mixes. In heavy rainfall regions, utilization of RCA might have negative impact on durability of BC

Table 7 Retained stability test results for DBM mixes

Aggregates used		Sample 1	Sample 2	Average value	Retained stability (%)
NA	Marshall stability value of unconditioned specimen (kN)	19.02	19.39	19.21	86.73
	Marshall stability value of conditioned specimen (kN)	16.87	16.45	16.66	
RCA	Marshall stability value of unconditioned specimen (kN)	14.09	13.73	13.91	75.77
	Marshall stability value of conditioned specimen (kN)	10.69	10.38	10.54	
TRCA	Marshall stability value of unconditioned specimen (kN)	12.06	12.39	12.23	83.16
	Marshall stability value of conditioned specimen (kN)	10.35	9.98	10.17	

Table 8 Compaction test results

		Test results
IS method	OMC_{mix}	11.40%
	MDD_{mix}	17.98 kN/m^3
Colorado procedure	OMC of material finer than 4.75 mm (M_f)	14.60%
	MDD of material finer than 4.75 mm (D_f)	17.14 kN/m^3
	Water absorption of material coarser than 4.75 mm (M_c)	4.09%
	Bulk specific gravity of material coarser than 4.75 mm (D_c)	23.48 kN/m^3
	OMC_{mix}	7.51%
	MDD_{mix}	19.83 kN/m^3

mixes. So, RCA mixes might be effectively used in normal rainfall conditions with provision of an adequate drainage system.

4.4 Compaction and CBR Test

Compaction and CBR test had been conducted on RCA as per WMM gradation recommended by MORTH, 2013 [19]. The results of compaction tests are given in Table 8.

A few limitations of IS method of compaction test were observed during the test. The workability of test specimens during compaction was observed as lower, as the mix contained material finer than 20 mm unlike the Colorado procedure, in which material taken for compaction test was finer than 4.75 mm. Another limitation observed was inadequate and non-uniform mixing of water in the mix. Clearly, it can be seen from the results which indicated higher value of OMC for WMM mix than the usual. In view of above, the values of OMC and MDD obtained by Colorado procedure were more reasonable. Therefore, OMC and MDD determined by Colorado procedure were adopted. Then, CBR test was conducted on RCA after four-day soaking of specimen considering the worst moisture conditions, the results of which are given in Fig. 7.

As per MORTH specifications, CBR value for subbase and base course material should be greater than 30% [19] and RCA were satisfying the permissible requirement as indicated in Fig. 7. Thus, these can be used even in worst moisture conditions.

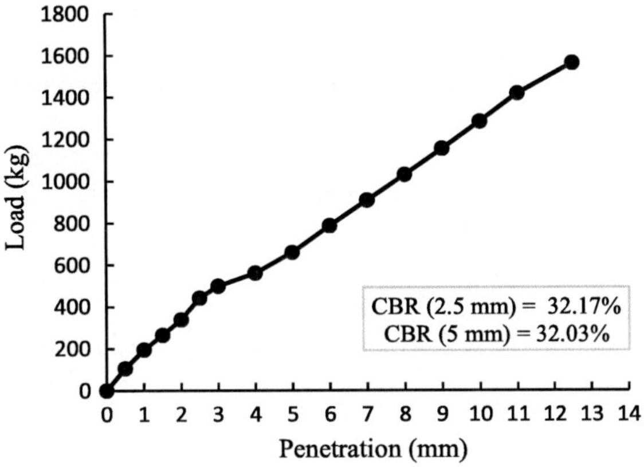

Fig. 7 CBR test results (RCA)

5 Conclusions

Based on the experimental results, the following conclusions have been made:

- The water absorption of recycled concrete aggregates (4.19%) and thermally treated concrete aggregates (2.91%) was more as compared to natural aggregates. All other properties of recycled concrete aggregates (RCA) and thermally treated concrete aggregates (TRCA) were satisfying the permissible limits specified by the MORTH for using in BC, DBM, and WMM layers.
- In case of DBM mixes, OBC for NA, RCA, and TRCA was 5.15%, 5.95%, and 5.83%, respectively. The higher value of optimum bitumen content in case of RCA and TRCA was primarily due to the higher bitumen content corresponding to 4% air voids.
- Thermal treatment decreased the absorption of RCA by removing the weak mortar from the aggregate surfaces. Hence, in case of DBM mixes prepared with TRCA, optimum bitumen content was lesser as compared to RCA.
- Although the thermal treatment of aggregates resulted in decrease in OBCas well as improved retained stability in comparison to RCA, but it affected the Marshall stability value which was lesser as compared to mixes containing NA and RCA for DBM mixes.
- There was significant reduction in Marshall stability value for mixes containing RCA (13.60 kN) in comparison to mixes containing NA (19.20 kN). As for VFB and VMA, all three types of aggregates were satisfying the permissible requirements.
- Retained stability value for RCA (75.77%) for DBM was marginally above minimum required value of 75% as per ASTM. In view of the above, it is recommended that use of RCA may be restricted to normal rainfall regions with effective drainage or it may be used with anti-stripping compounds in high rainfall regions.
- CBR value for RCA after four-day soaking (32.17%) was observed as more than the minimum CBR value (30%) required for using them in base or subbase course.

Colorado procedure was observed as a more suitable method for compaction test of aggregates for base/subbase course.

- Although thermal treatment helped in improving the properties of RCA, but the practical feasibility of this process needs to be determined. Considering the higher cost, higher energy consumption, and lesser influence of TRCA in comparison to RCA, the utilization of untreated RCA is recommended in DBM mixes. Other than environmental and economic benefits, there are various other indirect benefits of utilization of C&D waste such as effective use of land, energy conservation, etc.

Acknowledgements The authors of this paper are grateful to the Punjab Engineering College, Chandigarh (Deemed to be University) for providing all the facilities and funding for successful completion of the research work.

References

1. Ministry of Environment, Forests and Climate Change (MoEFCC) (2017) Guidelines on environmental management of construction and demolition (C & D) wastes. Central Pollution Control Board, New Delhi, India
2. Shaikh FUA, Nguyen HL (2013) Properties of concrete containing recycled construction and demolition wastes as coarse aggregates. J Sustain Cement-Based Mater 2(3–4):204–217
3. Shahidan S, Azmib MAM, Zuki SSM, Kupusamy K, Alie N (2017) Utilizing construction and demolition (C&D) waste as recycled aggregates (RA) in concrete. In: Global congress on manufacturing and management, Zhengzhou, China, 28–30 Nov 2016, pp 1028–1035
4. Wagih AM, El-Karmoty HZ, Ebid M, Okba SH (2012) Recycled construction and demolition concrete waste as aggregate for structural concrete. Housing Build Nat Res Centre 9(3):193–200
5. Yang J, Du Q, Bao Y (2010) Concrete with recycled concrete aggregate and crushed clay bricks. Constr Build Mater 25(4):1935–1945
6. IS 383 (2016) Coarse and fine aggregate for concrete—specifications. Bureau of Indian Standards, New Delhi, India
7. NBC CED 46 (2005) National building code of India. Bureau of Indian Standards, New Delhi, India
8. Arulrajah A, Piratheepan J, Bo MW, Disfani MM (2013) Geotechnical and geoenvironmental properties of recycled construction and demolition materials in pavement subbase applications. J Mater Civ Eng 25(8):1077–1088
9. Esfahani MA (2018) Evaluating the feasibility, usability, and strength of recycled construction and demolition waste in base and subbase courses. Road Mater Pavement Des 21(1):156–178
10. Leite FDC, Motta RD, Vasconcelos KL, Bernucci L (2011) Laboratory evaluation of recycled construction and demolition waste for pavements. Constr Build Mater 25(6):2972–2979
11. Tahmoorian F, Samali B (2017) Experimental and correlational study on the utilisation of RCA as an alternative coarse aggregate in asphalt mixtures. Australian J Civil Eng 15(2):80–92
12. Panda M, Giri JP, Sahoo UC (2018a) Performance of bituminous mixes containing treated recycled concrete aggregates and modified by waste polyethylene. J Mater Civ Eng 30(8):04018184–04018191
13. Panda M, Giri JP, Sahoo UC (2018b) Performance of bituminous mixes containing emulsion-treated recycled concrete aggregates. J Mater Civ Eng 30(4):04018052–04018061
14. Wu S, Zhong J, Zhu J, Wang D (2015) Influence of demolition waste used as recycled aggregate on performance of asphalt mixture. Road Mater Pavement Des 14(3):679–688

15. Bennert T, Papp WJ, Maher A, Gucunski N (2016) Utilization of construction and demolition debris under traffic-type loading in base and subbase applications. Transp Res Rec 1714(1):33–39
16. Barbudo A, Agrela F, Ayuso J, Jimenez JR, Poon CS (2012). Statistical analysis of recycled aggregates derived from different sources for sub-base applications. Constr Build Mater 28(1):129–138
17. Yang T, Daquan S, Guoqiang TS, Qi P, Fan Y, Xingyi Z (2018) Performance evaluation of asphalt mixtures containing recycled concrete aggregates. Int J Pavement Eng 19(5):422–428
18. Mahendra SP, Puneeth HC, Rohith M (2018) Characterization of construction and demolition waste aggregates for rigid pavement construction. Indian Highways 46(7):37–42
19. Ministry of Road Transport and Highways (MORTH) (2013) Specifications for roads and bridge works. In: Indian road congress, New Delhi, India
20. AASHTO T245–97 (2001) Resistance to plastic flow of bituminous mixtures using marshall apparatus. American Association of State and Highway Transportation Officials, Washington, DC
21. Asphalt Institute Manual Series No. 2 (2014) Asphalt mix design methods. Asphalt Institute, Lexington, USA
22. ASTM D1075 (2011) Standard test method for effect of water on compressive strength of compacted bituminous mixtures. American Association of State and Highway Transportation Officials, West Conshohocken, PA
23. IS 2720 (Part VII) (2011) Determination of water content-dry density relation using light compaction. Bureau of Indian Standards, New Delhi, India
24. Colorado Procedure 23-13 (2017) Determining maximum dry density and optimum moisture content of soil-rock mixtures. In: 2018 CDOT fields materials manual. Colorado Springs, CO, United States
25. AASHTO T85 (2001) Specific gravity of coarse aggregate. American Association of State and Highway Transportation Officials, Washington, DC
26. AASHTO T99 (2001) Moisture-Density relations of soils using a 2.5-kg (5.5-lb) rammer and a 305-mm (12-in.) drop. American Association of State and Highway Transportation Officials, Washington, DC
27. IS 2720 (Part 16) (2002) Laboratory determination of CBR. Bureau of Indian Standards, New Delhi, India

Application of Waste Bones in Civil Engineering Practices

Rajwinder Singh and Karanvir Singh Sohal

1 Introduction

The demand of constructional materials has increased rapidly over a small period because of the urbanization. The materials like cement, fine and coarse aggregates have been used in the excessive quantities to meet the demand of development rate. Out of these essential products, cement is the essential building material that has found its usage in almost every structure making process. In the year 2017, around 270 million tons of cement was produced by the cement industries, which is further expected to increase by about 290 million in next year [45]. With the increase in the cement production rates, the amount of CO_2 generation has also increased rapidly, which is an issue to concern about. During the production of 1 ton of cement by the industries, around 1 ton of carbon dioxide gas is emitted to the atmosphere [18]. However, a significant reduction in the amount of such materials has been noticed, which has triggered the researchers to find an alternative way to meet the growth rates [2]. The wastes from the agricultural and industrial fields have already been used by various researchers as partial replacement of cement, fine and coarse aggregates for the production of concrete. However, the positive influence on the properties of concrete has been caused by the integration of various wastes [4, 6, 15, 20, 24, 40]. It has also been reported that the addition of wastes in the concrete will not only stabilize the waste by-products but also has resulted into lowering the overall cost of construction [5, 8, 26, 41]. Apart from various waste materials generated from their respective fields, slaughterhouse waste is one of those wastes which are not handled/appropriately treated before disposing of thus causing air and

R. Singh (✉)
Dr. B.R. Ambedkar National Institute of Technology, Jalandhar, Punjab, India
e-mail: rajwinderbirdi2@gmail.com

K. S. Sohal
Guru Nanak Dev Engineering College, Ludhiana, Punjab, India

H. Singh et al. (eds.), *Sustainable Development Through Engineering Innovations*, Lecture Notes in Civil Engineering 113,
https://doi.org/10.1007/978-981-15-9554-7_28

water problems [16, 33]. According to a report, meat consumption in India is around 1.4 kg per capita per annum. In contrast, this amount goes up to 60–90 kg per capita per annum in the European countries, reported by CPHEEO. After the consumption of meat-based products, animal bones are directly dumped in landfills without any prior treatment, which creates various health and environmental problems [11, 35]. To stabilize/reduce the adverse impacts caused by animal bones, researchers have worked on the addition of waste bones to the concrete as various forms of replacements (cement, fine and coarse) [9, 36, 41, 46].

This review paper aims to gather the available information on the utilization of waste bones as a partial replacement of cement, fine and coarse aggregates in the concrete. Based on the observations, the authors will also suggest the new ideas which can be carried out by the readers in the future to explore the novel/hidden possibilities to use this waste in their respective fields.

2 Physical and Chemical Properties of Bone Powder

2.1 Physical Properties

This section comprises the introduction of various physical properties of bone powder used in the previous studies. The range or specific values used in such studies are also written and given in Table 1.

Table 1 Represents the values of specific gravity and density of bone powder found by various authors

Author(s)	Specific gravity	Density (kg/m^3)
Ogarekpe et al. [29]	1	2088
Wamanuel et al. [48]	2.9	–
Akinyele et al. [3]	1.83	–
Falade et al. [12]	2.22	–
Vispute et al. [47]	1.95	775
Otunyo et al. [36]	2.41	1022
Bhat et al. [9]	1.61	822
Navya [28]	1.4	–
Kotb et al. [21]	1.85	690
Subrahmanyam and Dhana [43]	2.18	1510
Falade et al. [14]	2.22	–

2.1.1 Appearance

Usually, the color of the bone powder is light whitish [3, 4], whereas, in some cases, the color can also be light yellow [21].

2.1.2 Water Absorption

Due to the porous nature, the particles absorb much water. The application of the grinding process will also lead to produce particles of smaller size, which will result in the more absorption of water on the surface. The range of water absorption lies from 2.5 to 4.4% [43, 47, 48].

2.1.3 Fineness Modulus

The rate of hydration is affected by the fineness of particles. The fineness modulus of bone powder particles ranges from 3.81 to 6.45 [37, 47].

2.2 Chemical Properties

The chemical composition of bone powder used by [3, 12, 14, 21, 22, 27, 31, 32, 48, 44] is given in Table 2.

3 Optimal Value for Replacement

This section covers the optimal values of replacement of essential material used for the production of concrete such as cement, fine and coarse aggregates with bones in crushed or powder form. The additional materials used along with bones are also given in Table 3.

4 Results and Discussions

This section includes the outcomes of various durability and mechanical properties of concrete made with the replacement of cement, fine and coarse aggregates. This section comprises of three subsections that cover the properties of each material in detail.

Table 2 Shows the percentage of chemical compound found in bone powder

Compound	% Range
Al_2O_3	0.01–13.21
CaO	47.77–76.31
SiO_2	0.01–19.02
.MgO	0.21–2.58
Fe_2O_3	0.01–0.52
K_2O	0.01–0.51
P_2O_5	4.96–36.85
Na_2O	0.36–1.67
BaO	–
MnO	0.03
SrO	0.083–0.166
TiO_2	0.01–0.049
CuO	–
CuO	0.015–0.28
P_2O_5	–
Cl	0.148–0.159
ZnO	0.007–0.024
H2O	0.75–1.62
MnO	0.01–8.52
MnO_2	0.086
SO_2	1.24
SO_3	0.059–0.41
LOI	0.37–5.89

4.1 Properties of Concrete Made with Replacement of Cement with Bone Powder

4.1.1 Setting Times

The setting times of cement increase as the content of bone powder is increased for the replacement of cement [31, 48, 12, 44]. This suggests that because of the addition of bone powder to the cement, the rate of hydration gets reduced [48]. The setting time of the paste containing bone powder does not fall under the range specified by [10]. Authors have also suggested the application of this type of mortar in the tropical areas to prevent the cracks [31].

Table 3 Shows the optimum values for the replacement of cement, fine and coarse aggregates suggested by various authors

Author(s)	Material used	Replacement criteria (%)	Optimum value (%)
For cement			
Mirzabagheri et al. [25]	Cuttlebone	0, 3, 5, 7	3
Falade et al. [13]	Pulverized bone	0, 5, 10, 15, 20	20
Adekunleand Daramola [4]	Sawdust and rice husk + bone powder	0, 5, 10, 15, 20	Saw dust 10 Rice husk 15
Singh [41]	Animal bone powder + polypropylene fiber	ABP 0, 5, 10, 15 PPF 1, 1.5, 2	BP 10 PPF 1.5
Anwar et al. [7]	Bone powder	0, 5, 10, 15, 20	5
Kalra et al. [19]	Fly ash + metakaolin + bone powder	FA 0, 10, 20 MK 0, 10, 20 BP 0,5	Fly ash (10) + metakaolin (10) + bone powder (5)
Falade et al. [12]	Bone powder	0, 10, 20, 30, 40, 50, 60, 70, 80, 90, 100	20
Olutaiwo et al. [32]	Cow bone ash	0, 5, 10, 15, 20, 25, 30, 40, 50	20
Ramya and Shanthi [37]	Banana leaf + cattle bone powder	20, 30, 40 (BLP + CBP)	30
Kotb et al. [21]	Bone powder	0, 5, 10, 15	5
Falade et al. [14]	Bone powder	0, 5, 10, 15, 20	20
Akinyele et al. [3]	Bone powder	0, 5, 10, 15, 20, 25	10
Opeyemi and Makinde [34]	Bone powder + rice husk ash	0, 5, 10, 15, 20	10
Tsado et al. [44]	Cow bone ash + egg shells	CBA 0, 5, 10, 15, 20 ESA 0, 5, 10, 15, 20	CBA (15) ESA (5)
Okoye and Odumodu [31]	Bone ash	0, 10, 20, 30, 40, 50, 60, 70, 80, 90, 100	20
Varma et al. [46]	Bone ash	0, 5, 10, 15, 20	15
Falade et al. [14]	Pulverized bone	0, 5, 10, 15, 20	20
Opeyemi and Makinde [34]	Rice husk and bone powder	0, 5, 10, 15, 20	10
Wamanuel et al. [48]	Bone powder	0, 5, 10, 15, 20, 25	10
Akinyele et al. [3]	Bone ash + wood ash	0, 5, 10, 15, 20, 25	10
Replacement of fine aggregates			
Rao et al. [38]	Bone char sludge	1.5, 3, 6, 9	3
Vispute et al. [47]	Bone powder + fly ash	Fine aggregates with BP (20) Cement by FA(10)	BP (20) FA (10)
Ogarekpe et al. [29]	Crushed cow bone	10, 20, 30, 40, 50	40–50

(continued)

Table 3 (continued)

Author(s)	Material used	Replacement criteria (%)	Optimum value (%)
Siddique et al. [39]	Ceramic waste	20, 40, 60, 80	60
Replacement of coarse aggregates			
Bhat et al. [9]	Crushed bone aggregates	0, 25, 35, 50, 75, 100	50
Subrahmanyam and Dhana [43]	Animal bones + fly ash + silica fumes	AB 10, 20, 30, 40 FA 20 Silica fumes 5	AB (20) FA (20)

4.1.2 Consistency

The water requirement for making normal consistency improves as the amount of bone powder is increased to replace the cement [48]. However, Falade et al. [12] revealed that a smaller amount of water is needed to maintain the required consistency of paste, and this requirement reduces as the amount of bone powder is increased.

4.1.3 Workability

The slump value of concrete gets reduced when the content of bone powder is increased in the mix [14, 22, 31, 44, 48]. Kalra et al. [19] and Anwar et al. [7] show that the slump value of concrete increases as the content of bone powder is increased. Significant improvement in the concrete mixes has been noticed.

4.1.4 Compressive Strength

Various authors have worked to examine the effect of utilization of bone powder as a replacement of cement for the production of concrete and found that the compressive strength of specimens gets decreased as the content of bone powder is increased after 28 days of curing when compared to control mix [4, 7, 13, 22, 31, 34, 41]. The reason behind this reduction might be the lowering the content of production of calcium silicate hydrate in the concrete mix due to decrease in the total amount of ordinary Portland cement [31]. As reported in the chemical analysis, bone powder consists of a higher amount of calcium rather than silica, which indicates that the production of di-calcium silicate and tri-calcium silicate would also be less during the hydration reaction [48]. The similar pattern is observed in case of mortar by [12]. In special type of concrete (foam concrete), the concrete made with replacement of 20% of bone powder qualifies the strength requirement for the foamed concrete [14].

Apart from these studies, some authors have also revealed that the strength of specimens containing bone powder increases [1] when replacement of 20% [44],

10% [3, 37, 46], 7.5% [46] and 5% [19] is done. Even, the mortar containing bone powder shows the best performance at 5% replacement of cement [21].

4.1.5 Split Tensile Strength

The specimens containing bone powder show the reduction in the split tensile strength when cured at a more prolonged period when compared to the control mix [13, 14, 41, 48].

Whereas, in some studies, the increase in the tensile strength is reported at shows that at replacement level of 5% [19] of concrete and [21] mortar. The authors suggested the reason behind it was due to the filer action of bone powder, which fills the pores of specimens, thus increasing the strength of paste [21].

4.1.6 Flexural Strength

There were variations in the flexural strength of the specimens containing different dosages of bone powder with the curing time. Wamanuel et al. [48] reported the increment behavior in the flexural strength with the curing time, but, even after 28 days, the strength was below than the control mix. However, Kalra et al. [19], have reported the increase in the strength at 5% replacement of bone powder.

4.1.7 Water Absorption

The minimum reduction in the water absorption of concrete specimens containing bone powder is reported at the 5% replacement of cement. The water absorption obtained at this percentage is 15% less than the control mix [19].

4.1.8 Drying Shrinkage

Because of the addition of bone powder in the concrete mix, the microstructure of the specimens gets denser due to the pore refinement property of this material, which ultimately reported to reduce the shrinkage [19].

4.2 Properties of Concrete Made with Replacement of Fine Aggregates with Bone Powder

4.2.1 Setting Times

The addition of bone powder has resulted in increasing the setting times of the specimens, which is because of the reduction in the cement content required for the hydration reaction [36].

4.2.2 Workability

The concrete mixes show the reduction in the slump values as the percentage of bone powder is used to replace fine aggregates. The obtained paste reported to have a harsh nature due to the more water absorption of bone particles [29, 36].

4.2.3 Compressive Strength

The effect of the addition of crushed bones as a replacement material of fine aggregates in concrete specimens was checked by various authors and found that with the strength of specimens was decreased considerably as the amount of crushed bones is increased [17, 29, 36], whereas Vispute et al. [47] reported that the concrete specimens made with the replacement of fine aggregates with crushed bones showed the improvement in the compressive strength of concrete at all curing ages when compared to control mixes.

4.2.4 Water Absorption Capacity

The water absorption capacity of specimens containing bone aggregates showed higher absorption when comparison was made with the control mix. This was because of the voids caused by the bones, which allows the water to pass through the concrete [17].

4.3 Properties of Concrete Made with Replacement of Coarse Aggregates with Crushed Bone Aggregates

4.3.1 Workability

Due to the higher water absorption capacity of crushed bones than normal coarse aggregates, the workability of concrete reduces. This can be because of the denser nature of natural coarse aggregates than crushed bone aggregates. Thus, the addition of crushed bones to the concrete results to increase the specific surface area, which is attributed to the lightweight of crushed bone aggregates. Therefore, more quantity of water is required to make a proper concrete mix to have denser packing of the mix [9, 43].

4.3.2 Compressive Strength

In most cases, the use of waste materials as a replacement material in the concrete has provided satisfactory results. The use of coarse sized bone aggregates is one of those materials. It has been reported that when waste bones are crushed to the size of coarse aggregates and added to a proportion in the concrete mix. The compressive strength is said to be much higher than the control specimens [23, 28, 43].

4.3.3 Split Tensile Strength

The integration of crushed waste bones has resulted to influence the split tensile strength of the concrete mix when curing is done up to 28 days [28, 43].

4.3.4 Flexural Strength

The flexural strength of concrete mix made with the partial replacement of coarse aggregates with crushed bones has also reported increasing when compared to control concrete after 28 days of curing [28, 43].

5 Conclusion

Based on the available literature, the findings of this review paper are made as follows:

The extensive work is done only on the replacement of cement with bone powder rather than on fine and coarse aggregates as crushed form. The overall amount of work that has been done to date on the addition of bones (as a powder or crushed form) in concrete is very less, and based on the thoroughly examined literature, it is concluded that bones can be used as a placement material for the cement and coarse aggregates. Better results have been obtained when crushed bones were used to replace coarse aggregates, whereas the concrete made with replacement of fine aggregates failed to show good performance. The concrete made with the addition of bones (in powder/crushed form) has resulted in producing lightweight and cost-effective concrete. The application of concrete made with the addition of this waste is suggested to be used for temporary constructions. By using this waste as an alternative material, the reduction in emissions of greenhouse gasses can be done.

Authors suggest the readers to use this waste material for the concrete production by lowering down the intervals of replacement percentage, i.e. (2, 4, 6, 8, 10 up to 50), to know the optimal value of replacement, which can be done without affecting the overall performance of concrete. The effect on concrete made with bones processed through burning and grinding or crushing at different temperatures and time, respectively, can be evaluated. The work on the impact on the properties of concrete by the addition of this waste with other supplementary wastes is also not available in depth. Readers are suggested to utilize the bones (in the form of powder) in different civil engineering practices such as stabilization of soil to reduce its disposal and hazard effect on the human health and environment.

Supplementary Info

Waste bones are one of those wastes whose generation cannot be restricted as in some of the countries; the meat obtained from the cow is considered as a primary source of protein. The generation of waste bones from slaughterhouses is a major hazard to human health and the environment. It has been reported that only in Lagos State (Nigeria) around 55,000 tons per annum. Still, this number does not include the unofficially slaughtering during the parties or any festivals. In India, an earning around 7000 crores has been done only through meat exportation in the year 2010–2011, and an extreme increase in the demand has been noticed in the past few years. Worldwide, Nigeria is the only region which has given the special importance to the utilization and safe disposal of waste bones in civil engineering practices.

Acknowledgements Authors would like to acknowledge Dr. Mahesh Patel, Civil Engineering, DR. B.R. Ambedkar National Institute of Technology, Jalandhar, Punjab, India, for providing his support and encouragement throughout the writing of this paper.

References

1. Abubakar M, AbdullahI M, Aguwa JI (2016) Probability based design of concrete mixes with cow-bone ash admixed cement. Leonardo J Sci 28:31–42
2. Adesanya DA, Raheem AA (2009) A study of the workability and compressive strength characteristics of corn cob ash blended cement concrete. Constr Build Mater 23(1):311–317
3. Akinyele JO, Adekunle AA, Ogundaini O (2016) The effect of partial replacement of cement with bone ash and wood ash in concrete. Ann Faculty Eng Hunedoara 14(4):199
4. Aliu Adekunle O, Daramola AS (2013) Optimisation for the use of rice husk ash and sawdust as alternative binder for concrete. Int J Eng Sci 2(10):39–42
5. Alizadeh R, Chini M, Ghods P, Hoseini M, Montazer S, Shekarchi M (2003, June) Utilization of electric arc furnace slag as aggregates in concrete–environmental issue. In: Proceedings of the 6th CANMET/ACI international conference on recent advances in concrete technology. Bucharest, Romania, pp 451–464
6. Al-Zaid RZ, Al-Sugair FH, Al-Negheimish AI (1997) Investigation of potential uses of electric-arc furnace dust (EAFD) in concrete. Cem Concr Res 27(2):267–278
7. Anwar FH, Malami SI, Baba ZB, Farouq MM, Labbo MS, Aliyu DS, Umar AB (2019) Compressive strength of lightweight concrete and benefit of partially replacing cement by animal bone ash (ABA). J Emerg Technol Innov Res 6(6):555–560

8. Aprianti E, Shafigh P, Bahri S, Farahani JN (2015) Supplementary cementitious materials origin from agricultural wastes–a review. Constr Build Mater 74:176–187
9. Bhat JA, Qasab RA, Dar AR (2012) Machine crushed animal bones as partial replacement of coarse aggregates in lightweight concrete. ARPN J Eng Appl Sci 7(9):1202–1207
10. EN B 196-3 (1995) Method of testing cement-3: determination of setting time and soundness. British Standard Institute, 389
11. Ezeoha SL, Ugwuishiwu BO (2011) Status of abattoir wastes research in Nigeria. Nigerian J Technol 30(2):143–148
12. Falade F, Ikponmwosa E, Fapohunda C (2012) Potential of pulverized bone as a pozzolanic material. Int J Sci Eng Res 3(7):1–6
13. Falade F, Ikponmwosa E, Fapohunda C (2013) Environmental benefits of the structural use of pulverised bone in the production of foamed aerated concrete. J Emerg Trends Eng Appl Sci 4(5):717–722
14. Falade F, Ikponmwosa E, Fapohunda C (2016) A study on the compressive and tensile strength of foamed concrete containing pulverized bone as a partial replacement of cement. Pakistan J Eng Appl Sci 13:82–93
15. Filipponi P, Polettini A, Pomi R, Sirini P (2003) Physical and mechanical properties of cement-based products containing incineration bottom ash. Waste Manage 23(2):145–156
16. Franke-Whittle IH, Insam H (2013) Treatment alternatives of slaughterhouse wastes, and their effect on the inactivation of different pathogens: a review. Crit Rev Microbiol 39(2):139–151
17. Geetha SD, Kavyashree MP, Krupaksha HK, Sthuthi KA, Viksheetha AR (2018) Utilization of bone aggregate and mangalore tiles waste as a partial replacement to natural sand and course aggregate in concrete. Int J Sci Res Dev 4(12):247–251
18. Hanle LJ, Jayaraman KR, Smith JS (2004) CO_2 emissions profile of the US cement industry. Environmental Protection Agency, Washington DC, pp 1–14
19. Kalra M, Singh NB, Kumar M (2016) Properties of concrete made from ternary blended cement in the presence of animal bone powder. Int J Civil Eng Technol 7(6)
20. Kaur G, Siddique R, Rajor A (2013) Micro-structural and metal leachate analysis of concrete made with fungal treated waste foundry sand. Constr Build Mater 38:94–100
21. Kotb M, Assas M, Abd-Elrahman H (2010) Effect of grounded bone powder addition on the mechanical properties of cement mortar. WIT Trans Ecol Environ 138:201–212
22. Lamidi IO, Olomo RO, Mujedu KA, Alao MO (2017) Evaluation of rice husk ash and bone ash mixed as partial replacement of cement in concrete. Technology (ICONSEET) 2(34):258–264
23. Makunza J (2014, June) Investigation on fired animal bones as aggregate replacement in concrete. In: RILEM international workshop on performance-based specification and control of concrete durability, pp 401–408
24. Mehta PK (1983) Pozzolanic and cementitious byproducts as mineral admixtures for concrete-a critical review. Special Publ 79:1–46
25. Mirzabagheri S, Derhamjani G, Maharati S, Ziaee Z, Vatankhah F, Mirzabagheri D (2018) Using cuttlebone powder to produce green concrete. J Appl Eng Sci 8(2):25–28
26. Mujedu KA, Adebara SA, Lamidi IO (2014) The use of corn cob ash and saw dust ash as cement replacement in concrete works. Int J Eng Sci 3(4):22–24
27. Naga SM, Awaad M, El-Mehalawy N, Antonious MS (2014) Recycling of fish bone ash in the preparation of stoneware tiles. Interceram-Int Ceram Rev 63(1–2):15–18
28. Navya D (2017) Evaluation of sustainable green concrete with partial replacement of cement by rice husk ash and partial replacement of coarse aggregate by crushed cattle bones. Int Res J Eng Technol 4(6):2602–2609
29. Ogarekpe NM, Agunwamba JC, Idagu FO, Bejor ES, Eteng OE, Ndem HO, Oloko EO (2017) Suitability of burnt and crushed cow bones as partial replacement for fine aggregate in concrete. Nigerian J Technol 36(3):686–690
31. Okoye FN, Odumodu OI (2016) Investigation into the possibility of partial replacement of cement with bone powder in concrete production. Int J Eng Res Dev 12(10):40–45
32. Olutaiwo AO, Sefiu YO, Ezegbunem II (2018) Utilizing cow bone ash (CBA) as partial replacement for cement in highway rigid pavement construction. SSRG Int J Civil Eng 5(2):13–19

33. Omole DO, Ogbiye SA (2013) An evaluation of slaughterhouse wastes in south-west Nigeria. Am J Environ Prot 2(3):85–89
34. Opeyemi DA, Makinde OO (2012) The suitability of partial replacement of cement with rice husk ash and bone powder in concrete structures. Int J Emerg Technol Adv Eng 2(9):261–265
35. Oruonye ED (2015) Challenges of abattoir waste management in Jalingo metropolis. Nigeria. Int J Res Geogr 1(2):22–31
36. Otunyo AW, Umueyo LE, Kingsley S (2014) Machine crushed cow bones as a partial replacement of fine aggregates in lightweight concrete. ARPN J Eng Appl Sci 9(12):2799–2806
37. Ramya M, Shanthi R (2016) Evolution of sustainable green concrete with partial replacement of cement by banana leaf powder and cattle bone powder. Ind J Appl Res 6(3):884–886
38. Rao SM, Reddy BV, Lakshmikanth S, Ambika NS (2009) Re-use of fluoride contaminated bone char sludge in concrete. J Hazard Mater 166(2–3):751–756
39. Siddique S, Shrivastava S, Chaudhary S (2018) Influence of ceramic waste as fine aggregate in concrete: Pozzolanic, XRD, FT-IR, and NMR investigations. J Mater Civ Eng 30(9):04018227
40. Sikalidis C, Mitrakas M (2006) Utilization of electric arc furnace dust as raw material for the production of ceramic and concrete building products. J Environ Sci Health Part A 41(9):1943–1954
41. Singh H (2019) Study on partial replacement of cement with animalbone powder along with the addition of polyproplyene fiber in concrete. Int J Eng Res Technol 8(1):118–123
43. Subrahmanyam M, Dhana Sri K (2016) Comparative study on effect of silica fume and animal bone on concrete by partial replacement of coarse aggregate with animal bones and cement with fly ash. Int J Eng Res Technol 5(6):273–277
44. Tsado T, Adejumo T, Adamu Y Performance evaluation of egg-shells and cow bone ashes as pozzolana in concrete production
45. US Geological Survey (2018) Major countries in worldwide cement production from 2012 to 2017 (in million metric tons). https://www.statista.com/statistics/267364/world-cement-productionby-country/. Accessed 3 Mar 2018
46. Varma S, Naidu M, Mohan S, Reddy D (2016) An Effective study on utilizing bone powder ash as partial replacement of construction material. Int J Innov Technol Res 4(3):3060–3062
47. Vispute G, Suryawad S, Adhore A, Tupsoundare P (2018) Performance of concrete by using animal bone powder and fly ash. Int J Adv Res Innov Ideas Educ 4(3):1608–1617
48. Wamanuel AH, Quezon ET, Busier M (2018) Effects of varying dosage replacement of cement content by animal bone powder in normal concrete mix production. Am J Civil Eng Arch 6(4):133–139

Impact of Fiber Hybridization on Performance of Engineered Cementitious Composite

Maninder Singh, Babita Saini, and H. D. Chalak

1 Introduction

In recent years, rapid advances have been made in the construction industry with the development of new cement-based products. According to present scenario high-performance fiber-reinforced cementitious composites (HPFRCC) is one of the best-suited cement-based products used for the construction purposes. The superior tensile and durability properties of HPFRCC products promote its use under aggressive environmental conditions. The HPFRCC products are designed with the optimized use of polymeric fibers and constituents. Engineered cementitious composite (ECC), reactive powder concrete (RPC) and strain-hardening cementitious composite (SHCC) products are the well-known examples of HPFRCC. From the above discussed products, ECC is the superb matrix in construction industry which is commonly known due to its bendable behavior and ductile performance. The design of ECC relies on micromechanical principle. Tight crack width (<100 μm) and excellent tensile strain capacity (1–8%) are the novel parameters of ECC which made it alter than other types of cementitious products [1]. The narrow crack width of ECC, controlled matrix fracture toughness and also exhibits excellent durability performance under susceptible conditions [2]. Sequential development of multiple tiny cracks in ECC results into strain hardening, which is similar to ductile nature of metal [3–5]. The

M. Singh (✉) · B. Saini · H. D. Chalak
Department of Civil Engineering, National Institute of Technology Kurukshetra, Kurukshetra, Haryana, India
e-mail: maninder_6160049@nitkkr.ac.in

B. Saini
e-mail: bsaini@nitkkr.ac.in

H. D. Chalak
e-mail: chalakhd@nitkkr.ac.in

H. Singh et al. (eds.), *Sustainable Development Through Engineering Innovations*, Lecture Notes in Civil Engineering 113,
https://doi.org/10.1007/978-981-15-9554-7_29

novel characteristics and ductile nature of ECC promote its use in various cement-based structural applications. The cement consumption and cost of fibers minimize the use of ECC in large-scale structural application. Numerous research investigations reported that the composition of ECC mainly consists of finer size constituents and low content of polymeric fibers [6–10]. Therefore, to reduce the cost of ECC finer size constituents can be replaced with supplementary cementitious materials (SCM) in high quantity. Literature reported that fly ash as SCM and polyvinyl alcohol (PVA) fiber as internal reinforcement are the most widely used constituents for the typical ECC mixtures [1–3]. Fly ash is produced during the combustion of coal in thermal power plants. However, in present scenario most of the countries around the globe are moving fast toward the sustainable and green power generation. On the other hand, cement-based product manufacturers association has shown concern about the shortage in supply for good quality of fly ash. The PVA fibers are most widely used to reinforce the ECC; but high cost of these fibers limit the use of ECC in large-scale structural applications. PVA fibers are manufactured by some limited industrial groups/countries in the world which also limit its availability. Therefore, after observing all the above-discussed parameters, authors in current research work have tried to evaluate the performance of ECC with the utilization of high quantity of slag (iron industry waste) as SCM. Two different types of low cost and indigenous fibers, i.e., metallic (steel) and non-metallic (polyester) fibers, were used to reinforce the cementitious product. The hybrid use of these fibers was also used at different percentages to reinforce the ECC. The present study emphasizes on the impact of steel and polyester fibers in hybridization on various parameters of ECC matrix. Therefore, utilization of these products can save the overall cost, environmental ecosystem and also provides better guidelines for future design of ECC matrix.

2 Constituents and Mix Proportions

The powdered constituents selected during this research work were Portland cement (PC) confirming to IS 8112 (2013) [11] and ASTM C150 [12], fine river sand (RS), and slag (SL) (iron industry waste) as supplementary cementitious material. The two different types of fibers such as polyester (PET) and steel (SE) fibers were used in cementitious mixtures as internal reinforcement. Potable water confirming to IS 456:2000 [13] and high range water reducer admixture (HRWRA) were added into mixed solid constituents to achieve the required fresh properties. In current investigation, 70% (of the total cementitious products) iron industry waste (SL) and non-degradable polyester fibers have been utilized in the cement matrix. The addition of high volume of waste product and low-cost fibers help in saving the natural resources, overall cost and contribute in making the eco-friendly sustainable cement matrix. To examine the influence of fiber hybridization on the performance of cementitious mixes, total six mixes were prepared. The proportion of

Table 1 Mix proportions of constituents used in ECC

Mix ID	Ingredients						
	PC	SL	RS	W/B	PET fiber (%)	SE fiber (%)	SP (%)
H1	1	2.2	0.8	0.29	2	0	0.6
H2	1	2.2	0.8	0.29	1.6	0.4	0.6
H3	1	2.2	0.8	0.29	1.2	0.8	0.6
H4	1	2.2	0.8	0.29	0.8	1.2	0.6
H5	1	2.2	0.8	0.29	0.4	1.6	0.6
H6	1	2.2	0.8	0.29	0	2	0.6

selected constituents and designation of various cementitious mixes have been given in Table 1.

3 Experimental Program

3.1 Mixing and Casting Process

The mixing process of cementitious composites was done by using power driven mortar mixer as described below:

- Firstly, all the selected powdered constituents (PC, RS, and SL) were added into the drum of mixer and thoroughly mixed at low speed for 2–3 min.
- Then approximately 65% quantity of selected water and HRWRA were poured into the drum of mixer in rotating condition and the rotating process continued for 2 min.
- After that, remaining part of liquid constituents was added into the mixed components and rotated the drum for another 2–3 min.
- At last, the quantity of selected fiber (as per mix design) was slowly dispersed into the drum in rotary condition and allowed to be mixed until the homogenous mix was not achieved.

In current investigation, cubical and prism-shaped specimens of size 70.6 mm × 70.6 mm × 70.6 mm and 500 mm × 100 mm × 100 mm, respectively, were casted to record the compressive, electrical resistivity and flexural response of various mixes. After 24 h, the casted specimens were taken out and then kept into water for required curing age before testing.

3.2 *Test Methods*

3.2.1 Compression Behavior

To study the impact of fibers combination on compression behavior of various cementitious mixes, cubical-shaped specimens were casted and tested in the universal testing machine as per IS 516:1959 [14] and BS-EN-12390 part 3 [15] specifications.

3.2.2 Flexural Response

The impact of fiber combination on flexural response of various cementitious mixes has been recorded on prism specimens by performing four-point bending test as per IS 516:1959 [14] and BS-EN-12390 part 5 [16] specifications.

3.2.3 Electrical Resistivity (ER)

ER performance of cement-based products indicates its durability [17, 18]. It can represent the movement and distribution of ions through cementitious products. The impact of fiber combination on ER of various mixes has been recorded on cubical-shaped specimens by using two-point method. The ER (ρ) of the cementitious specimens was determined by using the following relationship.

$$\rho = R\frac{A}{L}\ \text{k}\Omega\text{-cm}$$

where A is the area of cubical specimen (cm^2), L is the length of the cubical specimen (cm), and R (kΩ), resistance of the samples.

4 Results and Discussion

4.1 *Compressive Strength*

The impact of fiber combination on compression performance of ECC mixtures at different curing ages is given in Fig. 1. It has been noticed from Fig. 1 that the ECC mixture made with PET fiber had lowest compressive strength; while, ECC mixture made with steel fiber had highest compressive strength. Hybrid use of both fibers in ECC mixtures strongly influenced the compressive strength of the cementitious matrix. The substitution of PET with SE fiber at 20%, 40%, 60%, 80%, and 100% enhanced the compressive strength of the matrix by 1.31%, 4.5%, 7.24%, 11.31%,

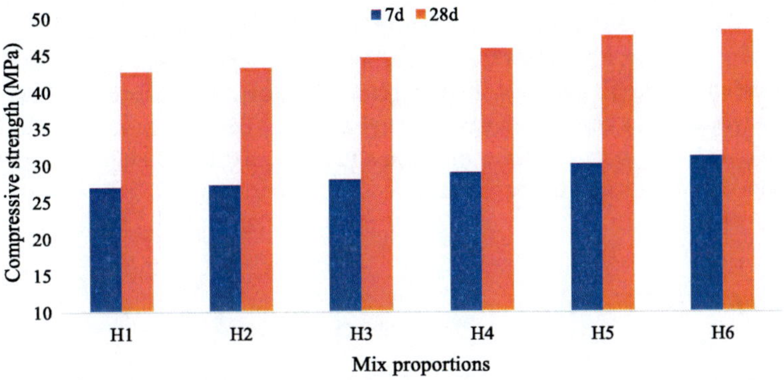

Fig. 1 Compressive strength of various ECC mixtures

and 13.01%, respectively, after 28d of water curing. It is observed from Fig. 1 that the matrix containing SE fiber performed better in terms of compressive strength as compared to mixes containing PET and hybrid fiber.

4.2 *Flexural Strength*

The influence of fiber combination on flexural strength of various ECC mixtures at different curing ages is given in Fig. 2. It has been observed from Fig. 2 that the ECC mixture made with PET fiber had lowest flexural strength; while, ECC mixture made with steel fiber had highest flexural strength. Hybrid use of both the fibers in ECC mixtures enhanced the flexural strength of the cementitious matrix as compared to H1 matrix. The substitution of PET with SE fiber at 20%, 40%, 60%, 80%, and 100%

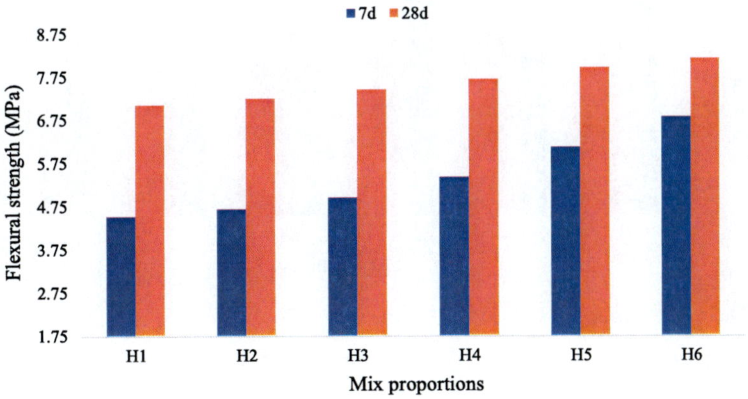

Fig. 2 Flexural strength of various ECC mixtures

enhanced the flexural strength of the matrix by 2.11%, 5.07%, 8.46%, 12.12%, and 15.09%, respectively, after 28d of water curing. It is observed from Figs. 1 and 2 that the mixes containing SE fiber gave better performance in terms of strength parameters as compared to mixes containing PET and hybrid fiber. The high modulus of SE fiber and strong bonding between fiber and cementitious mixtures were responsible for the enhancement in strength parameters.

4.3 Deflection Capacity

The impact of fiber hybridization on deflection capacity of various ECC mixtures at different curing ages is given in Fig. 3. It has been observed from Fig. 3 that the ECC mixture made with SE fiber had lowest deflection capacity; while, ECC mixture made with PET fiber had highest deflection capacity. Hybrid use of both the fibers enhanced the midspan deflection capacity of the cementitious matrix as compared to H6 matrix. The substitution of PET with SE fiber at 20%, 40%, 60%, 80%, and 100% decreased the deflection capacity of the cement matrix by 3.97%, 12.06%, 22.14%, 26.11%, and 34.21%, respectively, after 28d of water curing. It is observed from Fig. 3 that the mixes containing PET fiber gave better performance in terms of deflection capacity as compared to mixes containing SE and hybrid fiber. Deflection performance of all the ECC mixtures was opposite than strength parameters. The low modulus, better elongation capacity, and slip-hardening behavior between PET fiber and cement matrix were responsible for the enhancement in deflection capacity. Numerous research investigations reported that the inclusion of low modulus fiber improved the ductile performance of the cement matrix [19, 20].

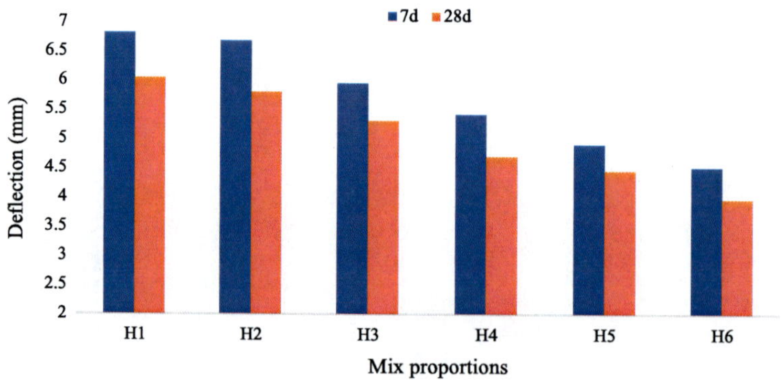

Fig. 3 Deflection capacity of various ECC mixtures

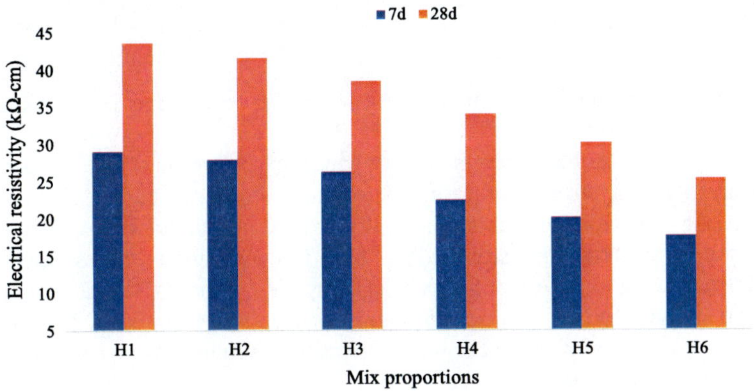

Fig. 4 ER of various ECC mixtures

4.4 Electrical Resistivity (ER)

The impact of fiber hybridization on ER of various ECC mixtures at different curing ages is given in Fig. 4. It has been noticed from Fig. 4 that the ECC mixture made with SE fiber had lowest ER; while, ECC mixture made with PET fiber had highest ER. Hybrid use of both the fibers enhanced the electrical resistivity of the cementitious mixes as compared to fully steel fiber containing ECC matrix (H6 mix). The substitution of PET with SE fiber at 20%, 40%, 60%, 80%, and 100% decreased the ER of the cement matrix by 4.69%, 11.97%, 22.16%, 31.06%, and 42.07%, respectively, after 28d of water curing. Matrix-containing PET fiber gave better performance in terms of ER as compared to mixes containing SE and hybrid fiber. The electrical conductivity of the PET fibers is very low and their inclusion makes a good bond with cement matrix, which was responsible for the better ER of ECC mixes. On the other hand, the steel is conductive material due to which the ER of ECC mixes containing SE fiber was found lower. Published literature reported that the inclusion of steel fiber decreased the ER of the cement matrix [21]. Therefore, in current investigation, the results obtained from mechanical and durability parameters revealed that the successful utilization of high volume of waste product with low cost and non-degradable fibers enhanced the overall performance along with promoting matrix sustainability.

5 Conclusions

In the current research investigation, the performance of ECC mixes containing fibers in hybridization was evaluated in terms of compressive strength, flexural strength, deflection capacity, and electrical resistivity. This study also emphasizes the impact of polyester (PET) and steel (SE) fiber as internal reinforcement in ECC mixes. The major key findings from the current research work have been mentioned below:

- Inclusion of SE fiber in ECC mixes gave better performance in terms of strength parameters as compared to PET and hybrid fiber containing cementitious mixes. The high modulus of SE fiber and strong bonding between steel and cementitious mixtures was responsible for the enhancement in strength parameters.
- Deflection capacity performance of all the ECC mixtures was opposite than strength parameters. Inclusion of PET fiber in cementitious mixes gave better performance in terms of deflection capacity as compared to SE and hybrid fiber containing mixes. The low modulus, better elongation capacity, and slip-hardening behavior of PET fiber with cementitious products were responsible for the enhancement in deflection capacity.
- Electrical resistivity of ECC mixture made with PET fiber was highest than other mixes. PET fibers are very low electrically conductive, and their inclusion reduced the porosity which was responsible for the better ER performance.
- Findings from current investigation depicted that the inclusion of metallic fibers improved the strength parameters while the inclusion of non-metallic fiber enhanced the ductile and durability parameters of ECC matrix made with slag. Therefore, the present investigation recommended that the hybridization of both the fibers was better option for the use in cementitious mixes.
- Examination from this study revealed that the addition of waste, non-degradable materials, low cost, and superior performance of cement matrix enhanced the durability and sustainability of the matrix.

References

1. Singh M, Saini B, Chalak HD (2019) Performance and composition analysis of engineered cementitious composite (ECC)—a review. J Build Eng 26:100851. https://doi.org/10.1016/j.jobe.2019.100851
2. Li VC (2007) Engineered cementitious composites (ECC)—material, structural, and durability performance. In: Nawy E (ed) Concrete construction engineering handbook. CRC Press
3. Singh M, Saini B, Chalak HD (2018) Properties of engineered cementitious composites: a review. ICSWMD 2018, LNCE 21, 2019, pp 1–11. https://doi.org/10.1007/978-3-030-02707-0_54
4. Li VC (1998) Engineered cementitious composites—tailored composites through micromechanical modeling. Fiber reinforced concrete: present and the future. In: Banthia N, Bentur A, Mufti AA (eds) Canadian society for civil engineering. Montreal, pp 64–97
5. Li J, Yang E-H (2017) Macroscopic and micro structural properties of engineered cementitious composites incorporating recycled concrete fines. Cem Concr Compos 78:33–42
6. Singh M, Saini B, Chalak HD (2019) Appraisal of hybrid fiber reinforced engineered cementitious composite. In: Proceedings of the 4th international conference on civil, structural and transportation engineering (ICCSTE 19) Ottawa, Canada, 11th and 12th June 2019. https://doi.org/10.11159/iccste19.192
7. Li VC, Mishra DK, Naaman AE, Wight JK, LaFave JM, Hwai-Chung W, Inada Y (1994) On the shear behavior of engineered cementitious composites. Adv Cem Based Mater 1:142–149

8. Li VC, Hwai-Chung W, Maalej M, Mishra DK (1996) Tensile behaviour of cement-based composites with random discontinuous steel fiber. J Am Ceram Soc 79:74–78
9. Singh M, Saini B, Chalak HD (2020) Performance characteristics of ECC using micro fibers in hybridization. Ind Concr J 94:66–73
10. Dan M, Ting H, Zhang YX, Lee CK (2017) Mechanical behaviour of a polyvinyl alcohol fiber reinforced engineered cementitious composite (PVA-ECC) using local ingredients. Constr Build Mater 141:259–270
11. IS-8112 (2013) Specification for 43 grade ordinary Portland cement. Bureau of Indian Standards. New Delhi
12. ASTM-C150/C150M-16e1 (2016) Standard specification for Portland cement. ASTM International, West Conshohocken, PA
13. IS 456-2000 Plain and reinforced concrete-code of practice, 4th Revision
14. IS 516-1959 (2006) Methods of tests for strength of concrete, 18th edn. Bureau of Indian Standards, New Delhi
15. BS-EN-12390–3, Testing hardened concrete, Compressive strength of test specimens, British Standard Institution, 2009.
16. BS-EN-12390-5 (2009) Testing hardened concrete, flexural strength of test specimens. British Standard Institution
17. Medeiros-Junior RA, Lima MG (2016) Electrical resistivity of unsaturated concrete using different types of cement. Constr Build Mater 107:11–16
18. Broomfield JP (2007) Corrosion of steel in concrete, 2nd edn. Taylor and Francis, New York
19. Soe KT, Zhang YX, Zhang LC (2013) Material properties of a new hybrid fibre-reinforced engineered cementitious composite. Constr Build Mater 43:399–407
20. Li Q, Zhao X, Xu S, Gao X (2016) Influence of steel fiber on dynamic compressive behavior of hybrid fiber ultra-high toughness cementitious composite at different strain rates. Constr Build Mater 125:490–500
21. Banthia N, Djeridane S, Pigeon M (1992) Electrical resistivity of carbon and steel micro-fiber reinforced cements. Cem Concr Res 22:804–814

Arresting the Heave of Black Cotton Soil Using Geogrid-Encased Granular Pile Anchor

Shweta Singh, Ravi Kumar Sharma, and Abhishek Sharma

1 Introduction

In India, the most commonly found expansive soil is black cotton soil and covers around 20% of land. Due to its swell-shrink characteristic, volumetric changes occur along with the variation in water content. Expansive soils absorb water during monsoon which results in swelling of the soil that leads to increment in the volume of the soil, while in dry season, the water evaporates resulting in shrinkage of the soil, and hence, the volume of the soil decreases [1–5]. In addition to the wetting of the expansive soil in monsoon, it swells or shows collapse compression which is totally based on the stress and suction history of the soil. The construction over expansive soils is very uncertain and challenging to the geotechnical engineers [4, 6]. Heaving is the most often problem that the geotechnical engineers have to deal with in case of expansive soils. Heave occurs when water enters within the clay minerals that cause increment in the volume of the soil, and the structures constructed over it lifted in upward direction. The damage to civil engineering structures constructed over these soils results in a huge financial loss. It has been predicted that the destruction to the structures constructed over swelling soils is remarkably high, equivalent or may even go beyond those caused due to floods, earthquakes, hurricanes, tornadoes, etc., combined [7]. However, many constructive technologies in the form of tension resistant foundations having wide bottom have been used in situ to reduce the heave in expansive soils such as under reamed piles, belled piers and drilled piers, friction

S. Singh · R. K. Sharma · A. Sharma (✉)
National Institute of Technology, Hamirpur, Himachal Pradesh, India
e-mail: abhishek1@nith.ac.in

S. Singh
e-mail: shwetasngh709@gmail.com

R. K. Sharma
e-mail: rksnithp61@gmail.com

H. Singh et al. (eds.), *Sustainable Development Through Engineering Innovations*, Lecture Notes in Civil Engineering 113,
https://doi.org/10.1007/978-981-15-9554-7_30

Table 1 Geotechnical properties of black cotton soil

Properties	Soil
Specific gravity, G	2.61
Liquid limit (%)	72.09
Plastic limit (%)	30
Plasticity index	42.09
Classification	CH
FSI (%)	87.5
MDD (g/cm^3)	1.435
OMC (%)	18.25

piers. Also, cohesive non-swelling (CNS) layer, sand cushion, physical alteration and chemical stabilization have been used so far to minimize the swelling problems of expansive soils. All the above-listed methods are costlier in nature and are not successful for small budget projects. The use of granular pile anchors to reduce the heave of poor soils has been done in the past two decades by some researcher [8–12]. It was revealed from the past studies that granular pile anchors are suitable for controlling the heave of the expansive soils. This paper is an attempt to improve the behaviour of granular pile anchors against heave by encasing piles with the help of a geogrid around its periphery.

2 Materials

2.1 *Black Cotton Soil*

For conducting the experimental tests, black cotton soil (BCS) was used and was collected from Jamnagar, Gujarat, India. According to the Unified Soil Classification System (USCS), the soil is classified as CH. Table 1 presents the geotechnical properties of the soil. The grain size distribution curve of black cotton soil is shown in Fig. 1.

2.2 *Granular Material*

The proportion of granular material plays a vital role in the analysis of heave, and hence, the mixture of gravel and sand in the proportion of 80:20 is used in the granular pile. The size of the gravels used was varied from 4.75 to 20 mm, and for the sand, the size was kept between 1.18 and 4.75 mm. Figure 1 shows the grain size distribution of aggregate and sand.

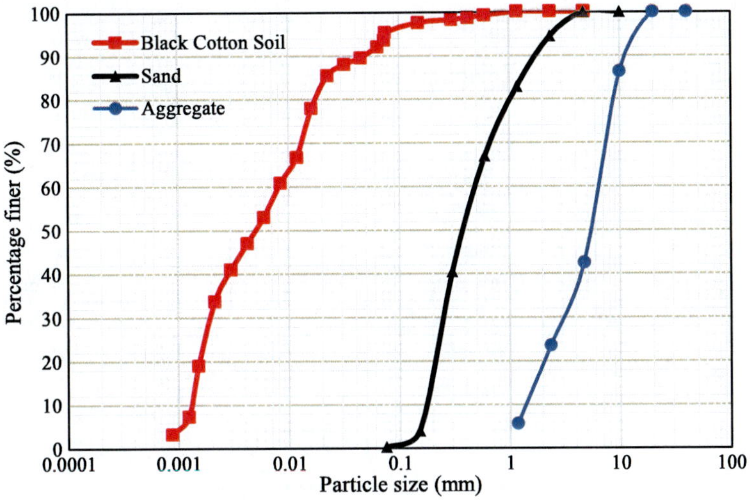

Fig. 1 Grain size distribution curve of material

2.3 *Anchor Plate and Model Footing*

A circular shape mild steel plate of diameter 20–60 mm, thickness 5 mm and modulus of elasticity 2.1 MPa $\times 10^5$ was used as anchor plate at the bottom and as model footing at the top of GPA.

2.4 *Geogrid*

The required size of geogrid to encase it around the periphery of the pile was cut from a biaxial geogrid roll. Various specifications of biaxial geogrid are given in Table 2.

3 Testing Programme

A cast iron tank having dimension 350 mm long, 350 mm width and 700 mm height and 4 mm thickness of the wall has been used for heave analysis (Fig. 2). For all set of tests, the soil was prepared at its optimum moisture content and maximum dry density. A total of 19 series of tests have been performed on various soil and pile parameters. The overall test plan is described in Table 3.

Table 2 Specifications of geogrid

Property	Value/type
Polymer type	HDPE
Mesh type	Square
Colour	Black
Packaging	Roll
Roll width	2 m
Roll length	30 m
Unit mass	650 g/m^2
Tensile strength	7.20 kN/m
Percentage elongation at maximum load	20.2%

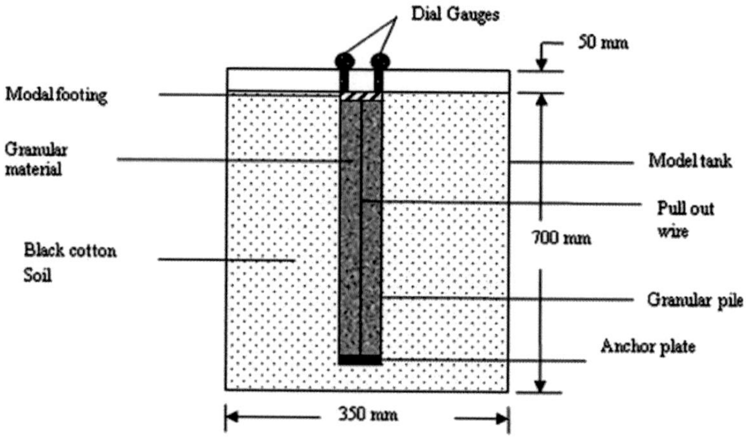

Fig. 2 Test setup for heave analysis

Table 3 Test plan

S. No.	Test condition	Length of GPA (mm)	Diameter of GPA (mm)	No. of tests
I	Expansive soil	–	–	1
II	GPA-reinforced soil	250, 350, 450	30, 40, 50	9
III	Geogrid-encased GPA in expansive soil	250, 350, 450	30, 40, 50	9
Total tests				19

4 Preparation of Test Setup

The schematic detail of test setup is shown in Fig. 2. The soil was filled in the tank by manual compaction at its MDD ensuring proper OMC. The soil was filled in layers of 50 mm each till the required height was achieved. The anchor was placed at the centre of the tank in the required position, and the casing for the granular pile was kept along with it. The granular material was filled in the casing after compaction of each layer, and simultaneously, the casing was withdrawn. The process was continued till the required length of the pile was achieved. When the geogrid was used to encase the granular pile, the geogrid was wrapped to the required dimensions and then was inserted into the casing and the same process (as followed above) was followed to construct the granular pile. At the top of the granular pile anchor, a model footing was placed having diameter equal to that of the GPA diameter, and two dial gauges were kept at its top to measure the heave. The saturation of soil was done with the help of water continuously, and the average heave was noticed from the two dial gauges. The heave was founded to complete when there was no change in dial gauge readings.

5 Results and Discussion

5.1 Effect of GPA Installation of Heave of Black Cotton Soil

The maximum value of heave was noticed for black cotton without any reinforcement. The final heave value of black cotton soil was 24.7 mm and required about 21,600 min to achieve. On reinforcing the black cotton soil with GPA of varying lengths of 250, 350 and 450 mm for constant diameters of 30, 40 and 50 mm, the heave was found to be decreased drastically with the large reduction in time to achieve the final heave. For 250 mm length GPA, the final heave was noticed as 18.76 mm (at time = 14,040 min), 17.86 mm (at time = 14,040 min) and 16.76 mm (at time = 14,040 min) for 30 mm, 40 mm and 50 mm diameter, respectively (Fig. 3). The similar rate of reduction of heave along with reduction in time was noticed for 350 mm and 450 mm length piles as shown in Figs. 4 and 5, respectively. This may be due the fact that the inclusion of granular pile anchor increased the permeability of the black cotton soil offering quick movement to passage of water. Further, it was noticed that, on increasing the length and diameter of the granular pile anchor, there was much reduction in the heave value along with time rate to reach the final heave. This may be due the fact that on increasing the length and diameter of GPA, the surface area increased resulting in high permeability, thus allowing passage of water more quickly. The similar results were obtained on installing GPA in expansive soils in the past by few researchers [8, 10].

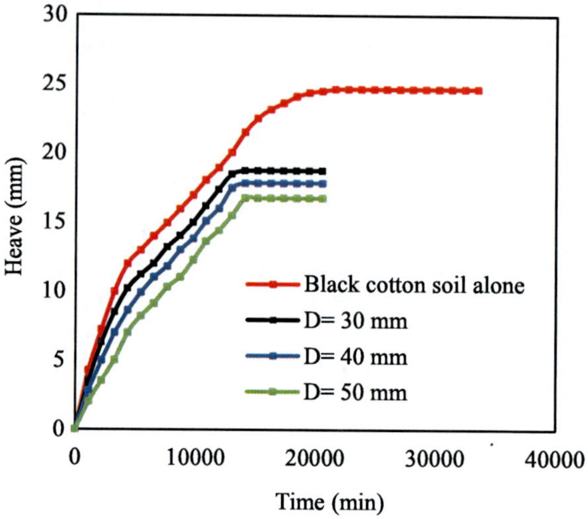

Fig. 3 Heave–time relationship for 250 mm length of GPA

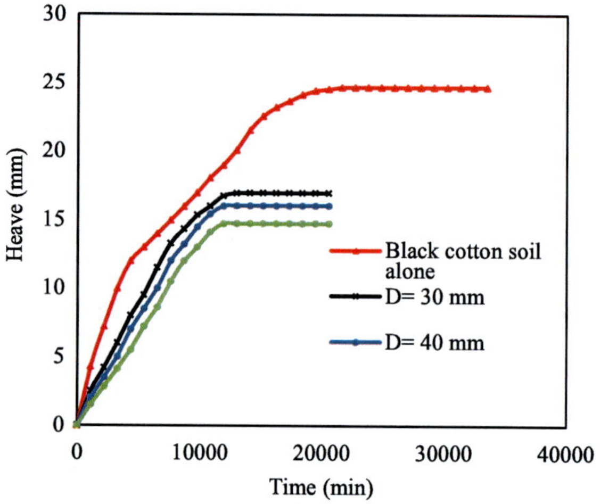

Fig. 4 Heave–time relationship for 350 mm GPA length

5.2 Effect of Geogrid-Encased GPA Installation of Heave of Black Cotton Soil

The effect of encasement of GPA with geogrid on amount of heave has been discussed in the current section and presented in Figs. 6, 7 and 8. For 250 mm length having diameter 30, 40 and 50 mm, the geogrid-encased GPA recorded final heave as 16.76, 15.76 and 14.55 mm along with time rate of 14,040, 12,960 and 12,960 min to reach the final heave; for 350 mm length having diameter 30, 40 and 50 mm, the geogrid-encased GPA recorded final heave as 14.86, 13.88 and 12.47 mm along with time

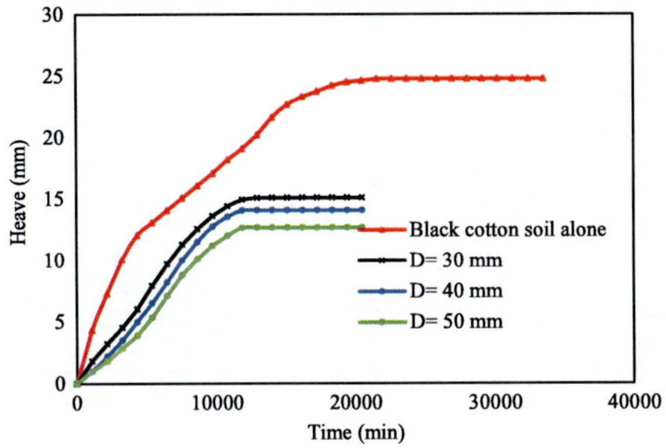

Fig. 5 Heave–time relationship for 450 mm GPA length

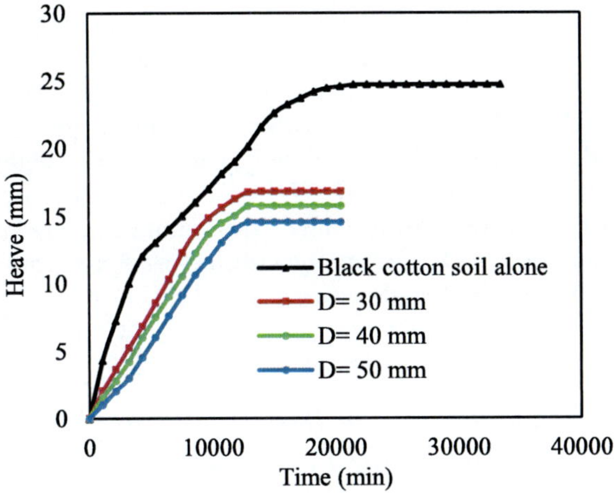

Fig. 6 Heave–time relationship for 250 mm length of geogrid-encased GPA

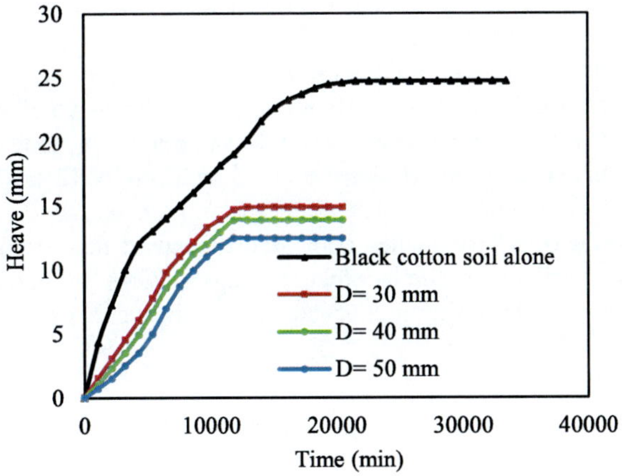

Fig. 7 Heave–time relationship for 350 mm length of geogrid-encased GPA

Fig. 8 Heave–time relationship for 450 mm length of geogrid-encased GPA

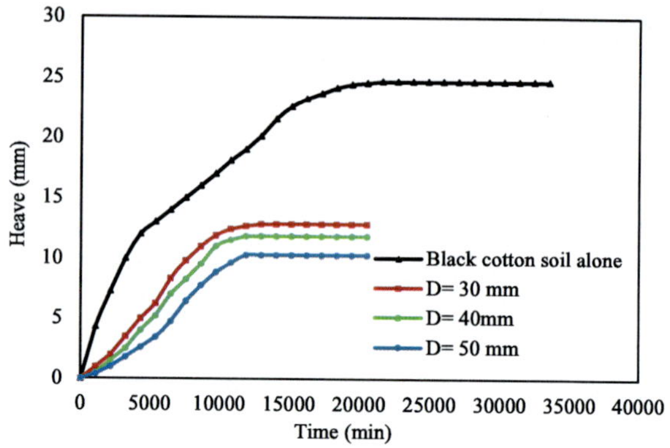

rate of 14,040, 12,960 and 12,960 min to reach the final heave; for 450 mm length having diameter 30, 40 and 50 mm, the geogrid-encased GPA recorded final heave as 12.88, 11.84 and 10.28 mm along with time rate of 14,040, 12,960 and 12,960 min to reach the final heave. It was observed from the graphical observations that the heave value reduced to higher extent on encasing GPA with geogrid. This may be due to the fact that on encasing GPA with geogrid, the friction interface increased along with increase in permeability which resulted into reduction of heave at larger scale. The similar results were obtained on installing geogrid-encased GPA in expansive soils in the past by few researchers [12].

5.3 *Comparison of Heave Value of GPA and Geogrid-Encased GPA*

The comparison of heave value of GPA and geogrid-encased GPA for various pile parameters has been presented in this section. It is very much clear from the graphs that on encasing GPA with geogrid, the final heave reduced to a larger extent. For 250 mm length having 30 mm diameter, the heave of geogrid-encased GPA reduced by 11% when compared to heave value of GPA alone (Fig. 9); for 40 mm diameter, the heave of geogrid-encased GPA reduced by 12% when compared to heave value of GPA alone (Fig. 10); for 50 mm diameter, the heave of geogrid-encased GPA reduced by 13% when compared to heave value of GPA alone (Fig. 11). This may be attributed to the fact that on encasing GPA with geogrid, the frictional resistance increased offering higher reduction of heave (Figs. 12 and 13).

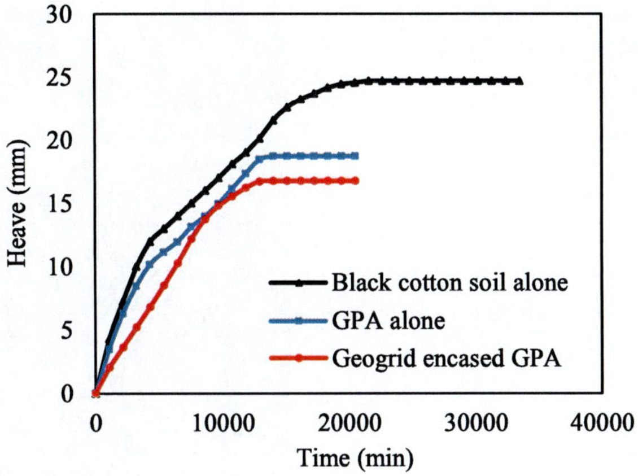

Fig. 9 Heave–time plot for 250 mm length at 30 mm diameter

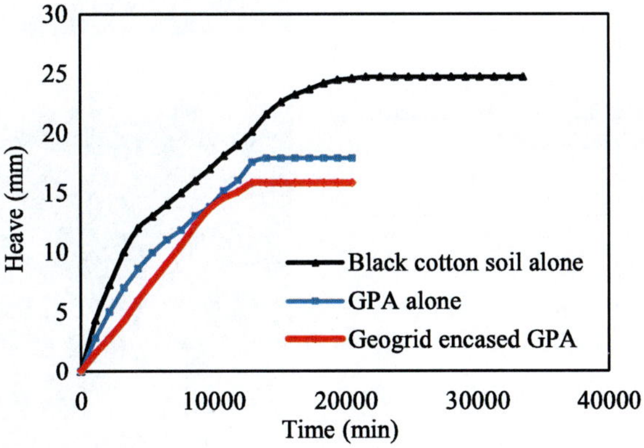

Fig. 10 Heave–time plot for 250 mm length at 40 mm diameter

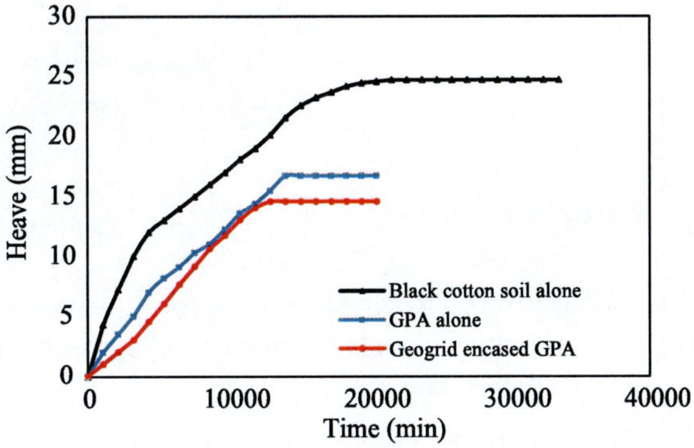

Fig. 11 Heave–time plot for 250 mm length at 50 mm diameter

Fig. 12 Complete saturation of soil without GPA

Fig. 13 Complete saturation of soil with GPA

6 Conclusions

A laboratory experiment has been done to reduce the amount of heave of black cotton soil by reinforcing it with GPA and geogrid-encased GPA. The following conclusion is drawn from the testing:

1. The heave of the black cotton soil becomes constant when there is a complete saturation of the soil.
2. The heave of the black cotton soil reduced on reinforcing it with granular pile anchor due to increased permeability.

3. The heave of the black cotton soil reaches to a minimum value on encasing granular pile anchor with geogrid around its periphery due to increased friction component.
4. The heave of the black cotton soil reduced on increasing the length and diameter of the GPA and geogrid-encased GPA due to increase in surface area.
5. The present study provides an economical solution of heave reduction of black cotton soil thus suggesting economical and sustainable design and construction.

References

1. Abhishek, Sharma RK (2018) A new foundation practice for predicting the behavior of granular pile anchor against uplift forces in expansive soils. In: International conference on sustainable waste management through design. Springer, Cham, pp 140–147
2. Abhishek, Sharma RK (2019) A numerical study of granular pile anchors subjected to uplift forces in expansive soils using PLAXIS 3D. Indian Geotech J 49(3):304–313
3. Sharma A, Sharma RK (2019) Effect of addition of construction–demolition waste on strength characteristics of high plastic clays. Innov Infrastr Solut 4(1):27
4. Sharma A, Sharma RK (2019) An experimental study on uplift behaviour of granular anchor pile in stabilized expansive soil. Int J Geotech Eng:1–14
5. Bhardwaj A, Sharma RK (2020) Effect of industrial wastes and lime on strength characteristics of clayey soil. J Eng Des Technol
6. Sharma A, Sharma RK (2020) Strength and drainage characteristics of poor soils stabilized with construction demolition waste
7. Jones DE, Holtz WG (1973) Expansive soils-the hidden disaster. Civil Eng 43(8)
8. Phanikumar BR, Srirama Rao A (1996) Granular pile anchors in expansive soils. In: Proceedings of a national seminar on partially saturated soils and expansive soils. Kakinada, India, pp 15–22
9. Rao AS, Phanikumar BR, Babu RD, Suresh K (2007) Pullout behavior of granular pile-anchors in expansive clay beds in situ. J Geotech Geoenviron Eng 133(5):531–538
10. Phanikumar BR, Srirama Rao A, Suresh K (2008) Field behaviour of granular pile-anchors in expansive soils. Proc Inst Civil Eng Ground Improv 161(4):199–206
11. Ibrahim SF, Aljornay AN, Aladly AI (2014) Heave behavior of granular pile anchor-foundation (GPA-foundation) system in expansive soil. J Civil Eng Urban 4(3):213–222
12. Raghuram ASS, Rao AS, Purnanandam K (2016) Pullout behavior of geogrid-encased granular pile anchor embedded in expansive clays. In: Proceedings of Indian geotechnical conference, Chennai

The Application of Low-Cost Natural Bio-Adsorbents for the Removal of Heavy Metals—A Review

Ankita Thanki, Arti Thanki, Rajwinder Singh, and Karanvir Singh Sohal

1 Introduction

Water is an utmost crucial and integral component on the earth for the necessary activities of living beings. Grievously, the quality of freshwater and water resources is the decline and deteriorate continuously due to industrialization, urbanization, civilization, boom population growth, domestic as well as agricultural activities, and other environmental activities. Shortage of water has become a perennial concern in the current scenario and therefore, adulteration of water becomes a grave challenge in the present scenario lead to affecting living creatures, recreation, transportation, commercial activities, and fishing. The Scientist, government bodies, and researcher facing a serious issue regarding the conservation of water. Expecting increases in population up to 7.9 billion in 2020 and because of it, the world may under a great scare city. Therefore, the elimination of toxic and hazardous impurities from contaminated wastewater is an urgent need for the better health of society [1]. The generation of wastewater containing excessive levels of toxicants has become a perennial

A. Thanki (✉) · A. Thanki
Department of Environmental Science and Engineering, Marwadi University, Rajkot, Gujrat, India
e-mail: thankiankita212@gmail.com

A. Thanki
e-mail: thankiarti111@gmail.com

R. Singh
Department of Civil Engineering, DR. B.R. Ambedkar National Institute of Technology, Jalandhar, Punjab, India
e-mail: rajwinderbirdi2@gmail.com

K. S. Sohal
Department of Civil Engineering, Guru Nanak Dev Engineering College, Ludhiana, Punjab, India
e-mail: karansohal875@gmail.com

H. Singh et al. (eds.), *Sustainable Development Through Engineering Innovations*, Lecture Notes in Civil Engineering 113,
https://doi.org/10.1007/978-981-15-9554-7_31

concern. Wastewater has become highly essential to treat for removal of contamination, as good quality water is a valuable thing and it is accessible in restricted amounts. The generation, handling, and disposal of toxic hazardous material and pollutants present in wastewater have intensely polluted our surrounding environment [1]. For soil and water, various pollutants have become growingly prevalent as the most dangerous pollutants. To reduce the amount of waste from the environment is one of the serious concerns. Wastewater treatment is a major issue nowadays. The waste generated can be either in liquid, solid and semi-solid forms.

Using commercially available adsorbent lead to a health hazard in some or the other way to the environment. This chemical or a heavy metal present in industrial wastewater or consumer waste, when coming in contact with water leads to groundwater pollution, pollution in lakes, rivers, and other water bodies. Both acute and chronic heavy metal toxicants may damage the nervous system, glands, gastrointestinal, cardiovascular, liver, and bones [2, 3]. Harmful effect with the intake of heavy metal has also been reported [4]. Few heavy metals with health effect and permissible limit are shown in Table 1.

The overall review study objective is to understand the effect of green adsorbent to treat the different wastewater by the removal of toxic material and heavy metal. Heavy metal removal from contaminated water plays a key role in addressing the environmental issues faced by developing countries. Mainly four adsorbents were evaluated as a capable low-cost bio-adsorbent for extraction of contaminants like heavy metals namely chromium, cadmium, mercury, zinc, lead, and many more. In this review study, a summary prepared of number of litreture studies on the utilization of natural adsorbent and its capacities related to the removal of pollutant from

Table 1 Effect of heavy metals present in wastewater on human health

S. No.	References	Heavy metals	Limits (WHO) (mg/l)	Health effects
1	[5]	Lead	0.05	Kidney damage, brain damage in fetal
2	[6]	Zinc	5.0	State of depression, increased thirst, complete or partial paralysis, neurological problem symptoms
3	[7]	Nickel	2.0	Nausea, coughing, carcinogenic, dermatitis, chronic asthma
4	[8]	Cadmium	0.003	Kidney failure, carcinogenic
5	[9]	Chromium	0.005	Headache, nausea, vomiting, diarrhea
6	[10]	Mercury	0.001	Kidney problem, a disorder related to the circulatory and nervous system
7	[11]	Iron	0.3	Tinnitus, depression, brittle nails, constipation

wastewater. The literature, showed the fruitfully used of adsorbent for the removal of heavy metal present in wastewater. Although this natural adsorbent in many cases are a source of pollutant so the aim is to contribute the idea about the relatively inexpensive natural adsorbent which are less expensive and its utilization in the treatment of wastewater.

2 Nature of Heavy Metal Ions and Its Sources

2.1 *Mercury*

Mercury (Hg) is considering as toxic heavy metal ions among all. Mercury is mainly found in form of Mercuric ion (Hg^{+2}) state. Mainly source of mercury is by both anthropogenic as well as natural activities. Natural activities like volcanic action and geological activities and anthropogenic activities like paper and pulp industries, pharmaceutical and cosmetic preparation, batteries and medical drugs, thermometers, etc. Because of its high harmfulness, European Union (EU) set the admissible furthest reaches of drinking water as well as wastewater as 0.001 and 0.005 mg/L, respectively [12].

2.2 *Chromium*

Mainly chromium exists in its trivalent (Cr^{+3}) and hexavalent (Cr^{+5}) oxidation states. Cr^{+5} is considered as highly toxic as well as carcinogenic in nature. The source of chromium in the water stream is mainly from industries like paints and pigments, ceramic industries, steel fabrication, and textile industries. Permissible limits of Cr^{+5} for drinking water are 0.1 mg/L according to the EPA. Along with this according to the FAO and WHO the limit of Cr^{+3} for drinking water is less than 5 mg/L [12].

2.3 *Arsenic*

Arsenic (As) is a metalloids element that is hazardous as well as toxic in nature. Arsenic is generally found in four states, i.e. arsenite (As^{-3}), arsenic (As), arsenite (As^{+3}), and arsenate (As^{+5}). From this arsenate (As^{+5}) is toxic compare to other forms in aerobic conditions. Sources like herbicides, mining, insecticides, phosphate fertilizers industries along with coal combustion which can cause arsenic contaminations. As reported by WHO, the permissible limit is 0.01 mg/l for drinking water [12].

2.4 Lead

Lead is the most common hazardous metal which is mainly found in industrial wastewater. Mainly it exits in divalent. Sources of lead are industries like pipes and products of bronze, battery manufacturing, ceramic, and glass industries. The permissible limit of lead is 0.05 mg/L according to WHO [12].

2.5 Cadmium

Cadmium is toxic in nature as well as a non-essential element that is used in daily life. Cadmium is commonly found in industries like nickel-cadmium batteries, metal plating, stabilizers, pigments, and alloy industries. According to the WHO the permissible limit of cadmium is 0.003 mg/l and with reference the EPA is 0.005 mg/L [12]. Few natural adsorbents with investigated wastewater and targeted pollutant are listed out in Table 2.

Table 2 List of natural adsorbent available and wastewater investigate for removal of heavy metals

S. No.	Adsorbent used	Type of wastewater	Heavy metals removal	References
1	Jackfruit Peel, *Moringa Oleifera*	Synthetic wastewater	Lead	[13, 14]
2	Bagasse, papaya wood, neem bark, rice husk ash, *M. oleifera*	cadmium and zinc bearing wastewater, milling agro-waste	Zinc	[15, 16, 13, 17]
3	Pine sawdust, rice husk, fly ash, *M. oleifera*	Multi-component aqueous solutions, synthetic wastewater	Nickel	[18, 19, 13]
4	Activated carbon, walnut shell, rise husk, oak bark, chesnut bur, banana peels, vermicompost, *M. oleifera*	Real wastewater sample, synthetic wastewater	Cadmium	[13, 20, 8]
5	Groundnut shell, *M. oleifera*	Aqueous solution	Chromium	[9, 13]
6	Rice husk, *M. oleifera*	Textile wastewater, water and wastewater	Mercury	[13, 21]
7	*M. oleifera*	Water and wastewater	Iron	[13]

3 Technologies Used for Removal of Heavy Metal Ions from Wastewater

From the last few years of research, many different conventional treatment technologies have been used to treat wastewater and have been applied for the removal of heavy metal and toxic pollutants from contaminated wastewater. Techniques used, experience the ill effects of certain deformities, for example, fragmented evacuation of contaminants and are portrayed by utilization of high reagent, low selectivity, the generation of other toxic wastes which need to be handled carefully, chemical and energy requirements, and relatively expensive. Many wastewater techniques are shown in Fig. 1 such as membrane filtration, electrolysis, chemical precipitation, ion exchange, solvent extraction, coagulation, reverse osmosis, and many more. Apart from various available treatment facilities for the disposal and use of by-products and waste material, use as adsorbent is the best option. As a peer, proper, and safe stabilization is not done properly. To do so, various researcher has conducted an experimental work based on some sustainable techniques.

3.1 Electrodialysis

In this, the compounds like heavy metals are separated with the help of semi-permeable ion-selective membranes. Electric potential between two electrodes can cause migration of anion and cation towards electrodes. The main objective of ED is to remove salts from aqueous solutions. A cell is divided into two equal compartments by placing cation and anion exchange membrane between electrodes. The current is applied directly. The advantages of using ED are no need for osmotic

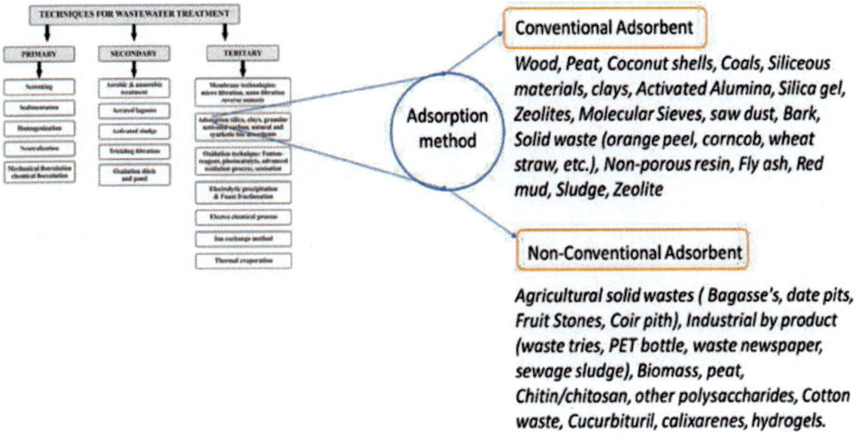

Fig. 1 List of wastewater treatment and adsorbent type

pressure, high-quality product, any additional chemical requirements, and environment friendly. With this, main disadvantage of this process is membrane fouling, which is responsible for decreases ions migration yield, flux decline, high level of polarization, and membrane resistance [22].

3.2 *Ultrafiltration*

In this process, the pressure is provided from porous membranes which helps for the removal of heavy metals. Mainly these pore size in membranes is between ranges 0.1–0.001 μ. Mechanism of this process is size exclusion, surface chemistry of particles along with electrical charge or membrane which affects the purification efficiency. Demerit of this process is the main generation of sludge.

3.3 *Ion Exchange*

In this, exchanges of ions take place between complex and electrolytes/between an electrolyte solution. This process of separation, purification, and decontamination of aqueous in this unwanted ion in the wastewater are exchanged for expedient ions as water pass through granular chemicals which are known as ion exchange resins. The advantage of this process is its cost-effective, very little amount of energy is required with the regeneration of resin is very economical.

3.4 *Chemical Precipitation*

Chemical precipitation is accomplished by the addition of coagulants such as iron salts, organic polymers, and alum. Main demerit of this process is an excess amount of sludge produced. Coagulants like iron salts, lime, organic polymers, and alum are used. Few studies also reported that 80% removal of Cu, Pb, and Zn along with this removal of oil (96.2%) from wastewater using precipitation and also air floatation system.

Among all techniques available, adsorption is the most promising, best choice compared to other techniques available for the treatment, due to its effectiveness with the removal of a metal ion at low concertation, and even it is cheap, simplicity in design, ease adaptation, easy availability, easy operation, and for its wider applicability [23, 24]. Adsorption has gained importance over the few decades in the removal of heavy metals and toxic materials because of its efficacy in the removal of inorganic as well as organic compounds from wastewater.

From an economic and circular economy point of view, environmentally friendly utilization of by-products/waste materials generated from agricultural, industries, residential, and domestic either as raw material or in production is the best option. It is a waste material/by-product from industries, households, and agriculture and found abundant in nature, due to ease availability, also requires less or no processing, has little or no economic values are the best choice to use [24].

3.5 *Adsorption*

Adsorption is the process in which material i.e. adsorbate migrate from liquid/gas and forms a layer on liquid/solid condensed phase(substrate). Adsorption is used to filter deodorize, detoxify, decolorize, separate along with concentrate to allow removal and recover harmful products from solution as well as gaseous mixtures (conventional). Adsorption is a physicochemical treatment with economically feasible and will treat the effluent standards as well as reuse the wastewater. Adsorption is the mass transfer process which includes cumulating of substances in an association of two different or same phases like gas–liquid, liquid–liquid, liquid–solid, and gas–solid. Adsorbate is the substance that is being adsorbed and adsorbent is the adsorbing material. Mainly surface affinity plays a driving force for the adsorption process [12]. Key parameters for adsorption are pH, chemical reactivity, surface area by volume for adsorption with a reduction in surface tension. Adsorption can be also defined as a phenomenon that intensifies chemical substances along with the layer of solid substances.

4 Type of Natural Adsorbent

Classification of adsorbents are as follows:

1. A natural material like wood, sawdust, bauxite, etc.;
2. Natural material which treated to develop their structures and properties (activated alumina, silica gel, and activated carbon);
3. few manufactured material like zeolites, polymeric resin, etc.;
4. Industrial by-product like fly ash, red mud as well as agricultural solid waste;
5. Few bio sorbents like fungi, bacterial biomass, and fungi [25].

Few kinds of literature give a different and simplified classification of adsorbent as shown in Table 3 [25].

These natural adsorbents are successfully reported to reduced pollutants and the positive result are observed. Due to relatively inexpensive, easy maintenance, available in large quantities, inexpensive, low energy consumption, low carbon footprint, abundant in nature, are the main advantage of using natural adsorbent for the treatment [12].

Because of these, the focus has turned on searching for another replaceable non-conventional adsorbent proposed as the green adsorbent, relatively inexpensive adsorbent, and efficient for reduction of pollutants [23]. Application of various

Table 3 Type of adsorbents with example

S. No.	Type of adsorbent	Examples
1	Commercially available adsorbent	Graphene, carbon nanotube, and activated carbon
2	Natural adsorbent	Clay, siliceous materials, and zeolites
3	Bio-adsorbent	Peat, fungal biomass, chitins and chitosan, yeast, and bacterial biomass
4	By-product from agriculture, industry, as well as waste material	Husk, bagasse's, peels, fly ash, barks, nutshell/hulls, straws, and brans
5	Animal waste material	Waste from animal processing industries, resultants, and houses (fish scales, eggshells, chicken feathers crab shells)
6	Industrial waste	Red mud

Source Chakraborty et al. [12]

natural materials to treat/reduce the concentration of heavy metals, toxic material, and organic matter is been done for the last few decades and shows a huge explosion in the development of new natural adsorbent [23].

Many natural adsorbents such as Mango peel, potato peel, Corncob, rice husk, Coconut husk, neem bark, fruit bagasse, palm fruit; sugarcane bagasse, *M. oleifera*, peanut husk, Almond shell, Barley straw, Apricot shell, Agave bagasse's, Corncob, and many more [26–37; Iqbal et al. 2009] have been investigated for the various pollutant from wastewater respectively. Few specific natural adsorbents shown in Fig. 2 such as *M. oleifera*, Peanut Husk, Rice Husk, and Sugarcane Bagasse are most frequently used in the treatment of various investigated wastewater. The use of the natural material term as "Green adsorption".

Rice Husk

Few studies reported that waste of agricultural and plants such as peels, maize, rice bran followed with rice husk was used as adsorbent. These types of adsorbents are generally referred to as bio-adsorbent. Rice husk is a by-product which is obtained from the rice mill, with a composition of cellulose (32.24), lignin (21.44%), mineral ash (15.05%), and hemicelluloses (21.34%). It has been used as an adsorbent for the removal of heavy metal ions in unmodified as well as modified form with good efficiency. It was observed that the highest adsorption capacities were 30.0 of Cr (III) and 22.5 mg/g Cu (II). The most extreme expulsion proficiency was 100% for metal particles at a contact time of 60 min and at pH 5–6, adsorbent portion 5 and 4 g, individually [12]. Table 4 shows the critical literature for rice husk as adsorbent.

Peanuts Husk

Peanut husk is a beneficial and important bio-waste to be used. Peanut husk is the biggest waste product of the food industry, it could be used to treat wastewater by extracting metal such as copper ions, cadmium, chromium, and many more heavy metals [47]. It was observed in the literature that efficient use of peanut husk for

Fig. 2 Natural adsorbents

the extratction of specific contaminants like Zn^{2+}, Pb^{2+} along with Mn^{2+} ions form the fertilizer industrial wastewater. This study reported peanut husk powder has a considerable high removal percentage for these three ions. Table 5 shows the critical literature for peanut husk as adsorbent.

Sugarcane Bagasse (SB)

SB is a by-product of the agricultural field. It's a fibrous material. After juice extracting from sugarcane thus material remains i.e. SB as relatively inexpensive adsorbent. This agriculture waste mainly consists of cellulose (40%), hemicellulose (24%) along with lignin (25%). Yu et al. reported that SB synthetically treated with pyromellitic dianhydride were utilized for the expulsion of Cd^{+2}, Pb^{+2}, Zn^{+2}, and Cu^{+2} particles from the watery arrangement. The most extreme adsorption limits of the adjusted sugarcane bagasse were 0.93, 1.06, 1.00, and 1.21 mmol/g for Cd^{+2}, Pb^{+2}, Zn^{+2}, and Cu^{+2} particles, separately, though, crude sugarcane bagasse displayed the adsorption limits of 0.13, 0.04, 0.07 and 0.10 mmol/g in a specific order [52]. Table 6 shows the critical literature for sugarcane bagasse as adsorbent.

Table 4 List of heavy metal with its capacities for investigated wastewater using rice husk as adsorbent

Heavy metal	Capacities (mg/g)	Type of wastewater used	References
(As^{+5})	615.11	Aqueous solution	[38]
	18.98	Aqueous solution	[39]
(Cd^{+2})	21.36	Aqueous solution	Roy et al. (38)
	0.16	Aqueous solution	[40]
	7	Lake water	[41]
(Cr^{+3})	0.32	Aqueous solution	[42]
(Cr^{+6})	14.2–31.5	Aqueous medium	[43, 44]
	164.31	Aqueous solution	[38]
	4.02	Aqueous solution	[40]
Hg	66.66	Aqueous solution	[45]
Pb	108	Aqueous solution	[46]
	45	Spiked water samples (mineral water and lake water) and spiked red wine samples	[41]
	129.48	Malachite green (MG) from aqueous medium	[46]
	11.40	Aqueous solution	[38]

Table 5 List of heavy metal with its capacities for investigated wastewater using peanut husk as adsorbent

Heavy metal	Efficiency (%)	Type of wastewater used	References
(Cd^{+2})	99.9	Aqueous solutions	[48]
	88.6	Aqueous solutions	[49]
(Pb^{+2})	98	Industrial wastewater	[50]
(As^{+3})	95	Surface water and groundwater streams	[51]
(As^{+5})	99	Surface water and groundwater streams	[51]

Table 6 List of heavy metal with its capacities for investigated wastewater using sugarcane bagasse as adsorbent

Heavy metal	Efficiency (%)	Type of wastewater used	References
(Ni^{+2})	66.4	Aqueous solution	[53]
(Cu^{+2})	90	Aqueous solution	[53]
(Pb^{+2})	99.9	Aqueous solution	[53]
(Hg^{+2})	97.58	Aqueous solution	[54]
(Cr^{+3})	73	Tannery wastewater	[55]
(Cr^{+6})	80.6	Aqueous solution	[55]

Table 7 List of heavy metal with its capacities for investigated wastewater using *M. oleifera* as adsorbent

Heavy metal	Efficiency (%)	Type of wastewater used	References
(As^{+5})	85.6	Aqueous solutions	[56]
	47	Drinking water	[57]
(Cd^{+2})	99.08	Synthetic wastewater	[58]
	3.8	Cadmium water	[59]
	72	Aqueous solutions	[60]
(Pb^{+2})	89.40 ± 0.00	Synthetic contaminated water	[61]
	99.68	Synthetic wastewater	[58]
	97.9	Contaminated water	[62]
(Cr^{+3})	97	Effluent for water and wastewater treatment	[63]
(Cr^{+6})	99	Aquatic pollutant	[64]
Hg	77.6	Polluted water	[65]

Moringa Oleifera (*M. oleifera*)
M.oleiferais a miracle tree. The pods of M.oleifera have been employed as an inexpensive, coagulant, disinfectant, and same time effective sorbent for the removal of organic matter present in wastewater. Vieira et al. [37] reported in its study about the potential used of *M. oleifera* pods as a natural adsorbent in the wastewater treatment of the dairy industry. It was found that the removal efficiencies are up to 98% for both turbidity and color. Also found, the pods of *M. oleifera* biomass have the great potential to be used in the dairy industry efficiently with the low-cost operation. It can be used for various wastewater such as industrial toxic dye, domestic wastewater, textile wastewater, and dairy industry. Table 7 shows the critical literature for *M. oleifera* as adsorbent.

5 Adsorbent Isotherms

Mainly there are four types of isotherm.

- The '*C*' Isotherm

The correlation between the concentration of substance holding in solution and the substance adsorbed on the surface of solid at the corresponding with the concentration. This correlation between is generally called as diffusion coefficients (kd, $L\ kg^{-1}$) also partition coefficient (kp, $L\ kg^{-1}$) [12].

- The '*L*' Isotherm

The correlation between the concentrations of the substance present in solution with adsorbed on the surface of solid which diminishes when the solute concentration

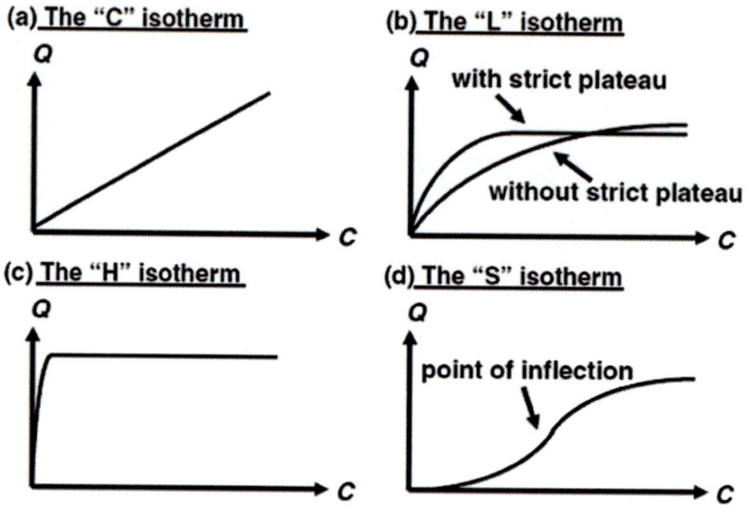

Fig. 3 Different type of isotherm curve

increases as shown in figure mainly two groups are present (1) curve which reaches to strict asymptotic plateau and (2) which do not reach the plateau [12].

- The '*H*' Isotherm

In 'L' isotherm the initial slope is very high. This type of isotherm is different from others as long as the substances which show a dominant affinity for the solid that the initial slope cannot be differentiated from infinity [12].

- The '*S*' Isotherm

Isotherm the curve is sigmoidal in shape. Due to this shape, isotherm got a point of inflection. Mainly this type of isotherm constantly results in two reverse mechanisms. This point of inflection explains the concentration for which this adsorption process overcomes the complexation [12]. Figure 3 shows the curve obtained from different types of isotherms.

6 Modeling of Concave Isotherm

About 70 years ago, Langmuir and Freundlich's isotherm was introduced. These are the most commonly applied adsorption isotherm equations. Liner forms of these isotherms are transformed so that their two adjustable parameters can be easily calculated with the help of linear regression as well as with graphical means.

- Langmuir models

Model is based on a few reaction hypotheses. In this model, it is assumed that solid contain limited adsorption capacity (Q_{max}). Assuming the sites for adsorption to be indistinguishable, each site contains at least one or more molecules of a particular compound at each site that is satirically and energetically independent.

Reaction: solute + free site → Surface complex.

- Freundlich models

Most widely isotherm, i.e., concave isotherm ('L' or 'H' Isotherm) are met. In this, the first model was empirical which is based on a reaction between the solute concentrations (C) with adsorbed quantity (Q).

Reaction: $Q = FC^n$

Where, (Dimensionless)$\{n < 1\}$

$FC\left(L\ \text{kg}^{-1}\right)$

With,

$\log Q = \log F + n \log C$

7 Conclusion

Quick industrialization has brought about expanded usage of overwhelming metal particles in the most recent couple of decades to make the genuine overall ecological issue. Because of its unsafe and bio-accumulative nature, there have been various endeavors to locate an appropriate healing activity. Various conventional technique has so far been used for the extraction of heavy metal ions, however, those methods could not be constructive due to some constraint. The adsorption process is very convenient and constructive. It has consequently been urgent to grow minimal effort and promptly accessible adsorbents for the adsorption of overwhelming metal particles from wastewater. The utilization of inexpensive adsorbents has some advantages like abundance, high adsorption capacity, relatively inexpensive, renewability, and easy separation. Adsorbents have practical gatherings that assume a consequential role in metal ions adsorption. Waste materials are related to utilitarian gatherings, to some degree, phenol gatherings, carboxyl, and hydroxyl which are oxygen- accommodate destinations and assume a significant role in metal ion removal. With the help of this analysis study, an assay has been made in the recent developments correlated with heavy metal removal by relatively inexpensive natural adsorbents. With the help of literature, it was demonstrated that the application of waste materials as relatively inexpensive natural adsorbents for eliminating heavy metals presents. Numerous enticing features particularly their endowment within the promoting environmental protection as well as limiting the costs for waste disposal. Following the adsorption process, the adsorbent can be disposed of without expensive regeneration due to their low-cost.

References

1. Ali I (2012) New generation adsorbents for water treatment. Chem Rev 112(10):5073–5091
2. Jaishankar M, Mathew BB, Shah MS, Gowda KRS (2014a) Biosorption of few heavy metal ions using agricultural wastes. J Environ Pollut Human Health 2(1):1–6
3. Jaishankar M, Tseten T, Anbalagan N, Mathew BB, Beeregowda KN (2014b) Toxicity, mechanism, and health effects of some heavy metals. Interdisc Toxicol 7(2):60–72
4. Fu F, Wang Q (2011) Removal of heavy metal ions from wastewaters: a review. J Environ Manage 92(3):407–418
5. Mahmoud AM, Ibrahim FA, Shaban SA, Youssef NA (2015) Adsorption of heavy metal ion from aqueous solution by nickel oxide nanocatalyst prepared by different methods. Egypt J Pet 24(1):27–35
6. Shobana R, Sahayaraj PA, Dharmalingam V, Soruba R (2014) Adsorption study on zinc (II) ions from aqueous solution using chemically activated fruit of kigelia pinnata (JACQ) DC carbon. Int Res J Environ Sci 3(9):65–69
7. Gupta, S., & Kumar, A. (2019). Removal of nickel (II) from aqueous solution by biosorption on A. barbadensis Miller waste leaves powder. Appl Water Sci 9(4):96.
8. Pyrzynska K (2019) Removal of cadmium from wastewaters with low-cost adsorbents. J Environ Chem Eng 7(1):102795
9. Bayuo J, Pelig-Ba KB, Abukari MA (2019) Adsorptive removal of chromium (VI) from aqueous solution unto groundnut shell. Appl Water Sci 9(4):107
10. Xia M, Chen Z, Li Y, Li C, Ahmad NM, Cheema WA, Zhu S (2019) Removal of Hg (II) in aqueous solutions through physical and chemical adsorption principles. RSC Adv 9(36):20941–20953
11. Ahamad KU, Jawed M (2010) Kinetics, equilibrium, and breakthrough studies for Fe (II) removal by wooden charcoal: a low-cost adsorbent. Desalination 251(1–3):137–145
12. Chakraborty R, Asthana A, Singh AK, Jain B, Susan ABH (2020) Adsorption of heavy metal ions by various low-cost adsorbents: a review. Int J Environ Anal Chem:1–38. https://doi.org/10.1080/03067319.2020
13. Kansal SK, Kumari A (2014) The potential of *M. oleifera* for the treatment of water and wastewater. Chem Rev 114(9):4993–5010
14. Ragadevan V, Kanchanabhan TE, Dayakar P, Mani A, Chockalingam MP (2019) Removal of heavy metal ions (lead) using natural adsorbent. Res J Pharm Technol 12(8):3693–3696
15. Bhattacharya AK, Mandal SN, Das SK (2006) Adsorption of Zn (II) from aqueous solution by using different adsorbents. Chem Eng J 123(1–2):43–51
16. Mohan D, Singh KP (2002) Single and multi-component adsorption of cadmium and zinc using activated carbon derived from bagasse—and agricultural waste. Water Res 36(9):2304–2318
17. Saeed A, Akhter MW, Iqbal M (2005) Removal and recovery of heavy metals from aqueous solution using papaya wood as a new biosorbent. Sep Purif Technol 45(1):25–31
18. Hegazi HA (2013) Removal of heavy metals from wastewater using agricultural and industrial wastes as adsorbents. HBRC J 9(3):276–282
19. Moodley K, Singh R, Musapatika ET, Onyango MS, Ochieng A (2011) Removal of nickel from wastewater using an agricultural adsorbent. Water SA 37(1)
20. Pereira MG, Arruda MAZ (2003) Vermicompost as a natural adsorbent material: characterization and potentialities for cadmium adsorption. J Brazil Chem Soc 14:39–47
21. Nhapi I, Banadda N, Murenzi R, Sekomo CB, Wali UG (2011) Removal of heavy metals from industrial wastewater using rice husks. Open Environ Eng J 4(1)
22. Akhter M, Habib G, Qamar SU (2018) Application of electrodialysis in wastewater treatment and the impact of fouling on process performance. J Membr Sci Technol 8(2):1–8
23. Crini G, Lichtfouse E, Wilson LD, Morin-Crini N (2018) Adsorption-oriented processes using conventional and non-conventional adsorbents for wastewater treatment. In: Green adsorbents for pollutant removal. Springer, Cham, pp 23–71

24. Sulyman M, Namiesnik J, Gierak A (2017) Low-cost adsorbents derived from agricultural by-products/wastes for enhancing contaminant uptakes from wastewater: a review. Polish J Environ Stud 26(3)
25. Crini G, Lichtfouse E, Wilson LD, Morin-Crini N (2019) Conventional and non-conventional adsorbents for wastewater treatment. Environ Chem Lett 17(1):195–213
26. Aman T, Kazi AA, Sabri MU, Bano Q (2008) Potato peels as solid waste for the removal of heavy metal copper (II) from wastewater/industrial effluent. Colloids Surf B 63(1):116–121
27. Annadurai G, Juang RS, Lee DJ (2002) Use of cellulose-based wastes for adsorption of dyes from aqueous solutions. J Hazard Mater 92(3):263–274
28. Ding Y, Su D (2012) Purifying native in-situ mastoid SiO_2 from rice husk. Energy Proc 16:1269–1274
29. Hosseinnia A, Hashtroudi MS, Pazouki M, Banifatemi M (2006) Removal of surfactants from wastewater by rice husk. Iranian Journal of Chemical Engineering 3(3):44–50
30. Idris MR, Alam MS, Rahman MW, Madani A, Bayazidi M, Alaskari MKG, Shanmugan K, Kannadasan E (2014) Use of renewable adsorbent (peanut husk) for the treatments of textile waste water. J Chem 4(4):156–163
31. Islamuddin G, Khalid MA, Ahmad SA (2019) Study of eco-friendly agricultural wastes as non-conventional low-cost adsorbents: a review. Ukrainian J Ecol 9(1):68–75
32. Lin SH, Juang RS (2009) Adsorption of phenol and its derivatives from water using synthetic resins and low-cost natural adsorbents: a review. J Environ Manage 90(3):1336–1349
33. Maheshwari U, Gupta S (2016) Performance evaluation of activated neem bark for the removal of Zn (II) and Cu (II) along with other metal ions from aqueous solution and synthetic pulp and paper industry effluent using a fixed-bed reactor. Process Saf Environ Prot 102:547–557
34. Pehlivan E, Tran HT, Ouédraogo WKI, Schmidt C, Zachmann D, Bahadir M (2013) Sugarcane bagasse treated with hydrous ferric oxide as a potential adsorbent for the removal of As (V) from aqueous solutions. Food Chem 138(1):133–138
35. Sarker TC, Azam SMGG, El-Gawad AMA, Gaglione SA, Bonanomi G (2017) Sugarcane bagasse: a potentially low-cost biosorbent for the removal of hazardous materials. Clean Technol Environ Policy 19(10):2343–2362
36. Velazquez-Jimenez LH, Pavlick A, Rangel-Mendez JR (2013) Chemical characterization of raw and treated agave bagasse and its potential as adsorbent of metal cations from water. Ind Crops Prod 43:200–206
37. Vieira AMS, Vieira MF, Silva GF, Araújo ÁA, Fagundes-Klen MR, Veit MT, Bergamasco R (2010) Use of *Moringa oleifera* seed as a natural adsorbent for wastewater treatment. Water Air Soil Pollut 206(1–4):273–281
38. Roy D, Greenlaw PN, Shane BS (1993) Adsorption of heavy metals by green algae and ground rice hulls. J Environ Sci Health Part A 28(1):37–50
39. Lee CK, Low KS, Liew SC, Choo CS (1999) Removal of arsenic (V) from aqueous solution by quaternized rice husk. Environ Technol 20(9):971–978
40. Munaf E, Zein R (1997) The use of rice husk for removal of toxic metals from waste water. Environ Technol 18(3):359–362
41. Tarley CRT, Ferreira SLC, Arruda MAZ (2004) Use of modified rice husks as a natural solid adsorbent of trace metals: characterisation and development of an on-line preconcentration system for cadmium and lead determination by FAAS. Microchem J 77(2):163–175
42. Marshall WE, Champagne ET, Evans WJ (1993) Use of rice milling byproducts (hulls & bran) to remove metal ions from aqueous solution. J Environ Sci Health Part A 28(9):1977–1992
43. Guo Y, Qi J, Yang S, Yu K, Wang Z, Xu H (2003a) Adsorption of Cr (VI) on micro-and mesoporous rice husk-based active carbon. Mater Chem Phys 78(1):132–137
44. Guo Y, Yang S, Fu W, Qi J, Li R, Wang Z, Xu H (2003b) Adsorption of malachite green on micro-and mesoporous rice husk-based active carbon. Dyes Pigm 56(3):219–229
45. Tiwari DP, Singh DK, Saksena DN (1995) Hg (II) adsorption from aqueous solutions using rice-husk ash. J Environ Eng 121(6):479–481
46. Wong KK, Lee CK, Low KS, Haron MJ (2003) Removal of Cu and Pb by tartaric acid modified rice husk from aqueous solutions. Chemosphere 50(1):23–28

47. Duc PA, Dharanipriya P, Velmurugan BK, Shanmugavadivu M (2019) Groundnut shell-a beneficial bio-waste. Biocatal Agric Biotechnol
48. Cheng Q, Huang Q, Khan S, Liu Y, Liao Z, Li G, Ok YS (2016) Adsorption of Cd by peanut husks and peanut husk biochar from aqueous solutions. Ecol Eng 87:240–245
49. Massie BJ, Sanders TH, Dean LL (2015) Removal of heavy metal contamination from peanut skin extracts by waste biomass adsorption. J Food Process Eng 38(6):555–561
50. Wierzba S (2015) Biosorption of lead (II), zinc (II) and nickel (II) from industrial wastewater by Stenotrophomonas maltophilia and Bacillus subtilis. Polish J Chem Technol 17(1):79–87
51. Sattar MS, Shakoor MB, Ali S, Rizwan M, Niazi NK, Jilani A (2019) Comparative efficiency of peanut shell and peanut shell biochar for removal of arsenic from water. Environ Sci Pollut Res 26(18):18624–18635
52. Yu JX, Wang LY, Chi RA, Zhang YF, Xu ZG, Guo J (2015) Adsorption of Pb 2+, Cd 2+, Cu 2+, and Zn 2+ from aqueous solution by modified sugarcane bagasse. Res Chem Intermed 41(3):1525–1541
53. Van Tran T, Bui QTP, Nguyen TD, Le NTH, Bach LG (2017) A comparative study on the removal efficiency of metal ions (Cu2+, Ni2+, and Pb2+) using sugarcane bagasse-derived ZnCl2-activated carbon by the response surface methodology. Adsorpt Sci Technol 35(1–2):72–85
54. Khoramzadeh E, Nasernejad B, Halladj R (2013) Mercury biosorption from aqueous solutions by sugarcane bagasse. J Taiwan Inst Chem Eng 44(2):266–269
55. Ullah I, Nadeem R, Iqbal M, Manzoor Q (2013) Biosorption of chromium onto native and immobilized sugarcane bagasse waste biomass. Ecol Eng 60:99–107
56. Kumari P, Sharma P, Srivastava S, Srivastava MM (2005) Arsenic removal from the aqueous system using plant biomass: a bioremedial approach. J Ind Microbiol Biotechnol 32(11–12):521–526
57. Pramanik BK, Pramanik SK, Suja F (2016) Removal of arsenic and iron removal from drinking water using coagulation and biological treatment. J Water Health 14(1):90–96
58. Kebede TG, Dube S, Mengistie AA, Nkambule TT, Nindi MM (2018) Moringa stenopetala bark: a novel green adsorbent for the removal of metal ions from industrial effluents. Phys Chem Earth Parts A/B/C 107:45–57
59. Masamba WRL, Mataka LM, Mwatseteza JF, Sajidu SMI (2010) Cadmium sorption by *Moringa stemopetala* and *Moringa oleifera* seed powders: batch, time, temperature, pH and adsorption isotherm studies
60. Jamal P, Muyibi SA, Syarif WM (2008) Optimization of process conditions for removal of cadmium using bioactive constituents of *Moringa oleifera* seeds. Med J Malays 63:105–106
61. Sajidu SM, Henry EMT, Kwamdera G, Mataka L (2005) Removal of lead, iron and cadmium ions by means of polyelectrolytes of the *Moringa oleifera* whole seed kernel. WIT Trans Ecol Environ 80
62. Tavares FO, Pinto LADM, Bassetti FDJ, Vieira MF, Bergamasco R, Vieira AMS (2017) Environmentally friendly biosorbents (husks, pods and seeds) from Moringa oleifera for Pb (II) removal from contaminated water. Environ Technol 38(24):3145–3155
63. Ghebremichael K, Gebremedhin N, Amy G (2010) Performance of *Moringa oliefera* as a biosorbent for chromium removal. Water Sci Technol 62(5):1106–1111
64. Shirani Z, Santhosh C, Iqbal J, Bhatnagar A (2018) Waste Moringa oleifera seed pods as green sorbent for efficient removal of toxic aquatic pollutants. J Environ Manage 227:95–106
65. Ranote S, Ram B, Kumar D, Chauhan GS, Joshi V (2018) Functionalization of *Moringa oleifera* gum for use as Hg^{2+} ions adsorbent. J Environ Chem Eng 6(2):1805–1813
66. Ali RM, Hamad HA, Hussein MM, Malash GF (2016) Potential of using green adsorbent of heavy metal removal from aqueous solutions: adsorption kinetics, isotherm, thermodynamic, mechanism and economic analysis. Ecol Eng 91:317–332
67. Chuah TG, Jumasiah A, Azni I, Katayon S, Choong ST (2005) Rice husk as a potentially low-cost biosorbent for heavy metal and dye removal: an overview. Desalination 175(3):305–316
68. De Gisi S, Lofrano G, Grassi M, Notarnicola M (2016) Characteristics and adsorption capacities of low-cost sorbents for wastewater treatment: a review. Sustain Mater Technol 9:10–40

69. Kyzas GZ, Kostoglou M (2014) Green adsorbents for wastewaters: a critical review. Materials 7(1):333–364
70. Mohammed MA, Shitu A, Tadda MA, Ngabura M (2014) Utilization of various agricultural waste materials in the treatment of Industrial wastewater containing heavy metals: a review. Int Res J Environ Sci 3(3):62–71
71. Pehlivan E, Altun T, Cetin S, Bhanger MI (2009) Lead sorption by waste biomass of hazelnut and almond shell. J Hazard Mater 167(1–3):1203–1208
72. Sarker N, Fakhruddin ANM (2017) Removal of phenol from aqueous solution using rice straw as adsorbent. Appl Water Sci 7(3):1459–1465
73. Tatah VS, Ibrahim KLC, Ezeonu CS, Otitoju O (2017) Biosorption kinetics of heavy metals from fertilizer industrial waste water using groundnut husk powder as an adsorbent. J Appl BioTechnol BioEng 2(6):00049
74. Zwain HM, Vakili M, Dahlan I (2014) Waste material adsorbents for zinc removal from wastewater: a comprehensive review. Int J Chem Eng

Self-Healing of Concrete Using Bacterial Solution

Navneet Singh, Harvinder Singh, and Ajitpal Singh

1 Introduction

Concrete is used widely because of its solitary characteristics, like high compressive strength, favorable fire resistance, and low cost when compared with other materials. But the problem regarding concrete is the formation of cracks which may appear during the initial period or in any phase of its life due to enormous reasons. There are various methods to repair the cracks like grouting, using mesh, sealing/healing of cracks has drawn attention over the past few years.The bacteria have also been used in various other works like cleaning of the weathered surface of concrete, restoring the deteriorated limestone, and improving the strength of mortar, all had shown promising results. The micro-organisms release calcium carbonate on coming in contact with moisture which heals the cracks. The micro-organism used in this research is "Bacillus Megaterium". Bacillus megaterium is a bacteria that is rod-shaped, gram-positive, mainly aerobic bacteria that need oxygen for their survival, the culture has spore-forming bacterium found in widely diverse habitats. They generally have a cell length of up to 4 μm and a diameter of 1.5 μm. The bacteria were in lyophilized form when received in a test tube from MTCC. The specimens were casted by adding micro-organisms during mixing of concrete to see the healing mechanism. The normal concrete specimens were casted, which were cracked under 80% of the ultimate load in which the healing solution consisting of bacteria and agar nutrients was injected to observe the formation of calcium carbonate.Various researchers had already assured about the benefits of adding micro-organisms into concrete-like Souidet al. [7] has mechanically cracked the concrete specimens containing bacteria added during mixing of mortar in a carrier calcium alginate. The cracks were first

N. Singh (✉) · H. Singh · A. Singh
Guru Nanak Dev Engineering College, Gill Road, Ludhiana, India
e-mail: nav.gndec@gmail.com

H. Singh et al. (eds.), *Sustainable Development Through Engineering Innovations*, Lecture Notes in Civil Engineering 113,
https://doi.org/10.1007/978-981-15-9554-7_32

visually inspected under the light microscope then it was further observed by Scanning Electron Microscope (SEM) to check the microstructure of the cracks any material precipitated (i.e. calcium carbonate). The results indicated that the precipitation of calcium carbonate indicated the healing of cracks. Restuccia and Reggio [2] research was based on enclosing the repairing agent is randomly distributed shells made up of glass and pharmaceutical capsules which were added into the material during its mixing. The material was chosen after investigating various factors like it does not react with the repair, the solution used was sodium silicate solution. Several beams and cubes were casted: the beams were subjected to three-point loading after 28 days of curing. When the opening of the crack reached a value of 0.25 mm, regardless of the load applied, the specimens were prolonged in a humid environment for 28 days (at 20 ± 2 °C and 90% relative humidity), and then all the specimens were tested again. The preliminary results of the experimented indicated that three-point bending tests reflected that sodium silicate was fastidiously liberated by the glass shells when the cracks obstructed them. Very small amount of solution infiltrated into the specimens which indicated that strength recovery has taken place. Chen et al. [1] have used ceramist carrier to immobilize bacteria, while substrate and nutrients were mixed evenly immobilized into another original carrier. The flexural strength and water permeation were observed, the result indicated that white precipitation was observed after 21 days of curing. The flexure strength was observed to be increased from 56 to 72% after crack was repaired by calcium carbonate released by bacteria. After curing of the cracked specimens, the water permeability of the cracked specimens also decreased indicating the healing of cracks. Sangadji et al. [5] demonstrated the attainable application of spore solution which acts as a healing agent, sprayed onto porous network concrete. Such type of concrete has a core that was porous in nature, used as a medium for transporting the agents which are going to heal the cracks in the fractured zone. The capability of formed solution to heal the cracks has been evaluated by supervising whether the bio-mineral precipitation is taking place in the cylindrical cores or not. Along with that, the water permeability test was conducted at several time intervals to check the self-healing mechanism. The specimens were encapsulated with bacteria-based solution. It was observed that specimens that were cured in water showed a faster rate of healing as compared to uncured specimens. The small cracks have a better chance to heal than deeper ones. Koster and Mors [4] coated the healing agents with geo-polymers with different techniques of making a mixture. The particles were coated by disintegrating in low shear granulating machine by operating the window for all activated liquids. To determine whether the application of particles is feasible in concrete mixture or not, various tests were performed on the coated particles which were integrated into cement paste-like strength and leaching test. Results showed the comparison among particles depicting that the coated particles with geo-polymers gave better protection from leaching than the particles which were not given geopolymer coating. Justs et al. [3] inspected about the process of curing done in ultra-high-performance concrete by the superabsorbent polymers which initiate internal curing. The study aimed to limit disintegration of concrete and also to reduce the shrinkage that takes place automatically leading to cracks formed at a very early age of ultra-high-performance concrete (UHPC).

Internal curing which is done by SAP is practical in decreasing the humidity which is inside the concrete which helps in reducing the shrinkage of concrete done automatically. In spite of the reduction in the cracks and self-healing, the mechanical properties like strength of concrete are largely distressed by the addition of SAP into the concrete mix, the attained strength was just 150 MPa after the curing of 28 days. Schlangen and Sangadji [6] researched on asphalt concrete self-healing mechanism and also on the specimens having cement as a raw material. The porous aggregate was used to fill bacteria and food with different cases like food and bacteria then only food containing concrete specimens (disks) were prepared. The concrete disks were cured for a period of 56 days and brought to partially crack in a tensile strength-controlled testing machine. After the specimens were cracked, they were kept for the permeability test in a setup in which the supply of water was maintained at one side for 24 h. The cracks were inspected after the healing mechanism took place under the microscope and the results indicated that the permeability of specimens decreased. In other projects, cementitious material embedded with hybrid fiber was studied which has the capability to mechanically repair cracks that occur by incorporated in microcapsules or steel fibers. SFRC previously has a high possibility for healing as their crack width is very small. The passage for ingress of chloride and other salt ions into the embedded steel is blocked, which enhanced the structure's durability.

1.1 Materials and Methodology

This section deals with the ingredients used for the casting of specimens as well as the method adopted for the casting of a specimen.

1.2 Bacillus Megaterium

There are various types of micro-organisms present in significant quantities (numbers) everywhere in sewage water, normal water, and surroundings. The most fundamental and the simplest wholly contained life systems are bacteria or prokaryotes. Their presence can be detected only by their reactions to different stains or cultures. Because after staining the various organelles of cells along with their plasma membrane becomes visible. The food of bacteria nutrient agar was added into the bacterial solution.

1.3 Preparation of Bacterial Culture Solution

The frozen bacterial culture carrying megaterium bacteria was received in a test tube that contained bacteria in a lyophilized form. The test tube was covered with a thick layer of cotton to protect the bacteria from contamination. The culture was received in the lyophilized form. The lyophilized bacteria were added into the media along with the distilled water and it was kept in the incubation chamber at 28 °C for 3 days, then the growth was observed and checked with the help of a haemocytometer under the microscope at 40X zoom as shown in Fig. 1. The bacterial culture prepared in which food of bacteria i.e. urea was also before casting as shown in Fig. 2. The count of bacterial cells was taken with the formula given as:

$$\text{Total Count} = \text{Average of total no. of Bacterial Cell in sub} - \text{square boxes} * 16 * 25 * 10^4 \text{ cells}$$

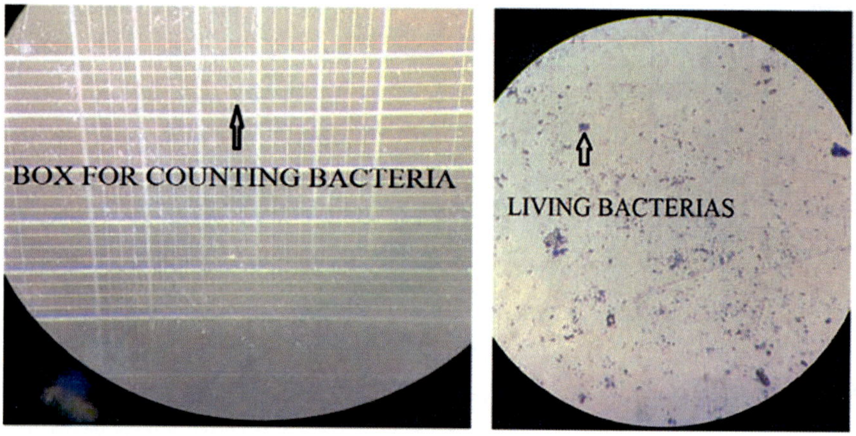

Fig. 1 Haemocytometer under the microscope 40× and on right bacteria slide stained with methylene blue

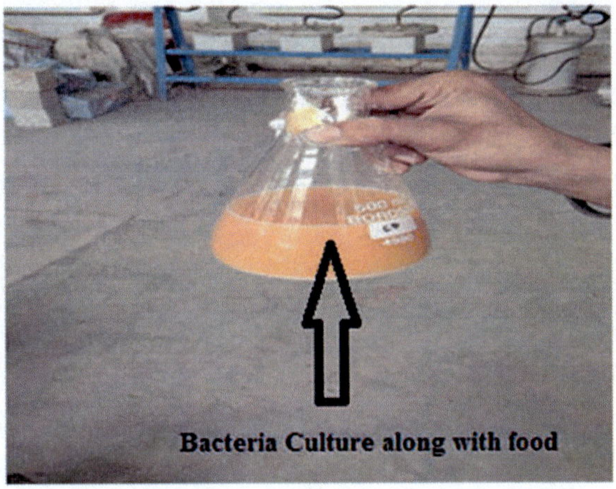

Fig. 2 Bacteria culture along with the food mixed

2 Casting of Specimens

2.1 Two Sets of Specimens Were Casted

2.1.1 1st Set of Normal Concrete (*N*)

The Coarse Aggregate and Fine Aggregate weighed and were added to the mixer and allowed to rotate for 1–2 min. After mixing of the Coarse Aggregate and Fine Aggregate, cement was added in dry stage into the mixer and rotated for 3–4 min for complete mixing. Required Amount of Water was added. The mixture was allowed to rotate for 3–4 min to completely mix all the ingredients and form a homogeneous mix.

2.1.2 2nd Set of Specimens Containing Normal Bacteria (*B*)

The Coarse Aggregate and Fine Aggregate weighed were added to the Mixer and allowed to rotate for 1–2 min. After mixing of the Coarse Aggregate and Fine Aggregate, cement was added in dry stage into the mixer and rotated for 8–9 min to completely mix, then the coated clay aggregates impregnated with bacteria added into dry mix. Required Amount of Water was added. The mixture was allowed to rotate for 3–4 min to completely mix all the ingredients and form a homogeneous mix (Table 1).

Table 1 Proportion of specimens

Set No.	Cement (in kg) for 1m^3	Coarse aggregate (in kg) for 1 m^3	Fine aggregate (in kg) for 1 m^3	Bacterial cells for 1 cm^3
1st set (*N*)	390.346	1100.90	680.02	–
2rd set (*B*)	390.346	1100.90	680.02	5.8×10^8

The water-cement ratio was kept as 0.45

2.2 Testing of Specimens

Compressive Strength Test:

The compressive strength of concrete is given in terms of the characteristic compressive strength of 150 mm size cubes tested at 28 days (f_{ck}). The characteristic strength is defined as the strength of the concrete below which not more than 5% of the test results are expected to fall. For M20 as per IS 456 it should be more than 20 MPa after 28 days. The compressive strength conducted on cube specimen as shown in Fig. 3 were loaded upto 80% of loading so that cracks which are less widened can be filled with solution.

Fig. 3 Compression strength test of M20

Fig. 4 Flexural testing of specimen

2.3 *Flexural Strength Test*

Flexural strength, also known as modulus of rupture, or bend strength, or transverse rupture strength is a material property, defined as the stress in a material just before it yields in a flexure test. The transverse bending test is most frequently employed, in which a specimen having either a circular or rectangular cross-section is bent until fracture or yielding using a three-point or four-point flexural test technique. The flexural strength represents the highest stress experienced within the material at its moment of yield. The UTM machine used for flexural testing is having a capacity of 100 kN. The load was applied gradually. The various observations like load–displacement graph stress–strain and various other graphs were prepared automatically. The beam specimen having bacteria in concrete to compare the result with normal concrete specimen as shown in Fig. 4.

2.4 *Addition of Micro-Organisms in Cracked Specimens*

The micro-organisms along with their food were injected into the cracked specimens with the help of a syringe as shown in Fig. 5 so as to observe the self-healing action of micro-organisms if externally added. It was done so that the technique of self-healing of concrete specimens by formation of calcium carbonate can be observed which will lead to an increase in the life of already made structures by reducing or healing the crack. The specimens in which the bacteria were injected were kept in dry state and wet state both so that the rate of self-healing can be observed and compared. The

Fig. 5 Syringe filled with bacteria culture and its food

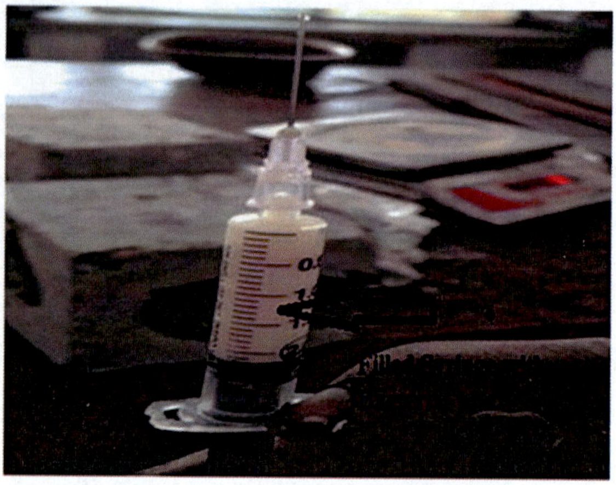

points were marked with the marker where the micro-organisms were injected into the specimens so that the formation of calcium carbonate can be observed in the specimens. Then the velocity at different time intervals of 7, 14, 28, 42 and 56 days was observed.

The minor and the major cracks were located on the surface of concrete and then the bacterial solution prepared was injected into the cracks with the help of a syringe. The size of needle was 0.55 mm × 2.5 mm, due to the smaller size of the needle it easily penetrated into minor cracks. The capacity of the syringe was 2.5 ml, filled many times so that all cracks can be filled with the bacterial solution made. All the cracks observed on the surface of specimens were filled as shown in Fig. 6. The two smooth surfaces were fixed so that the velocity of the ultrasonic waves can be calculated which was taken as the criteria for evaluating the phenomenon.

3 Results and Discussions

3.1 Compressive Strength of Concrete

Proportion of concrete = M20.

Cross-section Area = $0.15 \times 0.15 = 0.0225\ m^2$.

The load-carrying capacity of the cube specimens of different samples casted was calculated and it has been observed that the load-carrying capacity of the concrete decreases when we added urea into it a shown in Fig. 7 which is a major nutrient of bacteria.

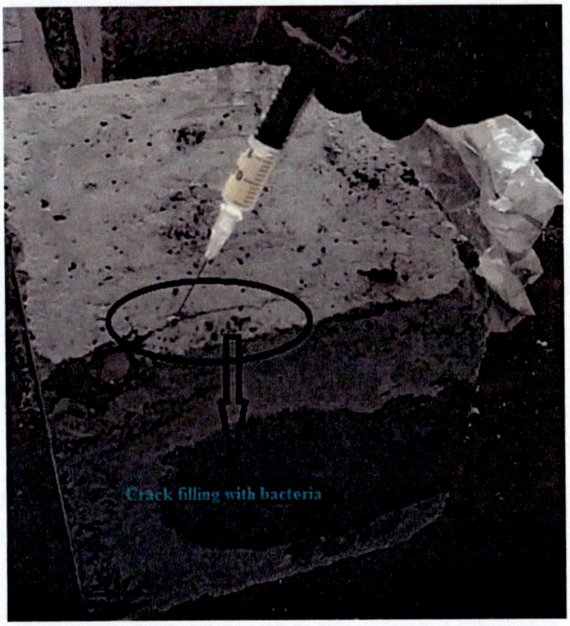

Fig. 6 Addition of bacteria along with its food into the cracks for self-healing action

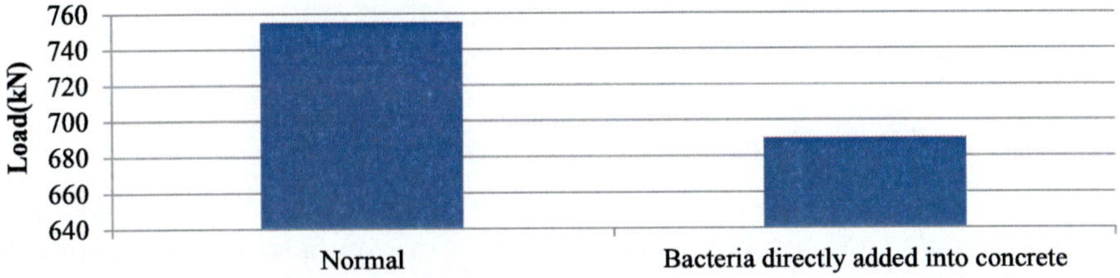

Fig. 7 Graph showing comparison of load-carrying capacity of cube

3.2 *Flexure Strength of Concrete*

Flexural strength of concrete is computed with the help of plain cement concrete beam calculated by Eq. (1) as per code (Table 2):

Table 2 Showing variation of flexural strength in different specimens

Specimens	Flexural strength (N/mm^2)
Normal M20	5.75
Bacteria directly added	5.2

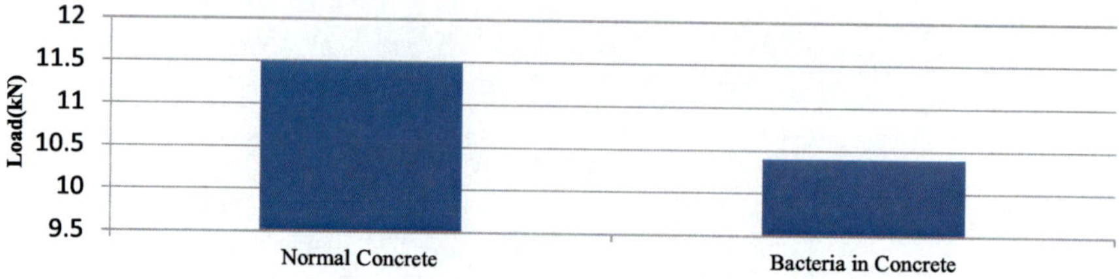

Fig. 8 Graph showing a comparison of the load-carrying capacity of beams

$$\begin{aligned} f_{\text{cr}} &= 0.7\sqrt{f_{\text{ck}}} \\ &= 0.7\sqrt{20} \\ &= 3.13\,\text{N/mm}^2 \end{aligned} \tag{1}$$

3.3 Variation of Load-Carrying Capacity of Beams

Similar type of behavior was observed in the case of beams as that of cubes. The overall maximum 10% decrease of load-carrying capacity is observed in all set of specimens when compared with the normal specimens because of the addition of urea into the mix as shown in Fig. 8

3.4 Stress–Strain Curve for Normal Concrete (Beam)

The above graph shown in Fig. 9 expresses the stress–strain results of the normal concrete beam M20 in which no bacterial culture was added and result of this specimen is used to compare them with the other sets of specimens carrying bacterial culture along with urea which reduces the strength of concrete.

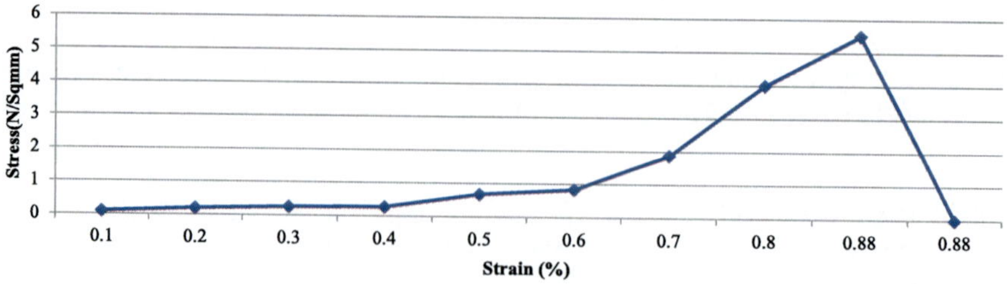

Fig. 9 Graph showing stress–strain curve for normal concrete beams

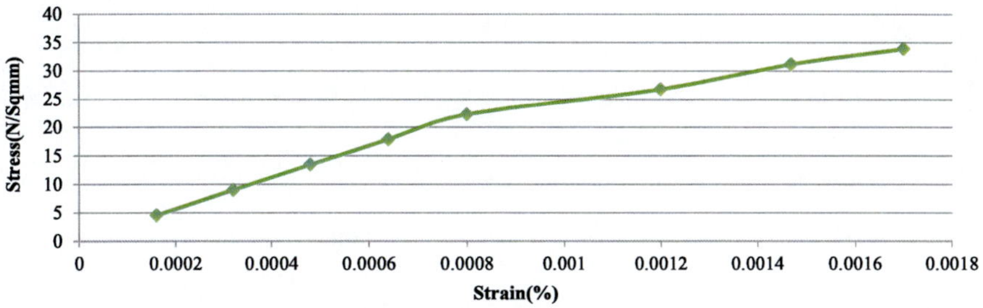

Fig. 10 Graph showing stress–strain curve for normal concrete cube

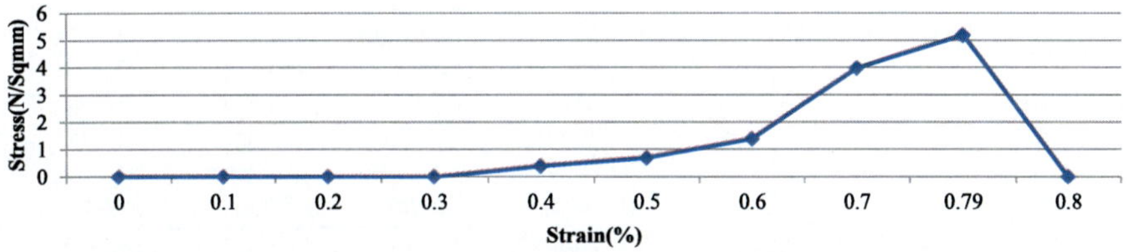

Fig. 11 Graph showing stress–strain curve for bacterial concrete beam

3.5 Stress–Strain Curve for Normal Concrete (Cube)

The behavior of stress–strain curve shown in Fig. 10 indicates the stress reaches a maximum value of 33.78 N/mm^2 and hook's law is valid up to 17.76 N/mm^2 as shown above. The stress taken by the cube is more than the standard value for M20.

3.6 Stress–Strain Curve for Bacterial Concrete (Beam)

The stress–strain curve for the bacterial culture concrete beam as shown in Fig. 11 has shown that the flexural strength is decreased due to the addition of urea into the specimens.

3.7 Stress–Strain Curve for Bacterial Concrete (Cube)

See Fig. 12.

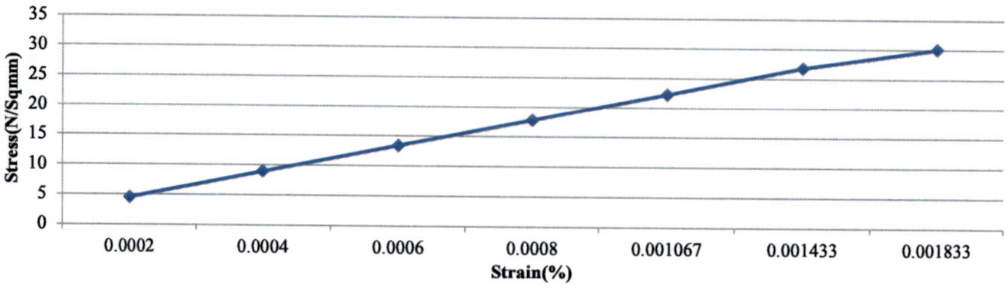

Figure12 Graph showing stress–strain curve for bacterial concrete cube

Table 3 Results for specimens in which bacteria were added after cracking of specimens

Specimens	Velocity after cracking (m/s)	Velocity after 7 days of adding bacteria (m/s)	Velocity after 14 days of adding bacteria (m/s)	Velocity after 28 days of adding bacteria (m/s)	Velocity after 42 days of adding bacteria (m/s)	Velocity after 56 days of adding bacteria (m/s)
Dry condition	3300	3343	3500	3620	3650	3704
Dry condition	3100	3134	3250	3284	3343	3404
Wet condition	3150	3300	3468	3600	3724	3946
Wet condition	3082	3202	3354	3486	3688	3806

3.8 Ultrasonic Pulse Velocity Test Results

To evaluate the self-healing mechanism the bacteria were added into the concrete during casting so that they can evenly spread in the homogeneous mixture and can heal the minor cracks formed during the shrinkage. The specimens casted were cured and subjected to 80% of the loading because most of the minor and major cracks are formed at this loading. The ultrasonic pulse velocity test was used to measure the velocity which is taken as a parameter to compare and observe the self-healing mechanism The velocity was measured at the different time intervals of 7, 14, 21, and 28 days after the cracking of specimens. Before cracking the normal velocity of the uncracked specimen lies between 6400 and 7000 m/s. The two specimens were kept in dry state after addition of bacteria, so that the moisture absorption from the atmosphere can take place and the water which was present in the voids can also be used for sealing of cracks. But the two specimens were kept in water because from the literature it was observed that the rate of self-healing accelerates due to the presence of water.

Table 4 Results for specimens in which bacteria were added during mixing

Specimens	After cracking (m/s)	7 Days after adding cracking (m/s)	14 Days after adding cracking (m/s)	21 Days after adding cracking (m/s)	28 Days after adding cracking (m/s)
Bacteria directly added	3845	4058	4664	4820	4900

From Table 3, it can be observed that Specimens kept in wet conditions showed better results as compared to the specimens kept in dry conditions. The velocity increases gradually in the specimens kept in wet condition and specimens in dry condition showed that the rate of self-healing in dry condition was very low but it also reflected gradual increase in velocity. Also, the self-healing in the concrete specimens in which bacteria were added was observed with the help of velocity measurement on the same point after an interval of 7, 14, 21 and 28 days as shown in Table 4.

4 Conclusions

- The specimens which were subjected to curing after the addition of micro-organisms into the cracks formed healed at faster rate than dry specimens as it can be observed from the change in velocity shown in Table 4 which was around 3150 m/s in 4th specimen and increased to 3806 m/s whereas if we observe in 1st specimen which was dry-cured the velocity was 3300 m/s in cracked specimen and reached 3704 m/s only. So, when micro-organism comes in contact with moisture the rate of healing increases.
- The decrease in strength is under limit as per IS 456 when bacteria were added into the concrete for both flexures as well as compression. Also, the decrease of strength when compared with normal concrete was appx. 8–10% only as shown Figs. 7 and 8.
- Maximum of crack healing occurs between age of 7–21 days which can be observed in Table 3 that the rate of healing slows down after 21 days. As soon as they absorb their food along with water it leads to formation of calcium carbonate which leads to increase in the velocity because cracks get healed with calcium carbonate.

Acknowledgement This project work could never have been possible without the permission of almighty GOD, who is always present by my side and has planned something best for me and all other people present in this world. I would like to pay my sincere thanks to my Guide **Dr. Harvinder Singh** and co guide **Er.Ajitpal Singh** for their guidance and encouragement in this research work. I would like to extend gratitude to Civil Engineering Department, Guru Nanak Dev Engineering College, Ludhiana for providing me best facilities and equipment in the laboratory.

References

1. Chen H, Qian C, Huang H (2016) Self-healing cementitious materials based on bacteria and nutrients immobilized respectively. Constr Build Mater 126:297–303
2. Restuccia L, Reggio A (2017) New self-healing technique for cement-based materials. XXIV Ital Group Fract Conf Struct Integr Proc 3:253–260
3. Justs J, Wyrzykowski M, Bajare D, Lura P (2015). Internal curing by superabsorbent polymers in ultra-high-performance concrete. Cem Concr Res 76:82–90
4. Koster SAL, Mors RM (2015) Geopolymer coating of bacteria containing granules for use in self-healing concrete. The 7th world congress on particle technology. Proc Eng 102:475–484
5. Sangadji S, Wiktor V, Jonkers H (2016) The use of alkaliphilic bacteria-based repair solution for porous network concrete healing mechanism sustainable civil engineering structures and construction materials. Proc Eng 171:606–613
6. Schlangen E, Sangadji S (2013) Addressing infrastructure durability and sustainabilty by self healing mechanism—recent advances in self healing concrete and asphalt. The 2nd international conference on rehabilitation and maintenance in civil engineering. Proc Eng 54:39–57
7. Souid A, Esaker M, Elliott D, Hamza O (2019) Experimental data of bio self-healing concrete incubated in saturated natural soil. Data Brief:10439

Flexural Strength Enhancement of Polypropylene Fiber-Reinforced Concrete Members

Lakhvir Kaur and Harvinder Singh

1 Introduction

Concrete is a popular binding material that is composed of cement, fine and coarse aggregates mixed with water which hardens with time. It offers more compressive strength as aggregates carry the compressive load, but weak in tension, having limited resistance against ductility and cracking. Cracks having greater width are more dangerous because there is possibility of corrosion due to corrosive environment, which leads to brittle fracture of concrete structures. The deterioration of such structures is of great concern because the repairing and rehabilitation of these structures are time consuming and costly. Its low tensile strength and brittle nature necessitate it to be reinforced with steel rods, fibers, etc. So, several attempts have been made to improve its low tensile strength and brittleness nature of concrete.

In addition to this, population growth and scarcity of land have raised the demand of construction of high rise buildings in the urban areas. This rapid development of tall buildings and long span structures demands low weight materials in cement composition which further demands sustainable materials.

Thus, sustainability of concrete can be achieved by reducing the raw materials of concrete and improving its strength when non-conventional and low-cost materials are added. So, selection of these materials is based on its durability, performance and environment preservation. With respect to this, many studies have been done on natural fibers and synthetic fibers. Thus, the waste of plastic polypropylene fibers

L. Kaur (✉)
Department of Civil Engineering, Punjabi University, Patiala, India
e-mail: lakhvir1177@gmail.com

H. Singh
Guru Nanak Dev Engineering College, Ludhiana, Punjab 141006, India

H. Singh et al. (eds.), *Sustainable Development Through Engineering Innovations*, Lecture Notes in Civil Engineering 113,
https://doi.org/10.1007/978-981-15-9554-7_33

is also considered in experimental work. The waste of PPF enhances the properties of concrete and has proven to reduce the environmental pollution, thus environment friendly [1].

Polypropylene is a thermoplastic material made by chemical industry and used to make fiber, bundling and materials like rope and warm clothes. It has low density and high wear resistance. Also, PP fibers are chemically inert, so will not corrode. This allows it to be used as an engineering material.

Polypropylene fiber-reinforced concrete contains short discrete fibers that are uniformly distributed and randomly oriented in concrete and act as an internal reinforcement so as to increase the structural integrity of concrete (delays the failure mechanism). The principal reason for using these fibers in concrete is to increase toughness and tensile strength of concrete and to improve post crack response of resultant composite. Its resistance again corrosion attack of environment is suited to minimize the erosion damage in structures and helped to prove it eco-friendly material.

The properties of FRC primarily depend upon the type of the fibers (geometry), orientation, distribution and volume fraction of fibers used in the concrete [2]. Fibrillated polypropylene fibers (mesh-type structure when filament is opened) are produced as continuous mono-filaments with circular cross section. These fibers can be chopped to required lengths and tapes that can be fibrillated to form the fibrils of rectangular cross section [3]. The use of fibrillated polypropylene fibers in concrete is to provide tensile strength, tensile strain capacity and the improved resistance to impact and fatigue. These fibers are most effective for controlling plastic shrinkage cracks; these are proven to deliver maximum post-peak flexural strength in concrete.

Mtasher Rana et al. [4] indicated that addition of fibers in concrete mix increases the compressive strength relative to conventional concrete. He examined that the polypropylene fiber inclusions in ranges of 0.4% and 1.5% increased the compressive strength up to 11% and 56%, respectively.

Mohod [5] has considered the effect of polypropylene fiber on the properties of concrete mix M30 and M 40. He found that flexure behavior of concrete of both mixes improved due to the addition of 0.5% fiber dosage. Govindasami et al. [6] have studied the effect of polypropylene fibers of 12 mm length on properties of M25 concrete. The flexural strength of polypropylene fiber-reinforced concrete at 28 days improved with increasing volume fraction. He observed a maximum increment of 33.61% in flexural strength of concrete at 1.5% dosage of polypropylene fiber relative to control specimen.

Yao [7] has examined the strength and deformation characteristics of polypropylene fiber-reinforced concrete (PFRC) beams by four-point bending. His investigation results showed that the strength of concrete containing drawing-wired polypropylene (DP) fibers was lower than that of concrete containing (FP) fibers. Due to 1.5% dosage of the fibre, an improvement of 27–114% in toughness for the concrete containing DP fibers and 25–163% for the concrete containing FP fiber were observed. It was found that the fibrillated polypropylene fibers were more beneficial for the improvement of flexural toughness.

Thus, polypropylene fibers are considered to serve these fibers for the purpose of sustainable use of materials. Polypropylene and plastic waste can be recycled by utilizing the waste for construction of fiber and also helped to overcome the disposal problems.

2 Materials Used

Cement: The cement used in this experimental investigation is OPC 43 grade.

Fine aggregate: Sand passing through IS 4.75 mm sieve as per IS: 383:1970 was used for all the specimens and having fineness modulus 2.8 and specific gravity 2.63.

Coarse Aggregates: Locally available coarse aggregates of 10 mm size are used for experimental work and having specific gravity of coarse aggregates 2.65.

Fibers Used: Fibrillated polypropylene fibers (PPF) of two different lengths (6 and 20 mm) were used in concrete at different volume percentages. The fibrillated polypropylene fiber (shown in Fig. 1) was provided by the *NINA Concrete Systems Pvt. Ltd.* And the physical properties of PPF were also shown in Table 1.

Concrete: Mix design of M 20 grade concrete was obtained as per provision of IS 10262 and mixture proportions of cement, fine aggregates and coarse aggregates were 1:1.5:2.8 and w/c ratio was 0.52 used [8].

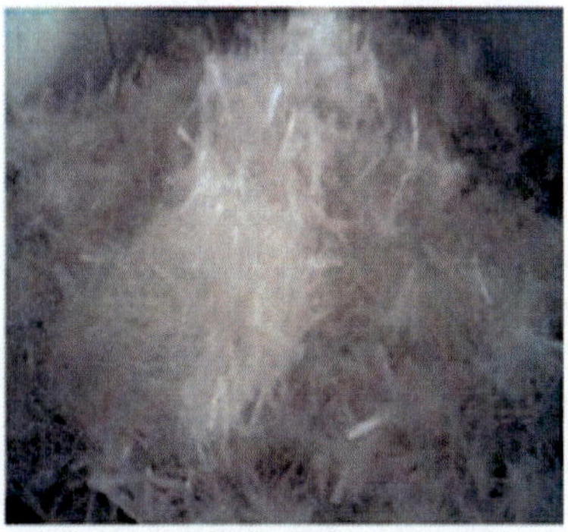

Fig. 1 Fibrillated polypropylene fiber

Table 1 Physical properties of PPF

S. No.	Property	Values
1	Fiber length	20, 6 mm
2	Type of fiber	Fibrillated/mesh type
3	Specific gravity	920 cc/gm
4	Water absorption	Nil
5	Alkali resistance	Alkali proof

3 Flexural Test

PPFRC beams of size 100 × 100 x 510 mm were tested after 28 days curing using a flexure testing machine under three-point loading. Prisms were placed on two supports with 480 mm length apart and a load at the midpoint of the samples applied. Specimen section and three-point bending test setup is shown in Figs. 2 and 3.

Fig. 2 Three-point bending test set

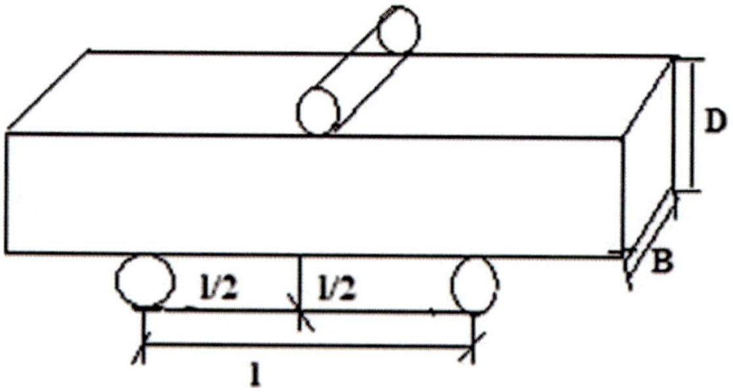

Fig. 3 View of beam for three-point bending test (for prismatic section B = D)

4 Flexural Capacity of Beams

Analytical Approach: Simple mechanics theory and equilibrium conditions have been used to calculate the flexural capacity in terms of ultimate moment of the beams containing randomly distributed fibers. The tensile forces carried by the added fibers create an additional internal moment. The analysis is based on the assumptions [9] such that plane section remains unchanged after bending, the internal moment is equal to external applied bending moment, and perfect bond is between reinforcement and surrounding material.

The rectangular beam section of width B and depth D was considered in the modeling. The polypropylene fibers are mixed randomly in the concrete, and these are assumed to present uniformly anywhere in the section. The tensile strength of PPFRC is assumed to be σ_t, whereas effect of length of fiber, type of fiber and V_f on compressive strength is ignored and stress block similar to conventional curves was used.

Flexural stress block for PPFRC was shown in this Fig. 4, where

x depth of neutral axis in compression zone;
b&d width and depth of rectangular ppfrc beam.
f_{ck} the characteristic strength of concrete;
$(d - x)$ depth of neutral axis in tensile zone;
σ_t the tensile strength.

Total compressive force C above the N.A. in PPRFC beam is given by Eq. 4.1

$$C = 0.36\ f_{ck}bx \tag{4.1}$$

And, total tensile force T below N.A. of a PPFRC beam is given by Eq. 4.2,

$$T = \sigma_t b(d - x) \tag{4.2}$$

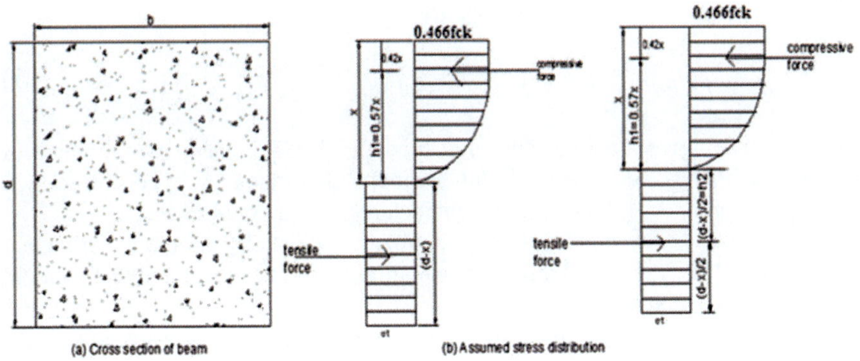

Fig. 4 Flexural stress block for PPFRC

For equilibrium of PPFRC beam, compressive force must be equal to tensile force below the N.A. of the beam. By equating both Eqs. 4.1 and 4.2, we got a simple expression in Eq. 4.3.

$$\frac{d}{x} = 1 + \frac{0.36 f_{\text{ck}}}{\sigma_{\text{t}}} \tag{4.3}$$

Here, $\frac{\sigma_t}{f_{\text{ck}}}$ in Eq. 4.3 can be replaced by a parameter called as fiber index (β) and put it in Eq. 4.3, and it is the ratio of residual tensile strength of PPFRC to its compressive strength.

$$\frac{d}{x} = 1 + \frac{0.36}{\beta}$$

Total moment of resistance (M) of PPFRC beam can be determined by taking moment of C and T about neutral axis and is given by in Eq. 4.4

$$M = M_1 + M_2 \tag{4.4}$$

where M_1 is the moment of compressive force (C) about neutral axis and M_2 is the moment of tensile force (T) about neutral axis. The magnitude of moment of resistance (M) can be worked out by substituting corresponding values of C, T in Eq. 4.4

$$M_1 = 0.36\ f_{\text{ck}} bx\ \times (0.57\times) \tag{4.5}$$

$$M_2 = \sigma_t b(d - x) \times \frac{(d - x)}{2} \tag{4.6}$$

Substituting the value of x in these above equations and after substituting M_1 and M_2 in Eq. 4.4, we get

$$\frac{M}{f_{\text{ck}}\ bd^2} = 0.2052\left(\frac{\beta}{\beta + 0.36}\right)^2 + 0.5\ \beta\left(\frac{0.36}{\beta + 0.36}\right)^2 \tag{4.7}$$

Equation 4.7 gives analytical flexural model of PPFRC beam. The moment of resistance of PPFRC beam depends only upon the fiber index (β) of concrete which is a function of V_{f}, length of fiber and characteristic strength of concrete.

In this above equation, the value of moment calculated from the experimental work, f_{ck}, b, d ($b = d = 100$ mm for prismatic section) is substituted and value of fiber index (β) is obtained.

From the value of fiber index, the flexural strength of concrete is calculated for different fiber lengths PPFRC at different volume fractions.

5 Results and Discussion

Values of tensile strength (flexural stress) obtained from above relation for PPFRC at different volume fraction and lengths of fiber are shown in Tables 2 and 3.

The results of flexural strength for various M 20 concrete PPF mixes are presented in Fig. 5. Bending strength of PPFRC of 20 mm fiber length shows better response, i.e., bending strength of PPFRC (20 mm fiber length) is increased from 0.5% V_f to 2% V_f. From experimental data, 0.5% V_f of 6 mm polypropylene fiber (PPF) concrete shows no significant response, but 1, 1.5 and 2% V_f of 6 mm PPFRC shows notable results as compared with plain concrete or 0.5% of PPFRC. During testing, it was observed that normal concrete beams suddenly failed after appearance of first crack, but PPFRC beams continue to resist load with increasing deformation and widening of cracks and failed at large deformations as fiber reach their maximum elongation.

The results of PPFRC having 20 mm fiber at different volume fractions show that flexural strength is increased about 48% from 0.5% to 2% V_f and 35% increase of flexural from 0.5% to 1.5% V_f.

The curves were developed between volume fraction (V_f) and flexural strength (f_b) for PPFRC of 6 and 20 mm, shown in Figs. 6 and 7. The regression equation generated from the flexural stress and volume fraction of fiber for L_f 6 and 20 mm curves can be used to estimate the flexural strength at various volume fractions. A theoretical correlation can be established between these two coefficients via a correlation constant, as follows.

Table 2 Calculation of bending stress parameters for PPFRC of L_f 6 mm

Volume fraction (V_f)	Length of fiber (mm)	Ultimate loads (kN)	Fiber index (β)	Stress (N/mm^2)	Average of stress, (N/mm^2)
0	–	6.0	0.0837	1.674	1.674
		6.0	0.0837	1.674	
		6.0	0.0837	1.674	
0.5	6	6.0	0.0830	1.674	1.834
		6.5	0.0957	1.914	
		6.5	0.0957	1.914	
1	6	7.5	0.1140	2.280	2.292
		7.0	0.1051	2.102	
		8.0	0.1247	2.494	
1.5	6	8.5	0.13511	2.702	2.77
		9.0	0.14581	2.916	
		8.5	0.13511	2.702	
2	6	9.0	0.14581	2.916	2.98
		9.0	0.1458	2.916	
		9.5	0.1570	3.14	

Table 3 Calculation of bending stress parameters for PPFRC of L_f 20 mm

Volume fraction (V_f)	Length of fiber (mm)	Ultimate loads (kN)	Fiber index (β)	Stress (N/mm^2)	Average of stress (N/mm^2)
0.5	20	8.0	0.1247	2.494	2.494
		8.0	0.1247	2.494	
		8.0	0.1247	2.494	
1	20	8.5	0.1351	2.70	2.844
		9.0	0.1458	2.916	
		9.0	0.1458	2.916	
1.5	20	9.5	0.1570	3.140	3.37
		10.0	0.1687	3.374	
		10.5	0.1808	3.616	
2	20	11.0	0.1934	3.868	3.70
		11.0	0.1934	3.868	
		10.5	0.1808	3.616	

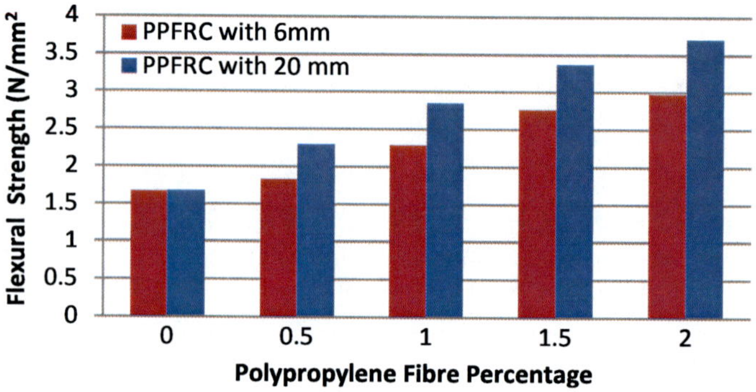

Fig. 5 Effect of polypropylene fiber on flexural strength of concrete

Fig. 6 Flexural strength versus volume fraction curve for PPFRC of 6 mm L_f

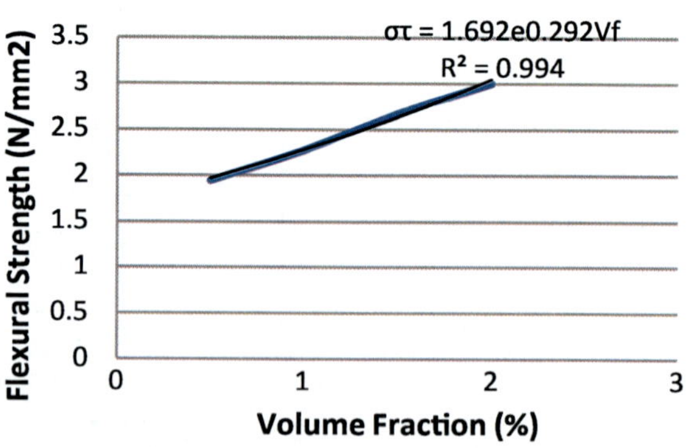

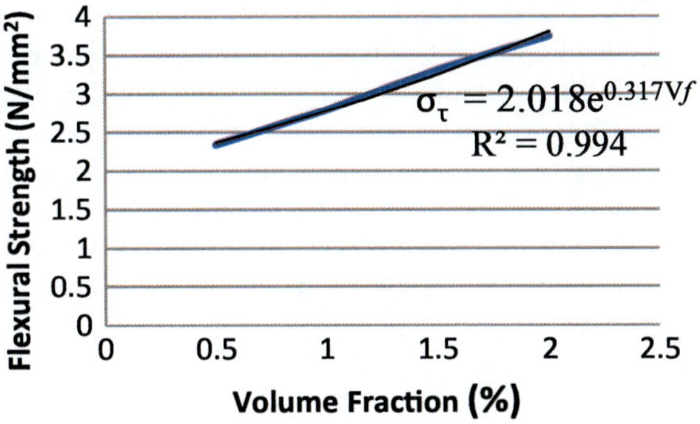

Fig. 7 Flexural strength versus volume fraction curve for PPFRC of 20 mm L_f

For PPFRC of 6 mm, regression equation is;

$$\sigma_\tau = 1.692\ e^{0.292V_f} \tag{5.1}$$

By substituting the values of V_f of fibrillated polypropylene fiber of 6 mm for the concrete mixture used in the flexural tests in Eq. 5.1, it gives the value of flexural strength of concrete.

Similarly, for PPFRC of 20 mm, regression Eq. 5.2 is;

$$\sigma_\tau = 2.018\ e^{0.317V_f} \tag{5.2}$$

By substituting the values of V_f of fibrillated polypropylene fiber of 20 mm for the concrete mixture used in the flexural tests in Eq. 5.2, the values of stress calculated from experimental results of PPFRC of 6–20 mm together show the significant increase in stress.

6 Conclusions

Based on the test results and predicted values of flexural strength, the following conclusions are drawn:-

- An analytical flexural model is proposed to predict the flexural response of PPFRC members, and it predicts the moment capacity of PPFRC members.
- It is observed from flexural test that on increasing the fiber dosage or percentage, there is significant improvement in the load carrying capacity and reduction in the crack opening and crack spacing. So, it reduces the repair cost and maintenance requirements. Thus, the use of polypropylene fibers in concrete is helped in making the sustainable structures/concrete.

- The flexural strength of concrete beams having 20 mm PPF was observed to increase by 48% when fiber dose V_f in concrete is increased from 0.5 to 2.0% than the conventional concrete. Effect of fiber lengths and volume fraction is shown in Fig. 6.
- The PPFRC having 20 mm long fibers is found to be more effective than the 6 mm PPFRC concrete in enhancing the flexural strength of concrete members. The addition of PPF enhances the cracking resistive strength of composite material due to the effect of the matrix bond between fiber and surrounding concrete.

References

1. Olaoye et al (2013) The use of fibre waste as complement in concrete for a sustainable environment. In: 2nd international conference on engineering and technology research, vol 4, no 9
2. Wu et al (2018) Improvement of mechanical properties in polypropylene and glass fibre reinforced peach shell light weight concrete. Adv Mater Sci Eng 625094:11
3. Daniel JI et al (1998) Fiber reinforced concrete. Portland Cement Association, Chapter 5, pp. 22–26
4. Mtasher A et al (2011) Strength prediction of polypropylene fiber reinforced concrete. Eng. Tech. J 29(2):305–311
5. Mohod MV (2015) Performance of polypropylene fibre reinforced concrete. J Mech Civil Eng 12:28–36
6. Govindasami S et al (2018) Strength assessment of polypropylene fibre reinforced concrete. Int J Eng Technol:436–438
7. Yao W (2000) Flexural strength and behavior of polypropylene fiber reinforced concrete beams. J Wuhan Univ Technol Mater Sci Ed 17(2):54–57
8. Indian Standard recommended guidelines for concrete mix design. IS: 10262-1982
9. Indian Standard Code of practices for plain and reinforced concrete IS: 456:2000

Sustainable Use of Sugarcane Bagasse Ash in Concrete Production

Karanvir Singh Sohal and Rajwinder Singh

1 Introduction

Sugarcane crop is mainly used for the production of sugar in the industries. Sugarcane bagasse is generally produced after juicing the sugarcane through various types of mechanical equipment. Usually, the amount of sugarcane bagasse generated depends upon the type and quality of sugarcane used. Over 110 countries produce approximately 1500 million tons of sugarcane, out of which 300 million tons per year is produced only in India [24]. Globally, in the year 2016, Brazil, India, and China produced approximately, 20.5%, 16.9%, and 6.3% of sugarcane, respectively [60]. The byproduct produced after the extraction of juice is normally used as a material for the cogeneration process for the production of steam energy, which is further used to generate the electricity. The cogeneration process results in generating the ash, which is the final product formed during the sugar production process. Approximately 25–50 kg of ash is generated during the burning of one ton of sugarcane bagasse [52]. The disposal of this waste exerts a hazardous effect on the soil, groundwater, and air quality when thrown directly to the landfills [27]. The generation of sugarcane bagasse ash (SCBA)after harvesting the sugarcane crop is represented in Fig. 1.

The obtained SCBA has typically higher silica content, which imparts the pozzolanic activity to the ash [42]. Therefore, many researchers have conducted studies by incorporating this ash for the production of concrete and found that its addition has affected the performance of concrete positively. As the silica content of ash reacts with the product formed at the time of the hydration reaction of cement and results in the production of C–S–H gel which enhances the durability and the

K. S. Sohal (✉)
Guru Nanak Dev Engineering College, Ludhiana, Punjab, India
e-mail: karansohal875@gmail.com

R. Singh
Dr. B.R. Ambedkar National Institute of Technology, Jalandhar, Punjab 141006, India

H. Singh et al. (eds.), *Sustainable Development Through Engineering Innovations*, Lecture Notes in Civil Engineering 113,
https://doi.org/10.1007/978-981-15-9554-7_34

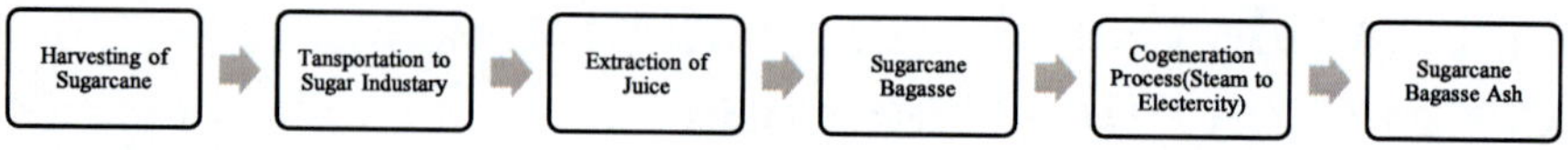

Fig. 1 The generation process of sugarcane bagasse ash

mechanical properties of the manufactured concrete [12, 30, 35, 37, 39, 43, 52, 57, 58].

A study revealed that the incorporation of SCBA being an alternative material for cement has caused a decrease in the emission of CO_2; around 519-kilo tons per year. The cost of the construction and energy has also been reduced as this waste is used as additional material in the production of concrete [3, 23, 25, 37].

This review paper aims to provide sufficient data associated with the physical and chemical properties of bagasse ash. Authors have also tried to gather information about the durability and mechanical properties of concrete made with the integration of SCBA as an alternative of fine aggregates and cement with suitable discussion. Authors have also suggested the work that can be done in the future by integrating SCBA as a partial substitution of cement for environment-friendly and sustainable concrete production.

2 Physical and Chemical Properties of Sugarcane Bagasse Ash

2.1 Physical Properties

The properties of SCBAsuch as texture, particle size, and the specific gravity are given in this section. Apart from these, other physical properties are written in Table 1.

2.1.1 Texture

The SCBA has an irregular rough surface [13, 15] and black color [47].

2.1.2 Fineness Modulus

The fineness modulus of the bagasse ash particle is approximately around 2.12 [11].

Table 1 Physical properties of bagasse ash used in previous studies

Author(s)	Particle size (μm)	Specific gravity	Density (kg/m^3)	Blaine fineness (m^2/kg)
Amin et al. [5]	5.1 (mean)	1.80	400 (compressed)	–
Cordeiro et al. [20]	76.3 (median)	–	2530	196
Rukzon and Chindaprasirt [51]	16.4 (median)	–	2240	1250
Cordeiro et al. [21]	2.7	–	2530	893
Somna et al. [56]	5.6	2.27	–	–
Bachtiar et al. [7]	5.4 (mean)	1.85	590 (compressed)	943
Srinivasan and Sathiya [57]	40.1 (mean)	–	2160	–
Lathamaheswari et al. [34]	75	2.65	–	–
Rao and Prabath [47]	–	2.20	1200	–
Singh et al. [54]	5.1 (mean)	1.8	–	–

2.1.3 Surface Area

The dried bagasse ash particles have a surface area of nearby 900 m^2/kg [5], whereas, Abdulkadir et al. [1], Srinivasan and Sathiya [57] reported that it is 400 and 471.6, respectively.

2.1.4 Temperature

In many studies, the emphasis is given on the temperature used to burn the sugarcane bagasse to get the ash, which is having maximum pozzolanic activity. Alyasri et al. [3] found that when sugarcane bagasse was burned at the temperature of 700 °C, a lower amount of carbon and higher pozzolanic activity is attained. It was also suggested that when sugarcane bagasse subjected to a temperature over 1000 °C, the ash obtained would have lower pozzolanic activity [43]. However, the loss of ignition of sugarcane bagasse gets reduced when it is burned at 550 °C for 45 min [15]. Whereas, Cordeiro et al. [18] revealed that when sugarcane bagasse is burned at 600 °C for 3 h at an increasing rate of 10 °C/min, ash has the most favorable pozzolanic properties to be used as a replacement material. However, in most of the studies, the ash is generated through the calcination process at the temperature range between 600 and 800 °C [16, 29, 32].

2.1.5 Sieving Process

The fineness of particles affects the pozzolanic activity a lot. Therefore, many authors grind the ash to provide the cementitious behavior and have used the material which was passed through the sieve no. 325 before using it as an alternative material for the replacement of cement [15, 17, 19, 32, 38, 39, 50, 56]. A few authors have used the material which was passed through sieve no. 50 [8, 33].

2.2 *Chemical Properties*

The composition of SCBA is reported to be rich in silica in most of the cases, and its pozzolanic activity depends upon the time and temperature used for its burning [9, 16, 33, 48]. By undergoing extensive research, the range of chemical compounds of SCBA is presented in Table 2.

Based on the Table 2, it is observed that the amount of silica is higher than the other compounds. However, it has also reported that the loss of ignition is too high, which highly depends upon the type of bagasse ash and temperature at which it is burned [9, 21, 35, 43, 48].

3 Research Findings of Previous Literature

The brief discussion about the properties of the earlier conducted studies on the incorporation of the SCBA for the initial and hardened state are presented in this section with appropriate reasoning.

Various authors have conducted different studies by varying the dosage of SCBA to replace some content of cement or sand or along with some other mineral admixtures. The positive effect of the replacement of the ingredients (cement and sand) has been observed when substitution is done up to a certain limit. It has been observed that the process of hydration of cement gets quicker as the content of bagasse ash is increased. The pattern of reduction of setting times has been reported for all of the replacement percentages of SCBA by Rauf et al. [49] and found that all samples were as per the standards. The opposite results were found by Oliveira et al. [29, 41], that if the amount of bagasse ash is increased, the setting times also gets increased. Similarly, the consistency of cement containing ash was also affected. However, Cordeiro et al. [20], have revealed that the addition of SCBA does not alter the consistency of mortar. The slump values of the concrete have also been increased by the addition of SCBA as a substitution material of cement, which indicates that there is no need to add chemical admixtures [20, 22, 53, 61].

Considering the mechanical performance of the manufactured concrete, it has been reported that by the increasing amount of SCBAfor the partial substitution of cement and fine aggregates up to a certain limit in the specimens, the compressive,

Table 2 Percentage of various chemical compounds present in bagasse ash

Compund		l_2O_3	CaO	SiO_2	MgO	Fe_2O_3	K_2O	P_2O_5	Na_2O	SrO	TiO_2	MnO	SO_3	CuO	L.O.I
%	Max.	29.5	12.6	96.2	6.77	14.34	12.16	3.01	3.82	0.01	3.1	0.27	4.33	0.096	19.6
	Min.	0.2	0.1	54.4	0.6	0.218	0.3	0.1	0.02	0.005	0.06	0.004	0.03	0.01	0.34

split tensile and flexural strength has reported increase for all the curing days [2, 4, 11, 12, 24, 28, 29, 37, 39, 48, 52, 55]. The reaction between the silica and alumina portion of SCBA and calcium content of cement results to form the C–S–H gel, which has the strong capacity to holds the particles together, thus, ultimately creates a dense structure [10, 15, 26, 40, 48, 56].

The concrete manufactured through the integration of SCBA represented the better results from the durability aspects in comparison to the control (reference) concrete. The properties such as water absorption capacity, sorptivity, porosity, sulphate attack, initial surface absorption, and water permeability have been reported to be enhanced by the substitution of SCBA [13, 26, 29, 36, 51, 56]. This is due to the finer particles of SCBA which behaves as a micro-filler material that improves the microstructure of concrete, thus ultimately, increases the density of concrete specimens and lowers the water permeability of concrete [14, 16].

4 Optimal Values of Replacement of Sugarcane Bagasse Ash Investigated by Earlier Studies

This section includes the conclusions given by various authors about the optimal percentage for the substitution of fine and cement with SCBA. The optimal values for replacement of cement and fine with SCBA are given in Tables 3 and 4, respectively.

As observed from Tables 3 and 4, the cement and fine aggregate replacement in the concrete can be successfully done by using SCBA up to a specific limit.

5 Conclusion

In this paper, various studies have been reviewed for the integration of SCBA as additional material for cement and fine aggregates in the sustainable development of concrete. The physical and chemical properties of SCBA have been thoroughly examined from the earlier available literature. The conclusion based on the reviewed literature are given below:

- Most studies have reported that the substitution of 20% cement with SCBA is the optimal value at which fresh and hardened properties of concrete gets improved in comparison to the control specimens.
- The incorporation of SCBA as a substitute material for fine aggregates and cement has enhanced the durability and mechanical properties of concrete when water curing is done up to 28 days. However, these properties keep on improving as the time duration for the curing of the specimens is extended. The improvement in the properties is concerned with the particle size of SCBA, which acts as a filler material. The reaction between the $Ca(OH)_2$formed because of the hydration reaction of cement and silicon dioxide portion of bagasse ash, which cause the

Table 3 Optimum values suggested by the studies on the replacement of cement with SCBA

Author(s)	Material(s)	Replacement criteria (%)	Optimum value (%)
Amin [4]	SCBA	0, 5, 10, 15, 20, 25, 30	20
Chi [13]	SCBA	0, 10, 20, 30	10
Cordeiro et al. [20]	SCBA	0, 10, 15, 20	20
Chusilp et al. [16]	SCBA	10, 20, 30	20
Abdulkadir et al. [1]	SCBA	0, 10, 20, 30	20
Bangar et al. [10]	SCBA	0, 2, 4, 6, 8, 10	10
Muangtong et al. [40]	SCBA	0, 10, 20, 40	20
Oliveira et al. [41]	SCBA	0, 10, 20, 30	20
Suvimol and Daungruedee [59]	SCBA	0, 10, 20, 30	20
Hussein et al. [31]	SCBA	0, 5, 10, 15, 20, 25, 30	20
Pratheba et al. [44]	SCBA	0, 5, 10, 15, 20, 25	15
Shafiq et al. [53]	SCBA	5, 10, 15, 20, 30, 35, 40, 45, 50	20
Batra and Kadam [11]	SCBA + steel fibers	5, 10, 15, 20	15
Praveenkumar and Vijayalakshmi [45]	SCBA + nano particles	20 (SCBA) 1, 2, 3, 4 (nano particles)	20
Pravisha and Anandhi [46]	SCBA + rubber tyre	5, 15, 25	15
Fauzi et al. [26]	SCBA + saw dust ash	5, 10, 15	10
Anand and Mishra [6]	SCBA + fly ash	10, 15, 20, 25	10

Table 4 Optimum values suggested by the studies on sand replacement with SCBA

Author(s)	Material(s)	Replacement criteria (%)	Optimum value (%)
Bhuvaneshwari and Tamilarasan [12]	SCBA + river sand/M-sand	0, 10, 20, 30, 40	20
Sales and Lima [52]	SCBA	10, 15, 20, 30, 50, 100	30
Moretti et al. [39]	SCBA + construction waste	30 (SCBA) 30, 50 (CW)	30 (SCBA) 30 (CW)
Sua and Makul [58]	SCBA + limestone	10, 20, 40, 60, 80, 100	20
Modani and Vyawahare [37]	SCBA	0, 10, 20, 30, 40	10–20
Dayo et al. [23]	SCBA	0, 10, 20, 30, 40	10
Akash et al. [2]	BA	10, 20, 30, 40, 50	40

production of calcium-silicate-hydrate. The formed product holds the cement particles closer, which provides the denser microstructure to the concrete.
- The application of SCBA as a mineral admixture for the replacement material of cement not only reduces health and environmental problems related to its disposal but also benefits in lowering the cost of construction and carbon dioxide emissions.

The available studies based on the addition of SCBA as a cement substitutional material have been reviewed thoroughly. Various authors have already worked in depth on the durability and mechanical performance aspects of mortar and concrete by incorporating SCBA at different percentages (from 5 to 50%) and found that 20% is optimum content. The readers are suggested to evaluate the performance of concrete at lower replacement levels (1–20%, with an interval of 1%) to have more accuracy about the optimal content. Less emphasis has been given on the substitution of fine aggregates with SCBA. Therefore, readers can also work in this approach to check the effect on the durability performance of the manufactured concrete. Authors also suggest the readers may explore the combination of various wastes along with SCBA at a different level of replacement, curing days, and temperature. Only a few authors have worked on the addition of sugarcane fibres in concrete, therefore, readers can also investigate the optimal value of this waste at which the best performance is observed.

Acknowledgements The authors would like to acknowledge Dr. H. S. Rai, H.O.D Civil Engineering Department, Guru Nanak Dev Engineering College, Ludhiana, Punjab, India for providing his support and encouragement throughout writing this paper.

References

1. Abdulkadir TS, Oyejobi DO, Lawal AA (2014) Evaluation of sugarcane bagasse ash as a replacement for cement in concrete works. Acta Technica Corviniensis-Bull Eng 7(3):71
2. Akash GS, Madhukar KK, Chetan T, Aishwarya T, Dinesh SM, Shivakumara B (2018) Experimental study on partial replacement of sand with sugarcane bagasse ash in concrete. Int Res J Eng Technol 5(5):3582–3584
3. Alyasri SAH, Alkroosh IS, Sarker PK (2017) Feasibility of producing nano cement in a traditional cement factory in Iraq. Case Stud Constr Mater 7:91–101
4. Amin NU (2011) Use of bagasse ash in concrete and its impact on the strength and chloride resistivity. J Mater Civ Eng 23(5):717–720
5. Amin N, Shah MT, Ali K (2009) Raw mix designing and clinkerization high strength portland cement from the raw material of DarukhulaNizampur, District Nowshera, NWFP, Pakistan. Mag Concr Res 61(10):779–785
6. Anand A, Mishra AK (2016) Comparative study of concrete strength by partially replacing cement with sugarcane bagasse ash and fly ash. Int J Sci Eng Technol Res 5(4):1063–1069
7. Bachtiar E, Marzuki I, Setiawan AM, Yunus AI, Gusty S (2019, March) Potency of Sugarcane bagasse ash partial subtitution of cement in concrete. In: First international conference on materials engineering and management-engineering section (ICMEMe 2018). Atlantis Press, pp 165, 27–31
8. Bahurudeen A, Kanraj D, Dev VG, Santhanam M (2015) Performance evaluation of sugarcane bagasse ash blended cement in concrete. Cem Concr Compos 59:77–88

9. Bahurudeen A, Marckson AV, Kishore A, Santhanam M (2014) Development of sugarcane bagasse ash based Portland pozzolana cement and evaluation of compatibility with superplasticizers. Constr Build Mater 68:465–475
10. BangarSayali S, Phalke Shubhangi N, Gawade Anjali Y, TambeRutuja S, Rahane AB (2017) A review paper on replacement of cement with bagasse ash. Int J Eng Sci Manage 4(6):127–131
11. Batra MJ, Kadam SS (2019) Experimental study of replacement of cement by bagasse ash with steel fibers. Int J Eng Res Gen Sci 7(1):10–15
12. Bhuvaneshwari M, Tamilarasan S (2016) Experimental studies on the effect of bagasse ash and m-sand on mechanical behavior of concrete. Int J Emerg Sci Eng 9(3):426–431
13. Chi MC (2012) Effects of sugar cane bagasse ash as a cement replacement on properties of mortars. Sci Eng Compos Mater 19(3):279–285
14. Chindaprasirt P, Homwuttiwong S, Jaturapitakkul C (2007) Strength and water permeability of concrete containing palm oil fuel ash and rice husk–bark ash. Constr Build Mater 21(7):1492–1499
15. Chusilp N, Jaturapitakkul C, Kiattikomol K (2009a) Effects of LOI of ground bagasse ash on the compressive strength and sulfate resistance of mortars. Constr Build Mater 23(12):3523–3531
16. Chusilp N, Jaturapitakkul C, Kiattikomol K (2009b) Utilization of bagasse ash as a pozzolanic material in concrete. Constr Build Mater 23(11):3352–3358
17. Cordeiro GC, Tavares LM, Toledo Filho RD (2016) Improved pozzolanic activity of sugar cane bagasse ash by selective grinding and classification. Cem Concr Res 89:269–275
18. Cordeiro GC, Toledo Filho RD, Fairbairn EMR (2009) Effect of calcination temperature on the pozzolanic activity of sugar cane bagasse ash. Constr Build Mater 23(10):3301–3303
19. Cordeiro GC, Toledo Filho RD, Fairbairn EMR (2010) Ultrafine sugar cane bagasse ash: high potential pozzolanic material for tropical countries Cinzaultrafina do bagaço de cana-de-açúcar: material pozolânico de alto potencial para paísestropicais. IBRACON Struct Mater 3:50–67
20. Cordeiro GC, Toledo Filho RD, Tavares LM, Fairbairn EDMR (2009b) Ultrafine grinding of sugar cane bagasse ash for application as pozzolanic admixture in concrete. Cem Concr Res 39(2):110–115
21. Cordeiro GC, Toledo Filho RD, Tavares LM, Fairbairn EMR (2012) Experimental characterization of binary and ternary blended-cement concretes containing ultrafine residual rice husk and sugar cane bagasse ashes. Constr Build Mater 29:641–646
22. da Cruz Sessa T, Silvoso MM, Vazquez EG, Qualharini EL, Haddad AN, Amaral Alves L (2016) Study of the technical capability of sugarcane bagasse ash in concrete production. Mater Sci Forum 866:53–57
23. Dayo AA, Kumar A, Raja A, Bheel N, Shaikh ZH (2019) Use of sugarcane bagasse ash as a fine aggregate in cement concrete. Eng Sci Technol Int Res J 3(3):8–11
24. Dhengare S, Amrodiya S, Shelote M, Asati A, Bandwaf N, Anand K, Jichkar R (2015) Utilization of sugarcane bagasse ash as a supplementary cementitious material in concrete and mortar—a review. J Impact Factor 6(4):94–106
25. Fairbairn EM, Americano BB, Cordeiro GC, Paula TP, Toledo Filho RD, Silvoso MM (2010) Cement replacement by sugar cane bagasse ash: CO_2 emissions reduction and potential for carbon credits. J Environ Manage 91(9):1864–1871
26. Fauzi MA, Rosseli SR, Seman MJA (2018) Utilisation of sugarcane bagasse ash and sawdust ash as cement replacement material in the production of structural concrete. In: Regional conference on science, technology and social sciences (RCSTSS 2016). Springer, Singapore, pp 455–463
27. Frías M, Villar E, Savastano H (2011) Brazilian sugar cane bagasse ashes from the cogeneration industry as active pozzolans for cement manufacture. Cem Concr Compos 33(4):490–496
28. Gandhare KU (2018) Effect on concrete properties after substituting fine aggregate by crushed groundnut shells and addition of sugarcane bagasse ash. Int J Trend Sci Res Dev 2(2):1490–1494
29. Ganesan K, Rajagopal K, Thangavel K (2007) Evaluation of bagasse ash as supplementary cementitious material. Cem Concr Compos 29(6):515–524
30. Gómez-Soberón JM (2002) Porosity of recycled concrete with substitution of recycled concrete aggregate: an experimental study. Cem Concr Res 32(8):1301–1311

31. Hussein AAE, Shafiq N, Nuruddin MF, Memon FA (2014) Compressive strength and microstructure of sugar cane bagasse ash concrete. Res J Appl Sci Eng Technol 7(12):2569–2577
32. Joshaghani A, Moeini MA (2017) Evaluating the effects of sugar cane bagasse ash (SCBA) and nanosilica on the mechanical and durability properties of mortar. Constr Build Mater 152:818–831
33. Kazmi SMS, Munir MJ, Patnaikuni I, Wu YF (2017) Pozzolanic reaction of sugarcane bagasse ash and its role in controlling alkali silica reaction. Constr Build Mater 148:231–240
34. Lathamaheswari R, Kalaiyarasan V, Mohankumar G (2017) Study on bagasse ash as partial replacement of cement in concrete. Int J Eng Res Dev 13(1):01–06
35. Martirena-Hernández JF, Betancourt-Rodriguez S, Middendorf B, Rubio A, Martínez-Fernández L, López IM, González-López R (2001) Pozzolanic properties of residues of sugar industries (second part). Materiales De Construccion 51(261):67–72
36. Mehta PK (1987, March) Studies on the mechanisms by which condensed silica fume improves the properties of concrete: durability aspects. In: Proceedings of international workshop on condensed silica fume in concrete, Ottawa, pp 1–17
37. Modani PO, Vyawahare MR (2013) Utilization of bagasse ash as a partial replacement of fine aggregate in concrete. Proc Eng 51:25–29
38. Montakarntiwong K, Chusilp N, Tangchirapat W, Jaturapitakkul C (2013) Strength and heat evolution of concretes containing bagasse ash from thermal power plants in sugar industry. Mater Des 49:414–420
39. Moretti JP, Sales A, Almeida FC, Rezende MA, Gromboni PP (2016) Joint use of construction waste (CW) and sugarcane bagasse ash sand (SBAS) in concrete. Constr Build Mater 113:317–323
40. Muangtong P, Sujjavanich S, Boonsalee S, Poomiapiradee S, Chaysuwan D (2013) Effects of fine bagasse ash on the workability and compressive strength of mortars. Chiang Mai J Sci 40(1):126–134
41. Oliveira de Paula M, Ferreira Tinôco IDF, de Souza Rodrigues C, Osorio Saraz JA (2010) Sugarcane bagasse ash as a partial-portland-cement-replacement material. Dyna 77(163):47–54
42. Patel JA, Raijiwala DB (2015) Use of sugar cane bagasse ash as partial replacement of cement in concrete—an experimental study. Glob J Res Eng 15(5):1–7
43. Payá J, Monzó J, Borrachero MV, Díaz-Pinzón L, Ordonez LM (2002) Sugar-cane bagasse ash (SCBA): studies on its properties for reusing in concrete production. J Chem Technol Biotechnol Int Res Process Environ Clean Technol 77(3):321–325
44. Pratheba S, Deepeka K, Kanimozhi A, Malathi J, Nandhini J (2018) An experimental study on bagasse ash as partial replacement for cement in concrete. Int Res J Eng Technol 5(3):771–774
45. Praveenkumar TR, Vijayalakshmi MM (2019) Mechanical characteristics of sugarcane bagasse ash along with nano particles in concrete. Int J Recent Technol Eng 8(1):2507–2510
46. Pravisha M, Anandhi V (2016) Experimental investigation on the effect of bagasse ash and rubber tyre waste in concrete. Int J Adv Res 4(4):1076–1081
47. Rao C, Prabath NVN (2015) Green concrete using agro industrial waste (sugarcane bagasse ash). Int J Soft Comput Eng 5(1):86–92
48. Rattanashotinunt C, Thairit P, Tangchirapat W, Jaturapitakkul C (2013) Use of calcium carbide residue and bagasse ash mixtures as a new cementitious material in concrete. Mater Des 46:106–111
49. Rauf N, Damayanti MC, Pratama SWI (2017, January) The influence of sugarcane bagasse ash as fly ash on cement quality. AIP Conf Proc 1801(1):040009
50. Rerkpiboon A, Tangchirapat W, Jaturapitakkul C (2015) Strength, chloride resistance, and expansion of concretes containing ground bagasse ash. Constr Build Mater 101:983–989
51. Rukzon S, Chindaprasirt P (2012) Utilization of bagasse ash in high-strength concrete. Mater Des 34:45–50
52. Sales A, Lima SA (2010) Use of Brazilian sugarcane bagasse ash in concrete as sand replacement. Waste Manage 30(6):1114–1122

53. Shafiq N, Hussein AAE, Nuruddin MF, Al Mattarneh H (2016, April) Effects of sugarcane bagasse ash on the properties of concrete. Proc Inst Civil Eng Eng Sustain 171(3):123–132
54. Singh G, Thakur N, Sharma NK (2019) Experimental research on the strength parameters of concrete using sugarcane bagasse ash, marble waste powder and recycled concrete. Int J Innov Technol Explor Eng 8(8):1378–1381
55. Sireesha G, Rao MK, Rao PK (2013) An experimental study on strength properties of concrete when cement is partially replaced with sugar-cane bagasse ash. IOSR J Mech Civil Eng:35–38
56. Somna R, Jaturapitakkul C, Rattanachu P, Chalee W (2012) Effect of ground bagasse ash on mechanical and durability properties of recycled aggregate concrete. Mater Des 1980–2015(36):597–603
57. Srinivasan R, Sathiya K (2010) Experimental study on bagasse ash in concrete. Int J Service Learn Eng Human Eng Social Entrepreneur 5(2):60–66
58. Sua-Iam G, Makul N (2013) Use of increasing amounts of bagasse ash waste to produce self-compacting concrete by adding limestone powder waste. J Clean Prod 57:308–319
59. Suvimol S, Daungruedee C (2008, May. Bagasse ash: effect of pozzolanic activity and application in cement use aspect. 3rd ACF Int Conf ACF/VCA 8:165–173
60. Xu Q, Ji T, Gao SJ, Yang Z, Wu N (2019) Characteristics and applications of sugar cane bagasse ash waste in cementitious materials. Materials 12(1):39
61. Yashwanth MK (2014) An experimental study on bagasse ash as replacement for cement in lightweight concrete. Int J Latest Trends Eng Technol 3(3):253–260

Image Steganography: An Inevitable Need for Data Security

Sneh Rachna and Rajesh Kumar

1 Introduction

While considering sustainability, it is said to be related to a kind of development wherein should strive to fulfill the need of future generations. In the modern world for sustainable development information has become inevitable. Information is a very valuable asset as it could be used to carry out policy and framework for sustainable development. But if the data integrity is lost and the data falls into wrong hands. This valuable source can become threat to nation. As the information which could be used to optimize various activities by surveying etc., if the coordinates for example falls into intruders they might use it in a negative way hence making the information vulnerable to society. Hence we need to think about security and prevention of data integrity. One of the problems which are inevitable in today's world is terrorism. We have been seeing significant growth in terrorism over the years. Also with the advent of modern computers and internet technology in the year, this terrorism has boosted even more. The death toll from terrorist activities around the world can be analyzed to have exponential growth over the years 2002–2017 [1].

The information which is being shared among the military personnel has become vulnerable to cyber-attacks. Hence it has become inevitable that the data need to be secured. So the option could be either sending the encrypted message or hiding the text in video, image, or audio file. This type of hiding is termed as image steganography. Steganography is not one of the novel techniques; it is being used for a very long time. The history of steganography can be traced back to ancient Greece, wherein the ruler used to shave the head of the slave, embed the message onto it, and

S. Rachna (✉) · R. Kumar
Modern Institute of Technology and Research Centre, Alwar, Rajasthan, India
e-mail: snehrachna@gmail.com

R. Kumar
e-mail: rjesh10391@gmail.com

H. Singh et al. (eds.), *Sustainable Development Through Engineering Innovations*, Lecture Notes in Civil Engineering 113,
https://doi.org/10.1007/978-981-15-9554-7_35

wait for the hair to grow [2–4]. Even during the times of war, steganography was used extensively by British and American forces. With the easy accessibility of digital images over the internet and the web, this becomes the most inevitable choice of medium for steganography [5]. For processing to be better, the image steganography needs to have a proper tradeoff between capacity, security, and robustness [6–8]. The basic requirement of stego system includes capacity, robustness, and security.

Capacity	Robustness	Security
Without perturbing the quality of cover medium and minimizing the hidden data perception, embedding maximum number of bits is termed as stenographic capacity [9]	Robustness is not the top priority, thus it is certain that stegno systems have limited robustness against technical modifications [10]	To avoid skeptics of attackers, there must be invisibility in the contents. Hence cover file attributes and characteristics are not altered and during the embedding process no distortion must be produced [11, 12]

There are three different approaches to hide the information in a cover object that is injection, substitution, and generation. In the case of injection, the end-user will be unable to realize that there is some additional information since these bytes do not affect the process. The file may look unusually large and may create skeptics as the technology changes the file size. The merit of the substitution technique is there is no changes in cover file size after execution of the algorithm [3, 13, 14]. Whereas demerit includes that the resulting image from steganography process may be dubiously affected by quality degradation. The amount of data that we can hide to the number of insignificant bits in the file is limited by the method of substitution. While considering the cases of injection and substitution method, the main demerit it is that we can compare pre-existing copy of the cover object with the stego object and may elucidate the comparative difference between the two. This problem can be deciphered with the application of the generation technique, as the result is original file and is thus resistant to comparison tests [3, 4]. Image steganography can be categorized into spatial domain approach and transform technique. Image is a constituent of different light intensities in different areas of the particular image for a computer. Pixels are usually a representation of individual points and grid formed of this numeric representation [8, 15]. The image's raster data is made by these pixels. Data hiding adheres to the limitation of the human visual system (HVS), particularly the area where there are a lower sensitivity and luminescence in pattern changes. Hence we utilize this margin of numerical value and visual perception of the multimedia carriers. It means applying slight distortions in the parts which are non- significant and invisible to human perception [7]. Apart from embedding information inside digital media, steganography also should be responsible to retrieve the information from the media. 8-bit and 24-bit per pixel image file are typical in image steganography [16–18]. Because of the comparatively lower size, 8-bit images are useful. Whereas 24-bit images offer much more flexibility and 24-bit images may use 16 million colors for RGB images [19].

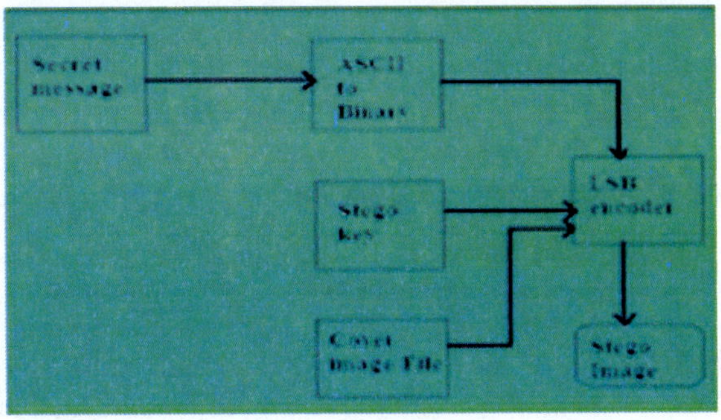

Fig. 1 LSB insertion mechanism

2 LSB Based Method of Image Steganography

Due to undeviating nature, this technique is a popular data hiding technique and is used widely. In this method remolding the least significant bit is done in pixels of stego-image, which results in variations in only the tone of the color [2, 4]. These modifications are as such that the human eye is unable to detect it. The hiding of the message to be transmitted by the LSB is performed either in a sequential or randomized fashion [10, 15]. A path is hence created for replacing the least significant bits of the image with the message bits.

2.1 Data Embedding

The process of embedding can be shown in the following steps wherein the input is the cover image, stego key, and the text file, and output is the resulting stego-image (Fig. 1).

Step 1 Extract the pixels of the cover image and then also of the stego key.
Step 2 Choose first pixel and pick characters of stego key and place it in first component of pixel.
Step 3 Insert characters of text in each component of the next pixels by replacing it.

2.2 Data Extraction

Here inputs are stego-image file and stego key and output is a secret message file.

Step 1 Based on the stego key a path is created at the receiver's side.
Step 2 By retrieving the LSB of the pixels' length of the secret message is recovered.
Step 3 LSB of each pixel is retrieved and based on the path pixels are traversed.

Step 4 Until the end of message length is reached traversing of all the pixels continues.

3 Performance Analysis

Most popular measurement used in steganography to evaluate quality of the stego-image is Peak Signal-to-Noise Ratio (PSNR). PSNR is used to evaluate the distortion between stego-image and cover image [20]. PSNR is defined by using Mean Square Error (MSE). MSE is measured between the original image and stego-image. Given width and height (m x n) of a cover image I and stego-image *K*, the MSE is defined as follows:

$$\text{MSE} = \frac{1}{mn}\sum_{i=0}^{m-1}\sum_{j=0}^{n-1}[I(i,j) - K(i,j)]^2 \tag{1}$$

where $I(i, j)$ is the pixel in original image, $K(i, j)$ is the pixel of stego-image m, n is number of rows and columns in the image.

$$\text{The PSNR is computed as: PSNR} = 10\log_{10}\left(\frac{\text{MAX}_I^2}{\text{MSE}}\right) \tag{2}$$

255 is the MAX of the cover image used for conducting experiments and testing. Increase in quality is testified with aggrandizement of PSNR. Hence for having a better image quality and minimum difference between stego-image and cover image, we need to have a higher PSNR. Low quality is indicated by PSNR values below 30 dB and hence distortion is higher. A PSNR of 40 dB or higher indicates high quality of stego-image [12, 20]. Considerably a lower error is indicated with lower value of MSE. Hence cumulative square error is represented by MSE. As MSE is in the denominator of PSNR, with the lower error we will have high PSNR.

3.1 Algorithm

Embedding Algorithm

i. Read character from text file that is to be hidden and convert the ASCII value of the character into equivalent binary value into an 8-bit integer array.
ii. Read the RGB color image into which the message is to be embedded and extract the red plane of the host image.
iii. Initialize the key which gives the random position of the pixel"s red plane to be processed for embedding. The key produces random sequence of pixel positions.

iv. Check the value of LSB (0 or 1) present in each pixel"s red plane.
v. Pick up the message bit. If the message bit is zero and pixel bit is one then decrement the pixel value. Else increment the pixel value. If message bit = pixel bit, do nothing.
vi. Repeat step (iv)–(v) for the remaining random sequences (pixel indices).
vii. Add the green and blue plane with the modified red plane
viii. Write the above pixel to stego-image file

Extraction algorithm

i. Open the stego-image, read the RGB color of each pixel.
ii. Extract the pixels red plane of the stego-image.
iii. Initialize the key that gives the position of the message bits in the random pixels red plane where the secret bits are embedded.
iv. For decoding, select the pixels using the same pseudo-random sequence. As the key knows the seed k, k_i can be reconstructed, and therefore the entire sequence of pixel indices y_i.
v. Read the LSB of the red plane of each pixel and put it directly in an array.
vi. Repeat step (v) for the entire sequence of pixel indices. vii.
vii. The content of the array converts into decimal value which is actually ASCII value of hidden character.

3.2 *Evaluation of Above Two Techniques*

With the aid of above-mentioned algorithms, these two techniques have been implemented in MATLAB and we can say that these are not void of weak and strong points with respect to image steganography. We need to culminate as to which approach needs to be applied. There could be the following parameters to measure the indices of steganographic systems. (i) **Perceptibility** helps in distorting the cover medium to an unacceptable level of embedded information. (ii) **Capacity** defines as to how much information could be hidden with reference to relative change in perceptibility. (iii) **Robustness to attacks** help surviving manipulation of embedded data of the stego medium in an attempt to destroy, remove, or change the embedded data [10] (Table 1).

Table 1 Comparison of characters of the above two techniques

Characteristics	Simple LSB	Random pixel selection
Imperceptibility	High	Higher
Robustness	Low	Low
Capacity	High	High
Tamper resistance	Low	High

4 Results

Now the time taken to encrypt or embed the text message or image in another image and the time taken to decrypt or extract the same message is also a measure for comparison. Table 4 shows the space taken and time taken by the two techniques (Figs. 2 and 3; Tables 2 and 3).

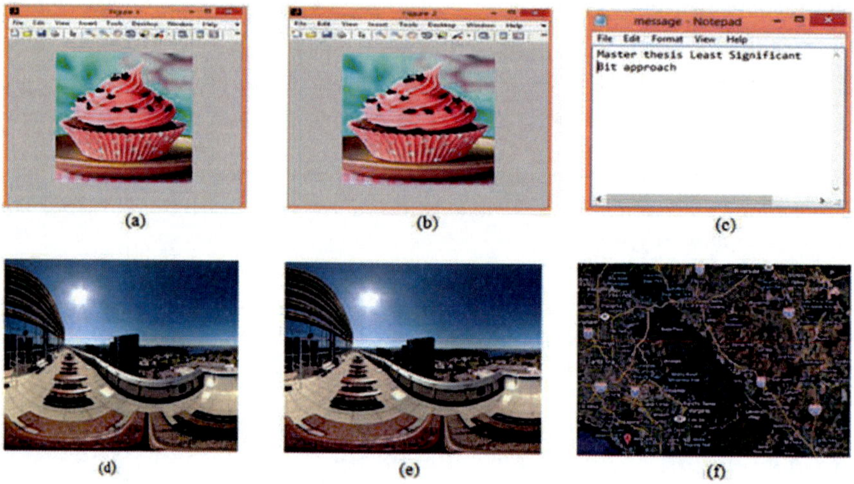

Fig. 2 **a** and **b** are the cover image and stego images, respectively for both the techniques, **c** is the hidden message file for both the techniques, **d** and **e** are the cover image and stego image respectively. **f** is the hidden message file

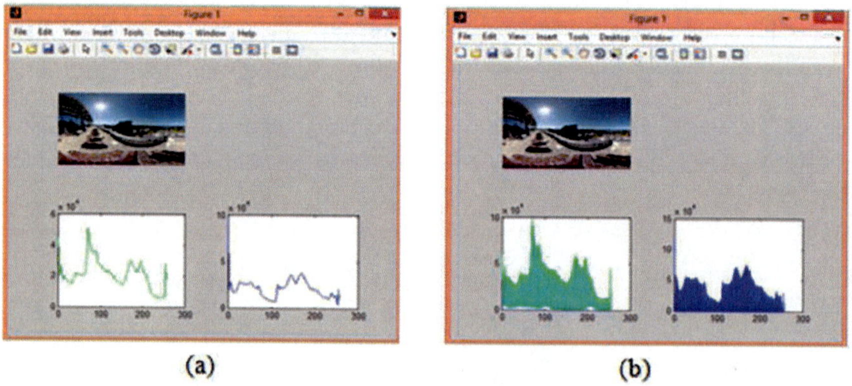

Fig. 3 **a** A cover image and the histogram of its G and B layers which are used to embed the message. **b** Histogram of Green (G) and Blue (B) layers of the cover image which are used to embed the image

Table 2 PSNR of random pixel selection technique

S. No.	Cover image	Secret message	MSE	PSNR
1	RGB image	Text message	0.0111	67.6835
2	RGB image	Image	0.0911	58.3346

Table 3 PSNR of least significant bit technique

S. No.	Cover image	Secret message	MSE	PSNR
	RGB image	Text message	0.0213	67.6697
2	RGB image	Image	0.0912	58.1311

Table 4 Comparison of space and time taken

Algorithm	Size of the cover image (kb)	Size of message (kb)	Time to embed (s)	Time to encrypt (s)
Least significant Bit	21	Text-1	3.385	2.805
	803	Image-30	10.045	9.495
Random pixel Selection	21	Text-1	4.035	2.937
	803	Image-30	10.936	10.003

As we can see the time taken is little bit more than the general least significant bit but it can be compromised with the improved PSNR.

5 Conclusions

As from [21] it is vindicated that terrorist activities are still prevailing and since most of activities in modern era are online. We need to think about security of the data, particularly the data which is of national interest. Hence it becomes inevitable to study image steganography. For securing information image steganography becomes one of the effective ways.

However, steganography is not only limited to the Military and Terrorism applications. Steganography can be used in supplementary to cryptography, watermarking, and fingerprinting. It has got wide applications as online banking, social media platforms, wide range of biomedical fields, smart identification cards. The usual application as in editing, enhancements, x-ray, scanning, etc. has always been there.

- The methodology carried out in the following work was with the aid of LSB technique and Random Pixel Selection Technique on the images to secure the information.

- Tables 2 and 3 show PSNR comparison of both the techniques. It could be culminated that LSB insertion using random key is better than simple LSB insertion in consideration with lossless compression.
- We could also observe that image resolution doesn't change in a significant manner and we can embed the message easily along with protection via personalized key. Hence unauthorized person cannot harm the integrity of the message.
- These algorithms are tested on both 8 Bit and 24 Bit images of same size of cover and secret image, hence it becomes simpler for implementation in both grayscale and color image.
- The major attributes of this work are to raise security via increasing PSNR and distortion rate reduction. But the application will also have effect on every algorithm. If more PSNR is demanded by the algorithm then the proposed algorithm can be used. However, while considering the overall aspect proposed algorithm can be said to be superior.

References

1. Zhou Z, Mu Y, Jonathan Wu QM (2019) Coverless Image Steganography using partial-duplicate image retrieval. J Soft Comput 23(13):4927–4038
2. Luo Y, Qin J, Xiang X, Tan Y, Liu Q (2020) Coverless Real-Time image information hiding based on image block matching and dense convolutional network. J Real-Time Image Process 17(1):125–135
3. Li C, Sun X, Zhou Z, Yang Y (2020) Real-time image carrier generation based on generative adversarial network and fast object detection. J Real-time Image Process:1–11
4. Parah SA, Sheikh JA, Akhoon JA, Loan NA, Bhat GM (2018) Information hiding in edges: a high capacity information hiding technique using hybrid edge detection. Multimedia Tools Appl 77(1):185–207
5. Deshmukh PU, Pattewar TM (2014) A novel approach for edge adaptive steganography on LSB insertion technique. In: IEEE international conference on information communication and embedded systems (ICICES), pp 1–5
6. Modi MR, Islam S, Gupta P (2013) Edge based steganography on colored images. In: 9th international conference on intelligent computing (ICIC), July 2013, pp 593–600
7. Qing X, Jianquan X, Yunhua X (2010) A high capacity information hiding algorithm in color image. In: Proceedings of 2nd IEEE international conference on e-business and information system security, May 2010, pp 1–4
8. Lee C-W, Tsai W-H (2010) A new steganographic method based on information sharing via PNG images. In: IEEE 2nd international conference on computer and automation engineering (ICCAE), Feb 2010, pp 807–811
9. Juneja M, Sandhu PS (2013) An improved LSB based steganography technique for RGB color images. In: Proceedings of second international conference on latest computational technologies, pp 10–14
10. Akhtar N, Khan S, Johri P (2014) An improved inverted LSB image steganography. In: IEEE international conference on issues and challenges in intelligent computing techniques (ICICT), Feb 2014, pp 749–755
11. Qazanfari K, Safabakhsh R (2014) A new steganography method which preserves histogram: generalization of LSB++. Int J Inf Sci:90–101
12. Gupta S, Gujral G, Aggarwal N (2012) Enhanced least significant bit algorithm for image steganography. Int J Comput Eng Manage, July 2012, pp 40–42

13. Prashanti G, Sandhyarani K (2015) A new approach for data hiding with LSB steganography, emerging ICT for bridging the future—proceedings of the 49th annual convention of the computer society of India CSI. Springer 2015, pp 423–430
14. Singh NJP, Biswas S, Sarkar D, Sarkar PP (2014) A huffman code based image steganography technique. In: 1st international conference on applied algorithm (ICAA), Jan 2014, pp 257–265
15. Feng B, Lu W, Sun W (2015) Secure binary image steganography based on minimizing the distortion on the texture. IEEE Trans Inf Forensics Sec
16. Baby D, Thomas J, Augustine G, George E, Michael NR (2015) A novel DWT based image securing method using steganography. In: International conference on information and communication technologies (ICICT), procedia computer science, pp 612–618
17. Mielikainen J (2006) LSB matching revisted. IEEE Signal Process Lett 13(5)
18. Juneja M, Sandhu PS (2009) Designing of robust image steganography technique based on LSB insertion and encryption. In: 2009 IEEE international conference on advances in recent technologies in communication and computing
19. Eyre W, Rogers M (2006) Steganography and terrorist communications: current information and trends—tools, analysis and future directions in steganalysis in context with terrorists and other criminals. In: Annual ADFSL conference on digital forensics, security and law. Springer, pp 567–579
20. Yang H, Sun X, Sun G (2009) A high-capacity image data hiding scheme using adaptive LSB substitution. J Radio Eng 18(4):509–516
21. Kumar R, Kumar N, Jung K-H (2020) Color image steganography scheme using gray invariant in AMBTC compression domain. In: Multidimensional systems and signal processing. Springer, pp 1–18
22. Mukherjee S, Sanyal G (2019) A multi level image steganography methodology based on adaptive PMS and block based pixel swapping. J Multimedia Tools Appl 78(13):17607–17622
23. Samidha D, Agrawal D (2013) Random image steganography in spatial domain. In: IEEE international conference on emerging trends in VLSI, embedded system, nano electronics and telecommunication system (ICEVENT), Jan 2013, pp 1–3
24. Karim SMM, Rahman MS, Hossain MI (2011) A new approach for LSB based image steganography using secret key. In: Proceedings of 14th IEEE international conference on computer and information technology, Dec 2011, pp 286–291
25. Yang C-H, Weng C-Y, Wang S-J (2008) Adaptive data hiding in edge areas of images with spatial LSB domain systems. IEEE Trans Inf Forensics Secur 3(3)
26. Zhang X, Wang S (2003) Vulnerability of pixel-value differencing steganography to histogram analysis and modification for enhanced security. Elsevier J, received 23 June 2003; received in revised form 13 Oct 2003
27. Wu H-C, Wu N-I, Tsai C-S, Hwang M-S (2005) Image steganographic scheme based on pixel-value differencing and LSB replacement methods. IEEE Proc Vis Image Signal Process 152(5)

A Taxonomy and Survey on Container Migration Techniques in Cloud Computing

Gursharan Singh and Parminder Singh

1 Introduction

The container is becoming increasingly popular with the continuous development of cloud computing. The migration of containers is regarded as the solution to these problems under the microservice characteristics. In general, container migration broadly categorized into two types: live migration and static migration. The container which is going to migrate will be stopped immediately during the static migration process and a different container will be created in the destination host. It will be better if we choose the way of the live migration process. In this approach, no migration cycle can be experienced by the consumer. This method is used to optimize the pre-copy live migration. It is possible to move a container between physical servers at different locations by making a container, which is not attached to a single hardware component. The type of migration is referred to as cold migration, and it is carried out as follows: stop a container at source machine, copy all its file system to another host, then start it at destination host Fig. 1. The whole state of a container can be saved into a disk file, which includes inter-process connections, network configuration, and other dependent parameters. It is then possible to restart a container from that file. The process of migration is referred to as moving a container along with its memory, storage, and network configuration from one system to another.

In this technique of container migration, the container will stop at source host and restart at destination host after complete transmission.

There are many applications for control and restart of a container:

G. Singh · P. Singh (✉)
School of Computer Science and Engineering, Lovely Professional University, Phagwara, Punjab 144411, India
e-mail: parminder.16479@lpu.co.in

G. Singh
e-mail: gursharan.dhot@gmail.com

H. Singh et al. (eds.), *Sustainable Development Through Engineering Innovations*, Lecture Notes in Civil Engineering 113,
https://doi.org/10.1007/978-981-15-9554-7_36

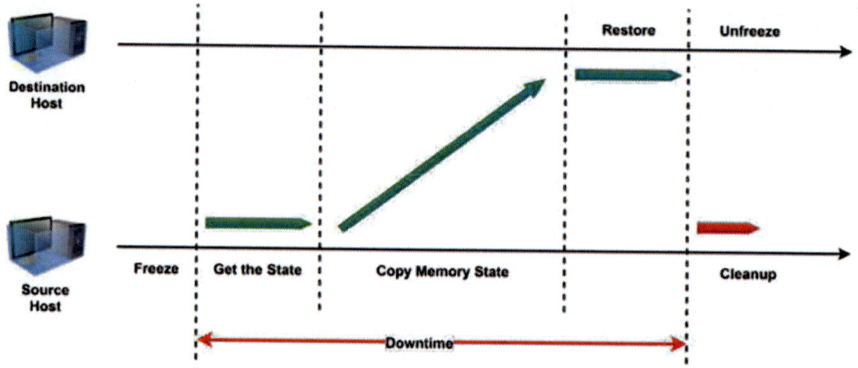

Fig. 1 Process of static container migration

- Maintenance or hardware upgrade.
- Server reboot or kernel upgrade.

Checkpoint and restart allow a running container without a reboot to be transferred from one host to another host. This feature of container migration is referred to as the live migration of containers. In this process, steps should be followed as:

1. Pass to the disk of the container file system.
2. The full state of the container including all resources and processes is stored into a disk file called dump.
3. The file will be transferred to a second host.
4. On another host from the dump, the container is restarted.

Using live migration is a better option as it provides high availability, dynamic load balancing, and fault tolerance.

2 Types of Containers

System containers such as LXC are similar to virtual machines, as they share the host operating system's kernel and offer isolation of user space Fig. 2. However, hypervisors are not used in system containers. It allows you to install various libraries, languages, databases, etc. Services running on each container use resources allocated to that container alone. System containers allow you to run multiple simultaneous processes, all under the same operating system. This improves efficiency and provides the advantages of VMs, such as carrying out several processes, along with modern container benefits, such as improved portability and fast startup times.

Configuration containers encapsulate the application's data, dependencies, and libraries to operate on an operating system 2. Application containers enable the consumer to build and operate a single container for several individual programs, or

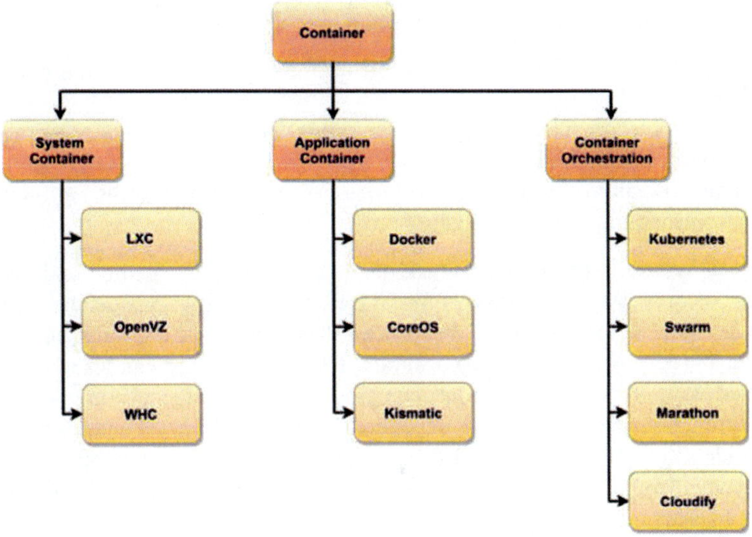

Fig. 2 Types of container

different resources that shape one program. For example, for a microservice app, a framework container will be well adapted, because each component that makes up the application operates independently of each other.

Nowadays, the number of container orchestration tools is available for the overall management of containers. Some of them are mentioned in 2. Google Kubernetes and Swarm are popular tools in this category.

3 Checkpointing and Restore in Userspace

A user-level checkpoint and restart process are started, and it is mostly implemented at OS kernel level, ensuring the complete transparency of the control process. Furthermore, implementation at kernel level requires no special interfaces for the recreation of resources. The following three steps are the basic checkpointing procedure:

1. The processes freeze—transfer processes to a previously recognized condition and deactivate the network.
2. Dump the container—collect the entire container operation state and save it to a storage server.
3. Stop the container—unmounted the file system and destroy all the processes of the container.

Checkpointing is the restart technique and vice-versa:

1. Container Restart: Build a container from the same state as in the dump file before.
2. Processes Restart: Create all processes within the frozen container and restore all resources from the dump file.
3. Container Resumes: Restart running processes, network activation, and container then continue to perform normally.

Process-freezer is the very first task of the checkpointing method and also the final task of the restarting process before the container's processes can be resumed. The freeze must be used to ensure that processes do not alter their status and the data from the stored processes remains accurate. Frozen processes can also be quickly replicated. It is very critical that all container processes are held in a consistent state. During restarting, all process dependencies should be saved and restored. The process identifiers (PGID, SID, TGID, and other identifiers) must be included in the dependency. Any parameters improperly restored result in termination of the process or even a kernel during restart.

4 Live Migration

Live migration is an advanced technique of container migration. The process of moving a running container along with its memory, storage, and network configurations between different hosts without disconnecting. It eliminates the drawback of offline migration. Types of live migration:

1. Post-copy live migration (Fig. 3)
2. Pre-copy live migration (Fig. 4)

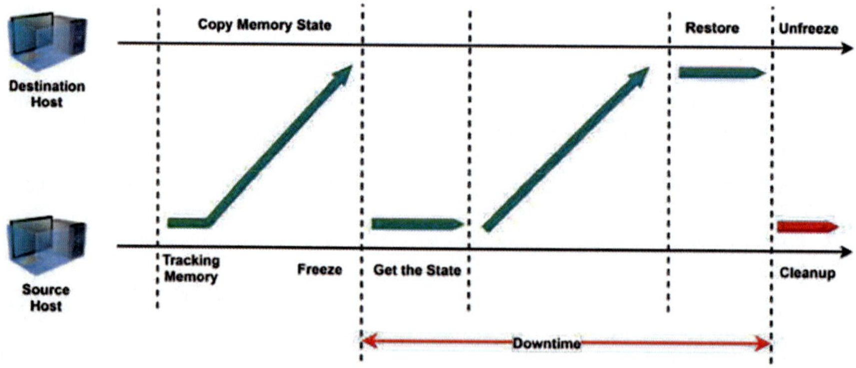

Fig. 3 Pre-copy live migration

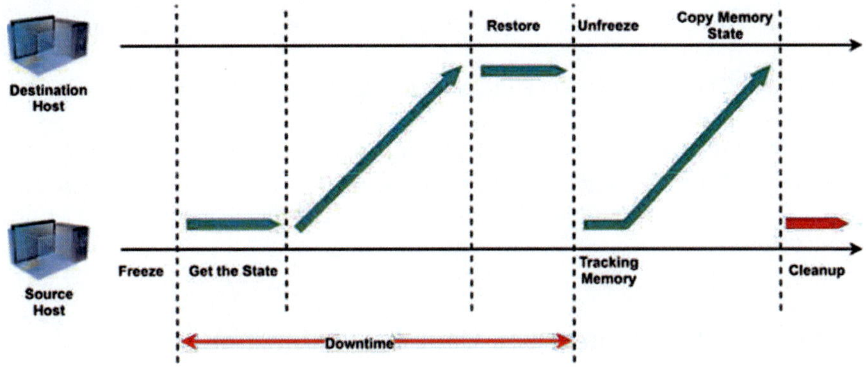

Fig. 4 Post-copy live migration

By using a checkpoint and restoration feature, live migration is easier to implement. A simple, hardware-free algorithm:

1. Synchronization of the container file system: Move the file system to the target host.
2. Freezing container: Freeze all the processes.
3. Dumping container: All collected resources will be saved on disk as a file.
4. Synchronization of the file system: A container is still running during the initial synchronization and some files can be non-updated at the destination server. That is why when a container is frozen, the second synchronization is carried out and its files are not changed.
5. Copy the dump: Upload the file to the server of the destination.
6. Reboot at the destination server: Now building a container and running processes inside the target server in the same condition as in the dump file. After this point, the processes are in the frozen state.
7. The container will resume again: Resume execution of the container on the server destination.
8. Stop the source server container: Kill the process of the container and unmount its system file.
9. Delete the root server container: Delete the file system of container and configure a source server files.

If anything goes wrong during restarting, then the migration of containers can be rolling back to the source host, and the container on the source side restarts as nothing had occurred. The migration process looks like a temporary network issue as external customers move to the container via the network. After a little delay, the container continues running normally. The only difference is that now, it is running on the destination host.

Stages 3–6 are responsible for the most interrupted operation in the above migration scheme. Let us have a close look at them:

1. Dump Time: Time required to pass through all the processes and resources and the data is saved into a disk file.
2. Synchronization of the second file system time: Time is taken to synchronize the second file system.
3. Dump copying of the file time: the time required to copy the dump from the source server to the target server over the network.
4. Time to spare: time needed to create a container from a dump file and all its processes.

5 Taxonomy and Survey on Container Migration Techniques

The checkpoint and restarting of a system have certain preconditions that the OS needs to provide. Container infrastructure contains: Namespaces (Process ID, User, group Id, Network configuration, Mount points, Inter-process communication, Host-name) and the cgroup subsystems (CPU cores, CPU time and utilization, Limit memory consumption, Limit and observe I/O accesses).

- PID virtualization—guarantees that a process can be allocated the same PID during restarting as before.
- Isolation of the process community—to ensure that the interaction between parents does not lead to external containers.
- Network separation and virtualization—to ensure the isolation of all network connections from the host operating system and all other containers.
- Virtualization of resources that is hardware-independent and can restart the container on another server.

During the design phase, other requirements must be considered:

1. A complete set of resources like delegated resources, register set, network connections, address space, and other private information of each process should be used by the system to test and restart a container.
2. Size of file dump should be reduced and all behavior between a freeze and a resume is streamlined so that the service is as late as possible.

For a better understanding of containers, we have studied various techniques and come up with a container taxonomy with six different parameters (Architecture, Tools, Purpose, Scope, migration technique, and Evaluation) Fig. 5.

Live migration service for container incorporates CRIU-based memory migration which reduces downtimes in migration [11]. Containers will re-operate on the target host immediately with a union view of the data between the source and the target hosts, while gradually transferring the disk state in the background. An optimized pre-copy algorithm introduces a gray-Markov prediction model. Memory pages are added to the Hot Workspace by the prediction model with high re-modified rates, and the stop copy is carried through on these memory pages and was designed by

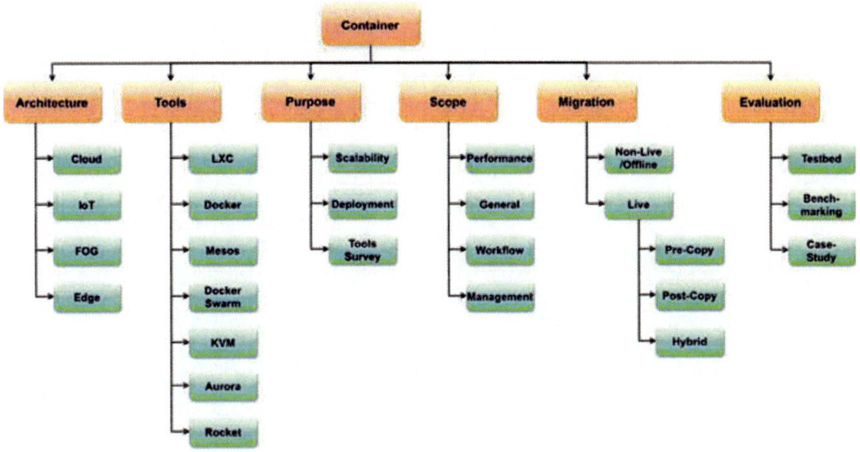

Fig. 5 A taxonomy on container in cloud computing

[12]. Experiments show that iterations and downtimes are reduced by OPCA. Mirkin [9] presented the checkpoint and restart functionality for OpenVZ containers.

Dump file size should be reduced, and all activities that occur between freeze and resume should be streamlined so that the service time is sustainable and as short as possible. Instead of initiating the migration based on thresholds, this migration is triggered by predictions, determining performance metrics, addresses device failure, and prevents excessive migration [14]. One of the methods to minimize extra overheads of migration is to identify the relatively stable memory pages. When the size is huge, stable pages should get the preference to reside in cache [2]. The detailed comparison of various techniques based on the proposed taxonomy is shown in Table 1.

1. Characterize the memory pages, actions on the basis of the page flags accessible on the proc file system of a Linux kernel.
2. Propose a forecast system that can be used to predict relatively stable memory pages.
3. Analysis of the memory and the proposed prediction method in two applications: memory deductions and live migration.

The pages that are updated in the last iteration but not modified in the current iteration are separated out in the first phase. Division of pages is done as per the second phase into two types, i.e., pages with high dirty rate and normal depending upon the updation in last some iterations. The filtered pages from the first phase are checked for the number of times the page modification has been done from history. The page that is modified more times is required not to be sent during the current iteration Jain and Agrawal [5].

Another container migration scheme is Application Oriented Docker Container (AODCs) to reduce Docker container application cost of deployment and to allow

Table 1 A taxonomy-based comparison on various container migration techniques

References	Archi tecture	Scope	Purpose	Tools	Evaluation	Migration	Technique
Qiu et al. [13]	Cloud	Performance	Deployment	LXD, CRIU	Tested	Live	Pre-copy
Bhardwaj and Krishna (2019)	Cloud	Performance	Deployment	LXD/CR	Tested	Live	Pre-copy
Kim et al. [7]	Cloud	Performance	Scalability	Kubernetes	Tested	Live	
Al-Dhuraibi et al. (2018)	Cloud	Performance	Scalability	Docker, RUBiS, KVM	Benchmarking		
Al-Dhuraibi et al. (2017)	Cloud	Performance	Scalability	Kubernetes, Graylog	Benchmarking		
Ma et al. [8]	IOT	Performance	Deployment	Docker	Tested	Live	Pre-copy
Nie et al. [12]	Cloud	General	Tools sur vey	Docker, Spark	Tested	Live	Pre-copy
Nadgowda et al. [11]	Cloud	General	Tools survey	LXC, CRIU	Tested	Live	Post-copy
Molt et al. [1]	Cloud	Perfor mance	Tools Survey	Docker, TOSCA	Tested	Offline	
Mirkin et al. [9]	Cloud	General	Deployment	OpenVZ, LXC	Tested	Live	Pre-copy
Dupont et al. [1]	IOT	General	Tools survey	OpenStack, Kubernetes	Tested	Live	Pre-copy
Raghunath and Annappa [14]	Cloud	General	Tools survey	XEN, RU-BiS	Tested	Live	
Wu et al. [15]	Cloud	General	Tools survey	Kubernetes	Tested	Live	Pre-copy
Guan et al. [4]	Cloud	General	Tools survey	CPLEX	Tested	Offline	
Elghamrawy et al. [2]	Cloud	Perfor mance	Scalability	LXC	Tested	Live	Pre-copy
Kan [6]	IOT	General	Deployment	Docker	Tested	Offline	
Yu and Huan [16]	Cloud	General	Deployment	OpenVZ, LXC	Tested	Offline	

automated scaling as the workload of cloud applications varies Guan et al. [4]. Pre-copy algorithms get rid of repeated transfer of dirty pages based on the probability of prediction of memory pages becoming dirty. When the dirty pages increase significantly, the probability will fail and raise sustainability issues. Fang et al. [3] propose an adaptive bandwidth technique which improves transmission bandwidth for decreasing transmission time when dirty pages rate suddenly increases. It decreases the iterative transfer total time as well as downtime.

Containers are very useful in edge computing. The edge computing platform is based on the concept of containers in which four key criteria were evaluated:

1. Deployment and termination.
2. Management of resources and services.
3. Fault tolerance.
4. Caching.

Docker container used in edge computing platform offers fast implementation, small footprints, and good performance based on our assessment and testing [10]. The containers are more compact, powerful, and easier to manage based on lightweight virtualization techniques. For live transportation of docker containers, both downtime and total migration time can be significantly reduced, helping to increase user experience with data center services. The docker container is migrated from the source host through logging and replaying in iterative phase [16]. An optimized pre-copying approach based on time series for the migration is suggested. It recognizes and transmits frequently updated dirty pages in the past and future era, in order to reduce unwanted, repetitive dirty pages transmission. This can reduce the overall migration time significantly.

Wu [15] propose pre-copy migration methods, including post-copying, pre-copy, pre-copy methods, etc., which are proposed to solve the problem of the too-long pre-copy migration period. Pre-copy approaches can reduce the total time of migration but can lead to longer downtime. In order to enhance live migration, this memory mechanism allows you to choose which pages you want to migrate in the iterative pre-copy stage or in the stop-and-copy process using the associated dirty rate. The suggested method can choose the best time to carry out memory migrations, according to the features of memory pages in order to reduce not only unnecessary migrations but also the total time of migration.

6 Future Directions

Pre-dump is the very first step to initialize the pre-copy container migration. In this phase, the complete set of a container memory is transferred to the destination host, and the container is still running on source host. If the source has a running container, it will be required to be shifted to destination. The following are the steps of the pre-copy migration technique.

1. First Phase: Only unmodified pages of memory (from the start of the container up to the current state) will be transferred to the destination. Pages that were modified have a strong probability of update in the next phase. We need not send them in the initial phase.
2. Second Phase: It will be decided on the basis of algorithmic design, which pages can be considered for transfer in the iterative phase. In the basic approach, all updated pages set to be sent in the same iteration.
3. Third Phase: If there is any page fault or any failure generated at the destination, we can get it from source host according to the proposed model.

Prediction of memory changed is the core component of pre-copy container migration. To improve prediction mechanism, a prediction scheme or set of multiple schemes should be applied to the memory pages to be migrated to the destination host. ARMA-based schemes, ARIMA-based schemes, regression-based schemes, Markov model, Bayesian scheme, and KNN-based scheme are some of the popular schemes available. It should be chosen wisely according to the pre-dump, iterative, and final dump phases of pre-copy migration.

7 Conclusion

A review of various container migration techniques is carried out in this research work. In order to achieve better performance and minimize the downtime, the important factor is the amount of memory transfer from source to the destination host and the repetition of the same in an iterative phase of memory migration. As discussed in related work, some of the research work is carried out to predict the change of memory. With the help of different prediction algorithms, researchers try to decrease the number of pages transferred during the iterative phase of migration. But the prediction of memory change will also help in the pre-dump phase of pre-copy container migration. As per the review of migrations techniques of containers, the pre-copy technique outperforms in most of the cases in Table 1. We have identified some research works related to container migration and record the following parameters: Architecture, Scope of the research, Purpose, Tools used, Evaluation, Kind of migration method opted, and migration techniques (pre-copy or post-copy). Pre-copy is the most commonly used migration technique, and it is best suited for containers. We will further extend this approach with time series prediction methods. It will help to identify the pages to be modified and provides a sustainable approach.

References

1. Dupont C, Gia Reda R, Capra L (2017) Edge computing in IOT context: Horizontal and vertical Linux container migration. In: 2017 global internet of things summit (GIoTS). IEEE, pp 1–4

2. Elghamrawy K, Franklin D, Chong FT (2017) Predicting memory page stability and its application to memory deduplication and live migration. In 2017 IEEE international symposium on performance analysis of systems and software (ISPASS). IEEE, pp 125–126
3. Fang Y, Chen Y, Ge J (2016) Improvement of live migration mechanism for virtual machine based on pre-copy. In: 2016 3rd international conference on materials engineering, manufacturing technology and control. Atlantis Press
4. Guan X, Wan X, Choi B-Y, Song S, Zhu J et al (2016) Application oriented dynamic resource allocation for data centers using docker containers. IEEE Commun Lett 21(3):504–507
5. Jain P, Agrawal R (2016) An improved pre-copy approach for transferring the VM data during the virtual machine migration for the cloud environment. Int J Eng Manuf 6(6):51–60
6. Kan C (2016) Docloud: an elastic cloud platform for web applications based on docker. In 2016 18th international conference on advanced communication technology (ICACT). IEEE, pp 478–483
7. Kim D, Muhammad H, Kim E, Helal S, Lee C et al (2019) Tosca-based and federation-aware cloud orchestration for kubernetes container platform. Appl Sci 9(1):191
8. Ma L, Yi S, Li Q (2017).E client service hando across edge servers via docker container migration. In: Proceedings of the second ACM/IEEE symposium on edge computing, pp 1–13
9. Mirkin A, Kuznetsov A, Kolyshkin K (2018) Containers checkpointing and live migration. Proc Linux Symp 2:85–90
10. Mostajeran E, Khalid MF, Mydin MNM, Ismail BI, Ong H (2016) Multifaceted trust assessment framework for container based edge computing platform. In: Fifth international conference on advances in computing, control and networking, ACCN
11. Nadgowda S, Suneja S, Bila N, Isci C (2017) Voyager: complete container state migration. In: 2017 IEEE 37th international conference on distributed computing systems (ICDCS). IEEE, pp 2137–2142
12. Nie H, Li P, Xu H, Dong L, Song J, Wang R (2017) Research on optimized pre-copy algorithm of live container migration in cloud environment. In: International symposium on parallel architecture, algorithm and programming. Springer, pp 554–565
13. Qiu Y, Lung C-H, Ajila S, Srivastava P et al (2019) Experimental evaluation of LXC container migration for cloudlets using multipath TCP. Comput Netw 164:106900
14. Raghunath BR, Annappa B (2017) Prediction based dynamic resource provisioning in virtualized environments. In: 2017 IEEE international conference on consumer electronics (ICCE). IEEE, pp 100–105
15. Wu T, Guizani N, Huang J (2017) Related dirty memory prediction mechanism for live migration enhancement in cloud computing environments. J Netw Comput Appl 3:1–14
16. Yu C, Huan F (2015) Live migration of docker containers through logging and replay. In: 2015 3rd international conference on mechatronics and industrial informatics (ICMII 2015). Atlantis Press

Sketching of EV Network: A Complete Roadmap

Amrit Pal Singh, Manuj Aggarwal, Harpuneet Singh, and Pankaj Bhambri

1 Introduction

This artifact review system signifies the various design approaches/strategies, DFDs, database design, and the User Interface design. It mainly highlights the up-keep of the current study in designing an innovative system to search for nearby available charging stations and allow verified users to make bookings according to their requirements at available charging stations. This section covers the actual implementation upshots, data taken from the user, and the way it is stored in the database. Brief: GATS India Ltd. "SCS (Smart Charging Station)" Online at https://www.gats.in/scs.html, accessed: Jan 28, 2020.

1.1 EVs Relation with Sustainable Development

Every nation is investigating to get an energy source that is renewable, contamination-free, and can be generated at faster rate with cheaper investments. Electric charge is the best suited pollution-free source of energy and since the batteries that store the charge are reusable, so they have a longer life. Moreover, electric charge has the capacity to do heavy work which would otherwise require a great quantity of fossil fuels in addition to their add-on transportation cost, where in contrast electricity is easily portable.

A. P. Singh · M. Aggarwal (✉) · H. Singh · P. Bhambri
Department of Information Technology, Guru Nanak Dev Engineering College, Ludhiana, Punjab 141006, India
e-mail: aggarwalmanuj17@gmail.com

A. P. Singh
e-mail: amrit220399@gmail.com

H. Singh et al. (eds.), *Sustainable Development Through Engineering Innovations*, Lecture Notes in Civil Engineering 113,
https://doi.org/10.1007/978-981-15-9554-7_37

Most vehicles in the whole world are largely dependent on fossil fuel consumption. This process not only contributes to their depletion but also contributes to environmental pollution. To save these resources for the upcoming generations, it is therefore primary concern to use alternatives for providing energy to ride vehicles. Hence, it is time to switch to electric-driven vehicles with 0% pollution. Additionally, they provide a greater run in an aspect of distance covered at a negligible cost which is the star feature of these EVs. It not only saves the energy resources but also protects other natural resources which are degraded due to their consumption.

Driving a vehicle on electricity eradicates a gigantic amount of Green House Gases (GHG) plus carbon-dioxide outrushes. EV driving assists to uphold milieu as it brings renewable energy into play. As result charging stations operate on this renewable energy eradicates exhalations and rate of electrical energy. EV's revamp atmosphere quality by consolidating exhalations at powerhouses and thus are more proficient than vehicles motorized by fossil fuels. Thus, EV's are precious facilities for customers. Driving on power is an enormous preference for both the planet and the wallet. This way EVs hold a great proportion in accomplishing sustainable development goals.

2 Approaches for Design

Design of software can be explained as a course of action that idealizes program necessities dependent on the program accomplishment. Design of software accepts consumer necessities as directive plus strives towards the discovery of optimal elucidation. Although software can be conceived, an arrangement can also be outlined for discovering an excellent and promising blueprint for executing proposed elucidation. In function-oriented design, a proposal consists of numerous minor small-systems called functions. These all functions can be proficient in carrying out a unique job within the proposal. A proposal can be observed like peak outlook for the entire task. Function centered design assumes a few properties for ordered design wherever split as well as surmount approach can be applied. This devised technique bisects the entire structure into minor tasks and it bestows a way for abstraction via hiding particulars as well as their working. Such type of operational segments distributes particulars among themselves by way of particular sharing and using particulars that are worldwide accessible. One more aspect of functions can be seen that when a program calls a function, the function changes the condition of the program, which occasionally is not satisfactory by other segments. Function centered design operates fine where the state of the system is not much significant and program or functions toil on the key input instead of depending on the state. Software devise course of action is alleged like sequence of precise ways. Despite the fact that it fluctuates as per the approach for design (function-centered or object-centered), hitherto this includes the subsequent ways:

- A resolution plan can be constructed commencing requisite otherwise previously applied structure plus the system sequence illustration.
- Objects are recognized and clustered into the classes on account of resemblance within quality features.
- Class chain of command and relation among them is determined.
- Application agenda is described.

3 Detailed App Design

This comprises the hope and the alleyway to reach the destination. The concern is with fabricating an android application design policy. This merely cannot build an app just for the reason that the contender also have. The contender may encompass a diverse business goal and mobile tactic which are quite diverse. Emerging a mobile policy that associates reverse to corporation policy involves 4 junctures:

- Know the commercial policy
- Commercial cell phone application policy
- Application policy
- Manufactured goods administration policy
- Know the Commercial policy

Grasping the general commercial policy should be supposed to outline the root of cell phone application plan. There are many advantages of fabricating and commencing a cell phone policy which is procured commencing the general corporation policy:

- Increases Return on Investment (ROI) as much as 74%.
- Decreases instructing desires.
- Leads to client pleasure.
- Reduces amalgamation necessities and errors.
- Enhancement in superiority, worth, efficiency, worker competence, and consumer interaction.

Overall a flourishing mobile policy is the juncture of business aims, mobile phone opportunities, and user's needs.

4 Design of System

System device is the course of action that identifies fundamentals of a system like the structural design, modules and parts, the distinctive coherence of those parts, and the data that goes all the way through the system. This is doomed to assure particular needs and necessities of an industry or association via engineering of an articulate and well-running system.

4.1 Data Flow Diagram (DFD)

Data flow diagram describes the surge of facts and figures among distinctive activities in commerce. DFD is a pictorial technique which displays particulars surge and shows transforms which are useful as facts budge from key into key out. It gives an uncomplicated, shrewd technique of defining commercial undertaking lacking focus on minutiae of PC systems. DFD's are smart practice that bestows what individuals do to a certain extent than what computers. Brief: Visual Paradigm "What is DFD? How to Draw DFD?" Online at https://www.visual-paradigm.com/tutorials/data-flow-diagram-dfd.jsp, accessed: March 10, 2020.

Depiction of Components:

DFD's engross four symbols:

- Process (Circle)
- Data Object (Arrow)
- Data Store (Open-ends Rectangle)
- External entity (Closed Square or Rectangle)

4.2 Relationship and Rules

DFD can be applied for any echelon of data abstraction. DFD may be screened in echelons. Each echelon has supplementary particulars surge and data purposeful minutiae than the prior echelon. Premier echelon is a Context Diagram. Some chief points are as follows:

- One gurgle (process) constitutes the complete system.
- Data arrows constitute I/O.
- Data Stores are not publicized. They are in the structure.

4.3 Importance of DFD

DFD picturesque represent operations and proceedings, that detains, maneuver, accumulate, and dispense facts amid structure also the surroundings and among elements of system. Optical depiction assembles this high-quality interaction mechanism among client and structure artificer. Composition of Data Flow Diagram authorizes beginning from an expansive outline and expands this on the way to pecking order of meticulous diagrams. DFD has frequently been applied owing to subsequent causes:

- Rational particulars surge of structure.
- Fortitude of corporeal structure edifice necessities.
- Straight forwardness of annotation.
- Inception of blue-collar and robotic system necessities.

Figure 1 depicts App Authentication Strategy which is based on the design that first user needs to self-authenticate by registering using Indian phone number instead of email-password authentication since phone authentication reduces the chance of hoax or multiple id creations by same single user on single device. Further for receiving invoices, offers or support mails only; one email-id gets attached to single profile and it is further verified. After authenticating using mobile number, if the user data is found to exist in the database then it is regarded as old user else he is directed for profile creation. After profile creation, it is compulsory for the user to enter the details of the electric vehicle mainly the manufacturer, model, and vehicle type using which the charger is automatically fetched matching the details.

For the authentication of the vehicle which belongs to the user and is a registered vehicle, the user is further requested to enter the RC (Registration Card) number of the vehicle.

In case vehicle is newly purchased and therefore RC creation is under process, then user is bound to enter Provisional RC and it will be valid for 30 days. If the user has entered Provisional RC during registration, then it needs to enter Original RC within 30 days as it will be issued within that time according to the Indian Vehicle

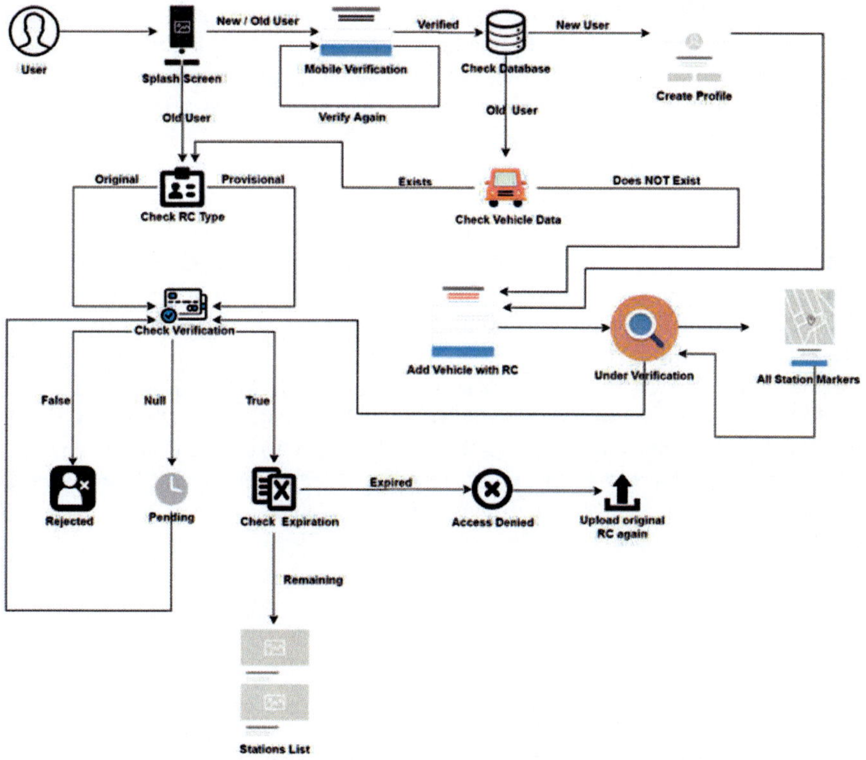

Fig. 1 Authenticating user identity using mobile and vehicle with original/provisional RC

Act, or else its access to the app will be denied automatically and will reopen upon original RC authentication. Either it is original or provisional RC; both are verified by the backend authority from Parivahan site of Indian Government only after which the app facilities get opened for the user. If there is any objection found in the RC details; then the RC details are rejected and the user is requested to re-enter the RC details.

In case of user with Provisional RC registration, opens the app then verified status, and the expiration date of Provisional RC is verified and user is given a notification too to upload the vehicle's original RC. If the RC verification is under process, then the user has the facility to see markers of all EV stations registered under the app on GMaps all over India. Upon clicking the marker, the user has the facility to open Google Maps app and thus finds its location, directions, open-hours, other info, etc.

If all the details are correct and verified, then the user has the privileges to observe the list of stations who matches w.r.t. charger type of vehicle registered, location of the user, and the slots availability at the station. After matching the user charger-type, user location, and available slots of stations; the user is offered a list of station choices to select from. Figure 2 shows the app booking process which includes the user's selection of the station, slot-time, and date to be reserved. By default, the selected date will be the current date of app usage. The slot-time for current date will automatically vanish if the time has elapsed with respect to current time of the app utilization and the engaged slots are automatically disabled in real-time due to booking of slots by other users. Upon slot selection by the user, a confirmation is required, and upon successful confirmation by the user; it is verified that user is doing the booking process first time for that particular day for a particular station. Multiple reservations for the same station on the same day are not allowed by single user in order to offer availability for all. Hence; if the verification is not objectionable, booking data is registered in the database in real-time. A booking-ID is generated along with 4-digit OTP which is required to authenticate user at the respective reserved station. Once the OTP gets confirmed only, then the EV charging access is made available to the user. Brief: J. Liu, "Research and Implementation of Electric Vehicle Fast Charging Station Parking Guidance System Based on Mobile Terminal," 2017 9th International Conference on Intelligent Human–Machine Systems and Cybernetics (IHMSC), Hangzhou, 2017, pp. 230–233, https://doi.org/10.1109/IHMSC.2017.60. After booking is confirmed, the next step that comes into action is the inbuilt routing to that reserved station and for this, it is essential to have access to either location permission/GPS or Network Access using mobile data/Wi-Fi, or maybe both if available for better accuracy and results.

5 Database Design

It is a gathering of procedures that smoothed the development, designing, accomplishment, and preservation of the venture data management systems. Appropriately

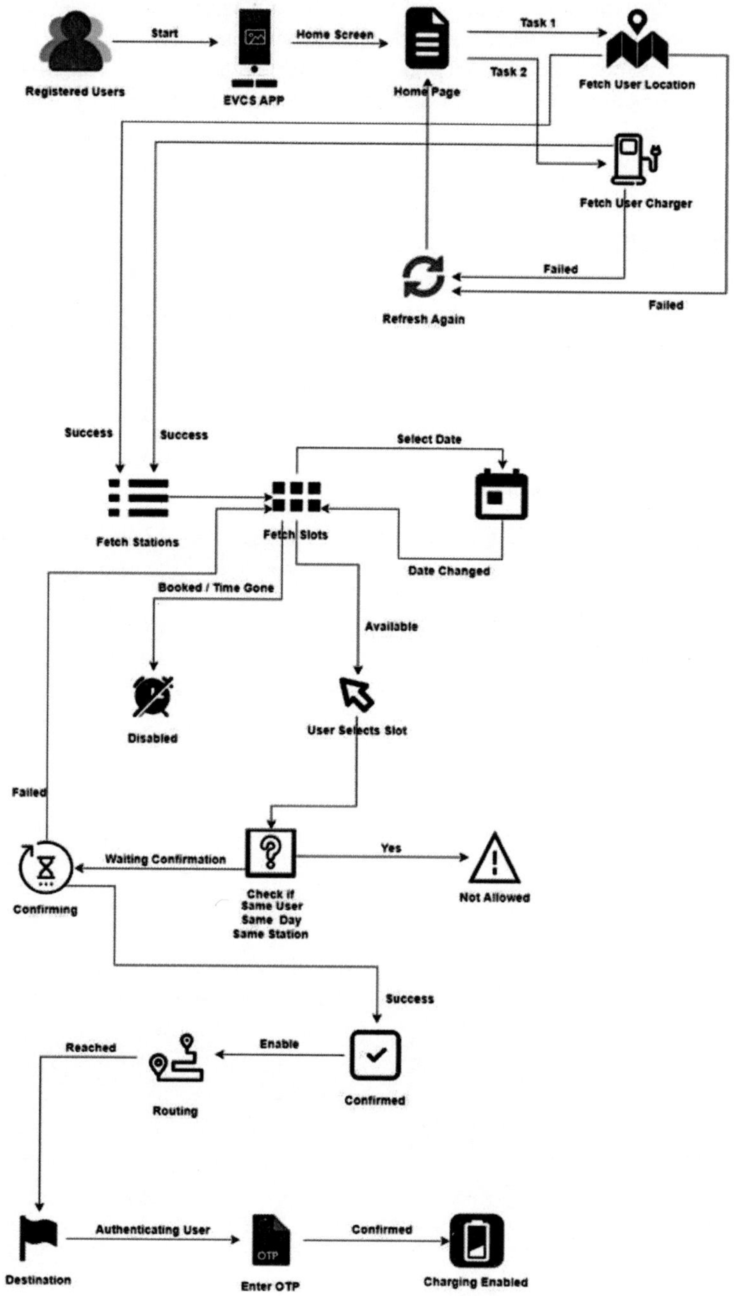

Fig. 2 Booking of the preferred time slot of an EV station by authenticated user

premeditated databases are simple to preserve, revamps data evenness, and are cost-efficient in provisions of disk storage space. The database designer decides how the data fundamental corresponds and what data must be piled. The significant goals of database devising are to produce rational and corporeal design models of the offered database structure.

5.1 ER Diagram

It is a graphic method for database devices. It is a top-level data model that defines data components and their connection for a number software system. An ER model is used to denote real-world objects. There are three basic models of ER Diagram:

- Objects-or-Entities
- Traits-or-Attributes
- Interactions-or-Relationships

Figure 3 shows the ER diagram for attributes of user, its vehicle, and the booked slot and Fig. 4 shows the ER diagram for all the essential attributes of an EV station

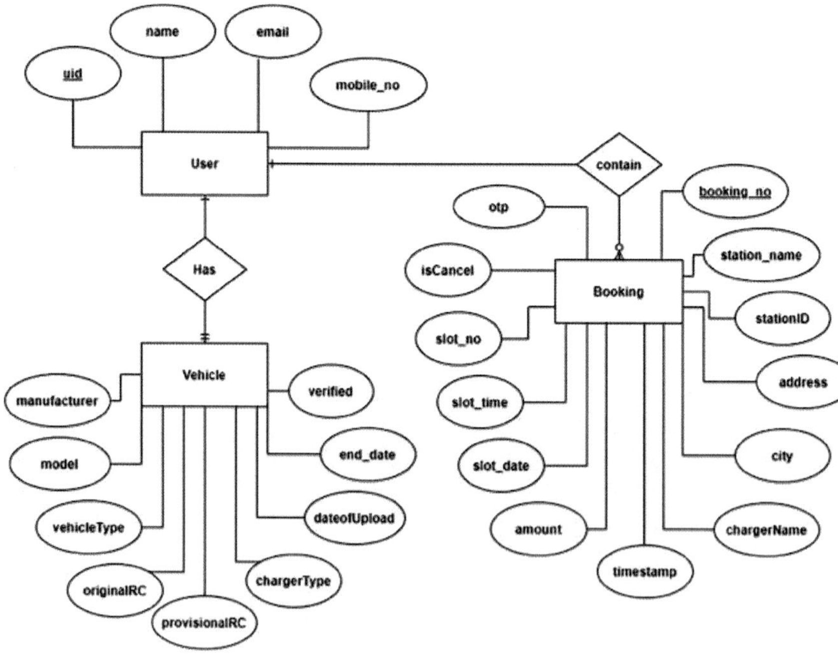

Fig. 3 ER diagram for user, vehicle and booking attributes

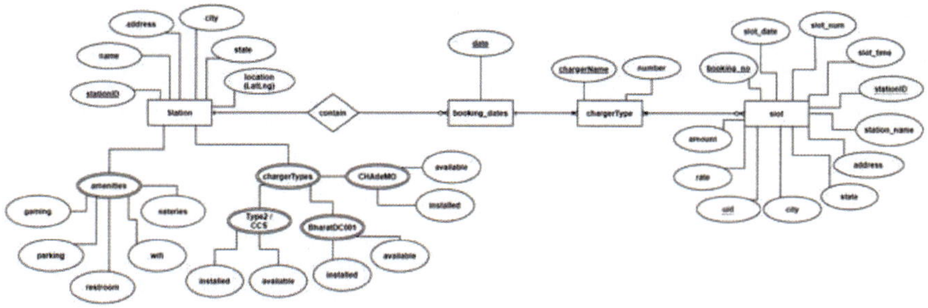

Fig. 4 ER diagram for EV station attributes necessary for database design

that an electric charging app must have and act as building blocks for designing the structure of database.

For instance, consider databank for a University/College, objects for Candidates, CourseNames, and Teachers/Professorsexist. Student's object contains features resembling Roll_No, Forename/Designation, and dept_ID. These may join connections through CourseNames and Professors.

6 Database Implications

The implementation of database and UI designing brings out the following results in the form of all required essential data for the app kept in Fire store database. The records are stored as per the attributes for respective entities drawn in the ER diagrams.

Figures 5 and 6 are the practical implementation of the ER diagram represented in Fig. 3 in which database used is the Firebase Fire-store provided by Google. The user bookings are stored as per the user preference of the slot-time at the station with the user's current location necessary for routing as shown in Fig. 6 with provision of slot cancellation before the reserved time. Figure 5 depicts the database implementation where it checks if the user has uploaded original RC or not and the 'verified' field shown in Fig. 5 is true because the bookings depicted in Fig. 6 can only be made after the user vehicle is verified by uploading the original RC. Also, the date of upload of RC is registered in the database with an end date after the month of upload so that if the user does not verify the RC details in the span of 30 days the account gets deactivated automatically, and the user is not allowed to access his/her profile and make any bookings further.

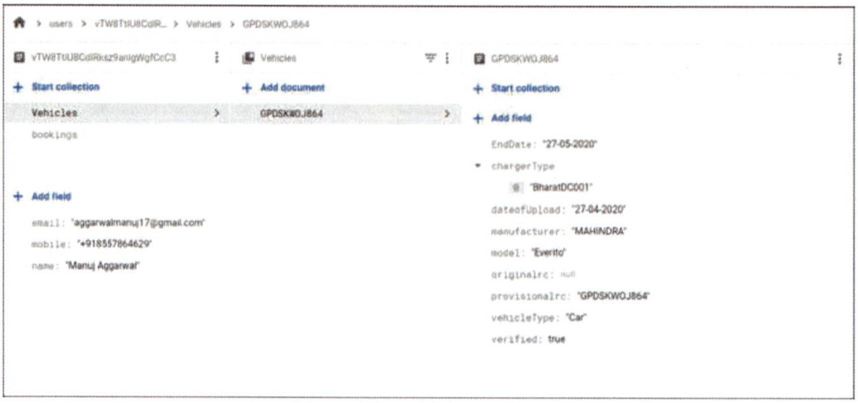

Fig. 5 Snapshot of firebase fire-store database containing user profile and vehicle data according to Fig. 3

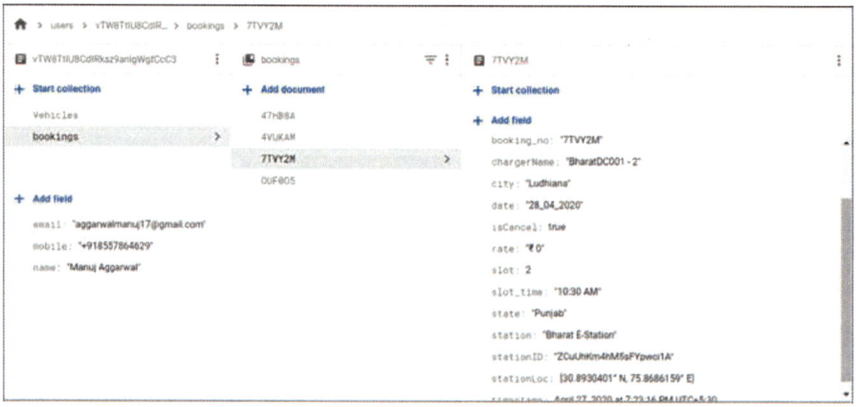

Fig. 6 Snapshot of firebase fire-store database containing user booking attributes according to ER diagram of Fig. 3

7 Results and Discussion

This topic of article specifies the results obtained from the project and the output screens that are obtained from **'Electrify'** application. Firstly, the user's current location from the Google map is fetched. Thereafter, a list of available charging stations matched with the vehicle details of the user along with the distance from the current location is displayed to the user as shown in Fig. 7. It shows only those stations which are having chargers compatible with the user's electric vehicle. It has some additional features like sort by according to the distance and number of charging options. This app allows users to pre-book charging slots not only for current day

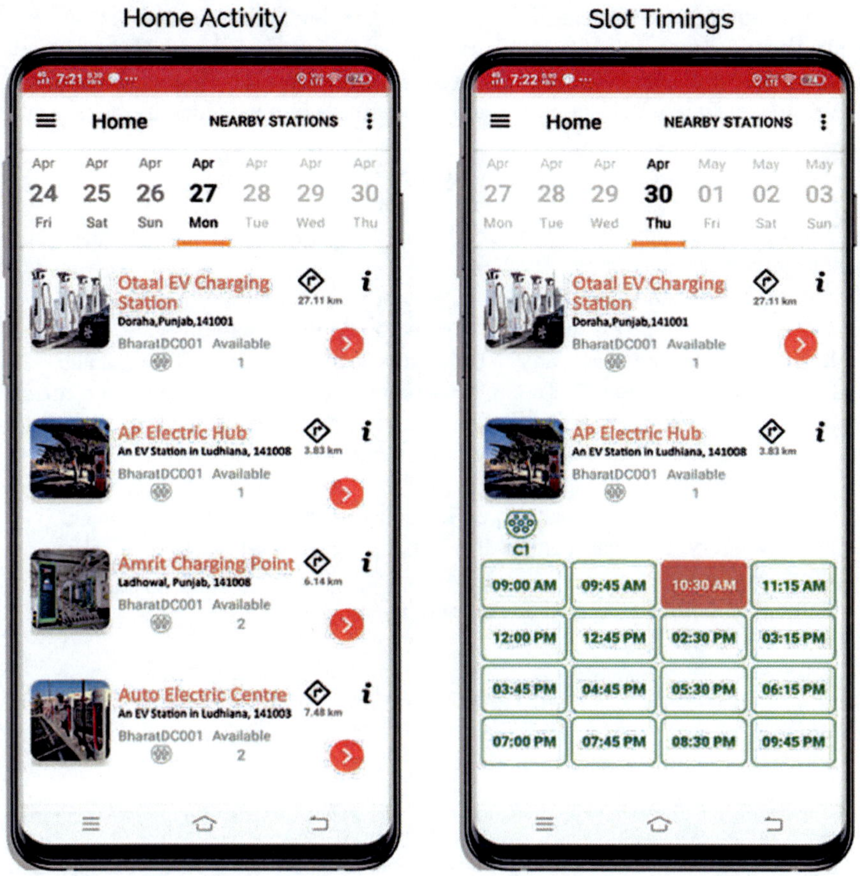

Fig. 7 Snapshots of electrify user app with list of nearby stations and its available slots

but also for future dates. Figure 7 shows an ELECTRIFY activity having built-in calendar with an option to traverse for future dates. The red filled block displays a booking made by another user and all other blocks show the available booking slots.

8 Conclusion

- Electric charge is the best-suited pollution-free source of energy and since the batteries that store the charge are reusable; so they have a longer life.
- EV driving assists to uphold milieu as it brings renewable energy into play. As a result, charging stations operate on this renewable energy. It eradicates exhalations and rate of electrical energy.
- EV's revamp atmosphere quality by consolidating exhalations at powerhouses.

- Driving on power is an enormous preference for both planet and wallet.
- The primary goal of the app is to retrieve and save user data, data of his/her vehicle and the booking data is efficiently saved in the database which makes it easier for the user as well as the station operator to verify each other's identity and ease the process of charging the EV.
- Register and login on the app with mobile OTP verification.
- On successful login, user will update his/her RC and vehicle details.
- The developed app displays nearby charging stations according to the user's current location.
- Sort charging stations with respect to distance and available chargers.
- The app reduces the manual searching and booking of charging stations.
- It supports a fast and easy way to book slots according to user's preference.
- Once the booking process is completed, route is displayed from the user's current location to the booked station that helps user to easily navigate to the station.
- User can review his/her past or upcoming bookings.
- It also allows users to update their profiles even after registration and cancel booking anywhere any time before the booked slot-time as per the demand.
- For any query, user can contact the responsible authority via email, phone, or WhatsApp.
- This app is both user as well as compiler friendly.

References

1. Murphy ML (2020) Android programming tutorials [E-book]. Available https://www.pdfdrive.com/android-programming-tutorials-e10189445.html. Accessed 20 Jan 2020
2. van Drongelen M (2020) Android studio cookbook: design, test, and debug your apps using android studio [E-book]. Available https://www.pdfdrive.com/android-studio-cookbook-design-test-and-debug-your-apps-using-android-studio-e185748368.html. Accessed 25 Jan 2020
3. Liu J (2017) Research and implementation of electric vehicle fast charging station parking guidance system based on mobile terminal. In: 2017 9th international conference on intelligent human-machine systems and cybernetics (IHMSC). Hangzhou, pp 230–233. https://doi.org/10.1109/IHMSC.2017.60
4. GATS India Ltd. (2020) GATS, SCS (smart charging station)—EV charging stations model [Online]. Available https://www.gats.in/scs.html. Accessed 28 Jan 2020
5. Google Firebase (2020) Add firebase to android project [Online]. Available https://firebase.google.com/docs/android/setup?authuser=0. Accessed 18 Jan 2020
6. Visual Paradigm (2020) What is data flow diagram (DFD)? How to draw DFD [Online]. Available https://www.visual-paradigm.com/tutorials/data-flow-diagram-dfd.jsp. Accessed 10 Mar 2020

A New Direction in TOPSIS Method for Evaluation of Alternates: R-TOPSIS

Raman Kumar and Harwinder Singh

1 Introduction

In this globalized era, individuals and organizations are focusing on better decision making which helps in the successful operation of the business. The justification of decision in evaluation processes is highly recommended prior to execution. The reliable ranking of alternates can help the organization for better sustainable development. The Technique for Order of Preference by Similarity to Ideal Solution (TOPSIS) technique for multi-criteria decision-making analysis was developed in 1981 [1]. The further notable development in the TOPSIS method was made in 1987 [2] and in 1993 [3]. The concept behind the TOPSIS is to evaluate the best alternate having the longest geometric distance from the positive ideal solution. Researchers have successfully proved the application of MCDM methods in logistic [4], advanced manufacturing technology [5, 6], material selection [7, 8], performance evaluation [9].

R. Kumar (✉)
Department of Mechanical Engineering, Chandigarh University, Gharuan, Mohali, Punjab 140413, India
e-mail: ramankakkar@gmail.com

H. Singh
Department of Mechanical Engineering, Guru Nanak Dev Engineering College, Ludhiana, Punjab 141006, India
e-mail: harwin75@rediffmail.com

H. Singh et al. (eds.), *Sustainable Development Through Engineering Innovations*, Lecture Notes in Civil Engineering 113,
https://doi.org/10.1007/978-981-15-9554-7_38

1.1 TOPSIS Method

As per TOPSIS method, the selected alternative needs to have probably the shortest distance from the positive ideal solution as well as the farthest from the negative ideal solution. The steps of TOPSIS method are presented below [10]:

Step 1. Develop decision matrix (D_{ij}): The decision table presents the value of each alternate with respect to every attribute.

Step 2. Compute the normalized matrix (ND_{ij}): The attributes values presented in decision table may have measured on the different scale and different unit. Therefore, normalization of decision table data is required. The following equation can be used to normalize the value of the attribute [1].

$$\mathrm{ND}_{ij} = D_{ij} / \left[\sum D2_{ij} \right)$$

Step 3. Develop weighted normalized decision matrix (W_{ij}): In this step, the normalized decision matrix is multiplied by its associated weights.

$$V_{ij} = W_j \times \mathrm{ND}_{ij}$$

Step 4. Determine the positive ideal (V^{+}) and negative ideal solutions (V^{-}): The values of positive ideal and negative ideal solutions are computed by using below-mentioned expressions.

$$V^{+} = \left\{ \left(\sum_{i}^{\max} V_{ij} / j \in J \right), \left(\sum_{i}^{\min} ij / j \in J' \right) \Big/ i = 1, 2, 3, \ldots\ldots, N \right\},$$
$$= \left\{ V_1^{+}, V_2^{+}, V_3^{+}, V_4^{+}, \ldots\ldots V_M^{+} \right\}$$

$$V^{-} = \left\{ \left(\sum_{i}^{\min} V_{ij} / j \in J \right), \left(\sum_{i}^{\max} V_{ij} / j \in J' \right) \Big/ i = 1, 2, 3, \ldots\ldots, N \right\},$$
$$= \left\{ V_1^{-}, V_2^{-}, V_3^{-}, V_4^{-}, \ldots\ldots V_M^{-} \right\}$$

where $J = (j = 1, 2, 3, \ldots, M)/j$ is associated with beneficial attributes, and $J' = (j = 1, 2, 3, \ldots, M)/j$ is associated with non-beneficial attributes.

Step 5. Calculate the positive and negative separation measures: The finding of separation measure is based on the n-dimensional Euclidean distance. The positive separation measures (S_1^{+}) and negative separation measures (S_1^{-}) are computed by using following equations.

$$S_1^+ = \left\{ \sum_{j=1}^{M} \left(V_{ij} - V_j^+ \right)^2 \right\}^{0.5} \quad S_1^- = \left\{ \sum_{j=1}^{M} \left(V_{ij} - V_j^- \right)^2 \right\}^{0.5} \quad (i = 1, 2, 3, \ldots, N)$$

Step 6. Compute the relative closeness: The relative closeness values (P_i) is computed by using following expression.

$$P_i = S_1^- \left(S_1^- + S_1^+ \right)$$

The alternate having highest P_i value should be placed at rank 1 and so on.

Researchers have successfully presented the modified TOPSIS method in different application. In this work, an attempt has been made to reduce the computational work of TOPSIS method. In present work, an effort has been made to modify the TOPSIS method in order to get the more reliable result. The condition-based TOPSIS is presented to make decision-making process more effective. The rest of paper is organized as follow. Section 2 shows the modification in TOPSIS, i.e., condition based R_1, R_2, R_3 TOPSIS. Section 3 demonstrates and discusses the validity of R_1-R_2 and R_3 TOPSIS. Section 4 presents the concluding remarks.

2 Condition-Based TOPSIS, I.E., R1, R2, and R3 TOPSIS

In this research work, three conditions have been put in TOPSIS work to improve the result of TOPSIS method. The first condition is based upon the values of positive separation measures only. The second condition and third condition are based on the condition whether the ranking provided by positive separation measure and negative separation measure is same or not.

2.1 R_1-TOPSIS

R_1-TOPSIS method reduces the steps and computational work involved in TOPSIS method. This method is suggested to find positive ideal valve and positive separation values. Following this, ranking of alternates is to be done based on positive separation, if there is significant difference between positive separation values. If the values of positive separation are much closer to each other, it is suggested to apply R_2-TOPSIS.

2.2 R_2-TOPSIS

This method suggests to compute the both positive separation and negative separation values. Following this, compare the ranking provided by positive separation and negative separation. If the ranking provided by both is almost same, then do ranking of alternates. However, if the ranking of positive and negative separation differs from each other, then is recommended to apply R_3-TOPSIS. There is no need to find relative closeness in R_2-TOPSIS, hence reduced the step involved in TOPSIS method.

2.3 R_3-TOPSIS

This method suggests to drop out the worst alternate based on ranking provided by both positive and negative separation and develops a new decision matrix. Now, repeat the whole procedure to find final ranking of alternates. The flowchart of R_1-R_2-R_3 TOPSIS is presented in Fig. 1.

3 Demonstration of R_1, R_2, and R_3 TOPSIS Method

The results of few published examples are discussed with R_1, R_2, and R_3 TOPSIS for the demonstration of condition-based TOPSIS method.

Example 1 Gumus [11] applied two steps methodology, i.e., fuzzy AHP and TOPSIS, to provide ranking to alternates. The fuzzy AHP was used to give weightage to attributes, and TOPSIS is provided to rank the hazardous waste transportation firms. The ranking is done on the basis of the relative closeness value as present in Table 1. The results are compared with R1 TOPSIS method as shown in Table 2.

As seen in example 1, there is significant difference between the positive separation measures; therefore, the ranking provided by positive measure is same as of relative closeness. The positive separation measure value of alternate E and alternate C is 0.110951 and 0.366725, respectively. This indicates that Alternate E must be placed at Rank 1. The positive separation measure of alternate C and alternate D is almost same, and there is a possibility the rank provided to alternate C and alternate D by R_1 and TOPSIS may differ to each other. In case, there is no significant difference between the positive separation measures, and it is recommended to use R_2 TOPSIS.

Example 2 Baykasoğlu and Gölcük [12] proposed an integrated MCDM approach for evaluation of alternates. The type-2 fuzzy DEMATEL and TOPSIS method were applied to find inter-dependencies among problem attributes and final ranking of alternatives, respectively. The SWOT-based strategy was demonstrated in decision making. The ranking of four alternatives, viz. SO, WO, ST, and WT strategies, was

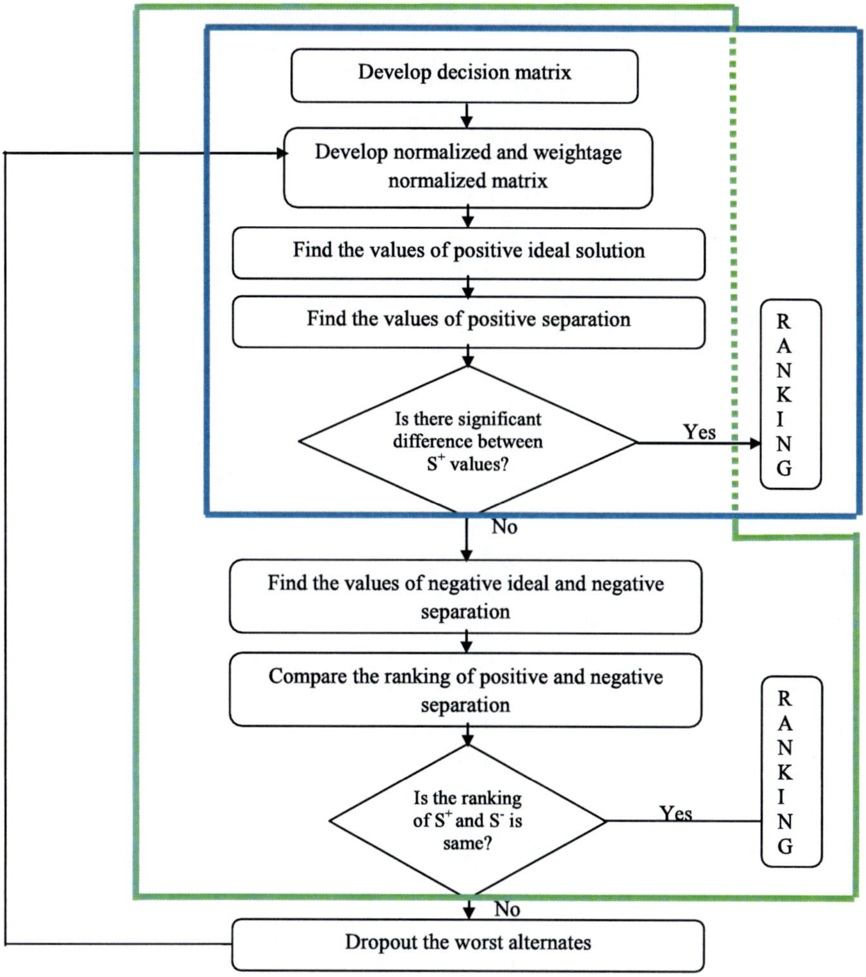

Fig. 1 Flowchart of R_1 -R_2-R_3 TOPSIS method

Table 1 Ranking on the basis of relative closeness value

Alternate	S^+	S^-	P_i	Ranking
A	0.431672	0.013506	0.030338	5
B	0.407930	0.108853	0.210636	4
C	0.366725	0.202463	0.355705	2
D	0.367593	0.142114	0.278815	3
E	0.110951	0.415603	0.789289	1

Table 2 Ranking on the basis of positive separation, i.e., R_1 TOPSIS

Alternate	S^+	Ranking
A	0.431672	5
B	0.407930	4
C	0.366725	2
D	0.367593	3
E	0.110951	1

Table 3 Ranking on the basis of relative closeness

Alternate	S^+	S^-	P_i	Ranking
SO	0.46	0.40	0.46	4
WO	0.38	0.52	0.58	1
ST	0.47	0.46	0.50	2
WT	0.42	0.38	0.47	3

Table 4 Ranking on the basis of positive separation and negative separation

Alternate	S^+	Ranking	S^-	Ranking	Ranking by R_2 TOPSIS
SO	0.46	4	0.40	4	4
WO	0.38	1	0.52	1	1
ST	0.47	2	0.46	2	2
WT	0.42	3	0.38	3	3

done on the basis of relative closeness as shown in Table 3. Table 4 presents the ranking of alternates on the basis of negative and positive separation measures.

As seen in Table 3, there is no significant difference between positive measure values; therefore, the values of negative separation measure are computed. It has been found the ranking provided by positive measure values is almost same as of negative separation measure. So, there is no need to find relative closeness and the ranking provided by positive separation measure is considered as final ranking as shown in Table 4.

Example 3 Başaran and Haruna [13] proposed the application of FAHP and TOPSIS for evaluating the mobile learning applications for mathematics. The FAHP and TOPSIS were used to compute the weightage of criteria and ranking of alternate, respectively. A four-level hierarchy diagram consists of 11 sub-criteria, and five alternate was proposed to solve the problem. The results are presented in Table 5. Table 6 and 7 present the results of R2 TOPSIS and R3 TOPSIS, respectively.

In this example, the ranking provided by positive separation measure and negative separation measure is not same; therefore, the R_3 TOPSIS is used. The worst alternate, i.e., alternate 5 is dropped and the whole procedure is repeated again.

Table 5 Ranking on the basis of relative closeness

Alternate	S^+	S^-	P_i	Ranking
A	0.0255	0.02203	0.4636	3
B	0.0307	0.02494	0.4486	4
C	0.0299	0.02866	0.4896	2
D	0.0260	0.03358	0.5632	1
E	0.0318	0.02488	0.4392	5

Table 6 Result of R2 TOPSIS

Alternate	S^+	Ranking	S^-	Ranking	Ranking by R_2 TOPSIS
A	0.0255	1	0.02203	5	Not applicable, it is suggested to use R_3 TOPSIS
B	0.0307	4	0.02494	3	
C	0.0299	3	0.02866	2	
D	0.026	2	0.03358	1	
E	0.0318	5	0.02488	4	

Table 7 Result of R3 TOPSIS

Alternate	S^+	Ranking	S^-	Ranking	Ranking by R_3 TOPSIS
A	0.0264	2	0.0219	4	3
B	0.0318	4	0.0226	3	4
C	0.0302	3	0.0239	2	2
D	0.0261	1	0.0349	1	1

Now the rank 1 provided by positive separation measure and negative separation is same. Therefore, alternate 4 is best alternate in this problem according to R_3 TOPSIS. Hence, the results are more reliable.

4 Conclusion

Effective decision making is essential for sustainable development of organization. TOPSIS method has received the attention of researcher and academician throughout the world.

- It is noticed that when the ranking of positive separation measure and negative separation measure is identical, then there is no need to find relative closeness value. This finding (R_2 TOPSIS) reduces the computational work involved in TOPSIS method.

- On the other side, when the ranking of positive separation measure and negative separation measure is not identical,and then it relies on the ranking provided by relative closeness, which is not a reliable attempt.
- So, new guidelines in the form of R_3 TOPSIS are presented. The use of R_3 TOPSIS increases the computational work, but the best alternate provided by R_3 TOPSIS is more realistic and reliable. The use of R_3 TOPSIS enhances the credibility of decision-making process and can be used effectively in any evaluation-based decision.

Acknowledgements The authors would like to thank copyright office, Government of India, for registering this concept under RoC No. L-75436/2018.

References

1. Hwang C, Yoon K (1981) Multiple attribute decision making: method and application, Multiple attribute decision making: method and application. Springer Publications
2. Yoon K (1987) A reconciliation among discrete compromise situations. J Oper Res Soc 38:277–286
3. Hwang CL, Lai YJ, Liu TY (1993) A new approach for multiple objective decision making. Comput Oper Res 20:889–899
4. Kumar R, Singh H, Dureja JS (2012) An approach to analyze logistic outsourcing problem in medium-scale organization by CFPR and VIKOR. J Manuf Technol Manag 23(7):885–898
5. Singh H, Kumar R (2011) Two-phase methodology for effective utilization of advanced manufacturing technologies using AHP And VIKOR. Mater Sci Eng 2(1–2)
6. Singh H, Kumar R (2013) Hybrid methodology for measuring the utilization of advanced manufacturing technologies using AHP and TOPSIS. Benchmarking: Int J 20(2):169–185
7. Singh H, Kumar R (2012) Selection of material for bicycle chain in Indian scenario using MADM approach. In: Proceedings of the World Congress on Engineering, vol 3, pp 1377–1381)
8. Singh H, Kumar R (2014) Selection of chain-material in automobile sector using multi attribute decision making approach. In: Annual meeting of the ISAHP. Grand Hyatt Hotel, Washington, DC
9. Chandel R, Singh H, Kumar R (2017) Performance evaluation of state-owned thermal power plants in northern India using DEA. Int J Global Energy Issues 40(6):380–399
10. Shyur HJ, Shih HS (2006) A hybrid MCDM model for strategic vendor selection. Math Comput Model 44(7–8):749–761
11. Gumus AT (2019) Evaluation of hazardous waste transportation firms by using a two step fuzzy-AHP and TOPSIS methodology. Expert Syst Appl 36(2):4067–4074
12. Baykasoğlu A, Gölcük L (2017) Development of an interval type-2 fuzzy sets based hierarchical MADM model by combining DEMATEL and TOPSIS. Expert Syst Appl 70:37–51
13. Başaran S, Haruna Y (2017) Integrating FAHP and TOPSIS to evaluate mobile learning applications for mathematics. Procedia Comput Sci 120:91–98

Thermohydraulic Performance of a Packed Bed Solar Energy Storage System

Harmeet Singh, R. P. Saini, and J. S. Saini

1 Introduction

The energy from sun has intermittent nature, often unpredictable and diffused; this makes the energy storage critically important. The storage of energy of a solar process is necessary in order to have uninterrupted supply of energy in the absence of availability of solar energy. Packed beds represent the most suitable storage units for air-based solar system. The heat transfer to and from a flowing fluid to a packed bed has been the subject of many theoretical and experimental investigations since Schumann's work [1]. The rate of heat transfer to or from the solid in a packed bed is a function of the physical and thermal properties of the fluid and solid. The temperature difference between the fluid and the solid, mass flow rate of the fluid and the geometric characteristics of the packed bed material are the governing parameters for performance of the system. Bed porosity is also an important parameter, which depends upon the shape and orientation of the packing material. The packing is usually random, where particles of apparently the same size and shape are packed in an arbitrary manner into the container. However, the material elements may also be arranged in a definite manner in the bed so as to obtain the desired bed porosity.

Major disadvantage of the packed bed storage system is considered to be the pressure drop in the bed, i.e. energy consumption by fan to propel the air through the

H. Singh (✉)
Guru Nanak Dev Engineering College Ludhiana, Punjab, India
e-mail: harmeetpawar@gmail.com

R. P. Saini
Department of Hydro and Renewable Energy, Indian Institute of Technology Roorkee, Roorkee 247667, India

J. S. Saini
Mechanical & Industrial Engineering Department, Indian Institute of Technology Roorkee, Roorkee 247667, India

H. Singh et al. (eds.), *Sustainable Development Through Engineering Innovations*, Lecture Notes in Civil Engineering 113,
https://doi.org/10.1007/978-981-15-9554-7_39

bed. This reduces the overall benefit of the solar energy storage system. Packing the large-sized elements of storage material could reduce pressure drop in the bed. The thermal performance of the system may deteriorate due to lesser area available for heat transfer. Kulakowski and Schimdt [2] emphasized that diameter of the storage elements and pressure drop through the bed are considered to be two parameters of primary importance in the design of the storage unit. Gauvin and Kutta [3] mentioned that the major operating cost in packed bed system is directly related to the pressure drop in the bed, which is difficult to predict with reliability when the elements in the bed depart from the spherical shape.

As the uses of large-sized elements reduce the frictional losses, Singh et al [4] proposed to investigate the heat transfer and friction characteristics of large-sized elements in low porosity range. An experimental study was conducted, and the results were reported for five different shapes of elements made of concrete. The storage elements investigated in the present study are shown in Fig. 1. The results reported by them were used to investigate the thermohydraulic performance.

2 Thermohydraulic Performance

The study of heat transfer and friction characteristics shows that an enhancement in heat transfer, in general, is accompanied with a corresponding increase in friction factor. Therefore, it is essential to determine the shape and void fraction, which will result in maximum enhancement in heat transfer with minimum friction penalty. In order to achieve the objective of thermal along with hydraulic performance, a thermohydraulic parameter was evaluated. This parameter evaluates the thermal enhancement over the pumping power. The thermohydraulic parameter has been defined as the recommendation given by Webb [5]:

$$\eta = \frac{\mathrm{St}}{(f)^{1/3}} \tag{1}$$

where,

St Stanton number
f Friction factor

The variation of the parameter with respect to sphericity is shown in Figs. 2 and 3 for void fractions 0.48 and 0.275, respectively. The thermohydraulic parameter increases gradually with an increase in the sphericity and is having a minimum value at sphericity of 0.65, i.e. the flattest rectangular element, and it has a maximum value for the spherical-shaped elements.

The point contacts in the spherical-shaped elements may be the reason for such a change as the heat transfer is enhanced by having a large surface of the elements in contact with the hot air, and since there are no flat surfaces and only point-to-point contacts are there with elements, the pumping power required is less and hence

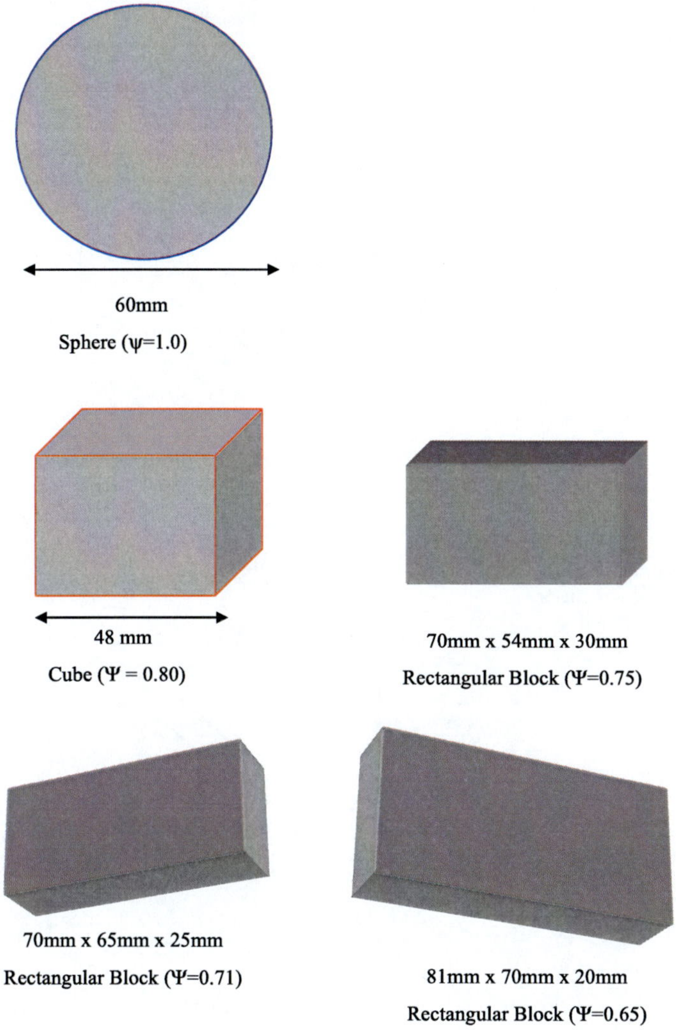

Fig. 1 Storage elements used in study by Singh et al. [4]

there is an overall increase in the thermohydraulic parameter. Whereas in the case of rectangular elements, there are surface contacts and due to this, the resistance to flow is increased with an increase in the flatness of the surface even though this causes an increase in turbulence and hence an increase in heat transfer, but along with an increase in heat transfer an increase in pumping power is also associated and hence the thermohydraulic parameter is minimum for the flattest surface, i.e. element having sphericity of 0.65. As the flatness decreases, the parameter also starts increasing gradually and finally has a maximum value corresponding to sphericity

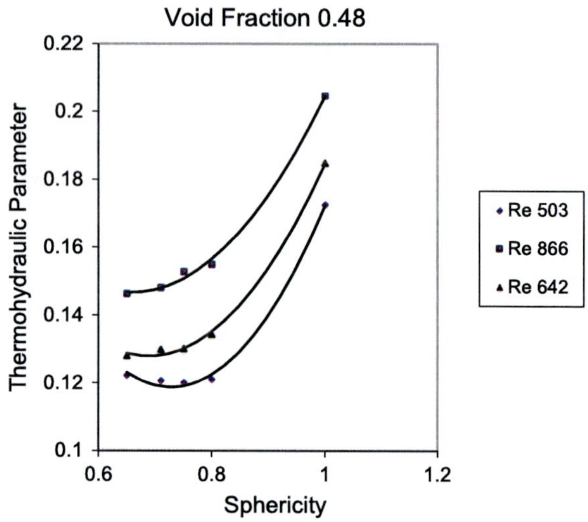

Fig. 2 Variation of thermohydraulic parameter with sphericity at void fraction 0.48

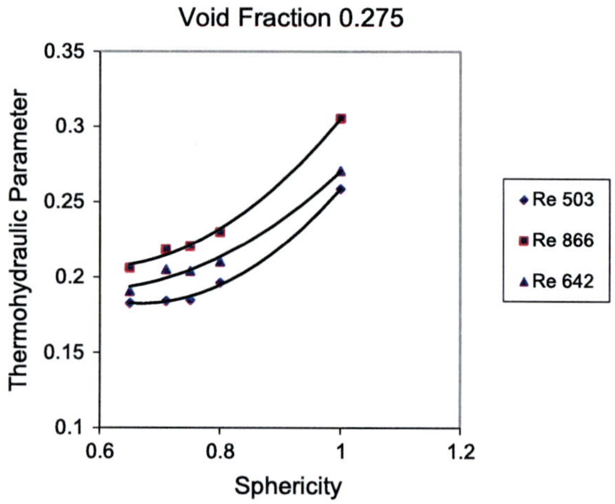

Fig. 3 Variation of thermohydraulic parameter with sphericity at void fraction 0.275

of 1.0. The thermohydraulic parameter increases with an increase in the Reynolds number. Similar trends have been obtained for other values of void fraction and Reynolds number.

Figures 4 and 5 show the variation of thermohydraulic parameter with void fraction for sphericity of 1.0 and 0.65, respectively. The results show that there is an increase in the value of thermohydraulic parameter with a decrease in the void fraction and is having a maximum value corresponding to the minimum value of void fraction, i.e. at 0.275 and the parameter decreases as the void fraction increase, and it has a minimum value corresponding to the maximum void fraction considered, i.e. 0.48. This may be due to the fact that the Nusselt number decreases sharply with an increase

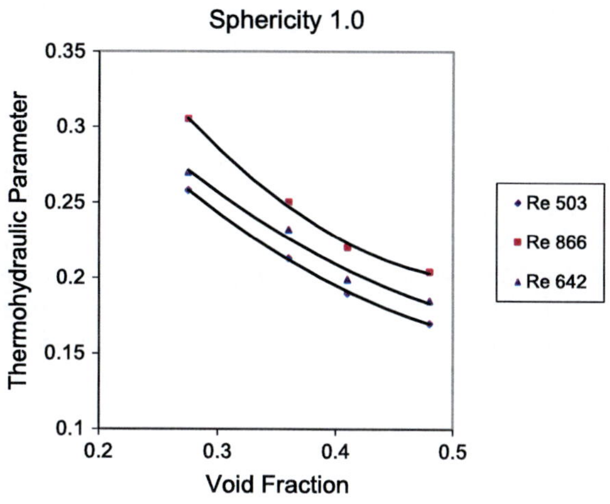

Fig. 4 Variation of thermohydraulic parameter with void fraction at sphericity 1.0

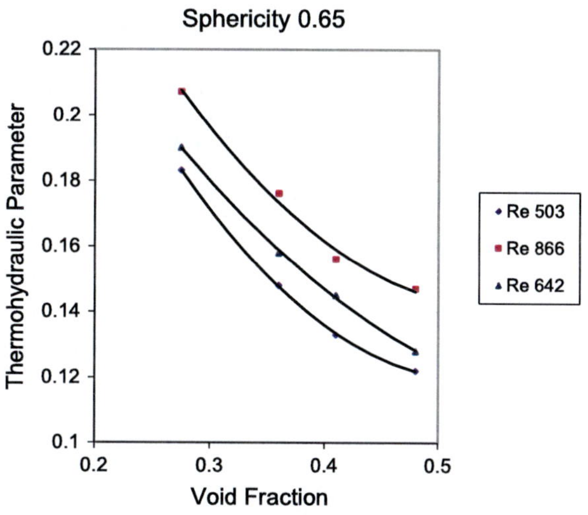

Fig. 5 Variation of thermohydraulic parameter with void fraction at sphericity 0.65

in void fraction, and hence, the heat transfer is reduced at a rapid rate; this decrease is enough to reduce the thermo hydraulic parameter even though the friction factor is also reducing along with void fraction. Similar trends have been observed for all other values of sphericity considered.

3 Conclusions

- Solar energy is one of the sustainable solutions available.

- An experimental study carried out on five different shaped storage elements in order to investigate the effect of sphericity and void fraction on heat transfer and friction characteristics in a packed bed was used to investigate thermo hydraulic performance.
- Thermohydraulic parameter has been evaluated and is found to have maximum value corresponding to the sphericity of 1.0, i.e. for spherical-shaped elements.
- Further as the sphericity decreases, the value of thermohydraulic parameter also decreases and is minimum for lowest sphericity of 0.65.
- The value of thermohydraulic parameter is maximum for the minimum value of void fraction of 0.275, and it decreases with an increase in void fraction.
- Therefore, it may be concluded that the thermohydraulic parameter is maximum for the spherical-shaped elements and void fraction of 0.275.

References

1. Schumann TEW (1929) Heat transfer: a liquid flowing through a porous prism. J Franklin Inst 208(3):405–416
2. Kulakowski BT, Schmidt FW (1982) Design of packed bed thermal storage unit for a solar system. Solar Energy Eng (ASME Trans) 104:223–228
3. Gauvin WH, Katta S (1973) Momentum transfer through packed beds of various particles in the turbulent flow regime. AIChE 19(4):775–783
4. Singh H, Saini RP, Saini JS (2013) Performance of a packed bed solar energy storage system having large sized elements with low void fraction. Solar Energy 87:22–34
5. Webb RL (1979) Towards a common understanding of the performance and selection of roughness for forced convection. In: Hartnett JP et al (eds) Handbook of studies in heat transfer: a festschrift for E.R.G. Eckert. Hemisphere, Washington, pp 257–272

Economically Sustainable Configuration Selection in Reconfigurable Manufacturing System

Prince Pal Singh, Jatinder Madan, and Harwinder Singh

1 Introduction

In recent years, manufacturers face challenges due to strict environmental regulations mandating them to focus on sustainable manufacturing (SM). According to US Department of Commerce [1], SM is the creation of cost-effective products that conserve energy as well as natural resources. To measure the progress in SM, manufacturers use sustainability indicators, which can be categorized into social, economic, and environmental. This paper focuses on the economic and environmental sustainability indicators by taking into account energy consumption and production costs.

Furthermore, the industry today needs to respond to variations in product variety and changing demand rates in a dynamic manner [2]. Therefore, to attain sustainable and competitive growth, manufacturers are attempting to adopt reconfigurable manufacturing system (RMS). RMS enables the system to produce new products and to change the production capacity of machines when required [3]. According to Koren et al. [4], "RMS is designed at the outset for rapid change in structure, as well as in hardware and software components, to quickly adjust production capacity and functionality within a part family in response to sudden changes in the market or

P. P. Singh (✉)
Department of Mechanical Engineering, I.K. Gujral Punjab Technical University, Kapurthala 144603, India
e-mail: princepal123@gmail.com

J. Madan
Department of Mechanical Engineering, Chandigarh College of Engineering and Technology (Degree Wing), Chandigarh 160019, India

H. Singh
Department of Mechanical Engineering, Guru Nanak Dev Engineering College, Ludhiana 141006, India

H. Singh et al. (eds.), *Sustainable Development Through Engineering Innovations*, Lecture Notes in Civil Engineering 113,
https://doi.org/10.1007/978-981-15-9554-7_40

regulatory requirements." RMS introduces flexibility in the manufacturing system using reconfigurable machine tools (RMTs). RMTs consist of basic and auxiliary modules. Where basic modules are structural in nature, such as base and columns and auxiliary modules are spindle units, work holding fixtures, tools, and tool holders [5]. In an RMS, for cost-effective reconfiguration at the system level, RMTs are added/removed to the system, whereas at the machine level, auxiliary modules are added/removed/readjusted to the basic module of an RMT. Thus, many machine configurations are possible using adjustable resources, such as basic and auxiliary modules. These machine configurations much depend on machine capabilities, such as production rate, operational ability, and machine availability. To measure RMS performance, manufacturers use RMS performance indicators, such as investment cost, system throughput, and changeover times.

Keeping in view the indispensability of RMS and sustainability challenges, it is highly desirable to develop a model that helps select a sustainable reconfigurable manufacturing system. This paper presents a single objective optimization problem that takes into account energy consumption cost as the environmental sustainability and investment cost, and maintenance cost as economic sustainability indicators to select RMTs in RMS. The proposed method would help process planners to determine a sustainable set of RMTs in a manufacturing line.

The rest of the paper is arranged as follows. Section 2 discusses the related literature to the RMT selection methodologies in RMS. Section 3 presents sustainability indicators for RMT selection. Section 4 presents a mathematical model for economically sustainable configuration design. Section 5 presents the implementation and results of the proposed work and the conclusions of this research work are given under Sect. 6.

2 Literature Review

This section discusses the literature related to RMS and SM performance indicators for configuration generation and optimization methodologies are used to select RMTs in an RMS.

Manufacturing system configuration is the arrangement of RMTs and their interconnections [6]. Xiaobo et al. [7] used investment cost of RMTs as a performance measure for configuration selection. Kimms [8] developed a mathematical model for the investment minimization of a manufacturing line. However, it focussed only on the investment cost of the manufacturing line. Spicer and Carlo [9] determined the minimum investment cost and reconfiguration cost using dynamic linear programming. Son [10] designed economic reconfigurable machining systems at an early stage of configuration design. However, they also consider cost of RMTs as an objective for configuration selection. Dou et al. [11] used nonlinear programming (NLP) to generate all the feasible arrangements of RMTs by considering the investment cost of RMTs. Furthermore, a genetic algorithm was used to generate optimized configurations. Yousuf and ElMaraghy [12] presented a methodology that helps to select the finest configuration considering reconfigurability cost and system availability using a discrete optimization procedure. Goyal et al. [5] selected the optimum set of RMT configurations based on machine reconfigurability, operational capability, and cost. It generates feasible machine configurations through a genetic algorithm. However, their study does not consider factors, such as machine availability and maintenance costs.

Huang et al. [13] explored SM performance of RMS using economic and environmental metrics. However, they used theoretical data to select a configuration and do not present an analytical method to determine economic and environmental indicators. Khezri et al. [14] presented RMS sustainability multi-objective mixed-integer programming model that selects process plans in RMS. They used completion time, production cost, and energy consumption as selection criteria. However, their study does not integrate energy consumption with total configuration cost.

The literature review has underlined the importance of RMTs in RMS and used different methodologies to select a set of RMTs for manufacturing configuration. However, there exist some limitations. First, most of the studies are focused on the investment cost of RMTs to make configuration selection decisions and factors such as maintenance costs are ignored. Second, most of the studies do not consider machine availability, which, however, directly affects the production rate. Third, a few studies considered sustainable manufacturing metrics, such as energy consumption, but does not integrate it with total configuration cost.

From the limitations mentioned above, it is desirable to develop a mathematical model that considers energy consumption cost, investment cost, and maintenance cost as sustainability indicators. Hence, the objective of the present study is to select an economically sustainable configuration using an optimized set of RMTs.

3 Sustainability Indicators

Reconfigurable machine tools perform many operations in their different machine configurations. These multiple configurations can handle various operational requirements of a part. However, these configurations vary in costs, such as investment, maintenance cost, and energy use cost. This section formulates three sustainability metrics, namely, machine investment cost, maintenance cost, and energy consumption cost. These sustainability indicators are discussed in the following paragraphs.

3.1 Machine Configuration Cost (MCC)

The combination of basic and auxiliary modules generates the machine configuration. Thus, the machine configuration cost is the sum of the investment cost of a basic module and auxiliary modules [15]. Equation (1) represents the economic sustainability indicator, namely, machine configuration cost in US$/hour.

$$\mathrm{MCC} = (\mathrm{BM}_c + \mathrm{AM}_c) / (\mathrm{PT} \times \mathrm{EHU}) \quad (1)$$

where BM_c is the basic module cost in US$,[1] AM_c is the cost of auxiliary modules in US$, PT is the payback period of 3 years, and EHU is the expected hour usage of the machine and equipment per year.

3.2 *Machine Maintenance Cost (MMC)*

Maintenance is a significant element of manufacturing lines. Machines undergo scheduled preventive maintenance and the cost incurred during that period is considered as machine maintenance cost [15]. Equation (2) determines the economic sustainability indicator, namely, machine maintenance cost in US$/hour.

$$\mathrm{MMC} = \frac{\mathrm{MYC}}{\mathrm{EHO}} \quad (2)$$

where MYC is the yearly maintenance cost in US$/year, EHO is an annual total estimated hour of operation.

3.3 *Energy Consumption Cost (ECC)*

It is the energy consumption cost of using a machine to perform operations on a part. Energy consumption cost is the product of machine power consumption by cost of electricity [16]. Equation (3) determines the environmental sustainability indicator, namely, energy consumption cost in US$/hour.

$$\mathrm{ECC} = \mathrm{PC} \times \mathrm{CE} \quad (3)$$

where PC represents power consumption an RMT in kWh, and CE is the cost of electricity in US$/kWh.

4 Mathematical Model for Economically Sustainable Configuration Design

This section formulates the total configuration cost (TCC), which combines the three sustainability indicators, namely, machine configuration cost, machine maintenance

[1] 1 US$ = INR 75.

cost, and energy consumption cost. Thus, the objective function is to minimize the TCC to select economically sustainable set of RMTs. The nomenclature of the mathematical model is: number of operations (O), number of machine configurations for a specific product (R), total configuration cost of rth RMT for carrying out oth operation on a product (TCC_{or}), production rate of rth RMT for carrying out oth operation on a product (PR_{or}), availability of rth RMT for carrying out oth operation on a product (A_{or}), demand rate required for oth operation (D_o), maximum number of machines allowed for operation o (N_o), and number of rth RMT for carrying out oth operation on a product (x_{or}). Equation (4) shows the objective function.

$$\text{minimize} \sum_{o=1}^{O} \sum_{r=1}^{R} \mathrm{TCC}_{or} \times x_{or} \tag{4}$$

4.1 Constraints

- Machine should satisfy the demand rate considering RMT availability for carrying out operations at each stage. Equation (5) ensures that the selected RMT meet the desired demand rate.

$$\sum_{o=1}^{O} \sum_{r=1}^{R} (\mathrm{PR}_{or} \times A_{or}) \times x_{or} \geq D_o \tag{5}$$

- Number of machines that can be installed at each stage is limited by Eq. (6), where x_{or} is an integer value.

$$0 \leq x_{or} \leq N_o \quad \text{and} \quad x_{or} \in Z^n \tag{6}$$

4.2 Optimal Configuration Design Using TLBO

Teacher-learner-based optimization (TLBO) is a metaheuristic technique based on knowledge transfer in a classroom environment [17]. It has two phases: teacher phase and the learner phase. In first phase, student acquires knowledge from teacher, and in the later phase, students interact between themselves and acquire knowledge from each other. This method is based on a greedy selection strategy. The pseudocode of the TLBO is as follows:

- User input: Objective function, lower and upper bounds, population size and number of iterations.
- Algorithm initialize with a random population (X)
- Using the above generated random population objective function is evaluated
- Teacher phase and learner phase loop starts in this step:

  for 1 to number of iterations (Teacher phase for loop)

    for 1 to number of population (Learner phase for loop)

    - X_{best} is selected
    - Determine mean of random population X_{mean}
    - $X_{new} = X_i + r\,(X_{best} - T_f \times X_{mean})$
    - Upper and lower bound applied to X_{new} and objective function is evaluated
    - Greedy selection: accept X_{new} if it is better than X_i
    - Select any solution randomly, Xp
    - Determine X_{new} as

      if objective function$_i$ < objective function$_p$

      $X^i_{new} = X_i + r\,(X_i - X_p)$

      else

      $X^i_{new} = X_i - r\,(Xi - X_p)$

      end
    - Upper and lower bounds applied to X_{new} and new objective function is evaluated
    - Accept X_{new} if it is better than X_i

    end

  end

Teacher Phase

Learner Phase

The next section implements the proposed mathematical model using TLBO algorithm using an industrial case study of an automotive piston.

5 Implementation and Results

The proposed method is implemented in MATLAB 2016a on a Microsoft Windows 10 platform with 8 GB RAM and 2 GHz processor. For this case study, an automotive piston is selected from a leading power train manufacturing unit located in Punjab (India). The company manufactures an extensive range of automotive pistons and exports to various countries, such as Japan and USA.

In the company, cast pistons are machined using a U-shaped cellular layout using CNC machine centers and RMTs. These RMTs can scale the production rate and/or change its functionality to adjust new products. Thus, there exist multiple configurations of RMTs with different machining capabilities, production rate, energy consumption, costs, and availability. The manufacturer wants to select an economically sustainable set of RMTs that has minimum total configuration cost and satisfies the demand rate. Table 1 shows the input data for the case study. This data is collected from the shop floor from the above-mentioned piston manufacturing unit.

The automotive piston is machined using specific operations. For the present case, the operation sequence is Op1-Op2-Op3-Op4-Op5. Figure 1 shows the automotive pistons and operations.

Table 1 Machine configurations, production rates, SM indicators, and machine availability

Configurations		Production rate per hour for operations					SM indicators			Machine availability
Machine	Conf	Op1	OP2	OP3	OP4	OP5	MCC/Hr	MMC/Hr	EUC/Hr	
L (broaching and lock-ring)	L11	90					0.435	0.317	0.234	0.85
	L12*	180					0.442	0.317	0.354	0.85
	L13			240	240		0.433	0.317	0.177	0.85
B (boring machine)	B11		180				0.146	0.023	0.265	0.90
	B12		360		200		0.153	0.023	0.38	0.90
	B13				360		0.119	0.023	0.152	0.90
	B14			180		144	0.122	0.023	0.152	0.90
	B15		360	300			0.132	0.023	0.259	0.90
C (center boring machine)	C11	180					0.223	0.317	0.379	0.80
	C12	240					0.227	0.317	0.506	0.80

* L12, where 'L' = broaching and locking machine, '1' = machine number, and '2' = machine configuration

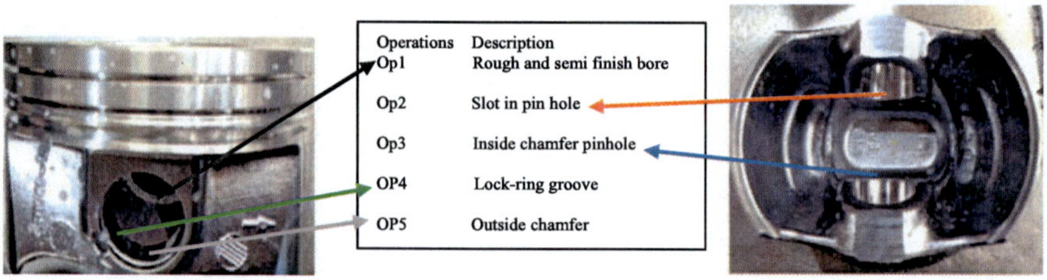

Fig. 1 Automotive piston for the case study

At each stage, only one operation is performed on the cast piston blank; therefore, the number of stages is five. Data from Table 1, which is computed using Eqs. (1)–(3), is input to the proposed methodology. The required demand rate is 400 pistons/hour. Due to space limitations, each stage can have a maximum of four machines. Next, there are specific parameters for TLBO algorithm: population size = 20 and number of iterations = 100. Afterward, the system determines sustainability indicators to assign RMTs at each stage. Table 2 shows the optimized total configuration cost for an automotive piston manufacturing line.

The results show that for carrying out operation sequence of OP1-OP2-OP3-OP4-OP5, three machines of type C with configuration C11, two machines of type B with configuration B15, two machines of type B with configuration B14 and B15 each, two machines of type B with configuration B13, and four machines of type B

Table 2 Total configuration cost for an automotive piston manufacturing line

Configurations		Number of machines required for each operation				
Machine	Conf	Op1	OP2	OP3	OP4	OP5
L	L11	–	–	–	–	–
	L12	–	–	–	–	–
	L13	–	–	–	–	–
B	B11	–	–	–	–	–
	B12	–	–	–	–	–
	B13	–	–	–	**2**	–
	B14	–	–	**1**	–	**4**
	B15	–	**2**	**1**	–	–
C	C11	**3**	–	–	–	–
	C12	–	–	–	–	–
MCC (US$/Hr.)		0.669	0.264	0.254	0.238	0.488
ECC (US$/Hr.)		1.137	0.518	0.411	0.304	0.608
MMC(US$/Hr.)		0.951	0.046	0.046	0.046	0.092
$TC_{operation}$(US$)/Hr		2.757	0.828	0.711	0.588	1.188
TCC (US$)/Hr		6.0720				

with configuration B14 are required at each stage. The total configuration cost for operation sequences OP1-OP2-OP3-OP4-OP5 is US$ 2.757, 0.828, 0.711, 0.588, and 1.188, respectively, which is the sum of economic and environmental sustainability indicators.

The economically sustainable configuration has a total configuration cost of US$ 6.0720. For the validation purpose, the proposed mathematical model is solved using mixed-integer linear programming using MATLAB 2016a inbuilt function, namely, *intlinprog*, and results are in line with the results from the proposed methodology using TLBO algorithm.

6 Conclusions

The proposed model combines three sustainability indicators, namely, machine configuration cost, machine maintenance cost, and energy consumption cost into one indicator, namely, total configuration cost. The proposed model is applied in an industrial environment and an automotive cast piston manufacturing configuration is selected for the implementation of the proposed model. Based on the proposed research work, the following points are concluded:

- The model selects an economically sustainable set of RMTs that has minimum total configuration cost.
- The proposed model helps the decision-makers in economically selecting number of RMTs during the reconfiguration period.
- The proposed model reduces time to make decisions, thus improves the efficiency of the RMS.
- The results from the proposed methodology are also validated using mixed-integer linear programming.

The proposed model can be extended by including more sustainable manufacturing indicators to improve the efficiency of the RMS.

References

1. US Department of Commerce (2010) How does commerce define sustainability. https://trade.gov/sustainability/.
2. Liu C, Ramirez-Serrano A, Yin G (2011) Customer-driven product design and evaluation method for collaborative design environments. J Intell Manuf 22(5):751–764
3. Shabaka AI, ElMaraghy HA (2008) A model for generating optimal process plans in RMS. Int J Comput Integr Manuf 21(2):180–194
4. Koren Y, Gu X, Guo W (2018) Reconfigurable manufacturing systems: Principles, design, and future trends. Frontiers Mech Eng 13(2):121–136
5. Goyal KK, Jain PK, Jain M (2012) Optimal configuration selection for reconfigurable manufacturing system using NSGA II and TOPSIS. Int J Prod Res 50(15):4175–4191

6. Koren Y, Hu SJ, Weber TW (1998) Impact of manufacturing system configuration on performance. CIRP Ann 47(1):369–372
7. Xiaobo Z, Jiancai W, Zhenbi L (2000) A stochastic model of a reconfigurable manufacturing system part 1: a framework. Int J Prod Res 38(10):2273–2285
8. Kimms A (2000) Minimal investment budgets for flow line configuration. IIE Trans 32(4):287–298
9. Spicer P, Carlo HJ (2007) Integrating reconfiguration cost into the design of multi-period scalable reconfigurable manufacturing systems
10. Son S (2001) Design principles and methodologies for reconfigurable machining systems.
11. Dou J, Dai X, Meng Z (2010) Optimisation for multi-part flow-line configuration of reconfigurable manufacturing system using GA. Int J Prod Res 48(14):4071–4100
12. Youssef AM, ElMaraghy HA (2007) Optimal configuration selection for reconfigurable manufacturing systems. Int J Flex Manuf Syst 19(2):67–106
13. Huang A, Badurdeen F, Jawahir IS (2018) Towards developing sustainable reconfigurable manufacturing systems. Procedia Manuf 17:1136–1143
14. Khezri A, Benderbal HH, Benyoucef L (2020) Sustainable Multi-objective Process Plan Generation in RMS Through Modelling Energy Consumption. Reconfigurable Manufacturing Systems: From Design to Implementation. Springer, Cham, pp 161–177
15. Pal Singh P, Madan J, Singh H (2020) Composite performance metric for product flow configuration selection of reconfigurable manufacturing system (RMS*)*. Int J Prod Res 1–21.
16. Singh PP, Madan J, Singh H (2019) Performance metrics for product flow configuration in a reconfigurable manufacturing system (RMS). In: International Manufacturing Science and Engineering Conference, vol 58745. American Society of Mechanical Engineers, pp V001T02A002
17. Rao RV, Savsani VJ, Vakharia DP (2011) Teaching–learning-based optimization: a novel method for constrained mechanical design optimization problems. Comput Aided Des 43(3):303–315

Day-Ahead Load Forecasting in PSPCL

Sukhjot Singh Sidhu and Arvind Dhingra

1 Introduction

In India, earlier SEBs were serving the consumers but there was problem of financial health of SEBs, inefficiency and poor Plant Load Factors (PLFs) and frequently SEBs were burdened with subsidies doled out by State Governments to the people. In order to bail out the SEBs, the Central Government came out with The Electricity Regulatory Commission Act, 1998 with independent State/Central Electricity Regulatory Commissions with power to determine tariff for SEBs/distribution utilities. With the initiation of reform process, some of the SEBs has been restructured leading to separation of generation, transmission and distribution. The enactment of The Electricity Act, 2003 led to the development of competitive electricity market in the country. The Act provides for non-discriminatory open access of the transmission and distribution system, de-licensing of generation, introducing captive power generation and recognized trading in power/electricity as a distinct activity. These provisions provide an enabling environment for development of bulk power market in India. Transmission has been recognized as separate activity and transmission utility is not allowed to trade in electricity which led to restructuring of all the SEBs into companies. The Act provides for non-discriminatory open access of the transmission and distribution system, de-licensing of generation, introducing captive power generation and recognized trading in power/electricity as a distinct activity. These provisions provide an enabling environment for development of bulk power market in India.

S. S. Sidhu
Guru Nanak Dev Engineering College, Ludhiana & Addl.SE, PSPCL, Ludhiana, India
e-mail: sidhusukhjot@gmail.com

A. Dhingra (✉)
EED, Guru Nanak Dev Engineering College, Ludhiana, India
e-mail: arvinddhingra@gmail.com

H. Singh et al. (eds.), *Sustainable Development Through Engineering Innovations*, Lecture Notes in Civil Engineering 113,
https://doi.org/10.1007/978-981-15-9554-7_42

Post-unbundling of the Punjab State Electricity Board, Generation and Distribution Utility, Punjab State Power Corporation Ltd (PSPCL) has been entrusted with the responsibility of Generation and Distribution functions for the State of Punjab. In line with the responsibilities assigned to the company, PSPCL has to undertake the following activities:

- Procuring power: Short-term, medium-term and long-term and also from renewable sources of energy to comply with Renewable Purchase Obligation (RPO)
- Arranging power during shortage periods from external sources within and outside the state, hence, maintaining the demand supply balance in the state (and minimize load shedding)
- Procuring and selling day-ahead power in exchange as per demand supply situation
- Executing necessary contracting agreements power purchase/sale agreements, transmission service agreements
- Complying with the regulatory requirement and maintaining merit order dispatch as mandated by the Punjab State Electricity Regulatory Commission
- Co-coordinating with generation, distribution and State Load Dispatch Center to ensure smooth operations of Transmission and Distribution System as per Grid Code.

2 Day-Ahead Forecasting

The need to accurately forecast day-ahead load arises on this account and effective market modeling is necessitated in Punjab on account of several reasons:

- **Huge intra-day variations**—Variation of over 2000 MW occurs during day and night in Punjab. With generation largely comprising base load plants and in the absence of sale of surplus power, PSPCL will face difficulty in managing its generation fleet.
- **Seasonality**—During summer when watering of paddy is done, the peak demand is of the order of 11,000 MW, whereas in winters during night hours the demand is of the order of 3000 MW.
- **Large cost outlay and high impact on the retail tariffs**—Power purchase cost is one of the largest cost components of PSPCL. Given the high impact of power purchase costs on utility finances, it is vital that these cost are controlled and optimized through a well-defined Power Procurement Plan.

- **Optimal Disposal of Surplus Power**—Day-ahead forecasting is very much essential for working out surplus and optimal disposal of same.

Based on the time scale used load forecasting techniques are of four types:

Very short-term load forecasting, short-term load forecasting, medium-term load forecasting and long-term load forecasting.

Generally, the load is considered block-wise for each 15 min time block which in the future is likely to be converted into 5 min time block. A five minute to thirty-minute ahead load forecast is called very short-term load forecast. This forecast is generally used for real-time control and security constraint. A one hour to one week ahead load forecasting is called short-term load forecasting (STLF) and is used for scheduling and day-ahead load forecasting comes under STLF. For maintenance scheduling, load forecasting ranges from a period of one week to few months, i.e., medium-term load forecasting (MTLF) is used. A few months to years ahead load forecasting is known as long-term load forecasting (LTLF) which is used for distribution/transmission planning.

Basic requirements for short-term load forecasting:

	Weather information	Economics	Land use	Updating cycle	Time scale
STLF	Essential	Non-essential	Non-essential	24 h	Fortnightly

3 Research Methodology

For the purpose of day-ahead forecasting, the maximum and the minimum temperature for major cities in Punjab was taken for the months of April and May 2018. The temperature maximum and minimum were given at different sections on the same Web page and the humidity was recorded at 7 equal intervals across each date so therefore an average of the seven readings was taken. Then, an average of the maximum and the minimum temperature was calculated, respectively, to get average maximum temperature and the average minimum temperature for that day. In the same manner, average humidity is obtained by taking the humidity of the major cities in Punjab. The impact of these four weather variable was studied with the help of regression analysis conducted on the average load (dependent variables) and the weather variables (independent variables).

4 Hypothesis

Taking load as the dependent variable and the weather variables of average maximum temperature, average minimum temperature, the average temperature and the average humidity as the independent variables the following null hypothesis and alternate hypothesis are developed:

H01: Average maximum temperature has no effect on the average load
Ha1: There is significant relation between average maximum temperature and average load
H02: Average minimum temperature has no effect on the average load
Ha2: There is significant relation between average minimum temperature and average load
H03 Average temperature has no effect on the average load
Ha3: There is significant relation between average temperature and average load
H04: Average humidity has no effect on the average load
Ha4: There is significant relation between average humidity and average load.

We shall be considering the analysis of average maximum temperature and average load only. All other analysis can be performed similarly.

5 Regression Analysis of Average Maximum Temperature and Average Load April 2018

Maximum temperature		Minimum temperature		Average temperature		Humidity		Avg. daily load	
Mean	35.47107	Mean	19.71786	Mean	27.45357	Mean	55.56071	Mean	4825.821
Standard error	0.55895	Standard error	0.611163	Standard error	0.467954	Standard error	2.693419	Standard error	97.85231
Median	35.05	Median	19.35	Median	27.15	Median	53.3	Median	4784.5
Mode	35.3	Mode	20.6	Mode	26.9	Mode	53	Mode	#N/A
Standard deviation	2.957683	Standard deviation	3.233969	Standard deviation	2.47618	Standard deviation	14.25223	Standard deviation	517.7857
Sample variance	8.747891	Sample variance	10.45856	Sample variance	6.131468	Sample variance	203.1262	Sample variance	268,102.1
Kurtosis	0.774988	Kurtosis	4.801105	Kurtosis	−0.0289	Kurtosis	−0.74845	Kurtosis	−0.16152
Skewness	0.001191	Skewness	1.516613	Skewness	0.176484	Skewness	0.268578	Skewness	0.631129
Range	13.6	Range	16.6	Range	9.8	Range	53.7	Range	1969
Minimum	27.7	Minimum	14.5	Minimum	22.3	Minimum	32.6	Minimum	3999
Maximum	41.3	Maximum	31.1	Maximum	32.1	Maximum	86.3	Maximum	5968
Sum	993.19	Sum	552.1	Sum	768.7	Sum	1555.7	Sum	135,123
Count	28	Count	28	Count	28	Count	28	Count	28

Summary output	
Regression statistics	
Multiple R	0.699741
R square	0.489637
Adjusted R square	0.47141
Standard error	379.7567
Observations	30
ANOVA	

	df	SS	MS	F	Significance F
Regression	1	3,874,045	3,874,045	26.863	1.68E-05
Residual	28	4,038,023	144,215		
Total	29	7,912,069			

	Coefficients	Standard Error	t Stat	P-value	Lower 95%	Upper 95%	Lower 95.0%	Upper 95.0%
Intercept	521.817	838.378	0.62241	0.53871	−1195.5	2239.16	−1195.5	2239.16
X Variable 1	122.064	23.5511	5.18295	1.68E-05	73.8218	170.306	73.8218	170.306

6 Interpretation of the Regression Analysis of Average Maximum Temperature and Average Load

Multiple R gives us the correlation coefficient between average maximum temperature and average load came out to be 0.69 for the month of April 2018.

The R squared value which is also known as the coefficient of determination is 0.48. It means that 48% of the change in average load is explained by the change in the average maximum temperature.

Significance The significance value of regression between average maximum temperature and average load the month of April 2018 is 0.000016. It is less than 0.05 Therefore, there is a significant relation between average maximum temperature and average load for the month of April; therefore, the null hypothesis is rejected in the alternate hypothesis is accepted.

7 Regression Analysis of Average Minimum Temperature and Average Load April 2018

Summary output	
Regression statistics	
Multiple R	0.757142
R square	0.573264
Adjusted R square	0.558023
Standard error	347.2528
Observations	30
ANOVA	

	df	SS	MS	F	Significance F
Regression	1	4,535,703	4,535,703	37.6143	1.28E-06
Residual	28	3,376,366	120,585		
Total	29	7,912,069			

	Coefficients	Standard error	t Stat	P-value	Lower 95%	Upper 95%	Lower 95.0%	Upper 95.0%
Intercept	2458.421	395.4236	6.217183	1.02E-06	1648.432	3268.41	1648.432	3268.41
X variable 1	121.1222	19.7491	6.133051	1.28E-06	80.66804	161.5764	80.66804	161.5764

8 Interpretation of the Regression Analysis of Average Minimum Temperature and Average Load

Multiple R gives us the correlation coefficient between average minimum temperature and average load came out to be 0.75 for the month of April 2018.

The R squared value which is also known as the coefficient of determination tells us how differences in one variable (here the average minimum temperature) explain the differences in another (here the average load). The R squared values is 0.57. It means that 57% of the change in average load is explained by the change in the average minimum temperature.

Significance The significance value of regression between average maximum temperature and average load the month of April 2018 is 0.0000012. It is less than 0.05 Therefore, there is a significant relation between average minimum temperature and average load for the month of April, therefore, the null hypothesis (H_{02}: Minimum temperature has no effect on the average daily load) is rejected in the alternate hypothesis (Ha2: There is significant relation between minimum temperature and load) is accepted. Therefore, it is concluded that the average minimum temperature has a significant impact on the average load.

9 Conclusions

The major conclusions from above are:

- The average maximum temperature significantly affects the average load
- The average minimum temperature significantly affects the average load
- The average temperature significantly affects the average load
- The average humidity temperature significantly affects the average load.

It is seen that accurate forecasting using the techniques mentioned in paper above not only helps in finding the relationship between various parameters but also helps in saving the money and making the power purchase more sustainable. This helps to make the electricity corporations sustainable in their business.

Bibliography

1. Calderon C, Serven L (2004) The effects of infrastructure development on growth and income distribution. https://elibrary.worldbank.org/doi/abs/https://doi.org/10.1596/1813-9450-3400
2. Chauhan R, Kuniyal JC, Pandey DC, Jamwal J, The worldwide historical facts behind the development of hydroelectric projects: a review https://www.eresearchco.com/ajabs/09%20RC%204_2_.pdf
3. AK Raja, Srivastava AP (2010) Power plant engineering, 4th edn

4. Singh A (2006) Power sector reform in India: current issues and prospects. https://www.sciencedirect.com/science/article/abs/pii/S030142150400254X
5. Power Sector in India: Growth, Policies and Challenges Semantic Scholar. Semantic Scholar, AI-Powered Research Tool. https://www.semanticscholar.org/paper/POWER-SECTOR-IN-INDIA%3A-GROWTH%2C-POLICIES-AND-Pal/a80f3ff2f19f444dd05affafcdb80302ab86f269
6. Dubash NK, Rajan SC (2001) Power politics: process of power sector reform in India. Econ polit wkly 36(35):3367–3390. doi:https://doi.org/10.2307/4411059. Accessed
7. Performance of Indian power sector during a decade under restructuring: a critique—science direct. ScienceDirect.Com, Science, Health and Medical Journals, Full Text Articles and Books. https://www.sciencedirect.com/science/article/abs/pii/S0301421503002830
8. Shahi RV Indian power sector: challenge and response : compilation of papers
9. Power sector reform in Orissa: an ex-post analysis of the causal factors—ScienceDirect. ScienceDirect.Com, Science, Health and Medical Journals, Full Text Articles and Books. https://www.sciencedirect.com/science/article/abs/pii/S030142150000063X
10. Impact Assessment of the Electricity Act 2003 on the Indian Power Sector—ScienceDirect. ScienceDirect.Com, Science, Health and Medical Journals, Full Text Articles and Books. https://www.sciencedirect.com/science/article/abs/pii/S0301421503003550
11. Status of Electricity Act, 2003: A Systematic Review of Literature—ScienceDirect. ScienceDirect.Com, Science, Health and Medical Journals, Full Text Articles and Books. https://www.sciencedirect.com/science/article/abs/pii/S0301421516306565
12. Assessment of Indian Power Sector Reform through Productivity Analysis: Pre and Post Electricity Act, 2003—IEEE Conference Publication. IEEE Xplore. https://ieeexplore.ieee.org/abstract/document/5484363
13. Willis HL (2002) Power distribution planning reference book. Revised and expanded, 2nd edn. Marcel Dekker, New York
14. Willis HL (2002) Spatial Electric Load Forecasting. Marcel Decker, New York
15. Papalexopoulas AD, Hesterberg TC (1990) A regression-based approach to short-term load forecasting. IEEE Trans Power Syst 5:1535–1547

The Future of Antenna Fabrication: 3D Printing Technology

Chahat Jain and Balwinder S. Dhaliwal

1 Introduction

In the course of recent decades, microstrip antennas have gathered a great deal of attention due to their low profile, low cost, ease of fabrication and conformability. 3D printed antennas that are fabricated using flexible/non-flexible substrates can even enhance the above-mentioned advantages to get a reliable, cleaner and faster technology for fabricating prototypes. A great zeal of interest and enthusiasm has been shown, both in industry as well as academia for such kind of electronics. Needless to say, that this research area now occupies the top of research pyramid of many national/international agencies. Moreover, flexible electronics, due to its ease of fabrication, low cost of manufacturing and generous availability of raw materials, have emerged as a perfect candidate for the future generation consumer electronics. Basically, 3D printing technology has emerged as a Silicon Valley start-up, Carbon Inc, which enables objects to rise from a liquid media continuously, thus, representing a fundamentally newer and a better approach to 3D printing [1]. A 3D object or prototype is the end product which builds up due to layering and piling up of many films. This technology therefore takes into account fast prototyping of an object with boundless outline, and the authority to plan and manufacture is left over to an individual user. 3D printing is also called as 'additive manufacturing' as it is the process of joining materials to make parts from 3D model data as a layer by layer technology in contrast to subtractive manufacturing or other formative manufacturing methodologies [2]. It can also reduce the usage of adhesives which are needed for

C. Jain (✉)
IKGPTU & Guru Nanak Dev Engineering College, Ludhiana, India
e-mail: chahatjain26@gmail.com

B. S. Dhaliwal
NITTTR, Chandigarh, India
e-mail: balwindersdhaliwal@nitttrchd.ac.in

H. Singh et al. (eds.), *Sustainable Development Through Engineering Innovations*, Lecture Notes in Civil Engineering 113,
https://doi.org/10.1007/978-981-15-9554-7_43

combing parts in order a get a finished product, thus reducing the wastage of materials, hence, aiding in a fast and speedy production which is of course not feasible with the conventional manufacturing processes which certainly involves expensive raw materials and costly machinery [3]. This technology has got impressive demonstrations and applications in different industries such as musical instruments, vehicles, housing usages and even in building materials. Although it has been commented that 3D printing technology holds the future and potential to create functional antennas and other electronic components (for creating functional antennas) at RF/microwave frequencies, however, its applicability is still to be explored [4].

The rapidly evolving 3D printing technology has some inherent advantages over the conventional methodologies that were followed for fabrication. Few of them include factors such as (1) this rapid prototyping topology has an ability to develop the prototypes same day, thus, saving the prototyping time. (2) This technology enables a designer with more degrees of freedom in terms of design and unit level customization. (3) Due to low cost production for initial prototyping, this technology provides edge over traditional ways in packaging as well as in the design of stacked interconnects, thereby producing less amount of waste. (4) Due to a wide range of materials available with unique properties, the designer gets more freedom to come up with materials which have good tensile strength. (5) As 3D printing enables low initial installation cost, it may pave way for processes delivering end use electronic packages. Apart from the positive side of 3D printing technique, there exists a darker side as well. Because 3D printing comes with plastic packaging of finished products, the results of it are obvious. Various issues observed with plastic packaging are: (1) Long-term usage and reliability are affected due to continuous exposure to environment thus resulting in moisture absorption and penetration, thereby deteriorating the quality of the product. (2) At higher frequencies, surface roughness plays a vital role challenging the reliability of the product. The design of antennas using 3D printing method has been progressively more required both in academic world and industry. The major advantage of using 3D printing technologies to fabricate microwave antennas include the rapid realization of designs without going through conventional processes such as machining and photolithography. Due to its evident advantages like time efficiency, environment friendliness and lighter weight as well as cost effective as compared to the conventional manufacturing techniques, it has become possible to develop almost any sort of complicated structure.

This article thus covers a brief design methodology of the antenna 3D printing process followed by the fabrication techniques that are used worldwide to develop such products. Last but not the least, the possible challenges and potential solutions are presented which provide sufficient knowledge to explore this field. A number of different materials and fabrication processes are reviewed in detail.

2 Design Methodology for 3D Printing Process

The methodology for designing a 3D printed antenna on flexible/non-flexible substrates is a systematic one. The complete procedure for designing, fabricating and testing of an antenna has been depicted in Fig. 1.

As drawn in the figure, the design methodology follows 5 vital stages, namely simulation of the designed antenna, 3D printing of the antenna, antenna curing, measurement of conductivity of the antenna fabricated and finally the measurement of the parameters of the antenna prototype produced [5]. The design and fabrication processes can be initialized with the help of simulation softwares available, viz., HFSS, CST, MWS, etc. After the simulation process is performed, the antenna's prototype is made by printing the simulated antenna using various kinds of 3D printers available. While printing, optimal parameters have to be taken care of. These parameters include number of printed layers, curing time, feed technique utilized and complexity of printed geometries over different substrates. After the optimal printing technique, curing has to be carried out using the high precision ovens. Curing is necessary as it helps achieving the required conductivity of the printed ink. To check the conductivity, probe technique is used in most of the cases [6]. The last step includes the measuring of antenna performance parameters such as resonant frequency, reflection coefficient, gain, VSWR and far-field pattern measurements.

The antenna design is uploaded to the printer software in Gerber format file. This file contains all the necessary related geometrical dimensions of the antenna

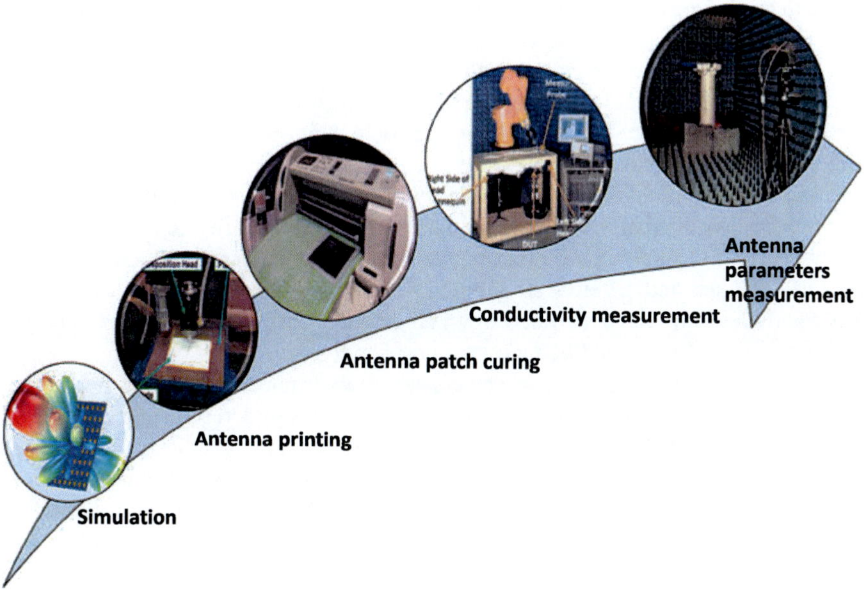

Fig. 1 Process depicting the design, fabrication and testing of a prototype

design. While the printing process proceeds, various things have to be taken care of, e.g., number of printed layers, heating temperature of the platen, number of jets used height of the cartridge from the platform, etc., even before actual printing takes place, many other steps need to be performed such as: (1) substrate must be cleaned and thoroughly dried, (2) the holder containing substrate should be heated to 50^0C so as to accelerate the evaporation of solvent. (3) Proper placement of drop from the nozzle should be taken care of, (4) precise distance between the nozzle tip and the substrate should be maintained and (5) droplet pitch values needed be decided accurately.

Curing process is necessary to ensure conductivity and physical continuity. It is usually done with the help of industrial ovens. A good cure is achieved after several trials and with controlled exposure times. Curing must be performed immediately after the completion of printing process, so as to avoid oxidation which may further lead to decreased conductivity.

3 Fabrication Techniques and Material

The development of 3D printing for manufacturing products has led to the emergence of a new market that not only affects the quality of end product but also presents products at a good price as well because of the usage of new materials. The evolution and success of product manufactured using 3D printed technology depends on the quality, strength and costs of material. Many options are available but one has to decide the right one depending on the required application. Materials like polylactic acid (PLA) and acrylonitrile butadiene styrene (ABS) can be used as they are soft and moldable and become rigid on cooling. The materials have to pass through three different tests, viz., initial extrusion through a plastic filament, then extrusion and trace binding and finally the end-use application during the process of 3D printing. ABS and PLA can be used as long-term materials, however, humidity exposure for a long time can have an effect on the quality of finished product [1]. Inkjet printing is a fast process that can be efficiently used in printing electronics on or in paper substrates. This technology will thus enable components such as antennas, memory, resistors, batteries and different sensors to be easily put in or on paper modules. [7]. This concludes that such a methodology will lead to a sustainable environment thus producing less pollution and being more eco-friendly.

The American society for testing and materials group, ASTM-F42 has given a set of standard methods that categorize additive manufacturing processes into various categories according to standard terminologies. The parameters considered for any typical AM process are speed of fabrication, resolution, cost, quality, surface finishing and strength of the part. Accordingly, the methods can be broadly classified as depicted in Fig. 2. The geometrical description for the same is shown in Figs. 3, 4, 5, 6 and 7

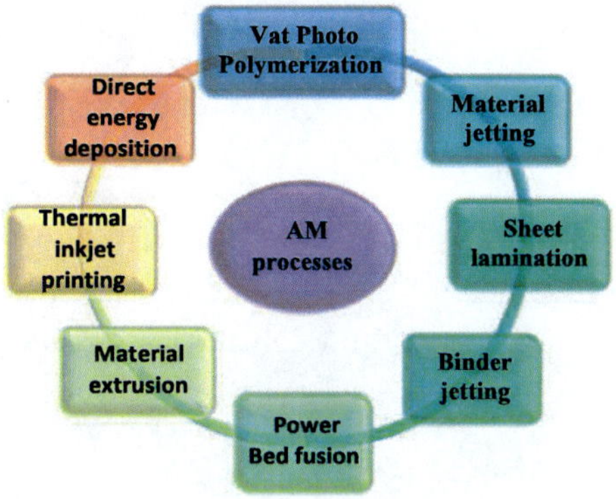

Fig. 2 Additive manufacturing processes

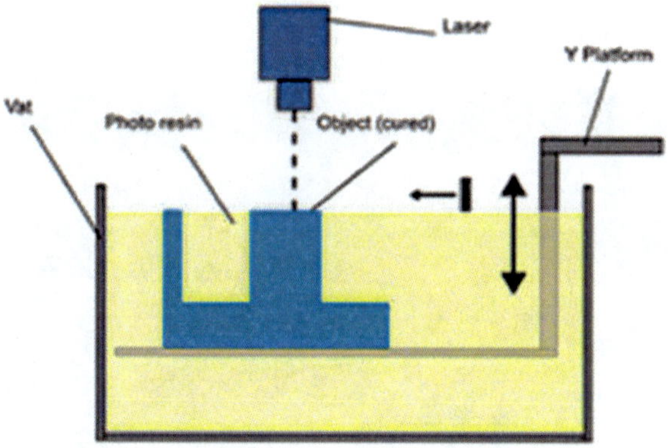

Fig. 3 Vat polymerization method [1]

3.1 Vat Photo-Polymerization

In this process, photo-curable resins are exposed to laser for the chemical reaction to happen. After which, they become solid. The chemical reaction is known as photo-polymerization. There must be sufficient cross-linking so that those molecules which are polymerized do not get redissolved into liquid monomers [8].

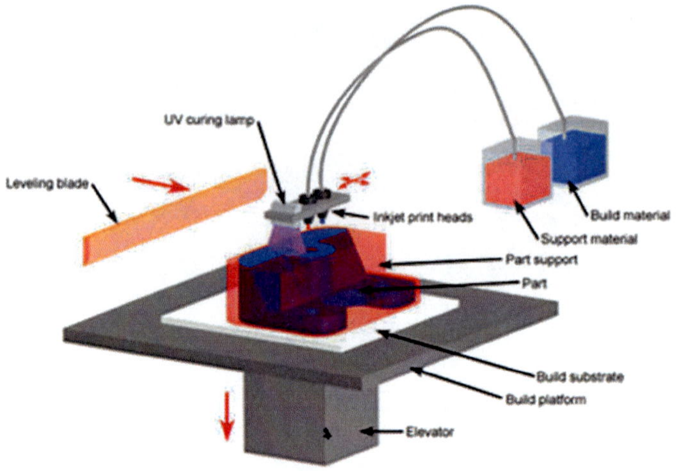

Fig. 4 Material jetting [1]

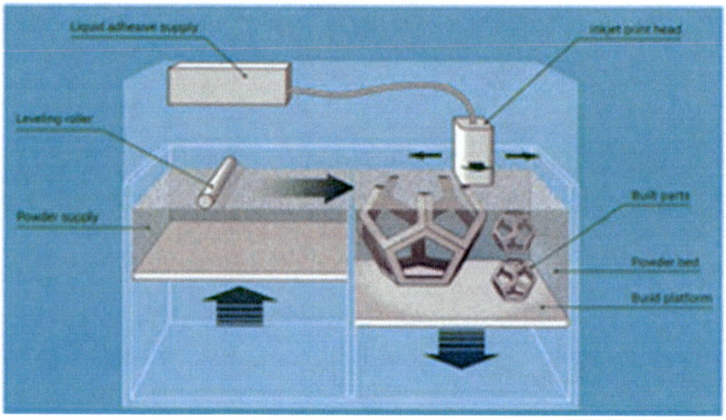

Fig. 5 Binder jetting [1]

Fig. 6 Fused deposition modeling [1]

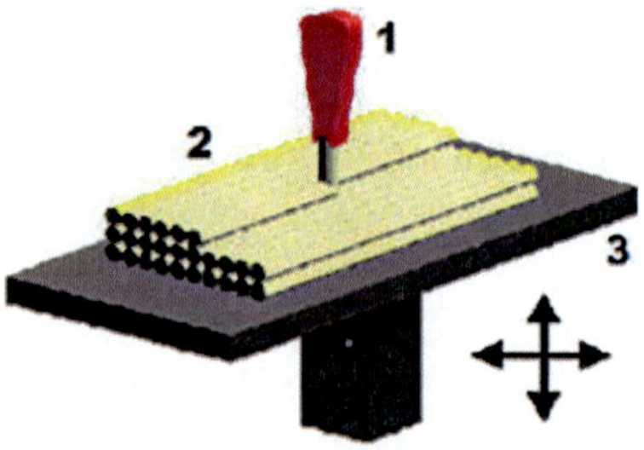

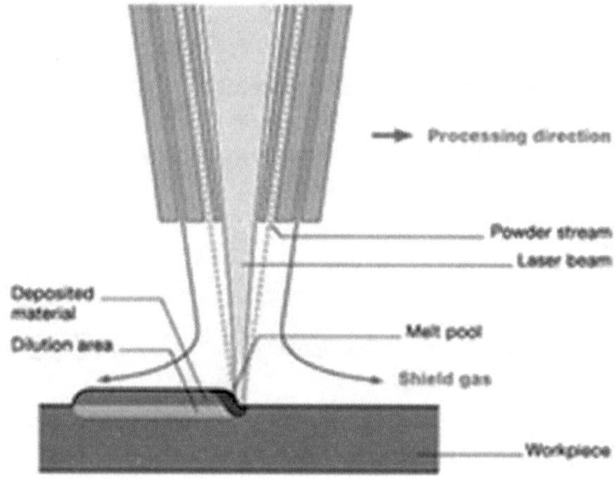

Fig. 7 Direct energy deposition [1]

3.2 Binder Jetting

In this AM technology, powder particles are joined together by selectively depositing a liquid bonding agent to develop a 3D part. This technology provides an edge in the sense that it has large build volume capacity, high print speed, design freedom and relatively low cost.

3.3 Material Jetting

In this, the previously deposited layers are softened by the deposition of liquid droplets on the working platform, which further turn into solid form as a single piece during the process of material jetting. Once the layers get deposited, the object is separated from the building platform for the removal of support material.

3.4 Sheet Lamination

Materials sheets are cut with the help of laser technique or by using ultrasound. Each layer is considered as a cross-sectional layer of solid object.

3.5 Power Bed Fusion

Thermal source like laser can be used to induce full/partial fusion between powder particles. The layer formed is then smoothened by moving a roller or blade recoater for adding another powder layer. The process of binding usually preferred for this technique is sintering and melting. Sintering used may be solid-state or liquid-state sintering.

3.6 Material Extrusion

In this technique, the printing material comes out through a nozzle on the application of constant pressure. After the material extrudes out of the nozzle, it gets deposited at a constant speed and fully turns into a solid form on the substrate. The new extruded material must bind with the previously deposited layer so that a solid part can be created and remain in that structure throughout the process.

3.7 Inkjet Printing

It is a 'non-contact' method that makes use of thermal, electromagnetic or piezo-electric technology to deposit droplets of ink onto the substrate. Size of the droplet can be varied by adjusting the applied temperature gradient, pulse frequency and ink viscosity. Because of its preciseness, this technology has been applied to the building of 2D/3D tissue and organs.

3.8 Direct Energy Deposition

The energy in this process is focussed onto a small region to heat up the substrate, and using a laser, melt the material that gets deposited. The laser technique has better resolution than an electron beam for melting of a substrate layer (Fig. 8).

4 Challenges and Potential Solutions

Although 3D printing is versatile and offers a variety of processes and techniques for structure building, but still it offers limited usage of materials for bio-printing.

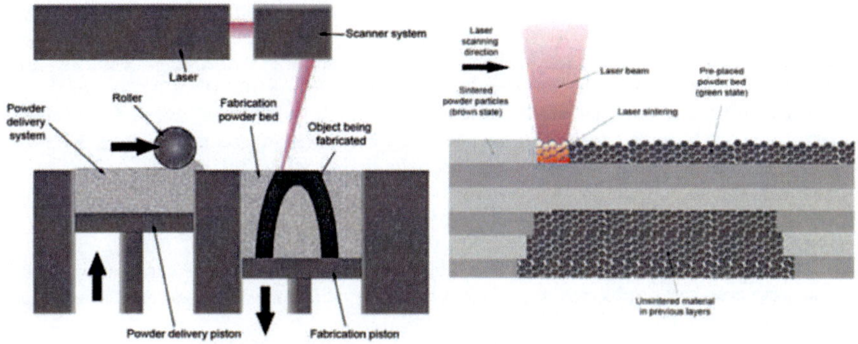

Fig. 8 SLS system schematic [1]

Thus, further development of novel materials is required. Major challenges include (1) lower energy consumption of AM processes with better speed and resolution, (2) development and utility of new printing materials with better chemical, mechanical and physical properties.

5 Conclusion and Scope in Future

3D printing is a captivating new technology that has inherently excited, enabled and engaged people of all ages. This technology has the potential to emerge as a breakthrough to meet the high expectations of the science and technology community. In coming future, 3D printed objects and antennas will be available for use by the researchers, educators and even the individuals in community. Future efforts may include the usage of hybrid AM processes for high dimensional accuracy, better finish and bigger build volumes. AM may be looked upon as a process to produce parts that can directly be used as end product. Thus, the future envisions an antenna which is integrated into 3D stacked package of digital, analog, components as a 3D printed package.

Acknowledgements Authors would like to thank Guru Nanak Dev Engineering College for providing financial assistance and facilities required for this work.

References

1. Srivastava VK (2017) A review on advances in rapid prototype 3D printing of multi functional applications. Sci Technol 7(1):4–24. https://doi.org/10.5923/j.scit.20170701.02
2. Lee JY, An J, Chua CK (2017) Fundamentals and applications of 3D printing for novel materials. Appl Mater Today 7:120–133. https://doi.org/10.1016/j.apmt.2017.02.004

3. Kaur A, Saini G (2018) 3D printed antennas: a review. Int J Eng Sci Comput 8(3)
4. Liang M, Shemelya C, Mac Donald E, Wicker R, Xin H (2015) 3D printed microwave patch antenna via fused deposition method and ultrasonic wire mesh embedding technique. IEEE Antennas Wireless Propag Lett 14. https://doi.org/https://doi.org/10.1155/2018/9815631
5. Al-Naiemy Y, Elwi TA, Khaleel HR, Al-Rizzo H (2012) A systematic approach for the design, fabrication and testing of microstrip antennas using inkjet printing technology. Int Sch Res Not (Hindawi) 2012(132465). https://doi.org/https://doi.org/10.5402/2012/132465
6. Monne MA, Lan X, Chen MY (2018) Material selection and fabrication processes for flexible conformal antennas. Int J Antennas Propag 2018(9815631). https://doi.org/https://doi.org/10.1155/2018/9815631
7. Rida A, Yang L, Vyas R, Tentzeris MM (2009) Conductive inkjet-printed antennas on flexible low-cost paper-based substrates for RFID and WSN applications. IEEE Antennas and Propag Mag 51(3). https://doi.org/10.1109/MAP.2009.5251188
8. Hoel KV, Ignatenko M, Kristoffersen S, Lier E, Filipovic DS (2020) 3D printed monolithic grin dielectric-loaded double-ridged horn antennas. IEEE Trans Antennas Propag 68(1). https://doi.org/10.1109/TAP.2019.2938563

Temperature Dependence of Multi-fin FinFET for Bulk and SOI Substrate at 20 nm Channel Length

Sarvesh Singh, Sandeep Singh Gill, and Navneet Kaur

1 Introduction

For sub-32 nm technology nodes, conventional metal oxide semiconductor (MOS) transistors will be difficult to survive due to various limitations posed by short channel transistors. These constraints known as short channel effects create a roadblock for the desired performance. To sustain the performance of nanoscale transistors, different types of gate or channel arrangements can present a viable solution. Placing gate on multiple sides of channel, in actual, increases the electrostatic control over charge carriers. Thus, in the upcoming future, trigate structure device, where gate is wrapped on three sides of channel, is going to be the most promising candidate to fulfill Moore's Law [1]. It has better gate controllability and lesser short channel effects that make it more reliable than CMOS. Latest research is focusing on this technology to serve IC industries more efficiently because multigate transistor structure offers sustainable solution while technological node scales down from 22 to 10 nm [2].

Fin-shaped field effect transistor, commonly referred to as FinFET is a transistor viewed as three-dimensional device in which gate electrode is placed on three sides of a thin conducting silicon channel. Gate controls the carrier movement inside the thin channel from top as well as sidewalls as shown in Fig. 1. The comparison between SOI and Bulk n-type FinFET has been studied to analyze the analog parameters for different temperature range. As multi-fin FinFET has multiple fins on single

S. Singh · N. Kaur (✉)
Guru Nanak Dev Engineering College, Ludhiana, India
e-mail: navneetkaur@gndec.ac.in

S. Singh
e-mail: singhsarvesh59@gmail.com

S. S. Gill
National Institute of Technical Teachers Training & Research, Chandigarh, India
e-mail: ssg270870@yahoo.co.in

H. Singh et al. (eds.), *Sustainable Development Through Engineering Innovations*, Lecture Notes in Civil Engineering 113,
https://doi.org/10.1007/978-981-15-9554-7_44

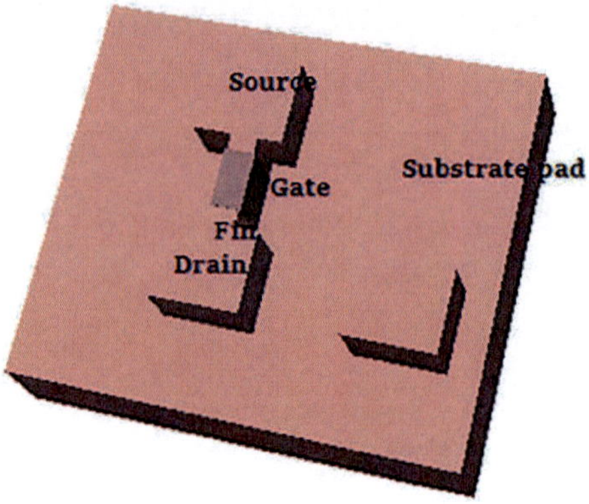

Fig. 1 Single fin rectangular FinFET

substrate, it gives more drain current than single fin FinFET. As temperature decides many characteristics of device, i.e., mobility, no of carriers, leakage current, etc., the study of effect of temperature is important to know the behavior of device [3, 4].

In brief, we propose a comparison Bulk nFinFET and SOI nFinFET. Study of different analog parameters, i.e., short channel effect (SCE), transconductance, early voltage, etc., have been carried out at temperature range from 200 to 400 K.

Device structure and result analysis are illustrated in Sects. 2 and 3, respectively. Further, the work is concluded in Sect. 4.

2 Device Structure

The Bulk nFinFET and SOI nFinFET have been designed on p-type Si substrate with 20 nm thick buried oxide. The device dimensions are as follows: channel length (L) of 20 nm each; fin width (W_{fin}) of 6.8 nm each; fin height (H_{fin}) of 30 nm each and gate material of hafnium oxide (HfO_2). The distance between consecutive fins is 24.8 nm. All these dimensions and other specifications of device like boundary labels along x-axis, y-axis and z-axis have been inputted in code of.inp file and then 3D designs of Bulk and SOI nFinFETs created are shown in Fig. 2. Figure 2a–b shows the 2-fin Bulk FinFET (fins are directly connected to substrate) and 2-fin SOI FinFET (buried oxide in between the fins and substrate), respectively. Figure 2c–d shows the 5-fin Bulk and SOI FinFETs, respectively. Besides that, the Bulk and SOI nFinFETs have 2 fins and 5 fins, respectively, in parallel. The doping concentrations used in substrate, fins, source and drain are 1×10^{15} cm^{-3}, 1×10^{15} cm^{-3}, 1×10^{19} cm^{-3}, 1×10^{19} cm^{-3}, respectively. These values are entered into code of.inp file using PROFILE command and setting the impurity type and peak concentration. The simulation is done on Cogenda TCAD [5].

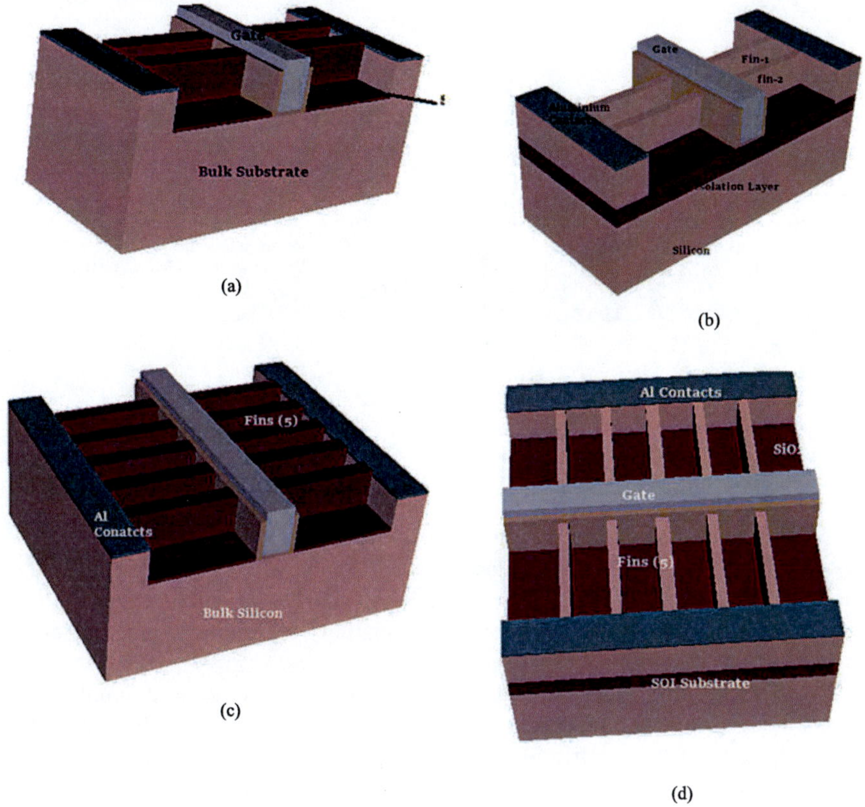

Fig. 2 Designed nFinFETs. **a** 2-fin bulk device. **b** 2-fin SOI device. **c** 5-fin bulk device. **d** 5-fin SOI device

3 Results and Discussion

Analog performance of designed Bulk and SOI n-type FinFETs has been studied for different temperature range, i.e., 200 to 400 K. Tungsten metal gate is used so as to avoid poly-depletion effects. Hafnium oxide (HfO_2) forms gate oxide to avoid gate tunneling currents. Source, drain and channel are of silicon material and insulator for SOI is silica (SiO_2). Metal contacts of aluminum have been provided to source/drain.

The simulation is carried with the variation of gate voltage (V_{GS}) from 0 to 1 V with the step size of 0.05 V. The drain voltage (V_{DS}) used is 50 mV for parameters like on-current, off-current, on/off-current ratio and SS. $V_{DS} = 1$ V & 50 mV is used for early voltage and transconductance.

The on-current (I_{ON}) as a function of temperature for different designed multi-fin FinFETs are shown in Fig. 3. I_{ON} is observed higher at room temperature for 2-fin and 5-fin SOI FinFETs. For Bulk nFinFETs, the on-current decreases with temperature increase which further follow decreasing trend with temperature. This

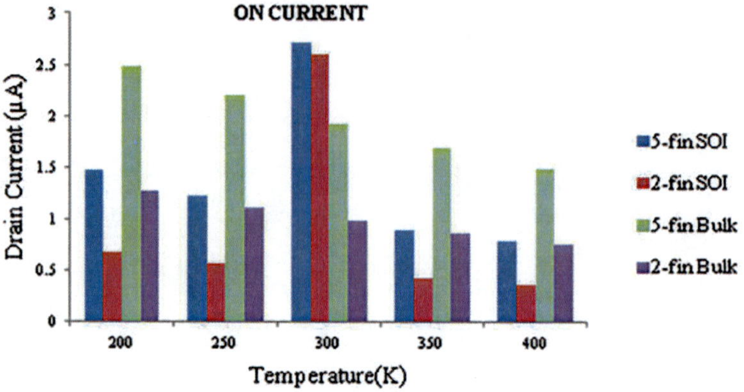

Fig. 3 On-current variations of designed bulk and SOI nFinFETs with respect to different temperatures

occurs due to the carrier mobility degradation with increasing temperature. While the on-current (I_{ON}) of SOI nFinFETs increases up to room temperature, and after that it decreases with increment in temperature because of dependence on temperature. Such demonstration of device proves its sustainability for below 32 nm gate length devices.

The leakage current is exponentially dependent on temperature. It doubles with every 10 degree rise in temperature [6]. The 5-fin and 2-fin Bulk FinFET show the maximum leakage current than SOI nFinFETs. Also, leakage current increases with increase in temperature in all designed multi-fin FinFETs as shown in Fig. 4, also observed in published studies [7].

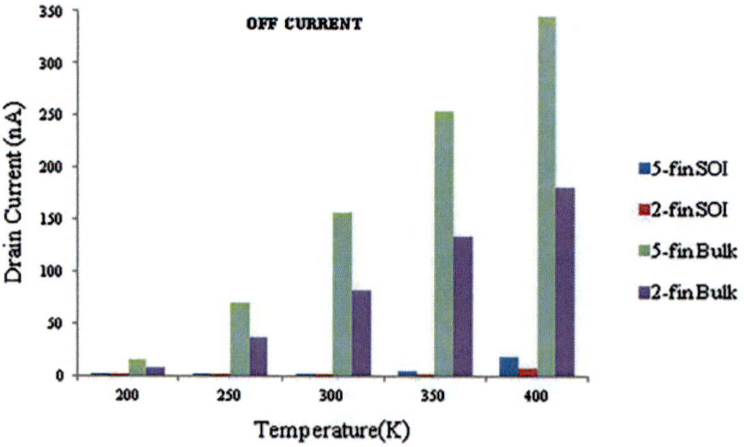

Fig. 4 Off-current variation of designed SOI and bulk nFinFETs

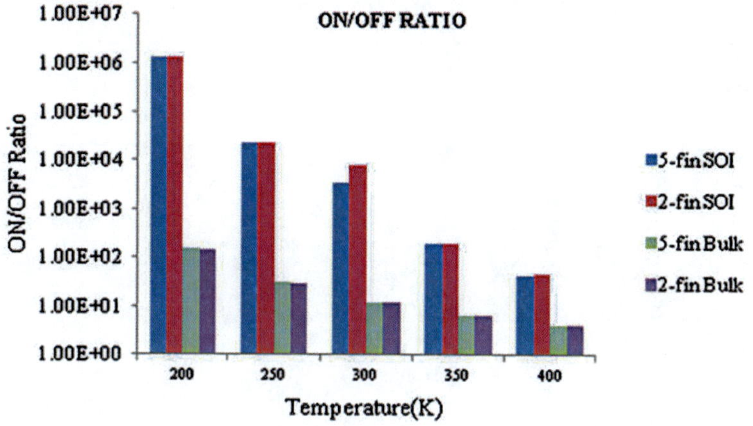

Fig. 5 On/off-current ratio of designed nFinFETs with respect to different temperature range

Due to large leakage current in Bulk nFinFET, its performance also gets decreased. 2-fin and 5-fin devices with oxide substrate show the minimum leakage current around room temperature and show the little increment after room temperature.

High on/off-current ratio indicates more switching speed of the device. Figure 5 shows the temperature effect on on/off-current ratio of designed devices. The on/off-current ratio is higher in case of SOI nFinFETs and lower in Bulk nFinFETs. On/off-current ratio inversely depends upon the temperature. Increment in temperature decreases the on/off-current ratio so the speed gets affected. It can be observed that the SOI nFinFETs have the highest switching rate due to their low leakage current, thus showing its ability to sustain at lower technology nodes and performing better. Bulk nFinFETs have the worst switching speed due to high leakage current.

Transconductance represents the change in output current for the change in input voltage. It represents the transfer of conductance. Figure 6 shows the transconductance changes of different designed SOI and Bulk nFinFETs. 5-fin SOI FinFET shows 2.72μS as the highest value of transconductance at room temperature because of high mobility of carriers at 300 K. Except at room temperature, bulk nFinFETs shows the highest transconductance.

The subthreshold swing (SS) is defined as how much change in gate voltage occurs corresponding to decade change in the drain current [8]. The short channel effects arise as the distance between the source and drain shortens. These effects are kept minimum to achieve the desired results. SS should be as low as possible. In Fig. 7, SS of designed nFinFETs is shown as a function of temperature. The SS is minimum at 200 K and increases with rise in temperature. 2- and 5-fin SOI FinFETs show the minimum SS. For 2-fin SOI FinFET, it starts with 40 mV/decade and ends at 110 mV/decade. For 5-fin SOI FinFET, it starts from 41 mV/decade and ends at 95 mV/decade. Bulk nFinFETs show the highest SS. 5-fin Bulk FinFET starts with 130 mV/decade and ends at 310 mV/decade which is very high than the normal

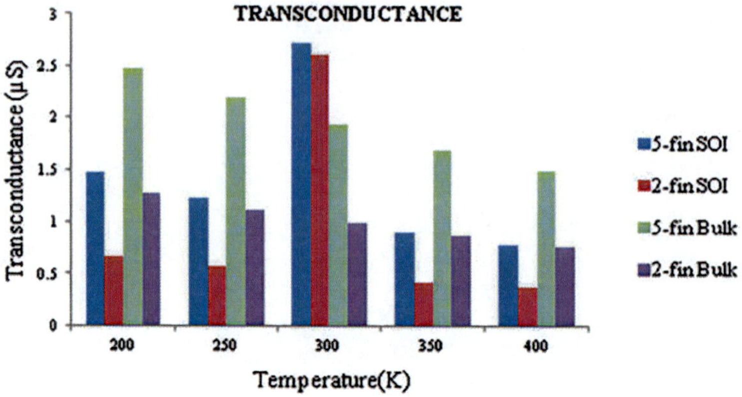

Fig. 6 Transconductance variations with respect to temperature of designed SOI and Bulk nFinFETs

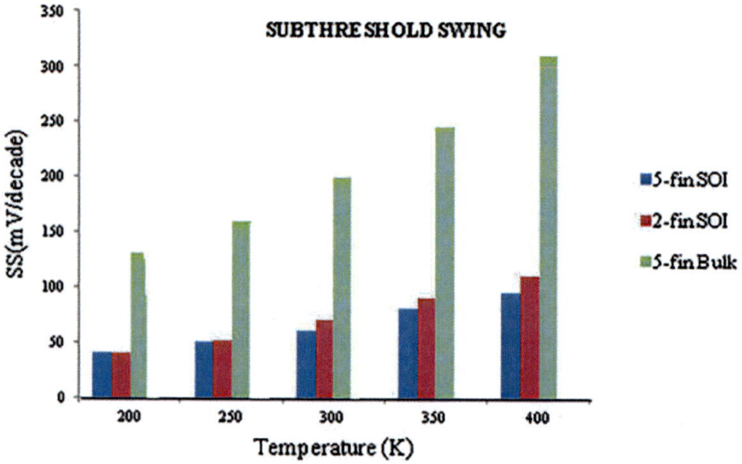

Fig. 7 Subthreshold swing variation with temperature of designed nFinFETs

value, i.e., 60 mV/decade. The SS of 2-fin Bulk FinFET starts with 100 mV/decade at 200 K and up to 300 K, it is at 200 mV/decade.

Early voltage (V_{EA}) measures the slope of drain current when plotted against drain voltage for fixed value of gate voltage. It measures how much channel length is modulated with scaling.

$$V_{\mathrm{EA}} = I_D / g_D \tag{1}$$

where g_D is transconductnce at constant V_{GS}.

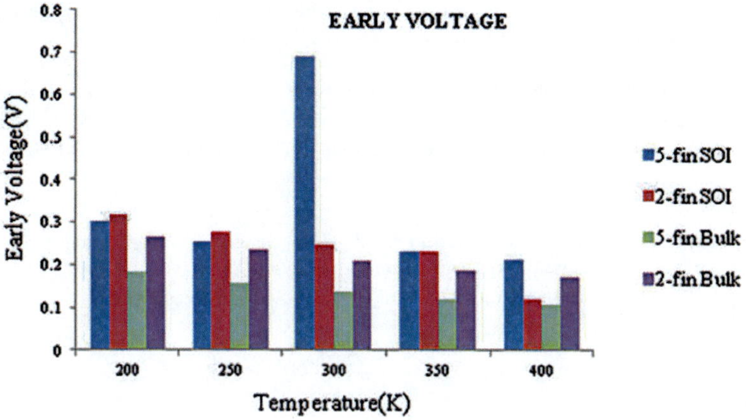

Fig. 8 Early voltage variation as a function temperature for designed nFinFETs

Degradation in early voltage can be seen with the increase in temperature. 5-fin SOI FinFET shows the highest V_{EA} (0.69 V) at room temperature. For 2-fin SOI, 5-fin Bulk and 2-fin SOI FinFETs, it decreases with increase in temperature as shown in Fig. 8.

4 Conclusion

In this work, FinFET device has been designed for channel length of 20 nm by placing two and five fins on a conducting channel and also the substrates on which channel formed were bulk silicon and SiO_2 insulator. Temperature under which the simulation was set up, varied from 200 to 400 K. Various performance metrics of designed multi-fin FinFET devices like subthreshold swing, on-current, off-current, on/off-current ratio, early voltage, transconductance were investigated for a particular set of input process parameters like fin height, fin width, doping concentration, channel length and fin pitch.

- It has been observed that subthreshold swing, on/off-current ratio and early voltage are improved for 2-fin and 5-fin devices when constructed on insulator substrate in comparison with bulk substrate. On the other side, these multi-fin transistors when built on bulk-Si resulted in more effective drain current, i.e., on-current but at the cost of leakage current. Reduced leakage current is obtained with inverse relation to temperature in 5-fin SOI device.
- Transconductance is more for 5-fin FinFET when formed over bulk silicon as device operates at lower value of temperature, but decreases with the increase in temperature. Thus, corresponding to a set of process parameters, fin-shaped transistors are providing sustainable performance even at nanometer scale,

where conventional MOS structures could offer only degraded performance and fabrication is also compatible with standard process.

As a future perspective, spacer engineering can be incorporated in multi-fin FinFETs. Changing the fin shape from rectangular to triangular can result in enhanced performance. Influence of gate stacking can be investigated in fin-shaped transistors with more than 5 fins.

Acknowledgements Authors would like to thank Guru Nanak Dev Engineering College for providing the facilities required to conduct this proposed work.

References

1. Chiarella T et al (2010) Benchmarking SOI and Bulk FinFET alternatives for planar CMOS scaling succession. Solid-State Electron 54:855–860
2. Simoen E, Claeys C, Oliveira AV, Agopian PGD, Martino JA (2015) High temperature influence on analog parameters of bulk and SOI nFinFETs. IEEE EUROSOI-Ultimate Integr Silicon 15:6911–6914
3. Pradhan KP, Priyanka S, Sahu PK (2016) Temperature dependency of double material gate oxide (DMGO) symmetric dual-k spacer (SDS) wavy FinFET. Superlattices Microstruct 89:355–361
4. Atalla Y, Hashim Y, Ghafar ANA, Jabbar WA (2019) A temperature characterization of (Si-FinFET) based on channel oxide thickness. TELKOMNIKA 17(5):2475–2480
5. Cogenda User's Guides. https://www.cogenda.com/article/downloads. Accessed 13 Dec 2019
6. Oxner ES (1988) FET technology and application. CRC Press, New York
7. Bhattacharya D, Jha NK (2014) FinFETs: from devices to architectures. Adv Electron 365689:1–21
8. Gaurav A, Gill SS, Kaur N (2015) Performance analysis of rectangular and trapezoidal TG bulk FinFETs for 20 nm gate length. In: IEEE INDICON Conference

Assessment of Crop Residues for Power Generation and Optimal Sites for Biomass Power Plants Using NDVI and Landsat8 Bands in Punjab, India

Harpreet Singh Dhaliwal, Yadwinder Singh Brar, and Gursewak Singh Brar

1 Introduction

Electrical energy is the most important part of today's life. Electrical energy is required for the social, economical, industrial and fundamental development of any country state or area [29]. Poor electricity supply affects the health, education, industrial and agricultural activities in rural as well as in urban areas. Demand of electrical energy is rising day by day throughout the world. India is also a big consumer of electrical power [17]. Most of electrical energy produced in India is dependent upon fossil fuels [4]. It is a fear that the conventional sources of generating electrical energy will be depleted soon [21]. According to the World Energy Forum, the conventional sources of energy in the country will deplete soon in the coming few decades [5]. The combustion of fossil fuels produces greenhouse gases, which further contribute to various environmental and ecological issues. Today we need to switch to non-conventional energy so that we may fulfil our energy needs [18]. Switching to non-conventional energy will also reduce the stress on fossil fuels. Biomass is an important non-conventional source of energy. Biomass is also considered environmental-friendly substance [24]. Carbon dioxide is emitted from the burning of crop residue. The same amount of CO_2 is balanced by the crops during

H. S. Dhaliwal (✉)
Inder Kumar Gujral Punjab Technical University, Kapurthala, Punjab, India
e-mail: dhaliwal361@gmail.com

Y. S. Brar
Department of Electrical Engineering, Inder Kumar Gujral Punjab Technical University, Kapurthala, Punjab, India
e-mail: ysbrar@ptu.ac.in

G. S. Brar
Department of Electrical Engineering, Baba Banda Singh Bahadur Engineering College, Fatehgarh Sahib, Punjab, India
e-mail: gursewak.singh@bbsbec.ac.in

H. Singh et al. (eds.), *Sustainable Development Through Engineering Innovations*, Lecture Notes in Civil Engineering 113,
https://doi.org/10.1007/978-981-15-9554-7_45

their growth. Agricultural crop residue is an important contributor of biomass. Agricultural crop residue is an important source of electrical energy generation. Many developing and developed countries have adopted the practice of electrical power generation from the agricultural waste [19]. Many researchers have calculated the agricultural biomass residues for the purpose of electrical energy generation using various techniques [13, 25, 28, 30]. Some of the researchers have also tried to find the locations of biomass power plants using various techniques [7, 27, 31, 33]. Although the collection of agricultural residues in the field and the transportation of this fuel to biomass power plant are two major issues for the successful and smooth running of these plants [26]. due to the problems of collection and transportation, the farmers are left with no other option rather than burning this crop residue in their fields. The burning of crop residual in Northern parts of India is the major issue nowadays [12]. Many environmental issues originate from the open field burning of these crop residues [22]. But if this crop residue is used to generate electrical power, it will not only eliminate the problem of open field crop residue burning, but it will also provide manifold benefits to the society. Also, the generation of decentralized electrical power will reduce stress on the grid and improve the efficiency of power system [3]. Generating electrical power from crop residues will also provide some extra income to the farmers as well as it will generate employment for the local people. The aim of this study is to calculate the surplus agricultural residues and to find optimal locations of biomass power plants in the Punjab state, which is the most agricultural rich state in the country.

1.1 Profile of Punjab State

Punjab is the land of five rivers situated in north India. The total geographical area of the state is 50,362 km^2, which is 1.5% of the total land area of India. The state is called 'The bread basket of India' [23]. There is diversity in terms of the terrain of the state. It includes the Himalayan foothills in the north-eastern part and the sandy deserts in the south-west part. Most of the land of the state is fertile and irrigated. The state bears the temperature up to 48 °C in summers to the chilling temperatures of 0 °C in the winters. The annual rainfall in the state is averaged between 460 to 960 mm. Punjab is one of the highest crop producers of India. Despite of very less geographical area, the state produces 10% rice, 22% wheat and 13% cotton of India [6]. As per census of 2011, the population of the state is approximately 280 million [15]. The cropping intensity of the state is around 189%. This cropping intensity is the highest in the country. The state has 22 districts, five divisions and 12,581 villages. Almost 65% of the total population of the state is residing in the villages. Agriculture is the chief profession of the people of Punjab. Agriculture in the state is extremely concentrated in terms of agricultural inputs.

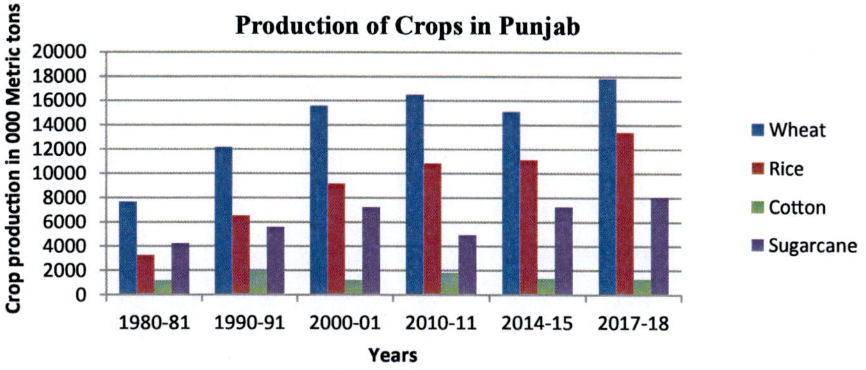

Fig. 1 Production of crops in Punjab [1]

1.2 Agricultural Structure of Punjab

Punjab is a recognized state for its agricultural activities in the country. Punjab is the largest contributor of food grains to the central pool, due to which it is called 'The bread basket of India'. Highly mechanised farms, adequate irrigation, use of fertilizers and hard work of farmers make this state to capable of producing various crops at very high yielding rates [32]. There are two main seasons of crops in the state, Rabi and Kharif season. Kharif is the crop which is grown in summer and harvested before the winters. Rabi crops are sown in winters and harvested before the summer season. The major crops grown in the state are wheat, rice, cotton, sugarcane, mustard, etc. (Fig. 1).

The green revolution leads manifold production of food grains in the state. The food grain production shoot up to 31.67 million tonnes in year 2017–2018, while the food grain production was 3.16 million tonnes in the year 1960–1961. The production of wheat escalated to 17.83 million tonnes in 2017–2018, which was 7.67 million tonnes in 1980–1981. The production of rice rose up to 13.38 million tonnes in 2017–2018 as compared to 3.23 million tonnes of 1980–1981. The production of sugar cane and cotton has shown changing scenario in past decades [1].

1.3 Crops and Crop Residues

The major crops grown in Punjab can be subdivided as per two main seasonal crops such as Rabi crops and Kharif crops. The main Rabi crop produced in Punjab is wheat. Wheat is the main source of food for the people of Punjab. Besides this cotton and mustard are also Rabi crops. The main Kharif crop of Punjab is rice. Sugarcane and maize are also grown in the Kharif season.

Crop residues are the leftover material which is the left behind in the fields after the harvesting of crops. Crop residues are generally left in the form of straw, husk, stalks, cobs, leaves, etc. The combine harvesters are used on the large scale for harvesting of crops in the state. These combine harvesters leave heavy amounts of crop residues in the fields [11].

1.4 Crop Residue Burning

The mechanised harvesting of wheat and rice crops produces heavy amounts of residue in the fields. The residue of wheat is used to make fodder for the cattle. But the residue of the rice crop cannot be used to make fodder due to high content of silica in it [14]. After harvesting of the wheat crop, there is enough time of 30–40 days to prepare the field for sowing of the rice crop. On the other hand, this time frame is limited to merely 10–15 days to prepare the field for sowing of the wheat crop, after harvesting of the rice crop [10]. Having no effective use of the residue of the rice crop, high labour cost for collection and transportation and the short time span between sowing of the wheat crop, farmers prefer to burn the rice residues in their fields [2] (Fig. 2).

25% of the wheat residue and 95% of the rice residue are burnt by the farmers of Punjab in their fields [16]. The burning of these crop residues create serious environmental, social, ecological issues like increased air pollution, rise in road accidents due to poor visibility, increased rates of pulmonary and cardiovascular diseases in people.

Fig. 2 Burning of crop residual in open fields

2 Methodology

The area of Punjab state is 50,362 km^2. The state has 22 districts. Pathankot is the smallest district of state with 929 km^2 area. Ludhiana is the largest district of state with 3767 km^2 of land area. The main purpose of this study is to calculate the crop residue and find the optimal locations of biomass power plants in the state. Hence, we need to know the vegetation area in the state. To enumerate the vegetation area, NDVI prediction is taken along with superpixeling of NDVI channels. For the projection of NDVI, we use landsat8 bands, which are taken from USGS server [8] (Fig. 3).

The NDVI-GIS model considered adherent to year 2017–2018. Eight bands of landsat8 are manipulated to get the NDVI model. Each band of landsat8 has a distinct range of wavelengths. To generate the NDVI channel band 4 having wavelength 0.632–0.680 μm and band 5 having wavelength 0.8452–0.885 μm are used. The NDVI channel is generated from Eq. (1).

$$\text{NDVI Channel} = \frac{\text{RED Channel} - \text{NIR Channel}}{\text{RED Channel} + \text{NIR Channel}} \tag{1}$$

Owing to a large area, superpixels of satellite images are created using the concept of simple linear iterative clustering (SLIC) [20]. The RGB channel is converted to superpixels model. This model is further converted to the NDVI heat map.

$$V_i = \left[\frac{\sum_{k=1}^{m} \text{pos}_m}{\sum_{p=1}^{n} \text{pos}_n}\right] \times \text{avg}(\text{NDVI}_r) \tag{2}$$

where V_i represents the vegetation normalization index of ith superpixel. m is the number of positions in the ith superpixels. n represents number of positions in the district where ith superpixel locates. NDVI_r shows the mean value of NDVI band of rth district.

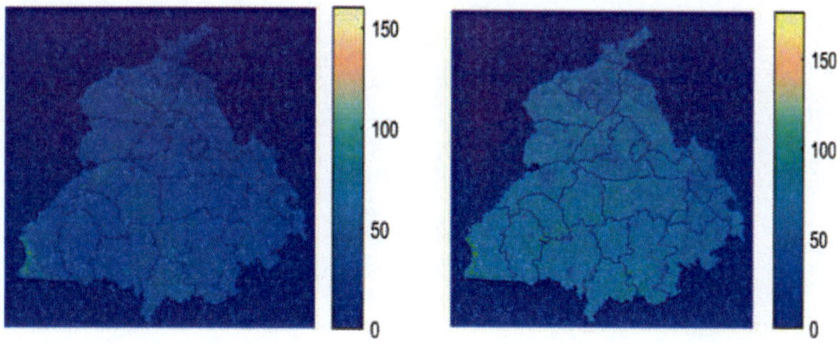

Band 4(0.632-0.680μm) Band 5(0.845-0.885μm)

Fig. 3 4th and 5th band of landsat8.

2.1 Assessment of Crop Residue

By using the superpixel region, the vegetation index is found and calculated. The crop data is taken from Punjab statistics report [9]. Main crops considered in this study are wheat and rice. Rest all crops are not considered in the study due to low production of crop residues from these crops. Figure 4a–b shows the production of wheat and rice crops in Kt per year by using the mapping of lansat8 bands with data. To find the amount of crop residue, we need to know the (CRR) crop residue ratios of these crops. It is the ratio of grain production of the crop to the ratio of its residue. Table 1 gives the information about the CRR values of different crops [6]. The landsat8 bands are mapped with the data and crop residue calculations are done for the state. Table 2 shows crop residues in all 22 districts of the state. It can be seen that the total crop residue generation comes to be 41,521 Kt for the year 2017–2018. It can be seen that Sangrur and Ludhiana districts are the topmost producers of crops as well as crop residues in the state. Pathankot and SAS Nagar are the districts with small production of grains as well as crop residues.

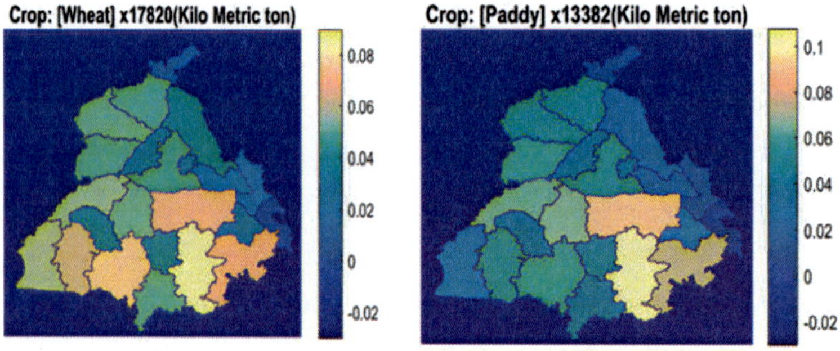

(a) (b)

Fig. 4 Crop production in Punjab (2017–2018).

Table 1 Crop residue ratios [6]

	Kharif	Rabi
CRR	Rice	Wheat
Straw	1.20	1.15
Husk	0.16	0.16
Stalk	0.00	0.00
Total	1.36	1.36

Table 2 Districtwise crop residue production in Punjab

Districts	Wheat		Rice		Total grain production (in 000 metric tons)	Total residue production (in 000 metric tons)
	Grain production	Residue production	Grain production	Residue production		
Amritsar	677	920	920	1205	1597	2125
Barnala	565	768	609	797	1174	1565
Bathinda	747	1015	1335	1748	2082	2763
Faridkot	511	694	619	810	1130	1504
Fatehgarh Sahib	395	537	454	594	849	1131
Fazilka	355	482	1007	1319	1362	1801
Ferozpur	842	1145	977	1279	1819	2424
Gurdaspur	643	874	876	1147	1519	2021
Hoshiarpur	292	397	620	812	912	1209
Jalandhar	715	972	861	1127	1576	2099
Kapurthala	513	697	523	685	1036	1382
Ludhiana	1224	1664	1296	1697	2520	3361
Mansa	499	678	890	1165	1389	1843
Moga	869	1181	908	1189	1777	2370
Pathankot	86	116	165	216	251	332
Patiala	1001	1361	1223	1602	2224	2963
Rupnagar	170	231	326	427	496	658
Sangrur	1442	1961	1599	2094	3041	4055
SAS Nagar	108	146	248	324	356	470
SBS Nagar	282	383	385	504	667	887
Sri Muktsar Sahib	687	934	1108	1451	1795	2385
Taran Tarn	759	1032	871	1141	1630	2173
Total	13,382	18,188	17,820	23,333	31,202	41,521

2.2 *Optimal Sites for Biomass Plants and Their Collection Centres*

To generate electrical power from crop residues, it is first important to collect crop residues at the collection centres and then to transport this biomass to the biomass power plants. For effortless conveying of crop residues to the biomass plants, optimal locations of collection centres and biomass plants need to be found. In the present work, the BAT optimization technique is adopted to find the optimal locations of collection centres and biomass power plants.

$$\text{dist}_{(c,i)} = \sqrt{(\text{cc\_}H_c - \text{SP\_}x_i)^2 + (\text{cc\_}y_c - \text{SP\_}y_i)^2} \tag{3}$$

$$\text{CR}_c = \text{if dist}_{(c,i)} = \min\{\text{dist(all}, i)\}$$

$$\text{then CR}_c = \text{CR}_c + \text{CR}_i$$

$$\text{else CR}_c = \text{CR}_c + 0 \tag{4}$$

$$\text{fitness\_ val} = \sum_{C=1}^{C=C_{\text{no}}} \left|\text{CR}_c - \text{avg}_{\text{CR}}\right| \tag{5}$$

where $\text{dist}_{(c,\, i)}$ denotes Euclidean distance between cth collection centre to ith super pixel centre position.

$\text{cc\_}x_{c,}$ $\text{cc\_}y_c$ shows x and y coordinates of cth collection centre.

$\text{SP\_}x_{i,}$ $\text{SP\_}y_i$ shows x and y coordinates of ith super pixel centre position.

CR_c represents crop residual of cth collection centre.

avg_CR = (total crop residual/number of optimal collection centres).

3 Results and Discussion

This study aimed on the assessment of crop residue and to find optimal sites of biomass power plants in the state along with the collection centres for the collection of crop residues. The experimental work was tested on computer the algorithm proposed for 6000 iterations and 20 bats. Locations of 20 biomass power plants and 50 collection centres were optimized using BAT optimization. Figure 5a shows the optimal location of all the 50 collection centres in the state. Figure 5b shows the optimal locations of 20 biomass power plants in the state. Total net surplus crop residue is 13.965 Mty^{-1}. This crop residue can be used as fuel in all the 20 biomass power plants. 20 biomass power plant locations were optimized and found that the total generating potential of available crop residue comes to be 1837.53 MW of electrical power. The largest power plant is of the capacity 331.35 MW, which is plant number 7. It is located in the Ferozepur district and it has four collection centres. The smallest power plant is of the capacity 20.08 MW, which is plant number 17. It is located in the Fazilka district and it has one collection centre.

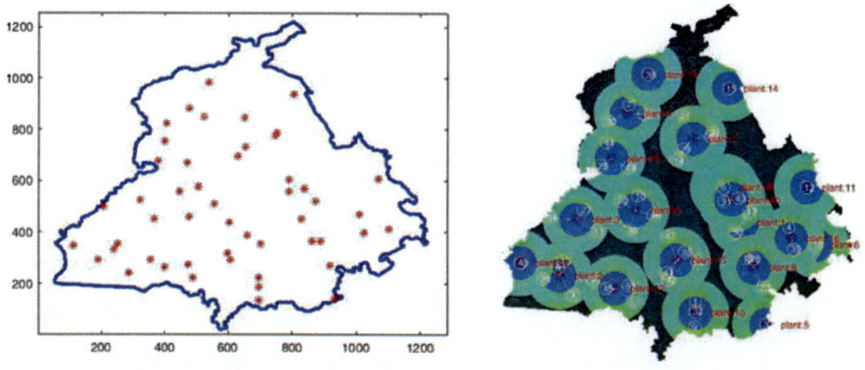

(a) Optimized locations of 50 Biomass Collection centres

(b) Optimized locations of 20 Biomass Power Plants

Fig. 5 Optimized locations of biomass collection centres and power plants

4 Conclusion

It is concluded from the study that:

- Crop residues produced annually in the state has a capacity to produce almost 1837.53 MW of electrical power.
- 20 biomass power plants can be set up in the state at various geographical locations.
- 50 biomass collection centres are optimized for the collection and storage of crop residues.
- Sustainability can be achieved both in terms of electrical energy production and in terms of environmental conditions.

5 Future Scope

The present study can further expand to find the other benefits of biomass power plants such as the reduction in environmental pollution, saving of coal in the power generation with crop residue, employment generation due to the set-up of biomass power plants and financial benefits to the farmers of state from the sale of crop residue. Similarly, the study can be further expanded to locate the sites of biomass power plants in other states of the country.

Acknowledgements The authors are grateful to the USGS for data and IKG Punjab Technical University, Kapurthala, Punjab for providing the required facilities and support for completion of this research work.

References

1. Agricultural Production and Cropping Pattern. (n.d.) Retrieved 17 May 2020, from https://punenvis.nic.in/index2.aspx?slid=5622&mid=1&langid=1&sublinkid=939
2. Ahmed T, Ahmad B, Ahmad W (2015) Why do farmers burn rice residue? Examining farmers' choices in Punjab, Pakistan. Land Use Policy 47:448–458
3. Bhattacharyya SC (2014) Viability of off-grid electricity supply using rice husk: a case study from South Asia. Biomass Bioenerg 68:44–54
4. Central Electricity Authority (2015) Government of India Ministry of Power Monthly Report on Broad Status of Thermal Power Projects in the Country, August 2015. Central Electricity Authority, pp 1–56
5. Channi HK (2015) Biomass Waste Potential Assessment in the Patiala District of Punjab. Int J Comput Appl 1–6.
6. Chauhan S (2012) District wise agriculture biomass resource assessment for power generation: a case study from an Indian state, Punjab. Biomass Bioenerg 37:205–212
7. Dhaliwal HS, Brar YS, Brar GS (2020) Optimization for biomass based plant localization using NDVI super pixels for Punjab State, India. Int J Adv Sci Tech 29(6s):2723–2733
8. Earth Explorer (n.d.) Retrieved 17 Feb 2020. From https://earthexplorer.usgs.gov/
9. Economic Adviser, Government of Punjab (2018) Statisticacal abstract of Punjab
10. Gupta N (2019) Paddy residue burning in Punjab understanding farmers' perspectives and rural air pollution. Ceew1–44
11. Gupta PK (2004) Residue burning in rice-wheat cropping system: causes and implications. Curr Sci 87(12):1713–1717
12. Gupta R (2012) Causes of emissions from agricultural residue burning in North-West India: evaluation of a technology policy response. SANDEE 66:1–25
13. Hiloidhari M, Baruah DC (2014) GIS mapping of rice straw residue for bioenergy purpose in a rural area of Assam, India. Biomass Bioenerg 71:125–133
14. IARI (2012) Crop residues management with conservation agriculture : potential, constraints and policy needs. Indian Agric Res Inst 7:32
15. Indian Administrative Service. (2011) Census of India 2011 provisional population totals. Accessed Apr 2014
16. Kaur A (2017) Crop residue in Punjab agriculture-status and constraints. J Krishi Vigyan 5(2):22–26
17. Kumar S, Bhattacharyya B, Gupta VK (2014) Present and future energy scenario in India. J Inst Eng (India): Series B 95(3):247–254
18. Maity SK (2016) Bio-fuels: green energy for development. Int Edu Res J 2(3):100–102
19. McKendry P (2002) Energy production from biomass (part 1): overview of biomass. Biores Technol 83(1):37–46
20. Achanta R, Shaji A, Smith K, Lucchi A, Fua P, Süsstrunk S (2012) SLIC superpixels compared to state-of-the-art superpixel methods. IEEE Trans Pattern Anal Mach Intell 34(11):2274–2281
21. Shafiee S, Topal E (2009) When will fossil fuel reserves be diminished? Energy Policy 37(1):181–189
22. Shaik DS, Kant Y, Mitra D, Singh A, Chandola HC, Sateesh M, Babu SS, Chauhan P (2019) Impact of biomass burning on regional aerosol optical properties: a case study over northern India. J Environ Manag 244(December 2018):328–343
23. Sharma AR, Kharol SK, Badarinath KVS, Singh D (2010) Impact of agriculture crop residue burning on atmospheric aerosol loading—A study over Punjab State, India. Annales Geophysicae 28(2):367–379
24. Shukla PR (1997) Biomass energy in India: policies and prospects. In: Biomass Energy: Key Issues and Priority Needs, Conference Proceedings, pp 357–376
25. Singh J, Panesar BS, Sharma SK (2008) Energy potential through agricultural biomass using geographical information system-A case study of Punjab. Biomass Bioenerg 32(4):301–307
26. Singh J, Panesar BS, Sharma SK (2010) A mathematical model for transporting the biomass to biomass based power plant. Biomass Bioenerg 34(4):483–488

27. Singh J, Panesar BS, Sharma SK (2011) Geographical distribution of agricultural residues and optimum sites of biomass based power plant in Bathinda, Punjab. Biomass and Bioenergy 35(10):4455–4460
28. Singh J (2013) Evaluation of crop residue potential for power generation for Indian State Punjab. IOSR J Agric Vet Sci 4(4):99–112
29. Singh J (2016) A roadmap for production of sustainable, consistent and reliable electric power from agricultural biomass—an Indian perspective. Energy Policy 92:246–254
30. Singh L, Singh J (2015) Assessment of crop residue potential for power generation using geographical information system. J Sci Ind Res 74(1):34–37
31. Singh R, Brar GS (2020) Optimized locations for biomass power plants using ant lion optimizer in District Bathinda, Punjab, India 1. 11(1):469–475
32. Singh Y, Humphreys E, Kukal SS, Singh B, Kaur A, Thaman S, Prashar A, Yadav S, Timsina J, Dhillon SS, Kaur N, Smith DJ, Gajri PR (2009) Crop performance in permanent raised bed rice-wheat cropping system in Punjab, India. Field Crops Res 110(1):1–20
33. Zhang F, Johnson DM, Sutherland JW (2011) A GIS-based method for identifying the optimal location for a facility to convert forest biomass to biofuel. Biomass Bioenerg 35(9):3951–3961

Location Optimization of Biomass-Based Power Projects

Ram Singh and Gursewak Singh Brar

Abbreviations

GIS	Geographic information system
NDVI	Normalized difference vegetation index
CRR	Crop-to-residue ratio
IISc	Indian Institute of Science
PEDA	Punjab Energy Development Agency
Kt	Kilo ton
MW	Mega Watt

1 Introduction

The human civilization started with the use of energy. It is assumed that with the start of human civilization, the use of energy started. With the development of society, the humans started to use energy in various day-to-day activities. Today, it is hard to imagine the human life without the use of energy during whole day for various personal and official activities. If we have a look around us, we can notice that there are a number of activities in which the human beings use energy, i.e., cooking, transportation, entertainment, official works, etc.

R. Singh (✉)
IKG Punjab Technical University, Kapurthala, Punjab, India
e-mail: sransingh008@gmail.com

G. S. Brar
Department of Electrical Engineering, Baba Banda Singh Bahadur Engineering College, Fatehgarh Sahib, Punjab, India
e-mail: gursewak.singh@bbsbec.ac.in

H. Singh et al. (eds.), *Sustainable Development Through Engineering Innovations*, Lecture Notes in Civil Engineering 113,
https://doi.org/10.1007/978-981-15-9554-7_46

1.1 Energy Demand

The use of energy is required by humans for the sake of the routine life. Energy is demanded in various forms such as electricity, gaseous, liquid, and solid fuels. The growth of energy demand is increasing globally. It is mainly due to the increasing population and with the change in living standard of citizens. The increase in global energy demand is due to the improved economic prosperity in the entire world and the increase of the world's population. These two factors are also expected to increase upward in the future, and hence, the global energy demand will also increase [11].

1.2 Energy Supply

The natural law about energy states that the energy neither be created nor be destroyed. Therefore, for the use of energy, it must be converted from one to another source. The main sources of energy existing in nature are fossil fuels, nuclear fuels, and renewable energy sources.

The fossil and nuclear fuels cannot be reproduced naturally in future, and these sources are limited sources of energy. These sources will be exhausted in the future. On the other hand, the renewable sources can be reproduced and are inexhaustible. These renewable sources are expected to deliver energy in nearby and far future.

The renewable energy sources still contribute less than 1% of the total energy supply. The world has to make significant technological progress in order to achieve a sustainable energy future that is based on renewable energy sources [11].

The demand of energy is mainly fulfilled by fossil fuels. As the consumption of these exhaustible energy sources is more than the replenishment, these sources will exhaust on some point of time in the future. Since the human civilization is dependent on the energy consumption, the adequate supply of renewable energy in the near future must be ensured.

1.3 Renewable Sources: Bright Future

The changing scenario throughout the past twenty-five years of development has presented the world meeting its rising needs of energy from the renewable sources [14]. The sources of renewable energy meet 40% of rise in the primary demand of energy and with their effective and explosive rise in the economy of power sector; it ends the booming period of coal. Figure 1 shows the source-wise installed power generation capacity (MW) as on 31.12.2019 in India. There is 23% share of renewable energy in total installed power generation capacity (in MW) in India.

Biomass as an energy source has multiple benefits. It is renewable, carbon-neutral, and has the potential to provide considerable employment in the rural areas of the

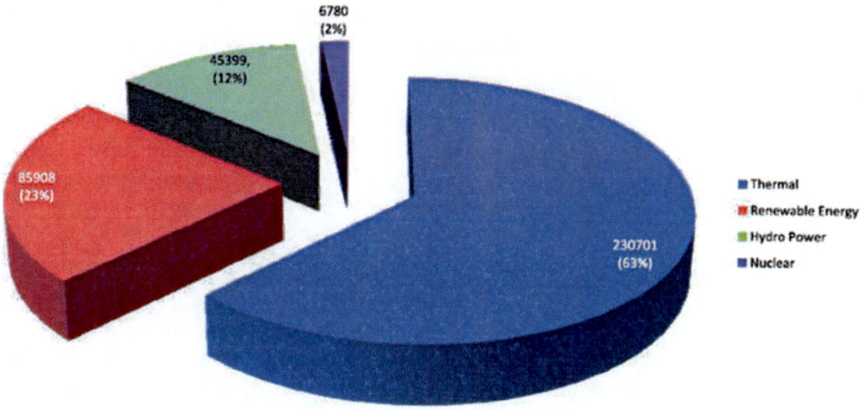

Fig. 1 India—source-wise installed power generation capacity (MW) as on 31.12.2019 [2]

country. About 32% of the total primary energy use in the country is still derived from biomass, and more than 70% of the country's population depends upon it for its energy needs [5].

Keeping in view, the scenario of demand of energy and its supply at global and India level, it is the need of the hour to develop renewable energy sources in India. Biomass is one of the important sources of renewable energy. The use of biomass for electricity is prominent in Europe and North America—predominantly produced from forestry products and residues. Cogeneration plants enable the use of biomass with increased efficiency, so much so that the combined efficiency of producing heat and electricity crosses 80%. The Europe and Americas continent contribute more than 70% of all consumption of biomass for electricity. In 2013, 462 TWh of electricity was produced globally from biomass. In the past few years, biomass is seeing increasing uptake in developing countries in Asia and Africa where significant population lacks access to electricity. Biogas and decentralized bio-energy systems are becoming more cost competitive. Already cogeneration plants using agricultural residues like Bagasse in India, Mauritius, Kenya, and Ethiopia are successful [21]. There is a large amount of biomass for the power generation in India [10]. Various crops such as rice, wheat, and cotton are categorized as the main sources of crop residue in India [9]. Many of the biomass-based power projects running successfully in rural areas of the country. There is sufficient amount of agricultural crop residual produced in the state, which can be used for power generation in the state [17]. The environmental effects of the burning of crop residues can be minimized with proper re-plantation of biomass. All biomass can be used to produce steam, and further, steam can be used for turbogenerators to produce electric power [12]. It is suggested that the remote and rural areas of the state can be benefitted from the decentralized power generation. There is enough surplus agricultural residue (biomass) available in the state [4]. Punjab is an agricultural-rich state which produces handsome amount of crops; hence, huge amount of crop residues are produced in the state. It results that the biomass from major crops can be used for the biomass power generation [16]. The available surplus crop residue for power generation is large in amount in Punjab, India. This study shows that there is a vast scope of biomass power generation in Punjab, India [7].

GIS has been used for planning a biomass power plant in the study area. Optimization of power plants along with collection centers has been performed [18]. In Punjab (India), agricultural biomass has immense potential for power generation. Punjab has sufficient amount of crop residue production every year, which is capable to generate 2000–3000 MW of electric power in the state. The establishment of new biomass power plants in the state will provide power to the energy deficit state as well as generate employment for local people and provide solution to the environmental issues raised due to crop residue burning [6]. The generation of power using agriculture waste is a better option in spite of burning it in open fields. This will not only reduce air pollution but will help to enhance the income of farmers. It will also provide the energy security for future generations [19]. Till date, there are total seven biomass power projects in Punjab, and three power projects are under implementation in various districts of Punjab [16]. It is the dire need of the hour to propose such type of projects in various locations of Punjab state. It would be favorable and beneficial for the society in general, power industry planners, Governments, and power engineers associated with this field. Also, it may get optimum revenue/returns out of it. In this study, the optimal locations of biomass-based power plants have been proposed in the study area of district Bathinda, Punjab (India).

2 NDVI Calculations

2.1 Identification of Geographical Area

Bathinda is a city and municipal corporation in southwestern part of Punjab, India. It is one of the oldest cities in Punjab, India, and the current administrative headquarters of Bathinda District. The proposed study considered the area encapsulated by geo-coordinates 30° 36′ 25″ N 74° 36′ 59″ E, 30° 36′ 25″ N 75° 26′ 35″ E, 29° 44′ 34″ N 75° 26′ 35″ E, and 29° 44′ 34″ N 74° 36′ 59″ E. The altitude of this focused area is 210 m above the sea level. The focused area of the district has seven blocks namely Sangat, Rampura Phool, Maur, Phool, Nathana, Bathinda, and Talwandi Sabo. Figure 2 shows the geographical representation of the study area.

2.2 NDVI Estimation with GIS Model

Normalized difference vegetation index (NDVI) has been used to calculate the density of vegetation in the region. Landsat collection has 11 bands, but none of them provides

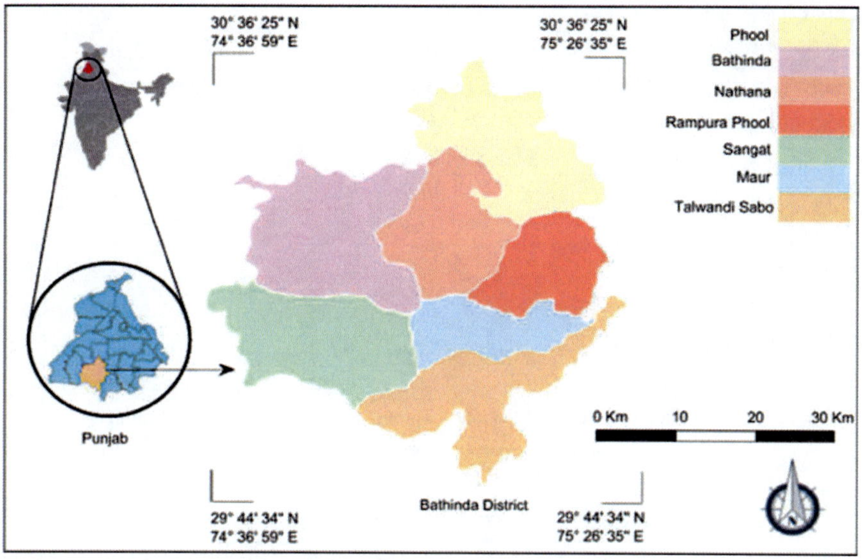

Fig. 2 Geographical area under study in district Bathinda

direct calculation of vegetation area without using NDVI [15]. Equation (1) shows NDVI prediction using 0.64–0.67 μm and 0.85–0.88 μm bands. The proposed work uses data from USGS for the above given coordinates for calculating NDVI of district Bathinda [8].

$$\text{NDVI} = \frac{\text{NIR}-\text{RED}}{\text{NIR}+\text{RED}} \tag{1}$$

3 Methodology

3.1 Assessment of Biomass Potential

The study area of district Bathinda produced total 2848.3 Kt of crop residual from major crops during 2017–18. With the use of Eq. (2), calculations of crop residuals for all blocks have been performed individually as given in Table 2 by considering CRR values given in Table 1.

Table 2 shows crop production for all blocks along with the crop residuals for the year 2017–18. Block Bathinda produced maximum crop residual of 605.53 Kt, whereas Block Maur 286.24 Kt. Equation (2) is given below [1, 3, 5, 6]):

Table 1 CRR values of crop residues

Season	Crop residue	Residue type	CRR value
Kharif	Paddy	Straw	1.20
		Husk	0.16
	Cotton	Stalk	1.00
Rabi	Wheat	Straw	1.15
		Husk	0.16

Source Chauhan [3]

Table 2 Crop production and crop residue production (2017–18)

Block name	Crop production (Kt)				Total crop residual generation (Kt)			
	Wheat	Paddy	Cotton	Total	Wheat	Paddy	Cotton	Total
Sangat	161.53	90.38	10.1	262.01	211.61	122.92	10.1	344.63
Bathinda	283.82	158.81	17.75	460.38	371.8	215.98	17.75	605.53
Talwandi Sabo	154.86	86.65	9.68	251.19	202.86	117.84	9.68	330.38
Maur	134.16	75.07	8.39	217.62	175.75	102.1	8.39	286.24
Nathana	206.25	115.41	12.9	334.56	270.19	156.95	12.9	440.04
Phool	199.58	111.67	12.48	323.73	261.45	151.88	12.48	425.81
Rampura Phool	194.58	108.98	12.18	315.74	255.15	148.22	12.18	415.55
Total	1335	747	83.5	2165.5	1748.9	1015.9	83.5	2848.3

Source Statistical Abstract of Punjab [20]

$$\text{Produced crop residual} = \text{CRR} \times [(A_i) \times (Y_i)] \quad (2)$$

where CRR_i is the crop-to-residue ratio of ith crop, A_i is the crop area covered by ith crop, and Y_i is crop yield of ith crop.

Figure 3 shows the graphical representation of crop production for all blocks along with the crop residuals for the year 2017–18.

3.2 *Location Optimization of Biomass-Based Power Plants*

The proposed work uses Antlion optimization technique to optimize locations for biomass power plants [13]. Geo-locations of biomass has already been concluded using GIS-NDVI model of year 2017–18 [8]. The next step is to optimize locations of biomass power plants accordingly. Biomass facilities require dependable fitness function to develop more accurate model. As it has been seen in Eq. (3), the summation of distances from biomass production sites to biomass power plants are used, this should be reduced using optimization as much as possible. In this study, 1000 iterations have been considered. It results the various locations in form of

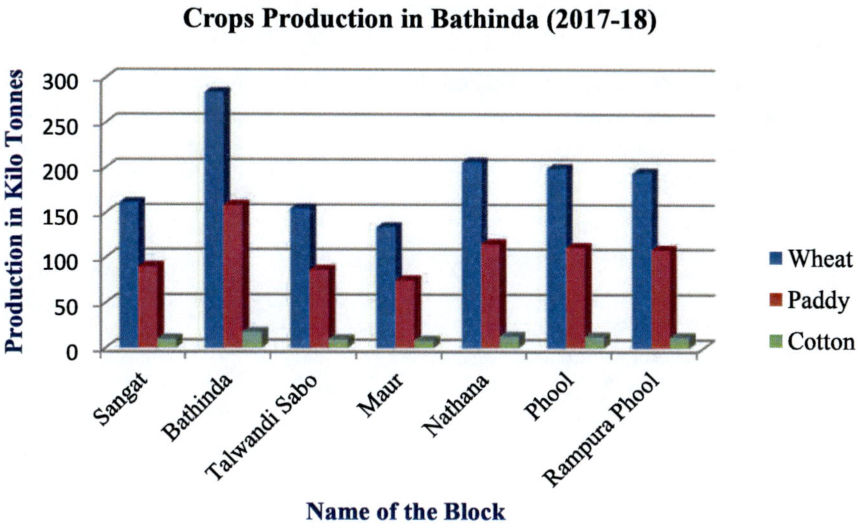

Fig. 3 Crop production in area under study

pixel coordinates in GIS-frame, and it has been changed into geo-coordinates using normalization and distance calculations.

One of the main contributions of this work is that the considered study area is not circular. This assumption can lead in false prediction of biomass sites at map boundaries

$$F(x) = \frac{\sum_{i=1}^{m} \sum_{j=1}^{n} \min\left\{\sqrt{(A_{x,j} - C_{x,i}) - (A_{y,j} - C_{y,i})}\right\}}{m} \tag{3}$$

where m is the biomass geographical locations and n is the number of optimal power plants.

$(A_{x,j}, A_{y,j})$ are coordinates of jth power plant. $(C_{x,i}, C_{y,i})$ are coordinates of ith biomass location.

Algorithm: Ant lion Optimization for Minimizing F(x)

1. *Population initialization of Ants and Antlions*
2. *Calculating fitness values of Ants and Antlions*
3. ***while** halting condition not satisfy **do***
4. ***for** i = 1 to i =β **do***
5. *Selection of Ant-Lion with Roulette Wheel*
6. *Update c_i and d_i equations given below*
7. *Random Walk Pattern for Ant-Lions*
8. *updation of fitness values*
9. *Calculation for best fitness values among Ants and Antlions*
10. *Replacing week Antlions with t Antlions*
11. ***end for***
12. ***end while***

where β= Number of Antlions. $Ci=c_i/I$, $di=d_i/I$

3.3 *Biomass Power Plant Calculations*

Total calculated surplus biomass for district Bathinda in year 2017–2018 is 928 Kt; this total surplus biomass is distributed to optimal locations for power generation. It is observed that if the study area of district Bathinda is optimized for seven biomass power plants, then surplus biomass and power generation capacity of the respective plants is calculated as given in Table 3. The calculations of power generation with crop residual, prediction of calorific value, and conversion factor have been done using a research work of IISc (Indian Institute of Science) Bangalore. Equation (4) is given as below which is used to calculate electric power generation capacity.

$$\text{Power Potential (MWe)} = \text{Available Biomass (kt)} \times \text{Factor of Power( )} \quad (4)$$

4 Results and Discussions

Total seven optimized locations of biomass power plants considering 1000 running iterations with 150 ant lions has been found in the study area. The results show that geo-locations of power plants mainly falls near the area having dense biomass potential as compared to study conducted by authors [18]. Figure 4a–b shows heat map of geographical areas covered by each biomass facility, and it is observed that area under each facility is nearly equal.

Table 3 Net surplus crop residue and plant-wise power generation capacity along with locations

S. no.	Plant no.	Net surplus crop residue (Kt)	Power potential (MW)
1.	Plant-1	161	21.18
2.	Plant-2	104	13.68
3.	Plant-3	131	17.23
4.	Plant-4	107	14.08
5.	Plant-5	190	25.00
6.	Plant-6	99	13.02
7.	Plant-7	136	17.89

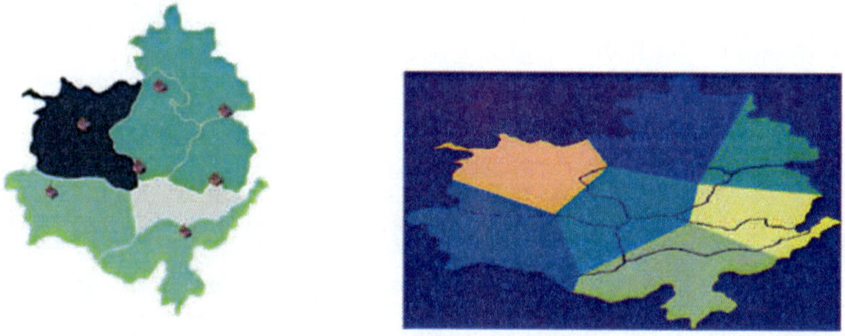

a) Seven Optimal Locations.

b) Coverage Area with seven Locations.

Fig. 4 Optimal locations and coverage area

The results also reflect that biomass potential is maximum for power plant number-5 (Bathinda block) which is 25 MW and minimum for power plant Number-6 (Maur block) having 13.02 MW out of seven optimized locations of biomass power plants.

5 Conclusion

This study shows that total electric power generation capacity using biomass of study area in district Bathinda is 122.08 MW, which is a significant electric power to meet the demand of area. It is observed that paddy and wheat are the major crops because they contribute 97% of total biomass produced from the considered crops in district Bathinda. The proposed research shows that generation of electric power using agriculture waste is a better option in spite of burning it in open fields. PEDA has not set up any biomass power project in district Bathinda of the state. This will not only reduce air pollution, but it will help to enhance the income of farmers. It will also provide the energy security for the future generations.

Acknowledgements The authors are grateful to the USGS for data and Dean, Research and Development, IKGPTU Kapurthala, Punjab, India, for facilities provided for completion of this research work.

References

1. Chahauan S (2010) Biomass resources assessment for power generation: a case study from Haryana state, India. Biomass Bioenerg 34(9):1300–1308
2. *Charts—Data & Statistics—IEA* (n.d.) Retrieved 30 May 2020, from https://www.iea.org/data-and-statistics/charts
3. Chauhan S (2011) District wise agriculture biomass resource assessment for power generation: a case study from an Indian state, Punjab. Biomass Bioenerg 37:205–212. https://doi.org/10.1016/j.biombioe.2011.12.011
4. Chauhan S, Silori CS (2003) Assessment of Biomass Availability for Power. Ecology 12(2):30–37
5. Current Status|Ministry of New and Renewable Energy, Government of India (n.d.) Retrieved 28 May 2020, from https://mnre.gov.in/bio-energy/current-status
6. Dhaliwal HS, Brar YS, Brar GS (2020a) Availability assessment of crop residue potential for electric power generation in Punjab, India: a review. Int J Emerg Technol 11(1):476–485. https://www.researchtrend.net/ijet/pdf/Availability Assessment of Crop residue Potential for Electric Power Generation in Punjab India A Review Harpreet Singh Dhaliwal 22051.pdf
7. Dhaliwal HS, Brar YS, Brar GS (2020b) Optimization for biomass based plant localization using NDVI super pixels for Punjab state, India. Int J Adv Sci Technol 29(6 Special Issue):2723–2733
8. Earth Explorer (n.d.) Retrieved 17 Feb 2020, from https://earthexplorer.usgs.gov
9. Hiloidhari M, Das D, Baruah DC (2014) Bioenergy potential from crop residue biomass in India. Renew Sustain Energy Rev 32:504–512. https://doi.org/10.1016/j.rser.2014.01.025
10. Kumar A, Kumar N, Baredar P, Shukla A (2015) A review on biomass energy resources, potential, conversion and policy in India. Renew Sustain Energy Rev 45:530–539. https://doi.org/10.1016/j.rser.2015.02.007
11. Machrafi H (2012) Green energy and technology. In: Green energy and technology. https://doi.org/10.2174/97816080528511120101
12. McKendry P (2002) Energy production from biomass (part 1): overview of biomass. Biores Technol 83(1):37–46. https://doi.org/10.1016/S0960-8524(01)00118-3
13. Mirjalili S (2015) The ant lion optimizer. Adv Eng Softw 83:80–98. https://doi.org/10.1016/j.advengsoft.2015.01.010
14. Owusu PA, Asumadu-Sarkodie S (2016) A review of renewable energy sources, sustainability issues and climate change mitigation. Cogent Eng 3(1). https://doi.org/10.1080/23311916.2016.1167990
15. Prabhakara K, Dean Hively W, McCarty GW (2015) Evaluating the relationship between biomass, percent groundcover and remote sensing indices across six winter cover crop fields in Maryland, United States. Int J Appl Earth Obs Geoinf 39:88–102. https://doi.org/10.1016/j.jag.2015.03.002
16. Punjab Energy Development Agency (PEDA) (n.d.) Retrieved 26 May 2020, from http://www.peda.gov.in/main/main_obj_resp.html
17. Singh J, Panesar BS, Sharma SK (2008) Energy potential through agricultural biomass using geographical information system-A case study of Punjab. Biomass Bioenerg 32(4):301–307. https://doi.org/10.1016/j.biombioe.2007.10.003
18. Singh J, Panesar BS, Sharma SK (2011) Geographical distribution of agricultural residues and optimum sites of biomass based power plant in Bathinda, Punjab. Biomass Bioenerg 35(10):4455–4460. https://doi.org/10.1016/j.biombioe.2011.09.004

19. Singh R, Brar GS (2020) Optimized locations for biomass power plants using ant lion optimizer in district Bathinda, Punjab, India 1. Int J Emerg Technol 11(1):469–475
20. Statistical Abstract of Punjab (2018) www.esopb.gov.in
21. World Energy Resources 2016 (2016) In World Energy Council 2016. https://www.worldenergy.org/wp-content/uploads/2016/10/World-Energy-Resources_SummaryReport_2016.10.03.pdf

Tuning PID Controller for Three Interacting Tanks Using ITAE Performance Criterion

Parminder Singh, Parag Nijhawan, and Arvind Dhingra

1 Introduction

In any industry, whether chemical or biological or electronics, almost all of the equipments are desired to operate well within suited range of physical constraints. These processes need to be controlled with well-tuned controllers so as to keep them in the bounded limits. Different control schemes like feedback, feedforward, inferential, cascade control, etc., are implemented in industries. Conventional negative feedback controller using proportional–integral–derivative (PID) algorithm is commonly used because of its quick and well-suited response.

In chemical industry, liquid-level tank is most commonly used process. Specific heads are maintained in the storage tanks due to safety or process constraints, in spite of the external disturbances. Liquid-level system can be a single storage tank or multiple tank system. The typical control systems that need to be designed for such complicated and cumbersome systems can prove to be tedious to work with. Khera et al. [1] and Basilio et al. [2] optimized PI and PID controller tuning parameters without and with time delay, respectively, for second-order system using Ziegler

P. Singh (✉)
Department of Chemical Engineering, Thapar Institute of Engineering and Technology, Patiala, Punjab 147004, India
e-mail: parminder.singh@thapar.edu

P. Nijhawan
Department of Electrical and Instrumentation Engineering, Thapar Institute of Engineering and Technology, Patiala, Punjab 147004, India
e-mail: parag.nijhawan@rediffmail.com

A. Dhingra
Department of Electrical Engineering, Guru Nanak Dev Engineering College, Ludhiana, Punjab 141006, India
e-mail: arvinddhingra@gmail.com

H. Singh et al. (eds.), *Sustainable Development Through Engineering Innovations*, Lecture Notes in Civil Engineering 113,
https://doi.org/10.1007/978-981-15-9554-7_47

Nichols method. Lather et al. [3] discussed internal model control (IMC) and IMC-based PID controller for various systems. Suresh et al. [4] presented the mathematical modelling of three tank systems. The transfer function for these systems was derived and optimized using Ziegler Nichols method.

2 Mathematical Modelling

Various three tank system configurations used in the work presented in this manuscript have been summarized as follows:

- ***Case 1***: In this case, interactive three tank system is considered as shown in Fig. 1. In this arrangement, all the three tanks are connected in cascade; i.e., the output of the previous tank is fed as input to its succeeding tank.

  Mass balance equation of Tank 1:
  Rate of accumulation in Tank 1 equals the difference of rate of mass into tank 1 and rate of mass out of Tank 1 as given below:

$$pf_1(t) - pf_2(t) = pA_1(\mathrm{d}h_1/\mathrm{d}t) \tag{1}$$

$$f_2(t) = [h_1(t) - h_2(t)]/R_1, \tag{2}$$

  Mass balance equation of Tank 2:
  Similarly, for Tank 2:

$$pf_2(t) - pf_3(t) = pA_2(\mathrm{d}h_2/\mathrm{d}t) \tag{3}$$

$$f_3(t) = [h_2(t) - h_3(t)]/R_2, \tag{4}$$

  Mass balance equation of Tank 3:
  Extending the same concept on Tank 3,

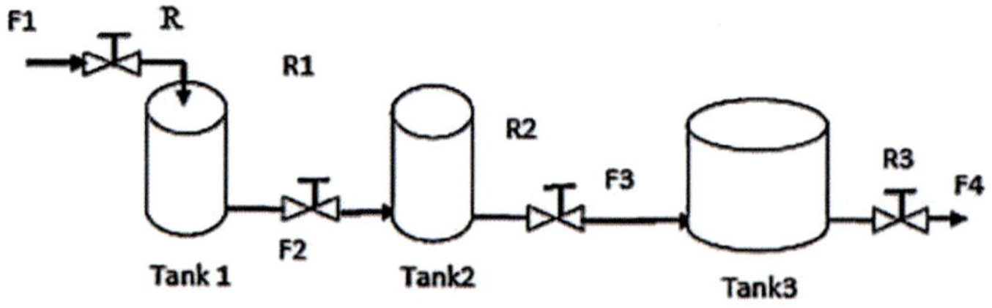

Fig. 1 Three interacting tank system (case 1)

$$pf_3(t) - pf_4(t) = pA_3(\mathrm{d}h_3/\mathrm{d}t) \tag{5}$$

$$f_4(t) = h_3(t)/R_3; \tag{6}$$

The transfer function of the system considered is obtained as

$$\frac{H_3(s)}{F_1(s)} = \frac{R_1R_2R_3}{Ps^3 + Qs^2 + Rs + A_1R_1R_2R_3} \tag{7}$$

where

$$P = A_1A_2^2R_1^2R_2^2R_3$$
$$Q = (A_2^2R_1R_2^2R_3 + A_1A_2R_1^2R_2^2 + 2A_1A_2R_1^2R_2^2R_3 + A_1A_2R_1R_2^2R_3)$$
$$R = (A_2R_1R_2^2 + 2A_2R_1R_2R_3 + A_1R_1^2R_2 + A_1R_2^2R_1)$$

Assuming $A_1 = A_2 = 2\ \mathrm{m}^2$; $A_3 = 1\ \mathrm{m}^2$ & $R_1 = R_2 = 2$; $R_3 = 4\ (\mathrm{m/m^3/s})$;

$$\frac{H_3(s)}{F_1(s)} = \frac{0.25}{16s^3 + 26s^2 + 3.5s + 1} \tag{8}$$

- ***Case 2***: The configuration of three tank system considered here is shown in Fig. 2 which consists of Tank 1 interacting with Tank 2 but Tank 3 non-interacting with Tank 2.

The system transfer function using Eqs. (2), (4) and (6) is obtained as

$$\frac{H_3(s)}{F_1(s)} = \frac{R_1R_2R_3}{P's^3 + Q's^2 + R's + R_1} \tag{9}$$

where

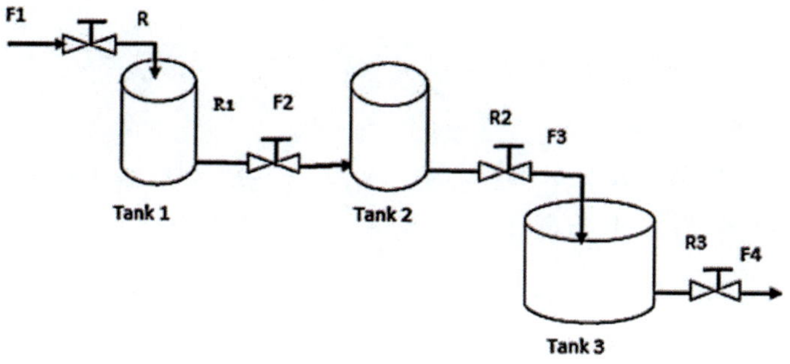

Fig. 2 Three interacting tanks system (case 2)

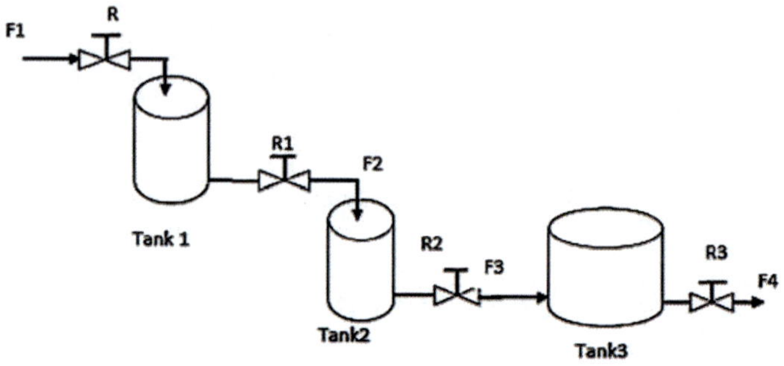

Fig. 3 Three interacting tanks system (case 3)

$$
\begin{aligned}
P' &= A_1A_2A_3R_1^2R_2R?_3 \\
Q' &= A_1A_2R_1^2R_2 + A_1A_3R_1^2R_3 \\
&\quad + A_1A_3R_1R_2R_3 + A_2A_3R_1R_2R_3 \\
R' &= (A_1R_1^2 + A_1R_1R_2 + A_2R_1R_2 + A_3R_1R_3)
\end{aligned}
$$

Taking $A1 = A_2 = 2\ \text{m}^2$; $A_3 = 1\ \text{m}^2$ & $R_1 = R_2 = 2(\text{m/m}^3/\text{s})$; $R_3 = 4\ (\text{m/m}^3/\text{s})$,

$$
\frac{H_3(s)}{F_1(s)} = \frac{4}{64s^3 + 64s^2 + 16s + 1} \tag{10}
$$

- ***Case 3***: The configuration of three tank system considered here is shown in Fig. 3 which consists of Tank 1 non-interacting with Tank 2 but Tank 3 interacting with Tank 2.

The system transfer function using Eqs. (2), (4) and (6) is obtained as.
The system transfer function is obtained as

$$
\frac{H_3(s)}{F_1(s)} = \frac{R_1R_2R_3}{P''s^3 + Q''s^2 + R''s + R_2} \tag{11}
$$

where

$$
\begin{aligned}
P'' &= A_1A_2A_3R_2^2R_1R_3 \\
Q'' &= \left(A_1A_2R_2^2R_1 + A_2A_3R_2^2R_3 + A_1A_2R_1R_2R_3 + A_1A_3R_1R_2R_3\right) \\
R'' &= \left(A_2R_2^2 + A_1R_1R_2 + A_2R_2R_3 + A_3R_2R_3\right)
\end{aligned}
$$

Considering $A_1 = A_2 = 2\ \text{m}^2$; $A_3 = 1\ \text{m}^2$ and $R_1 = R_2 = 2\ (\text{m/m}^3/\text{s})$; $R_3 = 4\ (\text{m/m}^3/\text{s})$,

$$\frac{H_3(s)}{F_1(s)} = \frac{4}{64s^3 + 80s^2 + 20s + 1} \tag{12}$$

3 PID Controller Tuning for Test System

The block diagram of conventional closed loop control system is shown in Fig. 4. The error signal is generated based on the difference of measured and set point. The controller takes the remedial action based on the error. Transfer function of the PID controller is linear combination of error, integral of error and its derivative as given [5] below:

$$G(s) = K_c\left(1 + \tau_D s + \frac{1}{\tau_I s}\right) \tag{13}$$

where τ_D and τ_I are the derivative and integral time constants, respectively.

Tuning of PID Controller

The purpose of PID controller is to keep the dynamic error signal $e(t)$ as small as possible. Integral performance indices to the error signal $e(t)$ are often good choices to do so. In particular, the following integral performance indices [6] are used in the interface

$$\text{ITAE} = \int_0^\infty t|e(t)|\mathrm{d}t \tag{14}$$

where ITAE is integral of time-weighted absolute error, which is also our objective function.

The tuning of PID of the above performance indices is further compared with conventional Zeigler Nichols method, in terms of overshoot, peak time and settling time.

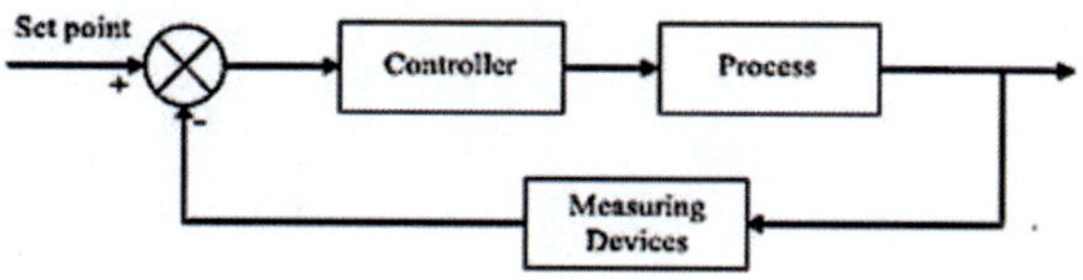

Fig. 4 Block diagram of closed loop system

4 Results and Discussion

Simulink diagram for the test system is shown Fig. 5. For the simplicity of the problem, the transfer function of the measuring element is taken as unity; i.e., there is no delay in the output signal of the measuring element.

MATLAB optimization toolbox is used in this work to minimize the objective function as defined by equations:

The performance index is calculated using Simpson's 1/3 rule at each evaluation step. Table 1 shows the optimized parameters of PID controller for all the three cases obtained using integral of time-weighted absolute error (ITAE) performance index.

Table 2 shows the Ziegler Nichols tuning parameters for all the three cases.

Figure 6a–c presents the response curves for all the three cases showing the comparison of Zeigler Nichols tuning with ITAE performance index tuning of PID controller. The performance parameters of the studied three tank systems are presented in Table 3.

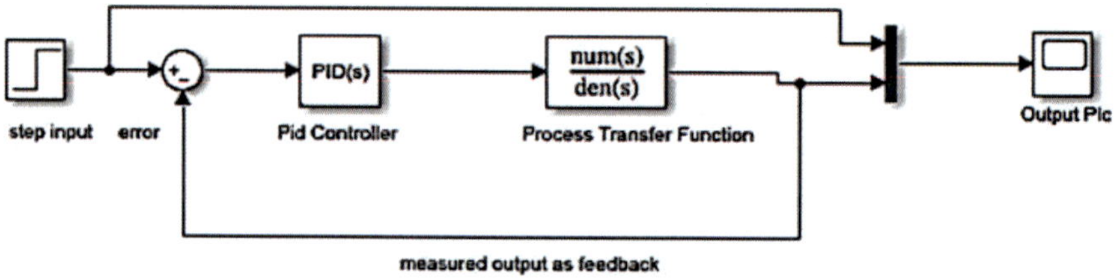

Fig. 5 Simulink block diagram of the feedback control system

Table 1 Optimized parameters of PID controller using integral time weighted absolute error (ITAE) performance index

	K_c	τ_I	τ_D
Case 1	2.031	4.76	9.522
Case 2	0.035	1.7521	1.366
Case 3	0.283	14.07	4.22

Table 2 PID controller tunning using Zeigler Nichols tunning parameters

	K_c	τ_I	τ_D
Case 1	8.05	3.25	0.56
Case 2	2.25	3	0.75
Case 3	1.5	3.6	1.02

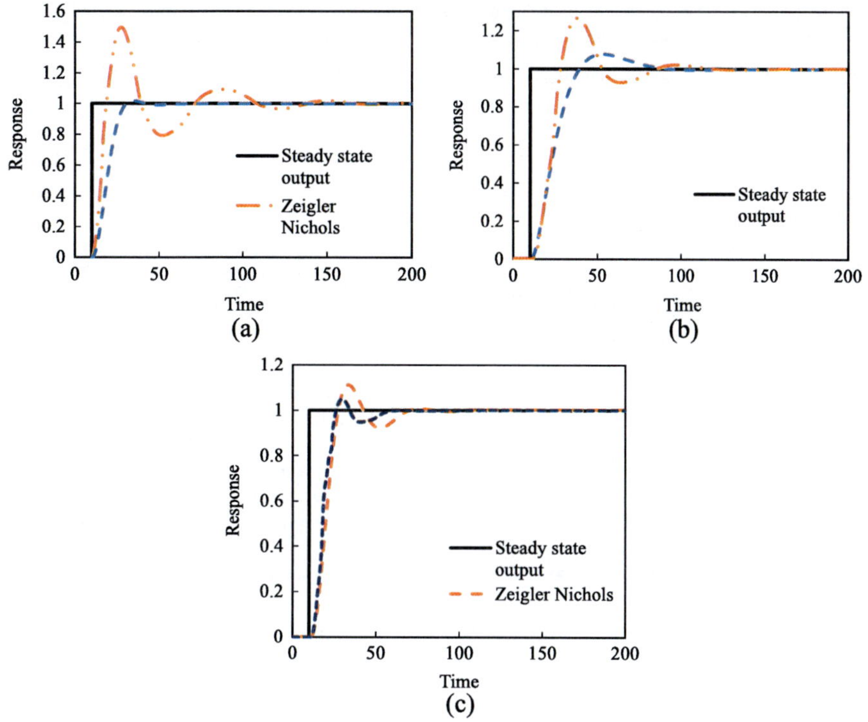

Fig. 6 **a** Response plot for case 1, **b** response plot for case 2, **c** response plot for case 3

Table 3 Performance parameter comparison of Zeigler Nichols tuning with ITAE tunning

Case	Controller tuning	Rise time (s)	Settling time (s)	Overshoot (%)
Case 1	Zeigler Nichols ITAE tuning	17.6 30.5	120 88	45.7 0.505
Case 2	Zeigler Nichols ITAE tuning	39.5 28.7	160 85	27.56 8.59
Case 3	Zeigler Nichols ITAE tuning	42.5 36.5	180 120	27.6 4.75

5 Conclusion

In this study, PID controller tuning using Zeigler Nichols and ITAE performance index is carried out for interacting three tanks in series. Three different arrangements of tanks were studied which therefore produced different process transfer functions. Simulation studies were carried out using MATLAB optimization toolbox in Simulink. The response curve for all the three cases showed that the controller tuning by ITAE performance index gave the faster steady state with lesser overshoot, in comparison to Zeigler Nichol tuning parameters. Also, rise time and settling time were also less for all the three cases for PID controller, tuned using performance index. Hence, ITAE performance index tuning method can be used for designing suitable control strategies for level processes.

References

1. Khera N, Balguvhar S, Shabarinath BB (2011) Analysis of PID controller for second order system using NI LabVIEW". Int J Emerg Technol Adv Eng
2. Basilio JC, Matos SR (2002) Design of PI and PID controllers with transient performance specification. IEEE Trans Educ 45(4):364–370
3. Lather JS, Priyadarshini L (2013) Design of IMC-PID controller for a higher order system and its comparison with conventional PID controller. Int J Innov Res Electr Electron Instrument Control Eng 1(3):108–112
4. Suresh M, Srinivasan GJ (2009) Integrated fuzzy logic based intelligent control of three tank system. Serbian J Electr Eng 6(1):1–14
5. Wayne Bequette B (2003) Process control: modelling, design and simulation. Prentice Hall. ISBN: 0-13-353640-8
6. Seborg DE, Edgar TF, Mellichamp DA (2004) Process dynamics and control. Wiley. ISBN: 81-265-0834-5

Sustainable Machining Aspects of Minimum Quantity Lubrication During Turning of EN-31 Steel Utilizing Singular and Dual Lubrication Nozzles

Gurpreet Singh, Sehijpal Singh, and Vivek Aggarwal

1 Introduction

Tool wear minimization is an essential objective for the researcher and industrial firm keeping in mind the end goal to deliver the required final product, so that it should be available for longer time, thus preventing any hazards to machine or decay of the surface finish. The tool wears out on the consequences of plastic deformation, mechanical failure, dull cutting edge, micro pits, and rise of interface temperatures [1]. It impacts the cutting force, product quality, tool life, and machining cost and hence require tool sharping or sometimes full replacement which deactivate the normal machining process, multiply tool cost, lower efficiency, and consequently, impacting the economic sustainability too. Metal cutting tools are subjected to extreme friction phenomenon due to physical contact between the chip and work piece and further on the account of compelling stress the high temperature developed in the proximity of tool surface [2, 3]. During cutting operation, the tools removes the material from the work piece to accomplish the required shape without compromising the product quality, tool life, frequent tool set up, and surface finish. As the tool worn out quickly, the indicators like surface finish, vibrations, cutting forces,

G. Singh (✉) · V. Aggarwal
Department of Mechanical Engineering, I. K. Gujral Punjab Technical University, Main Campus, Kapurthala, Punjab, India
e-mail: gssingh410@gmail.com

V. Aggarwal
e-mail: agarwalz_v@yahoo.com

S. Singh
Department of Mechanical Engineering, GNDEC, Ludhiana, Punjab, India
e-mail: mech@gndec.ac.in

G. Singh
Department of Mechanical Engineering, Chandigarh University, Gharuan, Mohali, Punjab, India

H. Singh et al. (eds.), *Sustainable Development Through Engineering Innovations*, Lecture Notes in Civil Engineering 113,
https://doi.org/10.1007/978-981-15-9554-7_48

and temperature appears more rapidly [4, 5]. Hence, to ensure the longer tool life and better product quality, the tool coating has been employed [6]. Since, the coating is expensive, hence the other alternatives to extend the tool life should be planned, which should be economical as well as sustainable [7, 8].

1.1 Types of Tool Wear

Crater wear: This type of wear occur on the rake face of tool due to continuous sliding and rubbing of chip and leading to elevated temperature and formation of localized pit and erosion of significant magnitude, thus impairing the product quality and weakening the tool strength.

Flank wear: This type of wear has been reported on the flank of tool due to friction between tool flank and work piece. It degrades the surface finish, dimensional accuracy and resulted in larger cutting power, higher temperature, and vibration. Consequently, the width of the flank wear land "V_B" is normally taken as a measure of the amount of wear, and an edge estimation of the width is characterized as tool reshape criterion.

1.2 Cause of Tool Wear

Tool wear systems in metal cutting incorporate abrasive wear, adhesive wear, diffusion wear, oxidation wear, electrochemical wear, and so on. Among them, abrasive wear, adhesive wear, diffusion wear, and oxidation wear are critical.

- **Abrasive wear** is chiefly brought on by the impurities influences inside the work-piece material, for example, carbon, nitride, and oxide mixes. This is a mechanical wear, and it is the main driver of the tool wear at low cutting speed.
- **Adhesive wear** Because of the high pressure and temperature, welding happens between the fresh surface of the chip and rake face. Serious wear is characterized by impressive welding and tearing of the gentler rubbing surface at high wear rate, and the arrangement of generally high wear particles. Adhesive wear happens for the most part at low machining temperatures on tool rake face, such developed edge.
- **Diffusion wear** is a procedure of molecular exchange at reaching high temp. Various experiments showed that the system of tool wear must include chemical activity and diffusion. They have shown welding and favored chemical assault of (W) tungsten carbide in (W-Ti) tungsten-titanium carbides. There are a few routes in which the wear might depend on the diffusion system.
- **Softening of the tool**: Diffusion of carbon in a moderately deep surface layer of the tool may bring about softening and resulting plastic flow of the tool. This

stream may deliver significant changes in the tool geometry, which bring about high powers and a sudden complete failure of the tool.

- **Diffusion of major tool constituents into the work (Chemical component loss)**: The tool grid or a major reinforcing constituent might be broken up into the work. In cast alloy, carbide or ceramic tools, this might be the prime wear phenomenon. With HSS tool, iron diffusion could happen.
- **Diffusion of a work-material part into the tool**: A constituent of the work material diffusing into the tool may modify the physical properties of a surface layer of the tool. For instance, the diffusion of lead into the cutting device may deliver a slim fragile surface layer, and this dainty layer can be expelled by crack or chipping.
- **Oxidation wear**: High temperatures and the nearness of air mean oxidation for generally metals. A slight oxidation of tool face is useful to decrease the tool wear. It decreases adhesion, diffusion, and current by segregating the tool and the work piece. In any case, at high temperature delicate oxide layers, for instance WO_3, TiO_2, are framed quickly, and after that taken away by the chip and the work piece. This outcomes in a fast tool material loss, which is oxidation wear.

1.3 Methods to Reduce Tool Wear

The following methods are used to reduce the tool wear

- Conventional flood lubrication (wet machining) [7]
- Cryogenic cooling [15]
- High-pressure jet cooling [10]
- Minimum quantity lubrication [3, 4, 7, 8, 12, 13]
- Near-dry machining [2]
- Spatter lubrication
- Mixed lubrication (solid + liquid lubricant) [14]
- Extreme pressure lubrication [4]
- Hot machining.

As per the literature survey, these techniques are used to cool the tool so as to reduce the tool wear. Some of the techniques are same but named differently by different authors although the mode of operation is same.

1.3.1 Conventional Flood Lubrication

It is the method of lubrication in which cutting fluid is applied in excess at the flow rate of 10–30 L per hours. It is also known as wet machining. But in present-day scenario of environmental consciences, strict environmental rule, competiveness, this technique is becoming less effective due to

- Degradation nature of cutting fluid to environment

- More health hazards
- Disposal problem of cutting fluid
- Uneconomical as compared to other technique due to cost of cutting fluid, cost of recycling, and cost of disposal which all known as cooling cost
- Non-biodegradability of cutting fluid (mineral oil).

Some of the developed countries have even banned on practicing this techniques for cooling and lubrication purpose. But still, this technique is used in industries [12, 13].

1.3.2 Cryogenic Lubrication

In this technique, the liquid nitrogen, carbon dioxide, and other gases are used for cooling the tool in which temperature of tool reduces to zero degree or even in negative scale. This technique is used where other techniques are not effective, such as during machining of aerospace and hard material. It is costly to use in daily production. So other alternatives for cooling should be investigated for cooling and lubrication in daily production [15].

1.3.3 High-Pressure Lubrication, Extreme Pressure Lubrication

In some cases, this techniques is used as an alternative to cryogenic cooling, and in this method of cooling, high-pressure jet of cutting fluid is targeted at tool–chip interface to reduce the cutting temperature. This technique has its own advantage and disadvantages of more operating power and machining set up cost [2, 6].

1.3.4 Minimum Quantity Lubrication (MQL)

This technique is gaining popularity for cooling and lubrication in present-day scenario of machining because in minimum quantity lubrication, the amount of lubricant is decreased. The maximum volume flow in MQL is less than 50 ml/hour. The air/oil mixture is supplied to the work station either directly via pressure. The optimum lubrication significantly reduces the frictional heat. Machining costs can be reduced with MQL. The environmental impact is also reduced, together with possible health hazards for the machine operator. The other name of this technique are near-dry machining, spatter lubrication, lean lubrication, etc. Some of the researcher has used MQL with solid lubrication. These days this technique has been applied in EDM.

In present research, MQL has been used for cooling and lubrication the tool using single nozzle and double because of following advantages [7–13]

- Less environmental pollution
- Reduced health hazards

- Low cooling cost
- Produce almost dry chips
- No disposal problem
- No recycling cost.

2 Sustainability Aspects of MQL

Sustainable machining is the requirement of present-day scenario which includes better utilization of man, machine, and working environment. The sustainable machining is the key to develop cleaner environment, economical machining, and environment friendly. Conventional machining is either performed dry or with the aid of wet cooling system that has numerous advantages as well as limitations which can be eliminated using of sustainable cooling techniques. The sustainable cooling is the need of hour to address the issues of environment safety, wastage disposal of cutting fluid, and higher coolant cost of mineral oil that is limited in nature. So, therefore, the role of quantity and quality of lubricant is much of importance to overcome all these problems [9, 12].

As per the literature survey, the possible solutions are dry machining, flood cooling, and minimum quantity lubrication. The dry cutting does not remain effective at high cutting speed; on the other hand, flood cooling is effective cooling technique, and besides this it pollutes the environment, thus lowering economic and ecological sustainability compared to dry machining. So, the suitable alternative to all these limiting scenario of stated processes lies in the application of minimum quantity lubrication [13].

As per literature review, most of the study has been done in MQL by changing the cutting parameters, materials, and the lubricants having different flow rates, but very limited literature has been available on the application of dual MQL for cooling. In present study, machining performance in terms of tool wear has been evaluated considering the effect multiple nozzles and comparison with dry and flood lubrication that has been done.

3 Experimentation

The experiment work has been carried out by turning of EN-31 material having 80 mm initial diameter and 800 mm length on lathe machine by using vegetable-based minimum quantity lubrication under different levels of cutting speed, feed rate, and depth of cut. EN-31 is a through hardening steel with a high degree of hardness. It has chromium and nickel as a main alloying element. Chromium increases hardness along with increase in toughness and wear resistance of steel. It is used in component

Table 1 Details of experimental data

S. No	Item	Description
1	Machine tool	Lathe machine, 5 H.P
2	Work specimen	EN-31 alloy steel, Size: Φ80 X 800 mm mm
3	Cutting tool (insert)	SNMG120408
4	Tool holder	PSDNN2020M12
5	Cutting velocity(rpm)	427, 646, and 1000
6	Feed rate	0.072, 0.088, and 0.138 mm/rev
7	Depth of cut	0.5, 1.0, 1.5, and 2.5 mm
8	MQL supply	Vegetable oil (soybean oil) at 4 bar
9	Air compressor	Single phase motor driven

such as gear, shaft, studs, and bolts. It has good abrasion resistance. The details of experimental set are given in Tables 1 and 2 (Fig. 1).

4 Methodology of Work

The full factorial design has been used for evaluating the performance at different cutting parameters in terms of tool wear. The research has been conducted in three phases. In first phase, the tool crater wear has been evaluated in different cutting environment like dry, flood, single, and dual nozzle by varying the cutting parameters on machining length of 30 mm. In second phase, tool wear has been measured in different machining time to evaluate the tool wear. Investigations have been conducted by varying the parameters like feed, speeds, depth of cut, and different cooling environments such as dry, flood, and MQL utilizing singular and dual MQL nozzles. On application of full factorial, design yields to 3*3*1*4 (Speed*Feed*Depth of cut*Environments) = 36 numbers of experiments. The tool wear has been measure with the help of surface roughness tester. The pressure of MQL lubricant has been kept constant to 4 bar during MQL cutting conditions.

5 Results and Discussion

In this section, the results obtained during machining of EN-31 steel have been expressed in terms of tool wear at different levels of process parameters so as to evaluate the fruitful parameter level, significance of utilizing singular and dual nozzle, and its sustainability prospectus with context to economic and ecological benefits. The graphs have been plotted as per the experimentation observation conducted in

Table 2 Experimental details

S. No	Speed (rpm)	Feed (mm/rev)	Doc (mm)	Cooling environment
1	427	0.072	0.5	Dry
2	427	0.072	0.5	Flood
3	427	0.072	0.5	MQL single nozzle
4	427	0.072	0.5	MQL double nozzle
5	646	0.072	0.5	Dry
6	646	0.072	0.5	Flood
7	646	0.072	0.5	MQL single nozzle
8	646	0.072	0.5	MQL double nozzle
9	1000	0.072	0.5	Dry
10	1000	0.072	0.5	Flood
11	1000	0.072	0.5	MQL single nozzle
12	1000	0.072	0.5	MQL double nozzle
13	427	0.088	0.5	Dry
14	427	0.088	0.5	Flood
15	427	0.088	0.5	MQL single nozzle
16	427	0.088	0.5	MQL double nozzle
17	646	0.088	0.5	Dry
18	646	0.088	0.5	Flood
19	646	0.088	0.5	MQL single nozzle
20	646	0.088	0.5	MQL double nozzle
21	1000	0.088	0.5	Dry
22	1000	0.088	0.5	Flood
23	1000	0.088	0.5	MQL single nozzle
24	1000	0.088	0.5	MQL double nozzle
25	427	0.138	0.5	Dry
26	427	0.138	0.5	Flood
27	427	0.138	0.5	MQL single nozzle
28	427	0.138	0.5	MQL double nozzle
29	646	0.138	0.5	Dry
30	646	0.138	0.5	Flood
31	646	0.138	0.5	MQL single nozzle
32	646	0.138	0.5	MQL double nozzle
33	1000	0.138	0.5	Dry
34	1000	0.138	0.5	Flood
35	1000	0.138	0.5	MQL single nozzle
36	1000	0.138	0.5	MQL double nozzle

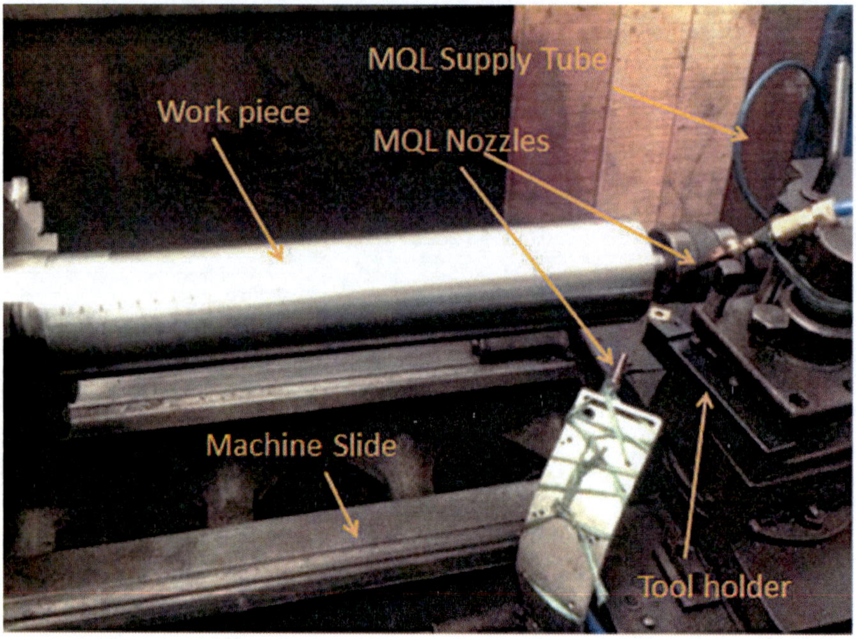

Fig. 1 Photographic view of setup

different phases to explore the viability of different MQL system for its suitability as sustainable cooling technology in present era of industry 4.0.

5.1 Analysis of Tool Wear

In this section, the crater wear of tool has been illustrated in the graphs as shown in Figs. 2, 3, 4, and 5. The graph indicates that tool wear is maximum during dry cutting and minimum in case of dual MQL nozzle. Further, tool wear increases as the depth of cut, and feed rate has been varied. In addition to this, Fig. 5 revealed that the crater wear increases as the machining time advanced which signifies that time have greater impact on the tool life and product quality.

The results cited above explored that utilizing single and dual MQL nozzle has impact on the tool wear as compared to dry and flood lubrication system, and more machining time has been reported in dual lubricating nozzle as shown in Fig. 5. Contray to this, less machining length has been found in case of dry machining due to rapid wear of tool. Hence, the MQL machining has maintained less tool wear and more tool life while incorporating single and dual nozzle system. As far flood lubrication has been concerned, the tool wear and machining time relevant to single and dual nozzle system is shown in Figs. 2, 3, and 4. Hence, it can be stated that

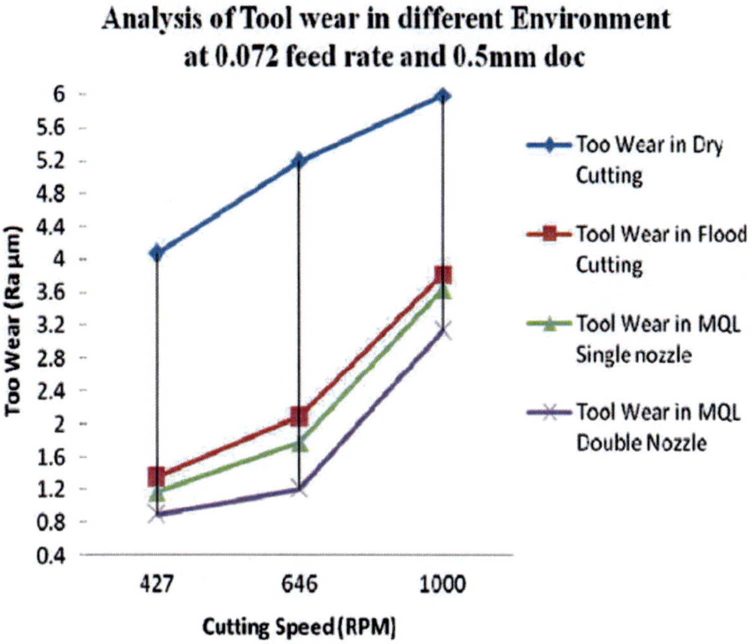

Fig. 2 Tool wear at 0.072 feed rate and 0.5 d.o.c

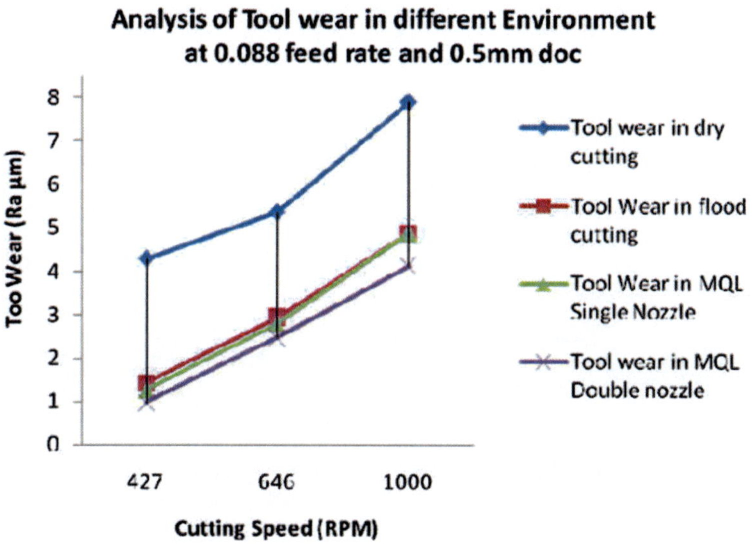

Fig. 3 Tool wear at 0.088 feed rate and 0.5 d.o.c

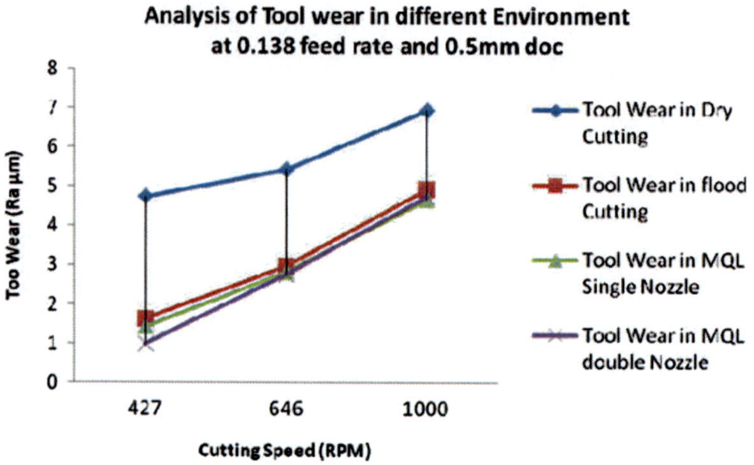

Fig. 4 Tool wear at 0.138 feed rate and 0.5 d.o.c

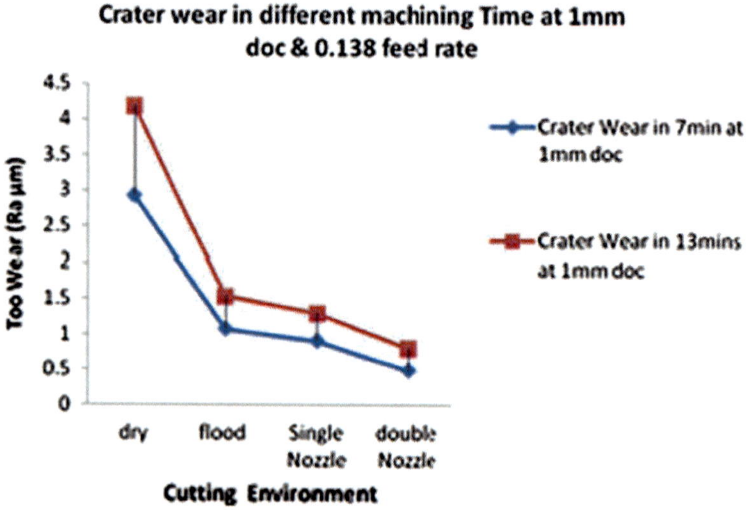

Fig. 5 Crater wear at 1 mm doc and 0.138 feed rate

it is a suitable candidate for cooling and lubrication along with less health hazards, reduced cooling cost, elimination of recycling system, exclusion of disposal cost. On the other hand, the advantages of MQL system are the drawbacks of flood cooling system. Therefore, on the observation of experimental results, it can be mentioned that MQL system has a potential to conserve economic and ecological sustainability with context to prolonged tool life due to stability of cutting edge and environmental friendly behavior.

6 Conclusions

As per the experimental results, the following conclusions have been drawn

- The number of nozzle affected the tool wear during turning operation.
- The position of nozzle in terms of stand-off distance and angle impact the tool wear.
- While placing the MQL nozzles away from cutting zone, lubricant droplets are not able to accommodate the cooling action in the cutting zone.
- Dual MQL system has reduced the tool wear as compared to other three cooling environment.
- The less amount of fumes, cleaner working environment, longer tool life are the indicators of sustainable machining aspects of MQL.
- MQL system also has potential for economic and ecological concern due to elimination of recycling and wastage disposal factors furthermore, pointing toward sustainable machining too.
- Machining with MQL reduced the environmental pollution, health hazards, tool wear as compared to dry and flood lubrication because of biodegradable, cooling, and lubrication nature of vegetable oil.

Acknowledgements The authors are highly thankful to I.K. Gujral Punjab Technical University, Kapurthala, Punjab, Chandigarh University, Mohali, Gegal industries Pvt. Limited, Mohali, and Sigma Freudenberg NOK Pvt. Ltd. Mohali, India, for providing research facilities to conduct this investigation.

References

1. Khan MMA, Dhar NR (2006) Performance evaluation of minimum quantity lubrication by vegetable oil in terms of cutting force, cutting zone temperature, tool wear, job dimension and surface finish in turning AISI-1060 steel. In: Proceeding of journal of Zhejiang University Science A. ISSN 1009-3095, pp 1790–1799
2. Carou D, Rubio EM, Lauro CH, Davim JP (2014) Experimental investigation on surface finish during intermittent turning of UNS M11917 magnesium alloy under dry and near dry machining conditions. Measurement 56:136–154
3. Dhar NR, Kamruzzaman M, Ahmed M (2006) Effect of minimum quantity lubrication (Mql) on tool wear and surface roughness in turning AISI-4340 Steel. Proc J Mater Process Technol 172:299–304
4. Ozcelik B, Kuram E, Cetin MH, Demirbas E (2011) Experimental investigations of vegetable based cutting fluids with extreme pressure during turning of AISI 304L. Proc Tribol Int 44:1864–1871
5. Autret R, Liang Y (2003) Minimum quantity lubrication in finish hard turning. In: Proceeding of Georgia Institute of Technology, Atlanta, Georgia, USA
6. Cantero JL, Dıaz-Alvarez J, Miguelez MH, Marın NC (2013) Analysis of tool wear patterns in finishing turning of Inconel 718. Wear 297:885–894

7. Kumar A, Singh G, Gill SS (2013) Impact of varying the nozzle stand-off distance on cutting temperature in turning of EN-31 steel with minimum quantity lubrication. Proc IJRET 2(6):931–936
8. Khan MMA, Mithu MAH, Dhar NR (2009) Effects of minimum quantity lubrication on turning AISI 9310 alloy steel using vegetable oil based cutting fluid. Proc J Mater Process Technol 209:5573–5583
9. Singh G, Aggarwal V, Singh S (2020) An outlook on the sustainable machining aspects of minimum quantity lubrication during processing of difficult to cut materials. Materials Today proceeding, May 2020
10. Kamruzzaman M, Dhar NR (2009) The influence of high pressure coolant on temperature tool wear and surface finish in turning 17crnimo6 and 42crmo4 steels. Proc ARPN J Eng Appl Sci 4(6):93–102
11. Dhar NR, Islam MW, Islam S, Mithu MAH (2006) The influence of minimum quantity of lubrication (MQL) on cutting temperature, chip and dimensional accuracy in turning AISI-1040 steel. Proc J Mater Process Technol 171:93–99
12. Singh G, Singh S, Kumar A (May–Oct 2013) Experimental evaluation of machining performance into turning of EN-31 steel with dry and vegetable based oil minimum quantity lubrication. Proc IJMERT 3(2):73–75
13. Singh G, Singh S, Singh M, Kumar A (2013) Experimental investigations of vegetable & mineral oil performance during machining of EN-31 steel with minimum quantity lubrication. Proc IJRET 2(6):1030–1037
14. Paturi UMR, Maddu YR, Maruri RR, Narala SKR (2016) Measurement and analysis of surface roughness in WS2 solid lubricant assisted minimum quantity lubrication (MQL) turning of Inconel 718. Proc CIRP 40:138–143
15. García Navas V, Fernández D, Sandá A, Sanz C, Suzonb S, Fernández deMendiola T (2014) Surface integrity of AISI 4150 (50CrMo4) steel turned with different types of cooling-lubrication. Proc CIRP 13:97–102

Use of Constriction Factor-Based Particle Swarm Optimization in Design of Reinforced Concrete Beams

Sonia Chutani and Jagbir Singh

Nomenclature

fy	Characteristic yield strength of steel in N/mm^2
fck	Characteristic compressive strength of concrete in N/mm^2
Ast	Area of steel in mm^2
Mu	Bending moment due to self weight and given loading in kN-m
M	Maximum moment capacity in kN-m
x_u	Actual position of neutral axis in mm
x_m	Limiting position of neutral axis in mm
d_B	Effective depth of beam in mm
b_B	Width of beam in mm
Con_Cover	Effective cover to reinforcement
δ_allowable	Allowable deflection in mm
δ_total	Total deflection in mm
P_t	Percentage of tension reinforcement
r	Effective depth to width ratio
$P_{\min}$	Minimum percentage of steel
$P_{\max}$	Maximum percentage of steel

S. Chutani (✉)
Department of Civil Engineering, DAV Institute of Engineering and Technology, Jalandhar, Punjab 144008, India
e-mail: soniachutani_annie@yahoo.co.in

J. Singh
Department of Civil Engineering, Guru Nanak Dev Engineering College, Ludhiana, Punjab, India
e-mail: jsdhillon26@rediffmail.com

H. Singh et al. (eds.), *Sustainable Development Through Engineering Innovations*, Lecture Notes in Civil Engineering 113,
https://doi.org/10.1007/978-981-15-9554-7_49

1 Introduction

During last few decades, the development in reinforced concrete (RC) construction has taken giant leaps across many countries. The popularity of RC construction owes it to many factors like high compressive strength, durability, good resistance to fire and water damage, low maintenance cost, and other aesthetic advantages of concrete. In addition to this, the lack of robust steel industry in different countries, which makes high-rise steel construction very expensive, is the major economic reason for the popularity of concrete. RC beams are the structural elements in RC frames for exhibiting excellent performance under vertical and gravity loads. Thus, they are an intricate and important part of a multi-storied and high-rise building system. Different techniques have been utilized in the previous literature to conduct optimization of RC beams and frames out of which most of the work has shown the application of classical techniques and genetic algorithm [2, 7, 14, 15, 18]. The use of particle swarm optimization (PSO), based on swarm intelligence, has been widely used in trusses and other steel structures ([4, 6, 16, 17], but has limited application for reinforced concrete structures [1, 5, 11, 13]. PSO has been reported computationally more efficient, fast convergent, and easier to implement than GA. Many other approaches, all based on the standard PSO algorithm, have been explored extensively in literature but their application in optimization problems of RC structures is still a subject need to be studied. Also, constantly increasing need for economical structures has enhanced the interests of designers in developing superior methodologies for optimum design of structural members which led to an attempt for optimization of RC beam structure using limit state method that fulfill the limitations (serviceability and strength requirements) of IS 456:2000 [8]. The utilization of constriction factor-based particle swarm optimization technique (CFPSO), a modified version of PSO with inertia weights [3, 10, 21, 22], is made to achieve optimum results. The remaining sections of this paper are organized in such a way that Sect. 2 includes the brief review of the PSO and CFPSO algorithm and Sect. 3 presents the problem formulation. Section 3.5 discusses general solution procedure in the form of flowchart. Section 4 contains convergence criteria, illustrative examples of beam design and their results are compared with the existing studies in Sect. 5. In Sect. 6, concluding remarks are discussed.

2 Overview of Techniques

2.1 *Standard Particle Swarm Optimization (SPSO)*

A population-based stochastic optimization method 'particle swarm optimization' has been developed by [12]. An optimal solution is explored from a swarm of moving particle vectors, based on a fitness function. Each ith particle vector represents a potential solution and has a position $x_i(t)$ and a velocity $v_i(t)$ at time t in the problem space. Each ith vector keeps a record of its individual best position $P_i(t)$, which is associated with its own best fitness achieved so far, at any time in the iteration process. This value is denoted as 'pbest'. Moreover, the optimum position among all the particles obtained so far in the swarm is stored as the global best position $P_g(t)$. This location is called 'gbest'. The new velocity of particle is updated as follows:

$$v_i(t+1) = w(t)v_i(t) + c_1r_1(P_i(t) - x_i(t)) + c_2r_2\big(P_g(t) - x_i(t)\big) \tag{1}$$

$$x_i(t+1) = x_i(t) + v_i(t+1) \tag{2}$$

Thus, $v_i(t)$ and $x_i(t)$ are the velocity and position of particle '*i*' at time '*t*', respectively. $w(t)$ is the inertia weight at time '*t*' which represents the memory of a particle during search. The inertia weighting function at each iteration is given as:

$$w(t) = w_{\max} - (w_{\max} - w_{\min}) \times \text{itr}/\text{itr}_{\max} \tag{3}$$

$w_{\max}$ and $w_{\min}$ are the maximum and minimum values of inertia weight, respectively, $\text{itr}_{\max}$ is the maximum number of iterations and itr is the current iteration number. The first right hand term in (1) enables each particle to perform a global search by exploring a new search space, whereas the last two terms represent cognitive and social parts, respectively, in which c_1 and c_2 are positive numbers illustrating the weights of the acceleration terms that guide each particle toward the individual best and the swarm best positions, respectively. r_1 and r_2 are uniformly distributed random numbers in (0, 1) and N is the number of particles in the swarm. Each particle changes its position based on the updated velocity according to the Eq. (2) which is known as flight formula. In this way, 'velocity updating' (1) and 'flight formula' (2) help the particles to locate an optimal solution in the search space. In order to keep the particles in a way so that they do not move too far beyond the search space, their velocities have been clamped by limiting the maximum velocity '$v_{\max}$' of each particle. Most of the time, value of maximum velocity is selected empirically, according to the characteristics of the problem. It is important to note that if the value of this parameter is too high, then the particles may move erratically, going beyond a good solution; on the other hand, if is too small, then the particle's movement is limited and the optimal solution may not be reached.

2.2 Constriction Factor Particle Swarm Optimization (CFPSO)

An improvement over standard PSO was introduced by [3] after [12] proposed the original particle swarm. He introduced an additional convergence agent known as constriction factor 'χ' to speed up convergence which is expressed as:

$$\chi = \frac{2}{2 - \varnothing - \left|\varnothing^2 - 4\varnothing\right|} \quad (4)$$

The characteristic of convergence for any system can be controlled by $\varnothing$, so it is called the convergence factor, $\varnothing = c_1 + c_2 > 4$. Generally $\varnothing$ is equal to 4.1, so χ is equal to 0.729. However, as $\varnothing$ increases, the constriction factor 'χ' decreases and diversification is reduced, yielding slower response. Unlike other evolutionary computation methods, the constriction factor approach ensures the convergence of the search procedure based on the mathematical theory. Therefore, the constriction factor approach generates higher-quality solutions than the basic PSO approach. These factors prevent particles to converge on local optima. Moreover, it is difficult to set an appropriate value for $v_{\max}$ due to its main effect on the convergence rate. Hence, to omit this obstacle, the constriction factor approach is applied. The velocity Eq. (1) will take a form of Eq. (5) in this case and new position of the particles will be determined as in Eq. (6) by same flight formula as in Eq. (2).

$$v_i(t+1) = \chi\{v_i(t) + c_1 r_1(P_i(t) - x_i(t)) + c_2 r_2\big(P_g(t) - x_i(t)\big)\} \quad (5)$$

$$x_i(t+1) = x_i(t) + v_i(t+1) \quad (6)$$

3 Formulation of Optimization Problem

In present optimization problem, some of the parameters are considered as pre-assigned while others are variable. The design variables are modified or adjusted in such a fashion that the cost (objective function) becomes minimum. Some restrictions called design constraints, limit the values of these design variables.

3.1 *Objective Function*

The total cost of material used is taken as objective function which includes the cost of concrete, cost of reinforcement, and cost of formwork. Since proposed algorithm is applicable for unconstrained and continuous optimization problem, the formulation of penalized objective function—including imposed penalties due to violation of constraints—is done to convert the constrained problem into an unconstrained one. This has been discussed in next section.

The cost of reinforced concrete beam is given as:

$$C = C_C V_C + C_{st} W_{st} + C_f A_f \quad (7)$$

where C is the total cost of structural element, i.e., beam; C_{st}—cost of steel per kg of steel; W_{st} —total weight of steel in the beam; C_C—cost of concrete per unit volume of concrete; V_C —total volume of concrete in the beam. C_f—cost of formwork per unit area; A_f—area of formwork.

Weight of steel W_{st} depends upon area of steel and its provided length. Similarly, gross volume of concrete depends upon cross-sectional area and length of beam.

3.2 *Constant Parameters and Design Variables*

Some input design parameters are considered fixed which include span of beam, grade of reinforcement and concrete, intensity of dead and live loads, effective cover of concrete, and unit cost of different components as given in Table 1. The independent design variables considered in the present beam design problem are effective depth (d) and percentage of longitudinal reinforcement (p_t). Design constraints for RC beam design consider not only Indian code provisions (IS 456: 2000) [8], but also few practical aspects as well.

Table. 1 Cost of different materials used

Grade of concrete	C_c (Rs/m^3)	Grade of steel	C_{st} (Rs/kg)	Formwork cost (Rs/m^2)
M20	4400	Fe250	40	100
M25	4550	Fe415	45	
M30	4750	Fe500	50	
M35	5000	Fe550	55	

3.3 Design Constraints

Condition for Moment of Resistance of the Section (Moment Capacity Constraint)

The cross-sectional dimensions and area of steel to be provided in the beam shall be such that moment of resistance of beam is greater than the actual moment applied on it.

$$M_r \geq M$$

where $M_r = 0.87\, f_y\, A_{st}\left(d_B - \frac{f_y\, A_{st}}{f_{ck}\, b_B}\right)$

Condition for Depth of Beam from Limit State of Serviceability (Deflection Constraint)

$$\delta\_\text{total} \leq \delta\_\text{allowable}$$

Total deflection includes short-term and long-term deflections both.

Condition for Limiting Depth of Neutral Axis (Ductility Constraint)

To avoid brittle failure of the section, maximum depth of neutral axis is restrained by ensuring that tensile steel does not reach its yield stress before concrete fails in compression. Mathematically, it can be expressed as

$$\frac{0.87\, f_y\, A_{st}}{0.36\, f_{ck}\, b_B\, d_B} < \frac{x_m}{d_B}$$

where x_m is the limiting depth of neutral axis, value of $\frac{x_m}{d_B}$ varies with the grade of steel and is given as $\frac{x_m}{d_B} = 0.53$, when $f_y = 250$ N/mm^2; $\frac{x_m}{d_B} = 0.48$, when $f_y = 415$ N/mm^2. $\frac{x_m}{d_B} = 0.46$, when $f_y = 500$ N/mms^2.

Condition for Minimum Tensile Steel (Minimum Flexural Strength Constraint)

The minimum area of tensile steel to be provided shall be taken as.

$$A_{st} \geq \frac{0.85\, b_B\, d_B}{f_y} \text{ and } = 100\, A_{st}/b_B\, d_B$$

Condition for Maximum Tensile Steel (Maximum Flexural Strength Constraint)

The maximum area of tensile steel shall not exceed $0.04 b_B D_B$. Mathematically, it can be expressed as.

$$A_{st} \geq \frac{0.85\, b_B\, d_B}{f_y} \text{ and } = 100\, A_{st}/b_B\, d_B$$

Condition for Nominal Cover (Durability Constraint)

$$\text{Con\_Cover} \geq 40 \text{ mm}$$

Depth-To-Width Ratio Condition (Practical Constraint)

$$1.5 \leq r \leq 4.0$$

Minimum Width of Beam (Practical Constraint)

$$b_B \geq b_{B\,\min}$$

From practical point of view, the beam shall be wide enough to accommodate at least two bars of tensile steel of given diameter or it should not be less than 200 as per (IS 13920-1993) [9].

3.4 Transformation of Constrained Problem into Unconstrained Optimization Problem

Since the algorithm is designed for unconstrained optimization problems, its application requires transformation of above-defined constrained optimization problem into an unconstrained problem. All previously defined constraints are normalized in such a way that they give zero or positive value for feasible designs. The penalty function approach is used to handle violation in constraints or infeasible designs. The penalty coefficients would be zero for all feasible designs and nonzero penalties are imposed for infeasible designs which deteriorate the quality of objective function. The normalized form of design constraints for optimal design of RC beam is given below:

$$g_1 = 1 - \frac{M}{M_r};\ g_2 = 1 - \frac{\delta\_\text{total}}{\delta\_\text{allowable}};\ g_3 = 1 - \frac{x_u}{x_m};\ g_4 = \frac{p_t}{p_{\min}} - 1;$$
$$g_5 = 1 - \frac{p_t}{p_{\max}};\ g_6 = 1 - \frac{\text{Con\_Cover}}{40};\ g_{7a} = 1 - \frac{r}{4};\ g_{7b} = \frac{r}{15} - 1;$$
$$g_8 = \frac{b_B}{b_{B\,\min}} - 1$$

Penalty coefficients are considered equal to absolute value of normalized constraint functions in case of any violation of above constraints.,

The penalized objective function $Z^{'}(x)$ has been considered as follows:

$$Z'(x) = C(1 + G)^r$$

where $G = Abs(g_1 + g_2 + g_3 + g_4 + g_5 + g_6 + g_7 + g_8)$

r is a parameter that depends on structure type. The value of r= 2 provided satisfactory results in this study.

The reinforced concrete beam design has been carried out for different material combinations of M20, M25, M30, M35 grades of concrete and Fe250, Fe415, Fe500, Fe550 grades of steel. The cost of formwork and materials of different grades are given in Table 1.

3.5 General Solution Procedure for Beam Design Problem Using CFPSO

The development of computer program to carry out the optimization process based on the formulation and optimization technique discussed earlier requires the input of constant parameters initially. The constant parameters of beam design include the load, span, grade of concrete, and steel. The output of decision variables includes the optimum percentage of tensile reinforcement and effective depth of the beam. The computer program for the optimum design procedure of RC beam using constriction factor-based PSO is explained in the form of flow chart in Fig. 1.

4 Convergence Criteria

The proposed algorithm CFPSO is terminated if any one of the predefined criteria is satisfied,

1. Average fitness of last five iterations in a single run does not change.
2. A pre-assigned number of maximum iterations are completed.
3. Standard deviation of minimum ten runs approaches to zero.

5 Illustrative Design Examples and Results

The effectiveness and efficiency of designing RC beams using CFPSO is presented with the help of three design examples. The scope of present paper is limited to simply supported singly reinforced beams. The first two problems of simply supported beams are taken from reference study of [19] and results obtained from them are compared with present study. In first example, beam carries uniformly distributed load including self-weight and in second example concentrated central point load is applied in addition to the self-weight of beam. The effective depth and percentage of reinforcement are the design variables required to reach at objective criteria. Grades of concrete and steel are taken as M30 and Fe415, respectively, for example 1 and as M25 and Fe500 for example 2. The effective depth-to-width ratio is designer's input parameter to avoid thin sections and kept equal to 2 for example 1 and 1.5 for example 2. The cost of materials, i.e., concrete, steel, and formwork is taken as mentioned in reference study for first two problems. Material cost is taken from Table 1 according to the

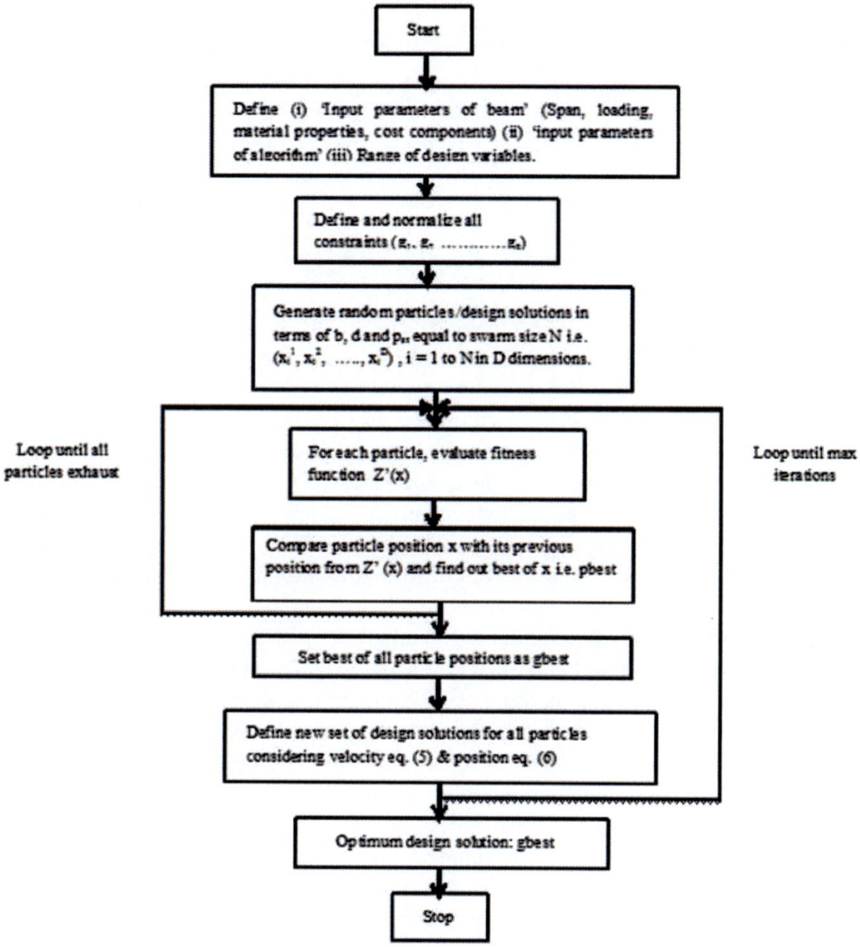

Fig. 1 Flowchart for design optimization of RC beam using CFPSO

grade of materials used in third design example. The unit weights of concrete and steel are taken as 25 and 78.5 kN/m^3, respectively. The cognitive and social coefficients equal to 2.05, swarm or population size as 20, and number of iterations as 500 are taken as CFPSO constant parameters for all examples.

The above design examples have been solved by limit state method and total deflection is calculated as per annexure C of IS 456:2000 [8] and a set of optimum solution is obtained by applying constriction factor particle swarm optimization technique. The constant parameters which have been found fine tuned to get best and consistent results from the algorithm are as given in Table 2. The convergence rate of the system is controlled by $\varnothing$. In current optimization problem, the search space

Table. 2 Constant parameters of CFPSO algorithm

Parameters	Values for CFPSO
Swarm size	20
Maximum iteration number	500
Maximum number of runs	10
χ	0.729
c_1	2.05
c_2	2.05

is bounded by $[x_{min}, x_{max}]$. The ranges of search space for different design variables, i.e., depth of beam (300 mm–1000 mm) and percentage of steel (0.12–4%) are considered as available in [20, 23].

The results obtained for simply supported singly reinforced beam using CFPSO method are given in Table 3 and compared with those obtained from genetically optimized artificial neural network-based approach of [19] in Table 4. As shown in Tables 3 and 4, the CFPSO has outperformed GA and has provided the better optimal solution at the cost of more deflection but within the permissible limit than those in reference study but overall cost of RC beam is found much less than the reference study.

All computations are carried out on a standard PC with an Intel® Core™ i3 CPU M350 @2.27 GHz frequency and 3 GB RAM memory. The algorithm has been coded in Turbo C++ installed in Window 7 at 32-bit operating system. Time required for single run is 1 s. Maximum ten runs of the program are made and their statistical analysis, i.e., mean and standard deviation have been performed to ensure that the obtained optimum solution from CFPSO is global or near global. Tables 3 and 4 present solution set in terms of independent design variables and derived quantities, respectively, to get at ready in hand solution.

Example 1. Perform optimum design of simply supported beam with following details of geometry, loading, and material properties:

Span = 4 m, live load = 35 kN/m, the grades of concrete and steel as M25 and Fe415 respectively. Maximum depth-to-width ratio = 2, clear cover to main reinforcement = 40 mm. The depth of beam section and percentage of steel are considered as independent variables bounded within upper and lower limit as defined earlier. The CFPSO parameters given in Table 2 and cost components as in reference study are used. The optimum design is governed by the constraints of limit state design as per IS 456:2000 [8] and solution details are shown in Table 3 and computed costs are given in Table 4.

Example 2. The geometry and loading details of singly reinforced concrete beam is given as: Span = 6 m, central concentrated live load on beam = 200 kN. The CFPSO parameters, maximum depth-to-width ratio, clear cover to main steel, grades and cost factors of materials for this problem are same as in Example 1. The optimum design follows the same code. The solution details and the computed costs are given in Tables 3 and 4.

Table. 3 Optimum design variables for RC beams

Design example	Design method	Design variables						
		D (mm)	B (mm)	p_t (%)	M_u (kN m)	M_c (kN m)	Δ_{all} (mm)	Δ_{tot} (mm)
1	Reference study	401.111	200.70	1.19412	111.645	111.655	16	9.09004
	Present study	401.500	200.75	0.481243	111.647	111.647	16	16
2	Reference study	606.30	404.2	0.943372	494.08	494.08	24	12.73
	Present study	606.32	404.21	0.365514	494.08	494.08	24	24
3	Present study	651.378	325.68	0.431168	476.748	476.748	24	24

Table. 4 Optimum quantities and cost components for RC beams

Design example	Design method	Quantity			Cost of components			Total cost (Rs/m)	% Reduction in cost
		Concrete (m^3)	Steel (kg)	Formwork (m^2)	Concrete (Rs.)	Steel (Rs.)	Formwork (Rs.)		
1	Reference study	0.350526	30.208292	4.334110	911.368307	604.165830	390.069900	476.40	18.58%
	Present study	0.352974	12.179713	4.335008	917.733320	243.594265	390.150752	387.87	
2	Reference study	1.553535	108.889942	10.18080	4039.192118	2395.578724	916.27200	1225.17	19.66%
	Present study	1.562124	42.192479	10.181093	4061.521617	928.234534	916.298376	984.342	
3	Present study	1.345558	43.082825	10.250675	6122.290697	1938.727136	1025.06752	1514.35	–

Example 3. An optimum design of singly reinforced concrete beam for which the design parameters are span = 6 m, live load = 65 kN/m, grades of materials M25 and Fe415, clear cover to main steel = 40 mm, is carried out as per IS 456-2000 [8] code provisions. The design variables are depth of beam and percentage of steel. Cost components given in Table 1 are used.

The results of optimum beam designs using CFPSO and reference study are summarized in Tables 3 and 4. Table 4 shows optimum cost of materials for above three problems and it is observed that present approach could reduce total cost for first two design examples by 18.28% and 19.66%, respectively. It is seen from solution detail given in Table 3 that optimum design is guided by deflection limit state causing reduction in steel.

It has been concluded from present study as shown in Table 4 that the overall cost of the beam will reduce by significant amount at the expense of deflection, i.e., total deflection (short-term and long-term deflection) will not reduce below maximum allowable deflection at optimal solution. Therefore, it can be stated that the optimum design is governed by deflection criteria. On the other hand, the design presented by [19] optimizer has reduced the total deflection by large amount considering the percentage of steel as dependent design variable which depends upon the depth of the section, but overall cost is more than the design proposed in present study. The cost of concrete and formwork has no significant role in achieving economy in beam design but the cost of steel is reduced considerably due to lesser percentage of steel.

Different convergence trends can be observed when different values of fixed parameters of the algorithm are used. However, the detailed parametric study for convergence trends is out of scope of this paper, still some observations are made for above examples. The convergence rate of CFPSO has been compared with particle swarm optimization with inertia weights (SPSO) for example 1 and example 3 with fixed values of CFPSO parameters as given in Table 2 and represented in Figs. 2 and 3. From the convergence trends, it is seen that CFPSO give optimum results faster, i.e., only a few number of iterations are required as compared to SPSO with

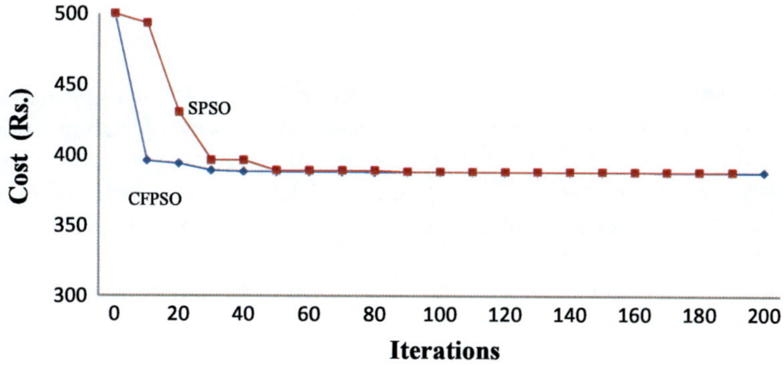

Fig. 2 Convergence trend for optimum beam design example 1

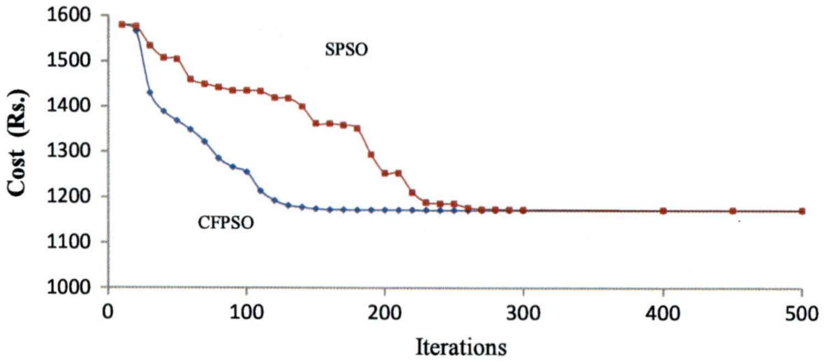

Fig. 3 Convergence trend for optimum beam design example 3

inertia weights. Therefore, it is proved to be a powerful tool capable of shortening the computation time.

6 Concluding Remarks

In this work, an application of constriction factor particle swarm optimization for design of reinforced concrete beams has been extended. Independent design variables taken during optimization process for beam are— 'depth' and 'percentage of reinforcement'. As the entire optimum design algorithm has been coded in C++, time taken to get the optimum design values has almost become an insignificant dimension. The limitations and restrictions of the Indian code IS 456: 2000 [8] have been used as constraints to formulate cost optimization problem and their violations are applied as penalties on the objective function of the CFPSO. Three design examples have been presented for successful implementation of the procedure. The lowest cost solution fulfilled deflection limit as per code provision has been achieved. Some other inferences which were drawn from the present study and are useful in design and construction of RC beams:

- Use of constriction factor along with standard particle swarm optimization technique has been found to expedite the convergence and has proved to be a relatively more efficient tool for exploring optimal solutions for reinforced concrete beams.
- A considerable percentage of saving in cost of RC beam has been found using proposed optimum design approach. Approximately 20% of the total cost will get reduced within the permissible deflection limit.
- Reduction in weight of steel contributes toward optimization of reinforced concrete beams.
- It is viewed that cost optimization is directly proportional to the ratio of depth to width of a beam.

- CFPSO has been proved as powerful tool for achieving the optimal design of beam by shortening the time of computation.

References

1. Ahmadi-Nedushan B, Varaee H (2011) Minimum cost design of concrete slabs using particle swarm optimization with time varying acceleration coefficients. World Appl Sci J 13(12):2484–2494
2. Camp CV, Pezeshk S (2003) Flexural design of reinforced concrete frames using a genetic algorithm. 129(1)
3. Clerc M (1999) The swarm and the queen: towards a deterministic and adaptive particle swarm optimization. In: Proceedings of the 1999 Congress on evolutionary computation, vol 3, pp 1951–1957
4. Flager F et al (2014) Fully constrained design: a general and scalable method for discrete member sizing optimization of steel truss structures. Comput Struct 140:55–65. Available at https://doi.org/10.1016/j.compstruc.2014.05.002.
5. Gharehbaghi S, Fadaee MJ (2012) Design optimization of RC frames under earthquake loads. Int J Optim Civil Eng 2(4):459–477
6. Gholizadeh S (2013) Layout optimization of truss structures by hybridizing cellular automata and particle swarm optimization. Comput Struct 125:86–99. Available at https://doi.org/10.1016/j.compstruc.2013.04.024.
7. Govindaraj V, Ramasamy JV (2005) Optimum design of reinforced continuous beams by genetic algorithms. Comput Struct 84:34–48
8. IS 456:2000, Plain and reinforced concrete code of practice, Bureau of Indian Standards, Manak Bhavan, New Delhi
9. IS 13920-1993 Ductile detailing of reinforced concrete structures subjected to seismic forces—code of practice
10. Iwamatsu M (2006) Locating all the global minima using multi-species particle swarm optimizer: the inertia weight and the constriction factor. pp 816–822
11. Kaveh A, Sabzi O (2011) Optimum design of reinforced concrete frames using a heuristic particle Swarm-Ant Colony optimization. pp 1–18
12. Kennedy J, Eberhart R (1995) Particle swarm optimization. In: IEEE international conference on neural networks, Perth, Australia, pp 1942–1948
13. Khajehzadeh M, Taha MR, El-Shafie A, Eslami M (2012) Modified particle swarm optimization of spread footing and retaining wall. J Zhejiang University-Sci A (Appl Phys Eng)
14. Kwak H, Kim J (2009) Computer-aided design an integrated genetic algorithm complemented with direct search for optimum design of RC frames. Comput Aided Des 41(7):490–500. Available at https://doi.org/10.1016/j.cad.2009.03.005.
15. Lee C, Asce M, Ahn J (2003) Flexural design of reinforced concrete frames by genetic algorithm. pp762–774
16. Luh G, Lin C (2011) Optimal design of truss-structures using particle swarm optimization. *Computers and Structures*, 89(23–24), pp.2221–2232. Available at https://doi.org/10.1016/j.compstruc.2011.08.013.
17. Poitras G, Lefrançois G, Cormier G (2011) Optimization of steel floor systems using particle swarm optimization. J Constru Steel Res 67(8):1225–1231. Available at https://doi.org/10.1016/j.jcsr.2011.02.016.
18. Rajeev S, Krishnamoorthy CS (1998) Genetic algorithm—based methodology for design optimization of reinforced concrete frames. Comput Aided Civil and Infrastructural Eng 13:63–74

19. Saini B, Sehgal VK, Gambhir ML (2006) Genetically optimized artificial neural networks based optimum design of singly and doubly reinforced concrete beams. Asian J Civ Eng (Build Housing) 7(6): 603–619
20. SP 16-1980, Design aids for reinforced concrete to IS: 456:1978, Bureau of Indian Standards, Manak Bhavan, New Delhi
21. Trelea IC (2003) The particle swarm optimization algorithm: convergence analysis and parameter selection. Inf Process Lett 85:317–325
22. Valle Y et al (2008) Particle swarm optimization: basic concepts. Variants Appl Power Syst 12(2):171–195
23. Vergese PC (2013) Limit state design of reinforced concrete. PHI Learning Private Limited, Delhi

Design and Performance Analysis of FinFET Based SRAM Cell Stability

Gurpurneet Kaur, Sandeep Singh Gill, and Munish Rattan

1 Introduction

Recently, several minute devices have been developed to satisfy the demand for low power, low cost, reduced area, and high performance. The fabrication and performance analysis of Double-gate (DG) Fin shaped field-effect transistor (FinFET) has been demonstrated for the smaller gate length [12]. The obtained results proved that a FinFET device has been considered as a strong competitor as compared to classical CMOS [3, 7]. The adoption of semiconductor on insulator (SOI) technology for fabricating of microprocessors has become adorable research. Therefore, it is a convenient device for the mobile industry with features of better switching performance to satisfy the need for efficient battery life. Moreover, low power circuit design is a basic requirement in portable devices [8]. A nanoscale FinFET based SRAM cells with appropriate read and write stability are required for satisfying the increasing demand for large data storage, low power dissipation, and high performance [5, 13, 14]. The SRAM cells stability has been greatly affected by process variations, voltage, and temperature fluctuations. The degradation of stabilities has been examined for scaled dimensions and low power behavior of circuits[10, 11].

The endeavor of designing an improved n-FinFET and p-FINFET devices have been taken. The proposed devices demonstrate great improvements for SCEs and

G. Kaur (✉) · S. S. Gill
Guru Nanak Dev Engineering College, Ludhiana, Punjab 141006, India
e-mail: gurpurneetkaur@gmail.com

S. S. Gill
e-mail: ssg270870@yahoo.co.in

M. Rattan
National Institute of Technical Teachers Training and Research, Chandigarh, Punjab 160019, India
e-mail: rattanmunish@gndec.ac.in

H. Singh et al. (eds.), *Sustainable Development Through Engineering Innovations*, Lecture Notes in Civil Engineering 113,
https://doi.org/10.1007/978-981-15-9554-7_50

transconductance. NanoscaleFinFET based SRAM cell devised with six transistors has been simulated using 3D cogenda Technology Computer-Aided Design (TCAD) simulator for low power applications. The stability parameters read static noise margin (RSNM) and write static noise margin (WSNM) have been evaluated using the butterfly method. The dependence of static noise margin (SNM) metrics on voltage and temperature has also been realized for SRAM circuit. This work is organized as follows: Sect. 2 explains the simulation methodology used for designing the device and its circuit; Sect. 3 illustrates the device and circuit performance; Sect. 4 concludes the work done.

2 Design Description and Simulation Methodology of Finfet and Sram Cell

The specifications of nanoscale FinFET device and the SRAM cell have been delineated in Tables 1 and 2, respectively. The three-dimensional simulator cogenda TCAD has been used for devising the device and its circuit. Figure 1 shows the 3D structure of designed n-FinFET and p-FinFETdevices. The gate electrodes, n-polysilicon (4.5 eV), and p-ploysilicon (4.85 eV)have been used for n-FinFET and p-FinFET respectively at 300 K. The SiO_2 is considered as interfacial oxide; aluminium& tungsten is used as metal contacts. Figure 2 outlines the schematic and layout structure of nanoscaleFinFET based SRAM cell. The flow chart of simulation procedure done in visual TCAD has been shown in Fig. 3 [2].

The set of performance metrics have been calculated for the designed FinFET device namely, drain current at maximal value of gate voltage (on-current, I_{ON}), drain current at minimum gate voltage (off-current, I_{OFF}), I_{ON}/I_{OFF} current ratio, drain induced barrier lowering (DIBL) and subthreshold swing (SS). These parameters have been computed with gate voltage (V_g)variation of 0 to 1.0 V and drain voltage (V_d) of 50 mV and 1.0 V as boundary conditions. SS indicates the gate potential requisite for altering the drain current by one decade. Equation (1) states the formula for the evaluation of the SS.

Table 1 Device parameters used in TCAD

Design parameters	n-FINFET	p-FINFET
Gate length, L_g (nm)	24	24
Oxide thickness, T_{ox} (nm)	1.1	1.1
Transistor Fin pitch (nm)	50	50
Transistor Fin width (nm)	12	12
Transistor Fin height (nm)	28	28
Supply voltage (V)	1	1

Table 2 Parameters for an SRAM design

Process parameters	Value
Design rule unit lambda (μm)	0.012
Substrate region thickness (μm)	0.03
Fin height (μm)	0.028
Gate oxide thickness (μm)	0.0011
S/D doping concentration (donor) for nMOS (cm^{-3})	3×10^{20}
S/D doping concentration (acceptor) for pMOS (cm^{-3})	3×10^{20}
Supply voltage, V_d (V)	0.9 V
Thickness of buried oxide (μm)	0.02
Poly-silicon gate thickness (μm)	0.002
ILD dielectric thickness (μm)	0.008
ILD Metal 1 thickness (μm)	0.008
Lateral characteristic length of S/D doping of nMOS (μm)	0.004
Vertical characteristic length of S/D doping of nMOS (μm)	0.003
Doping concentration in *p*-type substrate (cm^{-3})	1×10^{16}
Doping concentration in body (cm^{-3})	1×10^{17}

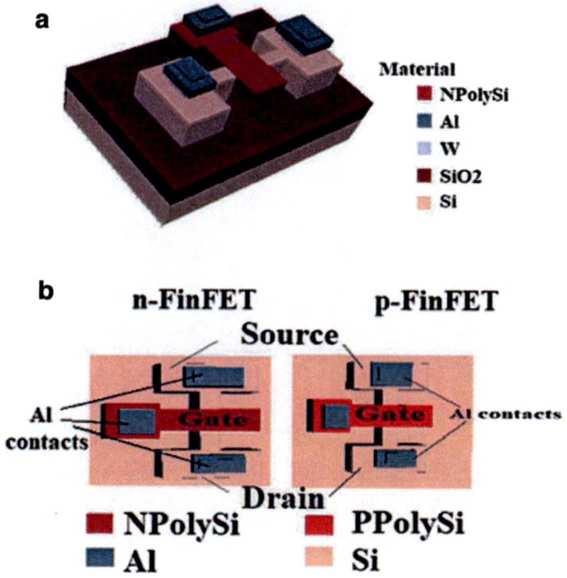

Fig. 1 Bird eye view of 3D SOI FinFET structure: **a** n-FinFET. **b** Top view of n-FinFET& p-FinFET

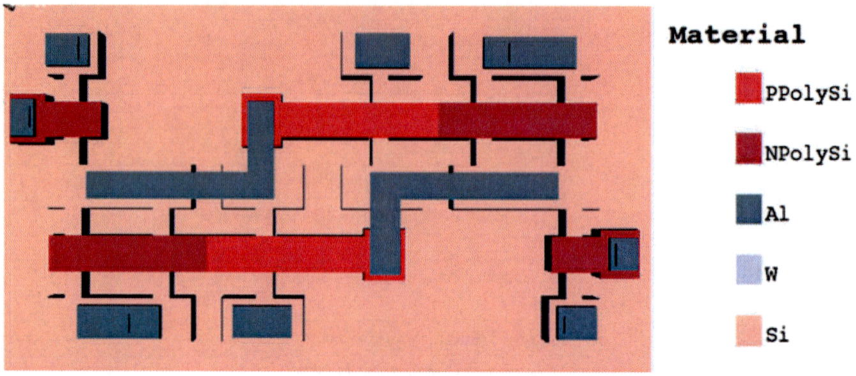

(a) Schematic

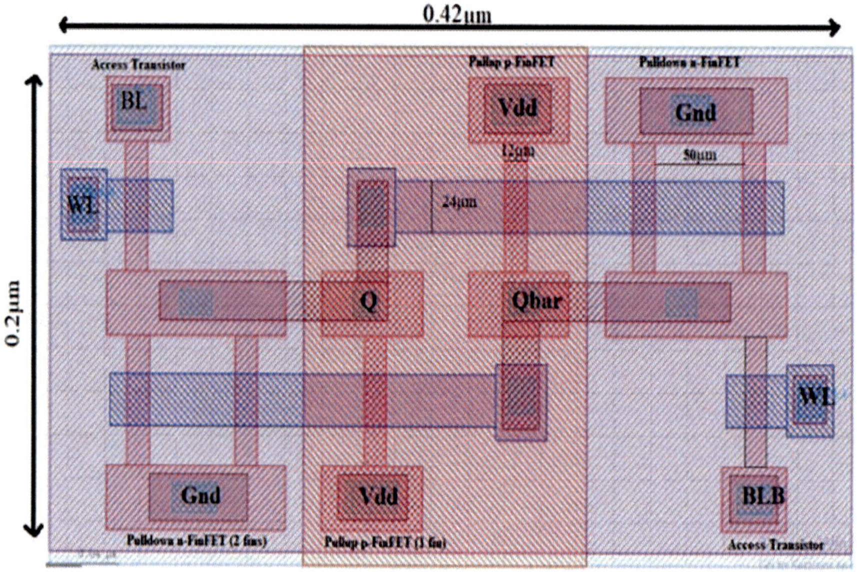

(b) Layout

Fig. 2 Nanoscale FinFET based SRAM cell implemented in TCAD environment. **a** Schematic. **b** Layout

$$\mathrm{SS} = \frac{\partial V_g}{\partial \log_{10} I_d} \tag{1}$$

where I_d indicates drain current in amperes. Subthreshold swing manifests the capability of the transistor to overcome the subthreshold regime. DIBL is calculated as the difference of gate voltage corresponding to both 50 Mv and 1 V drain voltage (V_d) at drain current (I_{DIBL}) of 2.83×10^{-7} A which is determined with the formulae

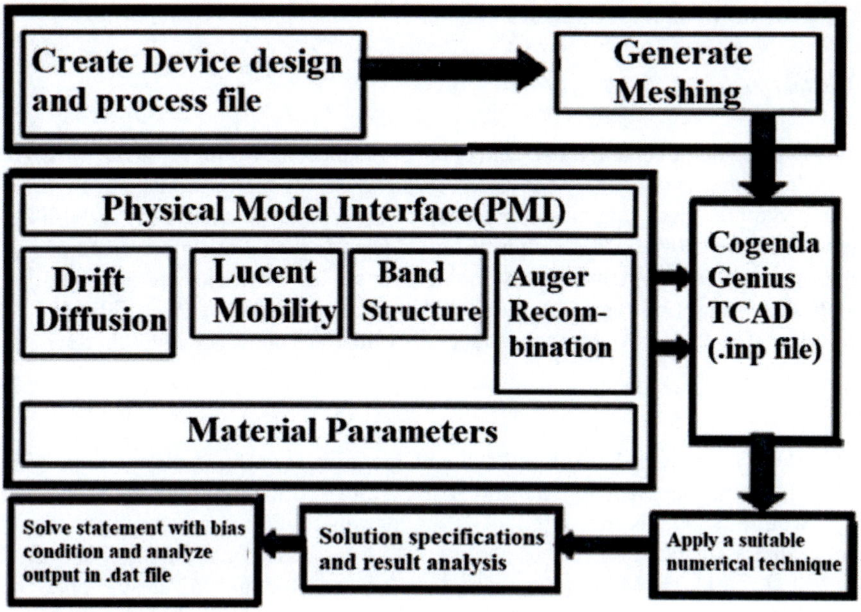

Fig. 3 Flow chart of simulation procedure of visual TCAD

stated in Eq. (2).

$$I_{\text{DIBL}} = \frac{\text{Weff}}{Lg} \times 10^{-7}\ \text{A} \tag{2}$$

$$\text{Weff} = 2H_{\text{Fin}} + W_{\text{Fin}} \tag{3}$$

where L_g is length of gate and Weff is the channel effective width which employs the fin height and fin width as in Eq. (3). Transconductance is evaluated by dividing the changing drain current and the changing gate voltage keeping drain voltage constant. It is expressed as $g_m = (\partial I_{d/}\partial V_g)$ Siemen [1, 6].

The memory circuit performance metrics based on voltage transfer curve include RSNM and WSNM. The ability to read data from SRAM circuit without flipping is characterized by RSNM. It also describes the robustness of SRAM. WSNM is another stability parameter that measures the writing ability of SRAM cells [5, 13, 14].

3 Results and Discussions

(A) *Device Performance*:

The input and output characteristic curves of SOI nanoscale n-FinFET and p-FINFET for different drain voltages are shown in Fig. 4a, b. The VI characteristic of n-FINFET is plotted on the right side of the figure where the similar VI curve for p-FINFET is shown on left side of the figure. It is observed from Fig. 4a that almost the same I_{ON} and I_{OFF} currents are obtained for two devices for similar dimensions in two operating regions i.e. saturation region at $V_d = 0.75$ and linear region at $V_d = 50$ mV. The extracted SCE metrics and transconductance values are mentioned in Table 3[1, 6].

(B) *Circuit Performance*:

(i) The schematic of SRAM cell for read operation is outlined in Fig. 5a. In read operation, the bit-lines (BL and BLB) are biased to supply voltage (V_d) and wordline (WL) is connected to V_d. In this situation, access transistors M5 and M6 are switched on. Let us assume, "1" data is saved at Q, and "0" is saved at Q_{bar}, M1 and M4 devices are switched off and M3 and M2 are switched on. Therefore, current will flow through BLB-M6-M2 as shown in Fig. 5a. Hence, voltage level of bit-line BLB discharges and voltage level of BL maintains at V_d. Figure 6 demonstrates the waveforms for read operation implemented

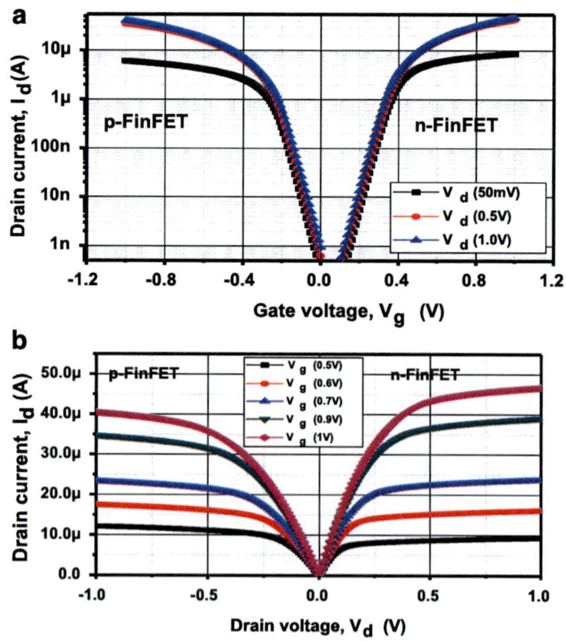

Fig. 4 VI characteristics of n-FINFET and p-FinFET. **a** Input. **b** Output

Table 3 Extracted electrical characteristics of FinFETs

Designed devices	Gate length (nm)	Supply voltage, V_d (V)	I_{ON} (μA)	I_{OFF}	I_{ON}/I_{OFF}	SS (mV/dec)	DIBL (mV/V)	g_m (S) [μA]
n-FINFET	24	1 V	46.8	13.4pA	3.49×10^6	66.5	40	78
		50 mV	8.64	4.71pA	1.83×10^6	65.5		27.3
p-FINFET	24	−1 V	−40.5	95.4nA	4.24×10^4	68.5	42	58
		−50 mV	6.04	33.1nA	1.82×10^4	67		13.2

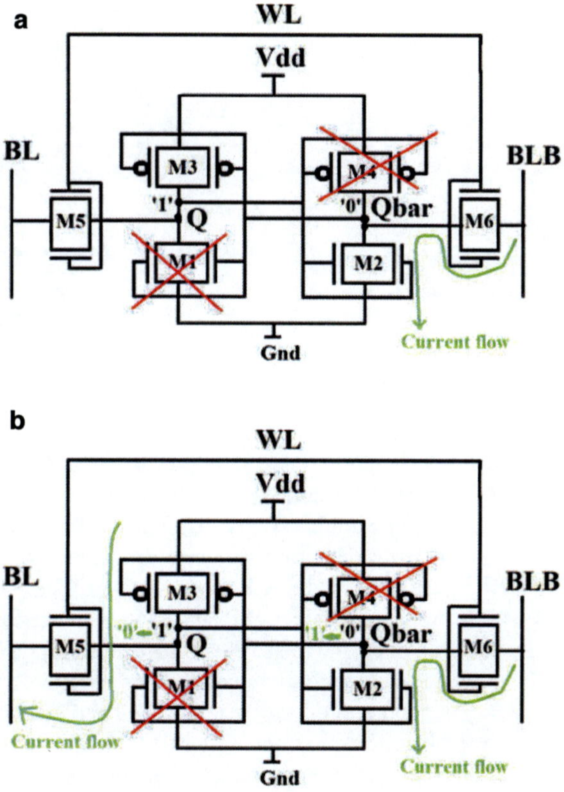

Fig. 5 Operations of 6T SRAM cell using nanoscale FinFET. **a** Schematic diagram for Read operation. **b** Schematic diagram for Write operation

Fig. 6 Waveforms for read operation performed in TCAD for FinFET based 6T SRAM cell

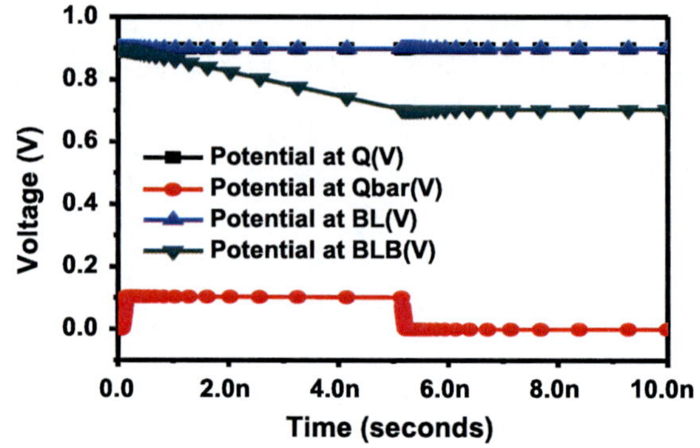

(a) Waveforms showing potential at Q and Qbar output

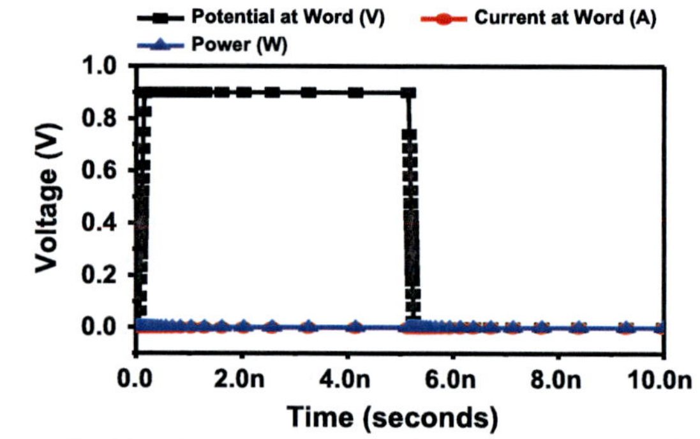

(b) Waveforms showing potential and current at wordline

in TCAD environment. Figure 6a displays the potential at different nodes of SRAM cell during read operation where Fig. 6b delineates the current, potential, and power at wordline (WL). Therefore, a successful read operation has been realized for simulated circuits. Figure 5b shows the schematic of designed cell for write operation. In write operation, the voltages of bit-lines (BL and BLB) are opposite to each other, and wordline (WL) is connected to V_d. In this situation, access devices M5 and M6 are switched on. This will drop the voltage level of Qnode and raises the voltage level of Q_{bar} node until the voltage level of Q node is enough to switch on M4 and switch off M2or the voltage level at node Q_{bar} is good enough to switch on M3 and switch off M1. Finally, the voltage level of nodes Qbar and Q will be turned over to '1' and '0', respectively. Figure 7 delineates the waveforms obtained after a successful write operation performed in TCAD for nanoscale FinFET based 6T SRAM cell [9].

(ii) The SRAM circuit stability has been measured by SNM. It is stated as the largest DC margin for which the cell condition does not flip during its access. The butterfly curve method is used for estimating SNM of a bit cell as shown in Fig. 8b corresponding to setup outlined in Fig. 8a. The square fitting method is utilized for determining RSNM. It is the largest square to be fitted in

Fig. 7 Waveforms for write operation performed in TCAD for FinFET based 6T SRAM cell

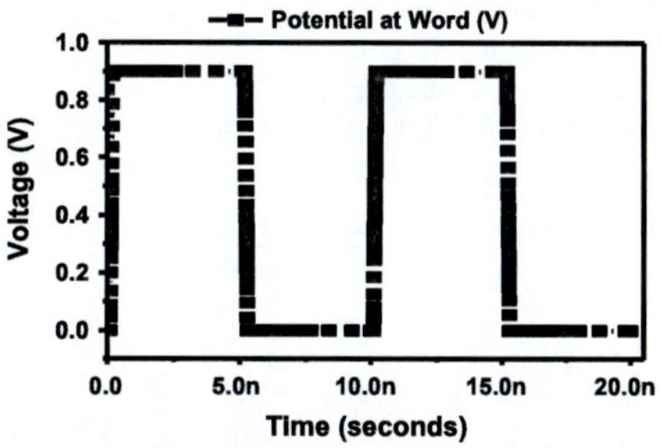

(a) Waveform shows Potential at wordline

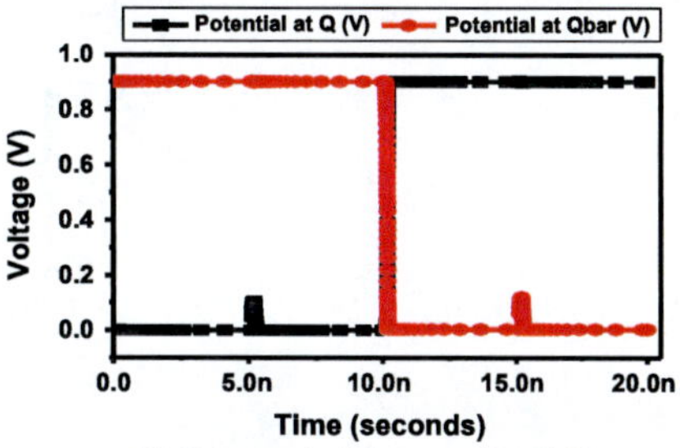

(b) Output potential at node Q and Qbar

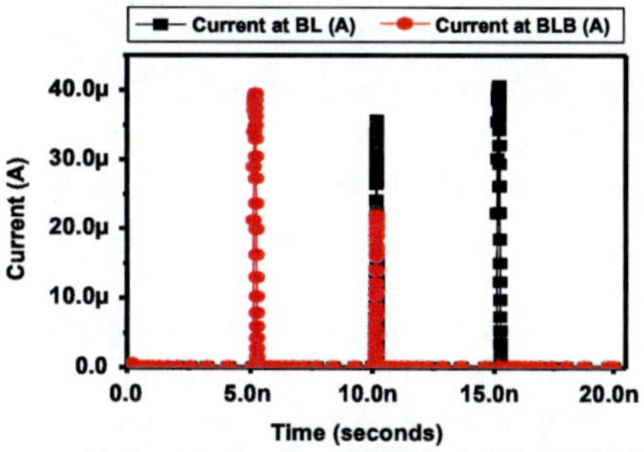

(c) Current waveforms at Bit-lines (BL and BLB)

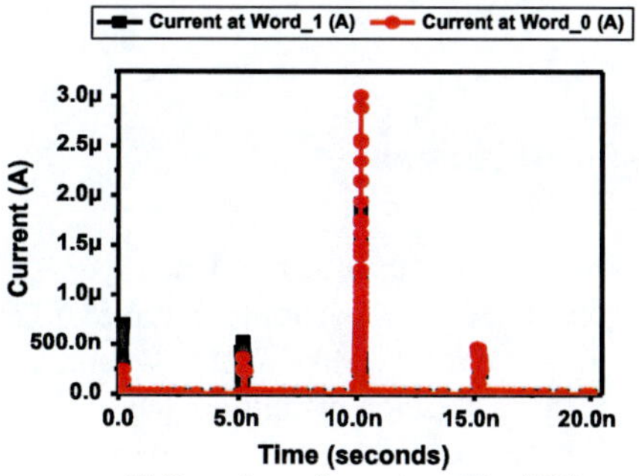

(d) Current waveforms at wordline (WL)

Fig. 8 Butterfly curve method of FinFET based 6T SRAM cell for measuring RSNM and WSNM

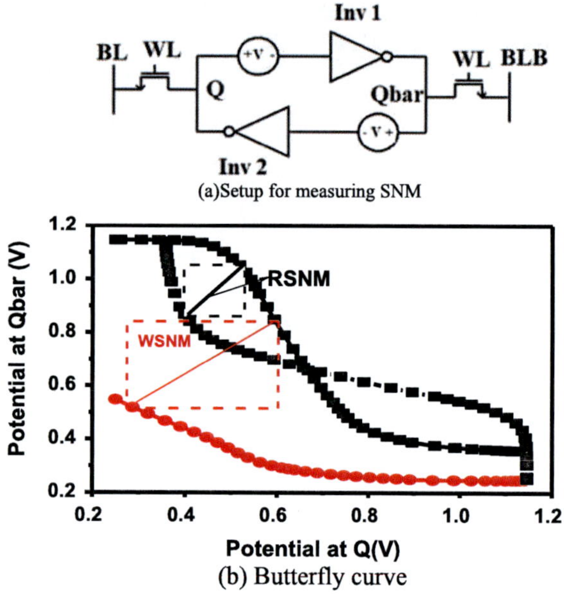

between overlapped plot of inverter transfer characteristics and its inverse characteristics. Similarly, WSNM is extracted as displayed in Fig. 8 [5, 13, 14].

(iii) The dependence of stability on temperature is shown in Fig. 9a. It is noticed that with the increase in temperature, RSNM and WSNM decreases due to increment of the random variation effects because of exponential dependence of current on sub-threshold operation. It is observed that RSNM and WSNM are reduced by 56% and as 16% as temperature varies from 253 to 450 K. The dependence of stability on voltage is illustrated in Fig. 9b. It can be clearly seen that both RSNM and WSNM are degrading for reduced supply voltage which explains the effects of voltage scaling [10, 11].

4 Conclusion

- In this work, the electrical characteristics of nanoscale FinFET devices have been discussed for low supply voltage. It has been noted that simulated devices show 47.4% improvement for DIBL and 7.32% enhancement for SS as compared to results extracted by authors [4].
- Further, low power and miniaturized FinFET based 6T SRAM cell have been demonstrated along with the calculation of SNM values using the butterfly curve method. The progress of WSNM for designed SRAM cell is 8.6% as compared to results demonstrated by researchers [4].

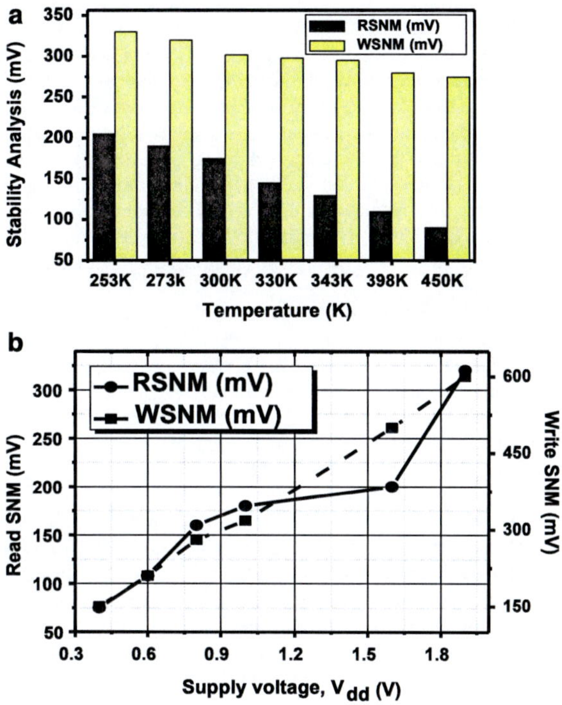

Fig. 9 Dependence of SRAM cell stabilities on temperature and supply voltage. **a** Impact of temperature on stability. **b** Impact of supply voltage on stability

- The impact of voltage and temperature variations on the stability of SRAM cell has also been discussed. The obtained values show improvement as compared to previous work.
- The miniaturized FinFET devices have lower leakage current and reduced power consumption as compared to the planar transistors. Therefore, nanoscaled FinFET technology has been considered as economical, reliable, sustainable, and energy-efficient for future generations. These devices could become an essential part of microprocessors in the future.

Acknowledgements The authors are grateful to MHRD, Govt. of India for sanctioning the grant for the purchase of software used for research work through TEQIP-III to Guru Nanak Dev Engineering College, Ludhiana. Authors would also like to extend gratitude to I. K. Gujral Punjab Technical University, Kapurthala for support in completion of this research work.

References

1. Aujla SK, Kaur N (2019) Optimization of dual-K gate dielectric and dual gate heterojunction SOI FinFET at 14 nm Gate Length. IETE J Res 1–9

2. Cogenda TCAD Tool Suite [Online] (2019). Available http://www.congendatcad.com
3. Colinge JP (ed) (2008) FinFETs and other multi-gate transistors, vol 73. Springer, New York
4. Farkhani H, Peiravi A, Kargaard JM, Moradi F (2014, September) Comparative study of FinFETs versus 22 nm bulk CMOS technologies: SRAM design perspective. In: 2014 27th IEEE International System-on-Chip Conference (SOCC). IEEE, pp 449–454
5. Grossar E, Stucchi M, Maex K, Dehaene W (2006) Read stability and write-ability analysis of SRAM cells for nanometer technologies. IEEE J Solid-State Circ 41(11):2577–2588
6. Kaur G, Gill SS, Rattan M (2020) Whale optimization algorithm for performance improvement of silicon-on-insulator FinFETs. Int J Artificial Intell 18(1):63–81
7. Kim TTH, Vaddi R, Agarwal RP, Dasgupta S (2011) Design and analysis of double-gate MOSFETs for ultra-low power radio frequency identification (RFID): device and circuit co-design
8. Kumar R, Babulu K (2018) Design and performance analysis of low power SRAM using modified MTCMOS G
9. Limachia M, Kothari N (2020) Characterization of various FinFET based 6T SRAM cell configurations in light of radiation effect. Sādhanā 45(1):31
10. Pattanaik M, Birla S, Singh RK (2012) Effect of temperature & supply voltage variation on stability of 9T SRAM Cell at 45 nm technology for various process corners
11. Reddy KN, Jayasree PVY (2019) Low power process, voltage, and temperature (PVT) variations aware improved tunnel FET on 6T SRAM cells. Sustain Comput Inf Syst 21:143–153
12. Solomon PM, Guarini KW, Zhang Y, Chan K, Jones EC, Cohen GM, Krasnoperova A, Ronay M, Dokumaci O, Hovel HJ, Bucchignano JJ (2003) Two gates are better than one [double-gate MOSFET process]. IEEE Circ Devices Mag 19(1):48–62
13. Saun S, Kumar H (2019, October) Design and performance analysis of 6T SRAM cell on different CMOS technologies with stability characterization. In: IOP conference series: materials science and engineering, vol 561, No. 1. IOP Publishing, p 012093
14. Takeda K, Hagihara Y, Aimoto Y, Nomura M, Nakazawa Y, Ishii T, Kobatake H (2005) A read-static-noise-margin-free SRAM cell for low-VDD and high-speed applications. IEEE J Solid-state Circ 41(1):113–121

Biomachining of Aluminum Alloy 46500 Using *Acidithiobacillus ferrooxidans*

Sonia, Amanpreet Kaur Sodhi, and Neeraj Bhanot

1 Introduction

Metal machining industries are important machining industries because they provide capital as well as intermediate metal goods and meet our infrastructure demands [1]. As per the estimation, half of the world population currently lives in urban areas, especially in developing countries [2]. By this logic, the demand for developing urban infrastructure has been raised two times. Being rigid and reliable, the various metals and alloys associated with the construction industry provide long-lasting service without failure [3]. Metal machining is a process of removing materials through chip formation by the application of mechanical energy on a sharp cutting tool. There are several production methods available with metal machining industries, and among them, conventional turning operation is most important and widely used at any scale of production.

Turning machining process is an important manufacturing process in aerospace and automotive industries. The machining process is cutting the metal workpiece as per the required dimension and shape at both intermediate and gives final products. Both rough and fine-tuning can be performed on a conventional turning machine [4]. As per the estimate, industries in developing countries like in India are established at

Neeraj Bhanot—The work has been guided during tenure at IIM Amritsar.

Sonia (✉)
Department of Civil Engineering, Guru Nanak Dev Engineering College, Ludhiana 141006, India
e-mail: karda.sonia9@gmail.com

A. K. Sodhi
Department of Applied Sciences, Guru Nanak Dev Engineering College, Ludhiana 141006, India

N. Bhanot
Quantitative Methods and Operations Management, Indian Institute of Management, Amritsar 143105, India

H. Singh et al. (eds.), *Sustainable Development Through Engineering Innovations*, Lecture Notes in Civil Engineering 113,
https://doi.org/10.1007/978-981-15-9554-7_51

small scale, and they do not have enough resources to follow protocol related to the disposal of cutting fluids, so they usually throw the unwanted used lubricants directly into the domestic sewage without giving any preliminary treatment. Several skin diseases are associated with it like skin cancer, wrinkle, cancer or eye irritation. While removing metals from the workpiece, the unwanted being removed from workpiece gets accumulated on the machine bed as metal curls. They are of grave concern from safety and economic point of view. The high density of electric consumption is also a concern because in India more than 50% of electricity has been generated from coal-based thermal power plants which, in turn, increases global carbon budget [5].

From the above discussion, it is widely clear that metal machining industry has been plagued with various discrepancies. Biomachining is an innovative and alternative metal machining process in which metal is machined using the metabolic activity of a microbial culture instead of the hard metal cutting tool used in the conventional machining process [6]. Biomachining is the organic process and can be achieved by exploitation different microorganisms like *Acidithiobacillus ferrooxidans, Bacillus circulans, Bacillus megaterium, Thiobacillus ferrooxidans*, etc., and their viability depends on various types of metals to achieve the good machining in the present situation. Various researches have carried out trails on the various materials like steel, copper and numerous alloys. Some examples are machining of pure copper by using *At. ferrooxidans* microbe to analyze the surface change and metal removal rate (MRR) [7], biomachining on EN-19 steel alloy using *At. ferrooxidans* and *Aspergillus niger* and to check the viability of biomachining on this metal [5] and machining on various type of metals such as copper, nickel and aluminum using different microbes [8, 9].

2 Methodology

Figure 1 gives an overview of the procedure and all the steps that are performed during the research.

2.1 *Culture of* Acidithiobacillus ferrooxidans

Acidithiobacillus ferrooxidans 5370 has been obtained from the National Collection of Industrial Microorganisms (NCIM), Pune, India. The 9 k medium used for the development of *At. ferrooxidans* (bacteria) was prepared by using two types of solutions: solution A and solution B.

1. Solution A: Mixing $(NH_4)_2SO_4$ 0.5g, K_2HPO_4 0.5g, $MgSO_4.7H_2O$ 0.5g, 1N H_2SO_4 5 min in one liter of distilled water.

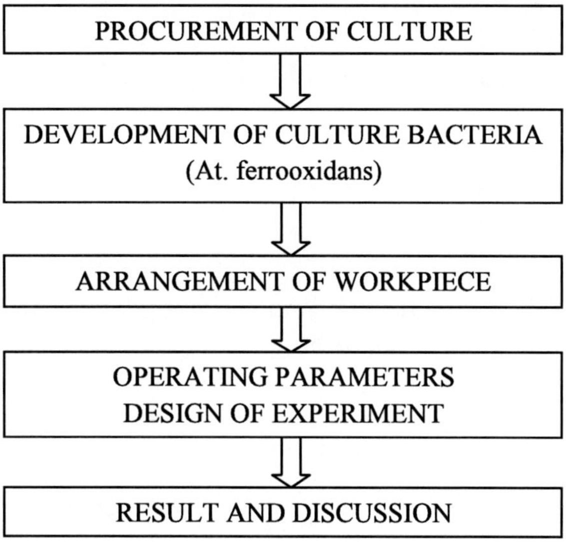

Fig. 1 Flow chart of methodology

2. Solution B: It was obtained by using one liter distilled water in which $FeSO_4.7H_2O$ 167 g and 1 N H_2SO_4 50 ml were added and filtered through the filter paper.
3. Half part of solution B in which added to one-fourth part of solution A.
4. The pH of the final solution (A + B) was maintained at 3.0.
5. Final Solution A + B was put in an incubator at 30 °C under the stationary condition for 2–5 days for the development of *At. ferrooxidans.*
6. The cells of bacteria have counted through the hemocytometer. In this study, the number of cells used is 10^4, 10^6,10^8 and 10^{10} (CFU/ml of H_2O) for machining aluminum alloy 46500.

2.2 Arrangement of Workpiece

Aluminum alloy 46500 has been obtained from P. S. Sura and Sons, Gill Road, Ludhiana. Upon characterization, it was found to be composed (%) of copper (2.37), magnesium (0.007), silicon (9.26), iron (1.29), manganese (0.174), nickel (0.077), zinc (2.65), lead (0.112), tin (0.049), titanium (0.033), chromium (0.033) and aluminum (83.67).

Table 1 Level and factors of L16 arrays

Level →	1	2	3	4
Factors↓				
pH	2	3	4	5
Shaking rate (rpm)	80	90	100	110
Cell concentration (CFU/ml of H_2O)	10^4	10^6	10^8	10^{10}
Time (h)	24	48	72	96

2.3 *Operating Parameters and Functioning Indicator for Biomachining*

The machining performance is analyzed by specific metal removal rate which is one of the major indicators.

$$\text{SMRR} = \frac{\text{Amount of metal removed (gm)}}{\text{Time (h)} \times \text{Area}\left(\text{cm}^2\right)}$$

The surface roughness is a functioning indicator used in this study to measure the surface roughness before and after machining. The surface roughness is checked by roughness tester machine.

2.4 *Experiments Using Taguchi Design*

The purpose of design is to make suitable combination and set the arrays according to Taguchi design and provides the arrangements of various arrays. Table 1 shows different levels and factors used in study.

Trials were performed according to the design of experiments as shown in Table 2.

The total of 16 trial was performed in duplicates to optimize the value of all functioning parameters for accurate results.

2.5 *Regression and Optimization*

In this work, the regression finds the various operating parameters like shaking rate (rpm), cell concentration (CFU/ml of H_2O), pH and Time (h) which are mention by X_1, X_2, X_3 and X_4 there-input parameters w.r.t output like specific metal removal rate (SMRR), and % change in surface roughness such as Y_1 and Y_2. This statistical tool (regression) is completed by with the help of Microsoft Excel software.

Table 2 Showing 16 arrays of Taguchi design

Arrays	pH	Shaking rate (rpm)	Time (h)	Cell concentration (CFU/ml of H_2O)
1	2	80	24	10^4
2	2	90	48	10^6
3	2	100	72	10^8
4	2	110	96	10^{10}
5	3	80	48	10^8
6	3	90	24	10^{10}
7	3	100	96	10^4
8	3	110	72	10^6
9	4	80	72	10^{10}
10	4	90	96	10^8
11	4	100	24	10^6
12	4	110	48	10^4
13	5	80	96	10^6
14	5	90	72	10^4
15	5	100	48	10^{10}
16	5	110	24	10^8

Optimization is done with the help of MATLAB software and using different objective of the genetic algorithm problem-solving tool. The various steps involved in the process were arranging the objective function, arranging the limits on variables, selecting the algorithm and finally evaluating the results.

3 Results and Discussion

In this work, 16 tests were conducted according to the Taguchi design using *At. ferrooxidans* bacteria. The test material was set over the shaker by keeping up their pH values and cell concentration (CFU/ml of water) in the medium at a particular shaking speed for different interval of time. Subsequent to this, the effects on specific material removal rate (SMRR) and percentage change in surface roughness (%ΔSR) were recorded to confirm the machining performance for removal of material.

3.1 Results of the Functional Indicators of At. ferrooxidans

The results of the functioning indicators for ferrooxidans have been given in Table 3.

Table 3 Results of SMRR and %ΔSR of various types of arrays

Arrays	SMRR (mg/h cm^2)	Δ SR (μm) final-initial	% age ΔSR
80 RPM			
1	0.235	−3.015	31.55
5	0.185	−2.253	47.45
9	0.106	−3.568	28.98
13	0.080	−1.665	19.36
90 RPM			
2	0.245	−2.587	38.10
6	0.175	−0.625	26.51
10	0.094	−2.411	24.67
14	0.096	−1.341	21.45
100 RPM			
3	0.212	−2.885	38.36
7	0.156	−2.268	26.87
11	0.145	−2.23	31.85
15	0.159	−1.969	24.73
110 RPM			
4	0.175	−2.376	31.66
8	0.163	−2.301	38.40
12	0.159	−1.788	34.18
16	0.157	−1.179	22.25

In Table 3, ΔSR shows the difference between the initial and final value of surface roughness, and SMRR shows the value of decreasing rate of material removal due to metabolic activity.

At 80 rpm, with increase in the value of pH, SMRR starts decreasing because bacteria perform the best at the pH value between 2 and 3 [10]. With the passing of time from 24 to 96 h, the value of SMRR decreases because the nutritional value of the solution starts decreasing, which also leads to the reduction of the metabolic activity of the microbe [5]. With the increase in cell concentration from 10^4 to 10^{10} CFU/ml of H_2O, SMRR starts decreasing. The addition of maximum concentration of polluted bacteria in solution informs about the presence of high Fe^{3+}in the solution. The maximum concentration of Fe^{3+}ion in the solution focuses on the maximum metal removal rate [6].

At 80 rpm, the surface roughness starts decreasing for every operating parameter such as pH, time and cell concentration. The surface roughness starts decreasing as pH value increases [7]. Similarly, the increase in time and cell concentration results in decrease in the value of surface roughness. But at array 9, surface roughness increases at pH value 4, time 72 h and cell concentration value of 10^{10} CFU/ml of H_2O, and this is the situation of indirect biomachining when maximum concentration of the microbes results in high metabolic activity within the solution. It implies maximum concentration of oxidant ion [6]. But upon employing low pH value or maximum machining time, surface roughness easily changes at minimum shaking rate of 80 rpm. The main reason for the distortion of surface roughness during the asymmetric transfer is the concentration of ions [5].

At 90 rpm, with increase in the value of pH, SMRR starts decreasing because bacteria perform the best at pH value between 2 and 3 [10]. With trespassing of time, the value of SMRR reduces because the nutritional value of the solution starts decreasing that further diminishes the metabolic activity of the microbe. It also results in decreasing cell concentration, which in turn affects the production of organic acid, thus impacting the acidly in process.

At 90 rpm, surface roughness starts decreasing as pH value increases. At 48 h, the surface roughness becomes maximum because the nutritional value of solution starts decreasing with the passing of time, which also decreases the metabolic activity of microbes [5]. Similarly, with increase in time and cell concentration, surface roughness reduces because of the decrease in the metabolic activity of *At. ferrooxidans* [11].

At 100 rpm, with increase in the value of pH, SMRR starts decreasing, but with the passing of time, the value of SMRR starts increasing from value 0.212 mg/h cm^2 to 0.156 mg/h cm^2. The value of SMRR is independent of the cell concentration, expect when cell concentration is too high [8, 9]. With increase in cell concentration, SMRR starts increasing owing to high concentration of Fe^{3+} ions in the solution, which would render a high removal rate.

At 100 rpm, the surface roughness becomes maximum at array 3 SR-2.885 μm ($pH = 2$, time = 72 h and cell concentration = 10^8 CFU/ml of H_2O). At 90 rpm, surface roughness starts decreasing with increase in pH value because ferrooxidans acts best at pH value in between 2 and 3. With the passing of time, the value of surface roughness starts decreasing owing to high metabolic activity, but it gradually slows down by passing of time which helps to increase the value of surface roughness.

At 110 rpm, with increase in the value of pH, SMRR starts decreasing because bacteria perform its best at pH value between 2 and 3 [10]. With the passing of time, the value of SMRR increases because of increase in bacteria and the amount of oxygen available and bacteria at the same pH and temperature, as bacterial metabolism is effected by oxygen concentration [12].

At 110 rpm, the surface roughness becomes maximum at array 4 SR-2.376 μm ($pH = 2$, time = 96 h and cell concentration = 10^{10} CFU/ml of H_2O). However, surface roughness starts decreasing as pH value increases because bacteria perform the best at pH value between 2 and 3 [10]. With increase in time, surface roughness starts increasing because either low pH value or high machining time can easily bring change in surface roughness [5]. With increase in cell concentration, the value of surface roughness also increases because of the maximum concentration of Fe^{3+} ion in the solution, which maximizes metal removal rate [6].

The maximum surface roughness is reached at array 9 having parameters (shaking rate = 80 rpm, $pH = 4$, time = 72 h and cell concentration = 10^{10} CFU/ml of H_2O).

3.2 *Effects of Different Parameters and Functioning Indicators by* **Acidithiobacillus ferrooxidans**

This main effect plot graph represents the effect of various parameters or indicators and also gives information about individual response of that parameter or indicator. It is done by MINITAB software. Figures 2 and 3 show the plotting of different operating parameters like pH value, shaking rate, time and cell concentration on SMRR value and %ΔSR.

Figure 2 shows that the SMRR values decrease as the value of pH increases because ferrooxidans works best at pH value between 2 and 3. The SMRR further decreases in case the shaking rate ranges between 80 and 90 rpm, which are demonstrated in Fig. 2. However, with increase in shaking rate from 90 to 100 rpm, the SMRR starts increasing and further decreases because increase in the shaking rate leads to increase in the oxygen concentration in this medium. The SMRR value increases with time but up to 48 h, after which its value decreases till 72 h or 96 h because with the passing of time, the metabolic activity of *At. ferrooxidans* decreases, which lead to the reduction in the concentration of H^+ ion present in solution.

Figure 3 shows that the value of %ΔSR independent on different operating parameters such as pH value, shaking rate, time and cell concentration. With increase in the pH value, there is decrease in the percentage change in surface roughness, because of %ΔSR ferrooxidans works the best for the range of pH value between 2 and 3. Initially, with increase in the shaking rate from 80 to 90 rpm, there is decrease in %ΔSR, but between 100 rpm and 110 rpm, the %ΔSR increases because of a

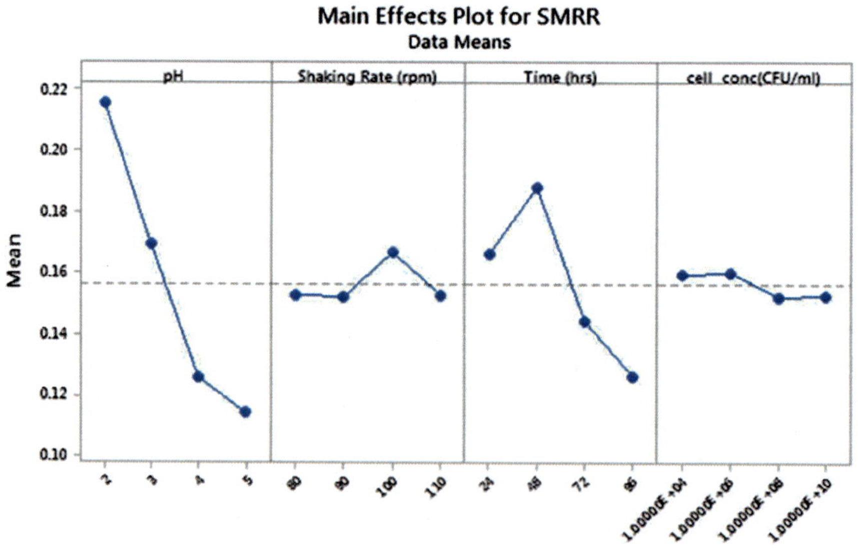

Fig. 2 SMRR main effect plot

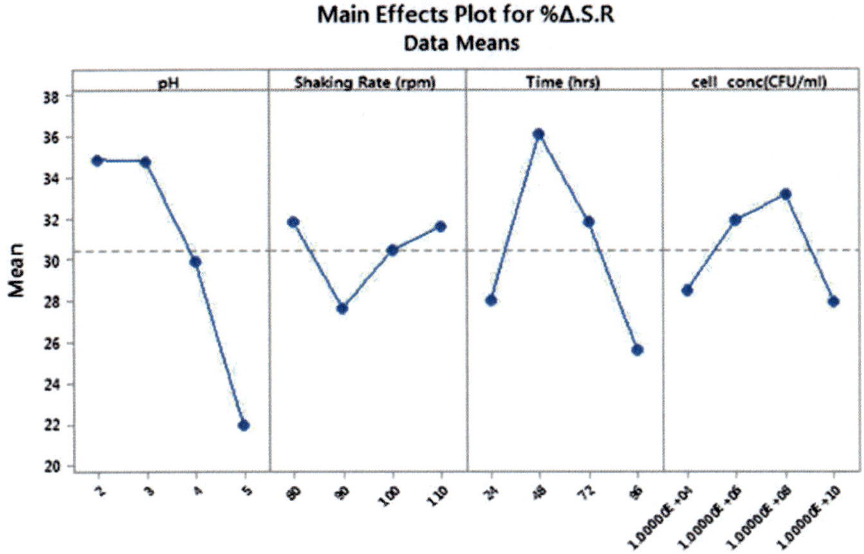

Fig. 3 % ΔSR main effect plot

Table 4 Operating parameters and maximum values of various functioning indicators according to main effects plots

Microorganism	Indicators	pH	Shaking rate (rpm)	Time (h)	Cell concentration (CFU/ml of H_2O)
Acidithiobacillus ferrooxidans	SMRR	2	100	48	10^6
	%ΔSR	3	80	48	10^{10}

uniform contact between the sample and metal at high shaking rate. Moreover, the %ΔSR value increases with time, but up to 48 h, after that the value of %ΔSR starts decreasing till time reaches 72 h or 96 h because with time, the metabolic activity of *At. ferrooxidans* decreases, which lead to the reduction in the concentration of H^+ ion present in solution (Table 4).

3.3 *Regression and Optimized Results*

The objective function of every output is obtained with respect to various input parameters. Regression analysis is used to reach the objective function, which is further used for optimization of input parameters. The output indicators are Y_1 for specific metal removal rate (SMRR) and Y_2 for percentage change in surface roughness, and X_1, X_2, X_3 and X_4 represent input parameters, i.e., pH value, shaking rate,

Table 5 Regression statistics of *Acidithiobacillus ferrooxidans*

Aspect	SMRR	% Δ SR
Regression equation results for various objectives	$Y_1 = 0.306 - 0.034X_1 + 0.00014X_2 - 0.0006X_3 - 4.502 \times 10^{-13}X_4$	$Y_2 = 47.39 - 4.379X_1 + 0.0213X_2 - 0.0482X_3 - 3.2 \times 10^{-10}X_4$
Multiple *R*	0.9169	0.71657
R-square	0.8407	0.51348
Adjusted *R*-square	0.782	0.33656
Standard error	0.0225	6.17109
Observations	16	16

Table 6 Optimized values of operating parameters

Culture	pH	Shaking rate (rpm)	Time (h)	Cell concentration (CFU/ml of H_2O)
Acidithiobacillus ferrooxidans	2	100	24	10^4

time and cell concentration CFU/ml of water. Table 5 shows the regression statistics of *Acidithiobacillus ferrooxidans,* and Table 6 presents the optimized parameter values.

At less pH value, the metal removal rate increases due to more H^+ ions, but with the passage of time, the concentration of H^+ particles decreases because of the utilization of H^+ for the metabolic action of microbe, which ultimately results in a decrease of metal removal rate [12].

The shaking rate 100 rpm and 110 rpm results in more metal removal because the uniform connection between the ferric particle and the workpiece does not permit the development of any layer over the workpiece.

3.4 Evaluation Between Empirical and Experimental Results

In SMRR and surface roughness, there is a minimum difference between both the experimental and empirical results that are depicted in Table 7.

Table 7 Difference between experimental and empirical results

Indicators	Experimental values	Empirical values	Differences (Exp.–Emp.)	Percentage change %
SMRR (mg/h cm^2)	0.2572	0.23429	0.02291	91
%ΔSR	41.3793	39.5405	1.8388	95

Fig. 4 SEM initial workpiece

3.5 Scanning Electron Microscopy (SEM Analysis)

SEM of sample is used to understand its topography, like the type of changes occurring on workpiece sample after biomachining. The changes occurring in the sample under the initial and final machining conditions are compared.

Figure 4 depicts the irregular and rough surface area of aluminum alloy 46500 samples. The irregularity of the surface of sample was determined before biomachining. The change in the sample of aluminum alloy 46500 after treatment by machining process is shown in Fig. 5. The image shows that the surface area becomes smooth after biomachining and oxidation by *At. ferrooxidans* (microorganism).

4 Conclusions

In the initial hours, the SMRR rate was maximum owing to the high metabolic activity of ferrooxidans, which helps in finding the actual trend of ferrooxidans. In addition, the results also provided better surface finish as also highlighted by scanning electron microscopy analysis. Therefore, it is not wrong to suggest that machining using ferrooxidans has shown favorable results based on current context of study.

1. In this study, the analysis shows that the greatest value of SMRR is 0.245 mg/h cm^2 at array 2 (SR = 90 rpm; pH = 2, T = 48 h or cell concentration = 10^6 CFU/ml of H_2O.), and the greatest value of percentage change in

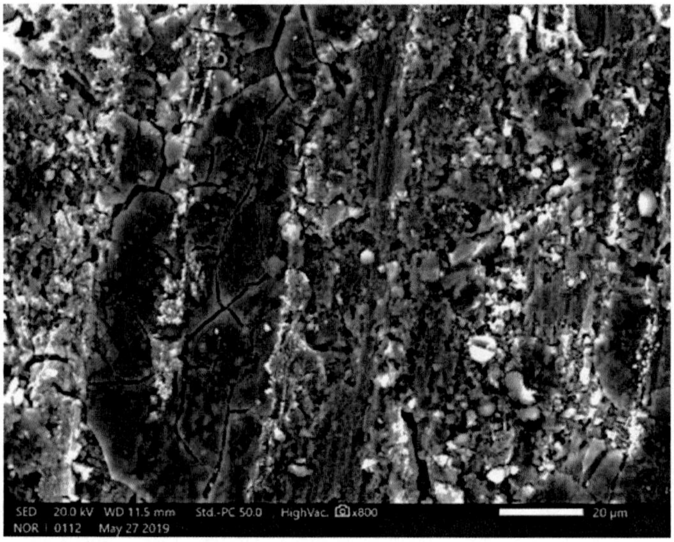

Fig. 5 SEM final workpiece

surface roughness is 47.4 at array 5 (pH = 3, SR = 80 rpm, $T = 48$ h, and cell concentration = 10^8 CFU/ml of H_2O).

2. Scanning electron microscopy (SEM) helps in identifying the trend, which has been followed by smoothness in surface roughness. From the image, it is clear that the surface has been smoothened through machining for 24 h and using different processing parameters.
3. The optimized values for SMRR are SR = 100 rpm, $T = 24$ h, pH = 2, and cell concentration = 10^4 CFU/ml of H_2O.

References

1. Iskanius P, Page T, Anbuudayasankar SP (2010) The traditional industry sector in the changing business environment—a case study of the finnish steel product industry. Int J Electron Customer Relat Manag 4(4):395
2. Zaĭtsev EV (1976) Nakoplenie i Raspredelenie Tseziia-137 v Organakh Karpov. Veterinariia 78(9):86
3. Williams JC, Starke EA (2003) Progress in structural materials for aerospace systems. Acta Mater 51(19):5775–5799
4. Fleischer J, Denkena B, Winfough B, Mori M (2006) Workpiece and tool handling in metal cutting machines. CIRP Ann Manuf Technol 55(2):817–839
5. Jhajj TS, Sodhi AK, Bhanot N (2019) A comparative study on application of *Acidithiobacillus Ferrooxidans* and *Aspergillus niger* for biomachining of EN-19 alloy steel. In: Proceedings of the 1st international conference on sustainable waste management through design, pp 323–335

6. Díaz-Tena E et al (2017) Biomachining: metal etching via microorganisms. Crit Rev Biotechnol 37(3):323–332
7. Johnson D, Warner R, Shih AJ (2007) Surface roughness and material removal rate in machining using microorganisms. J Manuf Sci Eng 129(1):223–227
8. Hocheng H et al (2012) Metal removal by *Acidithiobacillus Ferrooxidans* through cells and extra-cellular culture supernatant in biomachining. CIRP J Manuf Sci Technol 5(2):137–141
9. Hocheng H, Chang JH, Jadhav UU (2012) Micromachining of various metals by using *Acidithiobacillus Ferrooxidans* 13820 culture supernatant experiments. J Clean Prod 20(1):180–185
10. Uno Y, Kaneeda T, Yokomizo S (1996) NII-electronic library service. Fundamental study on biomachining: machining of metals by Thiobacillus ferrooxidans. JSME Int J Ser C Dyn Control Robot Des Manuf 39(4):837–842
11. Verma P, Sodhi AK, Bhanot N (2019) A study on biomachining of aluminium alloy 4004 using *Acidithiobacillus ferrooxidans*. In: Proceedings of the 1st international conference on sustainable waste management through design. Springer International Publishing, Berlin, pp 45–50. http://link.springer.com/10.1007/978-3-030-02707-0
12. Muhammad I, Ullah SMS, Han DS, Ko TJ (2015) Selection of optimum process parameters of biomachining for maximum metal removal rate. Int J Precis Eng Manuf Green Technol 2(4):307–313

Microplastics—A Review of Sources, Separation, Analysis and Removal Strategies

Ishfaq Showket Mir and Punnet Pal Singh Cheema

1 Introduction

Plastics find extensive use in daily life and industries due to their lightweight, resistance to environmental degradation and long life. The enormous growth in plastic industry from 1950 projects that plastic production will reach 33 billion by 2050. This exponential rate of plastic production and use has resulted deposition of 70–80% plastics in aquatic environments. The high accumulated volumes of plastics in the oceans are broken down due to wind and wave action, microbial degradation and UV radiations. This combined action of wind, waves, microbes and UV radiations results in formation of macroplastics, mesoplastics and microplastics. Microplastics have been defined as plastic particles below sizes 5 mm [2] resulting from fragmentation of larger particle. Microplastics may be formed by breakdown of large plastics and may be also found in textiles, wastewater effluents and paints [14]. These global contaminants are categorized as primary and secondary microplastics. Primary microplastics are directly released into the environment and are particularly leftovers of paints, personal care products, detergents or any other manufacturing wastes. These may be generated from industrial abrasives used to blast surfaces, modelling plastic powders, drilling fluid particles for gas, oil exploration and plastic fabrication [5]. Secondary microplastics are produced from weathering breakdown of plastic litter and fragments of post-use agriculture mulch films left in fields [11]. These are a result of photo-oxidation and mechanical degradation of primary microplastics. The sources may

I. S. Mir (✉) · P. P. S. Cheema
Department of Civil Engineering, Guru Nanak Dev Engineering Collage, Ludhiana, Punjab 141006, India
e-mail: Ishfaqshowketmir@gmail.com

P. P. S. Cheema
e-mail: ppsc390@gmail.com

H. Singh et al. (eds.), *Sustainable Development Through Engineering Innovations*, Lecture Notes in Civil Engineering 113,
https://doi.org/10.1007/978-981-15-9554-7_52

be degradation fragments of large plastic chunks, coastal markets, fishing nets, shipping and from natural disasters like floods, cyclones, etc. Other sources of secondary microplastics as listed are littering, dumping of plastics, natural disasters, plastic mulching, car tyres abrasion and paper recycling facilities. Microplastics have been reported in marine systems as well, e.g. the surface waters of north-western Pacific ocean are polluted with microplastic concentrations from 645 to 42500 items/km^2 [13]. Arabian Bay surface waters are stated with microplastic concentration ranging from 4.49×10^4 to 1.38×10^6 items/km^2. Arctic surface and subsurface water samples have been reported with microplastic concentrations ranging from 0–1.31 to 0–11.6 items/m^3. Microplastics have also been reported in freshwater environments. A study conducted on surface waters of Tamar estuary (England) revealed presence of microplastics with average concentration ranging from 0.0028 particles/m^3 [15]. Microplastic concentration ranging from 2427 to 7060 particles/m^3 has been reported from fresh water lakes of Changsha, China [19]. Saigon river in Vietnam has been reported with high concentration of microplastics ranging from 172400 to 419030 particles/m^3. Landfills are also considered to be a potential source of microplastics due to burial of large amount of plastics. Severe and harsh conditions such as PH, high salinity, fluctuating temperatures, stresses, enzymatic microbial degradation lead to breakdown of plastics to microplastics and get carried out by discharge of leachates [18].

2 Separation

In wastewater, microplastics can be collected by using various methods such as containers, autosampler, separate pumping and filtration and surface filtration. To concentrate microplastics, wastewater samples are filtered to an amount that depends on mesh and pore size of filters and sieves employed. Extraction and separation methods of microplastics vary for sediments and seawater. Large microplastics in water can be separated by using a net or by filtering using filters or sieves of appropriate size. A 10-L stainless steel bucket is used to collect sludge samples. Flotation is an appropriate method employed in case of sediments and soil with density ranging from 2.6 to 2.7 g cm^3. In flotation method, a high-density salt solution may be used for extracting microplastics. $CaCl_2$ (1.5 g cm^3) and NaCl (1.2 g cm^3) have been mainly used due to economic feasibility and lack of pollution potential. Acid and alkali solvents or oxidizing agents such as NaOH and H_2SO_4 are being used to digest organic matter from flotation. However, consideration should be given to disintegration of microplastics due to oxidising agents. Disintegration may affect weight, number and shape of microplastics. Minor changes in mass and size of microplastics with the use of Fenton's reagent under appropriate pH and temperature have been observed [7].

For purification and extraction of microplastics from sludge samples which may be enriched with high amount of organic matter, different methods may be employed. Some methods include wet peroxidation, enzymatic degradation and alkaline acid treatment. A new technique based on pressurized fluid extraction uses solvents such as dichloromethane and hexane to extract microplastics under high temperature and pressure (1500 PSI). Then a stream of nitrogen is used to remove solvents by evaporation [4]. For isolation of microplastics from marine sediments, elutriation technique has been found successful. In this technique, water is pumped up through the column (1.5 m) containing sediment sample and aeration is followed. Lighter particles separated from sand are collected on a sieve. Inorganic materials in sludge and wastewater samples are removed using density separation. Combined methods of density separation with 96% ethanol have been used in case of beach samples for extracting microplastics.

Enzymatic digestion and chemical degradation have been used for purification of microplastics from the environmental matrices. 10% KOH solution was found to be an optimum method for organic matter digestion without any change in microplastics except cellulose. Proteinase K has been reported as an efficient enzyme (97% recovery) for separating microplastics from biota-rich water samples. The main limitation of this method is its high cost. To eliminate this drawback, technical grade enzymes [9] in combination with oxidative purifying steps (35% H_2O_2) and a density separation with $ZnCl_2$ have been proven economical.

3 Analysis and Detection

Various detection and analytical techniques usually employed are visual detection, FTIR spectroscopy, Raman spectroscopy and thermal analysis. Depending on the two characteristics, physical (size, shape and colour) and chemical (polymer type) properties of microplastics, only a single method of analysis is not suitable. It is difficult to identify microplastic using a single analytical method, so a combination of various analytical methods is used. The critical factor for an identification method to be considered for analysis is the cut-off size of microplastics.

3.1 Visual Detection and Microscopy

The improved method in the visual identification of microplastics is the use of staining dye, e.g. Nile red that has the ability to bind with synthetic polymers. The microplastic particles are irradiated after staining and can be identified by microscopy [16]. Visual identification methods are used in combination with other appropriate methods. For example, scanning electron microscopy (SEM) technique coupled with spectroscopy makes use of X-rays to activate the sample and then identifying its composition and size. However, visual identification method cannot be used for high strandline samples usually found on beaches where the possibility of interference with high loads of organic and inorganic contaminants is more and where colours are same as interfering particles. Scanning electron microscopy can provide clear and high-resolution images of surface texture of plastic like particles that aids in distinguishing microplastics from other organic materials. Combining SEM and energy-dispersive X-ray spectroscopy (EDS) is effective in providing elemental composition of the same object. This elemental composition can be used to distinguish carbon-dominant plastics from inorganic particles. However, this combined method of SEM-EDS is expensive and takes substantial time and effort for examining and preparation of sample that can be a limitation in the number of samples to be handled. Other methods of microscopy such as polarized optical microscopy has been used successfully in specific cases, e.g. for identification of polyethylene particles in laboratory accumulation and toxicity tests. Crystalline nature of some polymers can be detected through the use of polarized light transmission that influences the crystal structure within that plastic. Biogenic materials that are left to be eliminated from sediment and neuston net samples make observation by microscopy difficult [1].

3.2 Fourier Transform Infra-Red Spectroscopy

In this method, microplastics may be exposed to IR radiation and obtaining spectrum whose characteristic peaks indicate chemical bonds between atoms. The obtained spectra are then compared with reference spectra. With the employment of FTIR spectroscopy, it was easier to identify carbon-based polymers. Production of unique spectra from different bond composition is used to establish a distinction between plastics and other materials. Micro-FTIR is used to observe micro-sized plastic particles by switching between object lens and IR probe. Different modes of FTIR for microplastic analysis are transmission, reflectance and attenuated total reflective mode (ATR). More stable spectra can be obtained for irregular surfaces of microplastic by using ATR mode. ATR-FTIR analysis is a sort of contact analysis. However, production of pressure in ATR probe can damage weaker microplastics and can be pulled from filter paper by electrostatic interaction with the tip of probe. Another limitation of using FTIR analysis is longer operating times required for mapping with single FTIR which can cover limited area of selected membranes for specific types of polymers. However, these drawbacks of FTIR can be removed in analysis by using focal plane array-based reflectance imaging technique. This technique is used to identify microplastics in the size range 150–250 μm with accuracy and less time.

3.3 Raman Spectroscopy

Raman microscopy uses submicron wavelength lasers as its source and can be used to identify particles down to 1 micron and less. This method can identify particles with aromatic bonds, C–H bonds and carbon double bonds. This method of analysis is not only used to identify microplastics but also provides information on composition of samples. For a complex identification of microplastics, Raman spectroscopy and FTIR can be used in compromised. As there is the use of small dia laser beam in Raman spectroscopy, microplastics of few micrometres size can be identified. Another advantage of Raman spectroscopy is that it does not damage the sample particles, so they can be used for further analysis. However, there can be interference in identification if some pigment and additive chemicals are present in samples.

3.4 Thermal Analysis

This is the recently tested method of identification for microplastics. This method measures the change in properties of microplastics and are identified on the basis of their thermal stability. Differential scanning calorimetry can be used to identify primary microplastics for which reference materials are available. Thermogravimetry in combination with differential scanning calorimetry can be used to identify polyethylene and polypropylene. For identification of polyvinylchloride, polyurethane, polyester, the combination of thermogravimetry and DSC cannot be used because of their overlapping phase transition signals. Thermal analysis however is categorized as a destructive method subsequent analysis cannot be done further on these microplastic samples. The combination of thermal analysis with GC-MS is used to analyse additive chemicals in microplastics. Weight basis microplastic concentration data can be obtained by employing the combined method of thermal analysis and GC-MS.

4 Treatment and Removal

A study revealed that microplastics in influent flow are mostly removed in primary treatment processes. From a comparative study in Australia, microplastic composition in primary, secondary and tertiary treatment zones was investigated. The study revealed 6.29, 0.48 and 1.55 microplastics per litre in respective steps of treatment. Spatial distribution study to observe the pattern of microplastics in a WWTP reported the major presence as secondary microplastics along treatment process and it was also revealed that primary microplastic concentration increased downstream of WWTP. A gravity operated filtration system with filter mode and back flush mode is used for removal of secondary effluents in WWTPs. With this gravity-operated filtration system, microplastics decreased from 0.8 to 0.002 MPs per litre. It has been proved that some treatment of wastewater may eradicate more than 95% of microplastics but it should be also noted that throughout the process of treatment, microplastics do not get thermally, chemically or biologically degraded but only a change in distribution occurs.

4.1 Application of Membrane Technology for Microplastics Removal

From last five years, there has been an increased attention towards removal of microplastics by employing membrane technology. Though there are some limitations associated with this technology, yet there is an increase in use of conventional membrane separation with membrane bioreactors. The removal of microplastics was found to be dependent on some parameters such as size, shape and mass. With employment of MBR technology, it was observed that microplastic concentration decreased from 6.7 to 0.004 MP per litre. With higher removal efficiency than tertiary treatment, MBR technology was used to treat primary clarified wastewater that consisted of more MP concentration than secondary effluent. The performances of membrane processes was evaluated by considering certain factors that influenced the operation such as material of membrane, pore size, thickness of membrane, properties of membrane surface, polluted water source. Other membrane process factors that influenced the operation of membranes were included as flux, trans-membrane pressure, polarization concentration, specific energy consumption, etc. Nowadays, ultrafiltration with coagulation is considered to be the main water treatment technology. However, this combined technology has not proven to be effective for removal of microplastics. So the worrying levels of microplastics in freshwater make it necessary to perform an in-depth research on behaviour of microplastics during treatment with coagulation and ultrafiltration. Some studies have focussed on behaviour of polyethylene in drinking water treatment by ultrafiltration and coagulation by the use of iron-based coagulant. However, these studies have reported removal efficiency of about 15% leading the process to be ineffective for microplastic removal. Instead to increase the coagulation performance when polyacrylamide was added, the removal efficiency of particles with size $d < 0.5$ mm increased from 15 to 91%. From these studies, we can conclude that ultrafiltration can be used to remove polyethylene, but still there should be more effort in order to understand cake layer formation and influence on membrane fouling due to presence of microplastics.

5 Impacts of Microplastics

Microplastic pollution effects have been experienced by aquatic life at each trophic level. Microplastics can pose impacts on environment physically, biologically and chemically. Physical impacts include entanglement and ingestion. Direct ingestion of plastic remains by marine species can be fatal as revealed by [8] on a study comprising of 200 species in terms of drowning, suffocating, strangulating or starving. Ingested plastic particles can lead to digestive blockage or internal damage due to abrasion. After ingestion, chemical and biological impacts imply that microplastics may cause poisonous impacts to living organisms through various passages and mechanisms. Microplastics can be incorporated into aquatic life through various means such as filter feeding, air-water surface inhalation and consumption of prey exposed to microplastics. Research has revealed physical damages or serious injuries caused by microplastics that abrade against the walls of internal organs. Hazardous chemicals that are added to polymers get attracted to plastic surfaces and may affect growth of organism, reproductive activities. The hydrophobic nature of some organic pollutants (e.g. PCBs, DDT and dioxins) makes them float on the surface of plastic particles [17]. Marine animals whenever feed on the microplastics ingest toxic chemicals and are transferred up the food chain [10]. Dicholoro diphenyl trichloroethane (DDT) leads to immune deficiency and neurological effects [12]. Some other effects on aquatic organisms may include change in faecal density, inflammatory effects, increase in death rate, decline in feeding efficiency and many other long-term impacts due to transfer of toxins in the underlying tissues of organisms body. Habitat destruction of microorganisms is a direct impact of microplastics sediment contamination resulting in alteration of water pathways and flow and the heat transfer.

6 Conclusion

The analysis of literature has revealed the lack of a reliable technology that efficiently retains microplastics at wastewater treatment facilities. Details on removal of microplastics at each step of treatment are unknown. Membrane bioreactors can efficiently remove microplastics during treatment but they are uneconomical and difficult to install in an existing treatment facility. MBBR can be only used when high standards are required. As per the current threat of microplastic pollution, there are some underlying issues associated that need to be considered for a better control on microplastic pollution. Priority concern that should be taken up are the sources, remediation and removal strategies. International level of encouragement is to be approached to avoid use of microbeads in personal care products. Instead use of biodegradable plastics like polyactide and polyhydroxyalkanates use should be encouraged. Even with considerable research on distribution and occurrence of microplastics in fresh water and aquatic ecosystems, there is however a dire need to learn about bioaccumulation and impacts of microplastics in both of aquatic environments. From various knowledge gaps identified after reviewing the literature on microplastics, there is need of appropriate evaluation of nano-sized particles and their distribution. Moreover, studying various interactions between microplastics and other pollutants while assessing broad range of chemical pollutants is needed so that more environmental implications associated with microplastics and their impacts on human health is evaluated. Multiple species exposure in varied environmental conditions should be assessed in order to find ecotoxicity of microplastics.

Sampling methods and techniques need further development in terms of their efficiency because sampling is important for correct. Spectroscopic analysis of microplastics can become difficult if polymer additions and biofilms are present infrared spectra of various microplastic types may vary with the variation of interaction with environment, which can produce new functional groups and changes at surface level. For emission prediction of microplastics, accidental release or littering is one of the great uncertainties to be quantified. Weathering and fragmentation issues related to microplastics needs to be extensively focussed. The microplastic occurrence data has shown that fibres (82%) are more dangerous than particles (18%) hence focus shifts towards microplastic fibres in wastewater and other microplastic sources. There is lack of knowledge considering various modifications that microplastics undergo once subjected to abrasions. Increasing demand of monitoring the pollution by microplastics at national and international level requires the development of novel methodology in order to reduce effort and time. Improving textile production reduces release of microfibres by (a) improving knitting techniques, (b) combining synthetic and natural textiles, (c) Applying textile coatings and (d) Eliminating loose fibres during production. For reducing pollution by microplastics, upstream sources need to be identified, because natural water streams make recovery of microplastic very different. Fate, fragmentation and distribution of polymer products need to be evaluated to study their short- and long-term effects. Effect of exposure experiments on microplastics needs to be evaluated out. Future research should be directed to develop new approaches in modelling to analyse microplastics transport in soil, water and sediments.

References

1. Anderson PJ, Warrack S, Langen V, Challis JK, Hanson ML, Rennie MD (2017) Microplastic contamination in Lake Winnipeg, Canada. Environ Pollut. https://doi.org/10.1016/j.envpol.2017.02.072
2. Browne MA, Crump P, Niven SJ, Teuten E, Tonkin A, Galloway T, Thompson R (2011) Accumulation of microplastic on shorelines woldwide: sources and sinks. Environ Sci Technol. https://doi.org/10.1021/es201811s
3. Erni-Cassola G, Gibson MI, Thompson RC, Christie-Oleza JA (2017) Lost, but found with Nile Red: a Novel method for detecting and quantifying small microplastics (1 mm to 20 μm) in environmental samples. Environ Sci Technol. https://doi.org/10.1021/acs.est.7b04512
4. Fuller S, Gautam A (2016) A procedure for measuring microplastics using pressurized fluid extraction. Environ Sci Technol. https://doi.org/10.1021/acs.est.6b00816
5. Hermabessiere L, Dehaut A, Paul-Pont I, Lacroix C, Jezequel R, Soudant P, Duflos G (2017) Occurrence and effects of plastic additives on marine environments and organisms: a review. Chemosphere. https://doi.org/10.1016/j.chemosphere.2017.05.096
6. Horton AA, Svendsen C, Williams RJ, Spurgeon DJ, Lahive E (2017) Large microplastic particles in sediments of tributaries of the River Thames, UK—abundance, sources and methods for effective quantification. Mar Pollut Bull. https://doi.org/10.1016/j.marpolbul.2016.09.004
7. Hurley RR, Lusher AL, Olsen M, Nizzetto L (2018) Validation of a method for extracting microplastics from complex, organic-rich, environmental matrices. Environ Sci Technol. https://doi.org/10.1021/acs.est.8b01517
8. Laist DW (1997) Impacts of marine debris: entanglement of marine life in marine debris including a comprehensive list of species with entanglement and ingestion records. https://doi.org/10.1007/978-1-4613-8486-1_10
9. Löder MGJ, Imhof HK, Ladehoff M, Löschel LA, Lorenz C, Mintenig S, Piehl S, Primpke S, Schrank I, Laforsch C, Gerdts G (2017) Enzymatic purification of microplastics in environmental samples. Environ Sci Technol. https://doi.org/10.1021/acs.est.7b03055

10. Lusher A (2015) Microplastics in the marine environment: distribution, interactions and effects. Mar Anthropogenic Litter. https://doi.org/10.1007/978-3-319-16510-3_10
11. Magni S, Binelli A, Pittura L, Avio CG, Della Torre C, Parenti CC, Gorbi S, Regoli F (2019) The fate of microplastics in an Italian wastewater treatment plant. Sci Total Environ. https://doi.org/10.1016/j.scitotenv.2018.10.269
12. Mansouri A, Cregut M, Abbes C, Durand MJ, Landoulsi A, Thouand G (2017) The environmental issues of DDT pollution and bioremediation: a multidisciplinary review. Appl Biochem Biotechnol. https://doi.org/10.1007/s12010-016-2214-5
13. Pan Z, Guo H, Chen H. Wang S, Sun X, Zou Q, Zhang Y, Lin H, Cai S, Huang J (2019) Microplastics in the Northwestern Pacific: abundance, distribution, and characteristics. Sci Total Environ. https://doi.org/10.1016/j.scitotenv.2018.09.244
14. Rezania S, Park J, Md Din MF, Mat Taib S, Talaiekhozani A, Kumar Yadav K, Kamyab H (2018) Microplastics pollution in different aquatic environments and biota: a review of recent studies. Mar Pollut Bull. https://doi.org/10.1016/j.marpolbul.2018.05.022
15. Sadri SS, Thompson RC (2014) On the quantity and composition of floating plastic debris entering and leaving the Tamar Estuary, Southwest England. Mar Pollut Bullet. https://doi.org/10.1016/j.marpolbul.2014.02.020
16. Shim WJ, Hong SH, Eo SE (2017) Identification methods in microplastic analysis: a review. Anal Methods. https://doi.org/10.1039/c6ay02558g
17. Teuten EL, Saquing JM, Knappe DRU, Barlaz MA, Jonsson S, Björn A, Rowland SJ, Thompson RC, Galloway TS, Yamashita R, Ochi, D., Takada H (2009) Transport and release of chemicals from plastics to the environment and to wildlife. Philos Trans R Soc B Biolog Sci. https://doi.org/10.1098/rstb.2008.0284
18. Triebskorn R, Braunbeck T, Grummt T, Hanslik L, Huppertsberg S, Jekel M, Knepper TP, Krais S, Müller YK, Pittroff, M. Köhler HR (2019) Relevance of nano- and microplastics for freshwater ecosystems: a critical review. TrAC Trends Anal Chem. https://doi.org/10.1016/j.trac.2018.11.023
19. Yin L, Jiang C, Wen X, Du C, Zhong W, Feng Z, Long Y, Ma Y (2019) Microplastic pollution in surface water of urban lakes in Changsha, China. Int J Environ Res Public Health 16(9). https://doi.org/10.3390/ijerph16091650

Minimizing Weight of Frames by Adopting Built-up Sections Over Hot-Rolled Sections for Sustainable Construction

Sukhwinder Singh and Harpal Singh

1 Introduction

Migration from rural to urban areas is increasing every year and thus, cities need to expand so as to accommodate everyone. For cities to grow, it requires more land and more buildings. Buildings are important part of urban infrastructure. With limited time and resources, building sustainable infrastructure is a challenge. Urban infrastructure mainly involves development of roads, buildings, electrical and water works. Urban sustainability can be achieved only if economic, social and environmental ramifications are kept in mind during the development of infrastructure. Major attributes related to the assessment of sustainable infrastructure are lower carbon footprints, protecting natural ecosystems, optimal use of natural resources, prompting industrial and technological modernization, escalate investment in education and research and development (R&D), raising employment and manifest financial viability.

Steel design promotes a cleaner environment by working together towards improving the environmental sustainability of the entire steel supply chain, from steel manufacturing and its usage in construction industry to its recycling and reuse after its design life is over through innovative technology. Steel structures in the past had been designed using conventional hot-rolled sections. Such buildings are called conventional steel building (CSB). Problem with the hot-rolled sections is that they are generalized sections and so, their design does not guarantee an optimized section that will have maximum utilization ratio and economy. Low utilization ratio means

S. Singh · H. Singh (✉)
Department of Civil Engineering, Guru Nanak Dev Engineering College, Ludhiana, Punjab 141006, India
e-mail: hps_bhoday@yahoo.com

S. Singh
e-mail: shahji1991@gmail.com

H. Singh et al. (eds.), *Sustainable Development Through Engineering Innovations*, Lecture Notes in Civil Engineering 113,
https://doi.org/10.1007/978-981-15-9554-7_53

excess steel is being used than required and this clearly indicates towards wastage of the material being done. For creating sustainable infrastructure, optimum utilization of the resource ought to happen. Moving in that direction, in the present study, hot-rolled steel I-sections are replaced with built-up steel I-sections so as to achieve maximum utilization ratio, thereby reducing quantity of steel and cost for a building.

The buildings that are built using built-up tapered steel I-sections are known as pre-engineered buildings (PEBs). The framework for construction of PEB is that the structural components are made in factories, then transported to the site and at site their assembling and erection is done. Pre-engineered design provides optimal usage of steel because the tapered I-sections being prepared in factories are as per the requirements of upcoming forces and bending moments on the structure. Construction of PEBs is faster and provides economic, environmental and social benefits in comparison to CSBs by optimizing steel weight, reducing carbon footprint of building and generating employment opportunities through opening up of factories.

The researchers, in the recent past, had been comparing the PEB steel frames with CSB frames. For instance, Saleem et al. analysed and designed steel industrial building by using conventional steel hot-rolled sections and pre-engineered tapered and cold formed sections. The study concluded that PEB buildings tapered sections used in primary frame is 32% cheaper and cold formed sections used as purlins and side girts had approximately 60% lesser weight as compared to conventional hot-rolled sections [1]. Pradeep and Rao presented their research on comparison between the steel weights of a 20 m CSB truss frame with that of a PEB frame and concluded that PEB frame is 26% lighter than the CSB truss frame [2]. Wakchaure and Dubey carried out comparative study of industrial PEB frame with CSB frame for steel requirement using two Indian standard design codes, IS 800:1984 and IS 800:2007. The results showed PEB structure required 30% lesser steel than CSB structure [3]. Shashank and Kulkarni modelled a 12.0 m wide steel frame in STAAD.Pro and found that PEB steel frames are 33% lighter than the CSB steel frames [4]. The research suggested that the 35% reduction in weight of steel can be achieved if PEB structure is built over CSB. Saleem and Qureshi had presented the study that compares 48 different 3-D frames in terms of their weights, lateral displacements (sway) and vertical displacements (deflection) by altering the spans, bay spacing and lateral forces of the frames [5]. It was observed that the reduction in the steel weight was in the range of 30–40%. Also, the PEB steel frames were found to be showing lesser lateral and vertical displacements as compared to CSB steel frames.

After reviewing the literature, it can be deduced that the percentage reduction in frame weights of PEB versus CSB is dependent on the frame span, spacing of bays and the lateral force effects coming on the frame. Moreover, not much research was found where built-up tapered I-sections were being compared with conventional hot-rolled I-sections using IS 800:2007 [6].

The present study discusses the effect of span and bay spacing on the economy achieved by the use of built-up steel I-sections. To achieve the objectives, 32 numbers of different frames were modelled in STAAD.Pro. Typical 3-D frames using built-up tapered I-sections and hot-rolled I-sections in STAAD are shown in Figs. 1 and 2, respectively. They were analysed and designed as per limit state method given in IS

Fig. 1 Typical 3-D frame consisting of built-up tapered I-sections

800:2007. 3-D regular steel frames consisting of three bays having different spans of 12, 15, 18 and 21 and different bay spacing of 4, 6, 8 and 10 m were taken up for the study. By changing the bay spacing, its effect on quantity of steel required by built-up sections and hot-rolled sections in the frames was observed.

2 Problem Formulation

3-D regular frames having fixed height (9 m) and three numbers of bays were modelled in STAAD.Pro software. The different spans taken up in study were 12, 15, 18 and 21 m. Four spacing for bays considered were 4, 6, 8, 10 m. For each regular

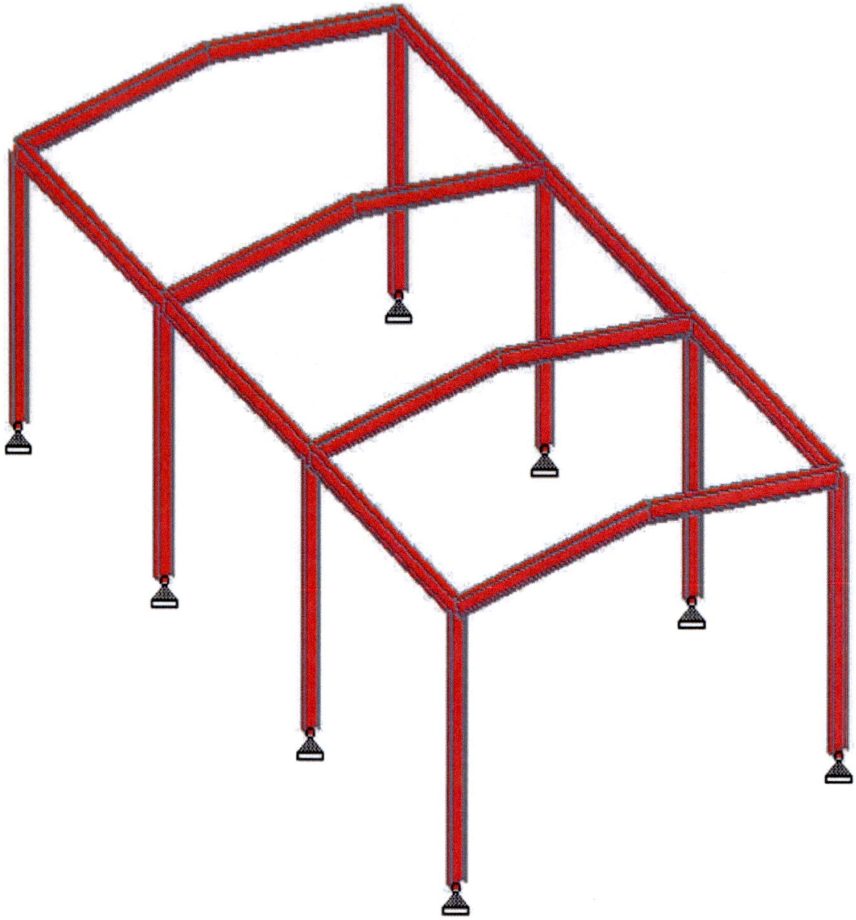

Fig. 2 Typical 3-D frame consisting of hot-rolled I-sections

frame with constant bay spacing, weight of steel required for different spans was taken from the software. The yield strength of the structural steel material considered in the study is 250 N/mm^2 for both PEB and CSB. The hot-rolled sections assigned were taken from IS 12778:2004 [7] and the tapered steel sections were built-up using steel plates of standard thickness mentioned in IS 1730:1989 [8]. The frames were analysed and designed using limit state design of Indian Standard code IS 800:2007.

2.1 Dead Load

The self-weight of primary frame members as shown in Figs. 1 and 2 is taken up by the software itself. Other imposed dead load due to secondary elements like roof sheeting, purlins, insulation, etc., on the roof is taken as 0.5 kN/m^2.

2.2 Roof Live Load

Roof live load is generated on the roof during construction and maintenance work. The value of the load adopted is 0.75 kN/m^2 as recommended in IS 875 (Part 2):1987 and no live load reduction was considered as roof slope in each frame was kept unchanged at 7.5 degrees [9].

2.3 Wind Load

The wind load is assessed with the help of IS 875 (Part 3):2015 [10]. The list of parameter and their values required for finding the design wind pressure, $P_{d,}$ are displayed in Table 1. The design wind pressure, $P_{d,}$ was calculated as 0.9277 kN/m^2 and remains same for all the frames being studied. Enclosure condition of all the frames was taken as 'Partially Enclosed having openings size ranging from 5 to 20 percent'. So, internal pressure coefficient (C_{pi}) was taken as ±0.50. External pressure coefficients (C_{pe}) for walls and roofs are based on geometry of the frame modelled and the direction of the wind. These values were selected from Table 55 and Table 6 of IS 875 (Part 3), respectively.

Table 1 Values of different parameters adopted for finding design wind pressure, P_d

Parameters	Values selected
Basic wind speed (V_b)	47 m/s
Probability factor (risk coefficient), K_1	1.0 (CLASS-A)
Terrain roughness and height factor, K_2	1.0 (terrain category-2)
Topography factor, K_3	1.0
Importance factor for the cyclonic region, K_4	1.0
AREA averaging factor, K_a	0.8
Wind directionality factor, K_d	0.8
Combination factor, K_c	0.9

Table 2 List of limit state design load combinations

S. No.	Load combinations
1.	1.5 (dead load ± live load)
2.	1.5 (dead load ± earthquake load)
3.	1.5 (dead load ± wind load)
4.	0.9 dead load ± 1.5 earthquake load
5.	0.9 dead load ± 1.5 wind load
6.	1.2 dead load ± 1.2 live load ± 1.2 earthquake load
7.	1.2 dead load ± 1.2 live load ± 1.2 wind load

2.4 Earthquake Load

The seismic analysis was performed by the STAAD.Pro software using IS 1893 (Part 1):2016 [11]. All the 3-D frames lies in the seismic zone IV, having response value of 5.0 and importance factor of 1.2.

2.5 Load Combinations

The list of limit state design load combinations used for analysis and design has been mentioned in Table 2. The earthquake and wind loads were applied on all the 3-D frames in longitudinal as well as in transverse direction.

3 Results and Discussions

For each regular 3-D frame with constant bay spacing, thirty-two numbers of models were prepared for different spans. Tapered steel I-sections were assigned to sixteen frames and the left ones were assigned narrow parallel flanged beams (NPB) or wide parallel flanged beams (WPB) hot-rolled steel sections taken from IS 12771:2004. The frames were analysed, designed and optimized. The results of the weight of steel utilized by the primary frame for all the cases considered in the research were tabulated in Table 3.

For drawing inferences precisely from the results in order to identify that whether by replacing the traditional hot-rolled I-sections with the tapered steel I-sections proved effective in material and cost saving and making structure more sustainable, Table 4 was prepared to study percentage variation in weight of steel consumed by PEB with respect to CSB corresponding to constant bay spacing and different spans. Also, it is required to understand the effect of changing bay spacing on the overall economy of the structure.

Table 3 Weight of steel utilized by primary frame for all the cases considered in the research

Bay spacing (m)	Weight in metric tonnes (MT) of 3-D frame using tapered I-sections (PEB)				Weight in metric tonnes (MT) of 3-D frame using hot-rolled I-sections (CSB)			
	Span (m)				Span (m)			
	12	15	18	21	12	15	18	21
4	15.4364	20.0150	25.4753	29.8234	19.0642	23.7844	32.0844	43.9813
6	18.1607	21.9724	28.1871	33.7869	22.3565	26.4759	35.7084	45.3965
8	20.6544	25.4653	31.7242	39.3497	28.5798	33.1751	38.2408	47.7976
10	23.5964	28.2319	35.1248	43.7787	31.9210	36.5864	42.2260	54.5436

Table 4 Percentage variation in weight of steel consumption by PEB with respect to CSB corresponding to constant bay spacing and different spans

Bay spacing (m)	Percentage (%) Variation in weight of steel consumption			
	Span (m)			
	12	15	18	21
4	−19.0	−15.8	−20.6	−32.2
6	−18.8	−17.0	−21.1	−25.6
8	−27.7	−23.2	−17.0	−17.7
10	−26.1	−22.8	−16.8	−19.7

The negative percentage variation means reduction in weight of the steel being used in PEBs from CSBs. The percentage reduction in steel quantity for frames designed using tapered I-sections lies nearly from 16 to 32%. Among the frames studied, all showed reduction in the quantity of steel when hot-rolled sections were replaced with built-up steel I-sections. It is due to the fact that the built-steel sections were tailored made sections that were selected as per the requirements of moments and forces as demanded by various frame members, whereas hot-rolled sections were generalized sections and had to be used as they were available. Furthermore, the reduction in the weight of the frame reduces the effect of seismic forces on it thereby, supporting better material utilization ratio.

The minimum and the maximum variation of 15.8% and 32.2% were found when 3-D frame was designed having constant bay spacing of 4 m and span of 15 m. All the frames designed were checked for allowable deflections as per IS 800:2007 and were found to be safe. Indeed, the results encourages to adopt built-up tapered steel I-sections over conventional hot-rolled I-sections as the former will definitely reduce the cost of building and will avoid excess usage of steel thus, making it more eco-friendly.

The effect of changing bay spacing on steel weight requirements has been projected in form of line graphs for both PEB frames and CSB frames individually, as shown in Figs. 3 and 4, respectively. Four line series are plotted representing steel weight required for different spans (12, 15, 18 and 21 m) against the bay spacing

Fig. 3 Effect of bay spacing on weight of steel required in PEB frames

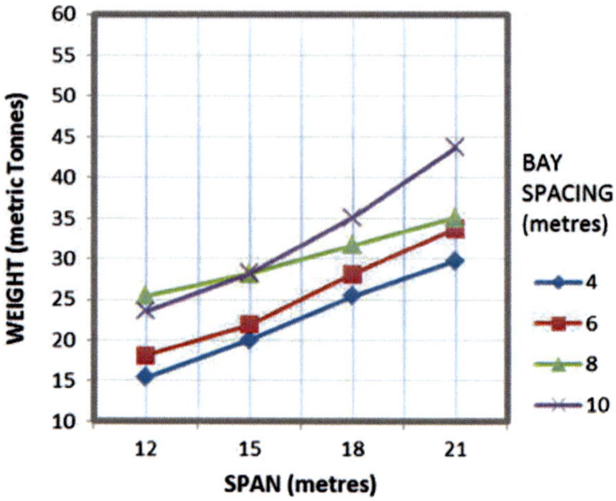

Fig. 4 Effect of bay spacing on weight of steel required in CSB frames

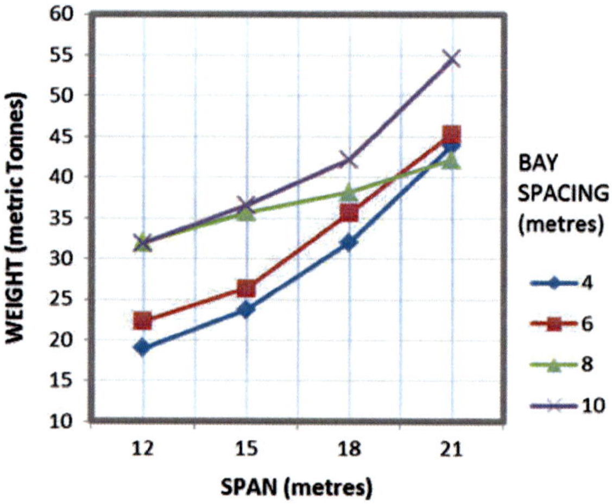

being considered for the regular frame. As four different bay spacing were taken up for study (4, 6, 8 and 10 m), the line in both the graphs corresponding to 8 m bay spacing showed a linear increase in quantity of steel consumed with the increase of span in the PEB as well as CSB frames. On the other hand, change in slope of the lines for other left bay spacing were found to be significant in case of CSB frames spacing when compared with PEB frames. The reason behind such remarkable finding was the fact that the tapered built-up steel I-sections assigned to the frame member were tailored made as per design requirements, whereas hot-rolled steel sections need to be used in CSB frames were to be adopted as available.

4 Conclusion

The present research was carried out with an aim of achieving more sustainable steel design for infrastructure development. Thirty-two numbers of 3-D regular frames with different spans and bay spacing were taken up for the study. The frames were equally divided among two categories, viz. PEB and CSB. The PEB frames were analysed and designed as per IS 800:2007 using built-up tapered steel I-sections, whereas the CSB frames adopted hot-rolled steel I-sections given in IS 12778:2000. Built-up steel sections being made in factory involves less wastage of material and energy due to cutting and welding as compared to fabrication of hot-rolled sections being done at site. Moreover, the fact that PEB is mostly fabricated inside the factory, it makes the assembling and erection of structural components swiftly on site, therefore requiring less completion time for construction when compared to CSB.

Modelling, analysis and design of steel frames were done on STAAD.Pro software. Comparison of PEB frames versus CSB frames in terms of quantity steel obtained was studied. Along with it, the effect of change in bay spacing on weight requirements of steel was reported and discussed. The inferences drawn from the research conducted are as follows:

- The built-up tapered steel I-sections used in the PEB frames lead to the reduction in weight of steel required by minimum 15.8% for a frame having span of 15 m and bay spacing of 4 m and maximum by 32.2% for the frame having the same span and bay spacing as above.
- The effect of bay spacing on quantity of steel utilized suggested that in case of PEB frames, the steel sections being used are optimized as per the forces and moments coming on the frames and therefore, showed more linearity in increments of weight of steel when compared with hot-rolled steel sections, which were to be used as available.
- The frames designed were checked for permissible deflections as specified by IS 800:2007 and were found to be falling under allowable limits.

Apparently, adopting built-up tapered steel I-sections instead of conventional hot-rolled steel I-sections resulted in better utilization of steel as a material will result in saving of material and cost. The carbon footprint generated by a building constructed using built-up tapered sections will be comparatively less than the building being built by using hot-rolled sections, thereby contributing less pollution in environment. Moreover, in order to provide built-up steel sections for construction of PEB, factories will be setting up that will lead to creation of jobs for the public. So, it can be concluded development of urban infrastructure using tapered steel section for construction of PEB is beneficial in terms of economic, social and environment aspects and therefore, supports sustainable infrastructure.

References

1. Saleem MU, Zahid AS, Hisham JQ (2013) Minimum weight design of pre-engineered steel structures using Built-up sections and cold formed sections. Adv Mater Res 684:125–129
2. Pradeep V, Rao PG (2014) Comparative study of pre-engineered and conventional industrial building. Int J Eng Trends Technol 9:1–6
3. Wakchaure S, Dubey NC (2016) Design and comparative study of pre-engineered building. Int J Eng Dev Res 4:2108–2113
4. Shashank P, Sachin MK (2017) Comparative study on the economy between pre-engineered and conventional steel buildings. Int Res J Eng Technol 4:2708–2711
5. Saleem M, Qureshi H (2018) Design solutions for sustainable construction of pre engineered steel buildings. Sustainability 10:1761
6. IS 800:2007 Indian standard general construction in steel-code of practice, third revision
7. IS 12778:2004 Indian standard hot rolled parallel flange steel sections for beams, columns and bearing piles-dimensions and section properties, first revision
8. IS 1730:1989 Indian standard steel plates, sheets, strips and flats for structural and general engineering purposes- dimensions, second revision
9. IS 875 (Part 2):1987 Indian standard code of practice for design loads (other than earthquake) for buildings and structures, second revision
10. IS 875 (Part 3):2015 Indian standard design loads (other than earthquake) for buildings and structures-code of practice, third revision
11. IS 1893 (Part 1):2016 Indian standard criteria for earthquake resistant design of structures, sixth revision

Improving Swelling and Strength Behavior of Black Cotton Soil Using Lime and Quarry Dust

Deepali Anand, Ravi Kumar Sharma, and Abhishek Sharma

1 Introduction

Black cotton soils (BCS) contain highly expansive montmorillonite clay mineral that predominantly belongs to the smectite group. These soils are prone to large volumetric changes on altering water content [11, 12]. The construction of structures is often avoided on these type of soils. However, in the recent years due to scarcity of good soil for construction purposes, engineers are forced to build structures on such soils by adopting suitable soil stabilization techniques. Lime is a common well-known stabilizer used in soil stabilization, and even a little amount of lime has a significant influence on soils comprising montmorillonite as compared to other expansive clay minerals such as kaolinite [3, 7]. A lot of past researchers have used lime as a chemical stabilizer for soil stabilization due to its effectiveness in gaining desirable engineering [3–5, 9, 10]. [3, 9] stated that the optimum lime content in clayey soil reduced maximum dry density (MDD) and plasticity index (PI) and increased optimum moisture content (OMC). Dash and Hussain [5] revealed that there was improvement in strength value and reduction in PI on adding optimum lime content to clay. Sharma and Hymavathi [10] presented that the addition of construction waste and rice husk ash with lime improved the strength and California bearing ratio (CBR) of the clayey soil. Bhardwaj and Sharma [4] showed that adding lime, waste foundry sand, and molasses alone and together in expansive soil increased

D. Anand · R. K. Sharma
National Institute of Technology, Hamirpur, Himachal Pradesh, India
e-mail: ce.deepali@gmail.com

R. K. Sharma
e-mail: rksnithp61@gmail.com

A. Sharma (✉)
Chandigarh University, Mohali, Punjab, India
e-mail: abhishek.e9490@cumail.in

H. Singh et al. (eds.), *Sustainable Development Through Engineering Innovations*, Lecture Notes in Civil Engineering 113,
https://doi.org/10.1007/978-981-15-9554-7_54

Table 1 Geotechnical characteristics of black cotton soil

Property	Value/description
Specific gravity, G	2.71
Soil type	CH
DFS (%)	40.19
Plasticity index (%)	40.60
Maximum dry density, MDD (gm/cc)	1.55
Optimum moisture content, OMC (%)	21.00
Unconfined compressive strength at 28 days, UCS (kPa)	149.06

optimum moisture content and unconfined compressive strength (UCS) value and reduced PI and MDD.

The dust produced due to quarrying activities is detrimental to the environmental pollutions concerning both air and land where it gets deposited. The gradation and physical properties of quarry dust are found analogous to that of natural sand [6, 13, 14]. Therefore, these can be used as an alternative to natural sand for soil stabilization, thus conserving sand as a natural resource. Soosan et al. [13] used quarry dust for stabilizing clay soils and found improvement in CBR value, plasticity, MDD, and compaction characteristics of the soil. Sridharan et al. [14] concluded that quarry dust is a good substitute for improving geotechnical characteristics of high plastic clays. Dixit and Patil [6] found that the gradation curve of quarry dust similar to sand and there was increase in MDD of soil on adding quarry dust to it.

2 Materials

2.1 Soil

The soil used in the study was brought from Rajkot, Gujrat, India, and was classified as clay of high plasticity (CH) according to unified soil classification system (USCS). The various geotechnical properties of clayey soil are given in Table 1.

2.2 Lime

Powdered lime was bought from a local hardware store from Hamirpur, Himachal Pradesh, India. The chemical composition of lime as supplied by manufacturer are given in Table 2.

Table 2 Chemical composition of lime

Chemical composition	Content (%)
Silicon dioxide (SiO_2)	45.40
Aluminum oxide (Al_2O_3)	18.50
Calcium oxide (CaO)	10.90
Ferric oxide (Fe_2O_3)	8.69
Potassium oxide (K_2O)	2.51
Sodium oxide (Na_2O)	2.30
Titanium dioxide (TiO_2)	2.10
Sulfur trioxide (SO_3)	1.86
Magnesium oxide (MgO)	1.42
Phosphorus trioxide (P_2O_3)	0.82
Loss of ignition	5.36

2.3 *Quarry Dust*

Quarry dust of sandstone was collected from a hillock in Charkhi Dadri, Haryana, India, having grayish-white appearance and sand-like texture. The chemical composition of quarry dust is given in Table 3.

3 Methods

Differential free swell (DFS) index, Atterberg's limits, and unconfined compressive strength (UCS) tests were performed for all mixes of lime and QD with BCS as per the standard codes. For experimentation, the soil was pulverized, so that no lump remains, and subsequently, the soil was oven-dried at 105 °C. The percentage amount of stabilizing materials and the volume of water used in all the experiments was taken corresponding to the dry weight of the soil. The stabilizing materials and BCS were first mixed in a dry state in measured quantity, and after that, required volume of water was added for each sample preparation.

Table 3 Chemical composition of quarry dust

Chemical composition	Content (%)
Silica dioxide (SiO_2)	85.81
Aluminum oxide (Al_2O_3)	10.57
Calcium oxide (CaO)	0.84
Ferric oxide (Fe_2O_3)	0.70
Potassium oxide (K_2O)	0.43
Magnesium oxide (MgO)	0.39

4 Results and Discussion

4.1 Differential Free Swell Index

The DFS test for BCS and various mixes of BCS along with lime and QD was conducted according to IS: 2720 (Part XL)-1977. The DFS value of BCS was 40.17% which reduced to 10% on adding 9% lime content and further increasing lime content beyond 9% increased the DFS value (Fig. 1). This may be due to the neutralization of clay ions by calcium ions of lime that decreases the moisture absorption potential of swelling minerals [3]. The addition of 24% quarry dust reduced the DFS to zero, and then, further addition of quarry dust did not show any change to DFS value (Fig. 2). The reduction in DFS value on adding quarry dust may be attributed the presence of few cementitious compounds such as Al_2O_3 and Fe_2O_3, and traces of CaO (Table 3)

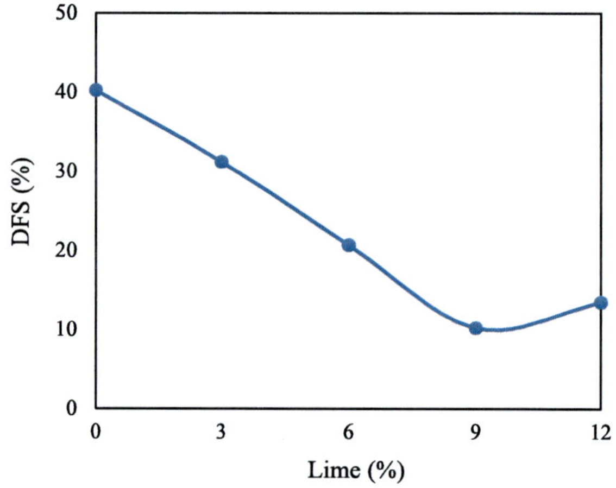

Fig. 1 DFS curve for lime-stabilized BCS

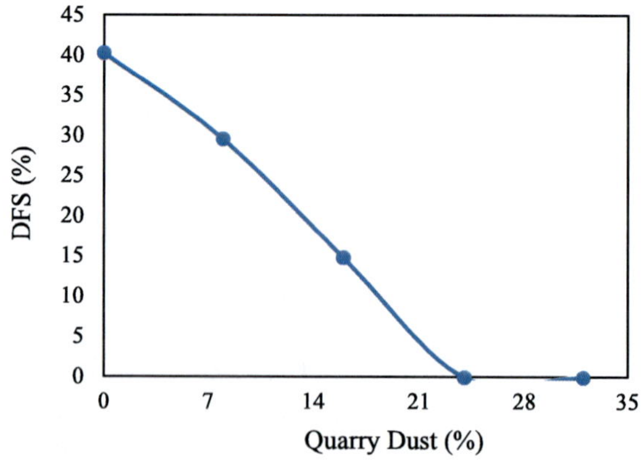

Fig. 2 DFS curve for QD stabilized BCS

which decreases the thickness of double layer between BCS and water molecules, thus helping in reducing the swelling.

4.2 *Atterberg's Limits*

The various Atterberg's limits (conforming to ASTM D4318-10) were evaluated for BSC and various mixes of lime and quarry dust alone and in combination to each other. The addition of 9% lime to BCS decreased the liquid limit of the composite from 71 to 56.8% and the overall plasticity index reduced from 40.6 to 22.3% (Fig. 3). The decrease in liquid limit on adding lime to BCS may be attributed to the filling of voids with calcium ions resulting to increase in electrolyte absorption which reduces the thickness of double layer helping in lowering the liquid limit [4]. On adding quarry dust to BCS in increasing amount, the liquid limit of the composite reduced from 71 to 50%, and plasticity index reduced from 40.6 to 25.5% with 24% quarry dust content (Fig. 4). The reduction in liquid limit on adding quarry dust may be attributed to the inclusion of bigger particles of quarry dust into BCS. The combined action of quarry dust (in varying amount) and lime (9%, chosen as optimum content based on DFS test) further reduced the liquid limit and plasticity index (Fig. 5). The liquid limit reached to 39.6%, and the plasticity index reaches to 7.6% for 24% quarry dust in BCS along with 9% lime composite. This may be due the fact that the combined action of lime and quarry dust further reduced the thickness of double diffuse layer.

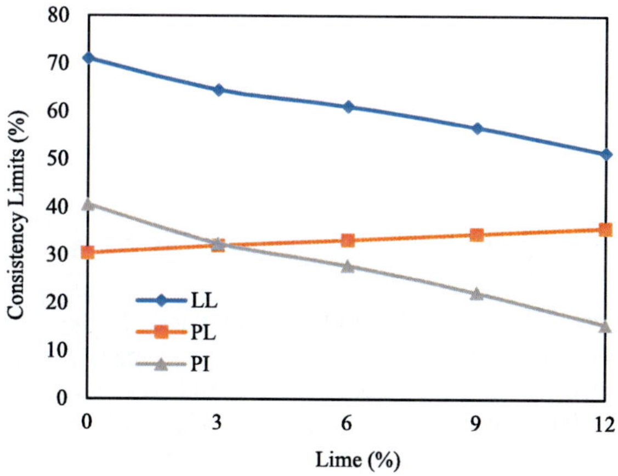

Fig. 3 Plot of LL, PL, and PI for lime-stabilized BCS

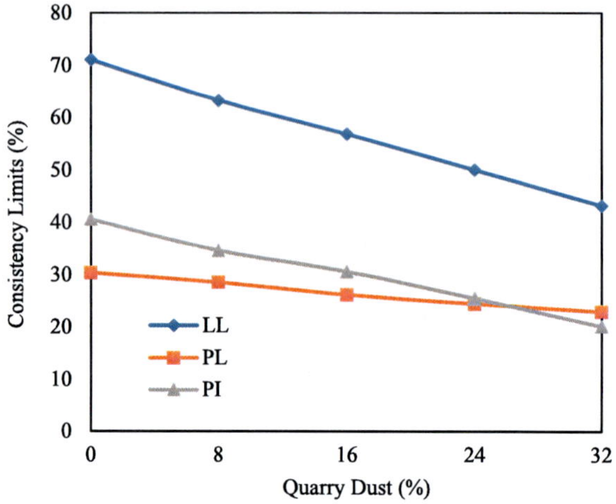

Fig. 4 Plot of LL, PL, and PI for QD stabilized BCS

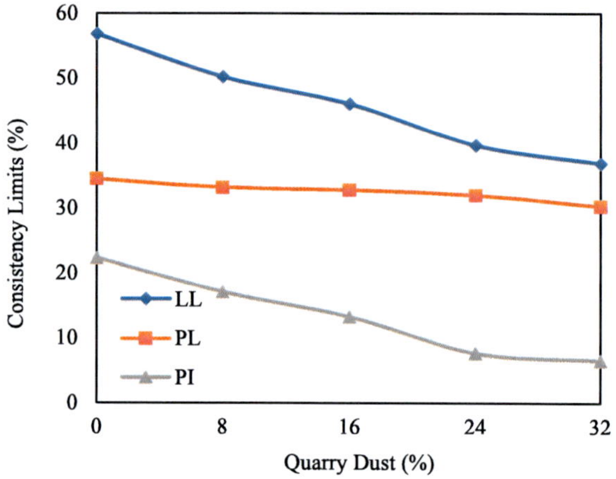

Fig. 5 Plot of LL, PL, and PI for varying QD stabilized BCS containing 9% lime

4.3 Unconfined Compressive Strength

The UCS test for BCS along combination of lime and quarry dust was conducted in accordance with ASTM D 2166: 2013. The results of UCS tests for various combinations for curing periods of 3, 7, and 28 days are presented in Figs. 6, 7, and 8. On increasing the lime and quarry dust content in BCS, the UCS value increased for all curing periods. It was revealed from Figs. 6 and 7 that on increasing lime content beyond 9% and quarry dust content beyond 24%, there was no appreciable increase in the UCS value for all curing periods, and hence, 9% lime and 24% quarry dust alone are sufficient for improving the strength of BCS. The UCS results of optimum content of lime (9%) in BCS along with varying percentages of quarry dust revealed that after 24% quarry dust content there was not much increase in strength

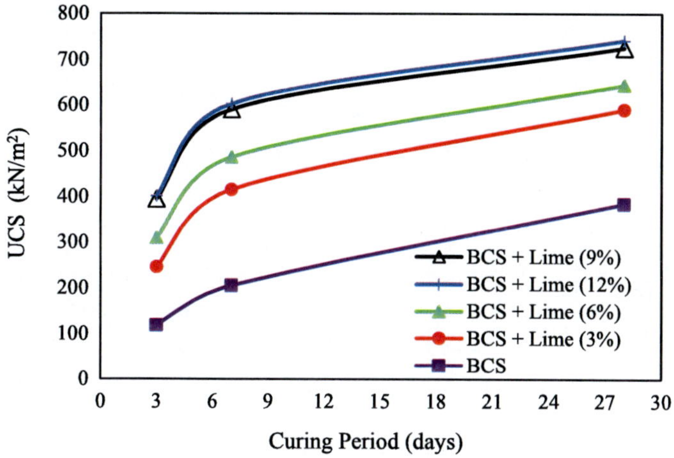

Fig. 6 Plot for UCS values with curing period for lime stabilized BCS

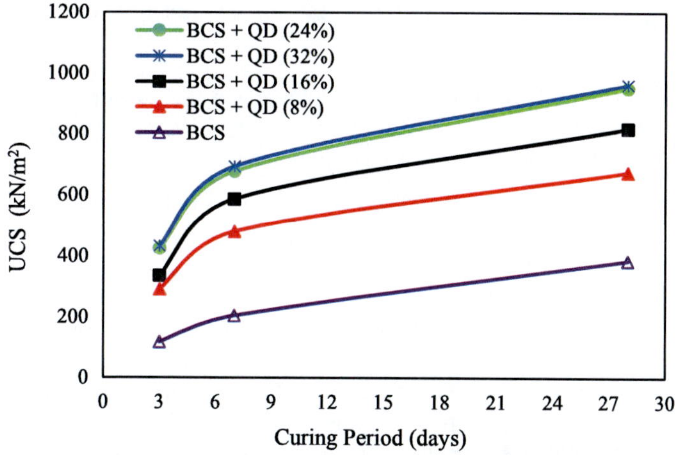

Fig. 7 Plot for UCS values with curing period for BCS for quarry dust-stabilized BCS

value (Fig. 8). The increase in UCS value on adding lime and quarry dust to clayey soil may be attributed the presence of cementitious compounds. Moreover, lime and quarry dust together in BCS increased the friction due to better gradation resulting in improved strength.

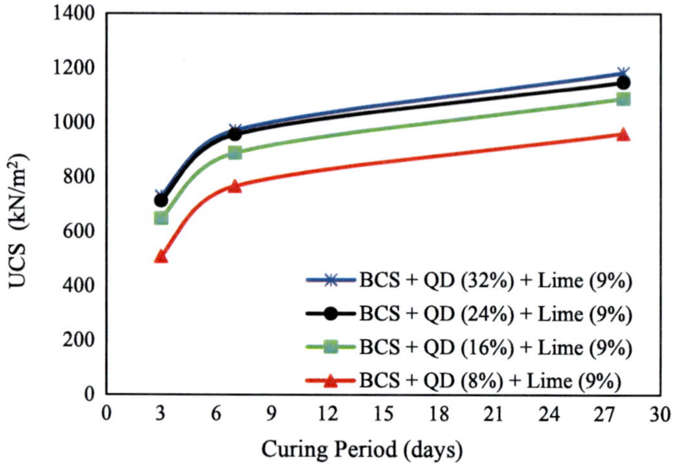

Fig. 8 Plot for UCS values with curing period for varying QD stabilized BCS containing 9% lime

5 Conclusions

This experimental study demonstrates the effect of addition of lime and quarry dust alone and in combination to each other to strengthen black cotton soil. Following conclusions have been withdrawn based on this study.

1. The addition of 9% lime and 24% quarry dust alone to black cotton soil reduces the differential free swell by 75 and 100%, respectively. This reveals that quarry dust is a best cheapest option to reduce the swelling of black cotton soils in place of lime which is also costlier.
2. The addition of 9% lime and 24% quarry dust alone to black cotton soil reduces the liquid limit by 20 and 30%, respectively, and decreases the plasticity index by 45 and 37%, respectively. On adding lime (9%) and quarry dust (24%) together to BCS, the liquid limit and plasticity index reduces by 31 and 66%, respectively.
3. The unconfined compressive strength tests reveal that the strength value is highest for optimum amount of lime and quarry dust alone and in combination for all curing periods; the highest value obtained for 9% lime and 24% quarry dust together mixed to clayey soil followed by mixture of 24% quarry dust and black cotton soil which is further followed by mixture of 9% lime and black cotton soil.

The study reveals that both lime and quarry dust may be used a stabilizer to reduce the swelling and increase the strength characteristics of black cotton soil.

Acknowledgements The author is thankful to NIT Hamirpur for proving all facilities to carry out this research.

References

1. ASTM D2166 (2013) Standard test method for unconfined compressive strength of cohesive soil. ASTM International, West Conshohocken, PA
2. ASTM D4318-10 (2013) Standard test methods for liquid limit, plasti limit, and plasticity index of soils. American Society for Testing of Materials Pennsylvania PA, USA
3. Bell FG (1996) Lime stabilization of clay minerals and soils. Eng Geol 42(4):223–237
4. Bhardwaj A, Sharma RK (2020) Effect of industrial wastes and lime on strength characteristics of clayey soil. J Eng Des Technol 16(6):1749–1772
5. Dash SK, Hussain M (2012) Lime stabilization of soils: reappraisal. J Mater Civ Eng 24(6):707–714
6. Dixit DM, Patil DK (2016) Utilization of stone dust to improve the properties of expansive soil. Int J Civil Eng Technol (IJCIET) 7(4):440–447
7. Hilt GH, Davidson DT (1960) Lime fixation in clayey soils. Highway Research Board Bulletin 262
8. IS 2720 Part 40 (1977) Methods of tests for soil, part 40, Determination of free swell index of soil. Bureau of Indian Standards, New Delhi
9. Kavak A, Akyarli A (2007) A field application for lime stabilization. Environ Geol 51(6):987–997
10. Sharma RK, Hymavathi J (2016) Effect of fly ash, construction demolition waste and lime on geotechnical characteristics of a clayey soil: a comparative study. Environ Earth Sci 75(5):377
11. Sharma A, Sharma RK (2019) Effect of addition of construction–demolition waste on strength characteristics of high plastic clays. Innovative Infrastruct Solutions 4(1):27
12. Sharma A, Sharma RK (2020) Strength and drainage characteristics of poor soils stabilized with construction demolition waste. Geotech Geol Eng J
13. Soosan TG, Sridharan A, Jose BT, Abraham BM (2005) Utilization of quarry dust to improve the geotechnical properties of soils in highway construction. Geotech Test J 28(4):391–400
14. Sridharan A, Soosan TG, Jose BT, Abraham BM (2006) Shear strength studies on soil-quarry dust mixtures. Geotech Geol Eng 24(5):1163–1179

Utilization of Pond Ash for Sustainable Construction

Gaurav Gupta, Hemant Sood, and Pardeep Kumar Gupta

1 Introduction

In excess of 70% of the total electricity produced in India is obtained from thermal power plants (TPPs) that use coal as the fuel [12, 24]. The firing of coal generates coal ash as a by-product that can be distinctly categorized in two types—fly ash and bottom ash. Fly ash refers to the lighter part of the coal ash rises with the flue gases and is eliminated from the flue gases with the use of mechanical ash collectors or through electrostatic precipitators. The leftover coal ash that is the coarser part settles down of the floor of the furnace or boiler is termed as bottom ash. The ashes are waste materials and need to be disposed of by their transportation, either in wet form or dry form. Due to lower operation cost, the majority of the coal-fired thermal power stations in India use a wet disposal method over the dry disposal method. In the wet disposal method, slurry is made by mixing water with both the wastes, and the slurry is dumped in open areas. This result in formation of ash ponds, and the ash from the ponds is termed as pond ash. During 2018–2019, 217.04 million tons of coal ash was generated by burning 667.4 million tonnes of coal at 195 coal-based TPPs in India, but only 68.7% of coal ash was utilized [9]. The unutilized coal ash is dumped in ponds that have already engaged approximately 113 million m^2 of useful land in India [2]. The requirement for coal-based power generation, total and domestic coal

G. Gupta (✉) · H. Sood
National Institute of Technical Teachers' Training & Research, Chandigarh, India
e-mail: gaurav007gupta@gmail.com

H. Sood
e-mail: sood_hemant@yahoo.co.in

P. K. Gupta
Panjab Engineering College (Deemed to be University), Chandigarh, India
e-mail: p_gupta_2000@yahoo.com

H. Singh et al. (eds.), *Sustainable Development Through Engineering Innovations*, Lecture Notes in Civil Engineering 113,
https://doi.org/10.1007/978-981-15-9554-7_55

Table 1 Predicted demand of coal for generation of electricity

Year	Coal-based generation (BU)	Total coal requirement (MT)	Domestic coal requirement (MT)
2017–18	958	630.0	584.0
2021–22	1118.8	735.0	685.0
2026–27	1319.7	877.0	827.0

requirement is envisaged to rise for the year 2021–22 and 2026–27, the values for which have been reported in Table 1 [8].

Table 2 shows the current stock of pond ash at various National Thermal Power Plant Stations in India as on 30.11.2019.

Table 2 Stock availability of different coal ashes at national TPPs in India

S. No.	Thermal power station	State	Availability of fly ash (MT)	Availability of bottom ash (MT)	Availability of pond ash (MT)
1.	Singrauli	Uttar Pradesh	600	–	72.99×10^6
2.	Korba	Chhattisgarh	700	–	51.12×10^6
3.	Ramagundam	Telangana	4969	–	72.46×10^6
4.	Farakka	West Bengal	2700	–	40.79×10^6
5.	Vindhyachal	Madhya Pradesh	4000	–	70.71×10^6
6.	Rihand	Uttar Pradesh	3000	–	36.21×10^6
7.	Kahalgaon	Bihar	3700	–	66.22×10^6
8.	Dadri	Uttar Pradesh	18,500	–	0.0×10^6
9.	Talcher Kaniha	Orissa	1770	–	79.63×10^6
10.	Feroze Gandhi, Unchahar	Uttar Pradesh	2000	–	8.33×10^6
11.	Talcher Thermal	Orissa	200	–	2.47×10^6
12.	Simhadri	Andhra Pradesh	5100	–	1.95×10^6
13.	Tanda	Uttar Pradesh	1450	–	4.72×10^6
14.	Sipat	Chhattisgarh	5200	–	33.51×10^6
15.	Mouda	Maharashtra	4000	10,000	0.86×10^6
16.	Barh	Bihar	875	–	6.65×10^6
17.	Bongaigaon	Assam	–	–	1.05×10^6
18.	Solapur	Maharashtra	500	–	0.04×10^6
19.	Kudgi	Karnataka	1250	–	0.55×10^6
Total (MT)			60,514	10,000	550.3×10^6

The release of leachate from the ash ponds contains toxic elements and poses a potential threat of soil, groundwater and surface water contamination [29]. Coal will persevere to be an ultimate component of the energy segment of India. Till the closing of 2018, India possesses 9.6% of the entire coal reserves on the earth, among which approximately 5% reserves are of lignite and sub-bituminous coal and 95% reserves are of anthracite and bituminous coal [13].

Table 3 Physical properties of pond ash acquired from different TPPs in India

Physical property	Location of thermal power plant					
	Panipat	Pankhi	Khedar	Panipat	Yamunanagar	Yamunanagar
	Mohanty and Patra [26]		Suthar and Aggarwal [29]			Gupta et al. [20]
Specific gravity	2.21	2.34	2.03–2.27	2.06–2.23	2.05–2.21	2
MDD (kN/m^3)	10.4	11.2	1.31–1.32	1.18–1.29	1.09–1.21	13.1
Coefficient of permeability (m/sec)	$3.8–4.2 \times 10^{-5}$	$2.4–2.9 \times 10^{-5}$	N.R.[a]	N.R.[a]	N.R.[a]	5.01×10^{-8}
Cohesion, c′ (kPa)	0	0	N.R.[a]	N.R.[a]	N.R.[a]	N.R.[a]
Angle of friction, ϕ' (°)	33.04–34.28	36.53–37.68	N.R.[a]	N.R.[a]	N.R.[a]	N.R.[a]
Grain size Distribution						
Clay (%)	0	0	0	0	0	0
Silt (%)	4.8	12	27.8–69.6	21.2–42.8	14.9–53.5	45.9
Sand (%)	95.2	87.2	72.1–30.3	78.7–56.9	85.1–46.4	54.1
Gravel (%)	0.024	0.743	0.08–0.17	0.12–0.35	0.0–0.06	0
Coefficient of Uniformity (C_U)	2.25	2.05	3.40–7.11	3.33–10.7	2.84–6.07	6.7
Coefficient of Curvature (C_c)	0.84	0.74	0.85–1.0	0.92–1.20	0.66–1.33	0.6
D_{10}, mm (%)	0.08	0.07	0.01–0.05	0.06–0.01	0.02–0.07	0.015
D_{30}, mm (%)	0.11	0.09	0.02–0.08	0.04–0.12	0.03–0.13	0.03
D_{60}, mm (%)	0.18	0.15	0.1–0.2	0.150–0.20	0.09–0.190	0.1

[a]N.R.: not reported

From Table 3, it is apparent that where the stock of pond ash is significantly larger than that of the fly ash available. Hence, there is an urgent need to utilize pond ash in various types of construction projects including road construction.

The coal ash may be classified as Class C or Class F, where the latter and the former are produced by combustion of bituminous coal and sub-bituminous coal, respectively, as per ASTM standards. Class C coal ash possesses pozzolanic along with cementing properties, while class F coal ash possesses only pozzolanic properties.

Though fly ash continues to be aggressively utilized in different industrial sectors of India, the use of pond ash is yet to pick up. This is evident from the fact that where only 20% of the total fly ash is provided free of cost, the complete pond ash is available free of cost.

2 Properties of Pond Ash and Its Utilization

For the current section, the literature related to the properties of pond ash has been presented along with its application in different fields of civil engineering. Important engineering properties such as physical, chemical, mineralogical, microstructural, geotechnical and toxicity are discussed below to get an insight of the governing mechanism prevalent in areas where pond ash shall be applied to attain effectual application such as construction projects.

2.1 *Physical Properties*

The physical properties of pond ash, as reported from samples acquired from different TPPs in India are reported in Table 3 [20, 26, 29]. Overall, it was seen that pond ash comprises sand and silt size particles, where usually the content of sand size particles was found to be higher than the silt size particles [20, 26, 29]. Pond ash was reported to be non-plastic in nature; therefore, it can be utilized for stabilization of soft soils for the subgrade.

2.2 *Chemical Properties*

Chemical composition of pond ashes from different TPPs in India is given in Table 4. The results show that pond ash mainly contains alumina and silica, with a low amount of calcium. Their chemical composition varies with the source of the coal [20, 26, 29]. Generally, pond ash reported was determined to be as class F ash, being low in calcium content pond ash is expected not to possess any cementitious properties. Loss on ignition values (LOI) was determined to be in the range of 1–2%. LOI being less than 5% ensures no peril of sudden self-ignition of the waste.

Table 4 Chemical composition of pond ash

Element	Thermal power plant location					
	Pankhi	Panipat	Khedar	Panipat	Yamunanagar	Yamunanagar
	Mohanty and Patra [26]		Suthar and Aggarwal [29]			Gupta et al. [20]
SiO_2 (%)	61.2	60.4	61.8–63.9	63.4–66.0	62.5–64.7	59.2
Al_2O_3(%)	24.1	30.0	21.6–26.7	20.6–25.0	21.3–27.7	30.9
Fe_2O_3(%)	9.8	3.8	4.4–11.5	6.0–8.8	4.5–8.7	4.2
CaO (%)	0.6	0.8	0.9–1.1	0.8–0.9	0.7–0.8	0.94
MgO (%)	0.6	0.5	0.5–0.7	0.5–0.6	0.5–0.6	0.48
TiO_2 (%)	1.5	1.9	1.4–1.7	1.2–1.6	1.2–1.8	1.93
P_2O_5 (%)	0.2	0.4	0.30–0.33	0.33–0.40	0.37–0.50	0.51
SO_3 (%)	0.1	0.2	0.11–0.19	0.15–0.16	0.08–0.09	0.12
MnO (%)	–	–	0.04–0.15	0.07–0.10	0.05–0.10	0.04
Na_2O(%)	0.1	0.1	0.13–0.26	0.06–0.07	0.08–0.09	0.17
K_2O (%)	1.8	1.9	0.9–1.0	1.3–1.6	1.4–1.8	1.34
LOI (%)	–	–	3.2–3.5	1.8–3.2	2.4–2.7	2.1

2.3 *Mineralogical Properties*

The XRD diffractogram of pond ash, reported by scholars, shows the existence of crystalline phases of quartz, alumina (in the form of mullite) and feldspar, accompanied by a minor quantity of iron oxide (in the form of magnetite and hematite) [20, 26, 29]. Presence of high amount of silica and alumina, and a low amount of iron was corroborated by the results of XRF analysis.

2.4 *Microstructural Properties*

The scanning electron microscopy (SEM) micrographs of pond ash reported by scholars show that pond ash comprises irregular shaped as well as spherical-shaped particles with a complex pore structure [20, 26, 29]. The SEM images of pond ash have been presented in Fig. 1a, b at 30x and 200x magnification, respectively Gupta et al. [20].

2.5 *Geotechnical Properties*

Geotechnical properties of pond ash namely compaction properties such as optimum moisture content (OMC) and maximum dry density (MDD), strength properties such

Fig. 1 SEM images of pond ash at a magnification of **a** 30x, and **b** 200x. Gupta et al. [20]

(a)

(b)

as bearing strength and shear strength properties of soils such as angle of internal friction and cohesion have been determined by various researchers. The following sections review these important properties of pond ash.

2.5.1 Compaction Properties

Bera et al. [4] determined the compaction properties of three samples of pond ash so that pond ash might be employed for filling in construction of structural fills and embankments. Two samples of pond ash were obtained from Kolaghat TPP, and one sample was obtained from the ash pond at Budge Budge TPP. The impact of different parameters, i.e. moulding moisture content, mould and tank area, specific gravity, layer thickness and compaction effort on compaction properties of pond ash was determined.

Results show that rise in compaction effort raises the MDD and lowers the OMC because of closer packing of the particles of the sample. The impact of mould area on MDD of pond ash was determined in Proctor mould and California bearing ratio (CBR) mould. Also, results show that with rise in the area of the mould, the MDD of pond ash increases. The reason cited by the authors was that with rise in the area of the mould, the consequence of bordering detention declines, resulting in improved organization of the particles. The influence of the size of tank on the MDD of pond ash was determined by compacting pond ash at OMC in two tanks having an area of 0.360 and 0.132 m^2. Results show that MDD ascents with rise in the area of tank, due to reduced lateral confinement in the larger area tank. The pond ash samples with higher specific gravity reported higher MDD. With the use of regression analysis, the authors developed an empirical formula, for prediction of dry density of pond ash at a known specific gravity, moisture content and compaction energy. The model is expected to be beneficial for field application of pond ash. A decrease in the thickness of layer of pond ash for compaction results in increased MDD, but only up to a certain limit. Linear empirical models were developed for estimation of MDD based on the specific gravity and compaction energy.

Singh and Saran [28] determined the compaction properties of pond ash samples, gathered from Rourkela Steel Plant at different compaction efforts. Results show that with rise in compactive energy MDD rises and the corresponding OMC falls. The MDD was reported to rise from 10.9 to 12.7 kN/m^3 on increasing the compaction energy from 0.36 to 3.5 MJ/m^3. On the other hand, the OMC was reported to drop from 38.8 to 28.1% on increasing the compaction energy from 0.36 to 3.5 MJ/m^3. The authors cited uniform gradation and round shape of pond ash particles as reasons for this trend.

2.5.2 Strength Properties

Strength properties of pond ash as in bearing strength and shear strength have been evaluated in the literature. Pond ash has been reported to have high shear strength, comparable to that of sand. Bearing strength of pond ash reportedly increased by stabilization with lime and phosphogypsum. Shear strength of pond ash reportedly increased by reinforcement with geotextiles. The impact of energy put in for compaction and saturation on shear strength, bearing strength and UCS of pond ash has been determined.

Jakka et al. [21] determined the shear strength of pond ash in loose and compacted state obtained from Indraprastha and Badarpur TPPs. Additionally, sand sample, from Yamuna river, was tested as a reference material to compare the results. Grain size analysis shows that the samples obtained from outflow points were found to be coarse-grained whereas the samples obtained from inflow points were found to be fine-grained. The samples were tested in dense state by preparing coarse-grained pond ash sample at 70% of the maximum Proctor density and fine-grained pond ash sample at 95% of the MDD. The samples were evaluated in loose state by preparing coarse-grained pond ash sample at 30% of the maximum Proctor density and fine-grained pond ash sample at 70% of the MDD. Typical consolidated undrained (CU) tests and consolidated drained (CD) tests were performed. CD test outcomes show that the effective cohesion for each sample was determined to be 0, whereas the residual angle of internal friction altered from 32.9° to 35.0° and the peak angle of internal friction altered from 32.9° to 41.7° for different pond ash samples, where higher angle of internal friction was recorded in the dense state. All the samples prepared in the dense state demonstrated dilative behaviour, and the amount of dilation was stated to reduce with a rise in the confining pressure. Every sample prepared in the loose state demonstrated contractive behaviour, under different confining pressures. CU test outcomes show that the effective cohesion for all the pond ash samples were determined to be 0, whereas the angle of internal friction, determined at maximum deviator stress, altered from 31.9° to 38.6° and the angle of internal friction, evaluated at maximum stress ratio, altered from 32.9° to 41.7° for different pond ash samples, where lower angle of internal friction was recorded in the loose state. All the samples prepared at dense state developed negative pore water pressure and its extent lowered with rise in confining pressure, whereas for all the samples prepared in the loose state developed positive pore water pressure and the magnitude of positive pore water pressure increases with increasing confining pressure. As compared with Yamuna sand, pond ash samples, under both, loose and dense state, depicted higher shear strength, due to irregular shape of pond ash particles that results in higher interlock friction among pond ash particles.

Bera et al. [3] determined shear strength of pond ash, obtained from three TPPs, through unconsolidated undrained triaxial tests and attempted to improve it, by reinforcing it with three types of geotextiles. The samples were prepared at standard Proctor effort. The effect of confining pressure, number of layers and type of geotextiles on the shear strength of pond ash was studied. Rise in confining pressure and number of layers of the geotextile resulted in rise in deviator stress at failure. The geotextile with lower stiffness reported a higher of deviator stress at failure. Overall, reinforcement of pond ash with the geotextiles resulted in a rise in the shear strength of pond ash.

Ghosh [15] increased the bearing strength of pond ash through its stabilization with lime alone and in combination of lime and phosphogypsum. In order to evaluate the bearing strength, CBR tests under unsoaked and four-day water soaked conditions were executed. Pond ash was treated with lime content in the range of 4–10% and phosphogypsum in the range of 0.5–1.0%, added by the dry weight of pond ash. The samples were exposed to curing for 7, 28 and 45 days. CBR of pond ash improved with rise in lime content, phosphogypsum content and curing duration. CBR of pond ash (seven-day cured), under unsoaked and soaked condition, was reported to be 34.0% and 5.0%, respectively. The CBR improves from 34.0 to 111.9% (unsoaked) and 5.0 to 99.9% (soaked) on addition of 10% lime alone, after seven days curing. Further, the CBR improves from 111.9 to 137.9% (unsoaked) and 5.0 to 25.7% (soaked) on treatment to pond ash sample stabilized with 10% lime, with 1% phophogypsum, added by the dry weight of the mix, after seven days curing.

Suthar and Aggarwal [30] increased the bearing strength of pond ash through its stabilization with lime and amalgamation of lime sludge and lime. In order to evaluate the bearing strength, CBR tests under unsoaked and four-day water soaked conditions were executed. Pond ash was treated with lime content in the range of 2–8% and lime sludge in the range of 5–15%, added by the dry weight of pond ash. The samples were exposed to curing for 728 and 45 days. CBR (seven-day cured) of pond ash improved with rise in lime content, lime sludge content and curing time. CBR of pond ash was reported to be 23.3% (unsoaked) and 4.1% (soaked), respectively. The CBR improves from 23.3 to 64.1% (unsoaked) and 4.1 to 53.3% (soaked) on addition of 8% lime alone, after seven days curing. Further, the seven-day cured CBR improves from 64.1 to 81.2% (unsoaked) and 53.3 to 63.1% (soaked) on addition of 15% lime sludge to pond ash sample containing 8% lime. For application as the sub-base course of flexible pavements, pond treated with an amalgamation of 2% lime and 10% lime sludge, cured for seven days or pond ash treated with an amalgamation of 2% lime and 5% lime sludge, cured for 28 days, are found suitable.

Singh and Saran [28] determined shear strength of pond ash, along with CBR and UCS of pond ash, at different compaction efforts and degree of saturation. Pond ash samples, obtained from Rourkela Steel Plant TPP, were tested at standard and modified Proctor density and different moisture contents. Results show that unit cohesion rises with rise in compaction energy. Also, unit cohesion rises with rise in moisture content, up to the moisture content equalizes with the OMC, beyond which unit cohesion decreases. The angle of internal friction rises with rise in compaction energy. Also, the angle of internal friction continuously decreases with rise in moisture content. The stresses at failure and preliminary stiffness of pond ash specimen prepared with higher compaction energies were reported to be higher than the specimens prepared at lower compaction energies. Unit cohesion values improve from 0.8 to 8.363 kN/m^2, and the angle of internal friction values rises from 37.48° to 44.47° with rise in energy, applied for compaction from 0.357 to 3.488 MJ/m^3. The samples depicted brittle failure with a low value of failure strains, in the range of 0.75–1.75%. Similarly, UCS values increase from 1.2 to 17.0 kN/m^2 with rise in energy, applied for compaction from 0.357 to 3.488 MJ/m^3. The UCS of pond ash was reported to be higher for samples prepared by compaction at modified effort than the samples prepared at standard effort. The consequence of saturation on the CBR of pond ash was studied by altering the moisture content of the specimen, in the range of 3.59–43.10% and 2.83–33.96% for samples prepared with standard and modified effort, respectively. CBR values first upsurge with rise in moisture content, reach maximum and then decrease with further rise in moisture content. The highest CBR of 45 and 7.5% was recorded at 8% and 7%, degree of saturation, for specimens prepared by compaction at modified and standard effort, respectively. Overall, it was determined that moisture content and compaction energy had a noteworthy consequence on the CBR of pond ash.

2.6 Toxicity

The leachate obtained from pond ash samples has been found to be rich in the concentration of toxic elements [29]. The literature highlights the potential threat of hazardous contamination of groundwater and surface water bodies by leaching of heavy metals due use of pond ash in the subgrade layer and embankments of pavements. Therefore, it is essential to assess the leachate of pond ash for concentration of toxic elements, before its utilization in any type of construction project. Suthar and Aggarwal [30] suggested lime sludge and lime stabilization for pond ash in order to lower the concentration of toxic elements, present in the leachate of pond ash.

Gupta et al. [20] performed toxicity characteristic leachate procedure (TCLP) to extract leachate of the pond ash and evaluate the concentration of toxic elements. Results as stated in Table 5 show that zinc and lead were recognized in the leachate of pond ash whereas erstwhile toxic elements were found below the detection limit of 0.1 mg/L of atomic absorption spectrophotometer.

Overall, the concentrations of heavy metals were discovered to be noteworthily under the permissible limits for hazardous waste, as prescribed by United States Environmental Protection Agency (USEPA). Therefore, it was determined that pond ash is safe for application in construction projects to attain sustainability.

The properties of pond ash reported in the literature show that pond ash possesses a number of properties, including high silt content and absence of plasticity, suitable for stabilization of weak soils for the subgrade of flexible pavements. Prediction of high demand of coal for generation of electricity in future necessitates the need for sustainable management of pond ash.

Table 5 Concentration of toxic elements in the pond ash

Heavy metals	Concentration (mg/L)	USEPA limits for hazardous wastes (mg/L)
Chromium	BDL[a]	5.0
Zinc	0.83	Not reported
Lead	0.28	5.0
Arsenic	BDL[a]	5.0
Cadmium	BDL[a]	1.0
Nickel	BDL[a]	Not reported
Copper	BDL[a]	Not reported
Mercury	BDL[a]	0.2

[a]BDL: Below Detection Limit; Detection Limit: 0.1 mg/L

2.7 Utilization of Pond Ash for Soil Stabilization

Pond ash has been utilized for stabilization of weak soils. The following section reviews the impact of pond ash stabilization on geotechnical properties of soils, namely strength properties such as unconfined compressive strength (UCS) and CBR, compaction properties such as OMC and MDD and plasticity. Consequently, utilization of pond ash for soil stabilization is expected to be a crucial solution for sustainable management of pond ash.

Bera and Kumar [6] considered the consequence of pond ash stabilization on the geotechnical properties of an inorganic soil (USCS classification: CH). The authors determined the impact of addition of pond ash on geotechnical properties of soils, namely strength properties such as CBR (unsoaked) and UCS, compaction properties such as OMC and MDD and plasticity. For the plasticity tests, samples were prepared by replacing soil with up to 55% content of pond ash. For the compaction tests, samples were prepared at standard Proctor effort by replacing soil with up to 70% content of pond ash. The authors reported that with increases in contents of pond ash, liquid limit, plasticity index, MDD and UCS of the stabilized soil decrease while OMC and CBR (unsoaked) increase. The reason for decreases in MDD, as cited by the authors, was lesser specific gravity of pond ash particles than the soil particles, whereas the reason for decreases in OMC, as cited by the authors, was higher fineness of pond ash particles than the soil particles. CBR (unsoaked) tests were performed on samples containing varying content of pond ash from 0 to 40%, where the CBR (unsoaked) increased from 8.5 to 16.5%, on addition of 40% contents of pond ash. UCS tests were performed on samples containing varying content of pond ash from 0 to 30% where the UCS decreased from 260 to 190 kPa on addition of 30% content of pond ash. The reason for decreases in UCS, as cited by the authors, was reduction of cohesive strength on addition of pond ash.

Kolay et al. [22] considered the consequence of pond ash stabilization on the geotechnical properties of peat soil. The peat soil was replaced by 5%, 10%, 15% and 20% content of pond ash. The authors determined the impact of addition of pond ash on geotechnical properties of soils, namely strength properties such as UCS and compaction properties such as OMC and MDD. The compaction properties were obtained at standard Proctor effort. The samples were exposed to curing for 7, 14, 28 and 120 days. The authors reported that with increases in contents of pond ash, UCS and MDD increased, whereas OMC decreased. The reason for rise in MDD, as cited by the authors, was higher specific gravity of pond ash as compared to that of the soil, whereas the reason for decreases in OMC was because of lesser fineness of pond ash as compared to that of the soil. The UCS also improved with rise in curing period, due to reduction in natural moisture content of soil, during curing. The reason cited for reduction in moisture content was filling of small air voids in the peat soil by pond ash and consumption of water during hydration of pond ash and soil particles. UCS, for 28 days cured samples, was reported to rise from 80 to 160 kPa on addition of 20% dosage of pond ash.

Bera [7] considered the consequence of pond ash stabilization on the geotechnical properties of a fine-grained soil (USCS classification: MH). The authors determined the impact of addition of pond ash on geotechnical properties of soils, namely strength properties such as UCS and compaction properties such as OMC and MDD. The compaction tests samples were prepared at standard Proctor effort by replacing soil with up to 100% content of pond ash. The authors reported that with increases in contents of pond ash, UCS and MDD of the stabilized soil decrease, while OMC increases. The reason for decreases in MDD, as cited by the authors, was lesser specific gravity of pond ash than whereas the reason for decreases in OMC, as cited by the authors, was higher fineness of pond ash particles than the soil particles. The reason for decreases in UCS, as cited by the authors, was reduction of cohesive strength on addition of pond ash. The effect of number of jute geotextile layer was evaluated on the UCS. Samples were made with a maximum of four layers of jute geotextile. UCS was reported to rise with upsurge in layers of geotextile for each of the samples, irrespective of the pond ash content in the soil.

Patil and Patil [31] considered the consequence of pond ash stabilization on the geotechnical properties of a fine-grained soil (USCS classification: CH), namely differential free swell, liquid limit, CBR (soaked), UCS, MDD and OMC. For the strength and plasticity tests, samples were made by substituting soil with up to 20% content of pond ash. On increasing the dosage of pond ash, differential free swell decreased from 60 to 50%, liquid limit decreased from 67 to 52%, and CBR increased from 2.56 to 4.56%. For the compaction tests, samples were prepared by replacing soil with up to 30% content of pond ash. On increasing the dosage of pond ash from 0 to 20%, MDD improved from 1.45 to 1.48 g/cc, but on further rise in content of pond ash from 20 to 30%, MDD decreased from 1.48 to 1.44 g/cc. On rise in dosage of pond ash from 0 to 10%, OMC decreased from 25.8 to 22.8%, but on further rise in dosage of pond ash from 10 to 30%, OMC increased from 22.8 to 26.3%. On rise in dosage of pond ash from 0 to 10%, UCS decreased from 4.35 to 0.58 kg/cm^2, but on further rise in content of pond ash from 10 to 20%, UCS increased from 0.58 to 1.62 kg/cm^2.

Gupta and Kumar [16] considered the consequence of soil stabilization using cement and pond ash, and soil reinforcement using polypropylene fibre. The effect of the proposed stabilization and reinforcement was studied on the geotechnical properties of a fine-grained soil (USCS classification: CL). The specimens were prepared with the content of cement varying from 0 to 6%, content of pond ash varying from 0 to 50% and content of polypropylene fibre of length 3 mm, 6 mm and 12 mm, varying from 0 to 2%, by the dry weight of the soil. The samples were exposed to curing for 7, 14 and 28 days. The authors determined the impact of addition of pond ash on geotechnical properties of soils, namely strength properties such as CBR (soaked), split tensile strength (STS) and UCS and compaction properties such as OMC and MDD. The results demonstrate that with rise in the dosage of pond ash or cement, the MDD falls, whereas the OMC rises. The reasons for lessening of MDD, as citied by the authors, were lesser specific gravity of pond ash, as compared to that of the soil and base-exchange aggregation and floc formation due to the hydration of cement. The reasons for rise in OMC, as citied by the authors, were higher OMC of pond ash as compared to that of the soil and consumption of higher volume of water during the hydration of cement. The increases in the contents of pond ash the CBR of the soil improves the CBR of the soil, up to 40% content of pond ash beyond which the CBR falls. Without curing and addition of cement, the CBR improves from 2.5 to 5.9% on addition of 40% content of pond ash. The CBR further increased with rise in the dosage of cement and period of curing. With the rise in the dosage and length of the fibre in the sample containing 40% pond ash and 6% cement, by the dry weight of the soil, the CBR, UCS and STS of the soil rise with escalation in the length and dosage of the fibre.

Kumar and Gupta [23] considered the consequence of fibre reinforcement using polypropylene fibre and soil stabilization using cement on the geotechnical properties of a soil stabilized with combination of pond ash and rice husk ash (RHA). Samples were made with varying content of pond ash from 30 to 45% of the dry weight of the soil, where the total dosage of pond ash and RHA was kept static to 50% of the dry weight of the soil. The cement dosage was altered from 0 to 4%. The samples were exposed to curing for 7, 14 and 28 days. The authors determined the impact of the proposed stabilization and reinforcement on geotechnical properties of the soil, namely strength properties such as CBR (soaked), STS and UCS and compaction properties such as OMC and MDD. The results show that addition of cement lowers the MDD and increases the OMC of all the soil–pond ash–RHA mixtures. Keeping the total stabilizer content constant, on increasing the contents of pond ash in the mix, the MDD increases because of lesser specific gravity of RHA than pond ash, and the OMC decreases due to higher OMC of RHA than pond ash. The authors reported that the optimal dosage of pond ash and RHA determined was 40% and 10%, respectively, for highest UCS and STS. The UCS improves with rise in cement content and curing period. Regarding fibre reinforcement, with the increases in the contents of polypropylene fibres, in the range of 0.5–1.5%, by the total dry weight of the mixture, further improves the UCS and STS. Soils reinforced with 12 mm length fibres showed better results than soils reinforced out with 6 mm length fibres. Addition of fibres resulted in conversion of brittle behaviour pond ash–RHA–soil mixtures and cement–pond ash–RHA–soil mixtures to a ductile behaviour.

Yadav and Tiwari [33] considered the consequence of soil stabilization with pond ash and cement, and coir fibre reinforcement on the geotechnical properties of a soil (USCS classification: CL). The soil samples were prepared at standard Proctor compaction effort, with the dosage of pond ash varying from 0 to 30%, dosage of cement altering from 0 to 4% and the dosage of coir fibre altering from 0 to 1.5%, by the dry weight of the soil. Samples were made with only pond ash, combinations of cement and pond ash and combinations of the fibre, cement and pond ash. The samples were exposed to curing for 7, 14 and 28 days. The specimens were evaluated to ascertain the compaction properties such as OMC and MDD and strength properties such as STS and UCS The results indicate that with rise in the dosage of pond ash, cement or fibre, the MDD falls, whereas the OMC rises. The reasons for lessening of MDD, as citied by the authors, were lesser specific gravity of pond ash and coir fibre, as compared to that of the soil and base-exchange aggregation and floc formation due to the hydration of cement. The reasons for rise in OMC, as citied by the authors, were higher fineness of pond ash than the soil, water absorption by the coir fibres and consumption of volume of water during the hydration of cement. On escalation of the dosage of pond ash, UCS and STS fall, the fall being gradual till 20% substitution of the soil with pond ash where after, a steep fall was observed in the values of UCS and STS at addition of 30% content of pond ash to the soil. The UCS of the soil falls from 121.7 to 102.6 kPa on 30% substitution of soil with pond ash. The reason, as stated by the authors, was slippage among the irregular particles of the soil and sub-rounded particles of pond ash. The STS and UCS rise with increases in cement content and curing period for any given mixture. The STS and UCS improve with increases in fibre content up to 1%, for any given mixture. Addition of fibre alters the brittle performance of cement or pond ash stabilized samples to a ductile one, where higher brittleness was reported for samples made with higher dosage of pond ash and cement. Overall, the authors recommended that up to 20% weight of the dry soil may be replaced by pond ash and up to 1% weight of the dry soil may be replaced by the fibre with the required cement content so as to achieve the target UCS and STS.

Gupta and Kumar [17] considered the consequence of soil stabilization with pond ash, RHA and cement on the CBR of a soil. The soil used in the study was classified as (USCS classification: CL). The specimens were prepared with varying content of pond ash ranging from 0 to 50%, RHA ranging from 0 to 20% and cement ranging from 0 to 4%, by the dry weight of the soil. Samples were also prepared with combinations of RHA and pond ash, where the total content of pond ash and RHA was kept fixed to 50%. Cement was added to the aforementioned samples. The samples were exposed to curing for seven days. The samples evaluated to ascertain the compaction properties such as OMC and MDD, and strength properties such as unsoaked and four-day soaked CBR. The results show that with increases in the contents of pond ash, RHA or cement, the MDD decreases, whereas the OMC increases. The reasons for decrease in MDD, as citied by the authors, were lower specific gravity of the wastes than the soil and base-exchange aggregation and floc formation due to the hydration of cement. The reasons for rise in OMC, as citied by the authors, were higher OMC of pond ash and RHA than the soil and consumption of higher volume of water during the hydration of cement. With the increases in the contents of pond ash, the soaked CBR of the soil increases up to 40% content of pond ash beyond which the CBR falls. Without curing and addition of cement, the unsoaked CBR increases from 2.1 to 10.5% on addition of 40% content of pond ash, whereas the soaked CBR increases from 1.1 to 7.1% on addition of 40% content of pond ash. The CBR further increased with increases in cement content. Similarly, with the rise in the contents of RHA, the soaked CBR of the soil upsurges up to 10% content of RHA beyond which the CBR falls. The CBR further increased with increases in cement content. Regarding the blend of soil–pond ash–RHA mixtures, highest improvement in the CBR was observed for the mixture containing 10% RHA and 40% pond ash. Without curing and addition of cement, the unsoaked CBR increases from 2.1 to 18.0%, whereas the soaked CBR increases from 1.1 to 13.1% on addition of 10% RHA and 40% pond ash by the dry weight of the soil. Overall, it was advocated that pond ash can be successfully used to escalate the CBR of the soil. Addition of RHA to pond ash results in higher bearing strength.

Gupta and Kumar [18] considered the consequence of soil stabilization with pond ash, RHA and cement on the UCS and STS of clay soil, following the same methodology as in their previous study Gupta and Kumar [17]. The samples evaluated to ascertain the compaction properties such as OMC and MDD, strength properties such as UCS and STS. The results related to the compaction properties were similar to that observed in their previous study Gupta and Kumar [17]. Other results show that with the rise in the dosage of pond ash, UCS and STS of the soil increase up to 40% dosage of pond ash beyond which the UCS and STS fall. With the increases in the contents of RHA, UCS and STS of the soil increase up to 10% content of RHA beyond which the falls. The UCS and STS further increased with increases in cement content for any given mixture. Regarding the blend of soil–pond ash–RHA mixtures, highest improvement in the UCS and STS was observed for the mixture containing 10% RHA and 40% pond ash. On comparison with the untreated soil sample, the STS and UCS of the soil treated with the mixture of 40% pond ash, 10% RHA and 4% cement, improved by 303% and 336%, respectively.

Yadav et al. [34] further extended his previous study. Yadav and Tiwari [33] studied the strength properties of soil (USCS classification: CL) stabilized with cement and pond ash and reinforced with polypropylene fibre of different lengths such as 6 and 12 mm, in the range 0–1%, by the dry weight of soil, instead of coir fibres. In this study, the dosage of cement was not altered and kept constant to 5%, by the dry weight of the soil. Unlike coir fibre, addition of polypropylene fibre did not have a noteworthy consequence on the compaction properties of the soil. The UCS and STS improved with rise in curing period and rise in the dosage of the fibre, for any given mixture. The authors reported that the magnitude of improvement in the UCS and STS was comparable for the samples reinforced with the 1% dosage of the fibre of 6 mm and the samples reinforced with the 0.5% dosage of the fibre of 12 mm length. Addition of fibres resulted in conversion of brittle performance of cement or pond ash stabilized samples to a ductile performance, where the brittleness augmented with upsurge in the dosage of pond ash. Overall, the authors recommended that up to 20% weight of the dry soil may be replaced by pond ash and up to 1% soil may be replaced by polypropylene fibre.

Gupta et al. [19] compared the efficacy of pond ash with RHA in increasing the UCS of a soil (USCS classification: CL). The soil samples were treated with varying content of pond ash ranging from 0 to 50% and varying content of RHA ranging from 0 to 20%, added by the dry weight of the soil. Results show that addition of pond ash or RHA both led to a noteworthy escalation of the UCS of the soil. RHA performed better than pond ash, as higher UCS was obtained with RHA stabilization. The optimum content of pond ash and RHA was determined to be 40 and 10%, respectively. On addition of pond ash or RHA, the MDD of the mixture lowers, and OMC rises with rise in contents of pond ash or RHA due to lower MDD and higher OMC as compared to that of the soil. Overall, it was observed that both pond ash and RHA can be utilized to rise the strength of the soil.

Gupta et al. [20] compared the efficiency of pond ash with brick kiln dust in improving stiffness and strength of a fine-grained soil. (USCS classification: CL). The soil samples were prepared with varying content of the wastes in the range of 0–40%, added by the dry weight of the soil. Samples were also prepared with combinations of brick kiln dust and pond ash. The samples were tested to establish MDD, OMC and CBR (unsoaked and four-day water soaked) and stiffness parameters such as resilient modulus. The results show that with rise in the contents of the wastes, the MDD lowers, whereas the OMC rises. The reason for decreases in MDD, as cited by the authors, was lesser specific gravity of the wastes particles than the soil particles, whereas the reason for decreases in OMC, as cited by the authors, was greater fineness of the wastes than that of the soil and existence of un-burnt carbon in the wastes. For soil stabilized individually with the wastes, with the increases in the contents of the wastes, the CBR of the soil increases up to 30% content of the wastes, beyond which the CBR falls. When added in combination, the maximum CBR was recorded on addition of 20% dosage of pond ash and 20% dosage of brick kiln dust. Resilient modulus of the soil stabilized with the optimum dosages of the wastes was determined through advanced cyclic triaxial testing. Resilient modulus of the soil stabilized with pond ash increased by 39.1% (unsoaked sample) and 44.3% (unsoaked sample). Resilient modulus of the soil stabilized with the optimum combined dosage of pond ash and brick kiln dust increased by 103.6% (soaked sample) and 62.8% (unsoaked sample).

2.8 *Utilization of Pond Ash in Erstwhile Engineering Purposes*

Various researchers have used pond ash in erstwhile engineering purposes such as stabilized/reinforced pond ash for its effective utilization as a safe foundation material, a fill material, material for building of the base and sub-base layers of flexible pavements and as a material for construction of liner of waste landfills.

Ghosh et al. [14] utilized pond ash as a safe foundation medium by reinforcing it with layers of jute geotextile. On varying the depth of placement of the geotextile, the authors reported that maximum bearing capacity was achieved when the depth of geotextile in a single layer system was placed at a depth (from the top of the tank) of 0.3125 times the depth of the footing. It was established that the bearing capacity improved with rise in the number of layers of the geotextile, till the optimum is reached. This optimum number was found dependent on the depth of the geotextile layer.

Bera et al. [5] reinforced pond ash with natural and synthetic geotextiles to utilize pond ash as a safe foundation medium. The authors successfully demonstrated and validated a power model for the prediction of bearing capacity of a foundation, square in shape, resting on pond ash reinforced with a geotextile.

Chand and Subbarao [10] increased the strength of in-place pond ash through hydrated lime columns. The in-place stabilization by lime column technique was established as an efficient technique in lowering permeability and raising the UCS. The technique was found to be effectual in lowering the potential of leaching of heavy metals by pond ash deposits, thereby decreasing the severely hazardous impact of pond ash dumps.

Chand and Subbarao [11] improved the strength and durability of pond ash by adding lime. The dosage of lime was 10 and 14% by the dry weight of pond ash for utilization of pond ash in the construction of base and sub-base course of flexible pavements. The specimens were subjected to curing for 28, 45, 90, and 180 days. Tests to determine strength were carried out like point load tests, UCS tests and Schmidt rebound hammer tests, and durability tests such as slake durability tests were performed. All the strength and durability properties increased with rise in content of lime and rise in curing time.

Mohanty and Patra [25] determined the liquefaction potential and cyclic behaviour of pond ash samples obtained from three different power plants such as Talcher, Panipat and Pankhi. Results show that Talcher pond ash was found to have higher in specific gravity. Overall, all the pond ash samples depict high resistance to liquefaction, and therefore, pond ash samples could be efficiently utilized as a fill material for construction of embankments.

Vijayasri et al. [32] determined the impact of reinforcing pond ash with a geotextile, on the liquefaction potential and cyclic behaviour of pond ash samples obtained from three different power plants such as Talcher, Panipatandand Pankhi. Results show that Talcher pond ash was found to have higher in specific gravity. Overall, all the pond ash samples depict high resistance to liquefaction, and therefore, pond ash samples could be efficiently utilized as a fill material.

Arora and Kumar [1] reinforced pond ash with polypropylene fibre. The reinforced material was used as a superimposing layer on soft clay for raising the bearing capacity of the foundation medium, resulting in a safe system. Pond ash was reinforced with 0.5%, 0.75% and 1.0% dosage of the fibre of different lengths such as 6, 12 and 18 mm. Model footing test with a circular plate of a diameter 0.1 m was performed where the depth of reinforced pond ash layer was altered. Results show that bearing capacity augmented with the rise in dosage and length of the fibre. The reason being that on addition of fibres to an optimum content in the weight of pond ash, bind the particles of the soil and reallocate stresses across a wider area. Addition of size (length) of the fibres raises the endurance (frictional) along the fibre–pond ash interface leading to added strength.

Rout and Singh [27] utilized pond ash stabilized with 20% commercial bentonite for application as a liner material in building of waste landfills. The stabilizer mixture was reinforced with polypropylene fibres dosage of 0.2%, 0.5% and 1.0% by the dry weight of the mixture. Addition of fibres had an insignificant effect on the compaction properties of pond ash-bentonite mixture. The UCS increased on inclusion of fibre where the maximum UCS was achieved on reinforcement with 1% fibre content. The permeability of the mixture got enhanced with a rise in the dosage of the fibre, though the permeability at 1% fibre content was discovered to be lesser than 10^{-7}cm/s, the upper limit of permeability for landfill liners, established by the USEPA. Shear strength of the mixture, determined by conducting direct shear tests, increased with a rise in the fibre content, being maximum at a dosage of 1%. On raising in the moulding moisture content, permeability decreases, whereas volumetric shrinkage strain increases, whereas the UCS and unit cohesion rise with a rise in the moulding moisture content till it dry of the optimum moisture content, peaking maximum at moisture content varying from 80 to 90% of the OMC.

3 Sustainable Construction with Pond Ash Stabilized Soil

Sustainable construction can be practiced by application of pond ash in activities and construction projects. Scholar has reported rise in CBR and UCS of soils on their stabilization with pond ash. Rise in the strength of soils on their stabilization with pond ash is expected to lower the required thickness of pavement layers, when fibre-reinforced soil is used in the subgrade layer of flexible pavement, thereby saving bitumen and aggregates, making the construction sustainable. Also, mitigation of greenhouse gas emissions, associated with pavement construction, is predictable because of reduction in consumption of bitumen and aggregates used in the building of pavements.

4 Conclusions

Pond ash was determined to be a noteworthy waste material that is generated in large quantities by TPPs and plays a significant role in polluting the environment. The literature highlights the potential threat of hazardous contamination of groundwater and surface water bodies by leaching of heavy metals due use of pond ash in engineering applications such as construction of roads. Consequently, there is an imperative need for effective utilization of pond ash in construction industry, and such a utilization of pond ash is expected to be a sustainable construction practice. Pond ash has been proved to be efficient in stabilization of weak soils and many other engineering applications. Following conclusions were drawn from the review of the literature.

1. As per the literature, physical properties of pond ash show that pond ash is made up of sand and silt size particles, where usually the content of sand was found to be higher than the silt. Pond ash was assessed to be non-plastic in nature.
2. As per the literature, chemical properties of pond ash (class F) show that pond ash mainly contains alumina and silica, with a low amount of calcium.
3. As per the mineralogical properties of pond ash reported in the literature, pond ash was found to be rich in quartz (SiO_2), alumina (Al_2O_3) and feldspar, whereas a minute quantity of iron oxide has been reported to be present in pond ash samples.
4. As per the microstructural properties of pond ash reported in the literature, pond ash was found to have a rough surface. Pond ash samples mostly comprise irregular and spherical and flaky-shaped particles.

5. As per the toxicity properties of pond ash reported in the literature, the leachate of pond ash possesses high concentrations of potential toxic elements such as zinc and lead. Overall, pond ash was found to be safe for use in construction as the concentrations of toxic elements in pond ash samples were found to lower than the permissible limits for hazardous waste, set by USEPA
6. Pond ash has been utilized in engineering applications other than soil stabilization such as a foundation medium. Strength of pond ash has been attempted to improve by stabilization with lime and reinforcement with fibres, geotextiles and geosynthetics.
7. Stabilization of soil with pond ash results in a gradual decrease of plasticity and MDD, and a gradual rise of OMC of the soil, with rise in the dosage of pond ash.
8. Stabilization of soils with pond ash results in increased bearing strength of soils of varying plasticity at an optimum dosage of 30% content of pond ash, added by the dry weight of the soil.
9. Stabilization of soils with pond ash results in increased stiffness, measured as resilient modulus.
10. Stabilization of soils with pond ash, for the subgrade layer of flexible pavements, results in reduction of the cost of pavement construction and thickness, and greenhouse gas emissions.
11. However, up till now, no study has been executed to find the effect of pond ash on permeability, split tensile strength, shear strength, unconfined compressive strength and stiffness of soils of soils of different plasticity.

References

1. Arora S, Kumar A (2020) Bearing capacity of circular footing resting on fiber-reinforced pond ash overlying soft clay. J Hazard Toxic Radioact Waste 24(1):04019037
2. Ahmaruzzaman M (2010) A review on the utilization of fly ash. Prog Energy Combust Sci 36(3):327–363
3. Bera AK, Ghosh A, Ghosh A (2009) Shear strength response of reinforced pond ash. Constr Build Mater 23(6):2386–2393
4. Bera AK, Ghosh A, Ghosh A (2007) Compaction properties of pond ash. J Mater Civ Eng 19(4):349–357
5. Bera AK, Ghosh A, Ghosh A (2008) Bearing capacity of square footing on reinforced pond ash. Proc Inst Civil Eng-Ground Improv 161(1):17–22
6. Bera AK, Kumar A (2010) Effect of pond ash content on engineering properties of fine grained soil. In: Indian geotechnical conference, pp 16–18
7. Bera AK (2013) Study on unconfined compressive strength of pond ash soil mixture reinforced with jute geotextiles. Emirates J Eng Res 18(1):59–65
8. Central Electricity Authority (2018) Ministry of Power, Government of India. National Electricity Plan, vol 1 Generation. Retrieved from http://www.cea.nic.in/reports/committee/nep/nep_jan_2018.pdf
9. Central Electricity Authority (2020) Ministry of Power, Government of India. Report on fly ash generation at coal/lignite based thermal power stations & its utilization in the country for

the Year 2018–19. Retrieved from http://www.cea.nic.in/reports/others/thermal/tcd/flyash_201819.pdf
10. Chand SK, Subbarao C (2007) In-place stabilization of pond ash deposits by hydrated lime columns. J Geotech Geoenviron Eng 133(12):1609–1616
11. Chand SK, Subbarao C (2007) Strength & slake durability of lime stabilized pond ash. J Mater Civ Eng 19(7):601–608
12. Chikkatur AP, Sagar AD, Sankar TL (2009) Sustainable development of the Indian coal sector. Energy 34(8):942–953
13. Dudhley B (2019) BP statistical review of world energy. British Petroleum, London, UK
14. Ghosh A, Ghosh A, Bera AK (2005) Bearing capacity of square footing on pond ash reinforced with jute-geotextile. Geotext Geomembr 23(2):144–173
15. Ghosh A (2010) Compaction properties & bearing ratio of pond ash stabilized with lime & phosphogypsum. J Mater Civ Eng 22(4):343–351
16. Gupta D, Kumar A (2016) Strength characterization of cement stabilized & fiber reinforced clay–pond ash mixes. Int J Geosynthetics Ground Eng 2(4):32
17. Gupta D, Kumar A (2017) Performance evaluation of cement-stabilized pond ash-rice husk ash-clay mixture as a highway construction material. J Rock Mech Geotech Eng 9(1):159–169
18. Gupta D, Kumar A (2017) Stabilized soil incorporating combinations of rice husk ash, pond ash & cement. Geomech Eng 12(1):85–109
19. Gupta D, Kumar A, Kumar V, Priyadarshee A, Sharma V (2019) Performance of pond ash & rice husk ash in clay: a comparative study. In: Recycled waste materials. Springer, Singapore, pp 145–153
20. Gupta G, Sood H, Gupta P (2020) Performance evaluation of pavement geomaterials stabilized with pond ash & brick kiln dust using advanced cyclic triaxial testing. Materials 13(3):553
21. Jakka RS, Ramana GV, Datta M (2010) Shear behaviour of loose & compacted pond ash. Geotech Geol Eng 28(6):763–778
22. Kolay PK, Sii HY, Taib SNL (2011) Tropical peat soil stabilization using class F pond ash from coal fired power plant. Int J Civil Environ Eng 3(2):79–83
23. Kumar A, Gupta D (2016) Behavior of cement-stabilized fiber-reinforced pond ash, rice husk ash–soil mixtures. Geotext Geomembr 44(3):466–474
24. Mishra UC (2004) Environmental impact of coal industry & thermal power plants in India. J Environ Radioact 72(1–2):35–40
25. Mohanty S, Patra NR (2014) Cyclic behavior & liquefaction potential of Indian pond ash located in seismic zones III & IV. J Mater Civ Eng 26(7):06014012
26. Mohanty S, Patra NR (2015) Geotechnical characterization of Panki & Panipat pond ash in India. Int J Geo-Eng 6(1):13
27. Rout S, Singh SP (2020) Influence of fibers on hydro-mechanical properties of bentonitic mixtures. Geotech Geol Eng 1–17
28. Singh SP, Sharan A (2014) Strength properties of compacted pond ash. Geomech Geoeng 9(1):9–17
29. Suthar M, Aggarwal P (2016) Environmental impact & physicochemical assessment of pond ash for its potential application as a fill material. Int J Geosynthetics Ground Eng 2(3):20
30. Suthar M, Aggarwal P (2018) Bearing ratio & leachate analysis of pond ash stabilized with lime & lime sludge. J Rock Mech Geotech Eng 10(4):769–777
31. Patil BM, Patil KA (2013) Effect of pond ash & RBI Grade 81 on properties of subgrade soil & base course of flexible pavement. Int J Civil Arch Struct Constr Eng 7(12):636–641
32. Vijayasri T, Raychowdhury P, Patra NR (2017) Seismic response analysis of Renusagar pond ash embankment in Northern India. Int J Geomech 17(6):04016141
33. Yadav JS, Tiwari SK (2016) Behaviour of cement stabilized treated coir fibre-reinforced clay-pond ash mixtures. J Build Eng 8:131–140
34. Yadav JS, Tiwari SK, Shekhwat P (2018) Strength behaviour of clayey soil mixed with pond ash, cement & randomly distributed fibres. Trans Infrastruct Geotech 5(3):191–209

Experimental Investigation on Bamboo as Structural Pile

Madhubala Mayilswamy, Anil Kumar Sharma, Surya Muthukumar, Jayakrishnan, Yamini Jayaprakash, Sakthipriya, and Amritha Velayudham

1 Introduction

Soil reinforcement techniques is used in the field of construction due to its versatile and flexible nature. Reinforced soil is used as a construction material by incorporating small percentage of material in the form of fibers, plates, solid rods, etc., to improve the tensile strength. The usage of natural inexpensive material as a reinforcement improves the sustainability in construction. This simple technique can cause a huge difference in the performance, cost, and sustainable nature of the project [1–3].

Bamboo is a perennial grass available in plenty in tropical, sub-tropical, and temperate zones [4]. It is a non-timber forest resource from large grass family [5]. There are around 1500 identified bamboo species in the world [6]. Bamboo is an easily available natural material used as a renewable resource in construction. It is a promising engineering composite material with fibers in one direction that gives high flexural strength to it [7–9]. Bamboo has wider application in the field of construction, as a building material for roofs, walls, also used for shoring and scaffolding [20]. It is used in stability of foundation pit excavation [10]. Studies show that it possesses high

M. Mayilswamy · S. Muthukumar (✉) · Jayakrishnan · Y. Jayaprakash · Sakthipriya · A. Velayudham
Department of Civil Engineering, Amrita School of Engineering, Amrita Vishwa Vidyapeetham, Coimbatore, India
e-mail: m_surya@cb.amrita.edu

M. Mayilswamy
e-mail: madhubalamayilswamy6@gmail.com

A. K. Sharma
Department of Civil Engineering, National Institute of Technology Patna, Patna, India
e-mail: sharma.kr.anil@gmail.com

H. Singh et al. (eds.), *Sustainable Development Through Engineering Innovations*, Lecture Notes in Civil Engineering 113,
https://doi.org/10.1007/978-981-15-9554-7_56

compressive and tensile strength and also been used as a reinforcement in concrete [11].

Bamboo is also used as a foundation member in soft soil. Instability and large settlement of peats and soft soil under the construction of embankment is improved by bamboo piles which also increases the bearing capacity of soil [1]. Having high compression and tensile strength [3], bamboo's strength can differ for different species based on the parts and moisture content.

Freshly cut bamboo possess high tensile strength, and it reduces gradually with its age. Selection of bamboo to be used in constructed is based on the seasoning and age. A well-seasoned bamboo should be around 3–6 years. The bottom portion of the seasoned bamboo possess high strength [12]. Slope reinforced with bamboo pile improves the stability by increasing the safety factor. The bearing capacity of a shallow footing placed on a sloped surface can be significantly increased by providing reinforcement through bamboo pile. The installation can improve the resistance to the lateral forces [13]. Bamboo piles can be widely used as a vertical reinforcement to improve the bearing capacity of sand in square footing.

Bamboo with branches is used as soil nails for hard soil and soft clay [14], which shows highest tensile capacity about 2.5–2.8 times than that of steel-pipe nails [15]. Bamboo pile performs better than geotextiles as a reinforcement in soil [16]. It is used as reinforcement in peat soil which has reduced settlement to a great extent [17]. Bamboo is used for stabilization and reinforcement to directly support the structure. From various studies, it is found that bamboo reinforcement can perform better than conventional steel piles, this is mainly due to the improved frictional resistance between the soil and the reinforcement and the inherent fibrous nature of the grass. The length and diameter should be decided based on the requirement [7].

The main problem with a natural reinforcement material is its durability. Bamboo is durable for 2 years without any treatment, and with preservatives, the durability can be improved to 4–7 years. But from the various techniques adopted over years, it is found that soaking bamboo in flowing water can improve the durability more than 10 years. Cutting and slicing of the natural bamboo trunk can reduce the strength, so it will be used in the original condition to retain the properties [7].

In the present study, it is aimed to understand load-settlement behavior of bamboo piles, to find the ultimate load-carrying capacity of bamboo piles and to compare the results of various dimensions bamboo piles.

2 Materials

2.1 Sand

River sand collected from Kaveri river basin near Kulithalai, Tamilnadu, is used for the test. The physical properties of sand are tested as per IS specifications and

Table 1 Properties of soil

Properties	Value
Coarse sand	2.9%
Medium sand	34%
Fine sand	61.3%
Silt and clay	1.8%
D_{10}	0.20 mm
D_{30}	0.32 mm
D_{60}	0.56 mm
Coefficient of uniformity, C_u	2.8
Coefficient of curvature, C_c	0.91
IS classification	SP

according to UCS the soil is classified as poorly graded sand (SP). The properties of sand are listed in Table 1.

2.2 Bamboo

Bamboo has a unique structure which is fiber reinforced in unidirectional with multiple nodes along its length [2, 18]. It is mostly used construction material because of its good mechanical properties compared with other materials. The density of bamboo is about 700–800 kg/m^3, and the fiber weight is around 60–70% of the total weight. It possess high tensile strength along the fiber direction [7]. Both treated and untreated bamboos possess good physical and mechanical properties [19]. The properties vary with respect to age and various parts of bamboo. Naturally available bamboo was collected from, bamboo market, Gandhi park in Coimbatore, Tamilnadu, for the model test.

Laboratory model tests were carried out to study the load-settlement behavior of individual bamboo piles which is hollow inside with a knot at the base of pile under vertical loading condition on medium dense sand. Tests were performed by varying the dimensions of bamboo piles, three different inner and outer diameters and three different lengths are shown in Fig. 1. The dimensions of the bamboo pile are selected based on the natural availability and to satisfy the model test condition.

Bamboo pile dimensions along with specifications are given in Table 2.

2.3 Pile Cap

Mild steel plate of circular in cross section with 150 mm diameter and 8 mm thickness is used as a pile cap kept at the top of all bamboo pile tested.

Fig. 1 Bamboo piles

Table 2 Bamboo pile dimensions

Bamboo piles	Inner diameter (ID) (mm)	Outer diameter (OD) (mm)	Length (mm)
C1-P1	60	70	300
C1-P2	50	65	300
C1-P3	24	34	300
C2-P1	60	70	200
C2-P2	50	65	200
C2-P3	24	34	200
C3-P1	60	70	100
C3-P2	50	65	100
C3-P3	24	34	100

3 Experimental Test Setup

The load-carrying capacity of bamboo pile was analyzed using a series of tests conducted on bamboos with varying inner and outer diameters and lengths. The vertical load tests were carried in a model tank setup of size 550 mm × 550 mm × 550 mm using a loading frame of capacity 50 kN. The dimensions of the tank were chosen, so that there will not be interference between the walls of the tank and the failure zone around the piles. It is also maintained to satisfy the condition to test end bearing pile. The tests were conducted for a single pile with a cap to analyze the improvement in bearing capacity. The foundation medium was prepared using river sand. The sand was filled in the tank setup by sand raining technique. The height of fall required for achieving a relative density of 50% was found by pouring sand from varying heights. This height of fall was found to be 280 mm to achieve 50% relative density [20], and to maintain uniformity in the series of tests performed, it was ensured that throughout the filling process this height was

Fig. 2 Vertical loading setup

constantly maintained. After filling up to the top of the pile, the bamboo was inserted to the soil to simulate the actual field placement condition. Pile was kept at the center of the tank. Experimental test setup for vertical loading is shown in Fig. 2.

4 Test Procedure

The pile cap was kept at the top and vertical compressive load was applied to bamboo pile through a hydraulic jack, which was supported centrally at the bottom flange of the steel girder made of channel sections. The vertical load tests were conducted on piles as per procedure recommended by IS 2911 (Part 4)-[21]. The proving ring and two LVDTs were placed in position over the pile cap. Axial load was applied in increments, and it was observed through 50 kN proving ring placed over the pile cap. Each increment of loading was maintained until the pile settlement become less than 0.02 mm/min. The vertical deflection of the pile was measured by LVDT corresponding to the load, and when the deflection of the pile ceases, the next load increment is applied. The test was proceeded until the pile achieves its ultimate axial capacity for the corresponding failure load. The load-settlement curve was plotted for vertical loading case, and safe load is calculated by minimum of two conditions for individual pile. Bamboo piles of the various inner and outer diameter and length were used as given in Table 2.

Table 3 Safe load of bamboo piles for vertical loading

Bamboo piles	Inner diameter (ID) (mm)	Outer diameter (OD) (mm)	Length (mm)	Safe load (kN)
C1-P1	60	70	300	0.511
C1-P2	50	65	300	0.367
C1-P3	24	34	300	0.188
C2-P1	60	70	200	0.325
C2-P2	50	65	200	0.164
C2-P3	24	34	200	0.064
C3-P1	60	70	100	0.195
C3-P2	50	65	100	0.13
C3-P3	24	34	100	0.044

5 Results and Discussions

5.1 Bamboo Piles

From the vertical pile load tests, the load-carrying capacity of bamboo piles for various conditions were obtained [22]. The safe load-carrying capacity of bamboo piles for different inner and outer diameters and lengths are reported in Table 3.

Bamboo pile of 300 mm length and ID60 mm-OD70 mm had larger load-carrying capacity than ID50 mm-OD65 mm, ID24 mm-OD34 mm diameters of length 300 mm. For length 200 mm, 100 mm with respect to same diameters, ID60 mm and OD70 mm had larger load-carrying capacity. Table 3 clearly shows that larger diameter bamboo pile shows higher load-carrying capacity and increase in length bamboo pile shows an increase in load-carrying capacity.

Load-settlement curves for bamboo piles shows ID60 mm-OD70 mm diameter piles that have higher peak than ID50 mm-OD65 mm, ID24 mm-OD34 mm diameters for the pile length of 300 mm, 200 mm, 100 mm, which concludes that load-carrying capacity is high for larger diameter bamboo piles.

Figure 3 shows that, for 300 mm length bamboo pile with diameter ID60 mm-OD70 mm, ID50 mm-OD65 mm, ID24 mm-OD34 mm, maximum diameter pile C1-P1 performs better with a greater load-carrying capacity, whose safe load is 28% greater than C1-P2 and 63% greater than C1-P3.

Figure 4 shows that, for 200 mm length bamboo pile with diameter ID60 mm-OD70 mm, ID50 mm-OD65 mm, ID24 mm-OD34 mm, maximum diameter pile C2-P1 performs better having greater load-carrying capacity, whose safe load is 50% greater than C2-P2 and 80% greater than C2-P3, here bamboo pile C2-P1 curve was high.

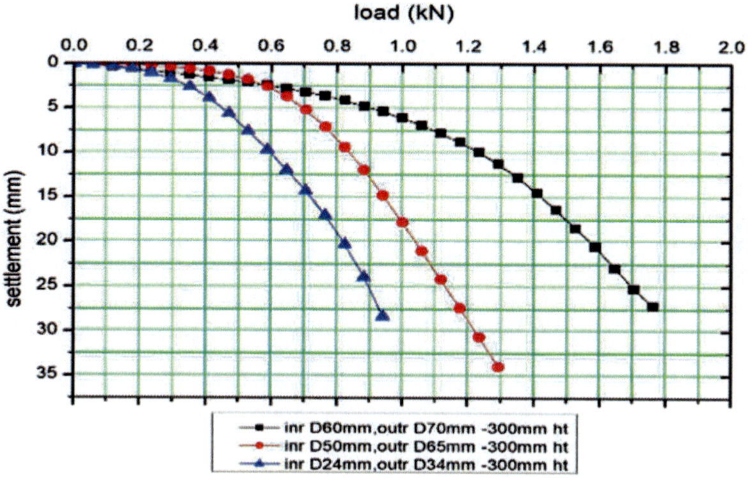

Fig. 3 Load versus settlement for vertical loading of bamboo pile with 300 mm length

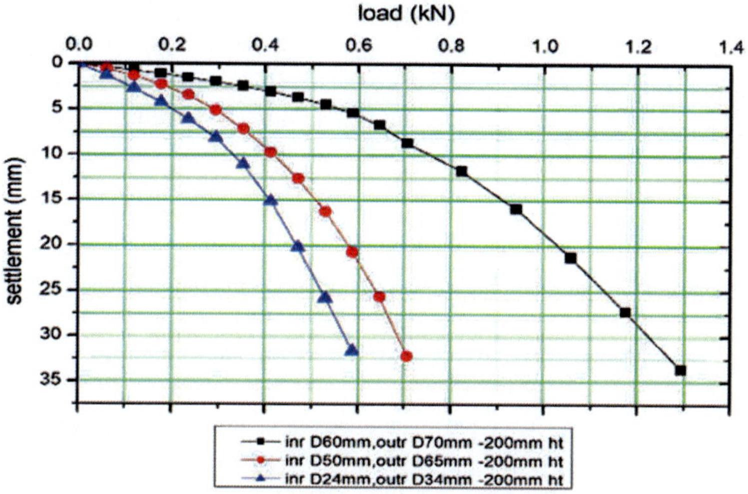

Fig. 4 Load versus settlement for vertical loading of bamboo pile with 200 mm length

Figure 5 shows that for 100 mm length bamboo pile with diameter ID60 mm-OD70 mm, ID50 mm-OD65 mm, ID24 mm-OD34 mm, maximum diameter pile C3-P1 performs better has greater load-carrying capacity, whose safe load is 33% greater than C3-P2 and 77% greater than C3-P3. Here bamboo pile C3-P1 curve shows high load-carrying capacity.

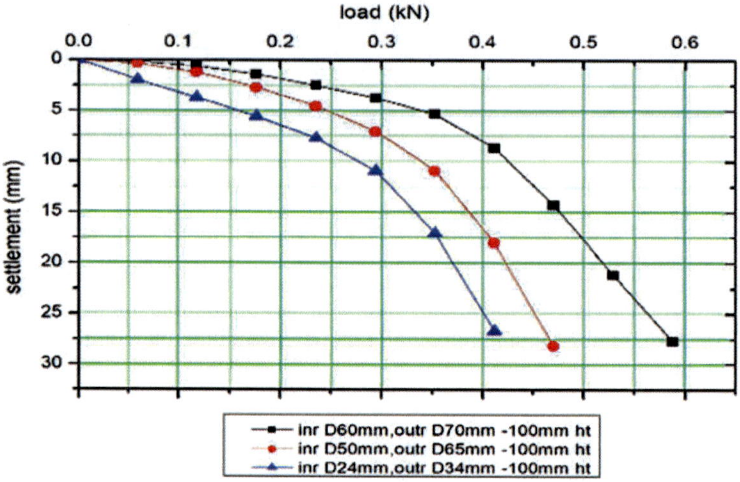

Fig. 5 Load versus settlement for vertical loading of bamboo pile with 100 mm length

Bamboo piles of ID60 mm-OD70 mm, ID50 mm-OD65 mm, ID24 mm-OD34 mm diameters with respect to 200 mm length piles (C2-P1, C2-P2, C2-P3) show maximum percentage increase in safe loads than 300 mm length and 100 mm length piles, change in length of bamboo piles influence the load-carrying capacity of piles, in 200 mm length piles D/B ratio was maintained as 2–3.5.

The durability of the untreated bamboo pile is around 2 years without any strength reduction. To improve the performance and durability, treatments like bitumen coating, oil coating, borax with water preservative can be performed.

6 Conclusion

From the experimental investigation on bamboo piles, the following conclusions can be made

- It is observed that the installation of bamboo piles as a vertical reinforcement in loose sand has considerably increased the bearing capacity of the footing.
- The series of tests showed that the soil confinement due to the vertical reinforcement has improved the bearing capacity and controlled the settlement to a greater extent.
- The pile condition C1-P1 has shown an improvement of around 63% compared to the C1-P3, and C2-P1 has shown an improvement of around 80% compared to the C2-P3, and C3-P1 has shown an improvement of 77% compared with C3-P3.
- The overall tests can be concluded that the bearing capacity increases with increase in diameter and length of the pile.

- Bamboo pile length influences the load-carrying capacity of piles, and percentage increase in safe load is maximum when D/B ratio is maintained between 2 and 3.5.

References

1. Rahardjo PP. The use of bamboo and bakau piles for soil improvements and application of pile raft system for the construction of embankment on peats and soft soils
2. Rogerson A, Edi Suhaimi B, Jegatheswaran R, Khairul Bin A (2016) Bamboo properties and suitability as a replacement for wood. PJSRR
3. Anusha R (2011) Behaviour of bamboo reinforced soils – state of art (2010), pp 15–18
4. Jha JN, Choudhary AK, Gill KS (2009) Bearing capacity improvement of soil using bamboo micropiles. In: ACSGE-2009, Oct 25–27, BITS Pilani, India
5. Xu Q, Harries K, Li X, Lui Q, Gottron J (2014) Mechanical properties of structural bamboo following immersion in water. Eng Struct 81:230–239
6. Chung KF, Yu WK (2002) Mechanical properties of structural bamboo for bamboo scaffoldings. Eng Struct 24:429–442
7. Farid M (2017) A review on the use of bamboo for earthwork construction. 04010. https://doi.org/10.1051/matecconf/201713804010
8. Das A, Sarkar S (2018) Importance of bamboo in building construction. Int Res J Eng Technol 5(6):389–392
9. Mahzuz HMA, Ahmed M, Dutta J, Ros RH (2013) Determination of several properties of a bamboo of Bangladesh. J Civil Eng Res 3(1):16–21
10. Mustapha AM (2008) Bamboo as soil reinforcement: a laboratory trial. Leonardo J Sci 13:69–77
11. IS 456:2000 Indian Standard—plain and reinforced concrete code of practice (Fourth Revision)
12. Jha AK, Sivapullaiah PV (2018) Potential of fly ash to suppress the susceptible behavior of lime-treated gypseous soil. Soils Found 58(3):654–665. https://doi.org/10.1016/j.sandf.2018.02.024
13. Munawir A, Murni Dewi S, Soehardjono A, Zaika Y (2013) Safety factor on slope modeling with composite bamboo pile reinforcement. Int J Eng Res Appl (IJERA) 3:150–154
14. Dai Z, Chen Y, Zheng G, Guo W (2014) Numerical analysis on mechanism of bamboo soil nails and bamboo piles in rows for retaining deep foundation pit. Tunneling and underground construction GSP 242 © ASCE 2014
15. Dai Z-H, Guo W-D, Zheng G-X, Ou Y, Chen Y-J (2016) Moso Bamboo Soil-nailed Wall and Its 3D Nonlinear Numerical Analysis. Int J Geomech 16(5):04016012
16. Marto A, Othman BA (2011) The potential use of bamboo as green material for soft clay reinforcement system. In: 2011 international conference on environment science and engineering
17. Maulana A, Susanti RD, Waruwu A (2018) Potential of bamboo pile as reinforcement of peat soil under embankment. ARPN J Eng Appl Sci 13(1):52–56
18. Structural design using Bamboo – code of practice CED 13(7702)WC, November 2009
19. Awalluddin D, Ariffin MAM, Osman MH, Hussin MW, Ismail MA, Lee H-S, Lim NHAS (2017) Mechanical properties of different bamboo species. In: MATEC web of conferences 138, 01024
20. IS I 2720 (Part 14) (1983) 'Methods of test for soils'-part 14 determination of density index (relative density) of cohesionless soils
21. IS I 2911 (Part 4)—1985-code of practice for design and construction of pile foundations-part 4 load test on piles. Properties of soils – appendix C
22. Singh NT (2016) Verifications of pile load capacity using static pile load test. CESDOC

Stabilization of Clayey Soil Using Waste Foundry Sand and Molasses

Avinash Bhardwaj, Ravi Kumar Sharma, and Abhishek Sharma

1 Introduction

Settlement problem in structures founded over expansive soils is most common as volumetric changes occur on altering water content [1–4]. These volumetric changes in expansive soils cost more than 9 billion dollars loss in the United States of America and 150 million dollars loss in the United Kingdom every year [5]. Numerous researchers found many techniques to solve the problems associated with expansive soil [6–8]. Soil stabilization is one of the best techniques to rectify the problems associated with the expansive soil by improving the engineering and index properties of poor soils [9, 10]. Various waste materials have been used by many researchers to stabilize the expansive soil [11, 12]; Verma and [1, 3, 4, 13, 14].

Metal and nonmetal casting industries blend sand with additives to make the mold. After the molding process, this foundry sand is of no use to the casting industry [15]. Numerous researchers found many ways to use waste foundry sand as a construction material [1, 11, 16]. Guney et al. [11] reported that CBR and UCS value of expansive soil increases with the addition of WFS. It is concluded that WFS can be used as a part of the subgrade in the pavement. Kumar et al. [16] performed UCS and CBR tests on clayey soil and reported that the addition of 40% WFS quantity gives higher MDD, UCS, and CBR than any other composite. Bhardwaj and Sharma [1] concluded that 20% quantity of WFS increases the UCS value of clayey soil.

A. Bhardwaj (✉) · R. K. Sharma · A. Sharma
Civil Engineering Department, NIT Hamirpur, Hamirpur, India
e-mail: 07bhardwaj@gmail.com

R. K. Sharma
e-mail: rksnithp61@gmail.com

A. Sharma
e-mail: abhishek1@nith.ac.in

H. Singh et al. (eds.), *Sustainable Development Through Engineering Innovations*, Lecture Notes in Civil Engineering 113,
https://doi.org/10.1007/978-981-15-9554-7_57

Soil stabilization using enzymes is yet another mode of improving the strength of clayey soil having poor geotechnical properties. Various enzymes such as terrazyme, permazyme, molasses, earthzyme, and eko soil have been used to enhance the low strength of poor soils [17–19]. In the cane sugar industry during the production of sugar, molasses is obtained as a by-product. As the demand for sugar is high, molasses is produced in bulk. A few researchers found that molasses has good geotechnical properties and it can be used in soil stabilization as a core additive [1, 12]. Taye and Araya [12] blended molasses with cement and found that the CBR value of soil enhanced from 1 to 64%. Bhardwaj and Sharma [1] blended molasses with lime and reported that a combination of lime 9% and molasses 10% attained a higher UCS value.

From the previous researches, it can be concluded that WFS and molasses showed good geotechnical properties. Very limited research has been done on combinations of waste foundry sand–molasses. So, for the current research, different combinations of waste foundry sand and molasses were made, and their effect on expansive soil is checked.

2 Materials

2.1 Soil

The soil used in the present study was collected from village Dangar, district Bilaspur, Himachal Pradesh (India). After removing foreign material (stones, grassroots, organic material, etc.), soil was taken from a depth of about 1–1.5 m and brought to geotechnical engineering laboratory, NIT Hamirpur, in sealed bags. To remove moisture from the soil, it was dried under sunlight, and after completely drying, soil was pulverized with the help of a pulverizer. To avoid the entry of natural moisture, pulverized soil was again sealed in airtight bags. Physical properties of the soil were determined using suitable ASTM standards and relevant Indian standards. ASTM D6913-04 and ASTM D422-63 test have been used to determine the gradation curve of the clayey soil. The various geotechnical properties of clayey soil used in the present study are tabulated in Table 1.

2.2 Waste Foundry Sand

Waste foundry sand (WFS) is a waste of foundry industries that finds no use elsewhere. For the present study, waste foundry sand is obtained from Shakti Foundries, Ludhiana (Punjab). Table 1 presents the various physical properties of the soil.

Table 1 Geotechnical properties of clayey soil and WFS used in present study

Properties	Value (soil)	Value (WFS)
Type	CH	–
Activity	1.20	–
Liquid limit	54%	–
Plastic limit	21%	–
Plasticity index	33%	–
Specific gravity	2.61	2.65
Optimum water content	16.8%	8.20%
Maximum dry density	1.70 g/cc	1.59 g/cc
Unconfined compressive strength	520 kPa	–

2.3 Molasses

Due to the high demand for sugar in the market, the production of molasses (a by-product from sugarcane industry) is high, and it creates a dumping problem to the cane sugar industries. For clarification processes in producing sugar, lime and sulfur dioxide are added, and these agents remain in molasses after the crystallization starts. The molasses used in the present study has been brought from Budhewal Co-Operative Sugar Mill Ltd., Ludhiana (Punjab).

3 Results and Discussions

3.1 Atterberg's Limits

The liquid limit and plasticity index of clayey soil as observed from Atterberg's limits test were 54% and 33%, respectively, clearly indicating that soil is of CH category and needs to be stabilized prior to use. On adding WFS to clayey soil in varying amount from 10 to 30%, both the liquid and plastic limit of the mixture reduced along with reduction in the plasticity index (Fig. 1). This may be due to the addition of coarser material (WFS) to finer material (clay). A comparable decrease in liquid and plastic limit by adding WFS to poor soils has been observed. The addition of molasses in varying percentages from 5 to 15% to clayey soil reduced liquid limit and plastic limit, thus reducing overall plasticity index (Fig. 2). This may be due to the reason that on adding molasses to clay, the thickness of double diffused layer decreased resulting in decreasing the liquid limit. Few past researchers have observed the similar results on adding molasses to poor soils [12]. The addition of 20% WFS in a varying mixture of clay: molasses, decreased the liquid limit and plasticity index further by 116 and 136%, respectively, on adding 20% molasses (Fig. 3).

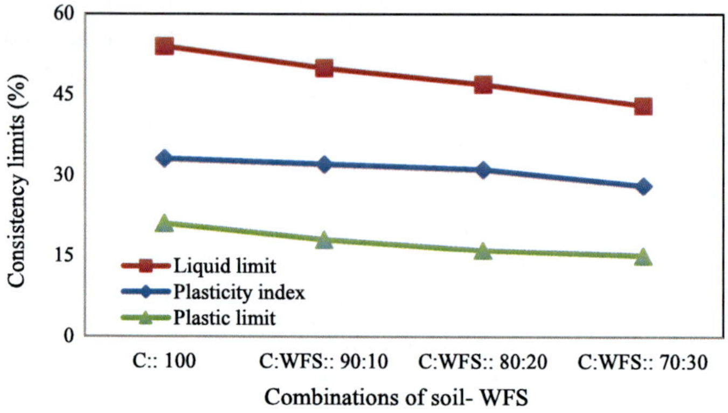

Fig. 1 Consistency limits of clay with varying percentages of WFS

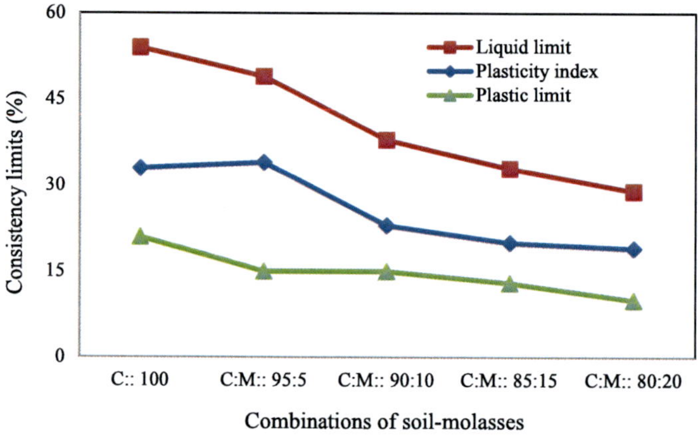

Fig. 2 Consistency limits of clay with varying percentages of molasses

3.2 Compaction Test

The optimum water content of the clayey soil was 16.8%, and maximum dry density was 1.70 g/cc. The WFS and molasses were blended with clayey soil alone and in combination with each other by varying percentages. Both the OMC and MDD increase on adding 10, 20, 30, and 40% WFS in clayey soil. The optimum water content of clayey soil increased from 16.8 to 18.2%, and MDD increased from 1.70 to 1.83 g/cc on adding 30% WFS (Fig. 4). This increase in MDD of clayey soil on addition of WFS may be due to the larger surface area of WFS particles and higher specific gravity of WFS [1]. The increase in optimum water content of clayey soil on addition of WFS may be attributed to the fact that the presence of bentonite (rich in montmorillonite mineral) in WFS increases water holding capacity. With the

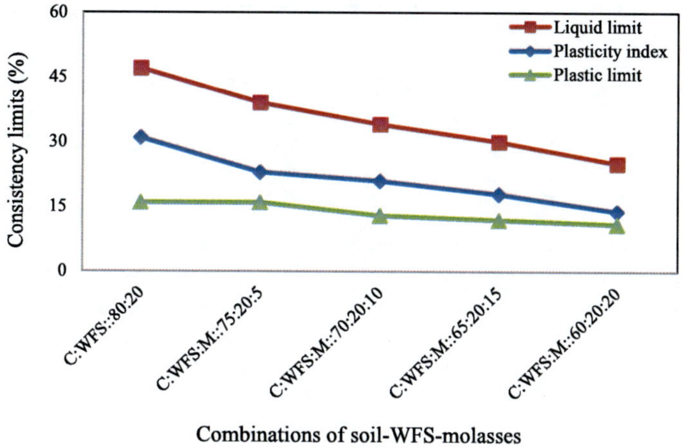

Fig. 3 Consistency limits of clay: WFS (20%) with varying percentages of molasses

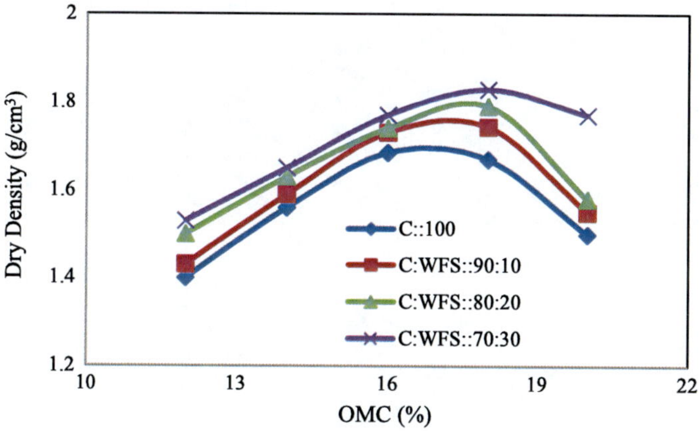

Fig. 4 Compaction curves for clay with varying percentages of WFS

addition of 5, 10, 15, and 20% molasses in clayey soil, the optimum water content reduces, and MDD increases. The optimum water content of clayey soil decreased from 16.8 to 10.0%, and MDD increased from 1.70 to 1.84 g/cc (Fig. 5). As molasses is having positively charged particles, so these positively charged particles attract toward negatively charged clayey soil particles, thus increasing the dry density of clayey soil. The reduction in OMC on adding molasses may be due to the transition of fine particles to coarse particles resulting in lower void ratio [12]. Further, addition of WFS and molasses in clayey soil by keeping WFS constant (20%) and varying molasses decreases the OMC and increases the dry density. The OMC of clayey soil decreased from 16.8 to 12.1%, and MDD increased from 1.70 to 1.93 g/cc (Fig. 6).

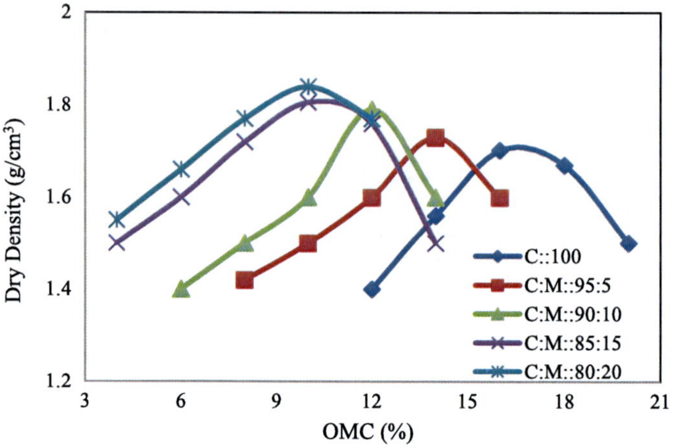

Fig. 5 Compaction curves for clay with varying percentages of molasses

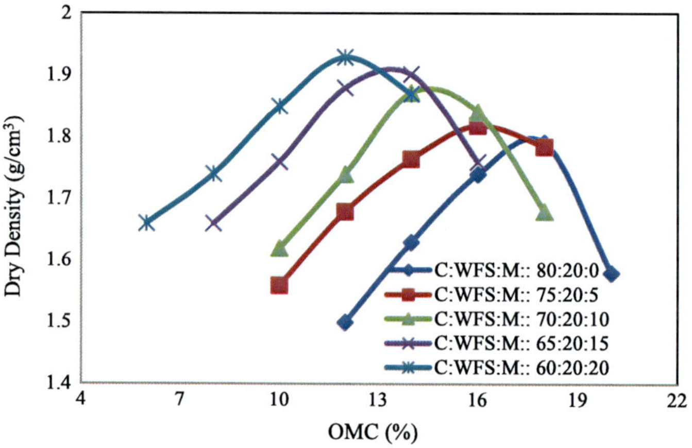

Fig. 6 Compaction curves for clay: WFS (20%) with varying percentages of molasses

3.3 UCS Tests

The UCS values of clayey soil after a curing period of 1, 3, 7, and 28 days were 246, 342, 433, and 520 kPa, respectively. On adding optimum percentage of WFS (20%) alone in clayey soil, the 28 days UCS value of clayey soil increased by 26% when compared with 28 days UCS value of composite. The unconfined compression strength (UCS) value for clay and WFS composite after 1, 3, 7, and 28 days obtained as 312, 406, 486, and 659 kPa, respectively (Fig. 7). This increase in UCS value of clayey soil on addition of WFS may be attributed to the compact structure of WFS. A similar trend of increase in UCS value on addition of optimum content of WFS in

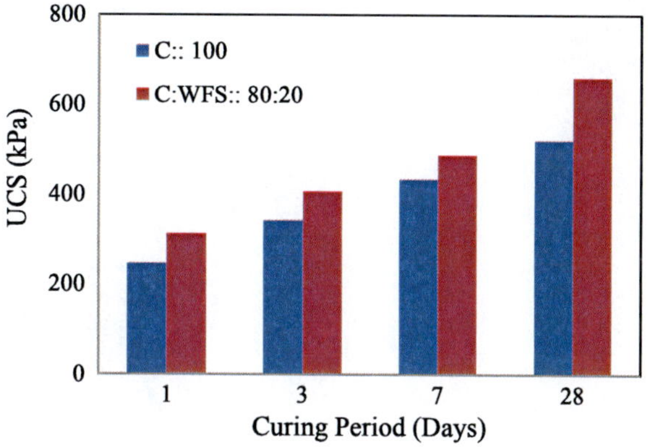

Fig. 7 UCS characteristics for clay with varying percentages of WFS

clayey soil has been reported in the past. On adding optimum percentage of molasses (10%) alone in clayey soil, the 28 days UCS value of clayey soil increased by 20% when compared with 28 days UCS value of composite. The unconfined compression strength (UCS) value for clay and molasses composite after 1, 3, 7, and 28 days obtained as 301, 387, 464, and 626 kPa, respectively (Fig. 8). This may be attributed to the fact that the addition of molasses in clayey soil attracts negatively charged fine clay particles (as molasses is positively charged) and bound fine clay particles, thus converting them to coarse particles and hence resisting higher compressive loads. Further, addition of WFS and molasses by keeping WFS constant (20%) and varying molasses in clayey soil the 28 days UCS value of clayey soil increased by 42% when compared with 28 days UCS value of the composite. The unconfined compression strength (UCS) value for clay, WFS, and molasses composite after 1, 3, 7, and 28 days obtained as 345, 454, 520, and 740 kPa, respectively (Fig. 9).

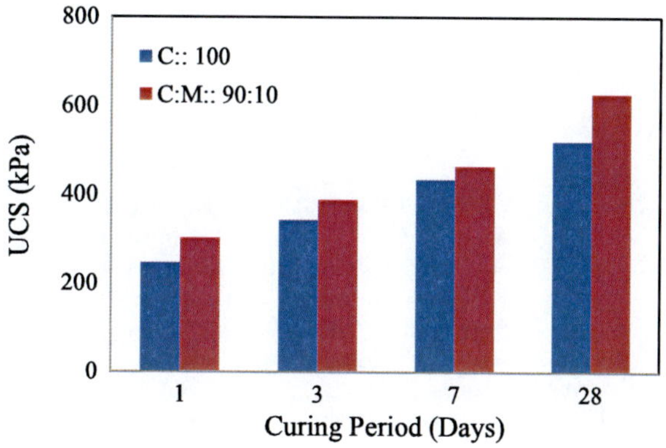

Fig. 8 UCS characteristics for clay with varying percentages of molasses

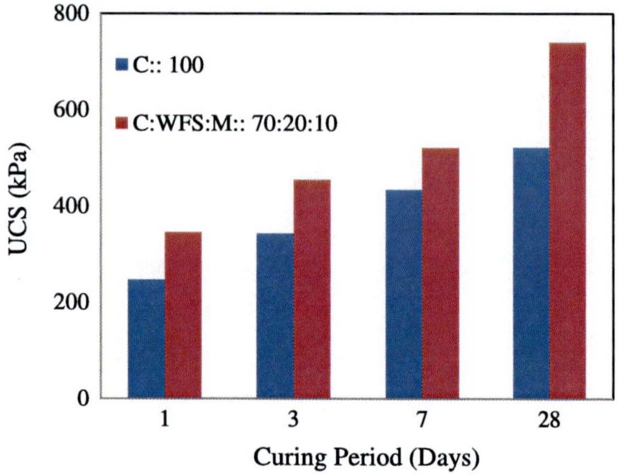

Fig. 9 UCS characteristics for clay: WFS (20%) with varying percentages of molasses

4 Conclusions

1. The addition of WFS and molasses alone and in combination to each other to clayey soil decreases both liquid limit and plastic limit, thereby decreasing the overall plasticity index.
2. The addition of WFS to clayey soil increases both maximum dry density and optimum moisture content. The addition of molasses to clayey soil alone and in combination to WFS increases the maximum dry density and decreases the optimum moisture content of the overall composite.
3. The addition of WFS and molasses alone and in combination to each other to clayey soil increases the unconfined compressive strength of the clayey soil for all the curing period. The highest unconfined strength is obtained for a curing period of 28 days for all the combinations.

References

1. Bhardwaj A, Sharma RK (2020) Effect of industrial wastes and lime on strength characteristics of clayey soil. J Eng Des Tech. https://doi.org/10.1108/JEDT-12-2019-0350
2. Lin B, Cerato AB (2012) Prediction of expansive soil swelling based on four micro-scale properties. Bull Eng Geo Environ 71(1):71–78
3. Sharma A, Sharma RK (2019) Effect of addition of construction–demolition waste on strength characteristics of high plastic clays. Innov Infrastru Solut 4(1):27
4. Sharma A, Sharma RK (2020) Strength and drainage characteristics of poor soils stabilized with construction demolition waste. Geotechn Geol Eng. 10.1007/s10706-020-01324-3
5. Nelson J, Miller DJ (1997) Expansive soils: problems and practice in foundation and pavement engineering. Wiley
6. Hunter D (1988) Lime-induced heave in sulfate-bearing clay soils. J Geotech Eng 114(2):150–167

7. Petry TM, Little DN (2002) Review of stabilization of clays and expansive soils in pavements and lightly loaded structures—history, practice, and future. J Mater Civ Eng 14(6):447–460
8. Sharma A, Sharma RK (2019) An experimental study on uplift behaviour of granular anchor pile in stabilized expansive soil. Int J Geotech Eng 2:1–4
9. Harichane K, Ghrici M, Kenai S, Grine K (2011) Use of natural pozzolana and lime for stabilization of cohesive soils. Geotech Geol Eng 29(5):759–769
10. Kolias S, Kasselouri-Rigopoulou V, Karahalios A (2005) Stabilisation of clayey soils with high calcium fly ash and cement. CemConcr Compos 27(2):301–313
11. Guney Y, Aydilek AH, Demirkan MM (2006) Geoenvironmental behavior of foundry sand amended mixtures for highway subbases. Waste Manag 26(9):932–945
12. Taye B, Araya AA (2015) Stabilization of expansive clay soil with sugar cane molasess and cement. Doctoral dissertation, Master's Thesis in Civil Engineering. Addis Ababa University, Addis Ababa Ethiopia
13. Abhishek RK, Sharma RK, Bhardwaj A (2018, November) Effect of construction demolition and glass waste on stabilization of clayey soil. In: International conference on sustainable waste management through design. Springer, Cham, pp 87–94
14. Abhishek VV (2018, November) Stabilization of Clayey soils using fly ash and RBI Grade 81. In: International conference on sustainable waste management through design. Springer, Cham, pp 95–102
15. Basar HM, Aksoy N (2013) Pre-treatment of Waste Foundry Sand via Solidification/Stabilization. Clean-Soil Air Water 41(1):94–101
16. Kumar A, Kumari S, Sharma RK (2016) Influence of use of additives on engineering properties of clayey soil. In: Proceedings of national conference: civil engineering conference-innovation for sustainability (CEC-2016)
17. Baby M, Gowshik A, Rajeshwar AK, Mohanasundram M (2016) Experimental study of expansive soil stabilized with TerraZyme. Int J Eng Res Technol 5:897–899
18. Eujine GN, Chandrakaran S, Sankar N (2017) Accelerated subgrade stabilization using enzymatic lime technique. J Mater Civ Eng 29(9):04017085
19. Pooni J, Giustozzi F, Robert D, Setunge S, O'Donnell B (2019) Durability of enzyme stabilized expansive soil in road pavements subjected to moisture degradation. Transp Geotech 100255

Energy Conservation for a Sustainable Future

Arvind Dhingra and Tejinder Singh Saggu

1 Introduction

Energy and its usage are major pointers to a country's status as developed or developing. Energy is available in various forms and is utilized as per the requirements of the individuals or groups. World at large is being run on conventional sources of energy with around 60% of the total energy consumption around the globe being met through them. These resources including oil, gas, coal, etc., are predicted to last a few more years. In addition to this, the growing population in developing countries is making the matters worse. The growth in population means growth in load and more burden on energy resources. To prolong the availability of these resources and increase the sustainability levels of mankind, it becomes necessary to conserve energy. Various routes for energy conservation are available and are discussed in the following sections of this paper.

2 Energy Conservation

Stated simply energy conservation is the judicious use of energy resources without compromising on the comfort levels. It is an effort to plug the wastage of energy and to utilize it properly. All sectors of economy wherein energy is consumed have immense potential for energy conservation. As per current estimates made by Bureau

A. Dhingra
EED, Guru Nanak Dev Engineering College, Ludhiana, India
e-mail: arvinddhingra@gmail.com

T. S. Saggu (✉)
EED, Punjab Engineering College, Chandigarh, India
e-mail: saggutejinder@gmail.com

H. Singh et al. (eds.), *Sustainable Development Through Engineering Innovations*, Lecture Notes in Civil Engineering 113,
https://doi.org/10.1007/978-981-15-9554-7_58

of Energy Efficiency, Government of India, Industrial sector has potential of 20–25% saving, domestic 15–20%, transport 15%, and so on. Various approaches for energy conservation are suggested viz

- Energy management
- Energy substitution
- Energy audit
- Use of energy efficient appliances
- Lifecycle analysis
- Use of simple energy conservation measures.

We shall discuss these approaches one by one.

2.1 *Energy Management*

Energy management is efficient utilization of energy resources. It is the judicious use of available energy resources to prolong their usage. Rising energy costs and scarce availability of resources to tap for use are major factors driving toward energy management. We discuss the example of electricity sector.

In most countries, electricity is now essentially an agency of Central Government. In India, the generation was fully under government control till very recently. But with new reforms in this sector, private generation has also been permitted. The total installed capacity in India as on 2016 was 200,000 MW and is expected to touch 400,000 MW by 2020 end. For electricity generation, majorly thermal route is being followed. With Indian coal not very good in quality and not expected to last long, we need to manage the load growth due to growing population. Load management techniques must be employed to keep the load factor toward unity as far as possible, i.e., load curve should be kept flat as much as possible, of course no control overload.

If load curve is flat, the same capacity can be used in a better way. Additional costly peak generation is not required. The loads can be staggered to shave off the peak and shift it to off peak hours. This would reduce the burden on the power system. In order to stagger the peak, due incentives would have to be extended to the major loads to shift their demands to off peak areas. Day-light saving is another method to manage the energy usage.

2.2 *Energy Substitution*

Replacing an existing energy source with another source of similar comparable values is called as energy substitution. Encouraging use of solar, wind, and other non-conventional sources of energy in place of electrical energy production using coal is a form of energy substitution. Also in the mobility sector, use of e-vehicles is energy substitution. Use of bio-diesel, ethanol mixed fuels, energy from geothermal sources are all examples wherein we are replacing the conventional energy sources with longer lasting easily available non-conventional energy sources.

2.3 Energy Audit

Energy audit is the process of finding out the utilization of available energy. Energy audit is the key to energy conservation. It is the process of validation of use of energy. It may be defined as verification, monitoring, and analysis of use of energy-containing recommendations for improving energy efficiency with cost benefit analysis and an action plan to reduce consumption of energy. It includes systematic analysis of energy usage in various facilities/departments of an organization. Energy audit is to be done by certified energy auditors. Energy audit can be a walk through audit where the aim is to just observe and point out the energy wastage points. In detailed energy audit, measuring and monitoring instruments such as energy analyzer, tong testers, digital oscilloscopes, lux meter, etc., are used to measure and monitor the usage of energy and to find out the wastage if any that is occurring. The first step here is to take the energy bill of the industry and establish a base line. A detailed audit is followed by a report that includes energy saving opportunities and their pay-back period and which is to be monitored for its implementation. A targeted audit may be conducted in case a particular department is showing unusual energy consumption patterns to find out the cause for aberrations. It is validated that any energy audit would lead to approximate up to 25% reduction in energy consumption in an industry.

2.4 Use of Energy Efficient Appliances

This is a technique which is valid in both industrial as well as domestic scenarios. Industries mainly use induction motors as their work horse. The new NEMA standards for energy efficient motors predict a saving of 5–7% of energy by use of IE3 or above standard motors in industry. This translates into attractive pay-back period of less than a year. In homes, the use of star rated appliances would promote energy efficiency. In India, the star ratings are issued by Bureau of Energy efficiency, Government of India, under standardization and labeling program and are revised from time to time to ensure that they remain in line with the world standards. All-star rated appliances are supposed to carrying the indication of energy efficacy and the possible savings incurred. Most of the appliances including window and split air conditioners, refrigerators, fans, and tube lights have all come under the labeling regime. It is found that the star-rated appliances in addition to saving the power offer attractive pay-back periods.

2.5 *Lifecycle Analysis*

Lifecycle analysis is a new concept that is gaining ground when we talk of sustainable energy consumption. Lifecycle analysis involves analyzing the energy usage right from the production of a product to its usage and finally discarding the product. Cradle to grave is the concept being followed in lifecycle analysis. For making a product, some energy is required. Energy is also an input in the raw material for the product. Once the product is produced, its usage also consumes energy. After its useful life, the way that we discard the product also demands some energy. All this cycle needs to be examined carefully to find out where and how energy can be saved to make this planet more sustainable.

2.6 *Use of Simple Energy Conservation Measures*

Use of simple energy conservation techniques could lead to substantial energy savings. Some of these are listed below:

- A simple electric switch is a powerful energy conservation device. Whenever we leave the room, switching off the lights and fans would help in conserving energy.
- Keeping the computer monitors off when not required would save enormous amounts of energy.
- In summer months, setting the air conditioners at 27 °C would save huge energy.
- Cleaning air filters of air conditioners would help increase their cooling efficiency and help in conserving energy.
- Switching off the televisions and others appliances from the switch and not remote would also help in energy savings
- Use of LED tube lights in place of ordinary fluorescent tube lights, incandescent lamps, CFLs would help in reducing energy consumption.
- For ceiling fan typically used in offices and homes, at least 1 foot space between the blades and ceiling should be maintained for ceiling fan to work efficiently.
- Use of star-rated appliances also helps in energy saving.
- Try using more day light to save energy.
- Defrost refrigerators regularly.
- Occupancy sensors can be installed to save power.
- Maintain proper power factor by using proper rating of capacitors.
- In industries, use motors of proper size.
- Use energy-efficient motors for continuous and near continuous systems.
- Do not use rewound motors.
- Use tight belts for proper coupling.
- For street lights, timers should be used.
- LED lights should be used for street lighting.

3 Case Studies

In this section, we present case studies of two different industries (names withheld as per owners insistence)

Case 1: Textile Industry
This case study was carried out on a 1200 spindle spinning mill located in Ludhiana. A detailed energy audit was conducted. The major point of focus was the lighting on the machines. The plant was using around 450 luminaires each with two 40 W fluorescent tube lights with average lux level of 280 lx and the exterior lighting included fifteen 250 W high-pressure sodium vapor lamp street lights with four 400 W metal halide fixtures as security lights. The following recommendations were made:

1. Replacing the 450 luminaires with two fluorescent tube lights with 2 × 20 W LED fixtures.
2. Replacing the exterior lights with 50 W LED street lights
3. Replacing 400 W metal halide fixture with 60 W LED flood lights.

The calculations are as shown:
With existing fixtures, power consumed: $450 \times 2 \times 50 = 45{,}000$ W
(50 W because each tube light 40 W and the ballast loss 5 W)
Replacement with 2 × 20 W LED fixture: $380 \times 2 \times 20 = 15{,}200$ W
The number of fixtures has reduced as the illumination level of 20 W LED fixture is more, and it does not require any ballast for operation.

So, total power saved: $45{,}000–15{,}200 = 29{,}800$ W

Assuming that the unit was working for single shift of 12 h, the amount of power saved per day $= 29{,}800 \times 12 = 357{,}600$ W h

Yearly power saved $357{,}600 \times 300 = 107{,}280{,}000$ W h = 107,280 units

Unit rate in Punjab is approximately Rs. 8/-, so amount saved = Rs. 858,240/-

Rate of 2 × 20 W LED fixture @ Rs. 1500/-, total investment $= 380 \times 1500 =$ Rs. 570,000/-

The simple pay-back period comes to be less than eight months.

Similarly for street lights and floodlights, the pay-back period is around one year.

In addition to the amount saved, valuable carbon credits were earned due to reduced electricity consumption.

Case 2: Steel Re-rolling Mills
The case pertains to steel re rolling mill located in Ludhiana. The study was carried out under a UNDP funded project for up gradation of steel rerolling mills in Ludhiana cluster wherein 20 odd units were taken into consideration. We shall discuss the case of only one of these units. The unit has scrap as the input and bars as output. The initial study revealed that the unit was using 120 kg of coal for 1 ton of output produced. The output was also having quality problems with scaling being reported. The following observations were made:

1. There was no temperature gauge installed in furnace.
2. The furnace output temperature was around 600 °C.
3. The furnace insulation was not proper.
4. Oxygen percentage inside the furnace was below 5% instead of required 20%.

The following recommendations were made and adopted by the industry:

1. Temperature gauge was installed to monitor furnace temperature.
2. Recuperator (tube type) to recover waste heat was installed. It brought down the outlet temperature to 200 °C and also the waste heat recovered was used to preheat the coal which made the heating more uniform thereby preventing scaling.
3. The furnace insulation was redone.

All these measures costed the industry around Rs. 6,50,000/-, and this brought down the coal consumption to around 58 kgs per ton of output. The payback period was around one year, and additional carbon credits were saved. The unit earned the tag of energy-efficient unit which helped in strengthening their market base.

4 Conclusion

Energy is a commodity which is required by everyone and everywhere. The costs of energy are also increasing, and the availability of energy is declining, so it becomes pertinent for us to save and conserve energy to make our planet more sustainable. As seen from the case studies above, the conservation of energy has led to saving of capital on one hand and has also contributed to sustainable development goals of United Nations promoting sustainable development.

References

1. Davis GR (1991) Energy for planet earth. Scientific American, pp 1–10
2. U.S. Department of Commerce (1998) Statistical Abstract of the United States, 114th edn.
3. Rahman S, Castro AD. Environmental impacts of electricity generation: a global perspective. IEEE Trans Energy Conv 10
4. Mohelnikova J (2008) Daylighting and energy savings with tubular light guides. WSEAS Trans Environ Develop 4(3):201–210. ISSN:1790-5079
5. Grigoriu M. Basis of energy efficiency economical and ecological approach method for pumping equipments and systems. WSEAS Trans Environ Develop 5(1):1–12. ISSN:1790-5079
6. Bukarica V, Tomsic Z (2007) Energy efficiency in croatian residential and service sector—analysis of potentials, barriers and policy instruments. WSEAS Trans Adv Eng Educ 4(12). ISSN:1790-1979
7. Sattari S, Avami A, Farahmandpour B (2007) Energy conservation opportunities: sugar industry in Iran. In: Proceedings of the WSEAS international conference on energy planning, energy saving, environmental education, Arcachon, France, 14–16 Oct 2007, pp 120–127

8. Arvind D, Tejinder S (2008) Energy efficient lighting. In: International conference on recent developments in mechanical engineering, ICRDME-2008, 23–25 Jan 2008, SUSCET, Tangori, pp 689–693
9. Arvind D, Singh T (2009) Energy saving in a spinning using energy efficient lighting. In: 5th WSEAS international conference on energy, environment and ecosystems, 27–29 Sept 2009, Vouliagmeni, Athens, Greece
10. Arvind D, Tejinder S (2011) Energy saving using energy auditing techniques : a case study for a steel power plant, electric power engineering & control systems 2011. In: Uzbekistan, International Conference, 24–26 Nov 2011, pp 64–68
11. Mikučionienė R, Martinaitis V, Keras E (2014) Evaluation of energy efficiency measures sustainability by decision tree method. Energy Build 76:64–71
12. Türkoğlu S, Kardogan OP (2018) The role and importance of energy efficiency for sustainable development of the countries. Lecture Notes in civil engineering

Sustainable Use of Plastic Waste and Crumb Rubber in Bituminous Concrete Production

Gurpreet Singh and Rajiv Chauhan

1 Introduction

India's rapid population growth has the world's second-largest population, and India's estimated total population amounted to around 1.33 billion people [1]. This massive crowd and industrial sector development adds an immense amount of non-biodegradable solid wastes to the environment. This growth involves not only livelihoods but also requires a broad network of mobility. Economic and social growth has a close connection in every nation's economy to the creation of its transport sector [2]. The nation has an overall road length of 5,603,293 km; of which 142,126 km are paved roads, as reported by the Transportation Ministry in 2013 [3].

Road network is one of the most widely used by people across all modes of transport. Hence, keeping the roads in good order is important, road performance is dependent upon the materials used. At the top layer of flexible pavement, the highest stress resistance occurs, which is why they are made primarily of bitumen from superior material. Bitumen has acted as a binding material for constructing roads and pavements for many years. Demand for bitumen has increased as a result of rapid urbanization. Bitumen is commonly used in constructing road, mainly due to its excellent binding and waterproofing properties. The appealing properties of bitumen depend on the form and composition of the mixture. Bitumen is a viscoelastic material; hence, its strength and adhesion mechanisms are determined by loading rate and temperature. If insufficient binder is used in the mix, then premature failure in the pavement can occur, such as cracks, and rutting, potholes. The bitumen thus

G. Singh
I. K. Gujral Punjab Technical University, Jalandhar, India
e-mail: gurpreet3622@gmail.com

R. Chauhan (✉)
Department of Civil Engineering, I. K. Gujral Punjab Technical University, Jalandhar, India
e-mail: dr.rajivchauhan@ptu.ac.in

H. Singh et al. (eds.), *Sustainable Development Through Engineering Innovations*, Lecture Notes in Civil Engineering 113,
https://doi.org/10.1007/978-981-15-9554-7_59

needs to be modified along with innovative sustainable pavement design techniques. Due to insufficient maintenance to the pavement, the situation is getting worse. One possible solution to this problem could be to improve the quality of bituminous mixtures. There are many binders available on the market to improve bituminous mix properties, but the use of these binders raises overall project cost. The escalating costs of bitumen and the lack of resources motivated highway engineers to explore alternatives for road construction [4].

A common method for improving bitumen quality is to change the engineering properties of bitumen by mixing it with polymers like rubber and plastics. In addition, bitumen properties are substantially improved when combined additives are used in bitumen mixtures as opposed to unmodified bitumen [5, 6]. Waste plastic and crumb rubber have been used in this study as an additive to base bitumen. The crumb rubber is made from shredding scrap tire which is a particular fiber- and steel-free material. Recycling these waste materials for the development of new roads and highways will help solve environmental problems by discarding those waste materials in dumpsites. Plastics and rubber remain on site for a long time as non-biodegradable materials and cause pollution for the environment [7]. Old methods of disposing of plastic waste and rubber tires, such as burning and land-filling, produce different types of air, water and soil pollution [8]. A recent survey by the Central Pollution Control Board in India estimated that about 33.7 million pounds of plastic waste are generated daily, around 13.2 million pounds of which remain uncollected and pollute environment [9]. Polymer use in modified bitumen improves the efficiency of funnel asphalt paving [10, 11]. The use of recycled plastic waste polythene bags enhanced the engineering properties of bituminous concrete mix, such as toughness or fatigue life [12]. On average there are more than 10 billion leftover tires per year [13]. Using crumb rubber in Superpave mix design was helpful in increasing mineral aggregate voids and enhancing asphalt mixture rutting resistance regardless of rubber size and type [14]. The improved performance of bituminous mixtures with crumb rubber has been demonstrated by many researchers over the past four decades [15–17]. Modified mix of crumb rubber has been cited as increased fatigue life, decreased reflective cracking and cracking at low temperatures and increased tensile strength [18]. This technique helps to develop green flexible flooring for future generations by using solid waste as a modifier to improve the engineering properties of bituminous concrete mixtures [19–22].

This study uses sustainable material economic mix developed with proper aggregate gradation and sufficient bitumen ratio to meet the desired mix properties of stability, toughness, resilience, skid resistance and workability. Using the Marshall stability mix design process on traditional bitumen mix samples, the optimum bitumen content is determined, and modified samples are prepared by incorporating waste plastic and crumb rubber into the mix in different percentages. The main objective of the study is to examine the behavior of waste plastic and crumb rubber tires as a modifier to determine stability, flow, and volumetric characteristics. These experimental findings have shown that overall performance of modified mixtures is more favorable than traditional mixtures.

2 Experimental Materials Used and Methodology

This research aims at achieving an environmentally friendly bitumen mix (waste plastic–crumb rubber modified bitumen Mix). This section deals with the methods used to measure the research work. The technique followed for carrying out the experimental program was demonstrated with the aid of a flowchart in Fig. 1, and the interpretation below is endorsed [23].

The materials used for experimental research include bitumen, aggregate, stone dust, plastic, and crumb rubber. Aggregate measurements were taken in compliance with specifications of the Ministry of Road Transport and Highways (MORTH). The bitumen used was degree 80/100 and was bought from a local contractor of Moga. The material used as plastic waste was polythene bags, waste plastic bottles, etc., collected from residential and municipal waste centers. Crumb rubber is purchased from tire shops in Ludhiana. The BC grade-2 was selected as it provides the aggregate mixture gradation given in Table 1.

In order to examine the mechanical properties of aggregates like strength, durability, hardness, water absorption ability, etc., various physical tests such as impact

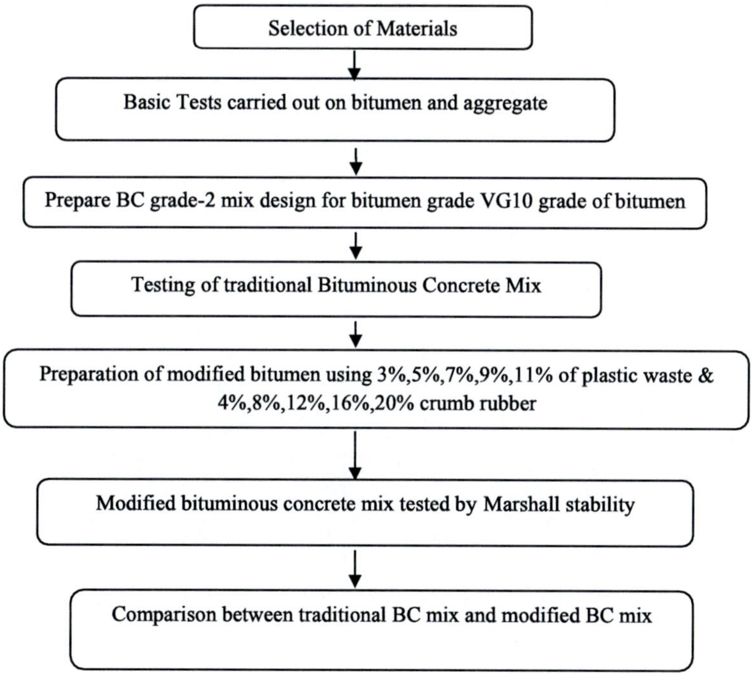

Fig. 1 Flowchart portray overall methodology

Table 1 Blending of aggregate for bituminous concrete

Sieve size (mm)	% Passing				% Passing blending ratio				MORT & H requirement
					13.2	11.2	SD	C	
	13.2	11.2	SD	Cement	0.10	0.35	0.53	0.02	
19.00	100	100	100	100	100.0				100
13.20	97	100	100	100	99.7				90–100
9.50	15	75.2	100	100	82.8				70–88
4.75	0.2	0.4	98.6	100	54.4				53–71
2.36	0	0.2	85.0	100	47.1				42–58
1.18	0	0	74.5	100	41.5				34–48
0.600			67.7	100	37.9				26–38
0.300			42.3	100	24.4				18–28
0.150			19.7	100	12.4				12–20
0.075			5.3	90	4.6				4–10

value, abrasion value, crushing test, specific gravity, flakiness, and elongation were performed. All the obtained results lies within the acceptable limits which were shown in Table 2.

Table 2 Mechanical properties of aggregates

Type of test	Test method	Result			MORT H (Fifth revision-2013 Specifications)
		13.2 mm	11.2 mm	Stone Dust	
Aggregate impact test	IS:2386-Part IV	13.5	14	–	<24%
Los angeles abrasion test	IS:2386-Part IV	21.8	22.2	–	<30%
Aggregate crushing test	IS:2386-Part IV	14.9	15	–	<24%
Combined flakiness and elongation test	IS:2386-Part I	31.7	32.6	–	Max 35%
Water absorption test	IS:2386-Part III	0.96	0.76	1.8	<2%
Specific gravity apparent	IS:2386-Part III	2.740	2.715	2.758	2–3
Specific gravity true	IS:2386-Part III	2.669	2.661	2.626	2–3

2.1 Modified Binder Preparation

Binders have been divided into two series: I and II. Series I and II represent the mixes, i.e., Bitumen (B) + Plastic (P) and Bitumen (B) + Rubber (R), respectively, with differing amounts of plastic and rubber in both bitumen. In order to alter BM, the waste plastic was added varied from a low modification of 3% to higher degrees of modification, up to 11%, with a rise of 2% by bitumen weight. The crumb rubber must be cut and scraped into small particles to create bituminous rubber and reduced to powder size; it will then be applied to traditional bitumen. The amount of crumb rubber ranged from 4 to 20%, with an improvement of 4%. Mixing of bitumen with Plastic waste and rubber were done at a temperature 200–220 °C.

2.2 Binder Testing

To compare the properties of modified and unmodified bitumen, various binder tests performed properties of both the binders have been tested under conditions of Indian Standard [24, 25]. Table 3 indicates results of the tests conducted on unmodified bitumen and modified bitumen.

2.3 Marshall Specimen Preparation for Traditional Mix

The bituminous concrete mixture has been formulated according to the standard Marshall stability test. To find out optimum binder content of the developed mix, Marshall specimens are casted with virgin bitumen and aggregate mix. First of all, aggregates and filler material were heated in a pan in required proportions and held for 2 h in an oven at a temperature of 160 °C. Bitumen (5.0% by weight of total aggregate) is heated to 150–160 °C and thoroughly mixed with heated aggregates at 154–160 °C temperatures. The mix is put in a preheated mold and compacted at a temperature of 138–149 °C by a rammer with 75 blows at either side A filter paper is placed on top of the sample and under the sample. After sample compaction, molds are held at 60 °C for 30 min in temperature-controlled water bath after 24 h. In the next trial, vary the bitumen content by +0.5% up to 7% and repeat the procedure above. In the Marshall test setup, the prepared mold is loaded. The optimum bitumen content for the unmodified bitumen mixes was found to be 5.5% by weight of the total mix. The Marshall stability and flow test provides the measure of output prediction for the Marshall mix design process. The test's portion of stability tests the average load that the test specimen bears at a load rate of 50.8 mm/minute. Load shall be added to the specimen before failure, and the full load shall be known as stability.

Table 3 Physiochemical properties of binders

	Composition	Penetration (mm)	Ductility (cm)	Softening point (°C)	Specific gravity (27 °C)	Kinematic viscosity (135 °C) cSt
Testing method		IS:1203	IS:1208	IS:1205	IS:1202	1206 (Part-3)
*TM	100% B	91.7	92.0	40.5	1.04	255
SeriesI						
BM1	97% B + 3% P	90.4	84.0	40.5	1.00	258
BM2	95% B + 5% P	90.1	78.0	41.7	1.07	260
BM3	**93% B + 7% P**	**89.7**	**76.0**	**42.2**	**1.10**	**259**
BM4	91% B + 9% P	88.0	62.0	43.0	1.12	262
BM5	89% B + 11% P	86.5	50.0	44.5	1.14	275
SeriesII						
BM6	96% B + 4% R	85.7	69.0	43.5	1.10	278
BM7	92% B + 8% R	80.3	77.0	50.4	1.02	320
BM8	**88% B + 12% R**	**76.3**	**79.0**	**57.7**	**0.93**	**400**
BM9	84% B + 16% R	40	58.0	55.0	0.95	450
BM 10	80% B + 20% R	26.3	58.0	85	0.97	432
**Specified limits (IS-73: 2013)*	*Paving bitumen VG-10*	Min 80	Min 75	Min 40	Not specified	Min 250
	Paving bitumen VG-30	Min 45	Min 40	Min 47	Not specified	Min 350

TM Traditional mix; *B* Bitumen; *P* Waste plastic; *R* Rubber
Bold indicates Optimzed mix in each series
Asterisk represents the Tradational mix without any modifier

2.4 Modified Marshall Sample Preparation

Modified Marshall samples were developed as per Marshall samples the standard procedure after deciding optimum binder content. A comparative analysis on BC mixes was performed using plastic waste and rubber crumb. For each aggregate gradation and content of asphalt, at least three specimens are needed. As noted above, plastic waste and adapted CR mixtures were used for the mechanical tests to prepare new bitumen mixes. The physical properties of modified mixtures such as air voids (Va), voids of mineral aggregate filled by bitumen (VFB), voids in mineral aggregate (VMA), and flow values are calculated.

3 Results and Discussions

The study shows that the softening point increased for both plastic and scrap rubber materials. It has been found that modified bitumen has ductility within specified limits and therefore proved to be stronger when used in pavements. As a result, it can be inferred that applying plastic wastes and crumb rubber to control mix will play a significant role in enhancing binder elastic behavior of pavements in terms of reduced rutting and cracking susceptibility. Marshall stability was also found to be higher for a changed mix. The study results showed a higher stability and VMA percentage of the modified mixture relative to the non-modified mixtures. This would have a positive influence on these mixtures' rutting resistance. Each compacted test specimen is subjected to the following tests, and obtained results are given in Table 4.

3.1 Marshall Stability and Flow Value

Analysis of Marshall stability on both unmodified and modified mixes was carried out. Stability value obtained from average of three samples prepared for each mix. Stability flow and volumetric analysis was performed on traditional mix and modified mixes (Series I and II) of bituminous concrete. Traditional mix stability value (5.5% by weight of total mix) is 10.2 kN, which meets the minimum BC mix stability requirement. The maximum stability value was obtained for by BM8 (88% B + 12% R) of Series II and BM3 (93% B + 7% P) of Series I. The stability value of mix BM8 (13.5) and BM3 (10.4) is considerably higher than the unmodified mix. These result values indicate that modified binder mix with crumb rubber and plastic waste provides better stability than traditional bituminous mix. For almost all the adjusted blends, the stability values obtained are higher than unmodified mix. The stability and flow tests are conducted after evaluating the bulk-specific gravity of the test specimens. In each sequence, the flow value of bituminous mixes is observed to grow with an increase in percentage replacement by waste additives. Unmodified mix flow value is 3.5 mm, while adjusted mix flow value, i.e., BM3 (3.3 mm) and BM8 (3.5 mm), within the specified limit of 2–4 mm. Also within defined limit, it is the flow value of the optimum adjusted mix. Exceptionally, high flow values are observed for mixes that substitute rubber material by more than 10%. Hence, plastic waste and crumb rubber can be used a sustainable material in pavement construction.

Table 4 Marshall stability flow and volumetric characteristics results of traditional and modified BC mix

	Composition	Marshall stability (kN)	Marshall flow (mm)	Bulk density (g/cc)	%Air voids	VMA(%)	VFB (%)
Testing method		Asphalt manual MS-2 seventh edition					
*TM	100% B	10.2	3.5	2.362	3.6	15.8	74.90
Series I							
BM1	97% B + 3% P	9.3	2.7	2.347	3.2	16.3	86.70
BM2	95% B + 5% P	9.7	2.9	2.356	3.3	16.4	80.00
BM3	**93% B + 7% P**	**10.4**	**3.3**	**2.369**	**3.7**	**16.5**	**77.20**
BM4	91% B + 9% P	9.2	3.4	2.351	3.8	16.6	77.00
BM5	89% B + 11% P	9.1	3.7	2.364	3.7	16.2	76.90
Series II							
BM6	96% B + 4% R	9.8	3.2	2.353	3.2	16.6	77.40
BM7	92% B + 8% R	11.43	3.5	2.361	3.7	17.0	81.00
BM8	**88% B + 12% R**	**13.5**	**3.5**	**2.374**	**3.9**	**17.2**	**87.30**
BM9	84% B + 16% R	12.9	3.6	2.323	4.2	18.7	76.30
BM 10	80% B + 20% R	13.9	3.7	2.278	5.5	19.5	71.90
**MORT H (V Revsion-2013 Specifications*	*BC Grade-2*	*>9*	*2–4*	–	*3–6*	*>12*	*65-75*

Bold indicates Optimzed mix in each series
Asterisk represents the Tradational mix without any modifier

3.2 *Density and Voids Analysis*

After the stability and flow check has been completed, a density and voids study for each collection of specimens is carried out. For each bitumen material, average the determinations of the bulk density. Unmodified mix bulk density are 2.362 g/cc. In series I, the maximum value of bulk density was obtain 2.369 g/cc for BM3, while in series II the maximum value of bulk density was obtain 2.374 g/cc for BM8. Density values of both the modified mix found to be higher than unmodified mix. Air voids were found to be 3.6% in the unmodified BC mix, well within the acceptable limits as per MORTH, 2013. In both modified mixes, the percentage of total air voids increased. In the unmodified mix, VMA are found to be 15.8%, that is within specified limits that depend on the aggregate size. The cumulative VMA increases with the partial percentage of waste additives. The result of density and voids analysis indicates that the sustainable use of waste plastic and crumb rubber is one of the best technique to be improved for the engineering properties of traditional bituminous mix.

It can be found that the stability value always increases up to certain limits with the addition of both modifiers, and further addition reduces the stability. This may be due to excessive quantities of additive which cannot be properly dissolved in bitumen. Thus, it was found in waste plastic and crumb rubber both maximum stability at 7% and 12%, respectively, at optimum bitumen content, varying content of modifiers. It is observed that with the rise in the flow value of the binder material but with the addition of modifier flow value decreases than that of traditional mixes. The experimental results from this analysis illustrate the benefit of using crumb rubber and plastic waste for modified bitumen. It has been determined that both waste plastic and modified bitumen CR exhibited reduced temperature sensitivity with increased additive content, according to traditional bitumen studies. Significant differences were found above 7% waste plastic content and 12% CR content in the adjustment indices. Such values can be seen as thresholds for the transition to higher performance properties.

4 Cost Analysis

Cost analysis of the modified and unmodified bituminous concrete mix which may serve as sustainable use of waste materials like plastic bottles, polythene bags, and crumb rubber for construction of green pavement in economical way was conducted on a pavement section as per IRC37:2012, and findings are presented in Tables 5 and 6.

The results showed that partial replacement of bitumen by crumb rubber and plastic waste contributes to improving strength and density of the mix itself. Higher value of density mixing gives better efficiency for build pavements. However, the estimated mix cost analysis revealed that the use of different waste materials as

Table 5 Cost analysis data for the concept

S. No.	Design data	Values
1.	Collective no. of standard axles	150 msa
2.	No. of lane	1 (one)
3.	Length of pavement	1 km
4.	Width of pavement	3.5 m
5.	Thickness of pavement	40 mm

Table 6 Cost reduction of traditional and modified mix developed

	Composition	Bitumen cost	BC layer thickness/length	Cost reduction
Units		Lakhs	40 mm/1 km	%
*CM	100% B	6.39		
Series I				
BM1	97% B + 3% P	6.20		2.97
BM2	95% B + 5% P	6.07		5.07
BM3	93% B + 7% P	5.95		6.66
BM4	91% B + 9% P	5.82		9.04
BM5	89% B + 11% P	5.69		11.11
Series II				
BM6	96% B + 4% R	6.14		3.96
BM7	92% B + 8% R	5.88		8.09
BM8	88% B + 12% R	5.63		12.06
BM9	84% B + 16% R	5.37		16.19
BM 10	80% B + 20% R	5.11		20.31

Bold indicates Optimzed mix in each series
Asterisk represents the Tradational mix without any modifier

additives in BC mix provide better solution to construct cheaper roads to the tune of 11–20%.

5 Conclusions

5.1 The Following Conclusions are Drawn on the Basis of the Experimental Investigation

- The use of rubber and waste plastic in roads can solve the problem of environmental pollution and their dumping issues by making sustainable use in bituminous pavement
- Crumb rubber provides satisfactory results by using it to substitute bitumen for various bitumen and bitumen mix tests in 12% of the proportion.
- Waste plastic provides satisfactory results by using it to substitute bitumen for various bitumen and bitumen mix tests in 7% of the proportion.
- Modified binder using plastic waste and crumb rubber increases the bitumen softening point and decreases the penetration and ductility value which improves the virgin bitumen quality.
- Single use materials such as rubber tires of vehicles, polythene bags, and plastic wrapper and bottles can be used as partial substitutes in bituminous

concrete mixtures that can help lower the increasing demand for bitumen in road construction thereby reducing the burden on economy.

Acknowledgements Authors would like to acknowledge the IKGPTU, Jalandhar, Punjab, for fulfilling their support during the research work.

References

1. Estimated total population from 2014 to 2024 (in millions) Published by H. Plecher, Nov 6 2019
2. Khan IM, Kabir S, Alhussain MA, Almansoor FF (2016) Asphalt design using recycled plastic and crumb-rubber waste for sustainable pavement construction. Procedia Eng
3. Basic Road Statistics of India 2015–16. Ministry of Road Transport and Highways. Retrieved 15 Jan 2018
4. Kok BV (2011) Laboratory comparison of the crumbrubber and SBS modified bitumen and hot mix asphalt. Constr Build Mater
5. Cardone F, Ferrotti G, Frigio F, Canestrari F (2014) Influence of polymer modification on asphalt binder dynamic and steady flow viscosities. Constr Build Mater 71:435–443
6. Baha VK, Mehmet Y, Mustafa A (2014) Evaluation of the conventional and rheological properties of SBS + Sasobit modified binder. Constr Build Mater 63:174–179
7. Bansal S, Misra AK, Bajpai P (2017) Evaluation of modified bituminous concrete mix developed using rubber and plastic waste materials. Int J Sustain Built Environ
8. Utibe JN, Johnson AJ, Feyisayo VA, Obioma UU (2017) Characterization of bitumen/plastic blends for flexiblepavement application. Int Conf Sustain Mater Process Manuf Sci Direct 7:490–496
9. Anonymous (2007) Waste and resources action programme (WRAP), annual local authorities plastics collection survey
10. King HW (1986) Polymer modified asphalts, an overview. Am Soc Civ Eng 240–254
11. Isacsson U, Lu X (1995) Testing and appraisal of polymer modifiedroad bitumen. Mater Struc 28:139–159
12. Zorrob SE, Suparama LB (2004) Laboratory design and investigationof proportion of bituminous composite containing waste recycledplastics aggregate replacement (Plastiphalt), In: CIB symposium on construction and environment theory into practice, vol 448
13. Alamo-NoleLuis A, Oscar P-P, Roman-Velazquez Felix R (2011) Sorption study of toluene and xylene in aqueous solutions byrecycled tires crumb rubber. J Hazard Mater 185:107–111
14. Xiao F, Amirkhanian SN, Shen J, Putman B (2009) Influences of crumb rubber size and type on reclaimed asphalt pavement (RAP) mixtures. Constr Build Mater 23:1028–1034
15. Lalwani S, Abushihada A, Halsa A (1982) Reclaimed rubber asphalt blends measurement of rheological properties to assess toughness, resiliency, consistency and temperature sensitivity. Proc J Assoc Asphalt Paving Technol Kansas City Kan 51:562–579
16. McGennis RB (1995) Evaluation of physical properties of fine crumb rubber-modified asphalt binders. J Transport Res Rec 1488:62–71. Transportation Research Board, Washington (DC)
17. Bahia HU, Davis R (1994) Effect of crumb rubber modifiers (CRM) on performance related properties of asphalt binders. J Assoc Asphalt Paving Technol 63:414–449
18. Oliver JWH (2000) Rutting and fatigue properties of crumb rubber hot mix asphalts. In: Proceedings of the asphalt rubber 2000 conference, Vilamoura, Portugal, pp 221–40
19. Modarres H, Hamidreza A (2014) Effect of waste plastic bottleson the stiffness and fatigue properties of modified asphalt mixes. Mater Des 61:8–15

20. Widojoko L, Purnamasari PE (2012) Study the use of cement andplastic bottle waste as ingredient added to the asphaltic concretewearing course. In: Proceedings of 8th international conference on traffic and transportation studies, vol 43, pp 832–841
21. EA MZ, Karim MoR, Shafigh P (2011) Using waste plastic bottles as additive for stone masticasphalt. Mater Des 32:4844–4849
22. Qadir A, Imam M (2006) Utilisation of waste plastic bags in bituminous mix for improved performance of roads. J Solid WasteTechnol Manage 32(3):185–195
23. Gupta K, Chopra T, Kumar M (2016) Laboratory inves-tigations of DBM (Grade 1) mix using different types of additives. Dissertation, Thapar University, Patiala, Punjab, India
24. Khanna SK, Justo CEG, Veeraragavan A (2009) Highway material and pavement testing (laboratory manual). King, G.N, Nemchand and Bros, Roorkee
25. Ministry of Road Transport and Highways (2013) Specifications for roadand bridge works. Fifth Revision, Published by Indian Road Congress on behalf of the Government of India

Analysis of Air Quality Index During Lockdown: A Case of Ludhiana District-Punjab

Pankaj Goel, Harpreet Kaur, Raman Kumar, Paramjit Singh Bilga, and Nidhi Aggarwal

1 Introduction

The year 2020 is the year of extensive pandemic corona virus disease of 2019 (COVID-19) caused by the SARS-CoV2 virus. First, the COVID-19 ailment was recognized in China's Wuhan city; over time, it rose in over one million cases globally. The COVID-19 impact led to lockdown in many nations around the world. In India, the first fourteen-hour nationwide lockdown was imposed on March 22, 2020, followed by 21 days beginning on March 24, 2020. The lockdown imposes restrictions and self-isolation measures, which diminishes transport and industrial emissions [11]. The trends in air emissions over this lockout era will provide an insight into the achievability of improving air quality as there are substantial emission limits from multiple sources, and improved proposals are being submitted to

P. Goel
Department of Business Management, Guru Nanak Institute of Management and Technology, Ludhiana, Punjab, India
e-mail: pankajgoel456@gmail.com

H. Kaur
Electrical Engineering Department, Chandigarh University, Mohali, Punjab, India
e-mail: harpreetchanni@yahoo.in

R. Kumar (✉) · P. S. Bilga
Mechanical Engineering Department, Guru Nanak Dev Engineering College, Ludhiana, Punjab, India
e-mail: sehgal91@yahoo.co.in

P. S. Bilga
e-mail: psbilga@gndec.ac.in

N. Aggarwal
Department of Chemistry, Arya College, Ludhiana, Punjab, India
e-mail: nidhiagg0506@gmail.com

H. Singh et al. (eds.), *Sustainable Development Through Engineering Innovations*, Lecture Notes in Civil Engineering 113,
https://doi.org/10.1007/978-981-15-9554-7_60

regulators to regulate air pollution. Air pollution has emerged as a growing issue worldwide, especially in developing nations such as India. In recent decades, India has perceived economic growth, quick urbanization, mechanization, and rapid infrastructure progress [3]. At the same time, India's air pollution level has enlarged to a chief health hazard and caused high premature mortality rates. Around one million people died in India in 2015 alone due to emissions from atmospheric particulate matter (PM) [2, 15]. Indian cities have always been the world's first 20 polluted cities. They have exceeded the pleasant air quality standards commended by the World Health Organization and Central Pollution Control Board. The PM, the most dominant pollutant, has significant influences from vehicles, domestic, energy, industrial, and dust in substantial parts of India [7]. To control the country's severe air pollution, the National Clean Air Program propelled a five-year action plan to reduce PM by 30% nationwide in 2019 [9].

Chay and Greenstone [4] have studied the importance of air quality and the role of government agencies in this direction. The enacting of laws like the Clean Air Act has resulted in a better atmosphere and reduced total suspended particulates air pollutants. However, Monks et al. [12] have discussed the relevance of quality air in the atmosphere and policies laid down by the various government in this direction globally. The study has shown that there is dire need on the part of different policymakers to lay down such options that ultimately reduce greenhouse impact. These policies should not be laid down on the cost of existing air quality, overall public health, and a healthy atmosphere. But this air quality, as per Jacob and Winner [8], depends upon the weather and global climate change. The study has made a relationship between the surface ozone and the present temperature in most of the polluted regions. It was observed that there are high adverse effects of global warming. A combined study of chemical transport models and general circulation models has found an increase in surface ozone in these polluted regions. Many studies have been published during the COVID-19 pandemic to study the influence of lockdown on the atmosphere. Gautam [6] has discussed that the COVID-19 has been first identified in Wuhan city of China and spread globally like a pandemic, where many countries of the world, including India, have to declare lockdown. This lockdown has many economic and social disadvantages in these countries. Despite these drawbacks, there were many positive impacts of this lockdown. The most noticeable effect has been noticed upon the air quality index.

The secondary data collected from NASA has shown that there is an almost 50% reduction in aerosol optical thickness in India. Xu et al. [16] have conducted an investigation of air quality index in three cities in China, viz. Anqing, Hefei, and Suzhou. The period of the inquiry was 2017–2019 in comparison to the COVID-19 period of 2020. The air pollutants $PM_{2.5}$, PM_{10}, SO_2, CO, and NO_2, have shown noteworthy change and lower concentration in the COVID-19 period as compared to the corresponding three years period. However, O_3 has not demonstrated significant change. The mean AQI air pollutants in the COVID-19 period were 45.1% lower than the 2017–2019 periods.

However, overall air quality improved during the COVID-19 period in China, primarily in these three cities under study. Mahato et al. [10] have studied a lockdown period of three weeks from March 24, 2020, to May 3, 2020. This period has shown a critical decline in the presence of air pollutants in the environment in India. The study has observed seven pollutants, viz. PM_{10}, $PM_{2.5}$, SO_2, NO_2, CO, O_3, and NH_3 for different 34 monitoring stations and secondary data have been collected from the Web page of the National Air Quality Index. The study has shown more than 50% reduction in PM_{10} and $PM_{2.5}$. The air quality has improved almost 40–50% in the respective areas. So the research has been demonstrated that lockdown has proved as an integrated alternative to pollution control measures carried out by authorities. A similar study accomplished by Sharma et al. [13] has shown that restricted human activities during Lockdown in COVID-19 period in India have helped in improving air quality. The study has been conducted in 22 cities for the period of March 16, 2020, to April 14, 2020. This period has been compared with the previous three corresponding years. The correlation has shown a significant improvement in air quality and suggested imposing such confident restrictions to improve air quality in the future.

2 Study Area

The current research focused on Ludhiana as it is located in the center of Punjab in the Malwa region, linking the Grand Trunk Road from Delhi to Amritsar at 30.55 North Latitude and 75.54 East Longitude in the northern territory. It is extended between north latitude 30°–34′ and 31°–01′ and east longitude 75°–18′ and 76°–20′. It is being touched the north by the river Satluj, which divides it from Jalandhar district. On other sides, it shares similar borders with the eastern area of Rupnagar, the western region of Moga, and the southeastern neighborhoods of Barnala, Sangrur, and Patiala. As per the 2011 census, the city has a population of 3,498,739. Ludhiana is the largest city of Punjab, with an area of 310 square kilometers with a dense population of 5200 per square kilometer. Population wise it is ranked six in north India with a massive population as per the 2011 census after Delhi, Jaipur, Lucknow, Kanpur, and Ghaziabad. Ludhiana is the commercial capital of Punjab and has a large number of Industrial units and branch offices of all major MNCs. The presence of a considerable number of the manufacturing hub, large population, heavy traffic, burden on all-natural and human-made resources is making the city severely polluted. Hence, the present study has focused upon understanding the extent of decline of pollution during the lockdown in this metro city.

3 Materials and Research Methods

The required data has been collected from the Web site of the Central Pollution Control Board (CPCB), Ministry of Environment, Forest and Climate Change, Government of India. The CPCB is maintaining an index to calculate environmental pollution, viz. National Air Quality Index (NAQI). The NAQI studies air quality index under six categories, namely good, satisfactory, moderately polluted, poor, very poor, and severe. Eight air pollutants, namely PM_{10}, PM2.5, NO_2, SO_2, CO, O_3, NH_3, Pb, and AQI, go under in-depth consideration. Continuous data monitoring is done at 20 state capitals and 46 cities with a population of more than 1 million by the State Pollution Control Board (SPCB). In Punjab, data monitoring is done at Amritsar, Bathinda, Jalandhar, Ludhiana, Khanna, Mandi Gobindgarh, Patiala, and Rupnagar. A case of Ludhiana district has been studied in the present study.

The study has adopted a descriptive research design where the sampling units were AQI of Ludhiana districts. The sampling design is a non-probability style. The research technique is based on time series analysis. The sample data collection period is pre- and post-COVID-19 lockdown declaration period in India. The sampling frame is restricted to the Ludhiana district only. The data was analyzed to investigate the improvements in air quality throughout the period from March 24, 2019, to May 21, 2019, and the lockdown period from March 24, 2020 to May 21, 2020. Concentrations of the different contaminants were analyzed for the period from March 24 to May 21, 2019, through 2020. The hourly quantities of seven air contaminants containing particulate matter ($PM_{2.5}$ and PM_{10}), nitrogen oxides (NOx, NO, and NO_2), sulfur dioxide (SO_2), ozone (O_3), and carbon monoxide (CO) have been collected from the CPCB Web-based database.

The statistical testing techniques used are Karl Pearson correlation and Paired sample t-test. Karl Pearson Test will help in analyzing the comparisons between previous year figures with the present data on AQI and hence establish a causal relationship. Paired sample t-test is a reliable method to determine whether various parameters understudy are statistically and significantly different from each other or not. The sampling analysis has been done with statistical software SPSS version 22. In a nutshell, to achieve the objective of this study, following methodology steps has been adopted:

1. The required data to calculate the AQI for Ludhiana City has been collected from the AQI station of PAU Ludhiana.
2. A descriptive statistics analysis was applied to check the acceptance level.
3. The sample data was then analyzed using correlation and t-test to find out the significant level of each parameter under consideration for the period from March 24, 2019, to May 21, 2019, and lockdown period of March 24, 2020, to May 21, 2020.

4 Results and Discussion

Based on the standards, actual values have been taken, as shown in Table 1. The averages have been calculated and displayed in the descriptive statistics of paired sample statistics in Table 2. The standard values as per Indian National Ambient Air Quality Standards (NAAQS) AQI Category, Pollutants, and Health Breakpoints [1], given by NAQI on CPCB Web site are shown in Table 1.

Table 1 Indian national ambient air quality standards (NAAQS) AQI category, pollutants, and health breakpoints [1]

AQI category (Range)	PM_{10} (24 h)	$PM_{2.5}$ (24 h)	NO_2 (24 h)	O_3 (8 h)	CO (8 h)	SO_2 (24 h)	NH_3 (24 h)	Pb (24 h)
Good (0–50)	0–50	0–30	0–40	0–50	0–1.0	0–40	0–200	0–0.5
Satisfactory (51–100)	51–100	31–60	41–80	51–100	1.1–2.0	41–80	201–400	0.5–1.0
Moderately polluted (101–200)	101–250	61–90	81–180	101–168	2.1–10	81–380	401–800	1.1–2.0
Poor (201–300)	251–350	91–120	181–280	169–208	10–17	381–800	801–1200	2.1–3.0
Very poor (301–400)	351–430	121–250	281–400	209–748	17–34	801–1600	1200–1800	3.1–3.5
Severe (401–500)	430+	250+	400+	748+	34+	1600+	1800+	3.5+

Table 2 Paired samples statistics

Parameter	Year	Mean	Acceptance level	N	Std. deviation	Std. error mean
$PM_{2.5}$	2020	40.93	Satisfactory	61	22.20	2.84
	2019	85.98	Moderately polluted	61	30.23	3.87
PM_{10}	2020	52.93	satisfactory	61	23.36	2.99
	2019	81.09	satisfactory	61	21.31	2.72
NH_3	2020	2.92	Good	61	1.02	0.13
	2019	9.73	Good	61	3.85	0.49
SO_2	2020	8.02	Good	61	4.70	0.60
	2019	18.31	Good	61	6.26	0.80
O_3	2020	40.03	Good	61	12.73	1.63
	2019	23.03	Good	61	8.37	1.07
AQI	2020	57.05	Good	61	21.34	2.73
	2019	88.98	Moderately polluted	61	29.99	3.84

Out of eight overall parameters, the study has investigated six significant parameters. These selected parameters have been tested against standard laid down by the national air quality index established by the Ministry of Environment and Forests, India. These parameters are particulate matters ($PM_{2.5}$ and PM_{10}), ammonia (NH_3) sulfur dioxide (SO_2), ozone (O_3), and overall air quality index [14]. The averages calculated for the current year and the corresponding previous year is given in Table 2 alongside a comparison with standards laid down as per Table 1.

The study has compared the average values of these parameters (for years 2019 and 2020), which have been further compared with standards as per the Indian National Ambient Air Quality Standards (NAAQS) AQI Category. The acceptance level (column four in Table 2) discusses the averages falling in the acceptance zone or not. Sixty-one days data has been taken for each year under study. Standard deviations have considered how much actual values in each parameter have deviated from average values in each year. Let us see how averages have moved from 2019 to 2020 for each parameter and falling in the acceptance zone or not.

Particulate Matter ($PM_{2.5}$): The accepted parameter for air pollutant particulate matter ($PM_{2.5}$) has shown that the average value has declined from 85.98 in 2019 to 40.93 in COVID-19 the year 2020. It indicates that lockdown in Ludhiana has a positive impact on reducing $PM_{2.5}$ and increasing visibility, improving the health and lives of masses.

Particulate Matter (PM_{10}): It has shown that the average value has declined from 81.09 in 2019 to 52.93 in COVID-19 the year 2020. It indicates that lockdown in Ludhiana has a positive impact on reducing PM_{10} and increasing visibility, improving the health and lives of masses. However, heavy particles like PM_{10} are easily washed by rain as compared to finer particles.

Ammonia (NH_3): It has shown that the average value has declined from 9.73 in 2019 to 2.92 in COVID-19 the year 2020. It shows that there are better chances of reducing health diseases, primarily related to lung and heart problems during the Lockdown in Ludhiana. The extent of poisonous gases, suffocation, and respiratory infections is prone to reduce in this lockdown period.

Sulfur Dioxide (SO_2): It has shown that the average value has declined from 18.31 in 2019 to 8.02 in COVID-19 the year 2020. It shows that the lockdown has resulted in the closure of manufacturing units and the burning of oil, coal, gas, and fossil fuel combustion, etc. There are better chances of reducing health diseases, especially corneal haze, breathing difficulty, airway inflammation, and high mortality rate during the lockdown in Ludhiana.

Ozone (O_3): It has shown that the average value has not declined from 23.03 in 2019 to 40.03 in COVID-19 the year 2020. It shows that the lockdown has not entirely resulted in dropping into several health-related problems like chest pain, throat irritation, etc. However, it may be very less period for ozone (O_3) to repair and leave an impact on the natural environment. Any source does not directly emit the Ozone. Still, it is formed by photochemical reactions between oxides of nitrogen (NOx) and other volatile organic compounds (VOCs) and gases in the air under the influence of sunlight and heat. Gases from all sources must be controlled to control ozone [5].

AQI: It has shown that the average value has declined from 88.98 in 2019 to 57.05 in COVID-19 the year 2020. It shows that the lockdown has resulted in a declining average pollution level during the lockdown period in Ludhiana.

Other tests have been conducted to find out if there exists any correlation between the samples collected for the lockdown period and the corresponding period during last year. The results so obtained are shown in Table 3, there exists a very low correlation between the pollutant levels in these two periods. Hence, it can be explained that the low correlation indicates figures have changed drastically, and AQI has improved in the period of study.

The next query was related to finding out if there exist significant differences in the figures of two periods. A paired sample t-test has been conducted, and the following output has been obtained, as shown in Table 4. The study has shown that there exists a significant change in the values of air pollutants for two periods. These have been summarized below:

1. There exists a significant statistical improvement in particulate matter ($PM_{2.5}$) during the lockdown period as compared to corresponding previous year figures

Table 3 Paired samples correlations

Air pollutants	N	Correlation	Sig.
$PM_{2.5}$ (2020) and $PM_{2.5}$ (2019)	61	0.316	0.013
PM_{10} (2020) and PM_{10} (2019)	61	0.148	0.256
NH_3 (2020) and NH_3 (2019)	61	0.049	0.705
SO_2 (2020) and SO_2 (2019)	61	−0.239	0.063
O_3 (2020) and O_3 (2019)	61	0.269	0.036
AQI (2020) and AQI (2019)	61	0.213	0.099

Table 4 t-Test results

Air pollutants	Paired differences					t	df	Sig. (2-tailed)	Ho significant or not
	Mean	Std. deviation	Std. error mean	95% confidence interval of the difference					
				Lower	Upper				
$PM_{2.5}$ 2020–2019	−45.05	31.34	4.01	−53.08	−37.02	−11.224	60	0	Yes
PM_{10} 2020–2019	−28.16	29.20	3.74	−35.64	−20.68	−7.532	60	0	Yes
NH_3 2020–2019	−6.82	3.94	0.50	−7.83	−5.81	−13.507	60	0	Yes
SO_2 2020–2019	−10.29	8.69	1.11	−12.52	−8.07	−9.255	60	0	Yes
O_3 2020–2019	17.00	13.22	1.69	13.61	20.39	10.039	60	0	Yes
AQI 2020–2019	−31.93	32.90	4.21	−40.36	−23.51	−7.581	60	0	Yes

with $t(60) = 11.224, p = 0.000$, and less than 0.005 with 40.93 ± 22.197 to 85.98 ± 30.22 with an improvement of 2.84–3.87.
2. There exists a significant statistical enhancement in particulate matter (PM_{10}) throughout the lockdown period as compared to corresponding previous year figures with $t(60) = -7.532, p = 0.000$, and less than 0.005 with 52.93 ± 23.37 to 81.09 ± 21.31 with an improvement of 2.99–2.72.
3. There exist a significant statistical improvement in ammonia (NH_3) during the lockdown period as compared to corresponding previous year figures with $t(60) = -13.507, p = 0.000$, and less than 0.005 with 2.92 ± 1.021 to 9.73 ± 3.86 with an improvement of 0.131–0.494.
4. There exist a significant statistical enhancement in sulfur dioxide (SO_2) during the lockdown period as compared to corresponding previous year figures with $t(60) = -9.255, p = 0.000$, and less than 0.005 with 8.02 ± 4.70 to 18.31 ± 6.26 with an improvement of 0.602–0.802.
5. There exist a significant statistical improvement ozone (O_3) during the lockdown period as compared to corresponding previous year figures with $t(60) = -7.581$, $p = 0.000$, and less than 0.005 with 40.03 ± 12.7345 to 23.02 ± 8.37 with an improvement of 1.63–1.07.
6. There exists a significant statistical improvement overall AQI throughout the lockdown period as compared to corresponding previous year figures with $t(60) = 10.039, p = 0.000$, and less than 0.005 with 57.05 ± 21.34 to 88.98 ± 29.99 with an improvement of 2.73–3.84.

4.1 Policy Implication and Suggestions

The study has shown that the country has faced significant economic and social setbacks in terms of declining GDP, GNP, output, employment, foreign reserves, exports, etc. There are huge losses to all industries, worsening conditions of the lower section of the society, and failure to combat COVID-19 rapidly. Despite these negative influences, there have been some positive impacts in terms of self-dependence, the fighting spirit of masses, and exposure of real strength of our medical facilities, government preparations, and the most crucial cleaner environment. Our country is dedicating enormous resources toward environment cleanliness, global warming, cleaning of rivers and oceans, etc. Lockdown has shown a self-framed mechanism to do all this. The environment has cleaned, and Ludhiana has become a city where one can breathe, less disease-prone, and a city where one can live smartly.

The study strongly suggests that the government should adopt such forced measures shortly to keep district pollution-free. Such actions can be designed with smart policies strategically laid down by the team of experts from a similar sector. Even a few legislations, in this regard, should be enacted in the parliament and assembly. Lockdown should be made a regular feature of environment cleaning where masses should abide it all costs.

5 Conclusion

The study has shown that the major air pollutants covered in air quality index worldwide, five pollutants under investigation, have shown a sharp decline in their aggregate percentage. The air environment has become cleaner than the previous corresponding period. The statistical tests have shown a significant improvement in the air quality parameters. The results, so obtained, viz. t(60) = 11.224 with $p = 0.000$ for $PM_{2.5}$, t(60) = −7.532, $p = 0.000$ for PM_{10}, t(60) = −13.507, $p = 0.000$ for NH_3, t(60) = −13.507, $p = 0.000$ for SO_2, t(60) = −7.581, $p = 0.000$ for O_3, and finally (60) = 10.039, $p = 0.000$ for overall AQI, have statistically proven the talk of town that environment has become more clean and AQI has improved significantly. This imposed lockdown has demonstrated that nature has its style of making major overhauling in the global atmosphere. Ludhiana is one of the top 10 air polluted cities in India. This forced lockdown has played an integral role in sweeping the pollution and making Ludhiana a less polluted city. This study can be extended to some more districts and regions of the country to get comparable results. A cross-nation study may bring some more concrete results for the policymakers. Even the timing of the survey can be increased to some more previous years to get better results.

References

1. AQICN (2020) https://aqicn.org/city/india/ludhiana/punjab-agricultural-university/. Accessed 30 April 2020
2. Awasthi A, Singh N, Mittal S, Gupta PK, Agarwal R (2010) Effects of agriculture crop residue burning on children and young on PFTs in North West India. Sci Total Environ 408(20):4440–4445
3. Beig G, Sahu SK, Singh V, Tikle S, Sobhana SB, Gargeva P, Murthy BS (2020) Objective evaluation of stubble emission of North India and quantifying its impact on air quality of Delhi. Sci Total Environ 709:136126
4. Chay KY, Greenstone M (2005) Does air quality matter? Evidence from the housing market. J Polit Econ 113(2):376–424
5. ED, Explained Desk (2020) https://indianexpress.com/article/explained/covid-19-lockdown-air-pollution-ozone-6479987/lite/, June 28 2020. Accessed 29 June 2020
6. Gautam S (2020) The influence of COVID-19 on air quality in India: a boon or inutile. Bull Environ Contam Toxicol 1:724–726
7. Guo H, Kota SH, Sahu SK, Hu J, Ying Q, Gao A, Zhang H (2017) Source apportionment of PM2. 5 in North India using source-oriented air quality models. Environ Pollut 231:426–436
8. Jacob DJ, Winner DA (2009) Effect of climate change on air quality. Atmos Environ 43(1):51–63
9. Kumar KS, Kumar KS, Srinivas N, Ziauddin A (2020) Atmospheric aerosols (PM 10 and PM 2.5) and their influence on air quality in Visakhapatnam City, Andhra Pradesh, India. In: Advances in air pollution profiling and control. Springer, Singapore, pp 143–166
10. Mahato S, Pal S, Ghosh KG (2020) Effect of lockdown amid COVID-19 pandemic on air quality of the megacity Delhi, India. Sci Total Environ 139086
11. Majumdar D, Purohit P, Bhanarkar AD, Rao PS, Rafaj P, Amann M, … Srivastava A (2020) Managing future air quality in megacities: emission inventory and scenario analysis for the Kolkata Metropolitan City, India. Atmos Environ 222:117135

12. Monks PS, Granier C, Fuzzi S, Stohl A, Williams ML, Akimoto H, ... Blake N (2009) Atmospheric composition change–global and regional air quality. Atmos Environ 43(33):5268–5350
13. Sharma S, Zhang M, Gao J, Zhang H, Kota SH (2020) Effect of restricted emissions during COVID-19 on air quality in India. Sci Total Environ 728:138878
14. Singh P, Dey S, Chowdhury S, Bali K (2019) Early life exposure to outdoor air pollution: effect on child health in India. Brookings Institute on India Center, vol 6, pp 1–46
15. Srivastava S, Kumar A, Bauddh K, Gautam AS, Kumar S (2020) 21-day lockdown in India dramatically reduced air pollution indices in Lucknow and New Delhi, India. Bull Environ Contam Toxicol 1:9–17
16. Xu K, Cui K, Young LH, Wang YF, Hsieh YK, Wan S, Zhang J (2020) Air quality index, indicatory air pollutants and impact of COVID-19 event on the air quality near central China. Aerosol Air Qual Res 20

Influence of Cement Treatment on Surface Deformation of Quarry Waste Bases Over Weak Subgrade

Ishfaq Rashid Sheikh, M. Y. Shah, and K. M. N. Saquib Wani

1 Introduction

Quarry waste is produced from crushing of rocks that can be used as an alternative material for pavement construction. Quarry waste is produced in huge quantiles that are stockpiled at quarry sites and the presence of it causes an environmental problem. The reuse of quarry waste as base course materials for pavement construction is a sustainable solution. Although the use of quarry waste is limited for a range of reasons, but in many cases would benefit the quarry operations. Recycled pavement materials have become viable alternative material for pavement construction, nearly 80% of materials produced are used for construction of pavement [1]. It is estimated that nearly 175 million tons of quarry waste are produced annually [2]. Thus this enormous amount of quarry waste needs disposal to preserve environment and lower the pavement construction cost [3]. The recycled materials are the best choice for pavement construction that allows the conservation of resources. The quarry waste is obtained after several stages the process consists of blasting, primary and secondary crushing, washing, screening, and stockpiling operations [4]. Quarry fines are generally angular and fine materials [5]. Depending upon the gradation, larger sizes can be replaced as sand in concrete mixtures and also in lateritic soil to improve its properties [6]. However, the variations gradation and compaction characteristics of the quarry fines obtained from different sources are marginal [7]. Sand occupies a larger

I. R. Sheikh (✉) · M. Y. Shah · K. M. N. S. Wani
Department of Civil Engineering, National Institute of Technology Srinagar, Srinagar 190006, India
e-mail: ishfaqrashid06@gmail.com

M. Y. Shah
e-mail: yousuf@nitsri.ac.in

K. M. N. S. Wani
e-mail: sakibwani_02phd17@nitsri.ac.in

H. Singh et al. (eds.), *Sustainable Development Through Engineering Innovations*, Lecture Notes in Civil Engineering 113,
https://doi.org/10.1007/978-981-15-9554-7_61

volume in the mixture and replacement by larger quarry fines can be considered as an economical, high-volume use option [8]. By varying the percentage of cement the surface deformation of quarry waste base decreases thus enhances the bearing capacity of unpaved test sections [9–12]. The increase in bearing capacity improvement factor of pavement depends upon the subgrade strength, base course strength, geocell stiffness, location, and thickness of base course [13–17].

The present paper discusses, Quarry waste, a residual deposit produced from limestone quarries, was used in this research for enhancing recycling efforts. Untreated quarry waste material exhibits low bearing capacity. To enhance these material properties 0, 2, and 4% of cement was added to the quarry waste. Cemented Quarry waste used as a pavement base was tested under static loading. The cement treatment is a sustainable and viable alternative to enhance the bearing capacity of low volume roads.

2 Objectives and Scope

The objective of the present study is to evaluate the reuse of quarry waste as base course material. Static plate load tests on cement-treated and untreated quarry waste bases over dredged soil were carried out. The parameters investigated in this study include the percentage of cement and base course thickness. The effectiveness of cement treatment in quarry waste bases was presented in terms of load versus deformation plots. A number of studies so far have focused on geosynthetic reinforced quarry waste base [18, 19]. Thus, an attempt has been made in this study to treat quarry waste bases with cement. The cement treatment increases the load-bearing capacity of base course material, thus decreasing rut depth in unpaved test sections.

3 Material

The various materials used in this study were cement, subgrade, quarry waste base course.

3.1 Cement

Ordinary portland cement grade 43 was procured from the structural lab National institute of Srinagar. The specific gravity of cement was 3.12.

3.2 *Subgrade*

The Soil used in this study was dredged soil from Shalimar Basin (34.143196 N, 74.861621 E) of Dal lake, Srinagar and was classified as CL according to the Unified Soil Classification System (USCS). The liquid limit and plastic limits of dredged soil were found to be 42 and 29%, respectively. Unconfined compressive strength was observed as 41 kN/m^2, angle of internal friction at OMC and Cohesion were found to be 37° and 27 kN/m^2 respectively. The subgrade had maximum dry density of 15.69 kN/m^3 and an optimum moisture content of 19% and a CBR value of 5%. Dredged soils are characterized by low shear strength and high compressibility and possess low bearing capacity [20–22].

3.3 *Base Course*

The quarry waste which was used as base course material was procured from the nearby quarry site in Srinagar. The properties of quarry waste are summarized in Table 1.

4 Test Equipment and Setup

Large scale model tests were performed in a steel tank of 1 m × 1 m in-plane and 1 m in height. Circular loading plate of 30 cm in diameter was used in this study. Manually operated jack of 147 kN capacity was used to apply footing load that simulates the wheel loads. The setup consisted of proving ring and linear differential transducers for displacement measurement. After every load increments, the deformation and vertical stresses were recorded using a data acquisition system [12]. The test setup is shown in Fig. 1.

Table 1 Properties of quarry waste used in this study

S. No.	Property	Quarry waste
1	Mean particle size (d50) (mm)	1.44
2	Coefficient of uniformity	26
3	Coefficient of curvature	2.5
4	Optimum moisture content (%)	13
5	Maximum dry density (kN/m^3)	22.16
6	CBR (%)	19

Fig. 1 Showing test setup and quarry waste material collection

4.1 Test Section Preparation

The thickness of the subgrade used in this study was 450 mm. Three different base course thickness 150 mm and 170 mm and 200 mm were used in test sections. The quarry waste base was treated by 0, 2, and 4%. The compaction of quarry waste (infill material) was done using hammer, the subgrade was compacted at 19% moisture content in three lifts with each lift of 150 mm thick. The moisture content of each layer was checked at three different locations and dry unit weight of each layer was checked. Quarry waste base was constructed on top of the subgrade. The quarry waste was compacted in a single lift at maximum dry density.

5 Results and Discussions

5.1 Permanent Deformation Versus Pressure

The static plate load tests were conducted on treated and untreated quarry waste bases. The varying cement percentage was used to stabilize the base course, and it was found that the cement treatment increased the bearing capacity of unpaved roads section. Furthermore, the varying thickness of the base course also increased the bearing capacity. The ultimate bearing capacity was obtained at 25 mm deformation from load versus deformation curve, as there is no clear failure [23]. Table 2 shows the improvement factors of cement-treated test sections.

The quarry waste bases were being structurally weak having less stiffness and bearing capacity, so to improve the structural properties of quarry waste bases a proper stabilization was required. It has been observed experimentally that without cement treatment the test sections undergo large surface deformation due to weak subgrade. The deformation replicates from the bottom to top layer of the pavement. During precipitation, the groundwater table also increases, thus the subgrade which

Table 2 Summary of unpaved test sections with varying base course thickness

Test no.	Base course thickness	Cement content (%)	25 mm settlement		
			q (kPa)	BCR	IF
1	150 mm	0	250	–	–
2		2	340	1.36	0.36
3		4	420	1.68	0.68
6	170 mm	0	410	–	–
7		2	500	1.22	0.22
8		4	570	1.39	0.39
11	200 mm	0	450	–	–
12		2	650	1.44	0.44
13		4	750	1.67	0.67

in direct touch with natural layer of soil gets saturated. The 150 mm thick base with 2% and 4% of cement shows 340 and 420 kPa of bearing capacity respectively as compared to untreated quarry waste base. As the bearing capacity increases the surface deformation decreases with each load increment as shown in Fig. 2.

The surface deformation decreases due to increase in percentage of cement. During the initial stage of loading each test section is showing less deformation, because during initial stage of loading, the load is taken by the base course material

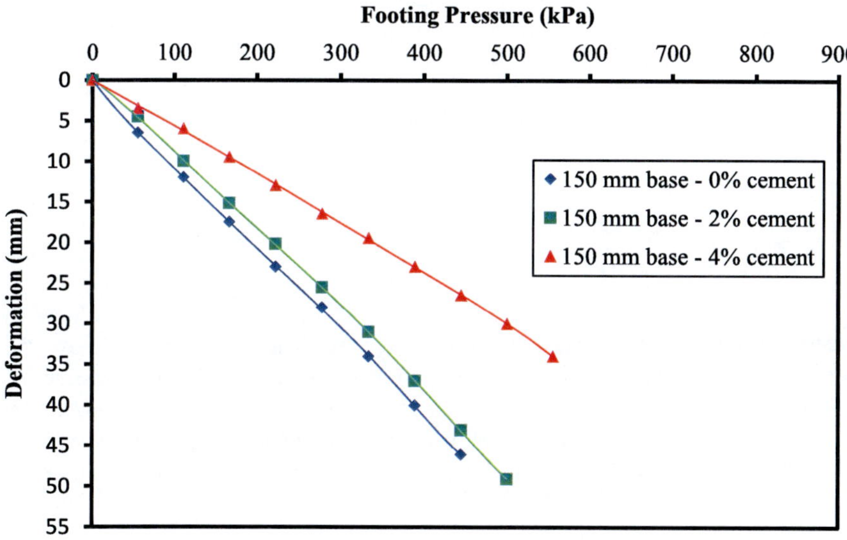

Fig. 2 Showing load versus deformation of 150 mm thick base

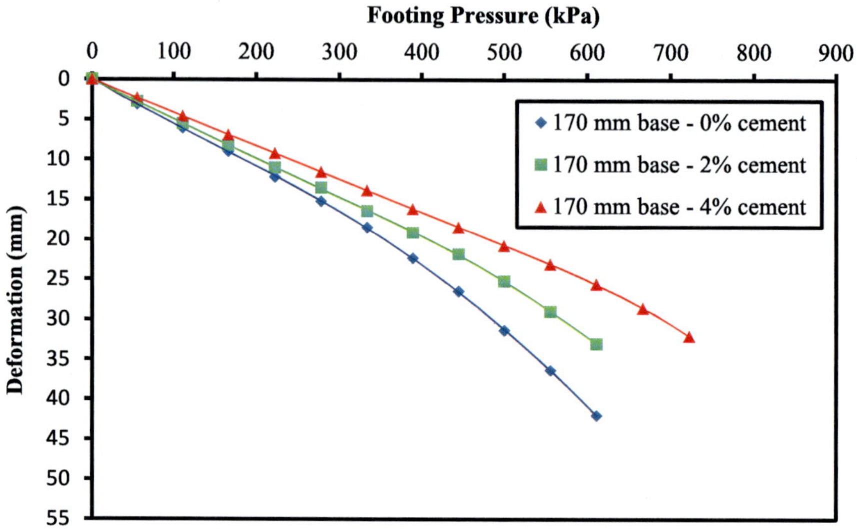

Fig. 3 Showing load versus deformation of 170 mm thick base

for densification. As the applied footing pressure increases the surface deformation also increases. Figure 3 shows the significant increase in bearing capacity with increase in the applied pressure, the 170 mm thick section shows that as the cement percentage increases the bearing capacity increases from 410 to 570 kPa. The quarry waste bases improved the performance of unpaved roads thus becoming a suitable material for construction of roads. 200 mm thick base shows a similar trend, and the bearing capacity increases by 300 kPa as compared to untreated bases as shown in Fig. 4. Figure 5, summarises the variation in bearing capacity.

5.2 *Effectiveness of Cement Treatment on Reducing Base Course Thickness*

The treated test sections show a notable increase in the bearing capacity ratio over the untreated test section. The bearing capacity improvement factor (BCIF) increases by 68% with the cement treatment in the case of 150 mm thick base. The improvement factor for 170 mm thickness base is 39% with 4% of cement. Similarly, 4% cement stabilization in case 200 mm base shows 69% of improvement factor as shown in Fig. 6. It should be noted that BCIF has been calculated with respect to corresponding unreinforced test sections. Hence it is evident from Fig. 6 that the bearing capacity of quarry waste bases increases significantly, provides adequate structural capacity for unpaved roads over weak subgrade, similar results were obtained by researchers [24, 25].

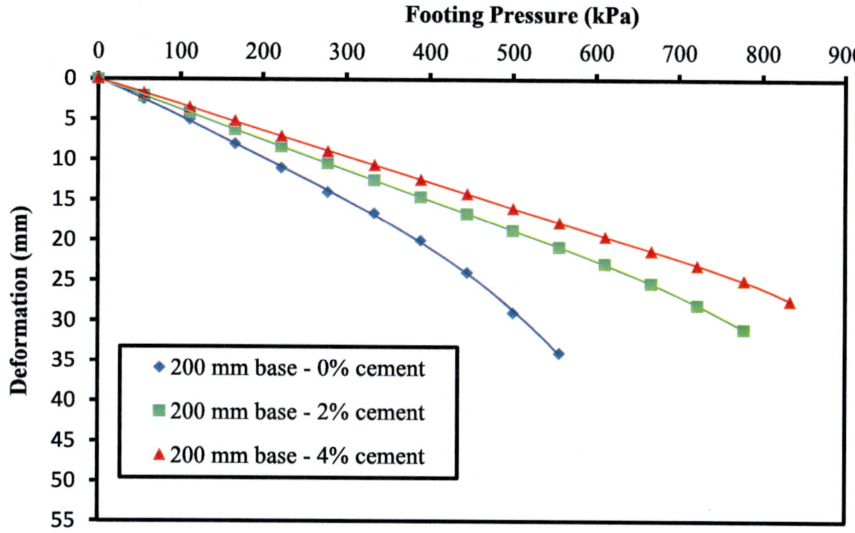

Fig. 4 Showing load versus deformation of 200 mm thick base

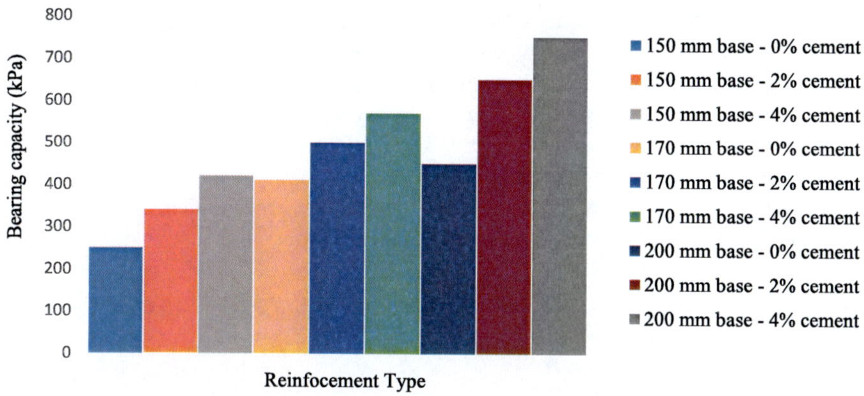

Fig. 5 Showing the bearing capacity with increase in cement percentage

6 Conclusions

Based on the results, following conclusion can be drawn:

- The bearing capacity improvement factor for 150 mm, 170 mm, and 200 mm base treated with cement was found to be 68%, 39%, and 69%, respectively in comparison to corresponding untreated sections.

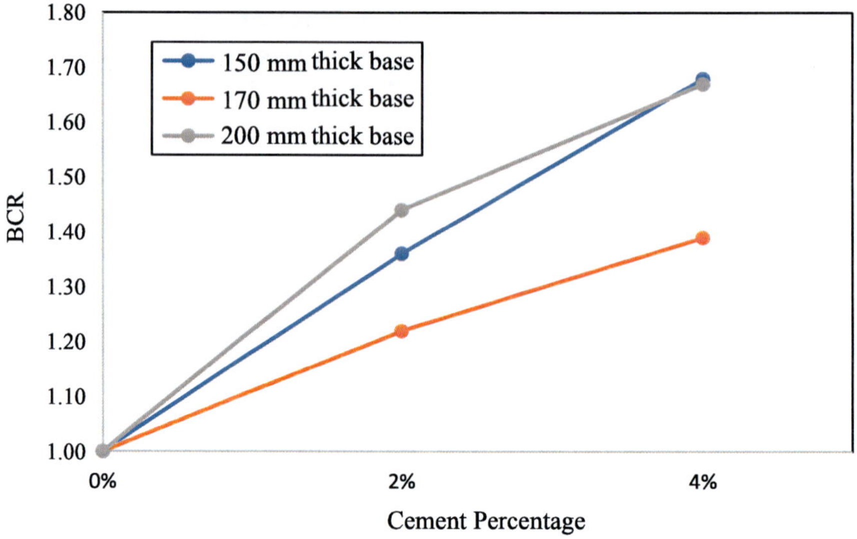

Fig. 6 Showing bearing capacity ratio versus cement percentage

- For each 2% addition of cement the average bearing capacity increased by 105 kPa. Maximum bearing capacity was obtained for 200 mm thick base. The increase in bearing capacity from 250 to 750 kN/m^2 is attributed to the addition of cement.
- The above findings suggest that the quarry waste can be used as a base course material, cement treatment distributes the footing load over a wider area thus preventing pavements from rutting failure.
- The use of quarry waste proved to be a sustainable, environment friendly solution reducing the use of natural aggregates worldwide.

Acknowledgments The authors would like to thank the Department of Civil Engineering, National Institute of Technology-Srinagar, India.

Funding
There was no direct funding available for this research. The first and the third author receive a doctoral fellowship from MHRD-India.

Conflict of Interest
The authors declare that there is no conflict of interest in this research.

References

1. Holtz K, Eighmy TT (2000) Scanning European advances in the use of recycled materials in highway construction. Public Roads 64(1):34–40

2. Collins RJ, Ciesielski SK, Mason LS (1994) Recycling and use of waste materials and by-products in highway construction: a synthesis of highway practice. Final report (No. PB-94-198140/XAB; TRB/NCHRP/SYN–199). National Research Council, Washington, DC (United States). Transportation Research Board
3. Taha R, Ali G, Basma A, Al-Turk O (1999) Evaluation of reclaimed asphalt pavement aggregate in road bases and subbases. Transp Res Rec 1652(1):264–269
4. Tepordei VV (1992) Construction sand and gravel: annual report. US Department of Interior, Bureau of Mines, Washington, DC
5. Okamura H (1997) Self-compacting high-performance concrete. Concr Int 19(7):50–54
6. Soosan TG, Jose BT, Abraham BM (2001, December). Use of quarry dust in embankment and highway construction. In: Proceedings of Indian Geo-Technical Conference, pp 274–277
7. Soosan TG, Sridharan A, Jose BT, Abraham BM (2005) Utilization of quarry dust to improve the geotechnical properties of soils in highway construction. Geotech Test J 28(4):391–400
8. Naik TR, Kraus RN, Chun YM, Canpolat F, Ramme BW (2005, October) Use of limestone quarry by-products for developing economical self-compacting concrete. In: Published at the CANMET/ACI international symposium on sustainable developments of cement and concrete, Toronto, Canada
9. Eze-Uzomaka OJ, Agbo D (2010) Suitability of quarry dust as improvement to cement stabilized-laterite for road bases. Electron J Geotech Eng 15:1053–1066
10. Mohammadinia A, Arulrajah A, Sanjayan J, Disfani MM, Bo MW, Darmawan S (2015) Laboratory evaluation of the use of cement-treated construction and demolition materials in pavement base and subbase applications. J Mater Civ Eng 27(6):04014186
11. Herrador R, Pérez P, Garach L, Ordóñez J (2012) Use of recycled construction and demolition waste aggregate for road course surfacing. J Transp Eng 138(2):182–190
12. Sheikh IR, Shah MY (2020a) Experimental study on geocell reinforced base over dredged soil using static plate load test. Int J Pavement Res Technol 13:286–295. https://doi.org/10.1007/s42947-020-0238-2
13. Shimizu M, Inui T (1990, May) Increase in bearing capacity of ground with geotextile wall frame. In: Proceedings of fourth international conference on geotextiles geomembranes and related products, vol 50, p 254
14. Mhaiskar SY, Mandal JN (1992) Soft clay subgrade stabilization using geocells. In: Grouting, soil improvement and geosynthetics. ASCE, pp 1092–1103
15. Pokharel SK, Han J, Leshchinsky D, Parsons RL, Halahmi I (2009) Behavior of geocell-reinforced granular bases under static and repeated loads. In: Contemporary topics in ground modification, problem soils, and geo-support, pp 409–416
16. Wani KS, Mir BA (2019) Influence of microbial geo-technology in the stabilization of dredged soils. Int J Geotech Eng https://doi.org/10.1080/19386362.2019.1643099
17. Wani KS, Mir BA (2019) Effect of biological cementation on the mechanical behaviour of dredged soils with emphasis on micro-structural analysis. Int J Geosynthetics Ground Eng 5(4):32. https://doi.org/10.1007/s40891-019-0183-9
18. Pokharel SK, Han J, Leshchinsky D, Parsons RL, Halahmi I (2010) Investigation of factors influencing behavior of single geocell-reinforced bases under static loading. Geotext Geomembr 28(6):570–578
19. Bortz BS, Hossain M, Halami I, Gisi A (2012) Innovative uses of quarry waste and reclaimed asphalt pavement. In: ICSDC 2011: integrating sustainability practices in the construction industry, pp 515–523
20. Wani KMNS, Mir BA (2020) Unconfined compressive strength testing of bio-cemented weak soils: a comparative upscale laboratory testing. Arab J SciEng. https://doi.org/10.1007/s13369-020-04647-8
21. Wani KMNS, Mir BA (2020) Effect of microbial stabilization on the unconfined compressive strength and bearing capacity of weak soils. Transp Infrastruct Geotechnol. https://doi.org/10.1007/s40515-020-00110-1
22. Sheikh IR, Shah MY (2020b) Experimental investigation on the reuse of reclaimed asphalt pavement over weak subgrade. Transp Infrastruct Geotechnol V7:634–650. https://doi.org/10.1007/s40515-020-00115-w

23. Abu-Farsakh MY, Akond I, Chen Q (2016) Evaluating the performance of geosynthetic-reinforced unpaved roads using plate load tests. Int J Pavement Eng 17(10):901–912
24. Subaida EA, Chandrakaran S, Sankar N (2009) Laboratory performance of unpaved roads reinforced with woven coir geotextiles. Geotext Geomembr 27(3):204–210
25. Fannin RJ, Sigurdsson O (1996) Field observations on stabilization of unpaved roads with geosynthetics. J Geotech Eng 122(7):544–553

Development of Sustainable Concrete Using Bacteria as Self-healing Agent

Krishna Murari and Pritpal Kaur

1 Introduction

In most of the infrastructures, concrete is used at large scale as construction material having advantage of high characteristic compressive strength, durability and cost economy. However, under the action of tensile forces it is susceptible to crack formation at micro- and macro-levels. Initially, micro-cracks are formed, which are unnoticeable, further leads to formation macro-cracks due to environmental and mechanical actions [26]. These cracks create a pathway for the oxygen, moisture and deleterious materials, which causes steel corrosion adversely affecting the durability. For this purpose, the inspection and maintenance techniques have gained increasing attention. Regular inspection and maintenance are difficult rather impossible for inaccessible large-scale infrastructures owing to capital investment required. Various other factors include labour and the location of the damage.

Bacterial concrete, eco-friendly in nature, is preferred nowadays as it helps in decreasing permeability, enhancing durability, easily apply to existing buildings in the form of liquids and has very less maintenance cost than conventional concrete.

Self-healing to repair cracks in cement-based matrix can be done by various techniques. Classification of self-healing for cement-based matrix is given in Fig. 1. Pros and cons of different self-healing techniques are shown in Table 1.

K. Murari (✉) · P. Kaur
Guru Nanak Dev Engineering College, Ludhiana 141006, India
e-mail: krishnamurarigne@gmail.com

H. Singh et al. (eds.), *Sustainable Development Through Engineering Innovations*, Lecture Notes in Civil Engineering 113,
https://doi.org/10.1007/978-981-15-9554-7_62

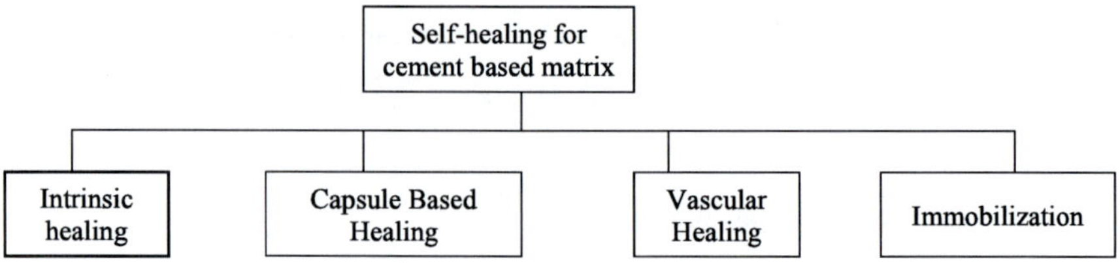

Fig. 1 Flowchart of various techniques of self-healing

Table 1 Pros and cons of different self-healing techniques

Technique	Benefits	Drawbacks	References
Vascular system	Healing agent discharge when required Adjustable high quantity of healing material	Casting difficulties Strength parameters of cement matrix affected in presence of too many tubes	[23]
Capsule-based healing	Healing agent discharge when required Medium quantity of healing agent Efficient under occurrence of multiple damage	Casting difficulties Strength parameters of cement matrix affected in presence of too many hollow fibres Difficulty in release of healing agent may be possible	[4, 13, 17, 19]
Immobilization	Response to multiple cracked areas at same time	Uncertainty in generation of healing products Need to meet several prerequisites	[16, 23]
Intrinsic healing	Excellent efficiency in healing Healing products are more compatible with cement matrix	Under ill-treatment, undesirable expansion Uncertainty in generation of healing products	[9, 18]

2 Self-healing Bacterial Concrete Methodology

In self-healing concrete, concrete mix is prepared by adding bacteria and its nourishing substances for preparing calcite. The main difficulty is to keep the bacteria and its food apart; for achieving it, bacteria are covered in capsules and will crack at the time of crack formation to release it. In many studies, capsules made of glass material are used for this purpose [14]. At most expected crack formation places, glass tubes are placed. Microcapsules carrying healing agent can be mixed in concrete without any effort, but due to impact of large aggregates, some of them get disintegrated.

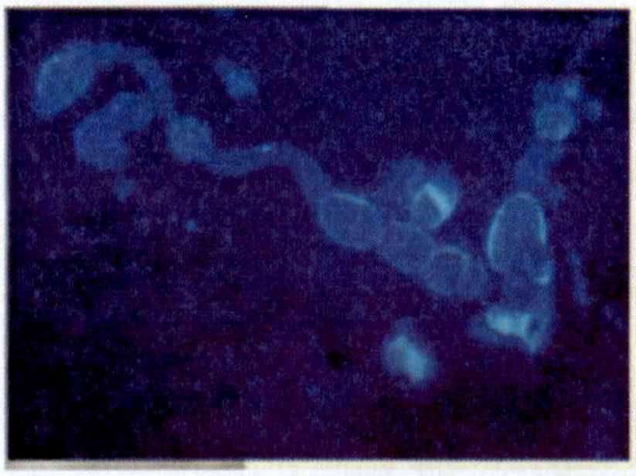

(a) Bacillus Fila

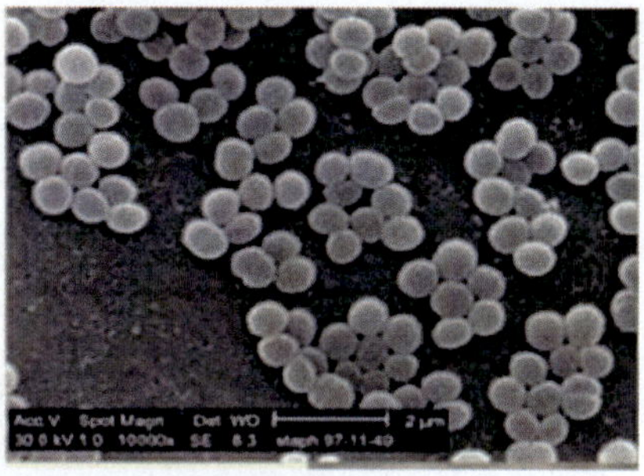

(b) Bacillus Cohnii

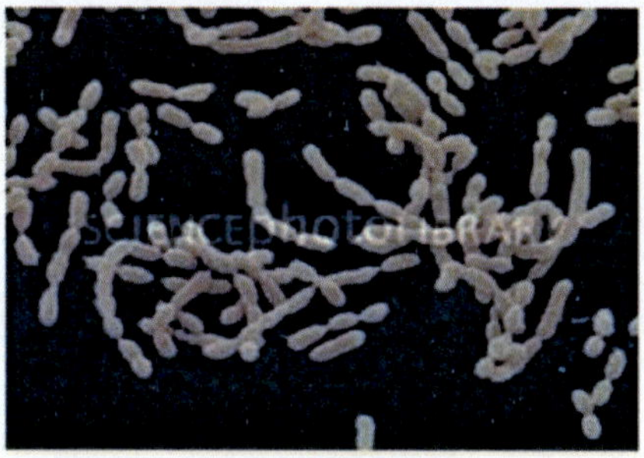

(c) Bacillus Parturri

Fig. 2 Bacteria used as self-healing agent [1, 21]

2.1 *Finding the Right Bacteria*

Concrete is highly alkaline; the genus bacillus group of bacteria (shown in Fig. 2) is capable of resisting to such alkaline conditions [1, 21]. Finding the right bacteria for making self-healing concrete is one of the important steps. These bacteria should have special characteristics, like capability of bearing mechanical stress and resisting high alkalinity. A source of nutrient is provided to the bacteria which undergo metabolic process, and precipitate of calcium carbonate is obtained [8, 22].

2.2 *Healing Mechanism*

Biological repair technique of bacterial introduction in the concrete dates back in 1990s in which environmentally friendly processes were suggested for crack repair [7]. In the precipitation pathway, urea is converted into NH_4^+ and HCO_3^- [15, 22]. In the starting of microbial urease mechanism, one mole of urea is hydrolyzed intracellularly to a mole of NH_3 and carbamate each. Further spontaneous hydrolysis of carbamate occurs leading to formation of NH_3 mol and H_2CO_3 mol [Eqs. (1–2)] [21].

$$CO(NH_2)_2 + H_2O \rightarrow NH_3COOH + NH_3 \tag{1}$$

$$NH_2COOH + H_2O \rightarrow NH_3 + H_2CO_3 \tag{2}$$

Following that, these products further produce unimolecular bicarbonate and bimolecular ammonium and hydroxide ions each [Eqs. (3–4)]. The rise in pH is observed which shifts bicarbonate equilibrium, and carbonate ion is formed (Eq. 5) [3].

$$H_2CO_3 \rightarrow HCO_3^- + H^+ \tag{3}$$

$$2NH_3 + 2H_2O \leftrightarrow 2NH_4^+ + 2OH^- \tag{4}$$

$$HCO_3^- + H^+ + 2NH_4^+ + 2OH^- \leftrightarrow CO_3^{2-} + 2NH_4^+ + 2H_2O \tag{5}$$

The formation of calcium carbonate precipitates due to availability of calcium ions after reaching a certain supersaturation level (Eq. 6).

$$CO_3^{2-} + Ca^{2+} \rightarrow CaCO_3 \tag{6}$$

Negative charge of the cell wall leads to attraction of calcium ions (Fig. 3) which results in crystal formation on bacterial cell [5].

Another pathway for production of $CaCO_3$ is aerobic oxidation of organic acids, which show less environmental effect in comparison to ureolytic pathway and through this metabolic pathway carbon dioxide is produced that leads to autogenous self-healing [10–12, 24].

3 Other Aspects of Self-healing Bacterial Concrete

3.1 Effect of Bacteria on Concrete Properties

Experimental study shows that addition of bacteria in cement has significant effect on mechanical properties of cement concrete. About 8–10%, loss in compressive and flexural strength was found by adding healing agent in concrete mix [25]. The spores of *Bacillus cohnii* and *Bacillus pseudofirmus* are dwindled approximately 10% after 28 days when added to cement stone specimen [11].

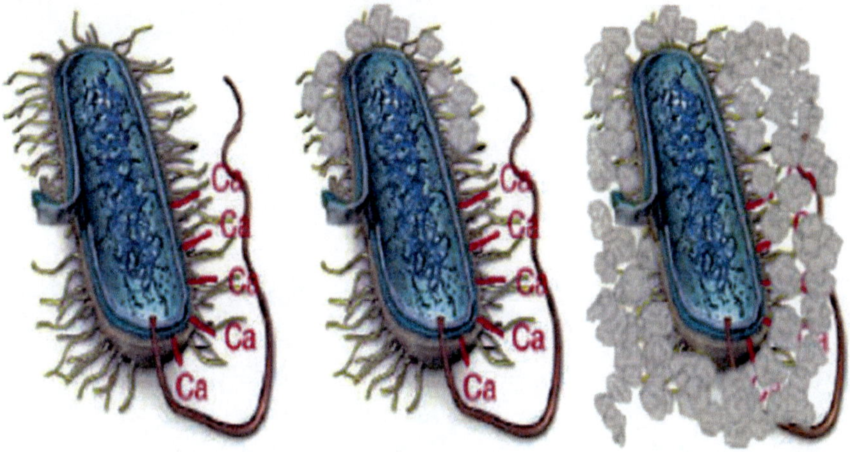

Fig. 3 Calcium carbonates formation on bacterial cell wall [22]

Study conducted by Naveen and Sivakamasundari (2016) showed that there is decrease in tensile strength. It has been found that by adding calcium lactate and calcium formate setting time is retarded, but it is accelerated by adding calcium nitrate as nutrient [20, 27]. It has been found that when *S. Pastteurii* bacteria with concentration of 105 cells/m is added in fly ash concrete then its water absorption capacity is reduced by four times [2].

3.2 *Cost*

Self-healing concrete mix is prepared by using bacteria and calcium lactate. Preparation of calcium lactate from milk is relatively expansive. Consequently, increases the cost of preparation of self-healing concrete than conventional concrete. To find the cost analysis different representatives like, design team members, client bodies and contractors had conducted survey for UK infrastructures [6]. It was reported that incorporating bio-agents present in self-healing concrete might lead to high initial cost but at the same time makes concrete more durable. Moreover, self-healing concrete can be made more sustainable and relatively cheaper by two possible ways—first is by decreasing the production cost of bio-agents, second is by designing the self-healing action for increased durability in order to get better behaviour under numerous cycles of loading and adverse environment condition.

(a) Marine Structure (b) Base Wall (c) Concrete flooring

(d) Tunnel Lining (e) Highway Bridge (f) Underground retaining wall

Fig. 4 Application of self-healing concrete

3.3 Applications

In the case of structures, those are built in a high water environments like underground basements and marine structures are particularly vulnerable to corrosion of steel reinforcement. Moreover, other structures such as motorway bridges are also vulnerable due to salts used to de-ice the roads, which percolates into the cracks of the structures and as a result accelerates the corrosion of steel reinforcement. Some major applications of self-healing bacterial concrete are shown in Fig. 4.

4 Conclusion

By observing the above-mentioned study on self-healing bacterial concrete, some points have been concluded below:

- Bacteria that can survive in highly alkaline medium represent promising candidates for application as self-healing agent in concrete.
- In comparison to other self-repairing concrete, bacterial concrete is eco-friendly in nature and preferred nowadays due to its less permeability and high durability.
- Bacteria have significant effect on concrete properties and depend on various factors like concentration of bacteria in concrete mix, curing agent.
- Incorporating bio-agent which is present in self-healing concrete has high initial cost, but it contributes in reduced maintenance and repair cost.

The overall conclusion is application of bacteria as healing agent leads to development of sustainable concrete and can contribute for increasing the durability of infrastructures. Moreover, in developed countries, self-healing concrete is used in infrastructures for crack healing purpose, but in the developing countries like India, it is still at research stage.

References

1. Bravo F, Silva D, Boon N, Verstraete W, De Belie N (2015) Screening of bacteria and concrete compatible protection materials. Constr Build Mater 88:196–203. https://doi.org/10.1016/j.conbuildmat.2015.04.027
2. Chahal N, Siddique R, Rajor A (2012) Influence of bacteria on the compressive strength, water absorption and rapid chloride permeability of fly ash concrete. Constr Build Mater 28(1):351–356. https://doi.org/10.1016/j.conbuildmat.2011.07.042
3. Dick J, De Windt W, De Graef B, Saveyn H, Van Der Meeren P, De Belie N, Verstraete W (2006) Bio-deposition of a calcium carbonate layer on degraded limestone by *Bacillus* species. Biodegradation 17:357–367. https://doi.org/10.1007/s10532-005-9006-x
4. Dry CM (2001) Three designs for the internal release of sealants, adhesives, and waterproofing chemicals into concrete to reduce permeability. Cem Concr Res 30(2000):1969–1977
5. De Muynck W, De Belie N, Verstraete W (2010) Microbial carbonate precipitation in construction materials: A review. Ecol Eng 36:118–136. https://doi.org/10.1016/j.ecoleng.2009.02.006
6. Gardner D, Lark R, Je T, Davies R (2018) A survey on problems encountered in current concrete construction and the potential benefits of self-healing cementitious materials. Case Stud Constr Mater 8(February):238–247. https://doi.org/10.1016/j.cscm.2018.02.002
7. Gollapudi UK, Knutson CL, Bang SS, Islam MR (1995) A new method for controlling leaching through permeable channels. Chemosphere 30(4):695–705
8. Hammes F, Boon N, Villiers J De, Verstraete W, Siciliano SD (2003) Strain-specific ureolytic microbial calcium carbonate precipitation. Appl Environ Microbiol 69(8):4901–4909. https://doi.org/10.1128/AEM.69.8.4901
9. Ish-Shalom M, Bentur A (1974) Properties of type K expansive cement of pure components I. Hydration of unrestrained paste of expansive component—Results. Cem Concr Res 4:519–532
10. Jonkers HM (2008) Self healing concrete: a biological approach. In: Zwaag S (ed) Self healing materials—an alternative approach to 20 centuries of materials science. Springer, The Netherlands, pp 195–204
11. Jonkers Henk M, Thijssen A, Muyzer G, Copuroglu O, Schlangen E (2010) Application of bacteria as self-healing agent for the development of sustainable concrete. Ecol Eng 36(2):230–235. https://doi.org/10.1016/j.ecoleng.2008.12.036
12. Jonkers HM, Schlangen E (2007) Crack repair by concrete-immobilized bacteria (April), pp 1–7
13. Kanellopoulos A, Qureshi TS (2015) Glass encapsulated minerals for self-healing in cement based composites. Constr Build Mater 98:780–791. https://doi.org/10.1016/j.conbuildmat.2015.08.127
14. Karaiskos G, Tsangouri E, Aggelis DG, Van Tittelboom K, De Belie N, Van Hemelrijck D (2016) Performance monitoring of large-scale autonomously healed concrete beams under four-point bending through multiple non- destructive testing methods. Smart Mater Struct 25(5). https://doi.org/10.1088/0964-1726/25/5/055003
15. Kaur M, Murari K, Kaur I (2020) Techniques and various efficiency evaluation tests for self-healing cement-based matrix: state-of-the-art. In: IOP conference series: materials science and engineering 814(1). https://doi.org/10.1088/1757-899X/814/1/012023

16. Khaliq W, Ehsan MB (2016) Crack healing in concrete using various bio influenced self-healing techniques. Constr Build Mater 102:349–357. https://doi.org/10.1016/j.conbuildmat.2015.11.006
17. Lv L, Yang Z, Chen G, Zhu G, Han N, Schlangen E, Xing F (2016) Synthesis and characterization of a new polymeric microcapsule and feasibility investigation in self-healing cementitious materials. Constr Build Mater 105:487–495. https://doi.org/10.1016/j.conbuildmat.2015.12.185
18. Nagataki S, Gem H (1998) Expansive admixtures (mainly ettringite). Cem Concr Compos 20:163–170
19. Pang JWC, Bond IP (2005) 'Bleeding composites'—damage detection and self-repair using a biomimetic approach. Compos Part A Appl Sci Manuf 36:183–188. https://doi.org/10.1016/j.compositesa.2004.06.016
20. Souradeep G, Kua HW (2016) Encapsulation technology and techniques in self-healing concrete. J Mater Civ Eng 28(2007):1–15. https://doi.org/10.1061/(ASCE)MT.1943-5533
21. Tittelboom K Van, Belie N De, Muynck W De, Verstraete W (2010) Use of bacteria to repair cracks in concrete. Cem Concr Res 40(1):157–166. https://doi.org/10.1016/j.cemconres.2009.08.025
22. Vijay K, Murmu M, Deo SV (2017) Bacteria based self healing concrete—a review. Constr Build Mater 152:1008–1014. https://doi.org/10.1016/j.conbuildmat.2017.07.040
23. Wang JY, Soens H, Verstraete W, Belie N De (2014) Cement and concrete research self-healing concrete by use of microencapsulated bacterial spores. Cem Concr Res 56:139–152. https://doi.org/10.1016/j.cemconres.2013.11.009
24. Wiktor V, Jonkers HM (2011) Quantification of crack-healing in novel bacteria-based self-healing concrete. Cement Concr Compos 33(7):763–770. https://doi.org/10.1016/j.cemconcomp.2011.03.012
25. Xu J, Yao W (2014) Multiscale mechanical quantification of self-healing concrete incorporating non-ureolytic bacteria-based healing agent. Cem Concr Res 64:1–10
26. Yang Z, Hollar J, He X, Shi X (2010) Laboratory assessment of a self-healing cementitious composite. Transp Res Rec 9–17. https://doi.org/10.3141/2142-02
27. Zhang Y, Guo HX, Cheng XH (2015) Role of calcium sources in the strength and microstructure of microbial mortar. Constr Build Mater 77:160–167

Sustainable Development of Prediction Model for Seismic Hazard Analysis

Sufyan Ghani and Sunita Kumari

1 Introduction

Earthquakes induced seismic hazard is always defined as an uncertain and disguised enemy of environment which has threatened and endangered humanity. The damages caused to the infrastructure and large number of human lives that are lost during strong earthquakes are still a problem all around the globe which cannot be stopped by mankind. One of the most disastrous phenomena that arise due to earthquakes is liquefaction which has always been a major concern for engineers due to the damages and devastation caused by it such as failures of earth structures, foundations and superstructures. Liquefaction occurs when soil transforms to liquefied state due to pore water pressure in the soil increases and effective stress reduces. Conditions required for liquefaction to occur are that the soils must be submerged below the water table, and the ground shaking must be intense along with duration of ground shaking sufficient enough for the soils to lose their shearing resistance. Seed and Idriss [20] proposed simplified procedures to determine the liquefaction potential of sandy soils. In numerous projects, the simplified Seed and Idriss criteria for assessing liquefaction potential based on standard penetration test (SPT) were used. In earlier years, the term liquefaction was referred for sandy soil deposits only but observations made during several earthquakes showed evidence of liquefaction in soil with fine content having medium to low plasticity. Materials like silt or clay are non-plastic or plastic in nature which tends to make an important and consistent

S. Ghani (✉) · S. Kumari
Department of Civil Engineering, National Institute of Technology Patna, Patna, Bihar 800005, India
e-mail: sufyan04@gmail.com

S. Kumari
e-mail: sunitafce@gmail.com

H. Singh et al. (eds.), *Sustainable Development Through Engineering Innovations*, Lecture Notes in Civil Engineering 113,
https://doi.org/10.1007/978-981-15-9554-7_63

difference in the cyclic strength of such soil deposits. Wang [25] noted liquefaction in silty sand to slightly sandy silt soils during Haicheng, 1975, and Tangshan, 1976, earthquakes. Also, criteria state that clayey soils could be susceptible to liquefaction only if all three of the following conditions are met: percent of particles less than 0.005 mm <15%, liquid limit (LL) <35% and ratio of water content and liquid limit (wc/LL) >0.9 [21, 25]. This standard came to be known as the Chinese criteria due to its origin. However, several cases were observed where ground failure caused considerable damage to buildings in silty and clayey soils containing more than 15% clay-size particles during Northridge (1994), Kocaeli (1999) and Chi-Chi (1999) earthquakes, thus questioning the efficiency and accuracy of Chinese criteria. Andrews and Martin [2] reviewed empirical approach and suggested a new assessment index which transformed the conventional Chinese criteria in accordance with US standards and renamed as modified Chinese criteria. Conclusion drawn from the work of various researchers suggests that interdependency exists in between plasticity and cyclic strength of soil which gives rise to use of fine content, liquid limit (LL) and plasticity index (PI) as key criteria for the evaluation of liquefaction susceptibility of soil deposits [3, 8–11, 13, 17–19, 22, 23].

Most of the relevant methods in evaluating the liquefaction potential are theoretical and empirical methods which are time-consuming, as well as there are many possible factors that may cause uncertainty in determining liquefaction potential. The uncertainty and lack of precision that occurs during testing and obtaining soil properties add up with uncertainties caused during the calculation of liquefaction potential using the abovementioned empirical methods and can lead to a false conclusion. In addition, these empirical methods need large number of parameters to determine the liquefaction susceptibility of soil due to earthquake loading which often makes it costly and time-consuming. A slight margin of error and mistakes are allowed in every aspect of engineering, but when it comes to safe designing against liquefaction especially in high seismic zones, a false conclusion can be extremely threatening for humanity. Therefore, simplified methods in evaluating soil liquefaction are popular among practicing engineers. These procedures are very useful at the initial designing stage to assess any risk of liquefaction, and only if the risk of liquefaction is high, a detailed analysis needs to be carried out to obtain the liquefaction potential.

Computational approach based on artificial neural networks (ANNs) is widely accepted by researchers to determine the liquefaction potential of soil deposits [1, 4–7, 12, 16, 24, 26, 27].

The accuracy of an ANN model can be highly improved by selecting appropriate inputs and providing large datasets for training and testing of the model. A powerful ANN model provides more precise and realistic results as compared to the conventional methods. In the present study, an updated semi-empirical approach developed by Idriss and Boulanger [9] is used as a conventional approach to evaluate liquefaction potential, as well as an ANN model has been developed to predict the liquefaction susceptibility of soil deposits considering fine content, liquid limit and normal moisture content as input parameters. A comparison has been established between the ANN and conventional Idriss and Boulanger method.

2 Methodology

Liquefaction susceptibility has been studied by many researchers based on different parameters. Various authors have presented the need for location-based study for seismic soil properties and analysis for liquefaction potential of soil deposits. In view of the above, for the present paper, the data are obtained for four investigation sites of Patna districts in Bihar, India, and are named as A, B, C and D, respectively. The data obtained are used to analyze the liquefaction behavior of soil deposits having a significant presence of fine content. Bihar is situated in the high seismic zone that falls on the boundary of the tectonic plate joining the Himalayan tectonic plate near the Bihar-Nepal border and has six sub-surface fault lines moving towards the Gangetic planes in four directions. The state has past experience of major earthquakes; the worst was the 1934, followed by 1988 earthquake and a recent earthquake in September 2011. The selected site lies in seismic Zone IV shown in Fig. 1. Figure 1 presents the seismic zone of Bihar developed by Building Materials and Technology Promotion Council (BMTPC).

The main scope of the study is to evaluate the liquefaction potential of soil deposits for the proposed sites A, B, C and D, respectively, using Idriss and Boulanger empirical method and developed Levenberg–Marquardt backpropagation algorithm-based

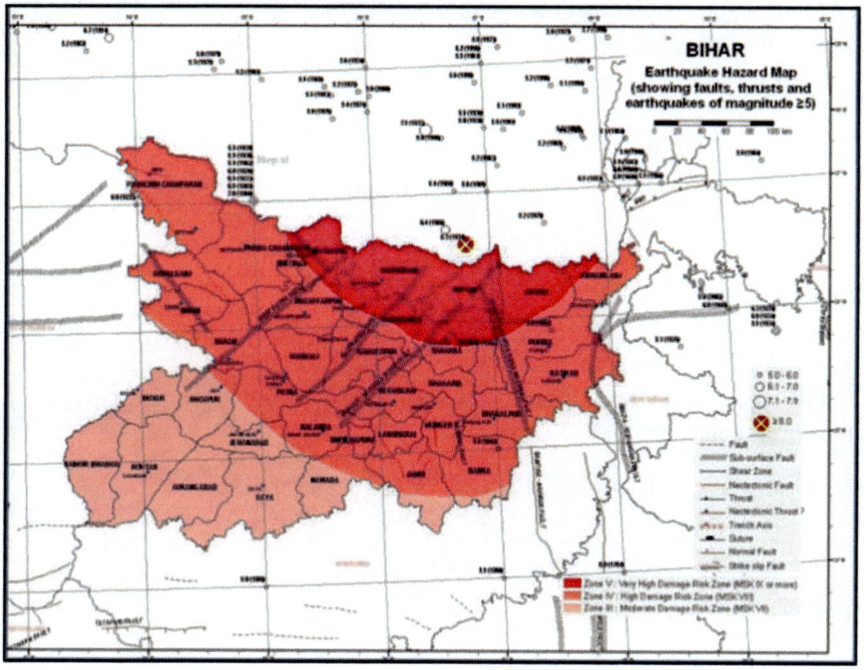

Fig. 1 Earthquake-prone areas of Bihar (http://bmtpc.org/topics.aspx?mid=56&Mid1=180). *Source* Vulnerability atlas of India

ANN model. Further comparison has been made between both the methods, and it has been indicated that the use of artificial intelligence is a robust approach as it eliminates the possibilities of human error and uncertainties. A suggestion has also been made for promoting the use of fine content and plasticity as a key criterion for determining liquefaction susceptibility.

2.1 Conventional Approach

The cyclic stress ratio (CSR) which illustrates the loading enforced by a seismic movement has been proposed by Idriss and Boulanger [9] as:

$$(\mathrm{CSR})_{7.5} = 0.65\left(\frac{\sigma_{vo} a\mathrm{max}}{\sigma'_{vo}}\right)\frac{r_d}{\mathrm{MSF}}\frac{1}{K_\sigma} \tag{1}$$

where σ_{vo} is the total vertical overburden stress, σ'_{vo} is the effective vertical overburden stress at specified depth z, $a_{\max}$ is the peak horizontal ground acceleration in g's, r_d is a stress reduction factor, MSF is the magnitude scaling factor, it is used for the adjustment of the induced CSR to the reference earthquake magnitude of 7.5, and K_σ is the correction factor for effective overburden.

The potential of soil deposits to resist liquefaction is termed as cyclic resistance ratio (CRR) can be evaluated from the corrected blow count $(N_1)_{60}$ using the following expression:

$$\mathrm{CRR} = \exp\left[\frac{(N_1)_{60cs}}{14.1} + \left\{\frac{(N_1)_{60cs}}{126}\right\}^2 - \left\{\frac{(N_1)_{60cs}}{23.6}\right\}^3 + \left\{\frac{(N_1)_{60cs}}{25.4}\right\}^4 - 2.8\right] \tag{2}$$

Where term $(N_1)_{60cs}$ depends on $(N_1)_{60}$ and given as:

$$(N_1)_{60CS} = (N_1)_{60} + \Delta(N_1)_{60} \tag{3}$$

Following equation is used to evaluate factor of safety (FOS) against liquefaction, which is described as:

$$\mathrm{FOS} = \frac{\mathrm{CRR}}{\mathrm{CSR}} \tag{4}$$

2.2 Artificial Neural Network (ANN) Based Approach

A multi-layer perception (MLP) is the most commonly used feedforward networks which usually comprise of three layers: an input layer, a hidden layer and an output layer. More [14] specified the theory and application of Levenberg–Marquardt backpropagation which has been widely accepted and used by various researchers for predicting liquefaction potential of soil deposits [5, 15, 16, 27, 28]. In this training process, the network errors are backpropagated into each neuron in the hidden layer and then transferred into the neuron in the input layer. The adjustments of the connection weights and biases depend on the distribution of error at each neuron. The global network error is reduced by continuously making adjustments of connection weights and biases. Before the training of a network, an error goal is set, and if the network error during the training becomes higher than the error goal, the training has to be stopped. The error backpropagation was the most popular learning algorithm which was proposed in 1969, but due to its challenging and demand of high-end computations, it was ignored. Again in the mid-1980s, the backpropagation learning algorithm was rediscovered and widely used (Negnevitsky 2002). The objective of the backpropagation method is to minimize the error signal of all the output neurons; therefore, it is a supervised learning method and the most widely used training method for the multilayer neural networks.

A multi-layer perception (MLP) trained with Levenberg–Marquardt backpropagation algorithm has been used in the present study to develop an artificial neural network (ANN) model using a neural network toolbox in MATLAB. The developed ANN model is based on test data obtained from the study of Bray and Sancio [10] from liquefaction sites of Adapazarı, Turkey, which liquefied during Kocaeli earthquake in 1999. It uses basic soil properties, i.e., liquid limit (LL), fine content (FC) and normal moisture content (NMC) as input parameters to predict liquefaction susceptibility of any given soil deposit.

3 Result and Discussion

Observation made from the literature suggested that the presence of fine content has a virtuous influence on liquefaction potential. Even some of the researcher suggested that liquid limit and moisture content also affect the liquefaction potential of the soil deposits. So the ANN model developed uses liquid limit (LL), normal moisture content (%) and fine content (%) as input parameter and predicts liquefaction susceptibility. The evaluation for liquefaction potential has been carried out using the conventional Idriss and Boulanger method as well as ANN model, and it has been shown in Tables 1, 2, 3, and 4 for all the four sites. Factor of safety has been evaluated using Idriss and Boulanger [9] empirical method, if FOS ≥ 1 the soil is said to be non-liquefiable and is denoted as '1', whereas for FOS ≤ 1, the soil is said to undergo liquefaction and is denoted as '0'. Similarly, the liquefaction susceptibility

Table 1 Liquefaction susceptibility of site A as per Idriss and Boulanger and ANN method

LL	NMC	FC	FOS = CRR/CSR	Liquefaction susceptibility as per Idriss and Boulanger	Liquefaction susceptibility as ANN
0	30.43478	31.9	0.553533	0	0
0	30.43478	31.9	0.971911	0	0
0	34.10138	40.2	0.448921	0	0
0	34.10138	40.2	0.493345	0	1
0	34.10138	40.2	0.342637	0	1
42.5	25.11078	89.2	0.513633	0	0
42.5	25.11078	89.2	1.067642	1	1
41.75	21.28936	90.3	0.768466	0	1
41.75	21.28936	90.3	1.118277	1	1
42	20.94259	84.6	1.195062	1	1
0	18.73467	84.6	1.518655	1	1
0	18.73467	43	1.97942	1	0

Table 2 Liquefaction susceptibility of site B as per Idriss and Boulanger and ANN method

LL	NMC	FC	FOS = CRR/CSR	Liquefaction susceptibility as per Idriss and Boulanger	Liquefaction susceptibility as ANN
0	42.1	40.2	0.242925	0	1
0	42.10069	40.2	0.245777	0	1
0	42.1	40.2	1.032964	1	1
41.25	26.2	86.9	0.732484	0	1
41.25	26.19647	86.9	4.654309	1	1
42.2	23.54	86.9	1.295994	1	1
42.2	23.5434	88.04	0.91503	0	1
42.2	23.54	88	0.703866	0	0
42.2	23.54	88	0.90524	0	0
0	15.55	22	1.211796	1	1
0	15.55248	22.15234	2.541637	1	1
0	19.03	9.073482	1.67194	1	0

as per ANN model has been represented, 0 denoting liquefaction and 1 denoting non-liquefaction.

The results obtained for the occurrence and non-occurrence of liquefaction by ANN method indicate dissimilar predictions as compared to Idriss and Boulanger method. The overall liquefaction susceptibility of all the sites when evaluated using

Table 3 Liquefaction susceptibility of site C as per Idriss and Boulanger and ANN method

LL	NMC	FC	FOS = CRR/CSR	Liquefaction susceptibility as per Idriss and Boulanger	Liquefaction susceptibility as ANN
0	36.64596	61.91	0.528431	0	1
0	36.64596	61.9	0.628616	0	1
0	36.64596	91.8	0.447802	0	1
42.5	29.20097	91.77	0.354429	0	1
42.5	29.20097	62.9	1.014233	1	0
35.5	22.61053	62.9	0.520122	0	0
35.5	22.61053	85.4	0.355602	0	0
38	26.43948	85.39	0.366889	0	1
38	26.43948	85.4	0.649645	0	1
38	26.43948	39.5	0.982301	0	0
0	23.3463	39.48	2.113911	1	0
0	23.3463	44.8	2.181767	1	0

Table 4 Liquefaction susceptibility of site D as per Idriss and Boulanger and ANN method

LL	NMC (%)	FC (%)	FOS = CRR/CSR	Liquefaction susceptibility as per Idriss and Boulanger	Liquefaction susceptibility as ANN
0	29.62038	65.8	0.429278	0	1
0	29.62038	65.8	0.356645	0	0
0	29.62038	68	0.36552	0	1
0	29.62038	68	0.420463	0	1
36.25	25.12799	68	1.105941	1	0
36.25	25.12799	72.6	0.864539	0	1
31	27.33813	72.6	0.844496	0	1
32.25	22.5834	96	0.580241	0	0
32.25	22.5834	66.2	0.774392	0	0
0	21.60584	66.2	0.848946	0	0
0	21.60584	40.9	1.35787	1	1
0	20.67682	40.9	1.119237	1	1

empirical method was 65%, whereas when the same sites were evaluated using ANN model, the liquefaction susceptibility is reduced to 40%.

Figure 2 represents the site liquefaction status when evaluated using Idriss and Boulanger [9] method, and it can be observed that the majority of soil layers lies close to zero indicating that the soil deposits are liquefiable. Figure 3 represents the site liquefaction status when evaluated using the trained ANN model. When the site

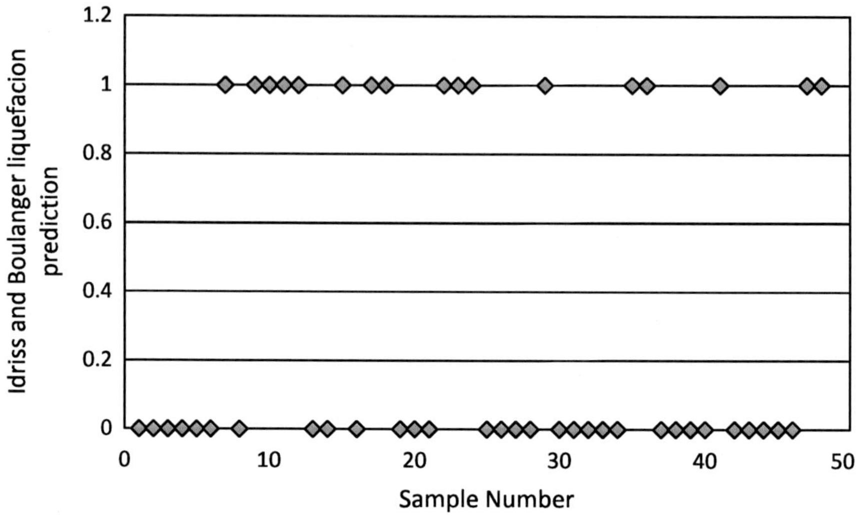

Fig. 2 Liquefaction prediction as per Idriss and Boulanger empirical method

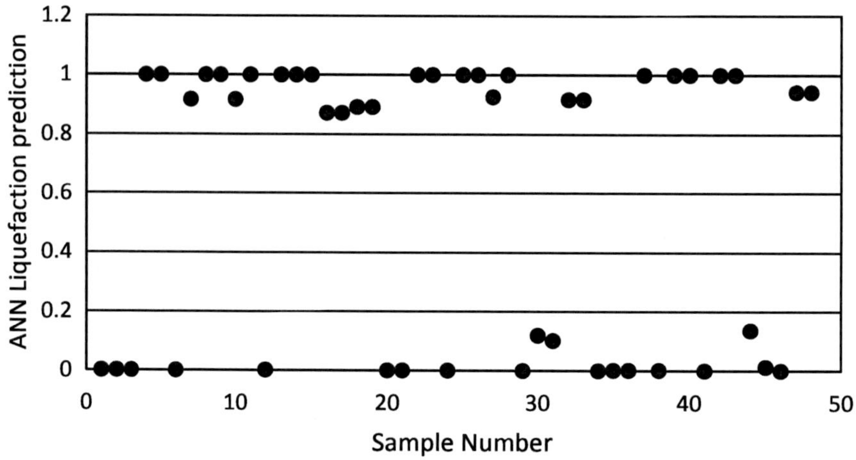

Fig. 3 Liquefaction prediction as per computer-based ANN model

was evaluated using ANN model, it was observed that the majority of soil layers lies close to one indicating that soil deposits are non-liquefiable. The difference in the results occurs due to the consideration of fine content as input parameters in evaluating liquefaction potential. The presence of fine content in the soil deposits is likely to increase the liquefaction resistance of a soil due to dilitative nature of clayey soil. It has been observed that an increment in the fine content causes a systematic decrease in the cyclic stress that is required to liquefy the soil deposits. Addition of

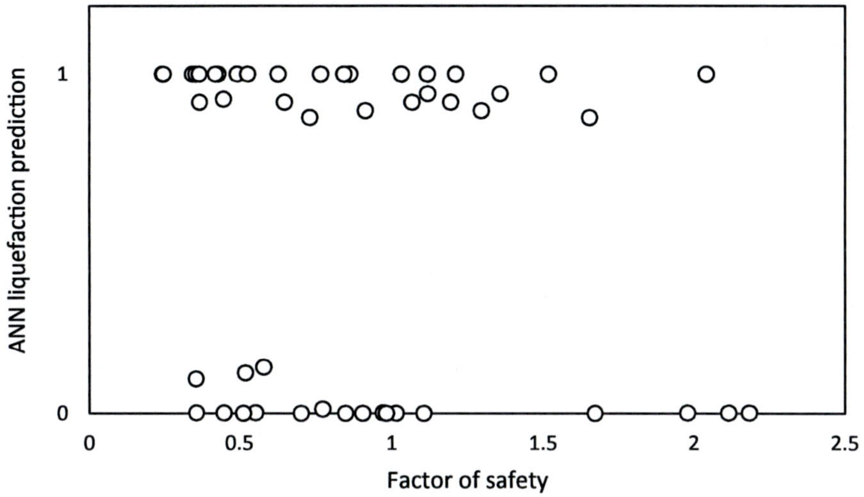

Fig. 4 FOS versus ANN liquefaction prediction for all the sites

fines in a soil leads to the filling up of void space up to a phase where the fines tend to displace the soil grains and starts to dominate the soil matrix; thus, the shear stress response becomes fines dominated.

Figure 4 depicts the relation between FOS evaluated using Idriss and Boulanger [9] method and liquefaction prediction by ANN model. The liquefaction prediction of the developed ANN model states that the soil deposits are non-liquefiable irrespective of the factor of safety of soil layers. Such contrast results are caused due to the consideration of plasticity and fine content properties of soil while evaluating liquefaction potential in computer-based approach. Equations (1), (2), and (4) mentioned above that are used evaluated liquefaction susceptibility use SPT *N* values to predict liquefaction, and this limits its predictability as observed in the literature. Figure 5 represents the relation between fine content and liquefaction susceptibility as per Idriss and Boulanger [9] empirical method and ANN model. It shows that for ANN model, soil layer having high percentage of fine content tends to be non-liquefiable indicating the influence of fine content of liquefaction prediction, whereas for the same values of fine content, soil layer tends to be liquefiable when computed using Idriss and Boulanger [9] method. From the abovementioned results, it can be observed that the ANN model developed concludes that the soil deposits of the proposed site are non-liquefiable. The computer-based approach can be considered as a crucial hazard mitigation technique as it eradicates most of the flaws of the conventional approaches, and at the same time, it is accurate and economic.

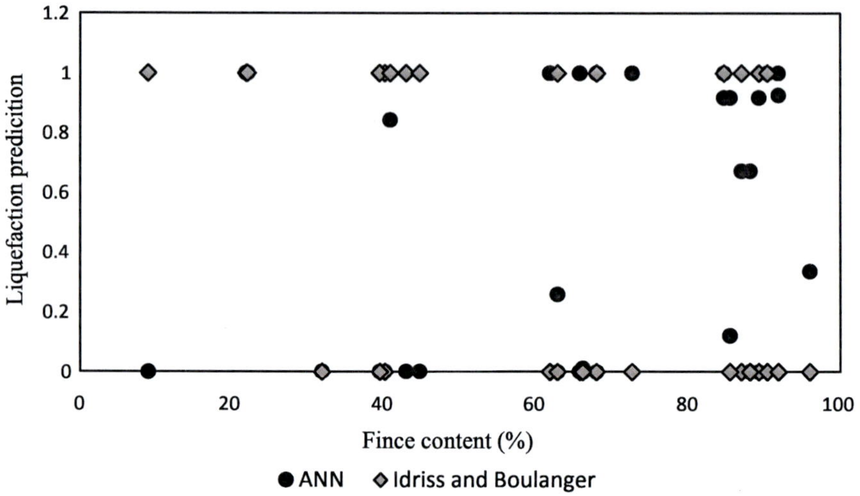

Fig. 5 Fine content versus liquefaction susceptibility as per Idriss and Boulanger and ANN model for all the sites

4 Conclusion

The importance of computer-aided liquefaction prediction in hazard mitigation has been well established based on the comparison between Idriss and Boulanger [9] method and developed ANN model. The artificial neural network has been developed using MATLAB for the assessment of liquefaction potential based on field and laboratory datasets. The results presented in the study following conclusions can be drawn:

1. The developed ANN model is a robust method in comparison with empirical method which consists of complex relationship between the soil and seismic parameters to evaluate liquefaction potential and thus raises the chances of uncertainty and errors.
2. Field and laboratory-based soil parameters may directly be used as input parameter for the developed ANN models, which is much simpler and responsive than the conventional methods to predict liquefaction potential.
3. The consideration of fine content as an input parameter in ANN model has significantly reduced the liquefaction potential of all the sites.
4. The results obtained from the ANN model states that the sites which pose the threat to undergo liquefaction when evaluated using Idriss and Boulanger [9] empirical method will fall in non-liquefiable zone, thus eliminating the long-term risk of life, property and environment.
5. Construction of any structure on liquefiable zone requires a great amount of capital and resources, so the developed ANN model reduces the liquefaction probability, thus contributing a huge saving of resource in the construction.

Therefore, the use of artificial intelligence for hazard mitigation can save us from incurring massive damages caused due to such hazards like liquefaction, and due to it cost efficiency and quick predictions, it should be categorized as a sustainable method for evaluating and predicting risk against any hazard.

Annexure

Liquefaction susceptibility as per Bray and Sancio [10] liquefaction criteria of soil in Adapazarı, Turkey, that liquefied in Kocaeli earthquake in 1999

Serial no.	Liquid limit (%)	Normal moisture content (%w)	Fine content (%)	Liquefaction susceptibility as per Bray and Sancio [10]	Liquefaction prediction as per Bray and Sancio [10]	Trained ANN model prediction
1	31	32	82	High	0	0
2	0	30	59	No	1	1
3	0	23	66	No	1	1
4	30	29	90	High	0	0
5	31	29	53	High	0	0
6	0	29	51	No	1	1
7	31	27	55	High	0	0
8	20	34	59	High	0	0
9	28	32	72	High	0	0
10	27	28	68	High	0	0
11	30	28	77	High	0	0
12	28	27	51	High	0	0
13	28	34	72	High	0	0
14	31	29	80	High	0	0
15	33	35	78	High	0	0
16	35	36	97	High	0	0
17	31	33	80	High	0	0
18	33	31	80	High	0	0
19	25	38	65	High	0	0
20	26	31	74	High	0	0
21	28	31	79	High	0	0
22	31	33	57	High	0	0
23	33	34	87	High	0	0
24	31	36	80	High	0	0

(continued)

(continued)

Serial no.	Liquid limit (%)	Normal moisture content (%w)	Fine content (%)	Liquefaction susceptibility as per Bray and Sancio [10]	Liquefaction prediction as per Bray and Sancio [10]	Trained ANN model prediction
25	22	30	77	High	0	0
26	0	27	58	No	1	1
27	32	34	86	High	0	0
28	32	32	99	High	0	0
29	0	28	88	No	1	1
30	0	22	54	No	1	1
31	0	34	71	No	1	1
32	19	22	59	High	0	0
33	0	34	67	No	1	1
34	0	37	76	No	1	1
35	25	29	54	High	0	0
36	25	31	69	High	0	0
37	32	30	96	High	0	0
38	30	30	82	High	0	0
39	33	30	58	High	0	0
40	29	32	78	High	0	0
41	30	33	88	High	0	0
42	0	30	81	No	1	1
43	0	28	93	No	1	1
44	30	28	87	High	0	0
45	38	37	94	High	0	0
46	25	29	54	High	0	0
47	28	37	82	High	0	0
48	29	28	82	High	0	0
49	34	39	85	High	0	0
50	24	32	64	High	0	0
51	28	27	77	High	0	0
52	0	26	65	No	1	1
53	25	32	67	High	0	0
54	28	29	92	High	0	0
55	29	28	93	High	0	0
56	34	38	85	High	0	0
57	26	35	97	High	0	0
58	33	40	95	High	0	0

(continued)

(continued)

Serial no.	Liquid limit (%)	Normal moisture content (%w)	Fine content (%)	Liquefaction susceptibility as per Bray and Sancio [10]	Liquefaction prediction as per Bray and Sancio [10]	Trained ANN model prediction
59	29	35	69	High	0	0
60	25	28	67	High	0	0
61	0	32	65	No	0	0
62	37	37	74	High	0	0
63	0	32	77	No	1	1
64	26	24	75	High	0	0
65	39	41	98	High	0	0
66	33	33	85	High	0	0
67	0	30	56	No	1	1
68	36	37	93	High	0	0
69	30	24	81	No	1	1
70	25	33	75	High	0	0
71	35	30	88	High	0	0
72	29	30	93	High	0	0
73	34	29	87	High	0	0
74	34	34	91	High	0	0
75	26	27	70	High	0	0
76	30	33	94	High	0	0
77	31	35	98	High	0	0
78	35	37	93	High	0	0
79	33	37	93	High	0	0
80	38	42	97	High	0	0
81	35	31	86	High	0	0
82	26	27	86	High	0	0
83	27	30	63	High	0	0
84	32	35	90	High	0	0
85	35	37	99	High	0	0
86	23	28	70	High	0	0
87	30	31	87	High	0	0
88	31	32	83	High	0	0
89	30	32	88	High	0	0
90	31	32	89	High	0	0
91	29	32	74	High	0	0
92	36	33	94	High	0	0

(continued)

(continued)

Serial no.	Liquid limit (%)	Normal moisture content (%w)	Fine content (%)	Liquefaction susceptibility as per Bray and Sancio [10]	Liquefaction prediction as per Bray and Sancio [10]	Trained ANN model prediction
93	33	37	84	High	0	0
94	30	28	69	High	0	0
95	23	30	74	High	0	0
96	31	26	73	Moderate	0	0
97	34	37	96	High	0	0
98	34	35	92	High	0	0
99	30	32	96	High	0	0
100	30	25	84	Moderate	0	0
101	32	31	91	High	0	0
102	45	39	86	High	0	0
103	25	25	64	High	0	0
104	30	36	86	High	0	0
105	30	32	93	High	0	0
106	28	32	87	High	0	0
107	25	32	58	High	0	0
108	22	26	93	High	0	0
109	31	38	83	High	0	0
110	34	34	54	High	0	0
111	25	23	58	High	0	0
112	31	32	73	High	0	0
113	33	35	94	High	0	0
114	29	24	65	Moderate	0	0
115	28	23	83	Moderate	0	0
116	34	35	97	High	0	0
117	31	34	92	High	0	0
118	31	34	92	High	0	0
119	35	38	93	High	0	0
120	33	33	84	High	0	0
121	33	29	87	High	0	0
122	31	37	89	High	0	0
123	38	29	97	No	1	1
124	36	28	97	No	1	1
125	31	35	90	High	0	0
126	28	31	84	High	0	0

(continued)

(continued)

Serial no.	Liquid limit (%)	Normal moisture content (%w)	Fine content (%)	Liquefaction susceptibility as per Bray and Sancio [10]	Liquefaction prediction as per Bray and Sancio [10]	Trained ANN model prediction
127	37	39	99	High	0	0
128	24	29	67	High	0	0
129	26	30	70	High	0	0
130	29	32	97	High	0	0

References

1. Agrawal G, Chameau JA, Bourdeau PL (1997) Assessing the liquefaction susceptibility at a site based on information from penetration testing In: Kartam N, Flood I, Garrett JH (eds) Artificial neural networks for civil engineers: fundamentals and applications. New York, pp 185–214
2. Andrews DCA, Martin GR (2000) Criteria for liquefaction of silty soils. In: Proceedings of 12th world conference on earthquake engineering, Auckland, New Zealand
3. Polito C (2001) Plasticity based liquefaction criteria. In: Proceedings of of the 4th international conference on recent advances in geotechnical earthquake engineering and soil dynamics. Paper No. 1.33
4. Chern SG, Hu RF, Chang YJ, Tsai IF (2002) Fuzz-ART neural networks for predicting CHI-CHI earthquake induced liquefaction in Yuan-Lin area. J Mar Sci Technol 10(1):21–32
5. Farrokhzad F, Choobbasti AJ, Barari A (2010) Artificial neural network model for prediction of liquefaction potential in soil deposits. In: International conferences on recent advances in geotechnical earthquake engineering and soil dynamics, p 4
6. Goh ATC (1996) Neural-network modeling of CPT seismic liquefaction data. J Geotech Eng ASCE 122(1):70–73
7. Goh ATC (2002) Probabilistic neural network for evaluating seismic liquefaction potential. Can Geotech J 39:219–232
8. Gratchev I, Sassa K, Fukuoka H (2006) How reliable is the plasticity index for estimating the liquefaction potential of clayey sands? J Geotech Geo-environ Eng 132(1):124–127
9. Idriss IM, Boulanger R (2006) Semi-empirical procedures for evaluating liquefaction potential during earthquakes. J Soil Dyn Earthquake Eng 26(2):115–130
10. Bray JD, Sancio RB (2006) Assessment of the liquefaction susceptibility of fine grained soil. J Geotech Eng 132(9):1165–1177
11. Bray JD, Sancio RB, Riemer M, Durgunoglu HT (2004) Liquefaction susceptibility of fine-grained soils. In: International conference on geotechnical earthquake engineering
12. Juang CH, Chen CJ (1999) CPT-based liquefaction evaluation using artificial neural networks. Computer-aided and infrastructure engineering. Blackwell Publishers, Oxford, pp 221–229
13. Marto A, Tan CS, Akhtar AM, Ung SW, Lim MY (2015) Effect of plasticity on liquefaction susceptibility of sand-fines mixtures. Appl Mech Mater 773-774:1407–1411
14. More JJ (1977) The Levenberg-Marquardt algorithm: implementation and theory. In: Watson GA (ed) Numerical analysis. Springer, Heidelberg, pp 105–116
15. Mughieda O, Bani-Hani K, Safieh B (2009) Liquefaction assessment by artificial neural networks based on CPT. Int J Geotech Eng 3(2):289–302. https://doi.org/10.3328/ijge.2009.03.02.289-302

16. Samuiand P, Sitharam TG (2011) Machine learning modelling for predicting soil liquefaction susceptibility. Nat. Hazards Earth Syst Sci 11:1–9
17. Paydar NA, Ahmadi MM (2016) Effect of fines type and content of sand on correlation between shear wave velocity and liquefaction resistance. Geotech Geol Eng 34(6):1857–1876
18. Prakash S, Sandoval JA (1992) Liquefaction of low plasticity silts. J Soil Dyn Earthquake Eng 71(7):373–397
19. Sandoval J (1989) Liquefaction and settlement characteristics of siltsoils. Ph.D. thesis, University of Missouri–Rolla, Mo
20. Seed HB, Idriss IM (1971) Simplified procedure for evaluating soil liquefaction potential. J Soil Mech Found Div. ASCE 97, SM8, pp 1249–1274
21. Seed HB, Idriss IM (1982) Ground motions and soil liquefaction during earthquakes. Earthquake Engineering Research Institute, Berkeley, Calif
22. Seed RB, Cetin KO, Moss RES, Kammerer AM, Wu J, Pestana JM, Riemer MF, Sancio RB, Bray JD, Kayen RE, Faris A (2003) Recent advances in soil liquefaction engineering: a unified and consistent framework, EERC-2003–06. Earthquake Engineering Research Institute, Berkeley, California
23. Seed RB, Cetin KO, Moss RES (2001) Recent advances in soil liquefaction hazard assessment. In: 15th ICSMGEG, TC4 satellite conference on lessons learned from recent strong earthquakes, Istanbul, Turkey
24. Vijay Kumar, Kumar Venkatesh, Tiwari RP, Yeetendra Kumar (2012) Application of ANN to predict liquefaction potential. Int J Comput Eng Res. ISSN: 2250–3005
25. Wang W (1979) Some findings in soil liquefaction. Report Water Conservancy and Hydroelectric Power Scientific Research Institute, Beijing, China, pp 1–17
26. Wang J, Rahman MS (1999) A neural network model for liquefaction-induced horizontal ground displacement. Soil Dyn Earthq 18(8):555–568
27. Omer M, Bani-Hani K, Safieh B (2009) Liquefaction assessment by artificial neural networks based on CPT. Int J Geotech Eng 3(2):289–302. https://doi.org/10.3328/IJGE.2009.03.02.289-302
28. Kumar V, Kumar V, Tiwari RP, Kumar Y (2012) Application of ann to predict liquefaction potential. Int J Comput Eng Res ISSN: 2250–3005

Design and Cost Analysis for the Rehabilitation of a Flexible Pavement Through the Study on Traffic Data

Sandeep Kaur and Prabhjot Singh

1 Introduction

In the world, India is having one of the prevalent road networks. Development of road network is considered as one of the most important infrastructure development for economic growth. India has a road network of over 5,472,144 km as on 31 March 2015, the second largest road network in the world [2]. India had completed and placed in use over 25,600 km of recently built 4- or 6-lane highways connecting many of its major manufacturing centres, commercial and cultural centres [2]. With the increase in the road traffic, the design service volume of two-lane road is exceeded due to which there is a need to update the road to a four-lane road. The widening of road also helps in improving road safety and making driving a hassle-free experience. The classification of pavements can be done into following types:

(i) Flexible pavement, (ii) Rigid pavement, (iii) Composite pavement

- **Flexible pavement**

The stress due to wheel loads over the flexible pavements is transferred at the bottom layers all the way throughout the granular structure due to grain-to-grain load transmittance under the point of contact. With increase in depth of the pavement, the stresses get reduced as the wheel load is scattered to a larger area. Because of this pro of stress scattering attribute, flexible pavement generally have various layers. Therefore, a concept of layered system is used in flexible pavement design. So, flexible roads are built in courses, and in order to withstand the maximum compressional

S. Kaur
Lemodular Design Consultants, Ludhiana, Punjab, India
e-mail: sandeeppannu2409@gmail.com

P. Singh (✉)
Civil Engineering Department, Guru Nanak Dev Engineering College, Ludhiana, Punjab, India
e-mail: prabhjotpannu58@gmail.com

H. Singh et al. (eds.), *Sustainable Development Through Engineering Innovations*, Lecture Notes in Civil Engineering 113,
https://doi.org/10.1007/978-981-15-9554-7_64

Fig. 1 Cross section of flexible pavement (www.googleimages.com)

stresses as well as wear and tear of the top surface, there must be a finest excellence in the top course. Low-quality materials can be used in other courses as these courses will face smaller amount of stresses [8] (Fig. 1).

2 Various Components of Pavement

A. Subgrade

The bottom most layer of the pavement just above the natural ground surface is a subgrade, as it handles the pavement from underneath, so it is an immanent part of flexible roads. While designing the flexible pavements, the properties of subgrade material are quite vital. The desirable properties of soil (subgrade) should be incompressibility, proper drainage, stability and ease of compaction.

B. Subbase

It is an optional layer between subgrade and base course and it normally consists of broken stones or aggregates mixed within soil. This layer is of different types such as stabilized soil subbase and water bound macadam subbase.

C. Base course

Beneath the surface course, a layer of different materials is laid that is known as base course. It helps in transmitting the forces or stress to the bottom layers. Base course can be following types: granular base, macadam base and cementitious base.

D. Binder course

This course is a suitable mixture of aggregates and bitumen material as per the specifications of Indian standards. This layer consists of the following materials:

- **Bitumen**: It should be paving bitumen of penetration grade S65 or A65 (60/70) as per IS stipulation [6]. On the contrary if above grade is unavailable then, by the consent of site engineer S90 (80/100) grade bitumen might be used [6].
- **Coarse aggregate**: These aggregates should consist of crushed gravel/shingle, crushed stone or other stones. The aggregate shall satisfy the different physical requirement set forth as per IS:2386-Part 1, 3, 4, and 5 [5].

- **Fine aggregates**: These are the naturally occurring or crushed aggregates or a mixture of these two, passing through 2.36 mm and retaining on 75 micron sieve [5].

3 Traffic Data Analysis

It is extremely essential to provide a valuable input for planning, financing, designing and operation of a project pavement, which can be done only through a precise estimation of traffic, i.e. going to use that pavement. Upcoming traffic inference is necessary and it is vital to gain the knowledge of traffic conditions or attributes along the major routes that are under the influence zone of study passageway, all along with the central project road [9]. Therefore, for knowing the current traffic conditions and its attributes on SH-11 Ludhiana to Malerkotla road and other major routes that are under its influence area, a thorough traffic surveys has to be conducted. For estimating the design traffic, following information is desired:

1. Initial traffic after construction in terms of number of commercial vehicles per day (CVPD).
2. During design life the rate of traffic growth in % age.
3. No. of years of design life.
4. Vehicle damage factor (VDF).
5. On the carriageway, commercial traffic distribution (Table 1).

So according to IRC: 73 [6], a four-lane road is required for the above calculated data.

4 Methodology: Centreline Readings

This project has been conducted on State Highway (SH)-11, from Ludhiana to Malerkotla for 1 km stretch of road. The following natural reduced levels of ground along the central line of the road longitudinally are noted with the help of surveying and the required formation level of the road are also shown in Table 2:

A. **Longitudinal section readings and graph** (Fig. 2)
B. **Cross section readings and graphs** (Figs. 3, 4, 5, 6, 7, 8, 9, and 10; Tables 3, 4, 5, 6, 7, 8, 9, and 10),

5 Earth Work Calculations

See Table 11.

Table 1 Traffic data analysis

Day	Car/jeep	2 Wheeler	Mini truck		Truck		Agricultural tractor		Bus	Cycle	Cycle rickshaw	Animal cart	Total
			Laden	Unladen	Laden	Unladen	Laden	Unladen					
1	7660	5326	536	369	886	287	80	124	214	500	78	42	
2	9066	6476	725	398	1058	247	137	130	236	449	115	47	
3	11,065	6323	891	652	1715	258	258	253	349	447	144	54	
4	8569	5799	507	451	865	332	129	93	238	441	139	42	
5	9115	6476	672	451	1014	296	188	76	236	337	173	46	
6	10,214	5665	855	665	1670	283	309	214	332	313	186	37	
7	8350	5557	444	483	815	376	170	51	223	273	186	42	
Total	64,039	41,622	4630	3469	8023	2079	1271	941	1828	2760	1021	310	
VPD	9148	5946	661	495	1146	297	182	134	261	394	146	44	
PCU(in terms of standard vehicle)	1	0.5	1.5	1.5	3	3	1.5	1.5	3	0.5	2	5	
CVPD	9148	2973	991	742	3438	891	273	201	783	197	292	220	20,149

Table 2 Longitudinal levels at particular intervals

Distance (m)	RL of ground (m)	Formation level (m)
0	98.625	99.500
150	98.975	99.250
300	99.045	99.000
450	99.110	98.750
600	99.015	98.500
750	98.560	98.250
900	98.250	98.000
1000	98.200	97.833

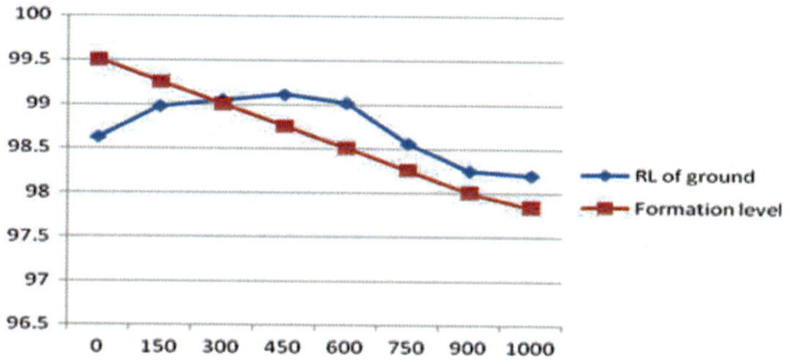

Fig. 2 Longitudinal section of road

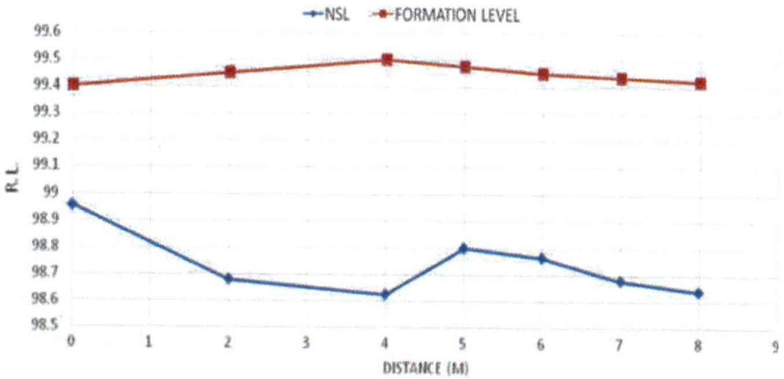

Fig. 3 Cross section at 0 m

6 Soil Exploration

A. Optimum Moisture Content test

The OMC is used to find the connection among dry density and moisture content of soil at particular compaction value.

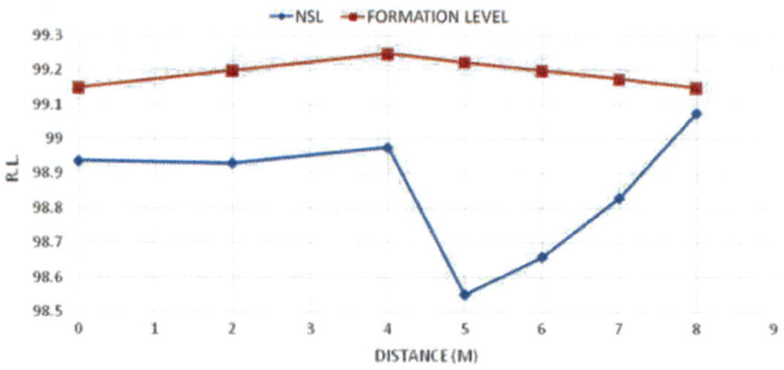

Fig. 4 Cross section at 150 m

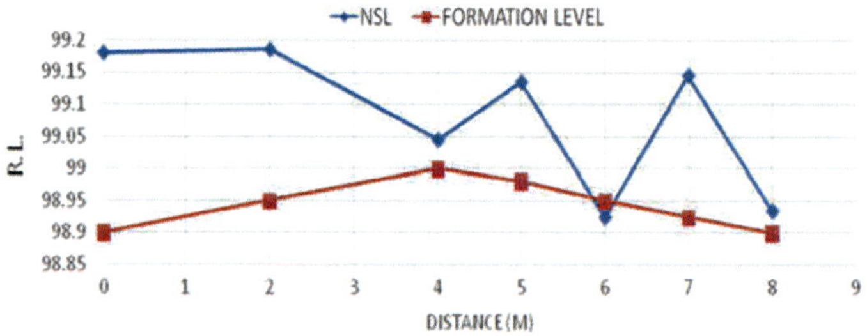

Fig. 5 Cross section at 300 m

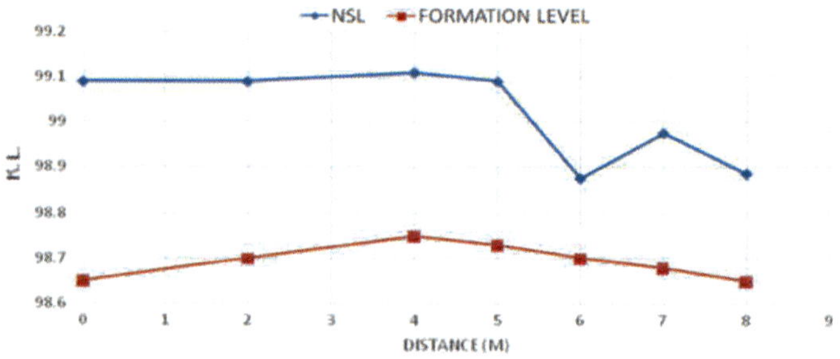

Fig. 6 Cross section at 450 m

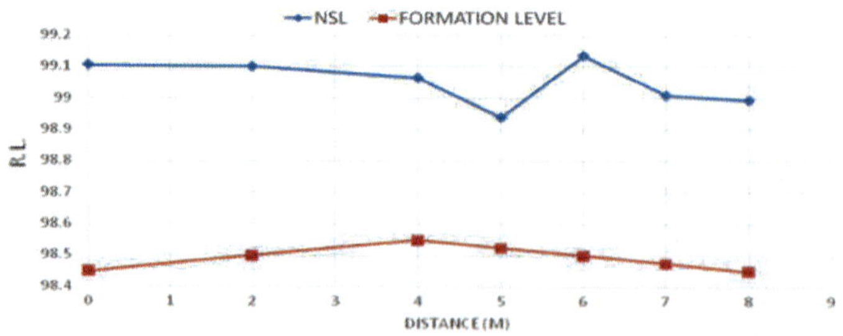

Fig. 7 Cross section at 600 m

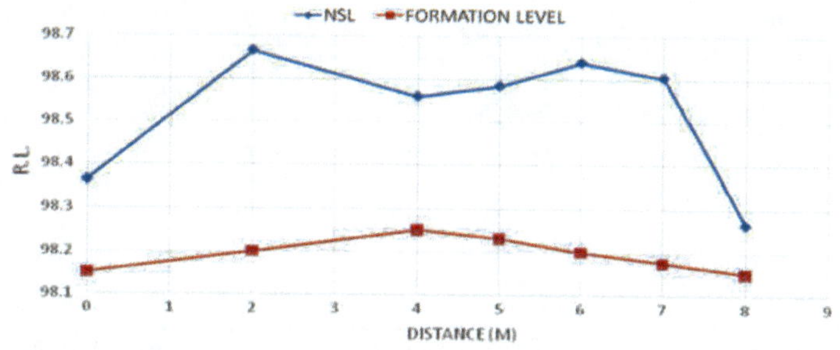

Fig. 8 Cross section at 750 m

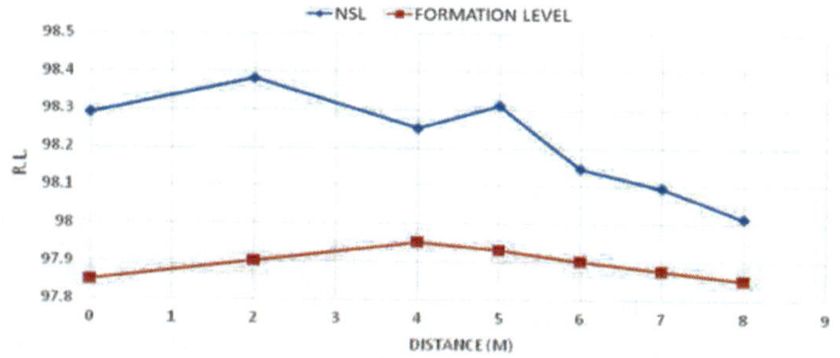

Fig. 9 Cross section at 900 m

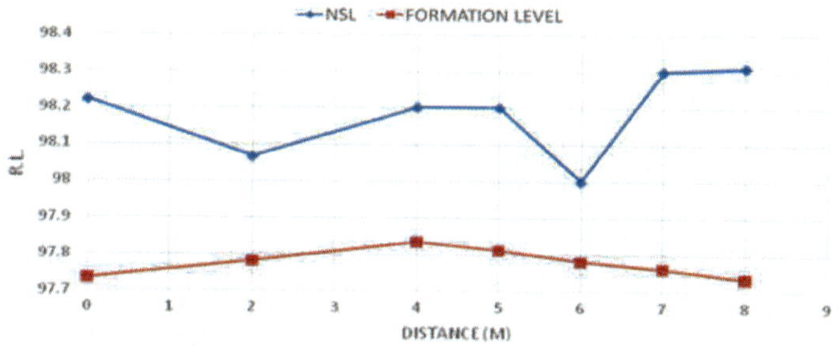

Fig. 10 Cross section at 1000 m

Table 3 RLs at distance 0 m

Distance	0	2	4	5	6	7	8
RL (g)	98.97	98.68	98.625	98.8	98.76	98.69	98.64
RL (f)	99.4	99.45	99.5	99.475	99.45	99.425	99.4

Table 4 RLs at distance 150 m

Distance	0	2	4	5	6	7	8
RL (g)	98.936	98.93	98.975	98.55	98.66	98.83	99.075
RL (f)	99.15	99.2	99.25	99.225	99.2	99.175	99.15

Table 5 RLs at distance 300 m

Distance	0	2	4	5	6	7	8
RL (g)	99.18	99.185	99.045	99.135	98.925	98.145	98.935
RL (f)	98.9	98.95	99	98.975	98.95	98.925	98.9

Table 6 RLs at distance 450 m

Distance	0	2	4	5	6	7	8
RL (g)	99.09	99.09	99.11	99.095	98.875	98.975	98.885
RL (f)	98.65	98.7	98.75	98.725	98.7	98.675	98.65

Table 7 RLs at distance 600 m

Distance	0	2	4	5	6	7	8
RL (g)	99.12	99.01	99.015	99.115	9.02	99.105	99.125
RL (f)	98.4	98.45	98.5	98.475	98.45	98.425	98.4

Table 8 RLs at distance 750 m

Distance	0	2	4	5	6	7	8
RL (g)	98.365	98.67	98.56	98.58	98.64	98.605	98.265
RL (f)	98.5	98.2	98.25	98.225	98.2	98.175	98.15

1. *OMC at the surface*

   Optimum moisture content = 15.2
   Maximum dry density = 1.733

2. *OMC at 500 mm depth*

   Optimum moisture content = 14

Table 9 RLs at distance 900 m

Distance	0	2	4	5	6	7	8
RL (g)	98.29	98.38	98.25	98.31	98.145	98.095	98.015
RL (f)	97.9	97.95	98	97.975	97.95	97.925	97.9

Table 10 RLs at distance 1000 m

Distance	0	2	4	5	6	7	8
RL (g)	98.22	98.07	98.2	98.2	98	98.3	98.31
RL (f)	97.973	97.8	97.83	97.805	97.78	97.755	97.773

Table 11 Earth work calculations

Distance in metres	Mean depth (d)	Central area (Bd)	Area of sides ($0.5\ sd^2$)	Total area ($Bd + 0.5\ sd^2$)	Length (L)	Banking ($Bd + 0.5\ sd^2$) L	Cutting ($Bd + 0.5\ sd^2$) L
0	0.714	7.14	0.51	7.65	150	1147.5	
150	0.342	3.42	0.11	3.53	150	529.5	
300	−0.135	−1.35	0.018	1.388	150		208.2
450	−0.322	−3.22	0.103	3.323	150		498.45
600	−0.556	−5.56	0.31	5.87	150		880.5
750	−0.333	−3.33	0.11	3.44	150		516
900	−0.32	−3.2	0.102	3.302	150		495.3
1000	−0.409	−4.09	0.167	4.257	100		425.7

Total excavation work in cutting = 3024.15 m^3
Total excavation work in banking = 1677.0 m^3

Maximum dry density = 1.8

B. **California Bearing Ratio test calculations:** [1]

$$\text{CBR value at 2.5 mm penetration} = \frac{\text{Load at 2.5 mm penetration}}{\text{Standard load at 2.5 mm penetration i.e. 1370 kg}} * 100$$

$$\text{CBR value at 5 mm penetration} = \frac{\text{Load at 5 mm penetration}}{\text{Standard load at 5 mm penetration i.e. 2055 kg}} * 100$$

CBR value is taken at 2.5 mm penetration if it is more than the value at 5 mm, but if the CBR value of 5 mm penetration is more than 2.5 mm then we have to repeat the test and if again CBR value at 5 mm is still more than we have to use that value [1].

- **Observations** (Fig. 11; Tables 12 and 13)
  - Effective CBR of compacted borrow material 500 mm thick = 11.5%
  - Effective CBR of compacted subgrade at surface = 6.81%
  - From graph, effective CBR of subgrade = 10%

7 Design of Flexible Pavement by IRC Method

The Indian Road Congress has established various standards and codes for the design and construction of pavements in India such as IRC: 37-2012 and IRC-SP-84.

The flexible pavements are designed by using the two input parameters:

1. Design traffic in terms of cumulative number of standard axles
2. CBR value of subgrade

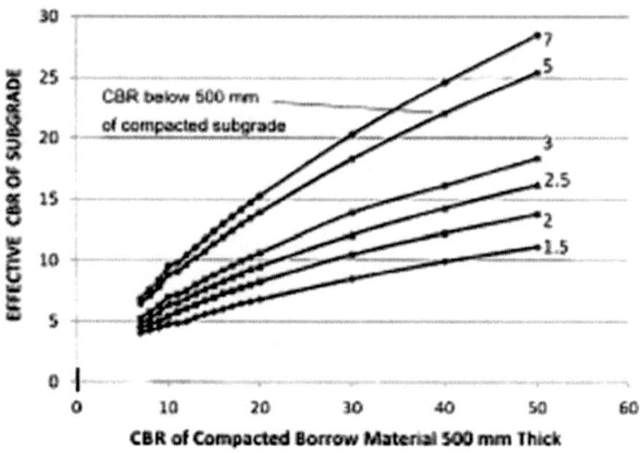

Fig. 11 Effective CBR graph of subgrade (IRC 37) [4]

Table 12 CBR test of compacted borrow material

Penetration (mm)	Standard load (kg)	Dial gauge divisions	Load (kg)
0		0	0
0.5		3.59	35.9
1		6.65	66.5
1.5		9.5	95
2		12.5	125
2.5	1370	15.75	158
3		17.1	171
5	2055	20.8	208
7.5		22.1	221
10		22.5	225
12.5		24.2	242

Table 13 CBR below 500 mm of compacted subgrade

Penetration (mm)	Standard load (kg)	Dial gauge divisions	Load (kg)
0		0	0
0.5		3.05	30.5
1		6.57	65.7
1.5		8.2	82
2		8.85	88.5
2.5	1370	9.34	93.4
3		9.53	95.3
5	2055	10.3	103
7.5		10.63	106
10		10.9	109
12.5		11	110

The design traffic is calculated by using the following formulae [1]: $N = \frac{[365*((1+r)n-1)*A*D*F]}{r}$

where N = standard axles cumulative no. to be catered for design in terms of msa, r = rate of growth commercial vehicles annually, F = vehicle damage factor, A = annual amount of traffic in the construction completion year in terms of the no. of the cvpd, n = design life (in years), D = lane distribution factor.

- **Design calculations**

  Design the pavement for upgradation to four lanes from two-lane road with the following data:

  - Dual carriage way with divider, i.e. four-lane divided carriageway
  - Traffic growth rate per annum, $r = 5\%$
  - Design life, $n = 15$ years
  - Lane distribution factor, $D = 0.75$
  - VDF, $F = 4.5$
  - Years in completion, $x = 1$ year
  - Average daily traffic (ADT) for commercial vehicles in both direction = 7118
  - Average daily traffic (ADT) in one direction = 3559

$$\begin{aligned} \mathbf{A} &= \mathbf{P\ (1 + r)^x} \\ &= 3559 * (1 + .05)^1 \\ &= 3737 \text{ cvpd} \end{aligned}$$

Where

A = initial traffic after construction in terms of cvpd, x = time interval in yrs between the last count and the construction completion year, P = no. of commercial vehicles as per last count, r = rate of growth of commercial vehicles annually.

$$\begin{aligned} \text{Now, standard axles no.}:\ \mathrm{N} &= \frac{(365 * 3737 * \{(1 + 0.05)15 - 1\} * 0.75 * 4.5)}{0.5} \\ &= 99 \times 10^6 \text{or 99 million standard axle (msa)} \end{aligned}$$

Now, according to effective CBR value as 10% and 99 msa thickness of the pavement is taken from PLATE 7 of IRC: 37-2012 (Figs. 12, 13)

- Total pavement thickness = 610 mm
- Thickness of granular subbase = 200 mm
- Thickness of granular base = 250 mm
- Thickness of dense bitumen macadam = 110 mm
- Thickness of bitumen concrete = 50 mm
- Width of shoulder for built-up area = 2 m

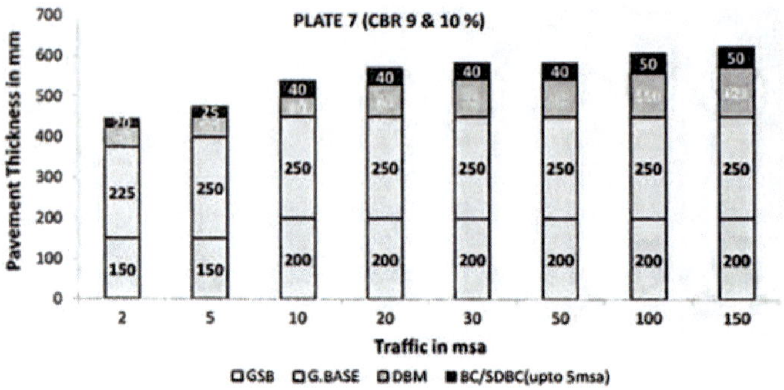

Fig. 12 Plate bar graph from IRC: 37-2012 [4]

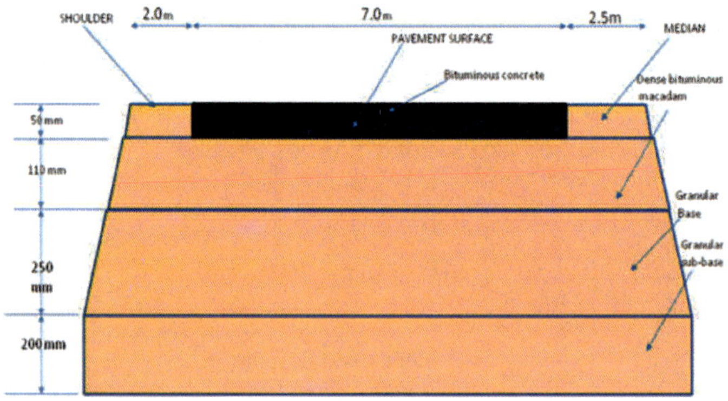

Fig. 13 Designed cross section of flexible pavement

8 Cost Analysis

Total length of road = 1.0 km

Metalled width of road = 7 m

Formation width of road = 11.5 m

Bitumen Present Rate (VG 10 Grade) = Rs. 42,020/metric ton (Table 14).

9 Conclusion

The flexible road is designed on the basis of CBR method according to IRC 37-2001 recommendations; as it is the utmost cost-effective and suitable method for small traffic amounts [7]. This method is much justified than other's because it depends upon material's strength parameters [3]. The top course thickness will be less for

Table 14 Cost analysis of flexible pavement

S. No	Description of item	No.	L (m)	B (m)	D/H (m)	Quantity (m^3)	Rate ($/m^3$)	Amount (Rupees)
1	2	3	4	5	6	7	8	9
1	Earth work from any lead including compaction and preparation of subgrade including loosening, leveling of earth 225 mm thick top layer, rough dressing of soil, final dressing of earth to give level, camber, watering, rolling with road roller, compacting the bed to achieve maximum dry density [As per C.S.R 24.5(A)]							
	Below the GSB	1	1000	11.5	0.225	2588		
					Total	2588	385.74	998,295.12
2	Providing and fixing brick on end edging as per CSR 24.15							
		2	1000	–	–	2000		
					TOTAL	2000	60.8	121,600
3	Granular subbase(coarse graded material grading 2)							
	Construction of granular sub-base by providing close graded material grading -II, spreading in uniform layers with motor grader on prepared surface, mixing by mix in place method with vibrator at OMC, and compacting with vibratory roller to achieve the desired density, complete as per technical clause 401 of MORT&H specifications							
		1	1000	11.5	0.2	2300		
					TOTAL	2300	1460.7	3,359,610
4	Water bound Macadam Grading II							
	Providing, laying, spreading and compacting stone aggregates of Grading II to water bound macadam specification including spreading in uniform thickness, hand packing, rolling with smooth wheeled roller 8–10 tones in stages to proper grade and camber, applying and brooming Water Bound Macadam Grading II requisite type of screening/binding Materials to fill up the interstices of coarse aggregate, watering and compacting to the required density complete as per technical clause 404 of MORT&H specifications							
		1	1000	7	0.25	1750		
					Total	1750	1741.8	3,048,150
5	Providing and laying tack coat for D.B.M. with bitumen @ 0.25 kg per sq m as per CSR 24.18 a							
		1	1000	7	–	7000		
					Total	7000	18.73	131,110

(continued)

Table 14 (continued)

S. No	Description of item	No.	L (m)	B (m)	D/H (m)	Quantity (m^3)	Rate ($/m^3$)	Amount (Rupees)
1	2	3	4	5	6	7	8	9
6	Providing and laying Dense Bituminous Macadam 110 mm thick using crushed aggregates of specified grading, premixed with bituminous binder G-30 @ 4%, carriage of mixed material to site of work, level and alignment, rolling with smooth wheeled, vibratory and tandem rollers to achieve the desired compaction complete as per technical clause 507 of MORT&H specifications							
		1	1000	7	0.11	770		
					Total	770	2970	2,286,900
7	Providing and laying tack coat for B.C. with bitumen @0.40 kg per sq m as per CSR 24.18b							
		1	1000	7	–	7000		
					Total	7000	20.79	145,530
8	Providing and laying Bituminous concrete 50 mm thick using crushed aggregates of specified grading, premixed with bituminous binder G-30 @ 5.5%, carriage of mixed material to site of work, level and alignment, rolling with smooth wheeled, vibratory and tandem rollers to achieve the desired compaction complete as per technical clause 509 of MORT&H specifications							
		1	1000	7	0.05	350		
					Total	350	4200	1,470,000
	Total amount							11,561,195.12
	Contingency charges @1.5%							173,417.93
	Grand total							11,734,613.05
	Amount in Lakhs							117.35 Lakh

higher value of CBR and this leads to more economical and sustainable roads as material consumptions will be less and vice versa. The subgrade material used is a black cotton soil. Construction cost of flexible roads is much less but its life span is not too long. So, it needs regular maintenance and its repairs cost could be more. So we need improvements in this regard.

Moreover, this study can be carried forward by using pozzolanic materials for stabilizing the diverse kinds of soils for obtaining better CBR value as well as we can take different soaking circumstances of soil w.r.t. time for obtaining more sustainable designs. Also, the design analysis can be done with some other suitable methods for betterment in the sustainable design and costing of flexible pavement.

Acknowledgements The authors are grateful to Dr. Prashant Garg (Associate Prof. at CE Department, GNDEC, Ludhiana) for their valuable guidance and assistance. Also, the authors appreciate the support from the undergraduate students in this project. We also acknowledge R&D department of Punjab Technical University, Jalandhar, CE departments of Guru Nanak Dev Engineering College and Gulzar Group of Institutes, Ludhiana.

References

1. Arora KR (2003) Soil mechanics and foundation engineering. Standard Publishers Distributors, Delhi
2. Chakroborty P, Das A (2017) Principles of transportation engineering, 2nd edn. PHI Learning Private Limited
3. Choudhary DK, Joshi YP (2014) A detailed study of CBR method for flexible pavement design. Int J Eng Res Appl 4(6) (Version 5):239–253. ISSN: 2248-9622
4. IRC: 37-2001 (2001) Code of guideline for the design of flexible pavement. Indian Road Congress, New Delhi
5. IS: 2386- 1963 Methods of test for aggregates for concrete. Bureau of Indian Standards, New Delhi
6. IS 73: 2013 Paving bitumen—specifications. Bureau of Indian Standards, New Delhi
7. Jain S, Joshi YP, Golia SS (2013) Design of rigid and flexible pavements by various methods and their cost analysis of each method. Int J Eng Res Appl 3(5):119–123
8. Khanna SK, Justo CEG (1993) Highway engineering, 7th edn. New Chand and Bros, New Delhi
9. Singh J, Singh P, Kaur S (2017) Upgradation of two lane road to four lane road. IJCRT 5(3). ISSN: 2320-2882

Relationship Between Compressive Strength and Split Tensile Strength for Sustainable Concrete—A Case Study

Anil Kumar Nanda and Jaspal Singh

1 Introduction

Building industry, a fast-growing sector, is one of the key areas of infrastructure development. To cater the requirements of building materials, we depend heavily on natural resources. As there is a limit up to which natural resources can be exploited, so it becomes imperative to find alternate sustainable substitute materials because there is a limit up to which natural resources can be exploited. By use of sustainable industrial wastes, we are not only conserving natural resources but also provide a solution to their disposal [1]. Sustainable materials like fly ash and silica fume are the promising industrial wastes which can be easily used in construction. With the increase in the number of coal-based thermal power plants in India, generation of fly ash has reached enormous proportions. Fly ash is an industrial by-product resulting from the combustion of pulverized coal. In India, about 100 million tonnes of fly ash is accumulated every year which is generated as waste from thermal plants. In spite of concerted efforts at national scale, only a very small fraction (around 6%) of the fly ash is put to use in India. Fly ash has many applications in the building industry. It is a better solution as a partial replacement of cement in concrete. Silica fume is also a waste by-product from the silicon metal and ferrosilicon alloy industries. The chief problems in using this material are associated with its extreme fineness and high water requirement when mixed with Portland cement. However, the availability of superplasticizers has opened up new possibilities for its use. The experimental results reveal that silica fume addition enhances strength development in geopolymer [2].

A. K. Nanda (✉)
Civil Engineering Department, RIMT University Mandi Gobindgarh, Mandi Gobindgarh, India
e-mail: aknanda4ap@gmail.com

J. Singh
Department of Civil Engineering, Punjab Agricultural University, Ludhiana, India
e-mail: jaspalsingh@pau.edu

H. Singh et al. (eds.), *Sustainable Development Through Engineering Innovations*, Lecture Notes in Civil Engineering 113,
https://doi.org/10.1007/978-981-15-9554-7_65

The experiment results indicate that concrete modification with silica fume can both increase and decrease the quality of bond conditions [3]. It is in this context that effort has been made to study the effect of addition of superplasticizer in addition to fly ash and silica fume on compressive strength of concrete.

It has been established that the waste materials like fly ash and silica fume can be converted into meaningful wealth as new construction materials having considerable cementitious properties which result in major saving of raw material of cement. The usage of waste materials such as fly ash and silica fume in construction sector is very important in terms of limiting the environmental pollution, cost reduction and slowing down the use of natural resources in the world.

2 Materials and Methods

Tests were conducted on concrete cube using fly ash, silica fume and superplasticizer to determine the effect of fly ash and silica fume on compressive and split tensile strength of concrete. The experimental programme consists of casting, curing and testing of silica fume and fly ash concrete specimens at different ages. The compressive and split tensile strength was investigated at the ages of 3, 7 and 28 days. Workability of various mixes was kept medium.

3 Properties of Material Used

The following materials were used in the present study.

3.1 Portland Cement

Ordinary Portland cement (OPC) of 43 grade (Ultratech) confirming to BIS: 8112:1989 was used for making concrete. It was fresh and without any lumps.

3.2 Coarse Aggregate

The coarse aggregate used was a mixture of two locally available crushed stone of 10 and 20 mm size in 50:50 proportion and confirming to BIS: 383:1970. Specific gravity and fineness modulus were determined 2.63 and 6.74, respectively [4].

3.3 Fine Aggregates

Fine aggregates were collected from Chakki River, Pathankot. It was coarse sand brown in colour. Specific gravity and fineness modulus of fine aggregates were experimentally determined as 2.62 and 2.925, respectively [4].

3.4 Fly Ash

Fly ash used in present work was obtained from Guru Hargobind Thermal Plant Lehra Mohabbat, Distt. Bathinda. It was F type fly ash, light brown in colour having specific gravity 2.09 [4].

3.5 Silica Fume

The silica fume used was obtained from Orkla India Private Limited (brand name: Elkem Microsilica 920-D), Navi Mumbai. It was grey in colour. The specific gravity and surface area were 2.26 and 20,500 mm^2/Kg [4].

3.6 Superplasticizer

The superplasticizer used in the study was Rheobuild SPI obtained from Basf Construction Chemicals (India) Pvt. Ltd., Navi Mumbai. It is based on naphthalene formaldehyde polymer. The physical and chemical properties of superplasticizer, which was obtained from the company, conform to BIS-9103-1979.

4 Experimental Programme

4.1 Mix Design by Indian Standard Recommendations

The guidelines given in BIS: 10262-1982 [5] have been adopted for mix design of concrete. The quantities determined by BIS design method are given in Table 1.

Table 1 Proportion of different materials [4]

Water	Cement	Fine aggregate	Coarse aggregate
185.4 l	579.375 kg	467.58 kg	1108.52 kg
0.32	1	0.81	1.91

Table 2 Test result of compressive and split tensile strength with different level of replacement

Sample	% age of replacement (fly ash + silica fume)	Compressive strength (MPa)			Split tensile strength (MPa)		
		3 days (MPa)	7 days (MPa)	28 days (MPa)	3 days (MPa)	7 days (MPa)	28 days (MPa)
N_0	0 + 0	27.14	34.85	54.71	2.71	3.14	4.60
N_1	0 + 4	45.33	54.22	67.19	4.10	4.40	5.18
N_2	0 + 8	40.94	50.47	67.52	3.80	4.34	5.00
N_3	0 + 12	40.74	51.78	69.57	3.51	4.44	5.78
N_4	10 + 4	35.60	50.29	76.83	3.85	4.42	6.10
N_5	10 + 8	27.59	40.77	63.75	3.10	3.77	4.55
N_6	10 + 12	26.18	38.24	66.61	2.58	3.68	5.02

4.2 Test of Strength

The compressive and split tensile strength tests were performed in accordance with BIS: 516 and BIS: 5816. For all the designated mixes and for each of the testing ages, three samples were cast and the average value of results is given in Table 2.

5 Results and Discussion

5.1 Compressive Strength

The compressive strength of all the mixes was determined at the ages of 3, 7 and 28 days for the various replacement levels of fly ash and silica fume. The results are reported in Table 2. The maximum/minimum value of compressive strength of concrete with the addition of 10% of fly ash to the various percentages of silica fume (i.e. 4, 8 and 12%) for 28 days was 76.83 N/mm^2/63.75 N/mm^2 as compared with compressive strength control mix 54.71 N/mm^2.

The increasing trends are due to the use of silica fume, which helps in filling concrete pores resulting in improved impermeability of concrete, lesser moisture diffusion, and increased strength may be attributed to the increased concrete density. Also secondary reaction takes place between silica and a by-product of cement hydration that is calcium hydroxide to produce a cementitious agent, calcium silicate hydrates. This additional cementitious agent, i.e. calcium silicate hydrates, adds strength to concrete in addition to increasing its density. The superfine size and high pozzolanic reactivity make silica fume a basic ingredient for concrete requiring high strength. The decreasing trends are due to alumina of fly ash, which is in the presence of lime forms gehlenite (C_2ASH_8)—an inert material. The low strength of fly ash—lime mix in spite of high reactivity of fly ash—is due to the formation of gehlenite. Further, the rate of growth of these pozzolanic products is small, and their influence on strength of the system is seen only after long periods of curing.

5.2 *Split Tensile Strength*

The split tensile strength of all the mixes was determined at the ages of 3, 7 and 28 days for the various replacement levels of fly ash and silica fume. The results are reported in Table 2. The maximum/minimum value of split tensile strength of concrete with the addition of 10% of fly ash to the various percentages of silica fume (i.e. 4, 8 and 12%) for 28 days was 6.10 N/mm^2/4.55 N/mm^2 as compared with split tensile strength control mix 4.60 N/mm^2. The reason for increase/decrease in strength was same as given in compressive strength.

The optimum value of compressive and split tensile strength is achieved with addition of 10% of fly ash and 4% of silica fume. But as the percentage of silica fume increases more than the optimum value for the same % age of fly ash, compressive and split tensile strength of concrete probably starts decreasing due to the excessive heat of hydration. Similarly, if percentage of dosage of fly ash increases more than the optimum value for the same % age of silica fume, compressive and split tensile strength of concrete starts decreasing due to the formation of gehlenite. It is recommended to use 10% fly ash and 4% of silica fume as replacement of cement for getting economical and better compressive and split tensile strength.

5.3 *Correlation Between Split Tensile and Compressive Strength Ratios*

The split tensile and compressive strength ratios (observed) are mentioned in Table 3. The correlation of split tensile and compressive strength depends upon the fine and coarse aggregate properties, w/b ratio, relative humidity, concrete compaction, temperature, curing age and ratio of coarse and fine aggregate. The predicted values of split tensile to compressive strength of concrete ratio were determined by equation derived by Artoglu et al. [6]. Integral absolute error of these two values was determined. The accuracy of the correlation is high if the value of error is small. The most of results show that the current study is in excellent and good agreement with Artoglu et al. [6] for 3, 7 and 28 days, respectively. It was also obtained that

Table 3 Split tensile and compressive strength ratios of concrete with and without silica fume and fly ash along with IAE

Samples	Split tensile and compressive strength ratios (%)								
	O_i after 3 days	P_i after 3 days	IAE after 3 days	O_i after 7 days	P_i after 7 days	IAE after 7 days	O_i after 28 days	P_i after 28 days	IAE after 28 days
N_0	9.80	11.41	9.64	9.01	10.40	8.72	8.41	8.80	4.15
N_1	9.04	9.44		8.11	8.83		7.71	8.16	
N_2	9.28	9.80		8.60	9.07		7.40	8.14	
N_3	8.61	9.82		8.57	8.98		8.31	8.05	
N_0	9.98	11.41	6.59	9.98	10.40	4.52	8.41	8.80	6.54
N_4	10.81	10.32		8.79	9.08		7.94	7.76	
N_5	11.23	11.34		9.25	9.81		7.14	8.32	
N_6	9.85	11.56		9.62	10.05		7.54	8.18	

Table 4 Relationship between compressive strength and split tensile strength of concrete

S. No.	Author's	Equations
1.	Nanda and Jaspal	$f_{\text{sts}} = 0.3019 f_c^{0.6782}$
2.	Gardner	$f_{\text{sts}} = 0.34 f_c^{0.66}$
3.	ACI318R-99	$f_{\text{sts}} = 0.56 f_c^{0.5}$
4.	CEB-FIB	$f_{\text{sts}} = 0.3 f_c^{2/3}$
5.	Oluokun et al.	$f_{\text{sts}} = 0.294 f_c^{0.69}$

f_{sts} = split tensile strength and f_c = compressive strength

the accuracy of the correlation is high, when the cement is replaced with silica fume along with fly ash as the value of IAE is near to 5 and best correlation was found for 28 days.

5.4 *Comparison of Relationship Developed by the Author(s) with Other Author(s)*

A total of 126 samples were casted, and based on the values determined experimentally, relationship was developed between compressive strength and split tensile strength. The relation derived from this current study is compared with those proposed by Gardner [7], given in ACI-318-99 [8], given in CEB-FIB [9] and by Oluokun et al. [10] as shown in Table 4.

On the basis of relations given by different authors, graphs were plotted in order to compare the relationships given by different authors as shown in Fig. 1. The relationship derived from the test results of the current study is very close to relationship

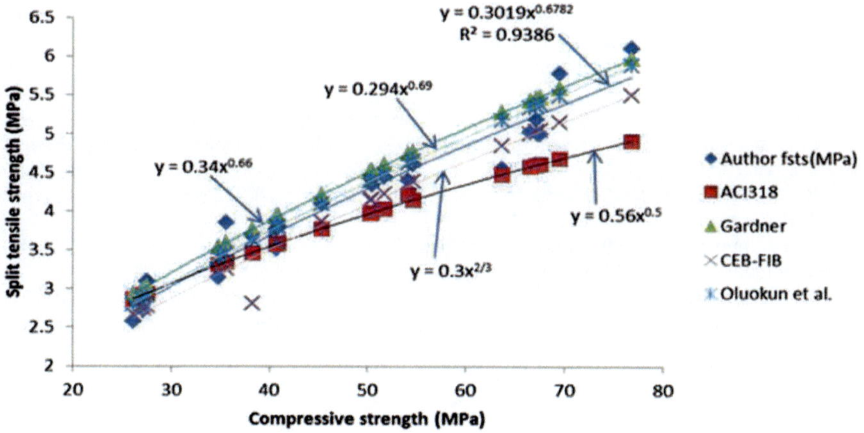

Fig. 1 Relationship between compressive strength and split tensile strength

derived by Oluokun et al. i.e. $y\,(f_{\text{sts}}) = 0.294 f_c^{0.69}$. The value of coefficient of determination ($R^2 = 0.9386$) attains good correlation between split tensile and compressive strength of concrete, and the relationship between these two is significant.

6 Conclusions

From the experimental investigation, the following conclusions can be drawn:

i. The partial replacement of cement with fly ash and silica fume results in formation of sustainable concrete. The optimum replacement levels of fly ash and silica fume are respectively 10% and 4%.
ii. The relationship derived between compressive and split tensile strength from the test results of the current study is very close to relationship derived by Oluokun et al. i.e. y $(f_{\text{sts}}) = 0.294_f^{c0.69}$.
iii. The split tensile and compressive strength ratio decreases with the increase in compressive strength and vice versa. Integral absolute error of experimental and predicted values of the split tensile and compressive strength ratio was found well within the specified range.

References

1. Dhariwal A Waste to wealth: flyash ijep.co.in/uploads/journal 1488542355199 pdf
2. Duan P, Yan C, Zhou W (2017) Compressive strength and microstructure of fly ash based geopolymer blended with silica fume under thermal cycle. Cement Concr Compos 78:108–119
3. Dybeł P, Furtak K (2017) Influence of silica fume content on the quality of bond conditions in high-performance concrete specimens. Arch Civ Mech Eng 17:795–805
4. Singh J, Nanda AK (2011) Potential benefits of flyash in attaining the workability of silica fume concrete. NBM&CW, New Delhi, pp 193–196
5. BIS: 10262-1982 (2004) Recommended guidelines for concrete mix design. Bureau of Indian Standard, New Delhi
6. Artoglu N, Girgin ZC, Artoglu E (2006) Evaluation of ratio between splitting tensile strength and compressive strength for concretes up to 120 MPa and its application in strength criterion. ACI Mater J 103(1):18–24
7. Gardner NJ (1990) Effect of temperature on the early age properties of type I, type III and type I/Flyash concrete. ACI Mater J 87(1):68–78
8. ACI Committee 318 (1992) Building code requirement for structural concrete (ACI 318-99) and commentary (ACI 318R-99). American Concrete Institution, Farmington Hills, p 391
9. CEB-FIP Model Code for Concrete Structures (1990) Evaluation of the time dependent behavior of concrete Bulleti d'Information No 199, Comite European du Beton/Federation International de la Precontrainte, Lausanne, 1991, 201
10. Oluokun FA, Burdette EG, Deatherage JH (1991) Split tensile strength and compressive strength relationships at early ages. ACI Mater J 88(2):115–121

Speech Parameters Extraction for Text-to-Speech Synthesis for Punjabi

Navdeep Kaur and Parminder Singh

1 Introduction

Humans communicate more effectively through the medium of speech. A system that translates written text into speech waveforms is called text-to-speech system or simply called TTS. Lately, statistical parametric techniques such as HMM and DNN which combine the speech synthesis techniques with vocoding to generate speech waveforms have gained popularity over concatenative speech synthesis techniques to generate synthetic speech from huge database of speech units [11]. In SPSS, various speech parameters such as excitation parameters and spectral coefficients are extracted by using vocoders from waveforms of training data, and then, synthesized speech is generated by using acoustic model at synthesized time [8]. To extract features, speech waveforms are transformed to its parametric representation which is more reliable and concise than the original signal. This makes it more suitable for subsequent data processing and analysis. This conversion into parametric forms is known as front-end signal processing. Quality of front-end affects the quality of synthesized waveforms [1].

Based on the known variation of human ear's critical frequency bandwidth, MFCC [3, 7] is the normally used in various speech processing methods among the various feature mining techniques such as linear predictive coefficients (LPC), perceptual linear predictive coefficients (PLP) and others.

N. Kaur (✉)
Government Polytechnic College for Girls, Amritsar, India
e-mail: nav_783@yahoo.com

P. Singh
Guru Nanak Dev Engineering College, Ludhiana, India
e-mail: parminder2u@gmail.com

H. Singh et al. (eds.), *Sustainable Development Through Engineering Innovations*, Lecture Notes in Civil Engineering 113,
https://doi.org/10.1007/978-981-15-9554-7_66

This paper is structured as Sect. 2 briefs the various speech parameters and their extraction methods; Sect. 3 is experimental setup and results; last section is for the conclusion.

2 Speech Parameters

Generally, a speech signal is presumed to be stationary for a short period of time typically in a window of 20–30 ms. Therefore, speech signal is represented by set of parameters between fixed small interval called frame. Each frame comprises 30–40 parameters to represent the spectral envelope of signal, and the value of fundamental frequency including their time derivatives is known as dynamic features of speech signal [11].

2.1 *Excitation Parameter (Fundamental Frequency)*

The fundamental frequency, or pitch, is the measure of lowest frequency of a periodic waveform, and it is associated with the period with which the signal repeats itself [5]. It is also denoted by $f0$. The excitations to the vocal cord caused by movement of vocal folds are reflected as high-energy areas in the speech signal. The time period between two excitation points is referred as pitch period. This vital information has abundance of applications. Some of them are speech recognition, voice assessment, speech perception, speech transformation including text-to-speech synthesis [4]. Several time and frequency analysis methods have been proposed in literature for pitch estimation such as autocorrelation method [6], robust algorithm for pitch tracking [9], harmonic product spectrum, subharmonic-to-harmonic ratio, single-frequency filtering (SFF) approach [2].

2.1.1 Harmonic Product Spectrum on Fast Fourier Transformed Data

The reason for selecting HPS for pitch estimation is that series of peaks in the speech spectrum correspond to the fundamental frequency as harmonic components lie at integer multiples of the fundamental frequency. In HPS, the original spectrum is multiplied with compressed/down-sampled spectra so that the peak at the fundamental frequency only gets enlarged. The various steps of algorithm are as:

Step 1. Read the time domain data, i.e. amplitude of speech signal into double array.

Step 2. Non-stationary speech signal is divided into small-sized frames typically of 20–30 ms for which speech signal is expected to be stationary. Signal is divided into overlapping frames where each frame shares the first part with the previous frame and last part with the next frame as shown in Fig. 1.

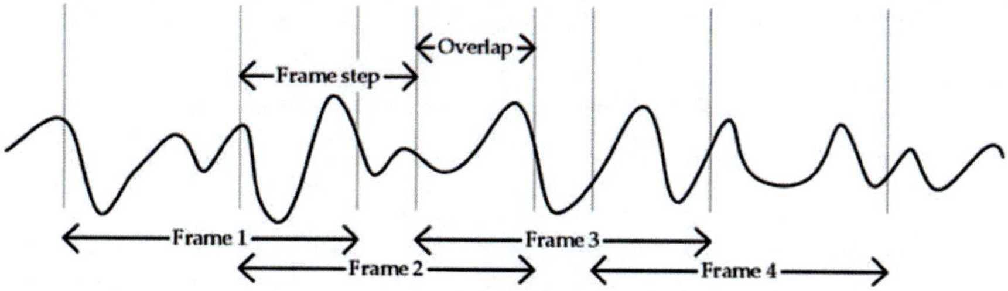

Fig. 1 Speech signal divided into four overlapping frames

The number of samples in each frame, i.e. frame length, depends upon the sampling rate of signal.

$$f_l = t * f_s \tag{1}$$

Here, f_l is the frame length, t is the duration of each frame and f_s is the sampling rate of signal.

Step 3. A speech signal may have frequency components that falloff at high frequencies. So to avoid overseeing the high frequencies, the high-frequency components are compensated using pre-emphasis filtering. A simple digital filter used for such compensation is given as [10]:

$$y(n) = s(n) - \alpha s(n-1) \tag{2}$$

Here, $y(n)$ is the output signal, $s(n)$ is the input signal and α is the pre-emphasis factor which takes value between 0.9 and 1.

Step 4. There are several types of window functions that can be applied to the signal. Hamming and Hanning window functions are commonly used, and both have a sinusoidal shape.

Hanning window function:

$$\begin{array}{ll} w(n) = 0.5 - 0.5\cos\left(\frac{2n\pi}{N} - 1\right), & 0 \le n \le N-1 \\ 0, & \text{otherwise} \end{array} \tag{3}$$

Hamming window function:

$$\begin{array}{ll} w(n) = 0.5 - 0.46\cos\left(\frac{2n\pi}{N} - 1\right), & 0 \le n \le N-1 \\ 0, & \text{otherwise} \end{array} \tag{4}$$

Step 5. To each frame of speech signal, fast Fourier transformation (FFT) is applied so that it is changed from time domain to frequency domain.

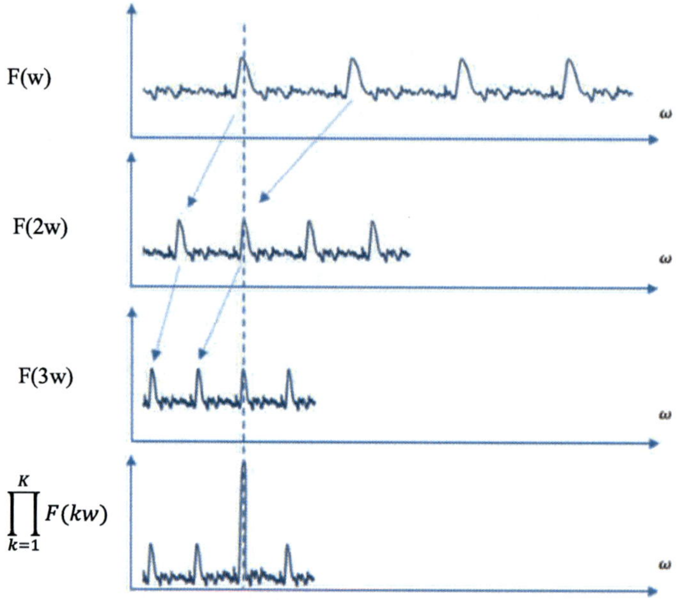

Fig. 2 Speech signal down-sampled three times to obtain the $f0$ value

Step 6. To compute the fundamental frequency of signal, the resultant FFT data is down-sampled N times as shown in Fig. 2. In Fig. 2, the $F(w)$ represents the amplitude spectrum of a time signal.

Step 7. FFT transformed data represents frequencies up to the Nyquist frequency which is half of the FFT data length. Frequencies lie in the second half are known as negative frequencies and removed.

Step 8. Resultant down-sampled arrays are combined into single array. The frequency index of maximum magnitude is found, and fundamental frequency of frame is then calculated as:

$$f0 = i * \mathrm{SR}/\mathrm{FFT}_l \tag{5}$$

Here, $f0$ is the fundamental frequency, SR is the sampling rate of signal and FFT_l is the length of FFT data.

2.2 *Spectral Parameters (MFCC)*

Spectral parameters such as Mel frequency cepstral coefficients (MFCCs) are extensively used in speech recognition and speech synthesis as these accurately represent the envelope of the short-time power spectrum. While calculating MFCCs coefficients, the performance of an algorithm may be affected by various parameters, such as (1) the number of the filters, (2) the shape of the filters, (3) the way in which the filters are spaced, overlapped or not, or (4) the way in which the power spectrum is wrapped [12]. The various steps to compute the MFCCs are as:

Step 1. Signal is divided into short overlapping frames to convert it into stationary signal.
Step 2. Each frame is transformed from time domain to frequency domain using FFT, and then, magnitude spectrum is computed from FFT data after removing negative frequencies from spectrum.
Step 3. Mel filter banks are applied to resultant spectrum, and energy is summed up.
Step 4. Discrete cosine transformation is applied to logarithm of filter bank energies.
Step 5. Generally, 2–13 DCT coefficients are retained that represents the MFCC values of a given frame.

2.2.1 Computation of Mel Filter Banks

To computer the Mel filter banks, lower and upper filter frequencies are chosen. Upper filter frequency is taken as half of the sampling rate of signal. These frequencies are converted into Mel frequencies using the formula:

$$M(f) = 2595 \log_{10}\left(1 + \frac{f}{700}\right) \tag{6}$$

If numFilter is the number of filter banks needed, then numFilter + 2 linearly spaced points between lower and upper frequencies are determined. These points are then converted back to Hz from Mel scale using formula:

$$f = 700\left(10^{\frac{M}{2595}} - 1\right) \tag{7}$$

These frequencies are converted to FFT bin number using formula:

$$f(i) = (\text{FFTSize}^* f(i))/\text{SamplingRate} \tag{8}$$

From the frequencies $f(i)$, triangular Mel filter banks are computed as:

$$H_m(k) = \begin{cases} 0, & k < f(m-1) \\ \frac{k-f(m-1)}{f(m)-f(m-1)}, & f(m-1) \le k \le f(m) \\ \frac{f(m+10)-k}{f(m+1)-f(m)}, & f(m) \le k \le f(m+1) \\ 0, & k > f(m+1) \end{cases} \tag{9}$$

Here, m is the number of Mel-filter bank and $f()$ is the list of $m + 2$ Mel spaced frequencies.

Starting from the first point, the first filter bank reaches to the peak at the second point, then coming back to zero at the third point. Similarly, staring from the second point, the second filter bank reaches to the peak at the third point, then returns to zero at fourth point. Each filter bank contains mostly zero values, but is nonzero for a definite segment of the spectrum.

2.3 *Deltas and Delta-Deltas*

The first- and second-order time derivatives of speech parameters are computed using:

$$\Delta = \frac{\sum_{k=-M}^{M} k * \text{coeffs}(k)}{\sum_{k=-M}^{M} k^2} \quad (10)$$

Here, k is the current frame for which delta components are computed using M frames before the current frame and M frames after the current frame. If M is taken as 1, then delta components are computed as:

$$\Delta = \text{currCoeffs} - \text{prevCoeffs} \quad (11)$$

3 Experimental Setup and Results

The Punjabi speech corpus from a female speaker at sampling rate of 44.1 kHz is used to perform experiment. The speech parameters are computed for frame size of 2048 with frame overlap of 1024 samples. The pre-emphasis factor is set to 0.95. The FFT size is set equal to frame size. Figure 3 shows a time domain representation of the phoneme/ ਜ਼ੀ/for an adult female speaker with a sampling frequency of 44.1 kHz. The signal is divided into ten overlapping frames. Figure 4 shows the fundamental frequency of each frame of the phoneme/ ਜ਼ੀ/calculated using Eq. (5).

For MFCC calculation, lower filter frequency value is taken as 30 Hz and upper frequency value is taken as 22,050 Hz which is half of the sampling frequency 44.1 kHz. A total of 17 Mel filter banks are used to compute the cepstral coefficients.

Following values represent the 19 (17 + 2) linearly spaced points between lower and upper filter frequencies in Mel scale which are computed using Eq. (6):

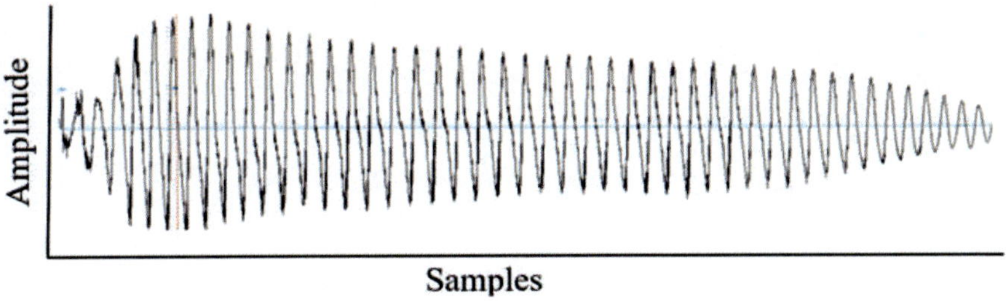

Fig. 3 Speech signal samples

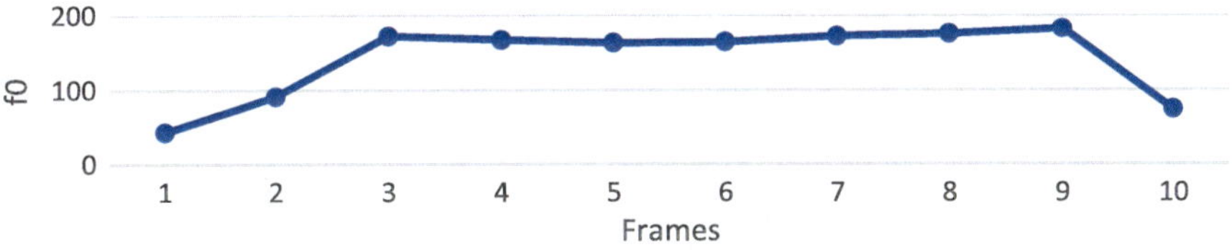

Fig. 4 $f0$ per frame of phoneme/ ਟੀ/

47.29341	262.6292	477.965	693.3007	908.6365	1123.972	1339.308	1554.644
1769.98	1985.315	2200.651	2415.987	2631.323	2846.658	3061.994	3277.33
3492.666	3708.002	3923.337					

Following values represent the Mel frequencies corresponding to above linearly spaced points which are computed using Eq. (7):

30	183.698	369.7565	594.9886	867.6423	1197.702	1597.254	2080.93
2666.442	3375.23	4233.25	5271.923	6529.283	8051.374	9893.934	12124.44
14824.56	18093.18	22050					

Following values represent the values of frequency bin to which Mel frequencies lie that are computed using Eq. (8):

1.393197	8.53092	17.17146	27.63122	40.29323	55.62117	74.17634	96.6382
123.8293	156.7454	196.5918	244.8276	303.2193	373.9051	459.4734	563.0578
688.4513	840.2458	1024					

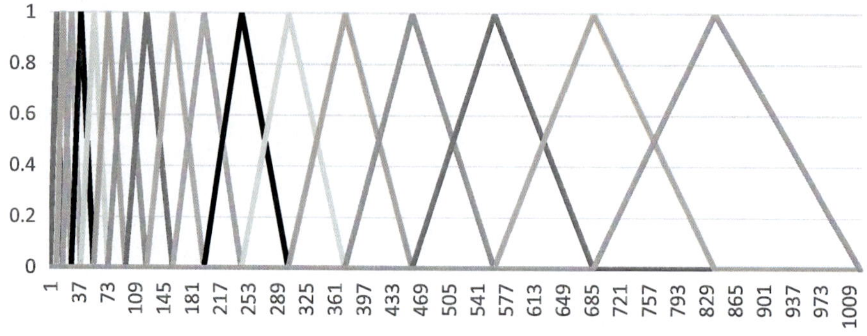

Fig. 5 17-Triangular Mel filter banks

Figure 5 shows the triangular Mel filter banks computed using Eq. (9) for the above-calculated values of frequency bin.

The computed Mel filter banks are applied to the magnitude spectrum of speech to compute the 13 MFCC coefficients of speech signal. In addition to this power of signal per frame is also calculated.

Along with the static features of speech, dynamic features called delta and delta-delta parameters are also computed by taking value if M as 2 in Eq. (10), which gives total 45 parameters per frame of speech signal.

4 Conclusion

- The statistical parametric speech synthesis approach uses the various speech parameters to generate speech waveforms from the written text.
- Speech signal is represented as set of excitation (fundamental frequency) and spectral parameters that are computed at fixed rate, typically at 5 ms frame rate.
- The harmonic product spectrum method computes the fundamental frequency from the series of peaks from the spectrum of speech which gives precise results as these peaks correspond to the integer multiples of fundamental frequency.
- The spectral parameters such as MFCCs of speech are used to determine the spectral envelope of aperiodic excitations as these mimic the human ear.

Acknowledgements This piece of research work is supported by I.K Gujral Punjab Technical University, Kapurthala.

References

1. Alim SA, Rashid NKA (2018) Some commonly used speech feature extraction algorithms. From natural to artificial intelligence—algorithms and applications, chapter 1, ISBN: 978-1-83881-548-6
2. Aneeja G, Yegnanarayana B (2017) Extraction of fundamental frequency from degraded speech using temporal envelopes at high SNR frequencies. IEEE/ACM Trans Audio Speech Lang Process 25(4):829–838. https://doi.org/10.1109/taslp.2017.2666425
3. Chauhan PM, Desai NP (2014) Mel Frequency Cepstral Coefficients (MFCC) based speaker identification in noisy environment using wiener filter. In: International Conference on Green Computing Communication and Electrical Engineering (ICGCCEE), Coimbatore, pp 1–5, https://doi.org/10.1109/icgccee.2014.6921394
4. Drugman T, Huybrechts G, Klimkov V, Moinet A (2018) Traditional machine learning for pitch detection. IEEE Signal Process Lett 25(11):1745–1749. https://doi.org/10.1109/lsp.2018.2874155
5. Hansen MW, Jensen JR, Christensen MG (2019) Estimation of fundamental frequencies in stereophonic music mixtures. IEEE/ACM Trans Audio Speech Lang Process 27(2):296–310. https://doi.org/10.1109/taslp.2018.2878384
6. Hernandez MJ (2016) A tutorial to extract the pitch in speech signals using autocorrelation. Open J Tech Eng Disciplines (OJTED) 2(1):1–10
7. Hossan MA, Memon M, Gregory MA (2010) A novel approach for MFCC feature extraction. In: 4th international conference on signal processing and communication systems, gold coast, QLD, pp 1–5. https://doi.org/10.1109/icspcs.2010.5709752
8. Ling ZH, Kang SY, Zen H, Senior A, Schuster M, Qian XJ, Meng HM, Deng L (2015) Deep learning for acoustic modelling in parametric speech generation: a systematic review of existing techniques and future trends. IEEE Signal Process Mag 32(3):35–52. https://doi.org/10.1109/msp.2014.2359987
9. Reddy MK, Rao KS (2017) Robust pitch extraction method for the HMM-based speech synthesis system. IEEE Signal Process Lett 24(8):1133–1137. https://doi.org/10.1109/lsp.2017.2712646
10. Tan L, Jiang J (2019) Digital signal processing: fundamentals and applications (3rd ed), chapter 6, pp 173–228, ISBN: 9780128150726
11. Tokuda K, Nankaku Y, Toda T, Zen H, Yamagishi H, Oura K (2013) Speech synthesis based on hidden Markov models. Proc IEEE 101(5):1234–1252. https://doi.org/10.1109/JPROC.2013.2251852
12. Zheng F, Zhang G, Song Z (2001) Comparison of different implementations of MFCC. J Comput Sci Technol 16:582–589. https://doi.org/10.1007/BF02943243

Application of Alternate Waste Materials as Barrier Material in Engineered Landfills

Jaskiran Sobti

1 Introduction

By 2050, the world is expected to generate 3.40 billion tons of waste annually, increasing drastically by nearly 75% from the figure of 2.01 billion tons as in the year 2016 [1]. A major chunk of the municipal solid waste generated is alleged to have exhausted the prospects of being reused or recycled necessitating adequate disposal. Landfilling is the most frequently adopted technique for waste containment by various countries. Compacted clay barriers have been used to line municipal waste landfills which essentially include low-permeability clays. In places where clays are not readily available, engineers face difficulties with economically sourcing clays for barrier applications. The sustainable solution to this problem lies in fabricating a barrier material using locally available soils and preferably some of the waste products with suitable admixtures fulfilling the functional requirements of low hydraulic conductivity for satisfactory performance. These alternate materials if put to beneficial use in such bulk quantities for fabricating the liner material would lead to conservation of the natural resources and promoting sustainable development by dispensing the need of borrowing the suitable soils from distant places. These waste products when used alone are considered to be deficient in meeting the requirements of a barrier material owing to high permeability. Mixing adequate quantities of suitable admixtures to these waste materials would help to enhance their performance ability satisfying the mandatory requirements of a barrier material. One such admixture is bentonite clay. Bentonite, which is a highly active clay belonging to the montmorillonite group of clays, with a high swelling and low permeability characteristics [2] has been used for the design of engineered barriers. This paper presents an exhaustive overview of the use of conventional soils and waste by-products generated from

J. Sobti (✉)
Department of Civil Engineering, Guru Nanak Dev University, Amritsar, Punjab 143005, India
e-mail: sobti.kiran@gmail.com

H. Singh et al. (eds.), *Sustainable Development Through Engineering Innovations*, Lecture Notes in Civil Engineering 113,
https://doi.org/10.1007/978-981-15-9554-7_67

various sources for probable utilization as a barrier material in liners and covers of engineered landfills.

2 Design Requirements of a Landfill Liner

Appropriately designed landfill liners must extend consistent performance and be efficient enough to refrain leakage of the generated leachate during the design life of the facility. There are certain essential requirements which should be given due consideration during selection of suitable materials as a landfill liner material. The points which need to be emphasized are low hydraulic conductivity, maximum shear strength, minimum shrinkage, minimum swelling, adequate suction property and good adsorption characteristics to impede the flow of contaminants into the subsoil or groundwater [3]. These vital characteristics need to be addressed to maximize the utility with minimum cost and result in obtaining a suitable and sustainable material with adequate strength for the entire service life of the landfill. A typical section of liner and cover material for municipal solid waste landfills is shown in Fig. 1. Liner system for composite clay liner consists of a 0.9-m-thick layer of compacted clay or amended soil, respectively, overlain by a 1.5–2-m-thick HDPE geomembrane along with leachate collection layer. The waste material is provided with a cover, 1.2–1.5 m thick, comprising of a 0.3-m-thick soil as gas collection layer, 0.6 m of clay or amended soil and 0.15 m of granular material for drainage purpose overlain by 0.45 m of natural soil supporting vegetation. A layer of geotextiles may also be provided at the interface between different component layers.

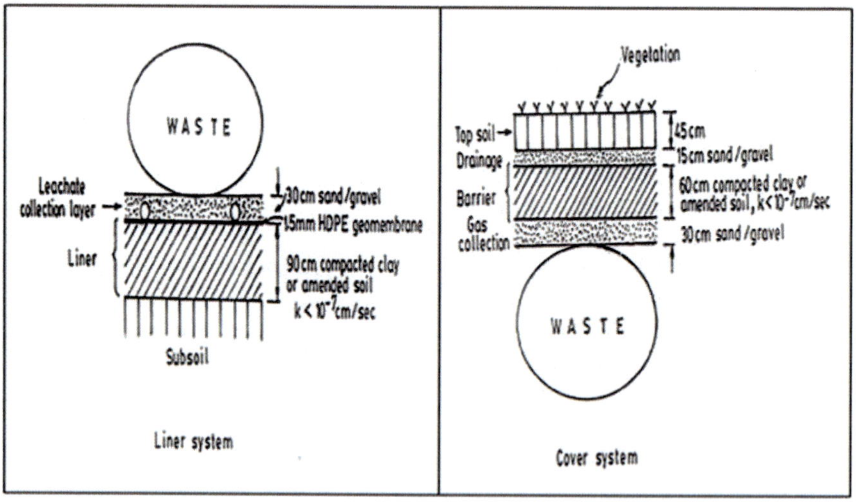

Fig. 1 Typical section of liner and cover material for MSW landfills [4]

3 Results and Discussion

3.1 Use of Sand–Bentonite as a Barrier Material

Sand–bentonite mixtures have been utilized as a barrier material in several engineering applications like landfills, cut-off walls, buffer and backfill materials of radioactive nuclear waste containments. The sand–bentonite mixture is a combination of two different materials in terms of particle size distribution and chemical activity to produce a material with low permeability, low compressibility and appropriate strength. The hydraulic conductivity of bentonite-enhanced soils can be controlled by adjusting the percentage of bentonite in the bentonite-enhanced mix. Chalermyanont and Arrykul [5] studied the impact of varying bentonite content on hydraulic conductivity, OMC and MDD of the sand–bentonite mixtures and suggested a range of 3–5% bentonite content to be mixed for sand–bentonite mixes. Zhang et al. [6] conducted an extensive testing programme to assess the effectiveness of bentonite-sand mixtures (varying the sand ratio in the test specimens from 0, 20, 30, 40 and 50%) as backfilling material for high-level radioactive waste. Low hydraulic conductivities of the mixtures were reported for the mixes that had sand ratio varying from 0 to 30%. In Fig. 2, the hydraulic conductivity is less than 1.0×10^{-10} m/s, even if the sand ratio is as high as 50%. Test results show that there is no significant change in hydraulic conductivity of bentonite-sand mixtures within the sand ratios ranging from 0 to 50%.

An elaborate experimental programme was conducted by the author on the use of sand–bentonite, silt–bentonite and coal ash–bentonite mixes for use as a barrier material by evaluating various properties like hydraulic conductivity, matric suction, adsorption characteristics, strength and swelling characteristics. The hydraulic conductivity varies for sand–bentonite, silt–bentonite and coal ash–bentonite mixes as shown in Fig. 3 according to Sobti and Singh [7]. It can be clearly seen that hydraulic conductivity decreases with increasing bentonite content.

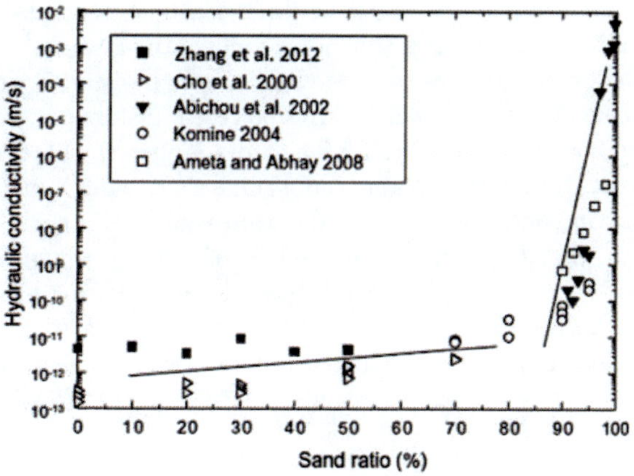

Fig. 2 Relationship between hydraulic conductivity and sand ratio [6]

Fig. 3 Effect of increased bentonite content on hydraulic conductivity [7]

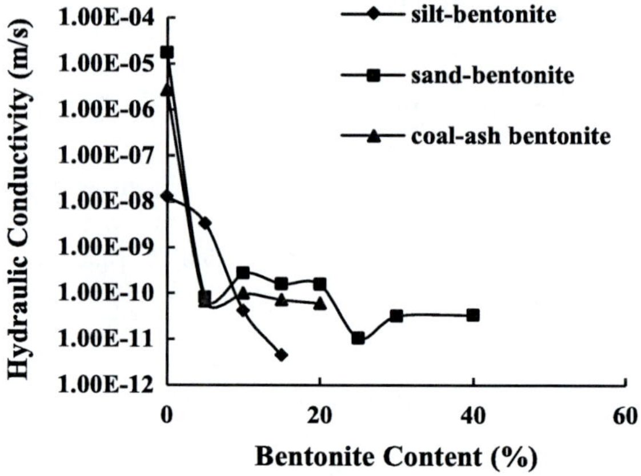

3.2 *Use of Alternate Waste Materials as Barrier Material*

3.2.1 Coal Combustion by-Products

A huge quantum of coal ashes are being produced by coal fired thermal power plants throughout the world. Coal ashes have advantageous properties such as low specific gravity, lower compressibility, high strength, high California bearing ratio and pozzolanic reactivity to be put to use as better alternative materials in various situations in the field of geotechnical engineering. The utilization of coal fly ash with clay in landfill bottom liners has attracted considerable attention in literature. Nhan et al. [8] investigated the hydraulic conductivity of class *F* fly ash mixed with lime kiln dust and bentonite. They found a reduced hydraulic conductivity with the addition of lime dust and bentonite in coal fly ash. Also, the dissolved metal concentration in the synthetic municipal solid waste leachate reduced. Cokca and Yilmaz [9] investigated the combination of rubber, bentonite and coal fly ash as a liner material. The composite material had a hydraulic conductivity of less than 10^{-9} m/s using water as permeant, and the effluent leachate was found to be non-hazardous. Kumar and Vaddu [10] experimentally studied the swelling behaviour of two bottom ash–bentonite mixtures compacted at varying moisture contents and showed a significant reduction in free swell index when the moisture content increased and the bentonite content decreased. In addition to this, Kim et al. [11] reasoned out that bottom and fly ashes have water soluble cations like Ca^{+2} and Na^{+} which can be exchanged for Na^{+} ion in bentonite resulting in a decreased swelling capacity of bentonite. According to Pal and Ghosh [12], the values of k of fly ash specimens of this study decrease to the range of 10^{-09}–10^{-10} m/s with the addition of 20% of montmorillonite clay, compared to that of 10^{-06}–10^{-07} m/s for fly ash samples alone. Reduction in the values of k of fly ash mixes continues with increase in percentages of montmorillonite clay from 20 to 50 but is not significant for addition of higher percentages as shown in Fig. 4.

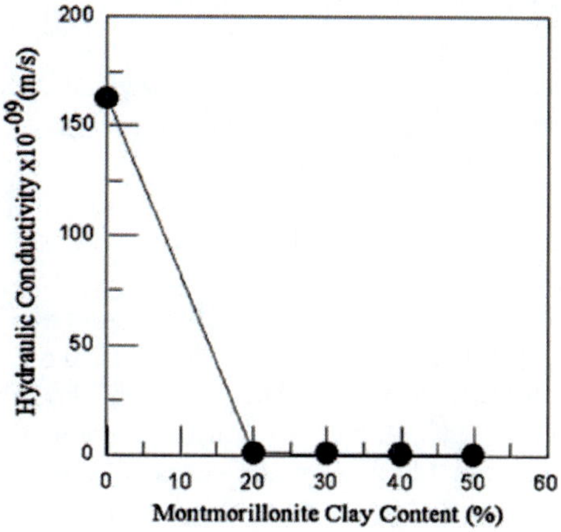

Fig. 4 Hydraulic conductivity versus montmorillonite %for Kolaghat fly ash [12]

Mishra and Ravindra [13] conducted the hydraulic conductivity and consolidation tests to evaluate the performance of fly ash and cement mixtures (0, 2, 5 and 10%) as a landfill liner material. A relatively lower value of compression index for all the mixtures indicating the settlement due to application of overburden pressure would be small. The mix with 90% fly ash and 10% cement compacted at 5% wet of OMC exhibited a hydraulic conductivity value less than 10^{-9} m/s. Thus, these studies demonstrate that coal ash mixed with adequate quantity of bentonite satisfies the hydraulic conductivity requirements, alleviates the swelling capacity of bentonite, provides favourable consolidation characteristics, indicating an affirmation of its usage as barrier material.

3.2.2 Waste Alum Sludge

Bashar et al. [14] advocated the use of waste alum sludge (WAS) generated from water treatment processes as alternative liner material fitting the sustainable landfill model for today's era. The properties of alum sludge were studied and found to match with kaolin clay to a considerable extent based on the XRD and SEM characteristics. The test results indicate that the hydraulic conductivity values are higher for alum sludge as compared to kaolin clay. WAS is considered as one of the major adsorbents for pollutant removal, and it has an ability to remove a large amount of heavy metals and organic constituents due to small particulate materials which provide available surface area for inter-particulate bonding [15]. Leaching tests showed that very low levels of Pb, Fe, Mn and Cr were leached out by distilled water from the WAS. The WAS contained a significant amount of organic matter as shown by FTIR spectra and TOC analysis. The results of the SEM and FESEM analysis concluded that the WAS is amorphous in nature. Good adsorption characteristics are a highly desirable characteristic for use as a liner material in addition to the permeability criterion.

3.2.3 Lime-Softening Sludge

The raw water is treated by adding slaked lime to remove the high carbonate hardness to acceptable limits before public supplies. The lime sludge chiefly consists of calcium carbonate and shows adsorption properties for oils, soluble compounds and various organic compounds, which is a characteristic looked out for a landfill liner material [16]. Dąbska [17] studied the long-term hydraulic conductivity of lime-softening sludge was with tap water, distilled water, sodium hydroxide solution, hydrochloric acid and municipal waste leachate as permeants. The results of the hydraulic conductivity tests showed that the k was largely affected by the compaction and the moulding moisture content. The short-term tap water permeation tests also showed that all k values at hydraulic gradients more than 30 were $<2.5 \times 10^{-8}$ m/s, when moulding water content range was 27.7–36.7% and the degree of compaction was from 0.95 to 1.05. The factors of permeating liquids and permeation time significantly affected the initial hydraulic conductivity. It was emphasized by the researchers that the investigated material met the essential requirements for use as a liner material in waste repositories.

3.2.4 Blast Furnace Slag and Other Waste from Iron and Steel Industry Blended with Bentonite

The amount of ground granulated blast furnace slag (GGBS) generated for every one ton of iron extraction is nearly estimated to be 400 kg [18]. Thus, disposal of this waste material on land poses huge environmental issues. Usage of GGBS within the landfill for fabricating the liner material would lead to bulk utilization of this waste solving dual purpose. Manikanta and Shankar [19] studied GGBS blended with bentonite and plain concrete mixtures for the replacement of conventional liners and conducted a detailed experimental programme by testing the material for Atterberg's limit, free swell index, compaction characteristics, unconfined compressive strength and hydraulic conductivity tests (k). The results from this investigation were encouraging, and the optimal percentage of GGBS to be used as a liner material was ascertained between 15 and 20%.

3.2.5 Red Mud

Red mud is a waste material generated during the production of alumina from bauxite ore using the Bayer process. Nearly forty percent of bauxite per ton treated using this method forms the red mud waste. The potential use of red mud as a barrier material has been examined by various researchers by performing the unconfined compressive strength, hydraulic conductivity tests and swell tests on the waste material. The results give the confirmation that red mud and cement–red mud samples have a high compressive strength and lower hydraulic conductivity values and swelling percentage as compared to natural clay samples, which is desirable to serve as a hydraulic barrier [20]. Rubinos et al. [21] studied the behaviour of red mud upon exposure to various organic and inorganic liquids along with various geotechnical characteristics and leaching column tests. It was observed that red mud displayed a considerable resistance to chemical attack against various chemicals which may be present in leachate in a real landfill. Ujaczki et al. [22] conducted a field study of various chemical, biological and ecotoxicity indicators to test the applicability of red mud and concluded that it could be used beneficially up to a 20% dose without any ecotoxic effect in the cover systems at landfill sites.

3.2.6 Waste Tyre Fibres

The waste tyres pose a critical disposal problem due to the huge amount of waste generated due to increasing number of vehicles circulating in the road network. Bekhiti et al. [23] studied the effect of addition of waste tyre rubber fibres (0–2%) on various geotechnical properties of cement stabilized bentonite clay (cement content 5, 7.5, 10%) including unconfined compressive strength, swelling behaviour, swell-consolidation and ductility. The results show that the liquid limits, dry density, swell potential decrease gradually when tyre rubber fibre content and cement increase indicating beneficial use of these fibres for reinforcing the soil for use as barrier material. Mukherjee and Mishra [24] conducted the consolidation and hydraulic conductivity tests on sand–bentonite mixes with bentonite percentage as 20% mixed with 0, 5, 10 and 15% of waste tyre chips. A reduction in the swell potential of the mixes was observed with reduction being the maximum at 15% tyre chip percentage. The authors suggested an optimal percentage of 10% of tyre chips for landfill liner and 15% for landfill cover. The benefit of incorporation of waste tyres in liner material is mainly seen as the improvement in shear strength and swelling characteristics and resistance to corrosive action of impurities present in leachate.

3.2.7 Waste Paper Sludge and Other Waste Generated During Wood Processing and Paper Industry

Kortnik et al. [25] examined various geotechnical properties of waste paper sludge, and test results indicated low-strength properties, which were enhanced using additives like fly ash, bentonite and cement, high compressibility and a low hydraulic conductivity of the order of 10^{-10} m/s. Slim et al. [26] tested mixtures of paper sludge and Class-C fly ash amended with three different polymers at varying percentages up to 5% to design an alternative liner for Cinder Lake landfill in Arizona, USA. It was found that the mixes exhibited favourable properties for use as a barrier material and the addition of polymers led to a densely packed soil structure leading to further reduction in hydraulic conductivity.

Such case studies increase the reliability quotient of the usage of these wastes in beneficial applications. The need of the hour is the sustainable use of materials to save onto the unscrupulous usage of various natural resources in the various construction activities.

4 Conclusions

- Each type of waste generated out of various processes possesses salient properties, and its utilization in beneficial use applications requires a thorough study of the physical, chemical and geotechnical characteristics and the environmental concerns posed by it.
- The study reveals that these wastes present pertinent desirable characteristics for consideration and bulk usage in design of hydraulic barriers of landfills in terms of their geotechnical and hydraulic performance, cost-effectiveness and environmental concerns in comparison/conjunction with the materials traditionally used as barrier materials in landfills.
- The usage of these waste by-products individually or in combination with each other as liner material serve a dual purpose of solving the problem of disposal of these wastes and saving onto our natural resources which is highly desirable considering the global context of exhaustion of natural resources and climatic change.
- These materials are not suitable for use as liner/cover material singly. But, in conjunction with certain admixtures like bentonite clay, cement, etc. desirable characteristics of hydraulic conductivity, shear strength, consolidation can be achieved.
- On the whole, protection of the environment needs to be prioritized, and thus, usage of particular waste material as a liner material should be thoroughly warranted by evaluating the site-specific characteristics, type of waste to be dumped in the landfill, geological and climatic conditions of the area, etc. in order to encapsulate the contaminants within and prevent the contamination of subsoil and groundwater by leaching in any case.

References

1. Kaza S, Yao LC, Bhada-Tata P, Van Woerden F (2018) What a waste 2.0: a global snapshot of solid waste management to 2050. Urban development. World Bank, Washington, DC. © World Bank. https://openknowledge.worldbank.org/handle/10986/30317 (License: CC BY 3.0 IGO)
2. Gleason MH, Daniel DE, Eykholt GR (1997) Calcium and sodium bentonite for hydraulic containment applications. J Geotech Geoenvironmental Eng ASCE 123(5):438–445
3. Sobti J, Singh SK (2019) A critical evaluation of the suction and swelling characteristics of sand–bentonite–coal ash mixes. Geotech GeolEng. https://doi.org/10.1007/s10706-019-00902-4

4. CPCB (2008) Guidelines and check-list for evaluation of MSW landfills proposals with information on existing landfills, PROBES/124/2008-2009. Central Pollution Control Board, Delhi
5. Chalermyanont T, Arrykul S (2005) Compacted sand-bentonite mixtures for hydraulic containment liners. Songklanakarin J Sci Technol 27(2):313–323
6. Zhang HY, Cui SL, Zhang M, Jia LY (2012) Swelling behaviors of GMZ bentonite–sand mixtures inundated in NaCl–Na_2SO_4 solutions. Nucl Eng Des 242:115–123
7. Sobti J, Singh SK (2017) Hydraulic conductivity and compressibility characteristics of bentonite enriched soils as a barrier material for landfills. Innov Infrastruct Solution 2:12. https://doi.org/10.1007/s41062-017-0060-0
8. Nhan CT, Graydon JW, Kirk DW (1996) Utilizing coal fly-ash as a landfill barrier material. Waste Manag 16:587–595
9. Cokca E, Yilmaz Z (2004) Use of rubber and bentonite added flyash as a liner material. Waste Manag 24(2):153–164
10. Kumar S, Vaddu P (2004) Swell potential of pulverized coal combustion bottom ash amended with sodium bentonite. J Energy Eng 130:54–65
11. Kim AG, Kazonich G, Dahlberg M (2003) Relative solubility of cations in class F fly ash. Environ Sci Technol 37(19):4507–4511. https://doi.org/10.1021/es0263691
12. Pal SK, Ghosh A (2013) Hydraulic conductivity of fly ash montmorillonite clay mixtures. Indian Geotech J 43(1):47–61
13. Mishra AK, Ravindra V (2015) On the utilization of fly ash and cement mixtures as a landfill liner material. Int J Geosynthetic Ground Eng 1:17. https://doi.org/10.1007/s40891-015-0019-1
14. Bashar NAM et al (2019) Feasibility study on the application of alum sludge (AS) as alternative landfill liner material in sustainable landfill infrastructure model: XRD and SEM analysis. J Phys Conf Ser 1349 012041
15. Awab H, Paramalinggam PTT, Yusoff ARM (2012) Characterization of alum sludge for reuse and disposal. Malays J Fundam Appl Sci 8(4):251–255
16. Ayoub GM, Merhebi F (2000) Characteristics and quantities of sludge produced by coagulating wastewater with seawater bittern, lime and caustic. Adv Environ Res. https://doi.org/10.1016/S1093-0191(01)00058-2
17. Dąbska A (2019) Hydraulic conductivity of compacted lime-softening sludge used as landfill liners. Water Air Soil Pollut 230:280
18. Sekhar DC, Nayak S (2017) SEM and XRD investigations on lithomargic clay stabilized using granulated blast furnace slag and cement. Int J Geotech Eng 6362:1–15. https://doi.org/10.1080/19386362.2017.1380355
19. Manikanta D, Shankar MU (2019) Use of ground granulated blast furnace slag blended with bentonite and cement mixtures as a liner in a landfill to retain diesel oil contaminants 7(5):103360
20. Kalkan E (2006) Utilization of red mud as a stabilization material for the preparation of clay liners. Eng Geol 87:220–229
21. Rubinos DA, Spagnoli G, Barral MT (2016) Chemical and environmental compatibility of red mud liners for hazardous waste containment. Int J Environ Sci Technol 13:773–792. https://doi.org/10.1007/s13762-015-0917-8
22. Ujaczki E, Feigl V, Molnar M, Vaszita E, Uzinger N, Erdelyi A, Gruiz K (2016) The potential application of red mud and soil mixture as additive to the surface layer of a landfill cover system. J Environ Sci 44:189–196. https://doi.org/10.1016/j.jes.2015.12.014
23. Bekhiti M, Trouzine H, Rabehi M (2019) Influence of waste tire rubber fibers on swelling behavior, unconfined compressive strength and ductility of cement stabilized bentonite clay Soil. Constr Build Mater 208:304–313
24. Mukherjee K, Mishra AK (2018) Hydraulic and mechanical characteristics of compacted sand–bentonite: tyre chips mix for its landfill application. Environ Dev Sustain. https://doi.org/10.1007/s10668-018-0094-2

25. Kortnik J, Cernec F, Hrast K (2008) Paper sludge layer as low permeability barrier on waste landfills. Soil Sediment Contam 17(4):381–392. https://doi.org/10.1080/15320380802146586
26. Slim GI, Morales M, Alrumaidhin L, Bridgman P, Gloor J, Hoff ST, Odem WI (2016) Optimization of polymer-amended fly ash and paper pulp millings mixture for alternative landfill liner. Procedia Eng 145:312–318

Various Approaches and Algorithms for Monitoring Energy Efficiency of Wireless Sensor Networks

Rachna, Yogesh Chhabra, and Pankaj Bhambri

1 Introduction

Wireless sensor networks (WSNs) are extensively applied in many social or army applications. Due to the poor capability of the fixed battery and the natural poor life of the sensors, it influences the operational time of the complete system badly. Consequently, the poor energy delivery is the major restraint to sustain the lengthy term and competent process of the system. As a result, falling energy utilization and humanizing energy competence are an essential requirement for planning a sustainable WSN. To manage this issue, many methods have been proposed. To direct the researchers for widely reorganization of various approaches and algorithms in this domain of exploration, categorization of the presenting energy-efficient polices for accomplishing sustainable WSNs has been followed. This article presents a number of essential thoughts and theories frequently implemented in energy-efficient WSN plans. Consequently, this article converses regarding present approaches considered for usual WSNs from different proprieties like clustering size, merits, clustering types, cluster head (CH) parameter selection, connectivity, device (nodes) capabilities, and device deployments. These plans are compared, and their significant features are extracted and elaborated [3].

Among the open research areas of WSNs, there are huge numbers of vital issues for the implementation of these networks. In many applications, WSNs are

Rachna (✉) · Y. Chhabra
CT University, Ferozepur Road, Ludhiana, Punjab, India
e-mail: rachnarana1981@gmail.com

Y. Chhabra
e-mail: yogeshfzr@gmail.com

P. Bhambri
Guru Nanak Dev Engineering College, Ludhiana, Punjab, India
e-mail: pkbhambri@gndec.ac.in

H. Singh et al. (eds.), *Sustainable Development Through Engineering Innovations*, Lecture Notes in Civil Engineering 113,
https://doi.org/10.1007/978-981-15-9554-7_68

used like tracing, supervising, scrutiny, edifice computerization, army applications, cultivation, home, health monitoring, and industry monitoring [19].

2 Wireless Sensor Networks

A wireless network is that network of devices which is utilized for communication of the information collected from an examined field without the wired links. Information is forwarded by many devices and with an entry for the place. This network can connect information with wireless local area network. The WSN architecture shows that there are hundreds to thousands sensor devices/nodes in a network. Sensor devices are responsible for gathering of surroundings' information and forwarding it to the sink device. The sink device receives this data from sensor devices through network and forwards it to end user. Wireless sensor network contains base stations and the number of wireless sensors. The monitoring of surroundings' limitations like voice, strain, warmth, and willingly transmit data to the key area is done by WSNs [2].

3 Sustainability Development

The idea of sustainability original came into view in the Brundtland Report, printed in 1987. This record is also referred to as Our Universal Forthcoming. It was sophisticated for the United Nations for notifying the relations of destructive ecological results of resourceful development and globalization. It was produced with the objective of contribution of explanation to the issues developing from industrialization and inhabitants' enlargement. In the present scenario, sustainability endeavors to safe existing requirements without bargaining the forthcoming invention. It becomes viable without allotting up any of the three necessary tower of strengths: environmental protection, social development, and economic growth [10]. Currently, various obstacles faced by human beings like environmental changes, limited water supply, and COVID-19 also need sustainable solutions. These issues can only be handled from a global point of view and by encouraging sustainable development [9].

According to World Health Organization (WHO), COVID-19 is a pandemic/deadly disease, and it covers almost whole world infected by this pandemic both economically and socially. Due to this virus, lockdown has been declared partial or full in all over the world for the safety of people from coronavirus. And at this time, people are connected with each other through wireless network. So it will not wrong to say that wireless sensor network is very crucial resource for removing distances between people due to COVID-19. The wireless sensor networks are useful to make possible mobile crowd-sensing for availing, distributing, as well as examining the data throughout the continuing deadly disease [10]. To meet up the Sustainable Development Goals (SDGs) by 2030, inhabitants must preserve the benefits already achieved as well as hasten hard work to attain reasonable, consistent, sustainable and current energy for all [12].

There are seventeen goals of sustainability development to get like no deficiency, zero starvation, high-quality healthiness and safety, femininity equal opportunity, pure water and hygiene, excellence edification, reasonable and pure energy, honest job and economic development, industriousness, improvement and transportation, lessen dissimilarity, sustainable urban and group of people, dependable utilization and manufacture, surroundings act, existence under water, existence on earth, harmony, reliability, and well-built organization, and association [11].

4 Parallel and Distributed Networks

In parallel computing, all processors are used for sharing memory to exchange the information among other processors. On the other hand, distributed computing does not share any memory; they use their self distributed memory. Information swaps between the processors are only giving a notice [1].

5 Clustering Techniques

Here, the entire network chooses the cluster head of each cluster with the help of main base station (BS). The main BS selects that CH which is based on modifying parameters like step compute, node kind, balancing energy, and slightest space from the BS. This CH is not dynamic for the whole network lifetime. For every cluster, there are predefined CHs that execute information collection and promotion of tasks. Numbers of nodes in a group are called as clusters. There are three types of clustering technique categories: 4.1.1 Centralized, 4.1.2 Distributed, and 4.1.3 Hybrid [1–3, 6].

6 Centralized Clustering

In this technique, first of all, the network chooses the cluster head of each cluster with the help of main base station. The main BS selects that CH which is based on modifying parameters like step compute, node kind, balancing energy, and slightest space from the BS. This CH is not dynamic for the whole network lifetime. For every cluster, there are predefined CHs that execute information collection and promotion tasks [1].

7 Distributed Clustering

It signifies to choose CH in each cluster by the distributed BS. The device that struggles above to the precise restriction turns into the CH. After a definite space of time when another member device in a cluster becomes more suitable to make CH, it shapes the CH on its hold. Additionally, the previously designated CH works as a regular member device. There is no static CH selection. Each device gets an equal facility to become the cluster head of each cluster [1].

8 Hybrid Clustering

It is a combination of centralized and distributed clustering [1]. Figure 1 explains the phraseology/taxonomy of clustering procedures.

The phraseology of clustering procedures are divided into four groups like cluster uniqueness, cluster supervisor uniqueness, clustering procedure, and entire progressing of algorithms. The clusters must be consistent. The characteristic of clusters specifies that cluster must be convex or linear. These must be connected to an area in data space with high density. The whole presentation of WSNs relies on the selection of cluster supervisor, and it is done by random method. The clustering process depends on which type of network is used because there are many types of network like centralized, distributed, and hybrid. There are many types of algorithms and approaches used for clustering like homogeneous, heterogeneous, static and dynamic clustering, centralized or distributed algorithms, probabilistic and non-probabilistic approaches, etc., and their comparison is explained in Table 1.

9 Literature Review

Nowadays, energy efficiency in WSNs with sustainable development is the most important issue. In order to sustainability development and implementation of WSNs in our daily life, energy efficiency and the life span of the network should be enhanced and energy consumption must be reduced. Literature in discussion targets to summarize that how WSNs are useful and how researchers are trying to improve energy efficiency with sustainable development and using clustering approaches in WSNs. Literature review reviewed the WSNs, equivalent and disseminated computing in WSNs, and then sustainable WSNs.

Charan et al. [5] presented DEAR protocol that always took efforts to minimize the power utilization with minimizing the expenses of packet in the network and giving the information to the supplicant with least amount stay with getting desired detail from the close by concealment of device which will available close to the desirer device. The results of simulation definitely display that the energy's utilization is

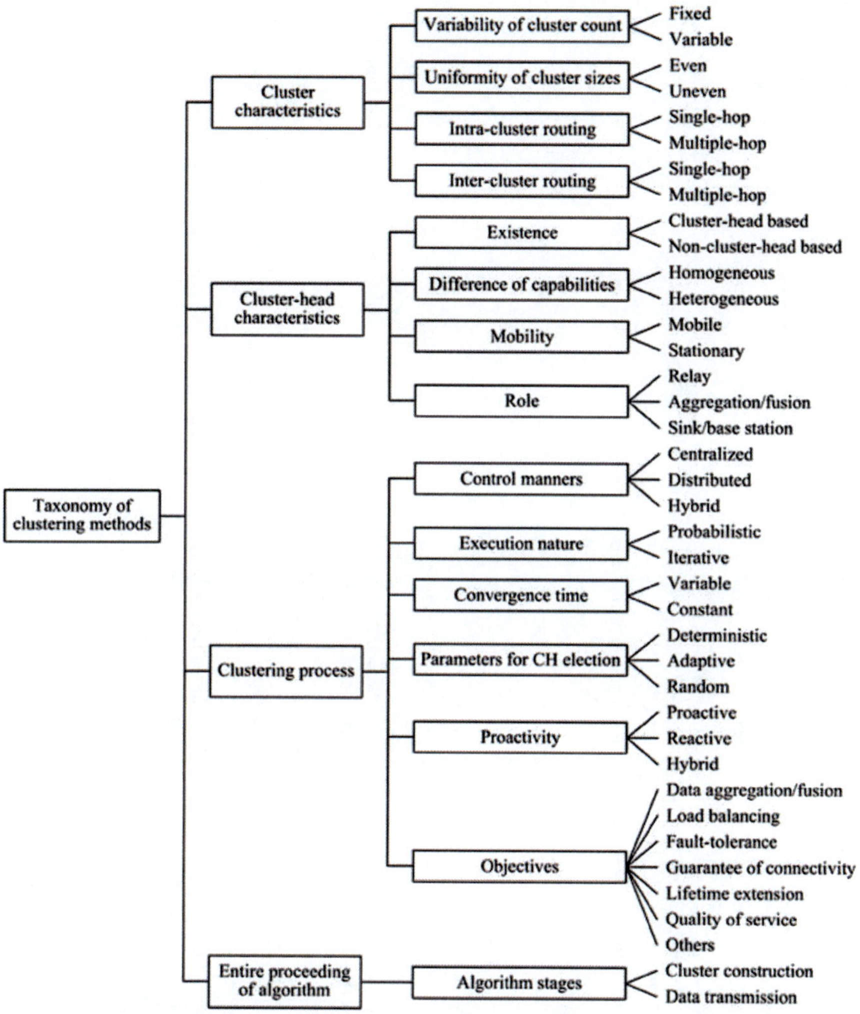

Fig. 1 Taxonomy of clustering methods [3]

minimum in cluster-depended model as compared to grid-based analytical model which is holding the same required symbols. García et al. [8] developed an ABS-SmartComAgri (agent-depended simulator of elegant announcement practices) for correctly supervising pesticide. This simulator is generously disseminated as unlock source from freely available study information to storehouse. Koriem and Bayoumi [14] presented the WHD algorithm to escalating the life of network and diminishing the transmission and calculation of outlay as well as fascinating the superiority of the system. Bello et al. [2] focused on layer to layer; physical and mass communication gain control approaches on inactive awaken radio depended on the previous

Table 1 Comparative study of various clustering approaches for WSNs

Clustering algorithms/Approaches	Clusters size	Merits	Clustering types	CH parameter selection	Connectivity	Node capabilities	Node deployments
EEDCA [18]	Equal size	Improves lifetime and efficiency	Distributed	Residual energy, density, and distance	Multihop	Homogenous	Random
OLE [4]	Unequal size	Improves the performance	Centralized	Location and residual energy	Multihop	Homogenous	Random
EEHMC [15]	Unequal size	Lengthen the system lifetime up to 27.63%, and executes two times superior than LEACH-C for first device expire	Hybrid	Remaining power, no. of neighbors and gap to BS	Multihop	Homogenous	Random
Hk-mean [16]	Unequal size	Prolongs the network lifetime	Hybrid	Residual energy, mean distance(intercommunication) and two hop coverage of the competing nodes	Multihop	Heterogeneous	Random
HADCC [1]	Unequal size	Prolongs the network lifetime	Hybrid	Minimum distance between nodes	Single hop	Heterogeneous	Random
H-CERP [17]	Unequal size	Reduces the energy consumption, and improve residual energy and lifetime	Hybrid	Minimum distance between nodes	Multihop	Homogenous	Random

(continued)

Table 1 (continued)

Clustering algorithms/Approaches	Clusters size	Merits	Clustering types	CH parameter selection	Connectivity	Node capabilities	Node deployments
JCR [19]	Unequal size	Increases the connectivity and efficiency of the network topology	Hybrid	Residual energy and no. of neighbor	Multihop	Homogenous	Random

works from the text. Here, the authors developed a MAC protocol for the novel inactive wake-up of radio technologies for coming Internet of Things' apps. Chen et al. [6] presented a state-of-the-art increasing miniature, and it is established on the real instrument of the non-line-of-sight dissemination and peaceful planning of the concave model into its convex envelope. The display of an indeterminate analysis and simulation is more for accessible procedures in both processing speed and localization precision. Zhang and Zhai [21] Showed that wireless sensor system depends on the Interweb which has robust steadiness as well as trustworthiness in information assortment. It be able to give efficient information sustaining the knowledge supervision survey of sustainable development of Winter Olympic Games. Yang et al. [20] presented a report of use of ZigBee-based WSNs within existing building duct system. This building duct method is developed for intellectual dissipate assortment in an manufacturing climate. The sustain sign concentration and route wastes are calculated for showing that the duct transmission medium works as a very successful waveguide. It provides a more trustworthy and standard network execution than previous practice's free area mediums. James and Erbacher [13] showed that University of California San Diego engineers are creating less cost, less energy wearable sensors which can be helped in measuring the vital signs of COVID-19. Due to coronavirus, the whole world is under the lockdown situation. There is major challenge for governments and health departments about solving this problem. Virus is spread under some extent control and finally enhances the survival rates by using wireless sensor networks. The engineers are working on those devices which are used for data transmission in smart phones through wireless sensor networks. These devices can be used for monitoring those patients which are coronavirus infected in real environments, and it influences degree of hotness and coldness as well as breathing. It is planned that this virus will be obsolete in the next year mid by creating a device and feasible developed process. Chu et al. [7] showed the ratio of the faithful Facebook admirer, those who are using Facebook daily, and sharing their pictures, important information on Facebook. The present and forthcoming fame has been evaluated straight forwarding by techniques of PSTIR as well as SFIBS.

10 Conclusion and Future Work

This article concluded that wireless sensor networks are extensively used by many domains like military applications, health applications, home applications, forest detection, and many other areas. Summarized siisue related to wireless sensor networks are elaborated as below:

- The constrained capability of the fixed battery and the lifespan of the wireless sensor devices are restricted, and it influences the functioning activities of the complete system.
- Restricted power supply is the major restriction for the sustainability of the extensive-time and proficient process of the system.

- Sinking energy utilization and enhancing energy efficacy is a crucial precondition for making a sustainable WSN.

To solve this problem, many approaches and algorithms for monitoring the energy efficacy of wireless sensor networks with sustainability development are needed to be designed and developed. The future works depend on the distributed clustering approach which is still work in progress. This article provides extensive study and analysis on various clustering approaches, algorithms, and approaches of WSNs with their different features like clustering types, and device deployment, clustering connectivity's with hops, clustering merits, etc. The sustainable WSNs enhance the lifespan of the network, energy efficiency, and energy consumption. Some current open issues in sustainable WSNs like energy efficiency, network lifetime, and energy saving are elaborated in detail.

References

1. Aslam M, Munir EUAA, Bilal M, Asad M (2014) HADCC: hybrid advanced distributed and centralized clustering path planning algorithm for WSNs. In: IEEE 28th international conference on advanced information networking and applications, pp 657–664
2. Bello H, Xiaoping Z, Nordin R, Xin J (2019) Advances and opportunities in passive wake-up radios with wireless energy harvesting for the internet of things applications. MDPI
3. Boukerche A, Wu Q, Sun P (2020) Efficient green protocols for sustainable wireless sensor networks. IEEE Trans Sustain Comput 5:61–80. https://doi.org/10.1109/tsusc.2019.2913374
4. Chakraborty KM, Chakraborty A, Mitra K, Naskar SK (2010) An optimized lifetime enhancement scheme for data gathering in wireless sensor networks (ArXiv preprint arXiv, 2010)
5. Charan P, Usmani T, Paulus R, Saeed HS (2019) Performance of Distributed Energy Aware Routing(DEAR) protocol cooperative caching for wireless sensor networks. Wireless Sens Netw 11:35–45
6. Chen S, Zhang J, Xu C (2019) Robust distributed cooperative localization with NLOS mitigation based on multiplicative convex model. IEEE Access
7. Chu W, Lee KT, Luo W, Bhambri P, Kautish S (2020) Predicting the security threats of internet rumors and spread of false information based on sociological principle. Comput Stand Interfaces. ISSN 0920-5489, https://doi.org/10.1016/j.csi.2020.103454
8. García MI, Lacuesta R, Lloret J (2018) ABS-Smartcomagri: an agent-based simulator of smart communication protocols in wireless sensor networks for debugging in precision agriculture. Sensors (Basel) 4:E998
9. https://www.activesustainability.com/sustainable-development/what-is-sustainability/
10. https://www.frontiersin.org/research-topics/14199/covid-19-role-of-wireless-communication-networking-sensing-technologies-for-rapid-pandemic-response#ov
11. https://www.un.org/sustainabledevelopment/sustainable-development-goals/
12. https://www.who.int/news-room/detail/28-05-2020-covid-19-intensifies-the-urgency-to-expand-sustainable-energy-solutions-worldwide
13. James L, Erbacher J (2020) Monitoring corona virus low power sensor developed for COVID-19 monitoring. Platform Power Electron
14. Koriem MS, Bayoumi AM (2018) Detecting and measuring holes in wireless sensor network. J King Saud Univ Comput Inform Sci Xxx
15. Kumar G, Mehra H, Seth RA, Radhakrishnan P, Hemavathi N, Sudha S (2013) Energy efficient hybrid multihop clustering algorithm in wireless sensor networks. In: IIEEE students' conference on electrical, electronics and computer science, pp 59–63

16. Kumar G, Mehra H, Seth RA, Radhakrishnan P, Hemavathi N, Sudha S (2014) An hybrid clustering algorithm for optimal clusters in wireless sensor networks. In: 2014 IEEE students' conference on electrical, electronics and computer science. https://doi.org/10.1109/sceecs.2014.6804442
17. Malarkkan S, Subha PC (2017) Energy efficient scheme for industrial WSN applications. Wireless Pers Commun 94(4):1937–1950
18. Meddah M, Haddad R, Ezzedine T (2017) An energy efficient and density control clustering algorithm for wireless sensor network. In Wireless Communications and Mobile Computing Conference (IWCMC), pp 357–364
19. Xu Z, Chen L, Chen C, Guan X (2016) Joint clustering and routing design for reliable and efficient data collection in large-scale wireless sensor networks. IEEE Internet Things J 3(4):520–532
20. Yang H, Kim B, Lee J, Ahn Y, Lee C (2018) Advanced wireless sensor networks for sustainable buildings using building ducts. MDPI Sustainability 10:2628. https://doi.org/10.3390/su10082628 www.mdpi.com/journal/sustainability
21. Zhang M, Zhai F (2019) The sustainable development information management of Winter Olympics based on Internet-based wireless sensor network. EURASIP J Wireless Commun Network 52. https://doi.org/10.1186/s13638-019-1360-1